TREATISE ON INVERTEBRATE PALEONTOLOGY

Prepared under the Guidance of the
Joint Committee on Invertebrate Paleontology

Paleontological
Society

Society of Economic
Paleontologists and
Mineralogists

Palaeontographical
Society

Directed and Edited by

RAYMOND C. MOORE

Assisted by CHARLES W. PITRAT

Part Q

ARTHROPODA 3

CRUSTACEA

OSTRACODA

R. H. BENSON, J. M. BERDAN, W. A. VAN DEN BOLD, TETSURO HANAI, IVAR HESSLAND, H. V. HOWE, R. V. KESLING, S. A. LEVINSON, R. A. REYMENT, R. C. MOORE, H. W. SCOTT, R. H. SHAVER, I. G. SOHN, L. E. STOVER, F. M. SWAIN, P. C. SYLVESTER-BRADLEY, and JOHN WAINWRIGHT

GEOLOGICAL SOCIETY OF AMERICA
and
UNIVERSITY OF KANSAS PRESS

1961

Library of Congress Catalogue Card
Number: 53-12913

Text Composed by
THE UNIVERSITY OF KANSAS PRESS
Lawrence, Kansas

Illustrations and Offset Lithography
MERIDEN GRAVURE COMPANY
Meriden, Connecticut

Binding
RUSSELL-RUTTER COMPANY
New York City

Address all communications to The Geological Society of America, 419 West 117 Street, New York 27, N.Y.

The *Treatise on Invertebrate Paleontology* has been made possible by (1) a grant of funds from The Geological Society of America through the bequest of Richard Alexander Fullerton Penrose, Jr., for preparation of illustrations and partial defrayment of organizational expense; (2) contribution of the knowledge and labor of specialists throughout the world, working in co-operation under sponsorship of The Palaeontographical Society, The Paleontological Society, and The Society of Economic Paleontologists and Mineralogists; and (3) acceptance by the University of Kansas Press of publication without cost to the Societies concerned and without any financial gain to the Press.

TREATISE ON INVERTEBRATE PALEONTOLOGY

Directed and Edited by
RAYMOND C. MOORE
Assisted by CHARLES W. PITRAT

PARTS

The indicated Parts (excepting the first and last) are to be published at whatever time each is ready. All may be assembled ultimately in bound volumes. In the following list, already published Parts are marked with a double asterisk (**) and those in press or nearing readiness for press are marked with a single asterisk (*). Each is cloth bound with title in gold on the cover. Copies are available on orders sent to the Geological Society of America at 419 West 117th Street, New York 27, N.Y., at prices quoted, which very incompletely cover costs of producing and distributing them but on receipt of payment the Society will ship copies without additional charge to any address in the world.

The list of contributing authors is subject to change.

(A)—INTRODUCTION.

B—PROTISTA 1 (chrysomonads, coccolithophorids, diatoms, etc.).

C—PROTISTA 2 (foraminifers).

**D—PROTISTA 3 (radiolarians, tintinnines) (xii+195 p., 1050 figs.). $3.00.

**E—ARCHAEOCYATHA, PORIFERA (xviii + 122 p., 728 figs.). $3.00.

**F—COELENTERATA (xvii + 498 p., 2,700 figs.). $8.00.

**G—BRYOZOA (xii + 253 p., 2,000 figs.). $3.00.

H—BRACHIOPODA.

**I—MOLLUSCA 1 (chitons, scaphopods, gastropods) (xxiii+351 p., 1,732 figs.). $7.50.

J—MOLLUSCA 2 (gastropods).

K—MOLLUSCA 3 (nautiloid cephalopods).

**L—MOLLUSCA 4 (ammonoid cephalopods) (xxii + 490 p., 3,800 figs.). $8.50.

M—MOLLUSCA 5 (dibranchiate cephalopods).

N—MOLLUSCA 6 (pelecypods).

**O—ARTHROPODA 1 (trilobitomorphs) (xix+560 p., 2880 figs.). $10.50.

**P—ARTHROPODA 2 (chelicerates, pycnogonids) (xvii + 181 p., 565 figs.). $3.50.

**Q—ARTHROPODA 3 (ostracodes) (this volume).

R—ARTHROPODA 4 (branchiopods, cirripeds, malacostracans, myriapods, insects).

S—ECHINODERMATA 1 (cystoids, blastoids, edrioasteroids, etc.).

T—ECHINODERMATA 2 (crinoids).

U—ECHINODERMATA 3 (echinozoans, asterozoans).

**V—GRAPTOLITHINA (xvii + 101 p., 358 figs.). $3.00.

*W—MISCELLANEA (worms, conodonts, problematical fossils). ADDENDA (index).

CONTRIBUTING AUTHORS

AGER, D. V., Imperial College of Science and Technology, London, England.

AMSDEN, T. W., Oklahoma Geological Survey, Norman, Okla.

†ARKELL, W. J., Sedgwick Museum, Cambridge University, Cambridge, Eng.

BAIRSTOW, LESLIE, British Museum (Natural History), London, Eng.

BARKER, R. WRIGHT, Shell Development Co., Houston, Tex.

BASSLER, R. S., U.S. National Museum, Washington, D.C.

BATTEN, ROGER, L., University of Wisconsin, Madison, Wisconsin.

BAYER, FREDERICK M., U.S. National Museum, Washington, D.C.

BEAVER, HAROLD H., Humble Oil & Refining Co., Houston, Texas.

BENSON, R. H., University of Kansas, Lawrence, Kans.

BERDAN, JEAN M., U.S. Geological Survey, Washington, D.C.

BOARDMAN, R. S., U.S. Geological Survey, Washington, D.C.

BOLD, W. A. VAN DEN, Louisiana State University, Baton Rouge, La.

BOSCHMA, H., Rijksmuseum van Natuurlijke Historie, Leiden, Netherlands.

BOWSHER, ARTHUR L., Sinclair Oil & Gas Co., Tulsa, Okla.

BRAMLETTE, M. N., Scripps Institution of Oceanography, La Jolla, California.

BRANSON, CARL C., University of Oklahoma, Norman, Okla.

BROWN, D. A., University of Otago, Dunedin, N.Z.

BULMAN, O. M. B., Sedgwick Museum, Cambridge University, Cambridge, Eng.

CAMPBELL, ARTHUR S., St. Mary's College, St. Mary's College, Calif.

CARPENTER, FRANK M., Biological Laboratories, Harvard University, Cambridge, Mass.

CASEY, RAYMOND, Geological Survey of Great Britain, London, Eng.

CASTER, K. E., University of Cincinnati, Cincinnati, Ohio

CHAVAN, ANDRÉ, L'École d'Anthropologie, Seyssel (Ain), France.

COLE, W. STORRS, Cornell University, Ithaca, N.Y.

COOPER, G. ARTHUR, U.S. National Museum, Washington, D.C.

COX, L. R., British Museum (Natural History), London, Eng.

CURRY, DENNIS, Pinner, Middlesex, Eng.

†DAVIES, L. M., Edinburgh, Scot.

DECHASEAUX, C., Laboratoire de Paléontologie à la Sorbonne, Paris, France.

DOUGLASS, R. C., U.S. National Museum, Washington, D.C.

DURHAM, J. WYATT, Museum of Palaeontology, University of California, Berkeley, Calif.

EAMES, F. E., Anglo-Iranian Oil Company, London, Eng.

ELLIOTT, GRAHAM F., Iraq Petroleum Company, London, Eng.

EMERSON, W. K., American Museum of Natural History, New York, N.Y.

ERBEN, H. K., Universität Bonn, Bonn, West Ger.

EXLINE, HARRIET, Rolla, Mo.

FAY, ROBERT O., Oklahoma Geological Survey, Norman, Okla.

FELL, H. BARRACLOUGH, Victoria University of Wellington, Wellington, N.Z.

FISCHER, ALFRED G., Princeton University, Princeton, N.J.

FISHER, D. W., New York State Museum, Albany, N.Y.

FRIZZELL, DONALD L., Missouri School of Mines, Rolla, Mo.

FURNISH, WILLIAM M., State University of Iowa, Iowa City, Iowa.

†GARDNER, JULIA, U.S. Geological Survey, Washington, D.C.

GEORGE, T. NEVILLE, Glasgow University, Glasgow, Scot.

GLENISTER, B. F., Iowa State University, Ames, Iowa.

HAAS, FRITZ, Chicago Natural History Museum, Chicago, Ill.

HANAI, TETSURO, University of Tokyo, Tokyo, Japan.

HANNA, G. DALLAS, California Academy of Sciences, San Francisco, Calif.

HÄNTZSCHEL, WALTER, Geologisches Staatsinstitut, Hamburg, Ger.

HARRINGTON, H. J., Tennessee Gas Transmission Company, Houston, Tex.

†HASS, WILBERT H., U.S. Geological Survey, Washington, D.C.

HATAI, KOTORA, Tohoku University, Sendai, Japan.

HAWKINS, H. L., Reading University, Reading, Eng.

HEDGPETH, JOEL, College of the Pacific, Dillon Beach, Calif.

HENNINGSMOEN, GUNNAR, Paleontologisk Museum, University of Oslo, Oslo, Norway.

HERTLEIN, L. G., California Academy of Sciences, San Francisco, Calif.

HESSLAND, IVAR, Geologiska Institutet, University of Stockholm, Stockholm, Swed.

HILL, DOROTHY, University of Queensland, Brisbane, Queensl.

HOLTHUIS, L. B., Rijksmuseum van Natuurlijke Historie, Leiden, Netherlands.

HOWE, HENRY V., Louisiana State University, Baton Rouge, La.

HOWELL, B. F., Princeton University, Princeton, N.J.

HYMAN, LIBBIE H., American Museum of Natural History, New York, N.Y.

JAANUSSON, VALDAR, Paleontological Inst., Uppsala, Sweden.

JELETZKY, J. A., Geological Survey of Canada, Ottawa, Can.

KAMPTNER, ERWIN, Naturhistorisches Museum, Wien, Aus.

KEEN, MYRA, Stanford University, Stanford, Calif.

KESLING, ROBERT V., Paleontological Museum, University of Michigan, Ann Arbor, Mich.

KIER, PORTER, U. S. National Museum, Washington, D.C.

†KNIGHT, J. BROOKES, Longboat Key, Fla.

KUMMEL, BERNHARD, Museum of Comparative Zoology, Harvard University, Cambridge, Mass.

LA ROCQUE, AURÈLE, Ohio State University, Columbus, Ohio.

†LAUBENFELS, M. W. DE, Oregon State College, Corvallis, Ore.

LECOMPTE, MARIUS, Institut Royal des Sciences Naturelles, Bruxelles, Belg.

LEONARD, A. BYRON, University of Kansas, Lawrence, Kans.

LEVINSON, S. A., Humble Oil & Refining Company, Houston, Tex.

LOCHMAN-BALK, CHRISTINA, New Mexico Institute of Mining and Geology, Socorro, N.Mex.

LOEBLICH, A. R., JR., California Research Corporation, La Habra, Calif.

LOEBLICH, HELEN TAPPAN, University of California, Los Angeles, Calif.

LOHMAN, KENNETH E., U.S. Geological Survey, Washington, D.C.

LOWENSTAM, HEINZ A., California Institute of Technology, Pasadena, Calif.

LUDBROOK, N. H., Department of Mines, Adelaide, S.Austral.

MARWICK, J., New Zealand Geological Survey, Wellington, N.Z.

MELVILLE, R. V., Geological Survey and Museum, London, Eng.

MILLER, ARTHUR K., State University of Iowa, Iowa City, Iowa.

MONTANARO-GALLITELLI, EUGENIA, Istituto di Geología e Paleontología, Università, Modena, Italy.

MOORE, RAYMOND C., University of Kansas, Lawrence, Kans.

MORRISON, J. P. E., U.S. National Museum, Washington, D.C.

MUIR-WOOD, HELEN M., British Museum (Natural History), London, Eng.

NEWELL, NORMAN D., American Museum of Natural History (and Columbia University), New York, N.Y.

OKULITCH, VLADIMIR J., University of British Columbia, Vancouver, B.C.

OLSSON, AXEL A., Coral Gables, Florida.

PALMER, KATHERINE VAN WINKLE, Paleontological Research Institution, Ithaca, N.Y.

PECK, RAYMOND E., University of Missouri, Columbia, Mo.

PETRUNKEVITCH, ALEXANDER, Osborn Zoological Laboratory, Yale University, New Haven, Conn.

POULSEN, CHR., Universitetets Mineralogisk-Geologiske Institut, København, Denm.

POWELL, A. W. B., Auckland Institute and Museum, Auckland, N.Z.

PURI, HARBANS, Florida State Geological Survey, Tallahassee, Fla.

RASETTI, FRANCO, Johns Hopkins University, Baltimore, Md.

RASMUSSEN, H. W., Universitets Mineralogisk — Geologiske Institut, København, Denm.

REGNÉLL, GERHARD, Paleontologiska Institution, Lunds Universitets, Lund, Swed.

REHDER, HARALD A., U.S. National Museum, Washington, D.C.

REICHEL, M., Bernouillanum, Basel University, Basel, Switz.

REYMENT, RICHARD A., University of Stockholm, Stockholm, Swed.

†RICHTER, EMMA, Senckenberg Natur-Museum, Frankfurt-a.-M., Ger.

†RICHTER, RUDOLF, Universität Frankfurt-a.-M., Frankfurt-a.-M., Ger.

ROBERTSON, ROBERT, Academy of Natural Sciences, Philadelphia, Pa.

ROWELL, A. J., Nottingham University, Nottingham, Eng.

SCHINDEWOLF, O. H. Geologisch-paläontologisches Institut der Universität Tübingen, Tübingen, Ger.

SCHMIDT, HERTA, Senckenbergische Naturforschende Gesellschaft, Frankfurt-a.-M., Ger.

SCOTT, HAROLD W., University of Illinois, Urbana, Ill.

SDZUY, KLAUS, Senckenbergische Naturforschende Gesellschaft, Frankfurt-a.-M., Ger.

SHAVER, ROBERT, University of Indiana and Indiana Geological Survey, Bloomington, Ind.

SIEVERTS-DORECK, HERTHA, Stuttgart-Möhringen, Ger.

SINCLAIR, G. W., Geological Survey of Canada, Ottawa, Can.

SMITH, ALLYN G., California Academy of Sciences, San Francisco, Calif.

SOHN, I. G., U.S. Geological Survey, Washington, D.C.

†SPENCER, W. K., Crane Hill, Ipswich, Suffolk, Eng.

†STAINBROOK, MERRILL A., Brandon, Iowa.

STEHLI, F. G., Western Reserve University, Cleveland, Ohio.

STENZEL, H. B., Shell Development Co., Houston, Tex.

STEPHENSON, LLOYD W., Dover, Ohio.

STØRMER, LEIF, Paleontologisk Institutt, University of Oslo, Oslo, Nor.

STOVER, LEWIS E., Tulsa, Okla.

STRUVE, WOLFGANG, Senckenbergische Naturforschende Gesellschaft, Frankfurt-a.-M., Ger.

STUBBLEFIELD, C. J., Geological Survey and Museum, London, Eng.

STUMM, ERWIN C., Museum of Paleontology, University of Michigan, Ann Arbor, Mich.

SWEET, WALTER C., Ohio State University, Columbus, Ohio.

SWAIN, FREDERICK M., University of Minnesota, Minneapolis, Minn.

SYLVESTER-BRADLEY, P. C., University of Leicester, Leicester, Eng.

TASCH, PAUL, University of Wichita, Wichita, Kans.

TEICHERT, CURT, U.S. Geological Survey, Federal Center, Denver, Colo.

THOMPSON, M. L., Illinois Geological Survey, Urbana, Ill.

THOMPSON, R. H., University of Kansas, Lawrence, Kans.

†TIEGS, O. W., University of Melbourne, Melbourne, Victoria, Austral.

TRIPP, RONALD P., Seven Oaks, Kent, Eng.

UBAGHS, G., Université de Liège, Liège, Belg.

VOKES, H. E., Tulane University, New Orleans, La.

WAINWRIGHT, JOHN, Shell Oil Company, Midland, Texas.

†WANNER, J., Scheidegg (Allgäu), Bayern, Ger.

WEIR, JOHN, Glasgow University, Glasgow, Scot.

WELLER, J. MARVIN, University of Chicago, Chicago, Ill.

WELLS, JOHN W., Cornell University, Ithaca, N.Y.

WHITTINGTON, H. B., Museum of Comparative Zoology, Harvard University, Cambridge, Mass.

WILLIAMS, ALWYN, Queens University of Belfast, Belfast, N.Ire.

WILLS, L. J., Romsley, Eng.

†WITHERS, T. H., Bournemouth, Eng.

WRIGHT, A. D., Queens University of Belfast, Belfast, N. Ire.

WRIGHT, C. W., London, Eng.

†WRIGLEY, ARTHUR, Norbury, London, Eng.

YOCHELSON, ELLIS L., U.S. Geological Survey, Washington, D.C.

YONGE, C. M., University of Glasgow, Glasgow, Scot.

†—Deceased.

EDITORIAL PREFACE

The aim of the *Treatise on Invertebrate Paleontology,* as originally conceived and consistently pursued, is to present the most comprehensive and authoritative, yet compact statement of knowledge concerning invertebrate fossil groups that can be formulated by collaboration of competent specialists in seeking to organize what has been learned of this subject up to the mid-point of the present century. Such work has value in providing a most useful summary of the collective results of multitudinous investigations and thus should constitute an indispensable text and reference book for all persons who wish to know about remains of invertebrate organisms preserved in rocks of the earth's crust. This applies to neozoologists as well as paleozoologists and to beginners in study of fossils as well as to thoroughly trained, long-experienced professional workers, including teachers, stratigraphical geologists, and individuals engaged in research on fossil invertebrates. The making of a reasonably complete inventory of present knowledge of invertebrate paleontology may be expected to yield needed foundation for future research and it is hoped that the *Treatise* will serve this end.

The *Treatise* is divided into parts which bear index letters, each except the initial and concluding ones being defined to include designated groups of invertebrates. The chief purpose of this arrangement is to provide for independence of the several parts as regards date of publication, because it is judged desirable to print and distribute each segment as soon as possible after it is ready for press. Pages in each part will bear the assigned index letter joined with numbers beginning with 1 and running consecutively to the end of the part. When the parts ultimately are assembled into volumes, no renumbering of pages and figures is required.

The outline of subjects to be treated in connection with each large group of invertebrates includes (1) description of morphological features, with special reference to hard parts, (2) ontogeny, (3) classification, (4) geological distribution, (5) evolutionary trends and phylogeny, and (6) systematic description of genera, subgenera, and higher taxonomic units. In general, paleoecological aspects of study are omitted or little emphasized because comprehensive treatment of this subject is being undertaken in a separate work, prepared under auspices of a committee of the United States National Research Council. A selected list of references is furnished in each part of the *Treatise.*

Features of style in the taxonomic portions of this work have been fixed by the Editor with aid furnished by advice from the Joint Committee on Invertebrate Paleontology representing the societies which have undertaken to sponsor the *Treatise.* It is the Editor's responsibility to consult with authors and co-ordinate their work, seeing that manuscript properly incorporates features of adopted style. Especially he has been called on to formulate policies in respect to many questions of nomenclature and procedure. The subject of family and subfamily names is reviewed briefly in a following section of this preface, and features of *Treatise* style in generic descriptions are explained.

A generous grant of $35,000 has been made by the Geological Society of America for the purpose of preparing *Treatise* illustrations. Administration of expenditures has been in charge of the Editor and most of the work by photographers and artists has been done under his direction at the University of Kansas, but sizable parts of this program have also been carried forward in Washington and London.

FORM OF ZOOLOGICAL NAMES

Many questions arise in connection with the form of zoological names. These include such matters as adherence to stipulations concerning Latin or Latinized nature of words accepted as zoological names, gender of generic and subgeneric names, nominative or adjectival form of specific names, required endings for some family-group names, and numerous others. Regulation extends to capitalization, treatment of particles belonging to modern patronymics, use of neo-Latin letters, and approved methods for converting diacritical marks. The magnitude and complexities of nomenclature problems surely are enough to warrant the

ix

complaint of those who hold that zoology is the study of animals rather than of names applied to them.

CLASSIFICATION OF ZOOLOGICAL NAMES

In accordance with the "Copenhagen Decisions on Zoological Nomenclature" (London, 135 p., 1953), zoological names may be classified usefully in various ways. The subject is summarized here with introduction of designations for some categories which the *Treatise* proposes to distinguish in systematic parts of the text for the purpose of giving readers comprehension of the nature of various names together with authorship and dates attributed to them.

CO-ORDINATE NAMES OF TAXA GROUPS

Five groups of different-rank taxonomic units (termed *taxa,* sing., *taxon*) are discriminated, within each of which names are treated as co-ordinate, being transferrable from one category to another without change of authorship or date. These are: (1) Species Group (subspecies, species); (2) Genus Group (subgenus, genus); (3) Family Group (tribe, subfamily, family, superfamily); (4) Order/Class Group (suborder, order, subclass, class); and (5) Phylum Group (subphylum, phylum). In the first 3 of these groups, but not others, the author of the first-published valid name for any taxon is held to be the author of all other taxa in the group which are based on the same nominate type and the date of publication for purposes of priority is that of the first-published name. Thus, if author A in 1800 introduces the family name X-idae to include 3 genera, one of which is *X-us;* and if author B in 1850 divides the 20 genera then included in X-idae into subfamilies called X-inae and Y-inae; and if author C in 1950 combines X-idae with other later-formed families to make a superfamily X-acea (or X-oidea, X-icae, etc.); the author of X-inae, X-idae and X-acea is A, 1800, under the Rules. Because taxonomic concepts introduced by authors B and C along with appropriate names surely are not attributable to author A, some means of recording responsibility of B and C are needed. This is discussed later in explaining proposed use of *"nom. transl."*

The co-ordinate status of zoological names belonging to the species group is stipulated in Art. 11 of the present Rules; genus group in Art. 6 of the present Rules; family group in paragraph 46 of the Copenhagen Decisions; order/class group and phylum group in paragraphs 65 and 66 of the Copenhagen Decisions.

ORIGINAL AND SUBSEQUENT FORMS OF NAMES

Zoological names may be classified according to form (spelling) given in original publication and employed by subsequent authors. In one group are names which are entirely identical in original and subsequent usage. Another group comprises names which include with the original subsequently published variants of one sort or another. In this second group, it is important to distinguish names which are inadvertent changes from those constituting intentional emendations, for they have quite different status in nomenclature. Also, among intentional emendations, some are acceptable and some quite unacceptable under the Rules.

VALID AND INVALID NAMES

Valid names. A valid zoological name is one that conforms to all mandatory provisions of the Rules (Copenhagen Decisions, p. 43-57) but names of this group are divisible into subgroups as follows: (1) *"inviolate names,"* which as originally published not only meet all mandatory requirements of the Rules but are not subject to any sort of alteration (most generic and subgeneric names); (2) *"perfect names,"* which as they appear in original publication (with or without precise duplication by subsequent authors) meet all mandatory requirements and need no correction of any kind but which nevertheless are legally alterable under present Rules (as in changing the form of ending of a published class/order-group name); (3) *"imperfect names,"* which as originally published and with or without subsequent duplication meet mandatory requirements but contain defects such as incorrect gender of an adjectival specific name (for example, *Spironema recta* instead of *Spironema rectum*) or incorrect stem or form of ending of a family-group name (for example, Spironemidae instead of Spironematidae); (4) *"transferred names,"* which are derived by valid emendation from either of the 2nd or 3rd subgroups or from a pre-existing transferred name (as illustrated by change of a family-group name from -inae to -idae or making of a superfamily name); (5) *"improved names,"* which include necessary as well as somewhat arbitrarily made emendations allowable under the Rules for taxonomic categories not now covered by regulations as to name form and alterations that are distinct from changes that distinguish the 4th subgroup (including names derived from the 2nd and 3rd subgroups and possibly some alterations of 4th subgroup names). In addition, some zoological names included among those recognized as valid are classifiable in special categories, while at the same time belonging to one or more of the above-listed subgroups. These chiefly include (7) *"substitute names,"* introduced to replace invalid names such as junior homonyms; and (8) *"conserved names,"* which are names that would have to be rejected by application of the Rules except for saving them in their original or an altered spelling by action of the International Commission on Zoological Nomenclature in exercising its plenary powers to this end. Whenever a name requires replacement, any individual may publish a "new name" for it and the first one so introduced has priority over any others; since newness is temporary and relative, the replacement designation is better called substitute name rather than new name. Whenever it is considered desirable to save for usage an otherwise necessarily rejectable name, an individual cannot by himself accomplish the preservation, except by unchallenged action taken in accordance with certain provisions of the Copenhagen Decisions; otherwise he must seek validation through ICZN.

It is useful for convenience and brevity of distinction in recording these subgroups of valid zoological names to introduce Latin designations, following the pattern of *nomen nudum, nomen novum,* etc. Accordingly, the subgroups are (1) *nomina inviolata* (sing., *nomen inviolatum,* abbr., *nom. inviol.*); (2) *nomina perfecta* (sing., *nomen perfectum,* abbr., *nom. perf.*); (3) *nomina imperfecta* (sing., *nomen imperfectum,* abbr., *nom. imperf.*); (4) *nomina translata* (sing., *nomen translatum,* abbr., *nom. transl.*); (5) *nomina correcta* (sing., *nomen correctum,* abbr., *nom. correct.*); (6) *nomina substituta* (sing., *nomen substitutum,* abbr., *nom. subst.*); (7) *nomina conservata* (sing., *nomen conservatum,* abbr., *nom. conserv.*).

Invalid names. Invalid zoological names consisting of originally published names that fail to comply with mandatory provisions of the Rules and consisting of inadvertent changes in spelling of names have no status in nomenclature. They are not available as replacement names and they do not preoccupy for purposes of the Law of Homonomy. In addition to *nomen nudum,* invalid names may be distinguished as follows: (1) *"denied names,"* which consist of originally published names (with or without subsequent duplication) that do not meet mandatory requirements of the Rules; (2) *"null names,"* which comprise unintentional alterations of names; and (3) *"vain or void names,"* which consist of invalid emendations of previously published valid or invalid names. Void names do have status in nomenclature, being classified as junior synonyms of valid names.

Proposed Latin designations for the indicated kinds of invalid names are as follows: (1) *nomina negata* (sing., *nomen negatum,* abbr., *nom. neg.*); (2) *nomina nulla* (sing., *nomen nullum,* abbr., *nom. null.*); (3) *nomina vana* (sing., *nomen vanum,* abbr., *nom. van.*). It is desirable in the *Treatise* to identify invalid names, particularly in view of the fact that many of these names (*nom. neg., nom null.*) have been considered incorrectly to be junior objective synonyms (like *nom. van.*), which have status in nomenclature.

SUMMARY OF NAME CLASSES

Partly because only in such publications as the *Treatise* is special attention to classes of zoological names called for and partly because new designations are now introduced as means of recording distinctions explicitly as well as compactly, a summary may be useful. In the following tabulation valid classes of names are indicated in boldface type, whereas invalid ones are printed in italics.

Definitions of Name Classes

nomen conservatum (nom. conserv.). Name otherwise unacceptable under application of the Rules which is made valid, either with original or altered spelling, through procedures specified by the Copenhagen Decisions or by action of ICZN exercising its plenary powers.

nomen correctum (nom. correct.). Name with intentionally altered spelling of sort required or allowable under the Rules but not dependent on transfer from one taxonomic category to another ("improved name"). (*See* Copenhagen Decisions, paragraphs 50, 71-2-a-i, 74, 75, 79, 80, 87, 101; in addition, change of endings for categories not now fixed by Rules.)

nomen imperfectum (nom. imperf.). Name that as originally published (with or without subsequent identical spelling) meets all mandatory requirements of the Rules but contains defect needing correction ("imperfect name"). (*See* Copenhagen Decisions, paragraphs 50-1-b, 71-1-b-i, 71-1-b-ii, 79, 80, 87, 101.)

nomen inviolatum (nom. inviol.). Name that as originally published meets all mandatory requirements of the Rules and also is uncorrectable or alterable in any way ("inviolate name"). (*See* Copenhagen Decisions, paragraphs 152, 153, 155-157).

nomen negatum (nom. neg.). Name that as originally published (with or without subsequent identical spelling) constitutes invalid original spelling and although possibly meeting all other mandatory requirements of the Rules, is not correctable to establish original authorship and date ("denied name"). (*See* Copenhagen Decisions, paragraph 71-1-b-iii.)

nomen nudum (nom. nud.). Name that as originally published (with or without subsequent identical spelling) fails to meet mandatory requirements of the Rules and having no status in nomenclature, is not correctable to establish original authorship and date ("naked name"). (*See* Copenhagen Decisions, paragraph 122.)

nomen nullum (nom. null.). Name consisting of an unintentional alteration in form (spelling) of a previously published name (either valid name, as *nom. inviol., nom. perf., nom. imperf.,* nom. *transl.;* or invalid name, as *nom. neg., nom. nud., nom. van.,* or another *nom. null.*) ("null name"). (*See* Copenhagen Decisions, paragraphs 71-2-b, 73-4.)

nomen perfectum (nom. perf.). Name that as originally published meets all mandatory requirements of the Rules and needs no correction of any kind but which nevertheless is validly alterable ("perfect name").

nomen substitutum (nom. subst.). Replacement name published as substitute for an invalid name, such as a junior homonym (equivalent to "new name").

nomen translatum (nom. transl.). Name that is derived by valid emendation of a previously published name as result of transfer from one taxonomic category to another within the group to which it belongs ("transferred name").

nomen vanum (nom. van.). Name consisting of an invalid intentional change in form (spelling) from a previously published name, such invalid emendations having status in nomenclature as junior objective synonyms ("vain or void name"). (*See* Copenhagen Decisions, paragraphs 71-2-a-ii, 73-3.)

Except as specified otherwise, zoological names accepted in the *Treatise* may be understood to be classifiable either as *nomina inviolata* or *nomina perfecta* (omitting from notice *nomina correcta* among specific names) and these are not discriminated. Names which are not accepted for one reason or another include junior homonyms, a few senior synonyms classifiable as *nomina negata* or *nomina nuda,* and numerous junior synonyms which include both objective *(nomina vana)* and subjective (all classes of valid names) types; effort to classify the invalid names as completely as possible is intended.

NAME CHANGES IN RELATION TO GROUP CATEGORIES

SPECIFIC AND SUBSPECIFIC NAMES

Detailed consideration of valid emendation of specific and subspecific names is unnecessary here because it is well understood and relatively inconsequential. When the form of adjectival specific names is changed to obtain agreement with the gender of a generic name in transferring a species from one genus to another, it is never needful to label the changed name as a *nom. transl.* Likewise, transliteration of a letter accompanied by a diacritical mark in manner now called for by the Rules (as in changing originally published *brŏggeri* to *broeggeri*) or elimination of a hyphen (as in changing originally published *cornuoryx* to *cornuoryx* does not require *"nom. correct."* with it. Revised provisions for emending specific and subspecific names are stated in the report on Copenhagen Decisions (p. 43-46, 51-57).

GENERIC AND SUBGENERIC NAMES

So rare are conditions warranting change of the originally published valid form of generic and subgeneric names that lengthy discussion may be omitted. Only elimination of diacritical marks of some names in this category seems to furnish basis for valid emendation. It is true that many changes of generic and subgeneric names have been published, but virtually all of these are either *nomina vana* or *nomina nulla*. Various names which formerly were classed as homonyms are not now, for two names that differ only by a single letter (or in original publication by presence or absence of a diacritical mark) are construed to be entirely distinct. Revised provisions for emendation of generic and subgeneric names also are given in the report on Copenhagen Decisions (p. 43-47).

Examples in use of classificatory designations for generic names as previously given are the following, which also illustrate designation of type species, as explained later.

Kurnatiophyllum Thomson, 1875 [*K. concentricum;* SD Gregory, 1917] [=*Kumatiophyllum* Thomson, 1876 *(nom. null.); Cymatophyllum* Thomson, 1901 *(nom. van.); Cymatophyllum* Lang, Smith & Thomas, 1940 *(nom. van.)*].

Stichophyma Pomel, 1872 [*Manon turbinatum* Römer, 1841; SD Rauff, 1893] [=*Stychophyma* Vosmaer, 1885 *(nom. null.); Sticophyma* Moret, 1924 *(nom. null.)*].

Stratophyllum Smyth, 1933 [*S. tenue*] [=*Ethmoplax* Smyth, 1939 *(nom. van. pro Stratophyllum); Stratiphyllum* Lang, Smith & Thomas, 1940 *(nom. van. pro Stratophyllum* Smyth) *(non Stratiphyllum* Scheffen, 1933)].

Placotelia Oppliger, 1907 [*Porostoma marconi* Fromentel, 1859; SD deLaubenfels, herein] [=*Plakotelia* Oppliger, 1907 *(nom. neg.)*].

Walcottella deLaub., *nom. subst.,* 1955 [*pro Rhopalicus* Schramm., 1936 *(non* Förster, 1856)].

Cyrtograptus Carruthers, 1867 [*nom. correct.* Lapworth, 1873 (*pro Cyrtograpsus* Carruthers, 1867), *nom. conserv.* proposed Bulman, 1955 (ICZN pend.)]

FAMILY-GROUP NAMES; USE OF "NOM. TRANSL."

The Rules now specify the form of endings only for subfamily (-inae) and family (-idae) but decisions of the Copenhagen Congress direct classification of all family-group assemblages (taxa) as co-ordinate, signifying that for purposes of priority a name published for a unit in any category and based on a particular type genus shall date from its original publication for a unit in any category, retaining this priority (and

authorship) when the unit is treated as belonging to a lower or higher category. By exclusion of -inae and -idae, respectively reserved for subfamily and family, the endings of names used for tribes and superfamilies must be unspecified different letter combinations. These, if introduced subsequent to designation of a subfamily or family based on the same nominate genus, are *nomina translata,* as is also a subfamily that is elevated to family rank or a family reduced to subfamily rank. In the *Treatise* it is desirable to distinguish the valid emendation comprised in the changed ending of each transferred family group name by the abbreviation *"nom. transl."* and record of the author and date belonging to this emendation. This is particularly important in the case of superfamilies, for it is the author who introduced this taxon that one wishes to know about rather than the author of the superfamily as defined by the Rules, for the latter is merely the individual who first defined some lower-rank family-group taxon that contains the nominate genus of the superfamily. The publication of the author containing introduction of the superfamily *nomen translatum* is likely to furnish the information on taxonomic considerations that support definition of the unit.

Examples of the use of *"nom. transl."* are the following.

Subfamily STYLININAE d'Orbigny, 1851

[*nom. transl.* Edwards & Haime, 1857 (*ex* Stylinidae d'Orbigny, 1851]

Superfamily ARCHAEOCTONOIDEA Petrunkevitch, 1949

[*nom. transl.* Petrunkevitch, herein (*ex* Archaeoctonidae Petrunkevitch, 1949)]

Superfamily CRIOCERATITACEAE Hyatt, 1900

[*nom. transl.* Wright, 1952 (*ex* Crioceratitidae Hyatt, 1900)]

FAMILY-GROUP NAMES; USE OF "NOM. CORRECT."

Valid emendations classed as *nomina correcta* do not depend on transfer from one category of family-group units to another but most commonly involve correction of the stem of the nominate genus; in addition, they include somewhat arbitrarily chosen modification of ending for names of tribe or superfamily. Examples of the use of "*nom. correct.*" are the following.

Family STREPTELASMATIDAE Nicholson, 1889

[*nom. correct.* WEDEKIND, 1927 (*ex* Streptelasmidae NICHOLSON, 1889, *nom. imperf.*)]

Family PALAEOSCORPIIDAE Lehmann, 1944

[*nom. correct.* PETRUNKEVITCH, herein (*ex* Palaeoscorpionidae LEHMANN, 1944, *nom. imperf.*)]

Family AGLASPIDIDAE Miller, 1877

[*nom. correct.* STØRMER, herein (*ex* Aglaspidae MILLER, 1877, *nom. imperf.*)]

Superfamily AGARICIICAE Gray, 1847

[*nom. correct.* WELLS, herein (*ex* Agaricioidae VAUGHAN & WELLS, 1943, *nom. transl. ex* Agariciidae GRAY, 1847)]

FAMILY-GROUP NAMES; USE OF "*NOM. CONSERV.*"

It may happen that long-used family-group names are invalid under strict application of the Rules. In order to retain the otherwise invalid name, appeal to ICZN is needful. Examples of use of *nom. conserv.* in this connection, as cited in the *Treatise,* are the following.

Family ARIETITIDAE Hyatt, 1874

[*nom. correct.* HAUG, 1885 (*pro* Arietidae HYATT, 1875), *nom. conserv.* proposed ARKELL, 1955 (ICZN pend.)]

Family STEPHANOCERATIDAE Neumayr, 1875

[*nom. correct.* FISCHER, 1882 (*pro* Stephanoceratinen NEUMAYR, 1875, invalid vernacular name), *nom conserv.* proposed ARKELL, 1955 (ICZN pend.)]

FAMILY-GROUP NAMES; REPLACEMENTS

Family-group names are formed by adding letter combinations (prescribed for family and subfamily but not now for others) to the stem of the name belonging to genus (nominate genus) first chosen as type of the assemblage. The type genus need not be the oldest in terms of receiving its name and definition, but it must be the first-published as name-giver of a family-group taxon among all those included. Once fixed, the family-group name remains tied to the nominate genus even if its name is changed by reason of status as a junior homonym or junior synonym, either objective or subjective. According to the Copenhagen Decisions, the family-group name requires replacement only in the event that the nominate genus is found to be a junior homonym, and then a substitute family-group name is accepted if it is formed from the oldest available substitute name for the nominate genus. Authorship and date attributed to the replacement family-group name are determined by first publication of the changed family-group name.

The aim of family-group nomenclature is greatest possible stability and uniformity, just as in case of other zoological names. Experience indicates the wisdom of sustaining family-group names based on junior subjective synonyms if they have priority of publication, for opinions of different workers as to the synonymy of generic names founded on different type species may not agree and opinions of the same worker may alter from time to time. The retention similarly of first-published family-group names which are found to be based on junior objective synonyms is less clearly desirable, especially if a replacement name derived from the senior objective synonym has been recognized very long and widely. To displace a much-used family-group name based on the senior objective synonym by disinterring a forgotten and virtually unused family-group name based on a junior objective synonym because the latter happens to have priority of publication is unsettling. Conversely, a long-used family-group name founded on a junior objective synonym and having priority of publication is better continued in nomenclature than a replacement name based on the senior objective synonym. The Copenhagen Decisions (paragraph 45) take account of these considerations by providing a relatively simple procedure for fixing the desired choice in stabilizing family-group names. In conformance with this, the *Treatise* assigns to contributing authors responsibility for adopting provisions of the Copenhagen Decisions.

Replacement of a family-group name may be needed if the former nominate genus is transferred to another family-group. Then the first-published name-giver of a family-group assemblage in the remnant taxon is to be recognized in forming a replacement name.

FAMILY-GROUP NAMES; AUTHORSHIP AND DATE

All family-group taxa having names

based on the same type genus are attributed to the author who first published the name for any of these assemblages, whether tribe, subfamily, or family (superfamily being almost inevitably a later-conceived taxon). Accordingly, if a family is divided into subfamilies or a subfamily into tribes, the name of no such subfamily or tribe can antedate the family name. Also, every family containing differentiated subfamilies must have a nominate *(sensu stricto)* subfamily, which is based on the same type genus as that for the family, and the author and date set down for the nominate subfamily invariably are identical with those of the family, without reference to whether the author of the family or some subsequent author introduced subdivisions.

Changes in the form of family-group names of the sort constituting *nomina correcta,* as previously discussed, do not affect authorship and date of the taxon concerned, but in publications such as the *Treatise* it is desirable to record the authorship and date of the correction.

ORDER/CLASS-GROUP NAMES; USE OF *"NOM. CORRECT."*

Because no stipulation concerning the form of order/class-group names is given yet by the Rules, emendation of all such names actually consists of arbitrarily devised changes in the form of endings. Nothing precludes substitution of a new name for an old one, but a change of this sort is not considered to be an emendation. Examples of the use of *"nom. correct."* as applied to order/class-group names are the following.

Order DISPARIDA Moore & Laudon, 1943

[*nom. correct.* MOORE, 1952 (*ex* Disparata MOORE & LAUDON, 1943)]

Suborder FAVIINA Vaughan & Wells, 1943

[*nom. correct.* WELLS, herein (*ex* Faviida VAUGHAN & WELLS, 1943)]

Suborder FUNGIINA Verrill, 1865

[*nom. correct.* WELLS, herein (*ex* Fungiida DUNCAN, 1884, *ex* Fungacea VERRILL, 1865)]

TAXONOMIC EMENDATION

Emendation has two measurably distinct aspects as regards zoological nomenclature. These embrace (1) alteration of a name itself in various ways for various reasons, as has been reviewed, and (2) alteration of

taxonomic scope or concept in application of a given zoological name, whatever its hierarchical rank. The latter type of emendation primarily concerns classification and inherently is not associated with change of name, whereas the other type introduces change of name without necessary expansion, restriction, or other modification in applying the name. Little attention generally has been paid to this distinction in spite of its significance.

Most zoologists, including paleozoologists, who have signified emendation of zoological names refer to what they consider a material change in application of the name such as may be expressed by an importantly altered diagnosis of the assemblage covered by the name. The abbreviation *"emend."* then may accompany the name, with statement of the author and date of the emendation. On the other hand, a multitude of workers concerned with systematic zoology think that publication of *"emend."* with a zoological name is valueless because more or less alteration of taxonomic sort is introduced whenever a subspecies, species, genus, or other assemblage of animals is incorporated under or removed from the coverage of a given zoological name. Inevitably associated with such classificatory expansions and restrictions is some degree of emendation affecting diagnosis. Granting this, still it is true that now and then somewhat radical revisions are put forward, generally with published statement of reasons for changing the application of a name. To erect a signpost at such points of most significant change is worth while, both as aid to subsequent workers in taking account of the altered nomenclatural usage and as indication that not-to-be-overlooked discussion may be found at a particular place in the literature. Authors of contributions to the *Treatise* are encouraged to include records of all specially noteworthy emendations of this nature, using the abbreviation *"emend."* with the name to which it refers and citing the author and date of the emendation.

In Part G (Bryozoa) and Part D (Protista 3) of the *Treatise,* the abbreviation *"emend."* is employed to record various sorts of name emendations, thus conflicting with usage of *"emend."* for change in taxonomic application of a name without alteration of the name itself. This is objectionable. In Part E (Archaeocyatha, Porifera) and later-issued divisions of the *Treatise,* use of *"emend."* is restricted to its customary sense, that is, significant alteration in taxonomic scope of a name such as calls for noteworthy modifications of a diagnosis. Other means of designating emendations that relate to form of a name are introduced.

STYLE IN GENERIC DESCRIPTIONS

DEFINITION OF NAMES

Most generic names are distinct from all others and are indicated without ambiguity by citing their originally published spelling accompanied by name of the author and date of first publication. If the same generic name has been applied to 2 or more distinct taxonomic units, however, it is necessary to differentiate such homonyms, and this calls for distinction between junior homonyms and senior homonyms. Because a junior homonym is invalid, it must be replaced by some other name. For example, *Callopora* HALL, 1851, introduced for Paleozoic trepostome bryozoans, is invalid because GRAY in 1848 published the same name for Cretaceous-to-Recent cheilostome bryozoans, and BASSLER in 1911 introduced the new name *Hallopora* to replace HALL's homonym. The *Treatise* style of entry is:

Hallopora BASSLER, 1911 [*pro Callopora* HALL, 1851 (*non* GRAY, 1848)].

In like manner, a needed replacement generic name may be introduced in the *Treatise* (even though first publication of generic names otherwise in this work is avoided). The requirement that an exact bibliographic reference must be given for the replaced name commonly can be met in the *Treatise* by citing a publication recorded in the list of references, using its assigned index number, as shown in the following example.

Mysterium DELAUBENFELS, *nom. subst.* [*pro Mystrium* SCHRAMMEN, 1936 (ref. 40, p. 60) (*non* ROGER, 1862)] [**Mystrium porosum* SCHRAMMEN, 1936].

For some replaced homonyms, a footnote reference to the literature is necessary. A senior homonym is valid, and in so far as

the *Treatise* is concerned, such names are handled according to whether the junior homonym belongs to the same major taxonomic division (class or phylum) as the senior homonym or to some other; in the former instance, the author and date of the junior homonym are cited as:

Diplophyllum HALL, 1851 [*non* SOSHKINA, 1939] [**D. caespitosum*].

Otherwise, no mention of the existence of a junior homonym is made.

CITATION OF TYPE SPECIES

The name of the type species of each genus and subgenus is given next following the generic name with its accompanying author and date, or after entries needed for definition of the name if it is involved in homonymy. The originally published combination of generic and trivial names for this species is cited, accompanied by an asterisk (*), with notation of the author and date of original publication. An exception in this procedure is made, however, if the species was first published in the same paper and by the same author as that containing definition of the genus which it serves as type; in such case, the initial letter of the generic name followed by the trivial name is given without repeating the name of the author and date, for this saves needed space. Examples of these 2 sorts of citations are as follows:

Diplotrypa NICHOLSON, 1879 [**Favosites petropolitanus* PANDER, 1830].

Chainodictyon FOERSTE, 1887 [**C. laxum*].

If the cited type species is a junior synonym of some other species, the name of this latter also is given, as follows:

Acervularia SCHWEIGGER, 1819 [**A. baltica* (=**Madrepora ananas* LINNÉ, 1758)].

It is judged desirable to record the manner of establishing the type species, whether by original designation or by subsequent designation, but various modes of original designation are not distinguished.

Original designation of type species. The Rules provide that the type species of a genus or subgenus may be recognized as an original designation if only a single species was assigned to the genus at the time of first publication (monotypy), if the author of a generic name employed this same name for one of the included species (tautonymy), if one of the species was named *"typus," "typicus,"* or the like, if the original author explicitly indicated the species chosen as the type, or if some other stipulations were met. According to convention adopted in the *Treatise,* the absence of any indication as to manner of fixing the type species is to be understood as signifying that it is established by original designation, the particular mode of original designation not being specified.

Subsequent designation of type species; use of "SD" and "SM." The type species of many genera are not determinable from the publication in which the generic name was introduced and therefore such genera can acquire a type species only by some manner of subsequent designation. Most commonly this is established by publishing a statement naming as type species one of the species originally included in the genus, and in the *Treatise* fixation of the type species in this manner is indicated by the letters "SD" accompanied by the name of the subsequent author (who may be the same person as the original author) and the date of publishing the subsequent designation. Some genera, as first described and named, included no mentioned species and these necessarily lack a type species until a date subsequent to that of the original publication when one or more species are assigned to such a genus. If only a single species is thus assigned, it automatically becomes the type species and in the *Treatise* this subsequent monotypy is indicated by the letters "SM." Of course, the first publication containing assignment of species to the genus which originally lacked any included species is the one concerned in fixation of the type species, and if this named 2 or more species as belonging to the genus but did not designate a type species, then a later "SD" designation is necessary. Examples of the use of "SD" and "SM" as employed in the *Treatise* follow.

Hexagonaria Gürich, 1896 [*Cyathophyllum hexagonum* Goldfuss, 1826; SD Lang, Smith & Thomas, 1940].

Muriceides Studer, 1887 [*M. fragilis* Wright & Studer, 1889; SM Wright & Studer, 1889].

SYNONYMS

Citation of synonyms is given next following record of the type species and if 2

or more synonyms of differing date are recognized, these are arranged in chronological order. Objective synonyms are indicated by accompanying designation "(obj.)," others being understood to constitute subjective synonyms. Examples showing *Treatise* style in listing synonyms follow.

Calapoecia Billings, 1865 [*C. anticostiensis;* SD Lindström, 1833] [=*Columnopora* Nicholson, 1874; *Houghtonia* Rominger, 1876].

Staurocyclia Haeckel, 1882 [*S. cruciata* Haeckel, 1887] [=*Coccostaurus* Haeckel, 1882 (obj.); *Phacostaurus* Haeckel, 1887 (obj.)].

A synonym which also constitutes a homonym is recorded as follows:

Lyopora Nicholson & Etheridge, 1878 [*Palaeopora? favosa* M'Coy, 1850] [=*Liopora* Lang, Smith & Thomas, 1940 (*non* Girty, 1915)].

Some junior synonyms of either objective or subjective sort may take precedence desirably over senior synonyms wherever uniformity and continuity of nomenclature are served by retaining a widely used but technically rejectable name for a generic assemblage. This requires action of ICZN using its plenary powers to set aside the unwanted name and validate the wanted one, with placement of the concerned names on appropriate official lists. In the *Treatise* citation of such a conserved generic name is given in the manner shown by the following example.

Tetragraptus Salter, 1863 [*nom. correct.* Hall, 1865 (*pro Tetragrapsus* Salter, 1863), *nom. conserv.* proposed Bulman, 1955, ICZN pend.] [*Fucoides serra* Brongniart, 1828 (=*Graptolithus bryonoides* Hall, 1858].

ABBREVIATIONS

A few author's names and most stratigraphic and geographic names are abbreviated in order to save space. General principles for guidance in determining what names should be abbreviated are frequency of repetition, length of name, and avoidance of ambiguity. Abbreviations used in this division of the *Treatise* are explained in the following alphabetically arranged list.

xvii

Abh., Abhandlungen
Abt., Abteilung, -en
Acad., Academia, Académie, Academy
aff., affinis
Afr., Africa, -an
Akad., Akademie
Ala., Alabama
Alb., Albian
Alba., Alberta
Alg., Algeria
Am., America, -n
Ann., Analen; Annal, -s; Annual
ant., anterior
Antarct., Antarctic
Apt., Aptian
Arbeit., Arbeiten
Arch., Archiv
Arct., Arctic
Arenig., Arenigian
Ark., Arkansas
Art., Article
Ashgill., Ashgillian
AsiaM., Asia Minor
Assoc., Association
Atl., Atlantic
Aus., Austria
Austral., Australia

Baikal., Baikalia
Barrem., Barremian
Barton, Bartonian
Bathon., Bathonian
Beil., Beilage
Belg., Belgique, Belgium
Bidr., Bidrag
Biol., Biology
Boh., Bohemia
Bot., Botany
Braz., Brazil
Br.I., British Isles
Brit., Britain, British
Bull., Bulletin
Burdigal., Burdigalian

C., Centigrade, Central
Calif., California
Callov., Callovian
Cam., Cambrian
Campan., Campanian
Can., Canada
Caradoc., Caradocian
Carb., Carboniferous
Carib., Caribbean
Cenom., Cenomanian
Centralbl., Centralblatt
cf., confero (compare)
Chatt., Chattian
Chazy., Chazyan
Clinton., Clintonian
cm., centimeter
Co., Company
Colo., Colorado
Comp., Comparative
Coniac., Coniacian
Conserv., Conservation
Contr., Contribution, -s

cosmop., cosmopolitan
Cret., Cretaceous
C.Z., Canal Zone
Czech., Czechoslovakia
Dept., Department
deutsch., deutschen
Dev., Devonian
Distr., District
Dol., Dolomite
dors., dorsal
E., East
ed., edition, editor
Eden., Edenian
e.g., exempli gratia (for example)
Ellesm., Ellesmereland
Eng., England
Eoc., Eocene
Est., Estonia
et al., et alii
Eu., Europe
ext., exterior
f., för, für, från
F., Formation
Fac., Facultad, Faculté, Faculty
fig., figure, -s
Fla., Florida
Förhandl., Förhandlingar, er
Fr., France, Francais, -e, French
G.Brit., Great Britain
Genoot., Genootschap
Geol., Geologiá, Geological, Geológico, Geologie, Geologisch, Geologiska, Geology
Géol., Géologie, Géologique
Ger., Germany
Gesell., Gesellschaft
Givet., Givetian
Gotl., Gotland
Gr., Group
Greenl., Greenland
Hauteriv., Hauterivian
Hemis., Hemisphere
Holl., Holland
Hung., Hungarica, Hungary
I.(Is.), Island, -s
ICZN, International Commission on Zoological Nomenclature
i.e., id est (that is)
Ill., Illinois
illus., illustration, -s
incl., including, inclusive
Ind., Indiana
Ind.O., Indian Ocean
Inst., Institut, Institute, Institutet, Institution, Instituto, Instituut
int., interior
Ire., Ireland
Jahrb., Jahrbuch
Jahrg., Jahrgang
Jour., Journal
Jur., Jurassic
juv., juvenile
Kans., Kansas

Kimm., Kimmeridgian
Kl., Klasse
Kolon., Kolonien
kon., koninklijk
kön., königlich
Ky., Kentucky
L., Lower, Land
La., Louisiana
lat., lateral
Led., Ledian
Lief., Lieferung, -en
Linn., Linnean, Linnéenne
Llandeil., Llandeilian
Llandov., Lladoverian
Llanvirn., Llanvirnian
long., longitudinal
Ls., Limestone
Ludov., Ludlovian
Lutet., Lutetian
m., meter
M., Middle
Maastricht., Maastrichtian
Mag., Magazine
Math., Mathematische
max., maximum
Md., Maryland
Me., Maine
Medit., Mediterranean
Meeresuntersuch., Meeresuntersuchungen
Mem., Memoir, -s, Memoria
Mém., Mémoire, -s
Mich., Michigan
Mijnb., Mijnbouw-kundig
Min., Mineralogie
Minn., Minnesota
Mio., Miocene
Miss., Mississippi, Mississippian
Mitteil., Mitteilungen
ml., milliliter
mm., millimeter
Mo., Missouri
Mon., Monograph
Mus., Museum
n, new
N., North
N.Am., North America
Nat., Natural; Naturale, -s; Naturali; Naturelle, -s
Naturf., Naturforschende
Naturv., Naturvetenskap, Naturvidenskapelig
Natuurwetensch., Natuurwetenschappen
N.B., New Brunswick
NC., North Central
N.Car., North Carolina
NE., Northeast
Neb., Nebraska
Ned., Nederland
Neocom., Neocomian
Neog., Neogene
Neth., Netherlands
N.J., New Jersey
no., number, -s; numéro, -s; número, -s

nom. correct., nomen correctum
nom. nov., nomen novum
nom. null., nomen nullum
nom. subst., nomen substitutum
nom. transl., nomen translatum
nom. van., nomen vanum
Nouv., Nouveau, Novelle
Nov., Novitates
NW., Northwest
N.Y., New York
N.Z., New Zealand
O., Ocean
obj., objective
Occas., Occasional
Okla., Oklahoma
Oligo., Oligocene
Onond., Onondagan
Ont., Ontario
Ord., Ordovician
p., page, -s
Pa., Pennsylvania
Pac., Pacific
Paläont., Paläontologie,
  Paläontologisch
Paleoc., Paleocene
Paleog., Paleogene
Paleont., Paleontologia, Paleon-
  tologica, Paleontological, Pale-
  ontologiese, Paleontology
Paleontgr., Paleontographica
Para., Paraguay
pend., pending
Penn., Pennsylvanian
Perm., Permian
Philos., Philosophical
pl., plural; plate, -s
Pleist., Pleistocene
Plio., Pliocene
Pol., Poland
post., posterior
Preuss., Preussische
Proc., Proceedings
Prof., Professional
Prov., Province
pt., part, -s
Pub., Publication

Quat., Quaternary
Rec., Recent
reconstr., reconstructed, -ion
Rept., Report, -s
Res., Research
Rev., Review, Revue
R.I., Rhode Island
Richmond., Richmondian
Roy., Royal, -e
Rupel., Rupelian
Russ., Russia
S., South, Sea
S.Am., South America
Santon., Santonian
Sarmat., Sarmatian
Sav., Savants
SC., South Central
Scand., Scandinavia
Schweiz., Schweizerische
Sci., Science, -s; Scientifique
Scot., Scotland
SD, subsequent designation
S.Dak., South Dakota
SE., Southeast
sec., section, -s
Senckenberg., Senckenbergischen
Senon., Senonian
ser., series, serial
Sh., Shale
Sib., Siberia
Sil., Silurian
Skr., Skrifter
Soc., Sociedad, Societâ, Société,
  Society
sp., species
Sp., Spain
Spec., Special
s.s., sensu stricto
Stud., Studies
subfam., subfamily
suppl., supplement
Surv., Survey
SW., Southwest
Swed., Sweden, Swedish
Switz., Switzerland
Tatar., Tatarian

Tenn., Tennessee
Tert., Tertiary
Tex., Texas
th., thoracic
Tidsskr., Tidsskrift
Tongr., Tongrian
Trans., Transactions
transl., translated, translation
transv., transverse
Tremadoc., Tremadocian
Trenton., Trentonian
Trias., Triassic
Turon., Turonian
Tyrrhen., Tyrrhenian
u., und
U., Upper
Univ., Universidad, Università,
  Université, Universitets, Uni-
  versity
USA, United States (America)
USSR, Union of Soviet Socialist
  Republics
v., volume, -s
Va., Virginia
var., variety
Venez., Venezuela
vent., ventral
Verh., Verhandlung, -en, Ver-
  handelingen
Verslag., Verslagen
Vidensk., Videnskab, Viden-
  skaberne
Visé., Viséan
Volg., Volgian
W., West
Wenlock., Wenlockian
Wiss., Wissenschaft, -en;
  Wissenschaftliche, -en
W.Va., West Virginia
Wyo., Wyoming
Ypres., Ypresian
Yugosl., Yugoslavia
Zeitschr., Zeitschrift
Zool., Zoologi, Zoologia, Zool-
  ogical, Zoologie, Zoologisch,
  Zoologiska, Zoology

REFERENCES TO LITERATURE

Each part of the *Treatise* is accompanied by a selected list of references to paleontological literature consisting primarily of recent and comprehensive monographs available but also including some older works recognized as outstanding in importance. The purpose of giving these references is to aid users of the *Treatise* in finding detailed descriptions and illustrations of morphological features of fossil groups, discussions of classifications and distribution, and especially citations of more or less voluminous literature. Generally speaking, publications listed in the *Treatise* are not original sources of inform- ation concerning taxonomic units of various rank but they tell the student where he may find them; otherwise it is necessary to turn to such aids as the *Zoological Record* or NEAVE's *Nomenclator Zoologicus*. References given in the *Treatise* are arranged alphabetically by authors and accompanied by index numbers which serve the purpose of permitting citation most concisely in various parts of the text; these citations of listed papers are inclosed invariably in parentheses and are distinguishable from dates because the index numbers comprise no more than 3 digits. Ordinarily, index numbers for literature references are given at the end of generic or family diagnoses.

SOURCES OF ILLUSTRATIONS

At the end of figure captions an index number is given to supply record of the author of illustrations used in the *Treatise,* reference being made to an alphabetically arranged list of authors' names which follows. The names of authors, but generally not individual publications, are cited. Previously unpublished illustrations are marked by the letter "n" (signifying "new") with the name of the author.

STRATIGRAPHIC DIVISIONS

Classification of rocks forming the geologic column as commonly cited in the *Treatise* in terms of units defined by concepts of time is reasonably uniform and firm throughout most of the world as regards major divisions (e.g., series, systems, and rocks representing eras) but it is variable and unfirm as regards smaller divisions (e.g., substages, stages, and subseries), which are provincial in application. Users of the *Treatise* have suggested the desirability of publishing reference lists showing the stratigraphic arrangement of at least the most commonly cited divisions. Accordingly, a tabulation of European and North American units, which broadly is applicable also to other continents, is given here.

Generally Recognized Divisions of Geologic Column

EUROPE	NORTH AMERICA
ROCKS OF CENOZOIC ERA	**ROCKS OF CENOZOIC ERA**
NEOGENE SYSTEM[1]	**NEOGENE SYSTEM**[1]
Pleistocene Series (including Recent)	Pleistocene Series (including Recent)
Pliocene Series	Pliocene Series
Miocene Series	Miocene Series
PALEOGENE SYSTEM	**PALEOGENE SYSTEM**
Oligocene Series	Oligocene Series
Eocene Series	Eocene Series
Paleocene Series	Paleocene Series
ROCKS OF MESOZOIC ERA	**ROCKS OF MESOZOIC ERA**
CRETACEOUS SYSTEM	**CRETACEOUS SYSTEM**
Upper Cretaceous Series	**Gulfian Series (Upper Cretaceous)**
Maastrichtian Stage[2]	Navarroan Stage
Campanian Stage[2]	Tayloran Stage
Santonian Stage[2]	Austinian Stage
Coniacian Stage[2]	
Turonian Stage	
Cenomanian Stage	Woodbinian (Tuscaloosan) Stage
	Comanchean Series (Lower Cretaceous)
	Washitan Stage
Lower Cretaceous Series	
Albian Stage	Fredericksburgian Stage
	Trinitian Stage
Aptian Stage	
	Coahuilan Series (Lower Cretaceous)

Barremian Stage[3]	Nuevoleonian Stage
Hauterivian Stage[3]	
Valanginian Stage[3]	Durangoan Stage
Berriasian Stage[3]	

JURASSIC SYSTEM	JURASSIC SYSTEM
Upper Jurassic Series	**Upper Jurassic Series**
Portlandian Stage[4]	Portlandian Stage
Kimmeridgian Stage	Kimmeridgian Stage
Oxfordian Stage	Oxfordian Stage
Middle Jurassic Series	**Middle Jurassic Series**
Callovian Stage	Callovian Stage
Bathonian Stage	Bathonian Stage
Bajocian Stage	Bajocian Stage
Lower Jurassic Series (Liassic)	**Lower Jurassic Series (Liassic)**
Toarcian Stage	Toarcian Stage
Pliensbachian Stage	Pliensbachian Stage
Sinemurian Stage	Sinemurian Stage
Hettangian Stage	Hettangian Stage

TRIASSIC SYSTEM	TRIASSIC SYSTEM
Upper Triassic Series	**Upper Triassic Series**
Rhaetian Stage[5]	(Not recognized)
Norian Stage	Norian Stage
Carnian Stage	Carnian Stage
Middle Triassic Series	**Middle Triassic Series**
Ladinian Stage	Ladinian Stage
Anisian Stage (Virglorian)	Anisian Stage
Lower Triassic Series	**Lower Triassic Series**
Scythian Series (Werfenian)	Scythian Stage

ROCKS OF PALEOZOIC ERA	ROCKS OF PALEOZOIC ERA
PERMIAN SYSTEM	PERMIAN SYSTEM
Upper Permian Series	**Upper Permian Series**
Tartarian Stage[6]	Ochoan Stage
Middle Permian Series	**Middle Permian Series**
Kazanian Stage [7]	Guadalupian Stage
Kungurian Stage	
Artinskian Stage[8]	Leonardian Stage
Lower Permian Series	**Lower Permian Series**
Sakmarian Stage	Wolfcampian Stage

CARBONIFEROUS SYSTEM	PENNSYLVANIAN SYSTEM
Upper Carboniferous Series	**Kawvian Series (Upper Pennsylvanian)**
Stephanian Stage	Virgilian Stage
	Missourian Stage
	Oklan Series (Middle Pennsylvanian)
	Desmoinesian Stage
	Bendian Stage
	Ardian Series (Lower Pennsylvanian)
Westphalian Stage	Morrowan Stage
	MISSISSIPPIAN SYSTEM
	Tennesseean Series (Upper Mississippian)
Namurian Stage	

Lower Carboniferous Series
Viséan Stage

Tournaisian Stage
Strunian Stage

DEVONIAN SYSTEM

Upper Devonian Series
Famennian Stage

Frasnian Stage

Middle Devonian Series
Givetian Stage

Couvinian Stage

Lower Devonian Series
Coblenzian Stage

Gedinnian Stage

SILURIAN SYSTEM

Upper Silurian Series
Ludlovian Stage

Middle Silurian Series
Wenlockian Stage

Llandoverian Stage (upper part)
Lower Silurian Series
Llandoverian Stage (lower part)

ORDOVICIAN SYSTEM

Upper Ordovician Series
Ashgillian Stage
Caradocian Stage (upper part)

Middle Ordovician
Caradocian Stage (lower part)

Llandeilian Stage
Llanvirnian Stage

Lower Ordovician Series
Arenigian Stage
Tremadocian Stage[9]

CAMBRIAN SYSTEM
Upper Cambrian Series

Chesteran Stage

Meramecian Stage
Waverlyan Series (Lower Mississippian)
Osagian Stage
Kinderhookian Stage

DEVONIAN SYSTEM
Chautauquan Series (Upper Devonian)
Conewangoan Stage
Cassadagan Stage
Senecan Series (Upper Devonian)
Chemungian Stage
Fingerlakesian Stage

Erian Series (Middle Devonian)
Taghanican Stage
Tioughniogan Stage
Cazenovian Stage

Ulsterian Series (Lower Devonian)
Onesquethawan Stage
Deerparkian Stage
Helderbergian Stage

SILURIAN SYSTEM
Cayugan Series (Upper Silurian)
Keyseran Stage
Tonolowayan Stage
Salinan Stage

Niagaran Series (Middle Silurian)
Lockportian Stage
Cliftonian Stage
Clintonian Stage

Medinan Series (Lower Silurian)
Alexandrian Stage

ORDOVICIAN SYSTEM
Cincinnatian Series (Upper Ordovician)
Richmondian Stage
Maysvillian Stage
Edenian Stage
Champlainian Series (Middle Ordovician)
Mohawkian Stage
Trentonian Substage
Blackriveran Substage
Chazyan Stage

Canadian Series (Lower Ordovician)

CAMBRIAN SYSTEM
Croixian Series (Upper Cambrian)
Trempealeauan Stage
Franconian Stage

Middle Cambrian Series
Lower Cambrian Series

EOCAMBRIAN SYSTEM
ROCKS OF PRECAMBRIAN AGE

Dresbachian Stage
Albertan Series (Middle Cambrian)
Waucoban Series (Lower Cambrian)

EOCAMBRIAN SYSTEM
ROCKS OF PRECAMBRIAN AGE

RAYMOND C. MOORE

---

[1] Considered by some to exclude post-Pliocene deposits.
[2] Classed as division of Senonian Subseries.
[3] Classed as division of Neocomian Subseries.
[4] Includes Purbeckian deposits.
[5] Interpreted as lowermost Jurassic in some areas.

[6] Includes some Lower Triassic and equivalent to upper Thuringian (Zechstein) deposits.
[7] Equivalent to lower Thuringian (Zechstein) deposits.
[8] Equivalent to upper Autunian and part of Rotliegend deposits.
[9] Classed as uppermost Cambrian by some geologists.

PART Q

ARTHROPODA 3

―――――

CRUSTACEA
OSTRACODA

By[1] R. H. Benson, J. M. Berdan, W. A. van den Bold, Tetsuro Hanai, Ivar Hessland, H. V. Howe, R. V. Kesling, S. A. Levinson, R. A. Reyment, R. C. Moore, H. W. Scott, R. H. Shaver, I. G. Sohn, L. E. Stover, F. M. Swain, P. C. Sylvester-Bradley, and John Wainwright

―――――

CONTENTS

――――――

[1] Sections of this volume prepared in collaboration by two or more contributors normally have indicated authorship with names arranged alphabetically, no seniority being indicated.

INTRODUCTION

By R. C. Moore, H. W. Scott, and P. C. Sylvester-Bradley

[University of Kansas, University of Illinois, and University of Leicester]

The Ostracoda are small crustaceans living in marine, brackish and fresh water. They are characterized by having a bivalved shell hinged along the dorsal margin. Most species are of microscopic size (0.4 to 1.5 mm.), though some fresh-water forms are rather larger (up to 5 mm.), and macroscopic, free-swimming, marine forms (up to 30 mm.) are known, both fossil and Recent. The shell in most species is calcareous, and may be smooth or highly ornamented, but it does not possess growth lines (except rarely when early molts are retained outside later ones), as ecdysis is complete at each instar.

In the existing oceans, ostracodes[1] live from shore line out to depths of about 1,500 fathoms (2,800 m.) or more. They are also found in most nonmarine aquatic habitats, and one terrestrial species has been recorded, living in association with myriapods and isopods in the damp leaf mold of a tropical forest. Ostracodes are also known parasitic or commensal on other Crustacea, on Amphibia, and on fish. Most aquatic Ostracoda are benthonic in habitat, though many belonging to the order Myodocopida are free-swimming during at least part of the life cycle, as are several of the fresh-water Podocopida.

As fossils, they are abundantly represented in limestones, shales, and marls from Cambrian times onward and are in many places so abundant that they form rock coquinas. In the Upper Jurassic of England such a rock (Cypris Freestones, Lower Purbeckian) has been used as a building stone.

The first ostracodes to be named were described by the great Danish naturalist, O. F. Müller, in 1776. Fossil representatives of the subclass were soon discovered, and by 1850 E. Forbes had already zoned the British Purbeck beds by means of Ostracoda. In 1866 G. O. Sars proposed subdivision of the Ostracoda into 4 groups classified as suborders (Myodocopa, Podocopa, Platycopa, Cladocopa) on the basis of their appendages. This classification has stood the test of time and with minor changes is now universally adopted for Recent ostracodes. Certain Paleozoic groups do not fit the scheme well, however, and in 1953 a separate division (Paleocopa) was proposed by Henningsmoen for their reception. Additional taxa are proposed herein, (Archaeocopida) for Cambrian forms with a flexible, partly calcified shell, thought to be ancestral to the other Ostracoda, and (Leperditicopida) for the distinctive, mostly large, thick-shelled Ordovician-to-Devonian forms included in the Leperditiidae.

The foremost student of fossil Ostracoda during the latter half of the 19th century was the Englishman, T. Rupert Jones. Thirty years after his death, in the 1920's, the greatly increased interest in micropaleontology that resulted from recognition of the value of fossil Ostracoda to explorations for oil led to a sudden revival of work on this group. Ulrich & Bassler proposed a new classification of Paleozoic ostracodes in 1923, and since then an ever increasing volume of papers has testified to the importance of Ostracoda as indices of stratigraphic horizons. In 1933 C. I. Alexander published an important paper on the finer shell structure of some post-Paleozoic ostracodes, and his work stimulated active research in details of shell morphology which previously had been overlooked. This has provided a firmer basis for taxonomic discrimination and has led to a great increase in the number of generic names proposed. The Ostracoda are now regarded as microfossils second only in importance to the Foraminifera as stratigraphic markers. At certain levels they have the advantage over Foraminifera of being more abundant and occurring in many environments of brackish- or fresh-water facies which are closed to Foraminifera.

Classification of the Arthropoda adopted in the *Treatise* is outlined in tabular form by Størmer (*Treatise*, Part O, p. O15-O16, 1959). This recognizes division of the true arthropods (Euarthropoda) into four main

---

[1] Although the spelling "ostracod" is employed by most British and some American writers, the *Treatise* adopts "ostracode" because this word is derived from the Greek ὀστρακῴδης (ostracōdes); Webster's New International Dictionary recognizes both spellings.—Editor.

groups (Trilobitomorpha, Chelicerata, Pycnogonida, and Mandibulata), which are ranked as subphyla. The Mandibulata comprises the classes Crustacea, Myriapoda, and Hexapoda. The subclass Ostracoda, described and illustrated in this volume, is not the most primitive or generalized major group of crustaceans but the most abundantly represented among fossils. It is convenient, therefore, to segregate the ostracodes in a separate volume, other crustacean assemblages and the remainder of mandibulates being assigned to *Treatise* Part R. Diagnosis of the subclass Ostracoda is given on page Q100.

Subphylum MANDIBULATA

Clairville, 1798

[*Emend.* from original application as major division of Insecta]

Euarthropods having mouthparts known as mandibles and 2 pairs of accessory feeding appendages called maxillae. Mandibles commonly modified to perform various feeding functions such as cutting, piercing, or sucking, but always present in some form during some stage of life. Either pair of maxillae sometimes absent or vestigial. *Cam.-Rec.*

Class CRUSTACEA Pennant, 1777

Highly diverse, mostly aquatic mandibulates bearing 2 pairs of antennae. Body usually consisting of 3 main regions, head (or cephalon), thorax, and abdomen; head and thorax commonly fused forming cephalothorax. Body generally covered with hard chitinous carapace impregnated with calcium salts. Respiration by means of gills. Nauplius larva characteristic. *Cam.-Rec.*

ACKNOWLEDGMENTS

Special thanks are expressed to J. R. Cornell and J. C. Kraft for permission to use photographs of ostracode genera made by them in connection with doctoral dissertations submitted to the University of Minnesota. Also, assistance in the loaning of type specimens has been given by C. W. Collinson, Illinois Geological Survey; G. A. Cooper, U. S. National Museum; John Imbrie, Columbia University; M. B. Marple, Ohio State University; and N. D. Newell, American Museum of Natural History. This aid is gratefully acknowledged.

MORPHOLOGY OF LIVING OSTRACODA

By H. V. Howe, R. V. Kesling, and H. W. Scott

[Louisiana State University, University of Michigan, and University of Illinois]

BODY SEGMENTATION AND APPENDAGES

The body of an ostracode is short, laterally compressed, and inclosed within a bivalved, calcareous carapace. The body shows no trace of segmentation, the boundary between head and thorax being represented merely by a slight constriction, but existence of ancestral segmentation is indicated by the nature and distribution of the appendages. A layer of soft tissue, the **epidermis** or **hypodermis,** hangs down on each side and secretes the shell.

The cephalic region of an ostracode is formed by a network of strong chitinous rods, which are connected by thin membranes of chitin. This framework includes the forehead and upper lip, fused together as a single structural unit. Attached to the framework are four pairs of appendages named (in order backward) **antennules, antennae, mandibles,** and **maxillae** (Fig. 1). The posteroventral edge of the upper lip forms the anterior margin of the mouth. This rim is roughened or serrate but not toothed. The **hypostome** is the lower lip, which is located on the ventral side of the body, forming the posterior part of the mouth. It is a somewhat canoe-shaped movable structure with the anterior, open end forming the mouth rim. The maxillae lie parallel to its sides and the first thoracic legs are attached to its posterior points. Many species have two paired structures at the front of the hypostome, chitinous **rake-shaped organs** embedded in the tissue and soft setiferous **paragnaths** at the sides. Some ostracodes have a sensory structure (**frontal organ**) attached to the forehead.

The thoracic region contains various organs of the digestive and reproductive sys-

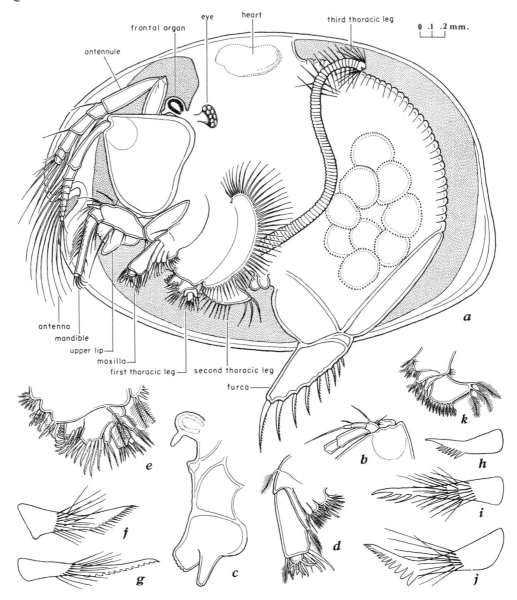

Fɪɢ. 1. Morphology of a representative myodocopid (myodocopine) ostracode, *Cypridina norvegica* Bᴀɪʀᴅ, Recent (Cypridinidae). *a*. Female with LV removed; eggs shown in one uterus at rear of body; genital lobes below base of 3rd thoracic leg.——*b*. Right antenna; inner face of endopodite and 1st podomere of exopodite.——*c*. Frontal organ and upper lip.——*d*. Left maxilla.——*e*. Left 1st thoracic leg; details of protopodite and endopodite.——*f-j*. Setae from protopodite of 1st thoracic leg.——*k*. Left 2nd thoracic leg. (Figs. *a, b, c,* and *k* to scale in upper right corner; from Kᴇsʟɪɴɢ, after Sᴀʀs and Sᴋᴏɢsʙᴇʀɢ.)

tems, and bears one to three pairs of **thoracic legs.** An abdomen is wanting. In the great majority of ostracodes, a pair of **furcal rami** forms the posterior end of the body.

All arthropod appendages are composed of a number of segments called **podomeres.** The typical crustacean appendage is biramous, consisting of a single basal branch, **(protopod)** composed of two podomeres **(coxa, basis).** The basis bears two branches, an inner called the **endopod** and an outer called the **exopod.** Very few appendages of ostracodes have exactly this arrangement. Some have the coxa and basis fused to form a single podomere; some have an extra podomere **(precoxa)** in the protopod; some lack an exopod; some, in addition to both endopod and exopod, have excess lobes such as the **epipodial plate** on the outer surface or various **endites** directed inward from the protopod.

The **antennules** (or "first antennae" of many European workers) are uniramous, the exopod being lost; they are attached to the forehead. Typically each antennule is composed of eight podomeres, but through fusion a condition may be approached in which there are only five. The protopod, composed of one or two podomeres, is much larger than the remainder, and houses a nervous ganglion. The antennules of some species have a locomotor function (swimming, climbing, or digging), or they may be sensory, or serve as balancing organs. In the Myodocopida they are sexually dimorphic and in some species are used in copulation, the distal setae of the male being equipped with suctorial structures (Fig. 2*b,c*); in the Halocyprididae (Fig. 3*b*) most of the setae of the male are longer than those of the female, and one is provided with a special sensory organ. In the Cladocopina they are used in swimming, being equipped with long natatory setae, which originate only on the small distal podomere (Fig. 4*a,c*). In the Platycopina the setae are strong and clawlike (Fig. 5*a*). In the Podocopina the antennules may bear long, feathered setae (Fig. 6) used for swimming or balancing, or they have clawlike spines that serve for digging or climbing (Fig. 7*a*).

The **antennae** ("second antennae" of some authors) are locomotor organs used for swimming, walking, or climbing. They are biramous, although the exopod is much reduced in some forms. They are attached to the sides of the head near the junction of the forehead and upper lip, from which they curve forward and downward. The protopod is large, strong and movable, the two podomeres being fused or separate. In the Myodocopida (Figs. 1*a*, 3*a*) the exopod is long and bears long natatory setae, but the endopod is shorter (Fig. 1*b*) and in many species dimorphic and developed as a clasping organ in the male (Fig. 3*b*). In the Cladocopina (Fig. 4*d*) both exopod and endopod bear long, stiff setae to aid swimming. In both Myodocopida and Cladocopina the protopod is long, strong and undivided, but in the Platycopina the coxa forms a knee with the basis, and both endopod and exopod are well developed, broad, flattened, and equipped with stiff setae to aid in walking (Fig. 5*a*). In the Podocopina the exopod is much reduced in most forms (Figs. 6*a,b*; 8*a*); in the Cytheracea it is devoped as a long, curved, hollow seta (Fig. 7*a*) serving as the duct for a powerful gland, and in some forms is dimorphic. The endopod in the Podocopina is leglike with four podomeres at most. Claws or natatory setae on the endopod are quite variously developed. Some are dimorphic, some sensory ("sense clubs"). Claws are developed on the terminal podomere.

The **mandibles** are situated at the sides of the mouth, and are very similar in all ostracodes. Each mandible consists of protopod, endopod, and exopod. The long coxa of the protopod is equipped at the ventral end with teeth which are used in mastication; those of the left and right mandibles meet in the center of the mouth. The basis and endopod together form the palp, which curves forward and downward; it is equipped with setae used for crawling and digging or for holding food fast and cutting off pieces (Fig. 1*a*,4*e*). In the Halocyprididae the proximoventral part of the basis is extended (Figs. 3*a,b*) and armed with cutting teeth. In the Platycopina the basis is long and provided with a comb of numerous long setae (Fig. 5*b*); a somewhat similar structure is found in the Darwinulidae of the Podocopina (Fig. 7*c*). In all forms the exopod is small and delicate, bearing a number of setae; in some species it functions as a **branchial plate,** accessory to that developed on the maxillae.

The **maxillae** work as supporting organs of the mandibles. They aid in carrying food to the mouth, in removing undesirable particles, and in creating water currents used in respiration and carrying in food particles in suspension. The maxillae lie at the sides of the hypostome, or, if a hypostome is ab-

sent, at the posteroventral sides of the head. The maxillae have a varied development in different groups of Ostracoda and the variations are regarded as important aids to taxonomy. The protopod in many species of all orders is equipped with two or three proximal endites bearing setae (Figs. 1*d;* 5*c;* 6*a,b;* 7*a*); in the Podocopina these endites are sometimes referred to as **masticatory processes.** In many forms the exopod is a

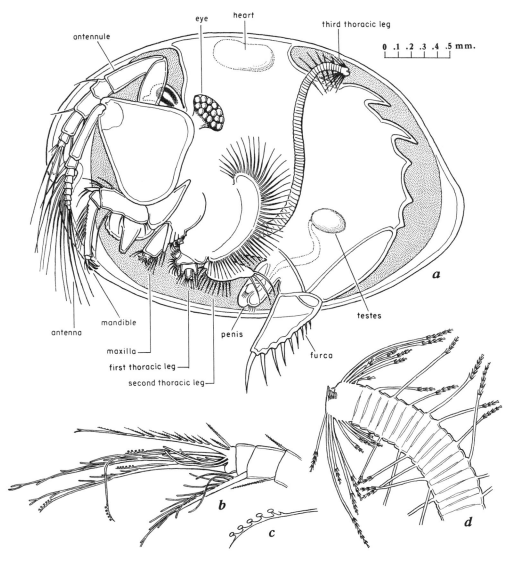

Fig. 2. Morphology of a representative myodocopid (myodocopine ostracode, *Cypridina norvegica* Baird, Recent (Cypridinidae). *a.* Male with LV removed; scale in upper right corner.——*b.* Distal end of male antennule.——*c.* Part of seta from male antennule showing suctorial structures.——*d.* Distal end of right 3rd thoracic leg. (From Kesling, after Sars and Skogsberg.)

large vibratory plate equipped with a comb of long setae, commonly feathered (Fig. 5c; 6a; 7a; 8d). Some Myodocopida, however, lack an exopod or this element is much reduced (Fig. 1d), and in the Cladocopina it forms a palp of one or two podomeres with several distal setae directed medially (Fig. 4a,b). The endopod in most species forms a strong palp equipped with distal setae (Figs. 1d; 3a,b), but in the Podocopina it may be lacking. In the Platycopina the first

podomere bears a comb of numerous long setae (Fig. 5c).

Three pairs of thoracic appendages occur in the Myodocopida and Podocopina, two pairs in the Platycopina, and only a single pair in the Cladocopina.

The **first thoracic legs** are attached to the body at the junction of head and thorax. In some families they are highly modified as accessories to the jaw apparatus, consequently being referred to by some authors

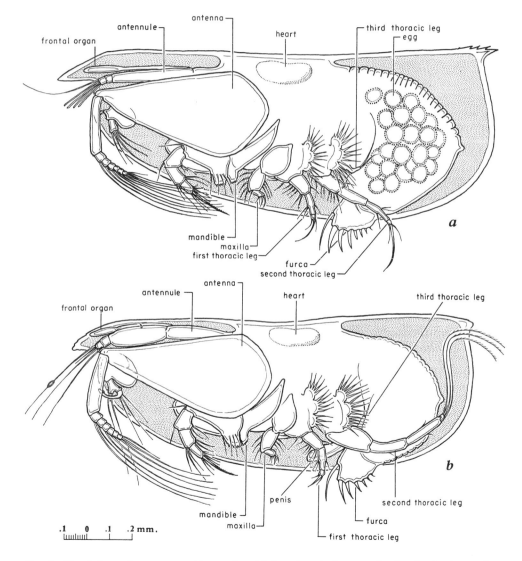

Fig. 3. Morphology of a representative myodocopid (myodocopine) ostracode, *Conchoecia elegans* Sars, Recent (Halocyprididae). *a.* Female with LV removed.——*b.* Male with LV removed. (Scale of both figures on lower left corner; from Kesling, after Sars.)

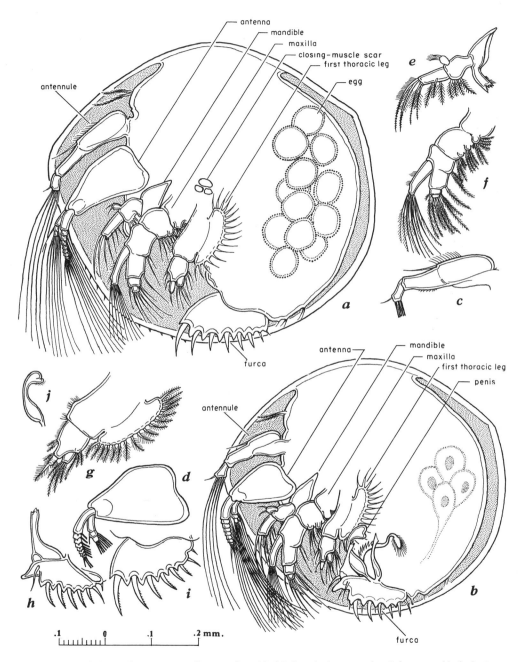

Fig. 4. Morphology of a representative myodocopid (cladocopine) ostracode, *Polycope orbicularis* Sars, Recent (Polycopidae). *a*. Female with LV removed.——*b*. Male with LV removed; testes shown in rear part of body.——*c*. Left antennule with distal setae broken.——*d*. Left antenna with setae broken on both exopodite (left) and endopodite (right).——*e*. Left mandible.——*f*. Left maxilla.——*g*. Left 1st thoracic leg.——*h*. Male left furca.——*i*. Female left furca.——*j*. Male penis. (All figures to scale in lower left corner; from Kesling, after Sars.)

as **maxillipeds** (or **second maxillae**). In other families they appear as unmodified leg structures, similar to the succeeding thoracic legs. In some species the first thoracic legs are dimorphic, and the male legs not only differ from the female, but the left and right

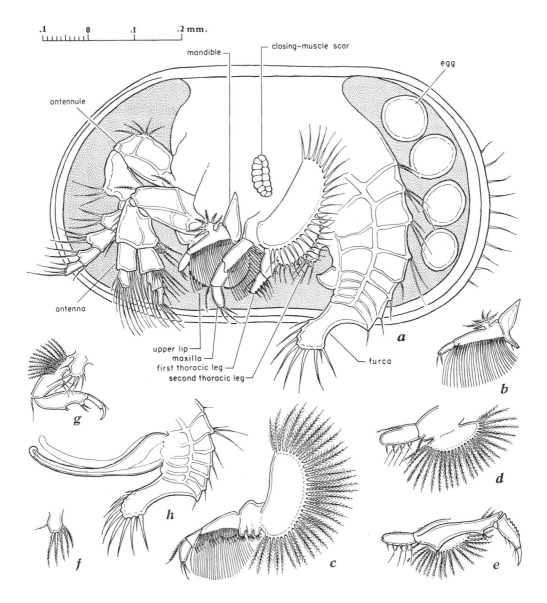

FIG. 5. Morphology of a representative podocopid (platycopine) ostracode, *Cytherella abyssorum* SARS, Recent (Cytherellidae). *a.* Female with LV removed; genital lobe shown above furca.——*b.* Left mandible. ——*c.* Left maxilla.——*d.* Female left 1st thoracic leg.——*e.* Male left 1st thoracic leg.——*f.* Female left 2nd thoracic leg.——*g.* Male left 2nd thoracic leg.——*h.* Male furca and penis. (All figures to scale in upper left corner from KESLING, after SARS.)

legs of the pair also may differ from each other. In the Myodocopida the protopod is large, vertical and unjointed, and provided with anteroventral endites (Fig. 1*e*) with clawlike setae (Fig. 1*f-j*). The exopod takes the form of a large vibratory plate that probably aids respiration (Figs. 1*a;* 2*a;* 3*a,b*). The endopod is either composed of short

podomeres on the posteroventral edge of the basis (Cypridinidae, Fig. 1*e*) or is leglike and directed backward (Halocyprididae, Fig. 3*a*). In the Cladocopina this is the only thoracic leg. Its homology is somewhat controversial. The large basal segment bears a vibratory plate which by some is regarded as the exopod and homologous with the

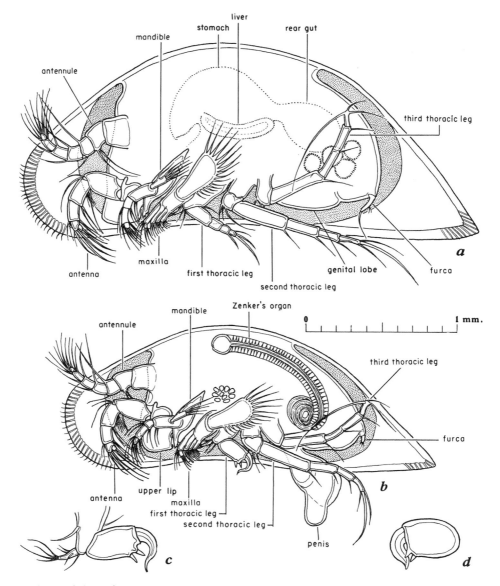

Fig. 6. Morphology of representative podocopid (podocopine) ostracode, *Macrocypris minna* (BAIRD), Recent (Macrocyprididae). *a.* Female with LV removed; 4 eggs shown in rear part of body.——*b.* Male with LV removed.——*c.* Male left 1st thoracic leg.——*d.* Palp of male right 1st thoracic leg. (Figs. *a* and *b* to scale at right; from KESLING, after SARS.)

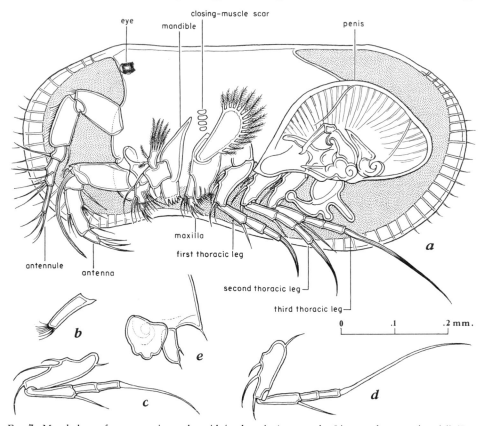

FIG. 7. Morphology of representative podocopid (podocopine) ostracode, *Limnocythere sanctipatricii* (BRADY & ROBERTSON), Recent (Limnocytheridae). *a*. Male with LV removed.——*b*. Male left brush-shaped organ. ——*c*. Female left 3rd thoracic leg.——*d*. Male left 3rd thoracic leg.——*e*. Female furca and genital lobe. (All figures to scale at right; from KESLING, after SARS.)

similar structure developed in the other orders and suborders; others regard it as an epidodial plate borne by the coxa (Fig. 4a,b). This basal segment bears a subtriangular podomere (?basis) to which are attached two setiferous lobes (?endopod and exopod). In the Platycopina the homology of the parts is likewise controversial. The appendage is dimorphic and in some species weakly developed or absent altogether in the females. That of males (Fig. 5e) bears a strong prehensile ramus of three podomeres directed backward (?endopod) used in copulation; this is absent in the female (Fig. 1d). In both sexes the basal protopod bears a vibratory plate with feathered setae (?epipod) and a distal lobe with a few short setae (?exopod). In the Podocopina this leg is variously developed and in many species is dimorphic. In the Cypridacea the protopod ends in a masticatory process bearing setae (Fig. 9); the endopod is modified as a palp, composed commonly of a single podomere in the female, but in the male further modified to form a prehensile claw of one or two podomeres; the exopod forms a small branchial plate. In the Cytheracea this pair of legs is pediform and lacks the branchial plate; the legs are dimorphic in some species however (Fig. 7a). In the Bairdiidae (Figs, 10, 11) they are pediform but possess a branchial plate; the legs are dimorphic in the Darwinulidae (Fig. 8e) both masticatory process and branchial plate are well developed but the endopod is pediform, as in the Cytheracea. In the Macrocyprididae there is no vibratory plate, the endopod in the female being subpediform, that of the male prehensile and asymmetric (Fig. 6c,d).

The **second thoracic legs** (not developed

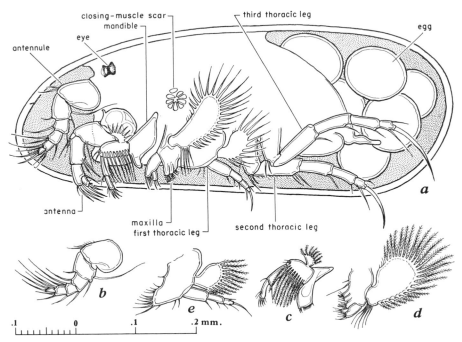

FIG. 8. Morphology of a representative podocopid (podocopine) ostracode, *Darwinula stevensoni* (BRADY & ROBERTSON), Recent (Darwinulidae). *a*. Female with LV removed.——*b*. Left antennule.——*c*. Left mandible.——*d*. Left maxilla from front.——*e*. Left 1st thoracic leg. (All figures to scale in lower left corner; from KESLING, after SARS.)

in the Cladocopina) closely resemble the first in the Halocyprididae, but in some species they are dimorphic (Fig. 3*a,b*). In the Cypridinidae (Fig. 1*k*) they take the form of a fixed lamelliform plate, never used in locomotion. In the Platycopina they are strongly dimorphic, that of the male (Fig. 5*g*) having an undivided protopod, an endopod of three backwardly directed podomeres, and a lamelliform exopod bearing setae. That of the female (Fig. 5*f*) is very small and consists of only the platelike exopod with setae. In the Podocopina this limb is uniramous and pediform, the protopod being strong and vertical and the endopod long and directed backward, with a strong terminal claw. In some species there is slight sexual dimorphy in the lengths of the podomeres and claws.

The **third thoracic leg** is developed only in the Myodocopida and the Podocopida. In the Cypridinidae this leg is long, mobile, vermiform and flexible, but lacks true joints (Figs. 1*a*; 2*a*). The distal end has long bristles with bell-shaped segments and a terminal comblike structure of setae (Fig. 2*d*).

This organ is used to clean the inside of the valves. In the Halocyprididae this limb is reduced to a small tapering stem of one or two podomeres and two setae (Fig. 3*a,b*). In some Podocopida this leg is also a cleaning organ of rather variable structure, upturned within the cavity of the shell and equipped with long cleaning setae (Fig. 6*a,b*), friction pads, a claw, and pincers, complete or in various combinations. In other Podocopida the limb is developed as a walking leg essentially similar to the two previous limbs (Fig. 7*a,c,d*; 8*a*); in some Cytheracea it is dimorphic in the proportions and lengths of podomeres and claws (Fig. 7*c,d*).

Male Cytheracea have paired **brush-shaped organs** (Fig. 7*b*) located in front of, between, or behind the thoracic legs. These are probably sexual sensory organs and not, as some suggest, rudiments of another pair of appendages.

The **furcae** ("furcal rami" of some authors) are appendage-like structures attached at the posterior end of the body. In the Myodocopida and Platycopina they con-

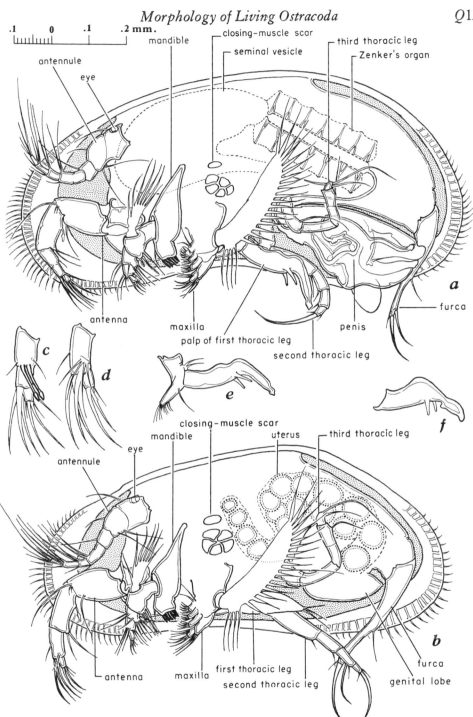

FIG. 9. Morphology of representative podocopid (podocopine) ostracode, *Candona suburbana* HOFF, Recent (Cyprididae). *a*. Male with LV removed; palp of maxilla turned backward to show "masticatory processes" or endites.——*b*. Female with LV removed.——*c*. Inner face of distal part of male right antenna, showing "male" setae.——*d*. Inner face of distal part of female right antenna.——*e*. Outer face of male left 1st thoracic leg.——*f*. Inner face of palp of male right 1st thoracic leg. (Figs. *a*, *b*, *e*, and *f* to scale in upper left corner, from KESLING.)

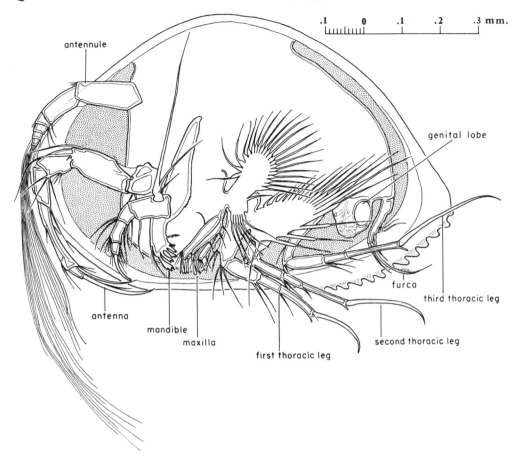

Fig. 10. Morphology of a representative podocopid (podocopine) ostracode, *Bairdia frequens* G. W. Müller, Recent (Bairdiidae). Female with LV removed (compare with male in Fig. 11); note dimorphism in terminal claws of antenna and setae of exopod of 1st thoracic leg. (From Kesling, compiled from Müller.)

sist of short, broad, lamelliform rami with several claws (Figs. 1*a*; 2*a*; 3*a,b*; 5*a*). In the Cladocopina they are similar but in some species are asymmetrical in the male, with the number of claws reduced in the left (Fig. 4*h,i*). In the Podocopina the furcae are extremely variable, and never lamelliform. In the Darwinulidae (Fig. 8*a*) they are lacking or represented by an unpaired reflexed process at the end of the thorax. In the Cytheracea they are dimorphic, in the female consisting of a small plate with two or more setae (Fig. 7*e*), in the males of most species fused with the penis (Fig. 7*a*). In the Bairdiidae they are small but well developed, variable, with at least three setae, of which one is long and strong (Figs. 10, 11). In the Macrocyprididae they are much

reduced and dimorphic (Figs. 6*a,b*). In the Cyprididae they are variable; in most genera, *Candona* for example, there is a long rodlike ramus and two powerful terminal claws and two setae (Fig. 12); this is reduced in *Cypridopsis* to a stumplike base and a long flagella-shaped terminal seta with or without basal spinule.

DIGESTIVE SYSTEM

The alimentary canal of ostracodes consist of the mouth, esophagus, stomach, and anus. Podocopidan Ostracoda have a short, narrow intestine and a voluminous rear gut between the stomach and anus. Some have livers at the sides of the stomach, which supply digestive fluids.

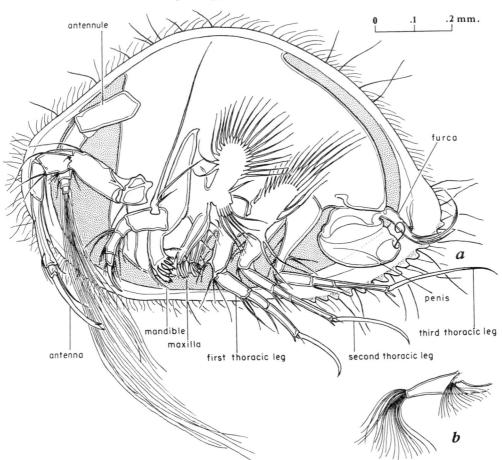

FIG. 11. Morphology of representative podocopid (podocopine) ostracode, *Bairdia frequens* G. W. MÜLLER, Recent (Bairdiidae). *a.* Male with LV removed; maxilla shown above its normal position and furca bent backward to show structures more clearly (at rest, exopod plate of maxilla lies outside that of 1st thoracic leg, and furcae lie between the penis).——*b.* Asymmetrical brush-shaped organs, which lie between bases of 2nd thoracic legs. (Both figures to scale in upper right corner; from KESLING, compiled from MÜLLER.)

GLANDULAR SYSTEM

The glandular system is not well understood in many families. The glands may be divided into secreting and excretory types, but glands of any sort have not been reported in some ostracodes. So-called "shell glands" (which have nothing to do with secreting of the shell) appear to be a combination of secreting and excretory glands. Other secreting glands are salivary (glands of the upper lip), livers (hepatopancreases), and those of the first thoracic legs. Certain marine ostracodes have glands in the upper lip that secrete light-producing substances. Some have glands that open at the borders of the valves. In some genera, at least, there are excretory glands opening near the antennules and maxillae.

RESPIRATORY AND CIRCULATORY SYSTEMS

Respiration is accomplished through the body wall, by gills on the rear part of the body, by vibratory plates of certain appendages, or by combinations of these three. A distinct heart is found only in the Myodocopida (Figs. 1*a*; 2*a*; 3*a,b*).

NERVOUS SYSTEM

Primary divisions of the nervous system

are the cerebrum, the ventral chain of ganglia, and the circumesophageal ganglion. Lesser ganglia lie in most of the appendages.

The cerebrum lies in the forehead area, from which nerves extend to the eye, antennules, and antennae, and to the epidermis of the valves. Motor nerves extend from the circumesophageal ganglion to the an-

tennae and forehead area. The ventral chain sends nerves to most of the appendages from the mandibles posteriorly, including the furcae and ventral area.

The eye, when developed, is a complex structure, composed of one median and two lateral divisions. It is situated dorsal to the basal podomeres of the antennules. The lat-

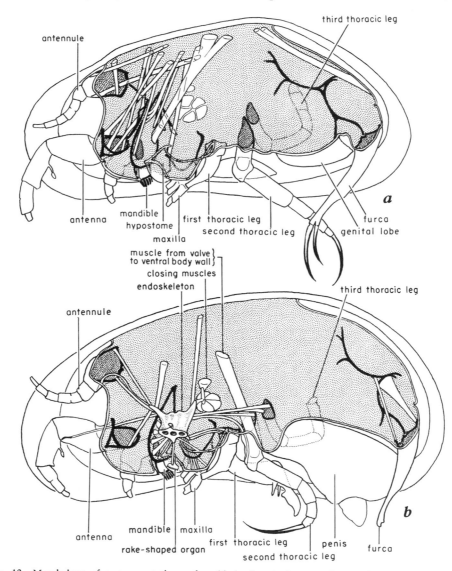

Fig. 12. Morphology of a representative podocopid (podocopine) ostracode, *Candona suburbana* Hoff, Recent (Cyprididae). *a*. Female with left half removed, showing muscles attached to dorsal part of valve and to appendages and rear part of body.——*b*. Male with left half removed, showing muscles from endoskeleton to appendages, rake-shaped organ, hypostome, closing muscles, and dorsal part of valve; muscles operating furca are attached to rear framework of body. Muscle from valve to ventral body wall is found only in males; it is contracted in copulation, unfolding the penis and swinging them out of the carapace.

eral divisions are reported to contain 10 to 15 cells and the median element 7 to 8. Many ostracodes have no eyes. In the Myodocopida (Cypridinidae) they are dimorphic, those of the male (Fig. 2a) being larger and with more lenses than those of the female (Fig. 1a). In the Podocopina (some Cytheracea) a glassy lenslike eye tubercle is observed in either valve at the anterior end of the hinge. An ocular sinus lies below the tubercle and serves to accommodate the eye. The frontal organ developed in some Ostracoda is believed to be sensory, but its use is not known certainly. Some authors believe that it is a kind of median eye.

MUSCULATURE

The body of an ostracode is provided with many muscles. Flexor and extensor muscles are well developed in all appendages. Separate muscles usually occur in each podomere so that each division is capable of independent action. The basal podomere of each cephalic appendage is attached to the inner dorsal area of the shell by flexor and extensor muscles (Fig. 12a). The bosses, at points of attachment, may be preserved in the fossil state.

Among the most important appendage muscles are those attaching the basal podomere of the mandibles to the shell. Several muscles extend from the mandible to the inner dorsal margin of the shell. Another set of mandible muscles is attached to the shell in an anteroventral position. They are attached slightly anterior and ventral to the closing muscles and serve as adjustor muscles. In most literature, the scars of the mandible-adjustor muscles have been erroneously included in the set of closing-muscle scars.

The adductors, or closing muscles, extend from valve to valve. Their distal ends are attached to the outer coating of the epidermis. In the calcareous portion of the shell, a raised boss is developed at the point of muscle attachment. This boss is the "muscle scar" referred to in fossil forms.

The adductor muscle fibers connect near the middle of the body in a chitinous rod. From the chitinous rod to the inner margin of the valves they may extend as a closely packed set of fibers, as in *Cytherella,* or may diverge to form several isolated bosses, as in the Cytheracea.

When the adductor muscles relax, the valves are opened by an elastic ligament that lies along the dorsal margin. Contraction of the adductors, while closing the valves, creates tension on the ligament. Relaxation of the adductors releases the tension and the ligament contracts, thus opening the valves.

In the mid-area of the ostracode body is a chitinous framework known as the **endoskeleton** (Fig. 12b). It is suspended by muscles which extend to other body parts. Muscles connect the endoskeleton with most of the appendages, several glands, and the alimentary tract.

SEX ORGANS

The sex system is paired and, with few exceptions, the left half is not connected with the right. Gonads of Cyprididae lie in the hypodermis, whereas those of other ostracodes are in the rear part of the body. The female system is made of ovaries, uteri, uterine openings, vaginae, and seminal receptacles. The seminal receptacles have adits and exits through the vaginae, and are not connected to the uteri inside the body. The male system includes the testes, vasa deferentia, and penes. Some genera have ejaculatory ducts or Zenker's organs, which pump sperm out through the penes (Fig. 6b). In some there are enlargements of the vasa deferentia that serve as seminal vesicles. Many species are parthenogenetic, but their females retain seminal receptacles. Of the syngamic species, a few have males with asymmetrical penes and many have dimorphic appendages. The penes in many Podocopina are extremely complicated (Fig. 7a) and accordingly some attempts have been made to use them in taxonomy, but the homology is imperfectly understood.

REPRODUCTION OF OSTRACODA

By R. V. KESLING
[University of Michigan]

PARTHENOGENESIS

Many species of fresh-water ostracodes lack males, as indicated both from observa-tions in nature and from cultures in aquaria. The females lay fertile eggs, which hatch out another generation of females.

Cyprinotus incongruens is parthenogenetic in one geographic range and syngamic in another. Furthermore, laboratory cultures of this species can be changed from syngamic to parthenogenetic by isolation of the females, and from parthenogenetic to syngamic by placing the females on a near-starvation diet. Such a reversal of reproductive processes does not occur in other species that have been investigated.

SYNGAMY

In species having males, the female lays fertile eggs only after copulation. The exact role of sperm in syngamic reproduction is questionable. The individual spermatozoa in many species are many times longer than the carapace, and absolutely, as well as relatively, larger than those of all other animals. LOWNDES, who studied the motility of ostracode spermatozoa, found that they advanced tail first and from this concluded that they were nonfunctional. If this is true, it does not explain why, in most syngamic species, impregnated and only impregnated female produce fertile eggs, nor why their eggs hatch out young individuals of both sexes.

COPULATION

Fresh-water species of ostracodes, which mature in about a month, live only a few months as adults. Although in some species young broods appear in the spring in great numbers and in others come forth in the summer, fresh-water ostracodes produce more than one or two generations per year. In each impregnated female, the seminal receptacles are distended with hundreds of spermatozoa, certainly more than enough to fertilize all eggs the animal could ever lay. Seemingly a single copulation is sufficient to impregnate a female for life.

Copulation has been observed in several Cyprididae, including *Candona fabaeformis, C. rostrata, Cyprois marginata, Candonopsis kingsleii, Cyprinotus incongruens,* and *Notodromas monacha.* In all except the last-named species, the procedure is the same. The male spreads his valves apart, clambers onto the posterodorsal part of the carapace of the female, clasps the edges of the female valves with the palps of the first thoracic legs, unfolds and extends the ends of the paired penes, and inserts them into the paired vaginae of the female. The female remains passive. The copulation is accomplished in minutes, and the accuracy of the male is attested by the fact that no female has ever been found to have spermatozoa in only one receptacle. ELOFSON recorded observations on the mating of many Cytheridae, including species of *Cythere, Leptocythere, Cythereis, Cytherura, Cytheropteron, Hirschmannia, Xestoleberis,* and *Paradoxostoma.* In these Cytheridae and in *Notodromas monacha,* the male approaches in a different way; the posteroventral edges of his valves are brought into contact with those of the female, so that the ventral sides of the two animals lie close together. No observations on the mating procedure of other families have been recorded.

Several marine species of Cypridinidae are known to have planktonic mating. One species, *Philomedes globosa,* is interesting because the adult male swims actively near the surface, whereas the adult female spends most of her life confined to the bottom, dragging her carapace through the mud. The adult females emerge from their final ecdysis with natatory setae; at certain times of the year, in the dark of night, flocks of females ascend over 100 fathoms from the bottom to copulate with the planktonic males. After mating, the females return to the bottom, shed their natatory setae by biting (or more properly "sawing") them off with the claws on the first thoracic legs, and by this self-mutilation lose for the rest of their lives the ability to soar upward from the mud. As for the males, they reach maturity with weak mouth parts, and probably die shortly after the nuptial swarming.

BROOD CARE

The Cyprididae, most Cytheridae, and presumably the Bairdiidae do not care for their young. After the eggs are laid (and in some species are attached to vegetation or bottom sediments), the female goes her way. Not all eggs are laid at one time, and broods of several stages may come from one mother.

The Darwinulidae, Cypridinidae, Halocyprididae, and Cytherellidae retain the eggs between the posterodorsal part of the body and the carapace. In many species, the young pass through more than one ecdysis before leaving the protection of the mother animal.

Nothing is known about the transfer of the eggs to the brood space, nor about the feeding of the young instars.

Eggs of most fresh-water ostracodes not only require no care, but can withstand desiccation for long periods of time. Sars described from various parts of the world ostracode faunas which were raised from dried mud shipped to him. Eggs have been reported viable after drying for 30 years. Eggs of marine ostracodes are not known to survive any drying out whatsoever.

ONTOGENY OF OSTRACODA

By R. V. Kesling
[University of Michigan]

HATCHING AND MOLTING

An ostracode hatches from the double-walled eggs as a bivalved nauplius. The egg splits along the equatorial line, thus freeing the young ostracode. In none of the species investigated does the animal have in its first instar, or growth stage, as many appendages as are developed in the adult.

Like other crustaceans, the ostracode grows by ecdysis, molting the old hard parts and secreting new and larger ones. This externally discontinuous growth habit is very similar to hatching again, for the individual that emerges after ecdysis differs from the one that inhabited the old carapace. The molting process begins with splitting of the inner lamella from the outer side along the front edge; the chitin breaks open along the median plane of the body, and in sequence the appendages are meticulously withdrawn from their old chitinous armor; then the expansion of the hypodermis springs the animal free from the old carapace. During the relatively brief ecdysis, the animal increases to about twice its former volume, adds new appendages and organs, and alters the form and function of the old ones before secreting calcareous valves and a chitinous coat on the body and appendages.

NUMBER OF INSTARS

Unfortunately, few authors have studied the immature instars of species, being content with examination and classification of adults only. In two suborders (Cladocopina, Platycopina) the number of instars has not been established for any species. Ostracodes belonging to the Cyprididae have nine instars, of which the last is the adult. Most of the Cytheracea also have nine instars, but three species of *Xestoleberis* have only eight. In the few species of Cypridinidae that have been investigated six instars are observed and in the Halocyprididae there are seven instars. Much additional work is needed to understand the immature instars of nearly all ostracodes.

CHANGES IN ANIMAL

The ostracode adds new appendages in the young instars. Each appendage, except for those present in the nauplius when hatched, begins as a simple lobe, variously called an anlage, primordium, or incipient appendage. In successive instars, the anlage assumes the definitive form of the appendage, adding podomeres and claws or setae. New organs are added in the same way, and

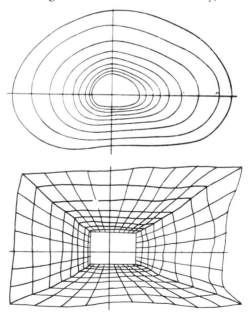

Fig. 13. Growth indicated by successive instars of *Cypridopsis vidua* (Müller). a. Outlines of instars in lateral view successive, ×5.——b. Plotted growth increments according to D'Arcy Thompson's system of Cartesian coördinates (44).

Fig. 14. Transverse section of adherent superposed instars of *Cryptophyllus obotoides* (Ulrich & Bassler), M.Ord., ×10 (49).

the animal becomes sexually mature only after the last ecdysis. In some species, however, there may be rare individuals which undergo further growth and ecdysis after sexual maturity is reached, for not uncommonly in a collection of a species rare individuals are found to be about double the volume of the majority.

The following chart summarizes the appearance of appendages in ontogeny, insofar as known. It is of particular interest that in all groups studied, the thoracic legs first reach their definitive form in the antepenultimate instar, regardless of the number of instars for the species. It seems that Cypridinidae hatch with the same stage of development as that achieved by the Cyprididae after three moltings. The three species of *Xestoleberis* with nine instars begin the first growth stage with the appendages that other Cytheracea have in the second instar.

CHANGES IN CARAPACE

The carapaces of successive instars of a growing ostracode commonly differ not only in size, but in proportions, growth being allometric. The allometric index is slight, however, and varies during growth (Fig. 13). The shell of young instars is in all species thinner in proportion to its size than in the adult. The duplicature, in species in which it is developed, is narrower in the young instars, and the hinge more fragile. In species which have an advanced hinge in the adult (e.g., amphidont), earlier instars show a more primitive type (e.g., merodont). This ontogenetic succession

therefore parallels the presumed phylogenetic history and is considered to be an example of palingenesis. In species that are highly ornamented as adults, the ornamentation increases during ontogeny, both in the number of ornamental elements and in their relative size.

In a few genera (e.g., *Eridoconcha, Cryptophyllus*) some, but not all, of the early molts are retained outside the later ones, and become cemented to them (Fig. 14), thus reproducing growth lines of the same nature as those in the Conchostraca.

TIME REQUIRED FOR MATURITY

The Cypridinidae reach maturity in about three years. The Cyprididae develop in about 30 days, and Cytheracea in 40 days to three years, according to the species. In general, fresh-water species grow and mature much faster than marine. These figures are based on scanty information, since few species have been studied.

Table. *Morphological Features of Ostracoda Occurring in Different Growth Stages*

Taxon					Appendage									
Cypridacea	Cytheracea	*Xestoleberis* sp.	Cyprididae	*Conchoecia* sp.	Antennule	Antenna	Mandible	Maxilla	First th. leg	Second th. leg	Third th. leg	Gonads	Genitalia	Furca
Instar number														
1	1	..	..	..	X	X	X	..	..	..	..	..	..	a
2	2	1	..	..	X	X	X	A	..	..	..	..	..	A
3	3	2	..	1	X	X	X	X	..	..	..	..	..	A
4	4	3	1	2	X	X	X	X	A	a	..	..	..	A
5	5	4	2	3	X	X	X	X	X	A	a	..	..	X
6	6	5	3	4	X	X	X	X	X	X	A	a	..	X
7	7	6	4	5	X	X	X	X	X	X	X	A	a	X
8	8	7	5	6	X	X	X	X	X	X	X	X	A	X
Adult					X	X	X	X	X	X	X	X	X	X

Explanation: X, structure present in definitive form; A, anlage always present; a, anlage present in some species, not in others; .., no trace of structure found.

SHELL MORPHOLOGY OF OSTRACODA

By H. W. Scott
[University of Illinois]

CONTENTS

GENERAL FEATURES OF CARAPACE

The ostracode carapace is bivalved, each valve being similar but not invariably a mirror image of the other. The two valves may be subequal or unequal in size. Inequality in size results in **overlap** of part of the free margin of the smaller valve by the larger, as in *Paraparchites* or *Kloedenella,* or in inclosure of the smaller valve by the larger all around the margin, as in *Cytherella.* A narrow edge of the free margin may be beveled in such a manner that the two valves fit without pronounced overlap, giving the appearance of equality as in *Amphissites.* Even in such shells, however, the beveled edge on one valve slightly overlaps the thin edge of the opposite valve. The two valves are articulated dorsally along the hinge and inclose the body of the ostracode. The carapace is composed of two parts: (1) a hard layer of calcium carbonate, and (2) a soft layer, the epidermis. The hard shell substance is preserved in fossils and therefore represents the portion commonly studied by paleontologists. During life the hard shell is coated with a chitinous layer and the epidermis is enclosed in chitin. The outer chitinous layer can be recognized in some thin sections of fossils.

The hard shell layer is usually composed of two parts, the **outer lamella** and **duplicature** (Fig. 15). The duplicature, not definitely recognized in the archaeocopids, leperditicopids, and palaeocopids, extends along the free margin of the valve and is welded to the outer lamella. Both parts are composed of crystalline calcium carbonate; the long axis of the crystals is arranged at right angles to the shell surface. The calcium carbonate is precipitated by epidermal cells, though no special lime-secreting cells have been recognized. Because the ostracode sheds its hard shell in a series of molt stages, it is obvious that calcium carbonate precipitation is not continuous but is rhythmic. It is known that the ostracode can store an excess amount of calcium carbonate in its body and can create a new shell after molting in a calcium-free environment. The amount of calcium in the diet may control some shell features. Unusually thin or thick shells, bizarre ornamentation, or other aberrant features may be environmentally controlled.

EXTERNAL FEATURES

SHAPE AND SIZE

Ostracodes vary greatly in size. They commonly measure 0.7 or 0.8 mm. in length,

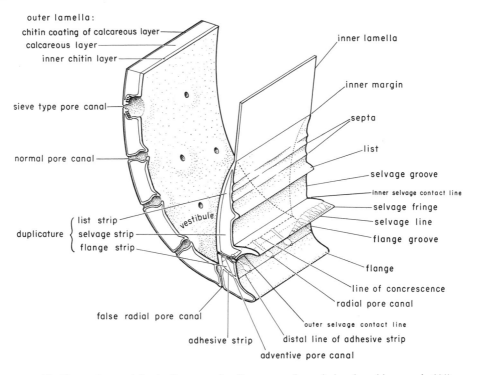

outer lamella:
chitin coating of calcareous layer
calcareous layer
inner chitin layer

sieve type pore canal

normal pore canal

duplicature { list strip
selvage strip
flange strip }

false radial pore canal

adhesive strip

adventive pore canal

inner lamella

inner margin

septa

list

selvage groove

inner selvage contact line

selvage fringe

selvage line

flange groove

flange

line of concrescence

radial pore canal

outer selvage contact line

distal line of adhesive strip

vestibule

Fig. 15. Nomenclature of the duplicature and wall structure of a typical podocopid ostracode (44).

but some species may greatly exceed the average. For example, *Eoleperditia fabulites* commonly attains a length of 3 to 5 mm.

The outline or shape of an ostracode is one of the most important criteria in classification. Many families and genera may be identified by general outline of the carapace. Regardless of great variation in shape, most ostracode carapaces may be described as ovate, elliptical, or quadrate. The prefix "sub-" may be added to any of these terms to indicate deviations. Rarely ostracode carapaces are subrectangular, trapezoidal, or rounded in outline.

The dorsal edge of the carapace may be convex or straight, and the ventral margin convex, straight, or concave. The ends are usually rounded, but in some species they may be extended into elongate structures such as found in *Bairdia* and *Cytherura*. Rounding of the ends may be symmetrical (e.g., *Cytherella*) but they are more commonly dissimilarly rounded. The greatest extension of curvature of the ends may be either above or below mid-height.

The area adjacent to the hinge, as seen in

dorsal view, is referred to as the **dorsum** (Fig. 16). It may be broadly or narrowly arched, flat (with **dorsal plica** or marginal ridge), or concave. The terminations of the hinge may be marked by a change in outline. These changes represent the juncture between the dorsal border and ends of the carapace, referred to as the **cardinal corners,** which are very important in classifying shape of the carapace (Fig. 17). The cardinal corners range from acute (e.g., *Kirkbya*) to obtuse (e.g., *Oepikella*); usually they are obtuse and unequal.

The position of greatest height also controls outline. If the greatest height is in front of the mid-length, the carapace may be referred to as **preplete,** if posterior to the mid-length as **postplete,** and if at or near the mid-length as **amplete.**

When the greatest height is near one end of the carapace it produces what is known as **swing.** Preplete and postplete shells possess swing. Usually the posteroventral border swings forward, resulting in a narrow posterior half and a high anterior half (Fig. 18). Some exceptions to the forward direc-

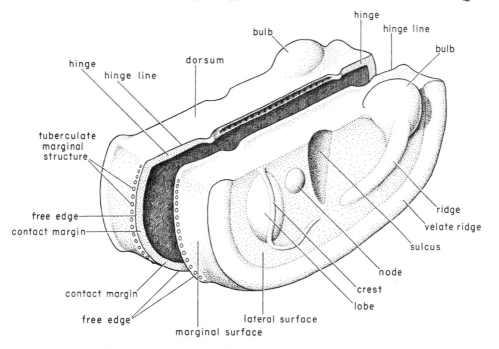

Fig. 16. Carapace nomenclature of a typical straight-backed ostracode (44).

tion of swing exist (e.g., *Leperditia* and a few other genera).

VENTRAL AREA

When the two valves of the carapace are opened they move outward around the **free edge** from the terminations of the hinge. The free edge extends around the anterior and posterior ends and along the ventral edge. The area adjacent to the ventral free edge, as seen from below, is the **venter.** Ventral structures, which may extend partly or entirely around the anterior and posterior ends, are referred to as **adventral.**

The closure of the carapace brings together a narrow portion of each valve. The area of contact is the **contact margin** (Fig. 16). The nature of contact is variable in detail but may be grouped into two types: (1) that in which one valve is distinctly larger than the other, the larger overlapping the smaller around all or a portion of the free edge (e.g., *Cytherella, Kloedenella*); and (2) that in which the valves are equal to subequal and overlap is wanting or slight (e.g., *Amphissites*).

The contact margin may be simple or complex. In the simplest type of closure a single ridge, termed the **selvage,** extends along the free margin. If more than one ridge is present, the selvage is considered to comprise the principal ridge. When the valves are closed, the selvage fits into a groove in the opposite valve. In some ostracodes the proximal covering of the epidermis *(inner lamella)* is calcified to form the **duplicature** (Fig. 15). In such shells the contact margin is complex and the selvage constitutes the principal ridge. In the living animal the space between the inner and outer lamellae, called the **vestibule,** is filled by part of the epidermis. Proximal to the selvage may be a minor second ridge, termed the **list.** The two ridges are separated by a groove. Additional secondary ridges may be proximal to the list. Distal to the selvage is the **flange,** a ridge that commonly is a part of the outer instead of inner lamella. When both selvage and flange are present they are separated by the **flange groove.** The duplicature is attached to the outer lamella by an adhesive strip of chitin; the proximal line of contact is referred to as the **line of concrescence** (Fig. 15).

The ventral area may be simple and show only the contact of the valves along the free

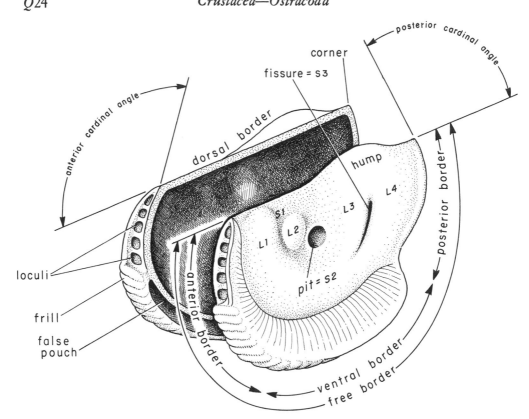

Fig. 17. Diagrammatic illustration showing some of the external features of a typical velate palaeocopid ostracode (44).

edge. The ventral area in many ostracodes, however, is modified by various ridges, frills, or flanges, which may be restricted to the venter or may extend to the cardinal corners. Jaanusson (1957) groups such features into types designed as velate, histial, and carinal structures. Experience has shown that these terms cannot be applied in all cases without confusion. Distinction of the histium is especially troublesome. The typical velum is a ventral ridge, flange, or frill that may extend around part or all of the anterior and posterior ends. If it is considered as the primary or main ventral frill, as typified by hollinids and eurychilinids, then histium cannot be used for such velate structures. In many ostracode carapaces, secondary ridges occur in addition to the primary velate structure. Kesling & Rogers (1957) refer to these secondary ridges in *Treposella* as "extra" and "marginal or submarginal."

The morphological significance of the **velum** is not clear. In some carapaces it is related to dimorphism, but in others it does not seem to function as a dimorphic structure. Sections of *Beyrichia* show that the velum is a downfold lined with the inner chitin layer. Even in *Hollinella* where the two sides of the velum are in contact, a dark line representing the infolded inner chitin layer is preserved. Perhaps in such shells the frill is a true dimorphic velum, whereas the marginal rim in such a form as *Zygobolba* is a **pseudovelum**.

The term "carinal structure" was proposed by Jaanusson (1957) for "different kinds of nondimorphic ornamental ridges situated lateroventrally and often occupying about the same position as the connecting lobe in quadrilobate valves." The term has been used to denote a kind of dimorphism, whereas it should refer only to prominent ornamental ribs on the lateral surface, not including low costae or velate ribs or the histial ridge. Many species of *Amphissites* are ornamented with typical carinae. Lateral or lateroventral ribs may or may not be superimposed on a dimorphic lobe.

Ribs and ridges varying in number and complexity commonly are found on the lateral surface of ostracode valves. These ribs or costae may be coarse (e.g., *Glyptopleura*) or fine (e.g., *Graphiadactyllis*). A simple ridge may occur where the lateral surface meets the venter. In some carapaces it is difficult to distinguish between a marginal costa and velum. If the rib is on the lateral surface, it is more appropriately termed a **costa,** whereas if it occurs on the venter or extends ventrally from the contact of the lateral surface with the venter, it is either a velum or a velate structure.

The velum commonly is a wide frill-like structure extending lateroventrally from the free edge (Fig. 19). It may be a smooth bladelike structure or its surface may be undulating, nodose, striate, or reticulate and its edge may be smooth, scalloped, or spinose. Velate structures are common in the Palaeocopida and previously have been applied only to this group, though velate-like structures occur in *Cythereis, Pterygocythereis, Kingmaina,* and other Mesozoic and Cenozoic genera.

LATERAL SURFACE

General features. The lateral surface of ostracode valves can be divided into posterior and anterior portions and into dorsal and ventral portions. For designation of specific points on the lateral surface, smaller subdivisions can be defined, as anteroventral, posteroventral, etc. (Fig. 18). The dividing line between anterior and posterior is placed at mid-length and between dorsal and ventral at mid-height. If the position of the adductor muscle scar is known, the surface in front of it is referred to as the preadductorial area and the posterior area as postadductorial.

Unfortunately, length and height are not always measured consistently, for such measurements vary according to methods of orienting the carapace. Students of the palaeocopids have oriented the carapace with the dorsal edge parallel to a horizontal line because in this group the hinge is fairly long and straight, producing a straight-backed dorsum in lateral view (Figs. 16, 18). However, students of Cenozoic and Mesozoic ostracodes mostly have oriented them with the ventral edge or long axis parallel to a horizontal line. This procedure has been adopted because of the great variation in the shape of the hinge or dorsum in platycopine, myodocopine, podocopine, and cladocopine groups. Because these different methods of orientation are well intrenched in practice it seems best to continue them.

The lateral surface of ostracode valves may be smooth or highly ornamented by granules, pustules, striae, costae, pits, spines, or reticula. **Granules** may be closely packed or sparsely distributed over part or all of the surface (Fig. 19). If the raised protuberances are distinct and larger than granules they are described as **pustulae** or **papillae. Striae** are fine furrows separated by very minute ridges. In *Entomis* they cover the entire surface and are nearly parallel to each other. In *Glyptopleura* the ribs are coarser and are described as **costae.** The lateral surface of *Pyxiprimitia* and similar forms is marked by depressions in the shell called **pits;** they are not the openings of normal pore canals. Larger pits such as found in *Thlipsura* are not common; their function, if any, is unknown.

One of the most distinctive of all ornamental features of ostracode carapaces is a **reticulate** surface, found in many genera from Paleozoic to Recent. Reticula are represented by a network of intersecting bars that produce a lacy pattern. In some species the connecting bars are narrow and the depressions large, whereas in others the depressions are small and the surrounding barlike portions are relatively wide and flat. In the latter type there is a gradation between reticulate and punctate surfaces (Fig. 19). If the depressions are wider than the bars, the reticulate pattern is obvious but if the barlike divisions are wider than the depressions, the appearance is more that of a pitted or punctate surface. One of the most common faults in early classification was the separation of smooth specimens representing internal molds from fossils showing reticulate exterior surfaces.

The shell may be pierced by **normal pore canals,** which represent the passageway for hairs or setae (Fig. 15). The canals may be few or many, widely distributed, closely

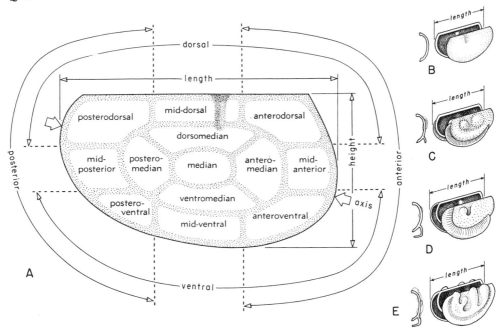

FIG. 18. Nomenclature of areas on the lateral surface and features relating to orientation and dimensions of typical straight-backed ostracodes. *A.* Diagram of RV. *B-E,* both valves and transverse sections of *Primitia, Hollinella, Piretella,* and *Tallinnella* (44).

packed, or entirely wanting. The projecting setae are sensitive to touch. In some ostracodes additional setae may extend through **radial pore canals** at the juncture of the duplicature with the outer lamella. Commonly they are concentrated at the anterior end but may extend along most of the free edge also.

Spines are distinctive features of many ostracode carapaces (Fig. 19). They are highly variable in number and size. Some are solid and ornamental, whereas others are hollow and apparently held some portion of the soft parts. Commonly but not invariably spines are located posteriorly. In *Pterygocythereis* and *Cytheropteron* the alate structure may extend backward into a spine in a posteroventral position; other spines may occur on both the posterior and anterior margins. In some genera (e.g., *Paracytheridea*) the caudal process in a posterodorsal position commonly is extended into a spine. Spines may develop along the dorsum (e.g., *Rakverella* and *Ctenonotella,* which may have several spines along the dorsal edge tilted slightly backward). *Dicranella* has two mid-dorsal spines and

Aechmina a single centrally placed dorsal spine; the lateral surface of *Cythereis* may bear many small spines, and spines are common on the edge of the velum in velate species.

Lobation and sulcation. Lobes and sulci are among the most distinctive features of many ostracode carapaces (Fig. 20). Early attempts to use them as a major basis of classification resulted in considerable confusion. **Lobes** represent elevations of the shell (domicilium) which are directly opposite internal depressions or troughs. This external lobation is a reflection of internal anatomy and therefore important. Unfortunately, the lobes cannot always be assigned to given internal organs. For convenience, lobes have been designated numerically from anterior to posterior parts of the valves, as L_1, L_2, L_3, and L_4. The presence of all lobes indicates a quadrilobate valve; a trilobate valve bears lobes designated as L_1, L_2, and L_3; a bilobate valve has only L_1 and L_2.

Sulci are elongate depressions of the domicilium labeled S_1, S_2, and S_3 from front to back; valves may be unisulcate, bisulcate, or trisulcate (Fig. 20). Of the sulci S_2 is the

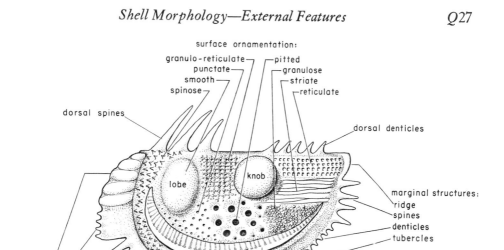

Fig. 19. Nomenclature of the lateral and marginal palaeocopid features of ostracode (composite) (44).

most significant because it marks the position of the adductor muscle. Sulci are expressed internally by corresponding elevations. Some shells possess elongate depressions which are not reflected internally; these may be referred to as **furrows or fissures** (Fig. 17). A typical sulcus opens dorsally and may open at both ends, whereas typical fissures do not open at either end and are contained within the shell.

In some forms all of the lobes merge ventrally into a ventral lobe. This is well illustrated in *Tetradella marchica*, a quadrilobate-trisulcate species in which L_1 and L_4 curve ventrally to form a long ventral lobe with L_2 and L_3 attached to it. A similar development is found in *Tallinnella dimorpha*.

The sulci S_1 and S_2 represent the points of attachment of muscles, S_2 denoting the position of the adductor muscle and S_1 the position of the muscles for one of the anterior appendages. S_2 is present in unisulcate species and S_1 and S_2 in bisulcate species. It has been observed that thick, short-hinged shells with a convex dorsum are seldom sulcate. S_2 has probably been developed due to the strain placed upon the carapace by the closing muscles. S_3 is com-

monly present in quadrilobate valves. The function of the lobes is not clear. In bilobate forms, where the two lobes are separated in part by S_2, the lobes may not have had any special function and are only an indirect product of S_2. The same may be true for some trilobate species wherin L_1, L_2, and L_3 are by-products of S_1 and S_2. This is not an explanation for all lobation. Certain organs, such as the liver and stomach, may have occupied some lobes; the posterior lobe may have been occupied by reproductive organs.

Unisulcate *Eu{loedenella*, bisulcate *K̄loedenella*, and trisulcate *Dizygopleura* are all closely related genera. In such forms the differences in sulcation are of only generic importance. There is closer relationship among these genera with different degrees of sulcation than exists between such unisulcate genera as *Bolbina*, *Dilobella*, *Euḱloedenella*, and *Plethobolbina*. It is common rather than exceptional that genera within a family show a variation in the number of sulci.

The number and character of sulci and lobes cannot be used generally for familial differentiation; they are aids to classification on all levels but are not by themselves gen-

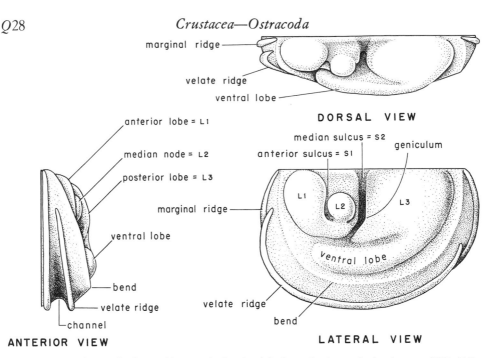

Fig. 20. Nomenclature of palaeocopid ostracode showing lobation, sulcation, and other features (LV) (44).

erally definitive in any category other than species. The number of sulci cannot be used solely for determining the affinities of genera. Phylogenetic lines based on sulcation have not been worked out, and much valuable information is expected to result from future studies of this subject. Classification based on lobation alone results in confusion.

DORSAL AREA

The general outline, nature of the hinge contact, form of lobation and sulcation, and presence of dorsal carinae are features that may be observed in dorsal view. Observation of the dorsum is essential to determine the position of greatest width of the ostracode carapace and character of the hinge line. The position of greatest width bears on dimorphism and orientation; the nature of the hinge bears on problems of classification and in some forms on orientation. The hinge line may be depressed in a channel (e.g., *Paraparchites*), comprise the highest, most dorsal edge of the carapace (e.g., *Welleria*), or be completely overlapped by the larger valve (e.g., *Cytherella*). The exterior terminations of the hinge fall into four general classes: (1) without cardinal angles and hinge contact not directly exposed (e.g.,

Cytherella, Bairdia); (2) with cardinal angles well developed, hinge line straight and uninterrupted by teeth or overlap, no channel (e.g., *Ctenobolbina, Primitia*); (3) with one end of hinge, usually anterior, interrupted by overlap of one valve, usually the larger, channeled or nonchanneled (e.g., *Eukloedenella*); and (4) with both ends of the hinge interrupted by toothlike structures of one valve fitting into socket-like notches in the other valve, channeled or nonchanneled (e.g., *Sansabella, Brachycythere*).

It should be pointed out that the internal characters of the hinge cannot be interpreted from the external characters and that hinge terminology can be used only when the internal features are known. However, the internal features of the hinge are known in only a few of the archaeocopids and palaeocopids; therefore, description of the carapace of such ostracodes specially calls for discussion of the external characters of the dorsum, including the hinge line.

INTERNAL FEATURES
MUSCLE SCARS

Muscle scars are very important shell features that commonly are preserved in Mesozoic and Cenozoic forms but less so in

Paleozoic ostracodes. Their value in classification and orientation has been neglected until recent years. Much is yet to be learned, but we know that orders and some superfamilies can be recognized on the basis of muscle patterns, as well as a few families and genera.

The presence of scars often has been overlooked because of the apparent opaqueness of the shell. However, coating the surface of many specimens with oil or water or converting calcareous specimens to fluorite may reveal excellently preserved scars. The use of stains and transmitted light also may aid in making muscle scars visible.

The best-known scars in Paleozoic ostracodes are found in the Leperditiidae (Fig. 21, *1,2*). In *Eoleperditia,* adductor, mandibular, antennal, and possibly stomach muscle-scars have been recognized. The large adductor scar is composed of as many as 100 small secondary scars. The over-all shape may be ovate (e.g., *Eoleperditia*) or chevron-shaped (e.g., *Herrmannina*). Muscle scars in many palaeocopids (Fig. 21,*3,4*) appear to be single circular structures about 100 microns in diameter. They are reflected exteriorly in several ways: as a "lucid spot," an unornamented, smooth circular spot, a raised, thickened area, or a pit; internally they may be marked by a circular depression or boss. In sulcate palaeocopids the adductor scar is most commonly opposite to the ventral and usually deepest part of S_2. Internally, this point may be rough or raised, but does not form a distinct pattern of multiple scars.

Only a few scars of myodocopids have been recorded. They are represented by large clusters of many secondary scars, commonly grouped in a distinctive pattern (Fig. 21,*8,13*). The shape and number of the secondary scars and the over-all pattern may vary in different genera. Certainly, the long thin bars of *Entomoconchus* differ strikingly from the subround secondary scars of *Cylindroleberis* or the ovate scars of *Cyclasterope*.

The cladocopine myodocopids have a closely set triangular group of three subround secondary scars placed near midcarapace (Fig. 21,*9*). They have not been reported often from fossils but are recorded from *Polycope.*

In the platycopines (Fig. 21,*7*) the adductor muscle scar is composed of a closely spaced set of secondary scars arranged in a double row. The secondary scars are usually 10 to 14 in number. Their arrangement in a biserial manner is typical of the suborder.

Scars in some of the metacopine podocopids are well known, in others unknown. In *Cavellina* and *Healdia* (Fig. 21,*5,6*) the adductor muscle scar consists of a cluster of as many as 40 closely spaced secondary scars. TRIEBEL (1941, pl. 5, fig. 50) shows mandibular, antennal and adductor scars in *Cavellina.* The large number of secondary scars in *Cavellina* clearly shows that such a form cannot be classified with an ostracode like *Cytherella,* which has a small number of secondary scars, even though the outline of the carapaces is almost identical. Muscle scars have not been reported from some families of Metacopina and, therefore, the variation of scar pattern is unknown.

HOWE & LAURENCICH (1958) show how muscle scars differ among the superfamilies of the podocopids. In the Bairdiacea seven or more secondary scars are closely packed into a subcircular group, very closely packed in *Bythocypris,* loosely packed in *Bairdia.* In *Darwinula* (Fig. 21,*20*) they are arranged in a rosette of nine or ten elongate-ovate scars. Scars in the Cypridacea (Fig. 21, *17,19*) are usually grouped into two sets, one large set of about six ovate scars of the adductor muscle and a second more anteroventral set of two discrete scars representing the mandibular muscles. In some valves antennal scars occur anterodorsally. None of the scars in the cyprids, which usually are visible through the translucent shell, are closely packed. Though not much work has been done on them it appears that the pattern is somewhat similar between such genera as *Candona* and *Cypria* and their value may be limited to family distinction. The Cytheracea (Fig. 21, *10,12, 14-16,18,21*) have a typical set of four discrete ovate scars in a vertical row, flanked anteriorly by one or more isolated mandibular or antennal scars. The axis of the four adductor scars is usually inclined anteroventrally. A few genera can be identified by the pattern or shape of scars. Variation is produced by fusion, fission, or change in shape of either the adductor or mandibular scars.

Very little can be said about the development of the muscle-scar pattern. In the

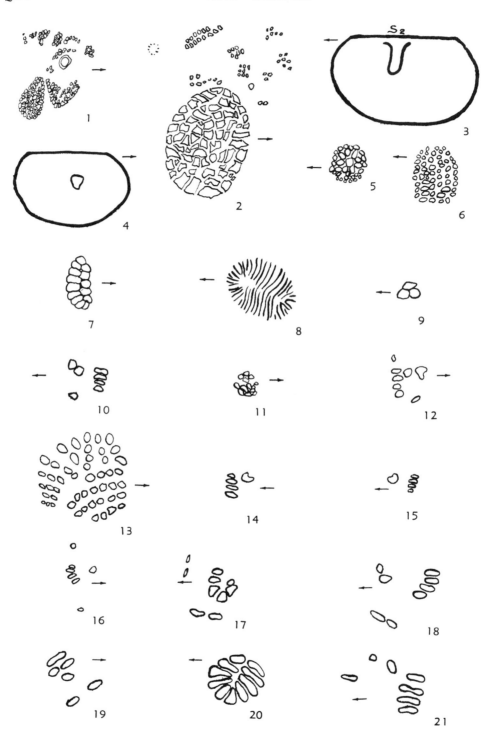

Ordovician *Eoleperditia* the adductor muscle was extremely large, consisting of bundled muscle fibers that produced a broad scar with many closely packed secondary scars. In late Paleozoic *Healdia* and *Cavellina* the adductor muscle is marked on each valve by as many as 40 closely grouped secondary scars. This is in sharp contrast to most post-Paleozoic forms in which the secondary scars are few and commonly spaced well apart.

HINGEMENT

Hingement is one of the most important features for use in classifying Mesozoic and Cenozoic ostracodes. Hinge characters are not so well known among the palaeocopids but are used wherever possible. The following discussion is based in part on the reports of LEVINSON (1950), HOWE & LAURENCICH (1958) and SYLVESTER-BRADLEY (1956). Most of the definitions and examples are based on their work.

The two valves of an ostracode are articulated in a variety of ways along the dorsal margin. Types of hingement may be divided into four broad groups: (1) smooth contact without interlocking devices; (2) straight or curved, smooth or denticulate hinge bar (tongue or ridge or apical list) that fits into a corresponding smooth or denticulate groove (Fig. 22*A*); (3) straight or curved hinge bar and groove supplemented by cardinal teeth and sockets, which are smooth or partially to wholly crenulate (Fig. 22*B*); and (4) a peripheral lock in genera in which one valve is larger than the other, the edge (commonly the selvage) of the smaller valve fitting into a groove in the larger. These major types have been subdivided for the most part on the basis of modifications such as degree of denticulation of the bar or cardinal teeth.

The nature of hingement in the archaeocopids is unknown. In the leperditicopids several kinds of hingement have been recognized. In *Eoleperditia* hingement appears to be of **adont** type (Fig. 23,*1*), consisting of a simple bar and groove. In *Herrmannina,* however, well-preserved vertical bars and slots occur along the hinge in manner that characterizes the **prionodont** type (Fig. 23, *2*). Thus, different types of hinges are recognized as occurring in one family (Leperditiidae) as early as the Ordovician.

The adont hinge was common among the palaeocopids, although by no means the only type of hinge in this group (LEVINSON, 1950). Some leperditellids appear to have valves that meet smoothly along the hinge contact as though they were held together only by an elastic band. In many genera of the Kloedenellacea the hinge is adont but may be modified by one or more faint cardinal teeth, approaching the **lophodont** type (Fig. 23,*3*), characterized by simple anterior and posterior cardinal teeth in one valve that fit into corresponding sockets in the opposite valve. In the Glyptopleuridae the hingement is distinctly lophodont and in the Miltonellidae it is reported as amphidont (see below). Hingement among the palaeocopids is not sufficiently known to make possible discrimination of significant types of articulation common to various genera or families and determination of evolutionary trends in the nature of hingement.

WAINWRIGHT (1959) reports that the hingement of *Dizygopleura swartzi* and *Eukloedenella sinuata* is characterized by the presence of a hinge bar in the left valve and a corresponding groove in the right (adont). In *Eukloedenella* the hinge bar of the left valve is bounded ventrally along its entire length by a distinct groove that engages a well-defined ridge bordering the

---

(See facing page)

FIG. 21. Muscle scars of representative ostracodes.——LEPERDITICOPIDA: *1. Herrmannia welleri,* L. Sil., ×8. *2. Eoleperditia fabulites,* M.Ord., ×11.——PALAEOCOPIDA. *3. Euprimites effusus,* M.Ord., ×37. *4. Tvaerenella carinata,* M.Ord., ×25.——PODOCOPIDA (PODOCOPINA). *10. Clithrocytheridea appendiculata,* Eoc., ×90. *11. Bythocypris arcuata,* Mio., ×60. *12. Cytheridea praesulcata,* Oligo., ×110. *14. Trachyleberis asperrima echinata,* Oligo., ×75. *15. Krithe papillosa,* Mio., ×75. *16. Loxoconcha subtriangularis,* Oligo., ×115. *17. Candona neglecta,* Rec., ×165. *18. Pseudocythere caudata,* Rec., ×250. *19. Cypridopsis vidua,* Rec., ×93. *20. Darwinula stevensoni,* Rec., ×190. *21. Paradoxostoma variabile,* Rec., ×190.——PODOCOPIDA (METACOPINA): *5. Healdia leguminoidea,* Penn., ×250. *6. Cavellina,* Penn., ×180.——PODOCOPIDA (PLATYCOPINA): *7. Cytherella abyssorum,* Rec., ×160.——MYODOCOPIDA (MYODOCOPINA): *8. Entomoconchus. 13. Cypridina homoedwardsiana,* Eoc., ×110.——MYODOCOPIDA (CLADOCOPINA): *9. Polycope orbicularis,* Rec. (*1* after 75; *2, 5, 6* after 70; *3, 4* after 36; *7, 9* after 68; *8* after 366a; *10-16* after 42; *17-21* after 88.)

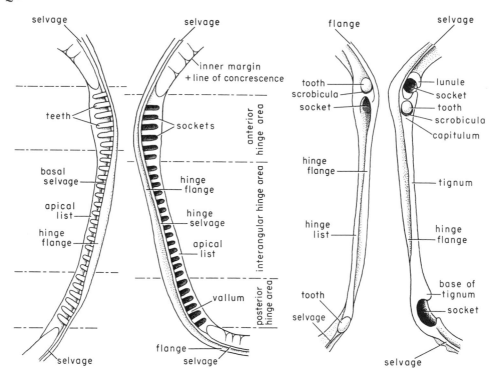

Fig. 22. Nomenclature of hinge elements of cytheracean ostracodes. *A, B, Cytheridea hungarica* ZALANYI, LV, RV (Cytherideidae); *C, D, Cythereis dentata* G. W. MÜLLER, LV, RV (Trachyleberididae) (44).

main hinge groove of the right valve. Such a groove is not present on the dorsal side of the main hinge element but a broad depression in the dorsum serves to accommodate the overhang of the right valve. In *Dizygopleura* the hinge bar of the left valve is not bounded either dorsally or ventrally by a supplementary groove; however, a depression of the dorsum serves to accommodate the overhanging right valve. The hinge bar shows no significant change in size throughout its length and no cardinal hinge teeth are present. As determined from the exposed hinge area of a right valve of *Kloedenella cornuta* no supplementary dentition occurs. The so-called anterior cardinal tooth of the kloedenellids is not part of the hingement but merely a peculiar structure of the overlapping left valve in the region of the anterodorsal free margin; this structure is accommodated by a depression in the right valve. The hinge devices proper terminate immediately behind the "tooth." This structure is best developed in dizygopleurid species; it is less definitive but nevertheless

present in species of *Kloedenella* and *Eukloedenella.*

In both *Dizygopleura swartzi* and *Eukloedenella sinuata* the left valve overlaps the right valve around the entire free margin, in most pronounced manner antero- and posteroventrally (Fig. 24). Overhang of the dorsum of the right valve with respect to that of the left is seen in both species but it is better developed in *Dizygopleura.* This overlap is due to a dorsally directed thickening of the shell, particularly in the region of L_3 and the main sulcus of *Dizygopleura* and in corresponding positions of *Eukloedenella.*

The anterior "cardinal tooth" is a very distinctive feature formed by a thickening of the shell and lateral projection of the margin of the left valve in the region of S_1 and L_2 in quadrilobate and trilobate species and in a corresponding position in bilobate forms. In dorsal view this feature appears as a large, triangular toothlike structure extending from the left valve and overlapping the right valve. The dorsal over-

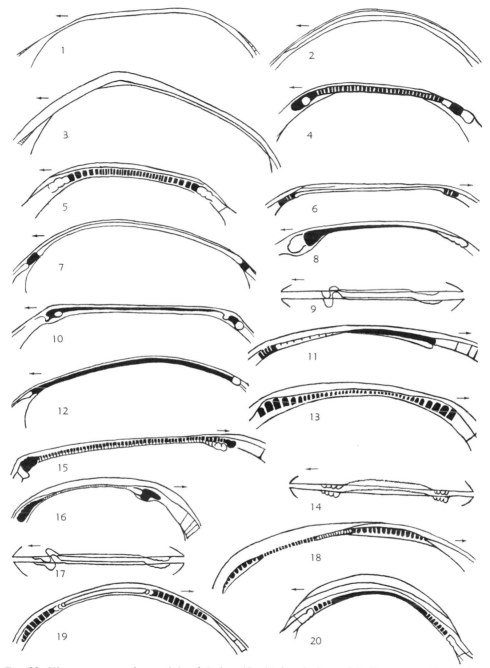

Fig. 23. Hinge structures characteristic of Podocopida (Podocopina).——*1-3.* CYPRIDACEA: *1, Candona compressa* (Cyprididae), ×84; *2, Cypria ophthalmica* (Cyclocyprididae), ×130; *3, Cypridopsis aculeata* (Cyprididae), ×125.——*4-20.* CYTHERACEA: *4, Loxoconcha rhomboidea* (Loxoconchidae), ×133; *5, Cytheropteron latissimum* (Cytheruridae), ×123; *6, Cytherura gibba* (Cytheruridae), ×137; *7, Paradoxostoma normani* (Paradoxostomatidae), ×182; *8-9, Heterocythereis albomaculata* (Hemicytheridae), ×95; *10, Cytheromorpha fuscata* (Paradoxostomatidae), ×130; *11, Hemicytherideis elongata* (Cytherideidae), ×90; *12, Paradoxostoma variabile* (Paradoxostomatidae), ×133; *13-14, Cythere lutea* (Cytheridae), ×137; *15, Leptocythere pellucida* (Leptocytheridae), ×137; *16-17, Hemicythere villosa* (Hemicytheridae), ×98; *18, Cyprideis torosa* (Cytherideidae), ×88; *19, Xestoleberis depressa* (Xestoleberididae), ×137; *20, Hemicytherura clathrata* (Cytheruridae), ×140 (88).

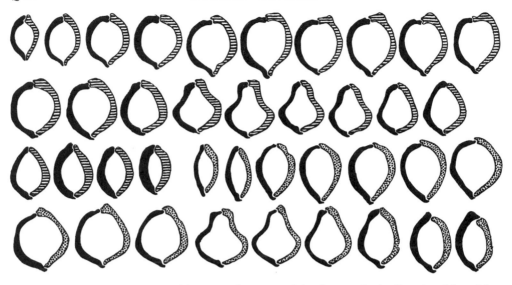

Fig. 24. Serial transverse sections of heteromorph carapaces belonging to Kloedenellocopina, left-to-right arrangement in each row indicating successive positions from rear toward front of carapace (camera lucida drawings). Sections in upper rows with left valve black and right valve ruled were drawn from a female of *Eukloedenella sinuata* ULRICH & BASSLER (M.Sil., Pa.), enlarged. Sections in lower rows with left valve black and right valve stippled were drawn from a female of *Dizygopleura swartzi* ULRICH & BASSLER (M.Sil., Pa.), enlarged (399).

hang of the right valve ends at the posterior margin of the tooth.

WAINWRIGHT also reports that in *Beyrichia moodeyi* and *Kloedenia normalis* the hinge element of the left valve consists of a bar extending the entire length of the hinge area without supplementary dentition or bounding grooves. The right valve is presumed to contain a corresponding groove. This agrees with the type of hinge structures reported by LEVINSON (1950, p. 66, fig. 1) for *Beyrichia fittsi* and *B. jonesi*. The manner of articulation of the valves along the free margin could not be deter-

mined, since only single valves were available.

In *Drepanellina clarki*, according to WAINWRIGHT, the left valve contains a hinge bar with no supplementary hinge structures. This element is formed by the tapered dorsal edge of the valve. The right valve is presumed to contain the corresponding hinge groove. *Mastigobolbina typus* possesses a very narrow, shallow groove extending across the length of the hinge area of the left valve. This is presumed to have accommodated a fine ridge in the right valve.

(See facing page)

Fig. 25. Diagrammatic illustrations of ostracode hinge structures; *1-14*, dorsal views of LV and RV; *15-42*, interior views.——*1*. Adont.——*2*. Prionodont.——*3*. Lophodont.——*4-7*. Merodont types: *4*, paleomerodont; *5*, holomerodont; *6*, antimerodont; *7*, hemimerodont.——*8*. Entomodont.——*9*. Lobodont.——*10-12*. Amphidont types: *10*, paramphidont; *11*, hemiamphidont; *12*, holamphidont.——*13*. Schizodont.——*14*. Gongylodont.——*15*. *Thlipsura furca*, ×40.——*16, 18*. *Thlipsurella fossata*, ×24.——*17*. *Ctenobolbina papillosa*, ×28.——*19*. *Milleratia cincinnatiensis*, ×52.——*20*. *Ceratopsis chambersi*, ×13.——*21, 23*. *Primitiopsis bassleri*, ×51.——*Bonneprimites bonnemai*, ×40.——*24*. *Eridoconcha rugosa*, ×57.——*25*. *Haplocytheridea curvata*, ×60.——*26*. *Paracytheridea brusselensis*, ×76.——*27*. *Cypridina homoedwardsiana*, ×33.——*28*. *Cyprideis apostolescui*, ×50.——*29*. *Bythocypris cuisensis*, ×40.——*30*. *Paracypris contracta*, ×40.——*31*. *Bairdoppilata gliberti*, ×30.——*32*. *Ruggieria micheliniana*, ×40.——*33*. *Paracytheridea grignonensis*, ×76.——*34*. *Cytheretta haimeana*, ×60.——*35*. *Costa edwardsi*, ×50.——*36*. *Cytheridea praesulcata*, ×13.——*37*. *Bradleya cornueliana*, ×40.——*38*. *Bradleya approximata*, ×40.——*39*. *Boldella deldenensis*, ×89.——*40*. *Haplocytheridea curta*, ×65.——*41*. *Cuneocythere foveolata*, ×60.——*42*. *Drepanella crassinoda*, ×27. BEYRICHICOPINA are represented by *17, 20, 21, 23, 42*; KLOEDENO-COPINA by *19, 22, 24*; PODOCOPINA by *25, 26, 28-41*; METACOPINA by *15, 16, 18*, and MYODOCOPINA by *27* (*1-14*, after 34; *15-24*, after 49; *25-39, 41*, after 42; *40*, after 7; *42*, after SCOTT, n.)

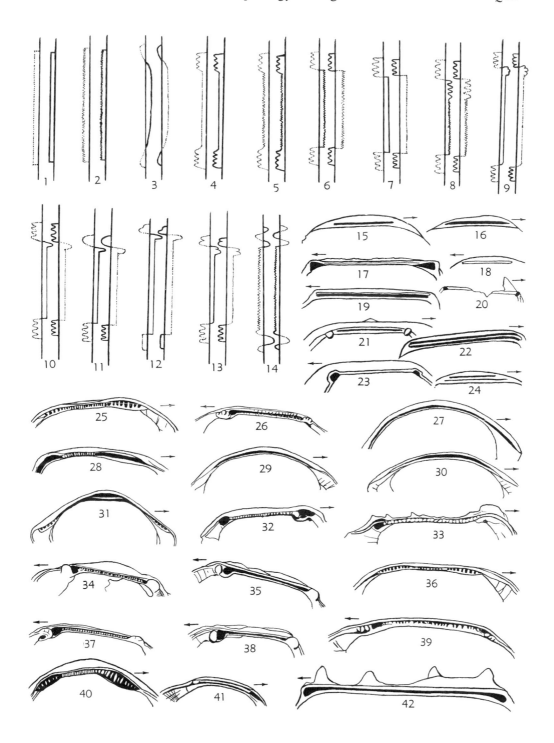

The ventral edge of the left valve of *Drepanellina clarki* contains a groove—very distinct in the region of the dimorphic inflation—that accommodated a projection of the ventral edge of the right valve.

The **peripheral** type of hinge lock is found in the Cytherellidae and among some of the Paleozoic Cavellinidae and a modified form in the Bairdiidae. This type is well represented in the modern *Cytherella* and *Cytherelloidea*.

Merodont hingement (Fig. 25,4,7) consists of a bar terminated by crenulate teeth in one valve opposed to a groove and sockets in the opposite valve. Four varieties are known and all are believed to be a development of adont or prionodont structures. The **paleomerodont** variety of merodont hingement (Fig. 25,4), characterized by simple bar and crenulate cardinal projections, is found in *Schuleridea* from the Jurassic; the **holomerodont** type (Fig. 25,5), with crenulate bar and terminal cusps, is recognized in *Haplocytheridea;* a partial reversal of holomerodont called **antimerodont** (Fig. 25,6), in which the terminal projections and groove are crenulate, is found in *Clithrocytheridea;* and **hemimerodont** hingement (Fig. 25,7), similar to antimerodont except for smooth median elements, is seen in *Palaeocytheridea*. Varieties of merodont articulation are common in Mesozoic ostracodes. Whether or not it occurs in the Paleozoic is unknown, but clearly it seems to have been derived by modification of Paleozoic types of hingement; in what part of geologic time the modification occurred is not precisely determined.

In **entomodont** hingement (Fig. 25,8) four elements are distinguished: a median bar (and groove) with (1) a coarsely crenulate anterior element and (2) smooth or crenuate remaining portion and (3) anterior and (4) posterior crenulate teeth (and sockets). A variety in which the anterior median element is lobate instead of crenulate is known as **lobodont** (Fig. 25,9). *Progonocythere* characterizes the former and *Acanthocythere* the latter.

Amphidont (heterodont of some authors) hingement (Fig. 25,10,12), characterized by a smooth tooth at the anterior end of the bar, is represented in the Cretaceous and becomes the most common type in the Tertiary. Its first reported appearance is in the Miltonellidae of the Permian. In the variety defined as **paramphidont** hingement (Fig. 25,10), as found in *Cythereis,* the median bar is crenulate or smooth, the anterior and posterior elements are notched or crenulate, and at the anterior end is also a smooth tooth or socket. In another variety called **hemiamphidont** (Fig. 25,11) the elements are like those of paramphidont hingement except that the posterior element is smooth or stepped, not crenulate (e.g., *Brachycythere, Alatacythere*). In a third variety, **holamphidont** (Fig. 25,12), the terminal elements are smooth or stepped and noncrenulate (e.g., *Amphicytherura, Trachyleberis, Pterygocythereis*).

The term **schizodont** (Fig. 25,13) has been applied to hingement observed in a small group having bifid anterior elements in both valves (e.g., *Paijenborchella*). This structure obviously is a minor modification of the amphidont type.

Loxoconcha has a rather complex hingement. Seen in lateral view, the hinge is somewhat convex, with the anterior element of one valve sharply downturned and the posteriormost element a knoblike tooth; in the opposite valve the reverse order of elements is found. This type has been referred to as *gongylodont* (Fig. 25,14).

Most of the named hingements may be found in the Cytheracea (Fig. 23,4,20). A highly varied set of hinge features are known to occur in the Cypridacea but they have not as yet been evaluated (Fig. 23,1,3).

REFERENCES

Howe, Henry V., & Laurencich, Laura
(1) 1958, *Introduction to the Study of Cretaceous Ostracoda:* Louisiana State Univ. Press, 536 p., text fig. (Baton Rouge).

Jaanusson, Valdar
(2) 1957, *Middle Ordovician ostracodes of central and southern Sweden:* Geol. Inst. Univ. Uppsala Bull., v. 37, p. 176-442, 15 pl., 46 fig.

Kesling, R. V., & Rogers, K. J.
(3) 1957, *Size, lobation, velate structures, and ornamentation in some beyrichiid ostracods:* Jour. Paleont. v. 31, no. 5, p. 997-1009, pl. 127-130 (Tulsa).

Levinson, S. A.
(4) 1950B, *The hingement of Paleozoic Ostracoda*

and its bearing on orientation: Jour. Paleont.,
v. 24, p. 63-75 (Tulsa).

Sylvester-Bradley, P. C.

(5) 1956, *The structure, evolution and nomen-clature of the ostracod hinge:* Brit. Mus. (Nat. Hist.), B, Geol., v. 3, no. 1, 21 p., illus. (London).

Triebel, Erich

(6) 1941, *Zur Morphologie und Ökologie der fossilen Ostracoden, mit Beschreibung einiger neuer Gattungen und Arten:* Senckenberg., v. 23, p. 294-400, 15 pl.

Wainwright, John

(7) 1959, *Morphology and taxonomy of some Middle Silurian Ostracoda:* Thesis, Univ. Illinois, p. 1-128 (Urbana).

DIMORPHISM OF OSTRACODA

By H. W. Scott and John Wainwright

[University of Illinois and Shell Oil Company]

CONTENTS

Syngamic and parthenogenetic methods of reproduction are known to occur in living ostracodes, somewhat rarely even within a single species that in part of its area of distribution may be syngamic and elsewhere parthenogenetic. Dimorphism in living or Mesozoic-Cenozoic ostracodes is not important in classification as it is among the Palaeocopida.

Dimorphic characters in the palaeocopids have been long recognized although much is yet to be learned about their morphology. Previous lack of adequate understanding often has resulted in misorientation of the carapace and confused classification. Various kinds of dimorphic structures have been described: (1) **kloedenellid** or **domiciliar,** characterized by slightly swollen to strongly inflated posterior portions of the carapace, as displayed in *Kloedenella* and many modern genera of the Podocopida; (2) **lobate,** distinguished by presence of a dimorphic lobe on the lateral surface in a lateroventral or anteroventral position, as found in *Zygobolbina* and *Bonnemaia;* (3) **beyrichiid,** marked by a strongly inflated portion of the domicilium that forms a pouch (crumina), as in *Beyrichia,* probably a special type of velar development; (4) **velate,** including structures classed as (a) **simple velum,** straight to slightly convex, as in *Hollinella,* (b) **closed velum,** forming a false pouch (dolon), as in *Uhakiella,* (c) **locular,** with compartment-like divisions, as in *Abditoloculina* and *Tetradella;* and (5) **histial,** defined by a flangelike structure continuous with the ventral ridge connecting the lobes parallel to the free margin and protruding ventrally, as in *Sigmoopsis* and *Glossomorphites.*

The sex of an ostracode cannot readily be determined in all cases. Adult females of species exhibiting dimorphic structures can be recognized without difficulty, whereas the adult males of such species lack carapace structures trustworthy for distinguishing them from juvenile stages of both sexes. Accordingly, Jaanusson (1956) has suggested that the adult females of dimorphic species be referred to as **heteromorphs** and the males and juveniles as **tecnomorphs.**

KLOEDENELLID DIMORPHISM

A common type of dimorphism seen both in the Kloedenellocopina and Recent ostracodes is characterized by inflation of the posterior portion of the domicilium of the female. In the palaeocopid group such inflation is referred to as kloedenellid. Kloedenellid dimorphism is observed most readily in dorsal view. The males and pre-adult instars are usually elongate-ovate, with the greatest width medial, whereas the adult female carapace is wedge-shaped, with the greatest width in posterior position. This has been observed in many genera (e.g., *Kloedenella, Dizygopleura, Glyptopleura, Geisina, Sansabella*) and is believed to be present in all Kloedenellacea.

The posterodorsal inflation of the domicilum in kloedenellid heteromorphs is on the whole so similar to the genital inflations of certain living ostracodes that their homologous relationship can hardly be doubted. Fossil forms possessing such inflations were distinguished as females by VAN VEEN (1922) in *Kloedenella*, and by SWARTZ (1933) in *Kloedenella, Eukloedenella,* and *Dizygopleura;* these authors interpreted the dimorphic inflation as posterior in position, agreeing with now accepted orientation. However, ULRICH & BASSLER (1923a,b) did not recognize the presence of dimorphism in Middle Silurian Kloedenellidae and identified the wider end of the carapace as anterior, thereby assigning the domiciliar inflation to a forward location and the main sulcus to a posterior position.

Among Middle Silurian Kloedenellidae the species of *Dizygopleura* best illustrate relationships of the dimorphic swelling to other lobal features of the domicilium. In *D. swartzi,* a typical representative of the genus, the inflation is in the posterodorsal region of the carapace and merges anteriorly with L_4 so as partially to obscure the posterior margin of this lobe; in effect it is an annex to L_4. TRIEBEL (1941, p. 356) points out that similar inflation in living ostracodes functions as a brood chamber, or that, in forms not offering protection for the young, it houses the enlarged oviduct. The exact function of the kloedenellid homologue is, of course, open to speculation.

In species of *Kloedenella* (e.g., *K. cornuta*) the dimorphic swelling tends to obscure the posterior border of the adjacent lobe L_3. In both *Dizygopleura* and *Kloedenella* the dimorphic feature is well defined laterally and is quite apparent in dorsal aspect. In species of *Eukloedenella,* how-

ever, the inflation is not very distinct laterally, though easily recognized in dorsal view.

Unfortunately, inadequate attention has been paid to effects of kloedenellid dimorphism on shell features, and accordingly different dimorphs have been described as different species. Inflation or swelling of the posterior region of the carapace not only changes its shape in dorsal view but also causes the hinge channel to widen backward, in some species altering the degree of overlap from the posterocardinal point around the posterior end.

LOBATE DIMORPHISM

In some palaeocopids the anterior, anteroventral, or ventral portion of the carapace is raised into a distinct lobe. In some species this ventral lobe may be a part of the anterior (L_1) and ventral lobes. Transverse sections through these lobes (Fig. 26), show no evidence of a remnant partition such as always is found in *Beyrichia* (Fig. 27); only an external swelling of the domicilium and a corresponding internal deepening of the carapace is seen. A form of domiciliar dimorphism is indicated, differing from typical kloedenellid only both in its location and its lobate nature. In kloedenellid dimorphism the posterior portion of the carapace is swollen to produce a wedge-shaped general outline to the carapace, whereas in lobate dimorphism it is a portion of the anterior, anteroventral, or ventral area of the carapace that is inflated to produce a distinct lobe. This dimorphic lobe may be short or long, straight or curved, high or low, and smooth or ornamented. The lobe of *Zygobolbina* is a typical example of lobate dimorphism.

In members of the Zygobolbidae, dimorphism is characterized by anterior to anteroventral inflation of the domicilium; such an orientation of the dimorphic feature places S_2 anteriorly. These inflations are not homologous to the posterodorsal swelling of kloedenellids. In *Mastigobolbina typus* dimorphism is characterized by inflation of part of the domicilium anterior and anteroventral to L_2 (Fig. 26); in *Drepanellina clarki* the dimorphic feature is an enlargement of the ventral lobe (Fig. 26).

The zygobolbid mode of dimorphism was

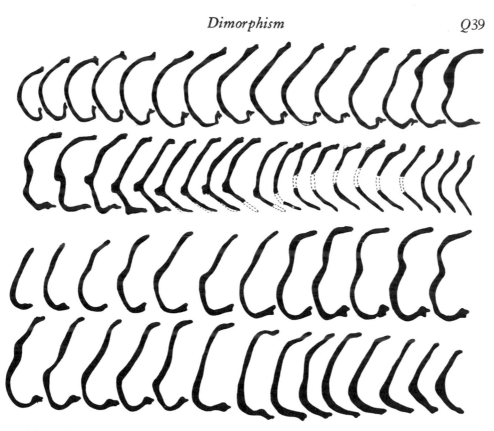

FIG. 26. Serial transverse sections of heteromorph left valves belonging to Beyrichicopina illustrating features of lobate dimorphism, left-to-right arrangement in each row indicating successive positions at approximate 100-micron intervals from rear toward front of valve. Upper two rows drawn from a specimen of *Mastigobolbina typa* ULRICH & BASSLER (M.Sil., Pa.), enlarged. Lower two rows drawn from a specimen of *Drepanellina clarki* ULRICH & BASSLER (M.Sil., Pa.), enlarged (399).

apparently initiated during early Middle Silurian time (lower part of Rose Hill Formation) in the many species of *Zygobolba, Zygobolbina,* and *Zygosella* (ULRICH & BASSLER, 1923b, pl. 39-45); it continued through Rose Hill time in the genera *Bonnemaia* and *Mastigobolbina* and culminated in the genus *Drepanellina* of the Rochester Shale.

BEYRICHIID DIMORPHISM

Dimorphism in *Beyrichia,* distinguished as the beyrichiid type (Fig. 27), was recognized long ago by the presence of a large "pouchlike" structure. This was considered to mark the posterior end of the carapace and only recently has it been realized that the prominent swelling is anteroventral in the adult female.

The pouch of *Beyrichia* seems to be directly related to expansion of the velum.

This type of dimorphism has been called cruminal, to distinguish it from the external pouch formed as a velate structure in the Eurychilinidae and Hollinidae.

RICHTER (1869) was first to recognize this type of dimorphism in several European species of *Beyrichia,* at the same time comparing the cruminae (small saclike projections) with the posterior swellings of the female of Recent *Cythere gibba.* ULRICH & BASSLER (1923a) suggested that cruminae were brood pouches in which eggs and larval forms could be protected. These authors (also KUMMEROW, 1931) interpreted the cruminae as belonging in a posterior position that would offer least resistance during movement of the animal through water and least impedance from obstacles encountered as the ostracode crawled along the bottom. BONNEMA (1930, 1932) placed the cruminae anteriorly, an orientation vigorously ob-

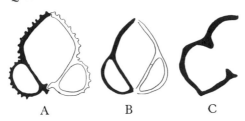

FIG. 27. Transverse sections of heteromorph carapaces and left valve belonging to Beyrichicopina, illustrating beyrichiid dimorphism.——*A. Hibbardia lacrimosa* (SWARTZ & ORIEL) (M.Dev., N.Y.), with complete inner partition in each valve (204).——*B. Phlyctiscapha rockportensis* KESLING (M.Dev., Mich.), with complete inner partition in each valve, enlarged (202).——*C. Beyrichia moodeyi* ULRICH & BASSLER (M.Sil., N.Y.), with incomplete inner partition enlarged (399).

jected to by KUMMEROW (1931). The currently accepted position of the cruminae is anterior to anteroventral, based on position of the adductor muscle scars.

The function of the cruminae has been the subject of much discussion among students of the Beyrichiidae. It has been suggested that this structure was used as a breeding room and should therefore occur posteriorly, a position in which eggs could most easily be transferred, by means of posterior appendages, from the region of the genitalia into the cruminae. However, the objection has been raised that currents produced by posterior natatory appendages would tend to flush the young out of the brood space. HESSLAND (1949), from a detailed study of the brood-pouch problem, reached judgment that the posterior location of the cruminae "would have been a serious hindrance during copulation or might even have made this process impossible." Such conclusions are based on observations made by ELOFSON (1941) on copulation in living ostracodes. HESSLAND (1949) points out that an anterior brood space would place the eggs and young in dangerous proximity to the oral appendages, and thus it is likely that they would suffer from indiscriminate eating habits of the adult animal.

SCHMIDT (1941) believed that cruminate specimens were males, and that the cruminae were used for storage of spermatozoa. TRIEBEL (1941) also suggested that the crumina was a male structure that served as a sperm vesicle. Conversely, it could be a structure of the female carapace into which

seminal fluids could be injected by the male animal during copulation, and subsequently utilized by the female when ovulation occurred. SCHMIDT (1941, p. 11) suggested that these chambers might be gas-filled cavities that could lend buoyancy to the carapace. The discovery by HESSLAND (1949, pl. 14, fig. 9) of larval carapaces in the cruminae of carapaces of *Beyrichia kloedeni* that were sectioned transversely, and similar observations by SPJELDNAES (1951) in *B. jonesi,* strongly suggest that the crumina was used as a brood chamber. In each of these reported occurrences, adventitious introductions of carapaces from without were not considered probable.

The possibility of transferring eggs from the region of the genitalia to anteroventral cruminae has been investigated by HESSLAND (1949).

Observation of lobate ostracodes in which orientation of the carapace is well established shows that the largest lobe of each valve (not to be confused with crumina) occurs posteriorly, and it is in this area that growth is most pronounced as the animal matures; we may therefore assume that the largest lobe of *Beyrichia moodeyi* occurs posteriorly and houses the genitalia. In the tecnomorphs of this species, the posterior lobe possesses an acuminate ventral extension that curves forward to its termination at a depression ventral to L_2; this depression occupies the same position as the crumina in heteromorphic valves. A slight inflation of the velum occurs subjacent to the extension of the posterior lobe. In the heteromorph, a similar extension of L_3 terminates at the posterodorsal margin of the crumina and is confluent with the subjacent inflation of the velum. This is indeed suggestive of a duct by means of which ova could be moved forward into the crumina by a bending of the abdomen.

The beyrichiid crumina has been judged by many workers to be an extension of the domicilium. KESLING & ROGERS (1957) conclude that ". . . the female brood pouch originated when a strongly convex part of the frill fused with the contact margin."

Transverse serial sections of *Beyrichia moodeyi* (Fig. 28) show that the crumina is not a swelling of the domicilium but a chamber within the velum formed by an inflation of the epidermal fold that produced

FIG. 28. Serial transverse sections of heteromorph left valve of *Beyrichia moodeyi* ULRICH & BASSLER (M. Sil., N.Y.) illustrating beyrichiid dimorphism; left-to-right arrangement in each row indicating successive positions at approximate 50-micron intervals from rear toward front of valve (two sections at left of top row hypothetical, showing coalescence of sides of velum), ×14 (399).

the velum. In this species, the bulbous crumina is continuous with the narrow, elongate cavity within the velum. The latter extends backward, becoming narrower as the sides of the velum converge, and is presumed to be completely eliminated in the posteroventral region of the valve where the open cavity is reduced to a mere dark line—vestige of the epidermal fold—bisecting the velum. A reflection of the crest of the velum can be traced across the external, ventral surface of the crumina; the position of the crest is marked internally by a depression on the lower floor of the crumina.

KESLING (1957) notes that in some beyrichiid heteromorphs the velate structure is completely interrupted by the crumina, in others it encroaches onto the sides of the crumina, and in still others it extends across the entire crumina. He suggests that during phylogeny the velate structure could either have retreated from the crumina or gradually grown across it, but favors the second explanation because "the frill does not extend onto the pouch in Lower Silurian ostracodes. It likewise seems significant that the oldest known ostracodes having a frill across the brood pouch are from Upper Silurian rocks." Study of the Middle Silurian *Beyrichia moodeyi* shows that a diminution in relief of the crest of the velum along the ventral surface of the crumina parallels the development of the crumina, the crest being most reduced where the crumina attains maximum inflation. This is only a short step beyond the point at which the reflection of the velar crest did not extend across the entire crumina. Originally, according to KESLING (1957), the crumina developed

as a specialized section of the velum, and encroachment of the unspecialized velate structure onto and across the crumina evolved later. He gives excellent illustrations of the relationship of the velate structure to the crumina in several members of the Beyrichiidae.

The relationship between the crumina and the velate structure in species in which the latter is complete and extends along the proximoventral side of the crumina is not clear. In beyrichiids with broad velate structures the crumina is contained within the velum, as in *Beyrichia moodeyi*, and does not restrict the lobes to any great extent. *B. salteriana* possesses a reduced velate structure and the crumina has encroached upon the domicilium to restrict the lobes (KESLING, 1957); in this species, the velate structure appears to be interrupted by the crumina. A further reduction of the velate structure occurs in *Phlyctiscapha rockportensis*, in which the crumina merges externally with the domicilium and the much-reduced velate structure occurs proximoventral to the crumina paralleling the ventral free margin. Internally, however, the crumina is quite separate from the domicilium, a partition being present between parts of the two features.

The beyrichiid crumina, although in some shells possessing an external resemblance to the dimorphic swelling of members of the Zygobolbidae, is distinguished from the latter by its internal relationship with the domicilium. A survey of published transverse sections of beyrichiid heteromorphic carapaces shows a prevailing tendency toward separation of the cruminal cavity from the domicilium, some being accentuated by the presence of a dividing wall (Fig. 27). Even where a partition is absent, the shell tends to be thickened and produced in a direction toward the contact margin of the valve in the area of the juncture between the domicilium and crumina; concurrently, the ventral marginal area of the valve is extended toward a union with this feature. On the other hand, no tendency toward separation of the domicilium and dimorphic inflation is observed in the Zygobolbidae. Further study of transverse serial sections of beyrichiid ostracodes is essential for proper understanding of structural relationships between the crumina and velate structure and determination of its phylogenetic significance.

VELATE DIMORPHISM

A fourth type of dimorphism is recognized in the development of certain velar adventral features. JAANUSSON (1957) has shown the relationship of the velum (frill) to dimorphism, but the problem has been complicated by introduction of the term "histium." We recognize the primary or main frill as developed in the Hollinidae as a velum. It may be associated with one or more secondary ribs in either a ventral or dorsal position; the secondary ribs may be continuous or broken into tubercles or spines. KESLING & ROGERS (1957) illustrate two subvelar ridges in *Treposella lyoni* and *T. stellata*. The velum is not just an external ornamental feature comparable to a costa but is a structure believed to have been formed by a fold of the inner chitin layer. As such, it shows a dark line of chitin between the outer and inner walls. This has been observed in *Hollinella* and *Hibbardia*.

The velum may be expressed as a broad frill that commonly is incurved. When curved strongly inward the free edges of the velum may approach each other closely or actually touch so as to form a pouch (dolon). The position of the dolon, if present, varies from anterior, anteroventral, ventral, to posterior. The function of the velum and velar dolon has been variously interpreted but it is generally accepted as an expression of dimorphism. Velate structures may have served different purposes in different genera.

In many species of *Hollinella* a splendid development of the velum is seen in the adult. In some species it is a frill that extends along the ventral and all or part of the anterior margin like sled runners; in others it is strongly convex and almost meets along a portion of the venter. A typical velar dolon is formed anteroventrally in the heteromorph of *Uhakiella*. In the Primitiopsidae the dolon occupies a posterior position and in the Eurychilinidae a ventral position.

In a few genera a very distinctive type of dimorphic feature has the form of successive closely spaced compartments called **loculi**. The development of this peculiar structure has been well illustrated by KESLING (1958). It is here considered that loculi are modifications of the velum. The locular type of dimorphism is well developed in *Ctenoloculina* where a row of locular cups open ventrally from an anteroventral and ventral position. Some loculi, as pointed out by KESLING (personal communication), are developed outside the frill (abvelar) (e.g., *Tetradella*) and others inside the frill (advelar) (e.g., *Ctenoloculina*).

HISTIAL DIMORPHISM

The term histium has been applied loosely to a dimorphic structure which has been confused with velum. Extensive studies of this problem indicate that the term velum should be retained for the main frill and that histium should be restricted to such marginal structures as are found in *Sigmoopsis*. KESLING (personal communication) says: "Histium is a ridge or flangelike structure parallel to the free edge generally protruding ventrally and lateroventrally, which, as seen in lateral view, (1) is continuous with a ventral ridge connecting lobes (e.g., *Glossomorphites*), (2) lies in the position of a connecting ventral ridge or lobe with no sharply defined boundary separating it from the lobes (e.g., *Sigmoopsis*), or (3) continues downward from the lobate area with sharp line of demarcation (e.g., *Ogmoopsis, Aulacopsis*). The base of a histium is broad and flared at its junction with the rest of the valve, which in most genera lies at the bend of the valve between the lateral and marginal surfaces. Histium is not the equivalent of velum or velar ridge, which in nearly all hollinids (e.g., *Hollinella*) has a sharp line of junction with the rest of the lateral surface. If a ventral lobe is present in the Hollinidae (e.g., *Hollinella, Hanaites, Abditoloculina, Falsipollex*), the velum or frill is below and distinctly set off from it. The broad structure around *Tetradella* is also a velum, and not continuous with the ventral connecting ridge. It is not certain that the ridge present in some sigmoopsids between the marginal ridge

and the histium is homologous to the velar structure in hollinids."

The primary character of a histium is that it is part of a ventral ridge connecting the lateral lobes or a ridge that continues ventrally from the lateral lobe. In this sense it does seem to characterize a distinct type of dimorphism.

REFERENCES

Bonnema, J. H.

(1) 1930, *Orientation of the carapaces of Paleozoic Ostracoda:* Jour. Paleont., v. 4, no. 2, p. 109-20, 14 fig. (Tulsa).

(2) 1932, *Orientation of the carapaces of Paleozoic Ostracoda:* Same, v. 6, no. 3, p. 288-295, 13 fig.

Elofson, Olof

(3) 1941, *Sur Kenntnis der marinen Ostracoden Schwedens mit besonderer Berücksichtigung des Skageraks:* Zool. Bidr. f. Uppsala, v. 19, p. 215-534, 52 fig., 42 maps.

Hessland, Ivar

(4) 1949, *Investigations of the Lower Ordovician of the Siljan District, Sweden. I. Lower Ordovician ostracods of the Siljan District, Sweden:* Uppsala Univ. Min-geol. Inst. Bull., v. 33, p. 97-408, pl. 1-18.

Jaanusson, Valdar

(5) 1957, *Middle Ordovician ostracodes of central and southern Sweden:* Geol. Inst. Univ. Uppsala, Bull., v. 37, p. 176-442, 15 pl., 46 fig.

——, **& Martinsson, Anders**

(6) 1956, *Two hollinid ostracodes from the Silurian Mulde Marl of Gotland:* Same, v. 36, p. 401-10, 1 pl., 2 fig.

Kesling, R. V.

(7) 1957, *Origin of beyrichiid ostracods:* Univ. Michigan Mus. Paleont. Contr., v. 14, no. 6, p. 57-80, pl. 1-17, fig. 1-5 (Ann Arbor).

——, **et al.**

(8) 1958, *Bolbineossia, a new beyrichiid ostracod genus:* Jour. Paleont., v. 32, no. 1, p. 147-151, pl. 24, 7 fig. (Tulsa).

——, **& Rogers, K. J.**

(9) 1957, *Size, lobation, velate structures, and ornamentation in some beyrichiid ostracods:* Same, v. 31, no. 5, p. 997-1009, pl. 127-130.

Kummerow, E. H. E.

(10) 1931, *Über die Unterschiede zwischen Phyl-*
locariden und Ostracoden: Centralbl. f. Min. Geol. Pal., Jahrg. 1931, Abt. B., no. 5, p. 242-257, 18 fig. (Stuttgart).

Richter, Reinhard

(11) 1869, *Devonische Entomostraceen in Thuringen:* Zeitschr. Deutsch. Geol. Gesell., v. 21, p. 757-776, pl. 20, 21 (Berlin).

Schmidt, E. A.

(12) 1941, *Studien in böhmischen Caradoc (Zahoran-Stufe). I. Ostrakoden aus den Bohdalecschichten und über die Taxonomie der Beyrichiacea:* Senckenberg. Naturf. Ges. Abh. 454, 87 p., 2 figs., 5 pl. (Frankfurt a-M.).

Spjeldnaes, Nils

(13) 1951, *Ontogeny of Beyrichia jonesi Boll:* Jour. Paleont., v. 25, no. 6, p. 745-755, pl. 103-104, 3 fig. (Tulsa).

Swartz, F. M.

(14) 1933, *Dimorphism and orientation in ostracodes of the family Kloedenellidae from the Silurian of Pennsylvania:* Same, v. 7, p. 231-260, pl. 28-30.

Triebel, Erich

(15) 1941, *Zur Morphologie und Ökologie der fossilen Ostracoden, mit Beschreibung einiger neuer Gattungen und Arten:* Senckenbergiana, v. 23, p. 294-400, 15 pl. (Frankfurt a-M.).

Ulrich, E. O., & Bassler, R. S.

(16) 1923a, *Paleozoic Ostracoda: Their morphology, classification, and occurrence:* Maryland Geol. Survey, Silurian vol., p. 271-391, fig. 11-26 (Baltimore).

(17) 1923b, *Systematic paleontology of Silurian deposits (Ostracoda):* Same, p. 500-704, fig. 27, pl. 36-65.

Veen, J. E. van

(18) 1922, *The identity of the genera Poloniella and Kloedenella:* Roy. Acad. Amsterdam, Pr. 23, no. 7, p. 993-996, 1 pl.

ORIENTATION OF OSTRACODE SHELLS

By H. W. Scott
[University of Illinois]

The question of which are anterior, posterior, dorsal, and ventral parts of fossil ostracode carapaces has been attacked in various ways by paleontologists. Of course, students of living ostracodes have not been concerned with such matters because the soft parts inside the carapace furnish definitive information. The problem of orientation has been most serious for those working with the palaeocopids, as evidenced by the fact that a given species has been oppositely oriented by workers who interpreted the same morphological feature as representing different functions. In fact, confusion has been so great that specimens of the same species have been oriented differently by the same worker.

ANTERIOR-POSTERIOR

The position of the brood pouch remained in doubt for a long time. At first it was considered to be invariably posterior, but later this was found to conflict with evidence furnished by other criteria. Seemingly the pouch in Beyrichiacea does not correspond to the dimorphic swelling in Kloedenellacea as indicated by the position of such features as major sulcus (S_2), adductor muscle scar, and "eye-spots," by the curvature of lobes or sulci, greatest width and height, narrowest and most extended ends, by size of the cardinal angles, and by the direction of spines. There has been no consistency in evaluation of these characters. Some have judged that dimorphic structures should be considered to outrank all other criteria in importance, but when the anteroventral position of the brood pouch in Beyrichiacea came to be recognized most problems in orientation were resolved.

The adductor muscle scars now universally are recognized to have primary value in orientation. Muscle scars in the great majority of Recent ostracodes are located in front of the mid-length.

In addition to impressions on the carapace made by the adductor muscle, scars of the mandibular, antennular, and antennar muscles may be preserved. In some shells the adductor muscle scar occurs near mid-length of the valves, but usually most of the muscle-scar pattern is anterior. In a few forms the scar may appear to be behind the mid-length, but if its position is compared to mid-length of the hinge it will be classifiable as anterior.

The position of the adductor muscle scar may be determined in several ways: (1) by direct observation of the scar on the interior or on molds and casts of the interior; (2) by looking through transparent valves; (3) by covering opaque shells with water or a clear oil so that the scar becomes visible through the shell; (4) by detecting one or more dark spots on the shell where the carapace is either thicker or thinner than average; (5) by noting the position of S_2, which usually is associated with the adductor muscle; (6) by determining the position on the surface of an interruption of reticula, punctae, or other ornamental features; (7) by observing the central position of a radiate surface pattern of lines; (8) by determining the position of a pit, as in the Kirkbyidae; and (9) by marking the position of a raised smooth central boss. Any of these found to be placed distinctly or slightly away from the valve mid-length or hinge mid-length may indicate the position of the adductor muscle and thus aid in orientation.

The greatest height of the carapace commonly is anterior (Fig. 18). A striking exception to this rule is found in the Leperditicopida where the greatest height is definitely posterior. Large and excellently preserved muscle scars of the adductor muscle and muscles of anterior appendages are found abundantly in several genera of the order. They are located in the narrow anterior portion of the leperditicopids (Fig. 29). In some genera of palaeocopids, the greatest height is nearly medial, and in a few (e.g., some species of *Kloedenella* and *Eukloedenella*) variations within a given species may shift the position of greatest height from posterior to anterior, but in spite of exceptions the position of the greatest height is a valuable clue to orientation of ostracodes.

The greatest width of the carapace is usually in the posterior half, corresponding to position of the genital organs, but in some

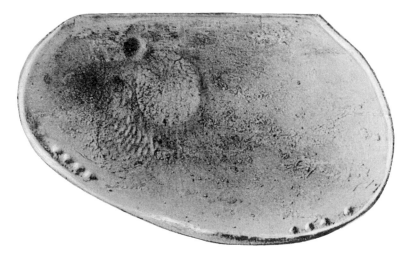

Fig. 29. *Eoleperditia fabulites* (CONRAD), M.Ord., USA(Ill.); RV int., showing eye spot, muscle scar, and tubercles along inner ventral margin (function of tubercles unknown but they may have acted as "door-stops" for smaller valve when valves were closed), ×10 (70).

species it is medial or anterior. In forms having a kloedenellid type of dimorphism and in all modern dimorphic groups the greatest width is posterior in females and medial to posterior in tecnomorphs. In some palaeocopids consisting possibly of partho-genetic genera, the greatest width is medial, but careful measurements often will show that even in these the greatest width is slightly posterior. In the Healdiidae a fur-row parallels the border of the widest end and is considered posterior. The Beyrichia-cea are exceptional in that greatest width measured through the dimorphic structures is anterior. Greatest width of most ostra-codes is in the end opposite to that possess-ing the greatest height and opposite to the end possessing the muscle scars. Exceptions are known but they do not lessen the value of greatest width as a criterion of orienta-tion.

Dimorphic features, except in the Bey-richiidae and Zygobolbidae, are found in the posterior half of the carapace. They are important both for orientation and classifi-cation of many ostracodes ranging from Ordovician to Recent.

Some smooth ostracodes without sulci, lobes, or dimorphic structures may be diffi-cult to orient, especially if the muscle scar is not discernible. In many shells, however, the cardinal angles differ, the more obtuse cardinal angle being usually anterior.

Though the problems of orientation have been most varied among the palaeocopids, some difficulty has been experienced with other ostracodes. Semiround carapaces, such as those found among the Polycopidae and some Entomozoidae, may present special problems. Only a few of these are found as fossils. In the Mississippian *Discoidella* the reticulate pattern is a little off-center and the adductor scar is assumed to be off-center in the same direction. The hinge may be slightly closer to the anterior end. Some symmetrical fossil cypridids are difficult to orient, but commonly living representa-tives are available for comparison.

Detection of muscle scars in most Cythera-cea is not difficult. Carapaces belonging to this group commonly are preserved in Meso-zoic and Cenozoic sediments as single valves, so that the interiors may be ex-amined in a translucent or transparent con-dition. Even closed valves may be oriented by the more acuminate posterior extremity (e.g., *Cythereis*), higher anterior end (e.g., *Cythereis*), or smoothly raised muscle swell-ing (e.g., *Cythereis dallasensis*). Generally, in the Podocopina the posterior end is more pointed than the anterior, and greatest height is in front of the mid-length.

Study of hingement aids orientation. LEVINSON (1950) found that the hinge, as a primary structural feature of the carapace, has more highly developed components at the anterior end.

In approximate order of importance for

orientation of fossil ostracodes the following criteria may be cited: (1) fossils with living representatives may be oriented accurately on the basis of observed internal morphology (e.g., *Bairdia, Cytherella*); (2) adductor muscle scars are anterior in position (e.g., *Healdia, Eoleperditia, Cythereis*); (3) in sulcate ostracodes S_2 marks position of adductor muscle scars and thus anterior position (e.g., Cytheridae, palaeocopids); (4) among dimorphic forms the posterior end is widest (e.g., *Kloedenella*) except carapaces exhibiting special types of dimorphism (e.g., *Beyrichia, Zygobolba*); (5) the anterior element of hinge structures usually is more complex than the posterior (e.g., *Kloedenella*); (6) in outline the posterior extremity of ostracode carapaces usually is more acuminate than the anterior (e.g., *Cythereis, Oligocythereis*); (7) ontogenetic development shows that the posterior half of the carapace is more acuminate than the anterior in instar stages (e.g., *Cytherella*); (8) among ornamental features it is observed that major spines are directed backward (e.g., *Cythereis*); and (9) the posterior portion of the hinge channel commonly is widest (e.g., *Kloedenella*).

In the past many new genera have been defined on the basis of "reversal of overlap" of the valves. Now it is known that reversal of overlap may occur within a population of a given species and therefore it cannot be used as a generic or specific character. Reversal has been reported in *Sansabella, Aurikirkbya,* and *Paraparchites*. Orientation should not be based on direction of overlap, but the direction of overlap must be determined after the orientation has been established by other criteria.

DORSAL-VENTRAL

To distinguish dorsal from ventral seemingly should be a simple matter of observation, but unfortunately, many published illustrations show the dorsal margin of ostracode carapaces in a ventral position. The problem is to determine the position of the hinge (dorsal) or the free margin (ventral). The question is readily resolved in single valves in which the hinge elements can be seen.

The long straight back of most leperditicopids and palaeocopids make recognition of the dorsum easily determined (Fig. 18).

In the straight-backed ostracodes of these groups the ventral margin is usually convex or rarely sinuate (e.g., some kloedenellids).

The position of the hinge may be determined by the presence of hinge elements, often discernible exteriorly as well as interiorly, and in many genera by the presence of a hinge channel. Cardinal angles denote the terminal points of the hinge but are lacking in some shells. Eye spots, such as are developed excellently in *Cythereis*, denote anterior as well as dorsal position. Alate structures distinguish the ventral portion of a valve or carapace (e.g., *Brachycythere, Cytheropteron*). Radial pore canals and velate and histial structures are along the free margin (ventral). Short sulci are in the dorsal half of the carapace. If a sulcus is open at one end only, it opens dorsally.

The problem of dorsal-ventral orientation is most acute in such groups as the Darwinulacea, Cypridacea, Thlipsuracea, Quasillitacea, and the cytherellids-cavellinids. In some of these (e.g., *Cytherella*) hinge elements are not developed; in others the surface may be smooth and the outline may appear symmetrical. In such shells the dorsal margin is usually more convex than the ventral, or the ventral margin is straight to gently concave (e.g., *Darwinula, Cavellina, Healdia, Cypris, Candona*).

The criteria applicable to shape in outline of the carapace when viewed laterally vary from group to group. In the Beyrichicopina the dorsal margin is straight and the free margin strongly to gently convex. In the Kloedenellocopina the dorsal margin is straight and the ventral convex to concave (sinuate). The dorsal outline of some Platycopina (e.g., *Cytherella*) is strongly convex and the ventral outline less convex, almost straight or slightly concave. However, in *Cytherelloidea* the dorsal margin varies from straight to gently concave or convex; the ventral margin is subparallel to the dorsal but is likely to be more concave.

Criteria other than outline, such as position of the pit, hinge elements, and costae, must be used in questionable forms. In the Metacopina great variation in outline is found; the Healdiacea have strongly convex dorsal outlines when oriented with the ventral margin parallel to the horizontal, and gently convex to straight or gently concave ventral outlines; the Thlipsuracea have dorsal outlines more convex than the ventral, but the ventral varies from gently convex to straight or gently concave; the Quasillitacea have a straight dorsal margin. When the straight dorsal margin of the Quasillitacea is oriented parallel to the horizontal, the ventral margin is usually broadly convex with an anterior swing or it is medially concave. The mid-ventral incurvature is similar to that found in the Podocopina.

A characteristic feature of the Podocopina is the medial concavity or incurvature of the ventral margin. This is sharply contrasted to the straight or convex dorsal outline.

ORIENTATION FOR ILLUSTRATION

The position of greatest height, magnitude of ventral swing, and line of greatest length may vary according to the manner of orienting the specimen for illustrative purposes. The palaeocopids are oriented with the hinge parallel to a horizontal line. The hinge, fairly long and straight in most of the palaeocopids and archaeocopids, is used as the basic reference line; on the other hand, students of the podocopids usually orient specimens with a line through the greatest length parallel to the horizontal, though some are oriented with the venter parallel to the horizontal. Each method has been in use so long that continuation of common practice is recommended.

REFERENCE

Levinson, S. A.
(1) 1950, *The hingement of Paleozoic Ostracoda and its bearing on orientation:* Jour. Paleont. v. 24, p. 63-75 (Tulsa).

GLOSSARY OF MORPHOLOGICAL TERMS APPLIED TO OSTRACODA

By R. C. Moore
[University of Kansas]

[Relative importance of terms is indicated in accordance with the collective judgment of contributing authors, first rank by boldface capital letters, second rank by boldface small letters, and third rank (including some obsolete terms) by italic small letters. Annotation or remarks by some authors are given, inclosed by square brackets.]

accommodation groove. Furrow above median element of hinge for reception of dorsal edge of opposite valve when carapace is open.

adanterior. Toward anterior extremity, forward.

ADDUCTOR MUSCLE SCAR. Impression on valve interior of muscle serving for closure of valves, generally located in front of mid-length.

adhesive strip. Thin chitinous layer between duplicature and outer lamella.

ADONT. Simple hinge of ridge-and-groove type, lacking teeth, with ridge or bar in one valve, which fits into groove along dorsal edge of other valve.

adplenate. Direction toward plenate end of carapace, referring to features in part of valve or carapace distinguished as plenate end.

adposterior. Toward posterior extremity, backward.

adventive pore canal. Tubule extending through duplicature.

adventral. Direction toward venter or location adjacent to venter.

ALA (pl., ALAE). Ventrally placed winglike lateral extension of valve, commonly directed backward (characteristic of some Cytheracea and Beyrichiacea).

alar process. Posteriorly attenuated ventrolateral ridge resembling but not associated with velate forms, common to more ornate cytherid ostracodes, particularly as development of one longitudinal ridge of the reticulate system of ridges; usually precedes dorsolateral or mid-lateral ridges in evolution of highly ornate, reticulate cytherids.

ALATE EXTENSION. Any outward lateral extension in ventral half of valve, usually increasing in width backward and terminating abruptly, tending to have triangular shape (e.g., *Pterygocythereis*).

AMPHIDONT. Four-element hingement resembling entomodont and lobodont in general features, one valve having well-defined toothlike projections at extremities, separated by median furrow with deep smooth socket at anterior end, opposite valve with reverse arrangement of projections and depressions. [All descriptions of complicated hingements, such as variations of amphidont and merodont, should refer to tripartite division (posterior and anterior terminal elements and median element), for each element is modified *in toto* in merodont hinges or separately and differently in amphidont hinges. For example, in paramphidont hinges the anterior and posterior elements (teeth or sockets) are notched or crenulate, the median element (groove or bar) may be smooth or finely crenulate but modified at the anterior end adjacent to the anterior element to form a smooth tooth or socket (e.g., *Cythereis*). Benson]

amplete. With greatest height at or near mid-length of valve or carapace.

ANTENNAL MUSCLE SCAR. Impression on valve interior of attachment for muscle joined to antenna, located in front of adductor muscle scar, generally above and in some podocopids behind mandibular muscle scar.

ANTERIOR. Part of carapace in which antennules, antennae, and upper lip are located, front region.

anterior area. Part of either valve surface adjacent to front border (or margin), extending backward not farther than mid-length, divisible into anterodorsal, mid-anterior, and anteroventral areas.

anterior cardinal angle. Junction between anterior margin and hinge line.

anterior corner. Area immediately adjacent to anterior cardinal angle of either valve.

anterior hinge area. Front part of hinge in either valve.

anterior horn. Projection of anterior lobe above hinge line (e.g., *Ceratopsis*).

ANTERIOR LOBE. Rounded elevation adjacent to front of valve, best developed dorsally (commonly designated by symbol L_1).

anterior longitudinal point. Front extremity of longitudinal section of carapace where valves meet.

ANTERIOR MARGIN (OR BORDER). Front part of outline of either valve, forming part of free margin.

ANTERIOR SULCUS. Depression of valve surface extending adventrally from dorsum and located nearest to front (commonly designated by symbol S_1).

ANTERIOR VIEW. Appearance of carapace or valve as seen from front in line parallel to hinge line or axis.

anterodorsal angle. Generally obtuse angulation where relatively straight dorsal margin of valve meets rounded anterior margin (*see* anterior cardinal angle).

anterodorsal area. Surface of either valve adjacent to and including anterior corner.

anteromedian area. Front part of median surface of either valve intermediate between median and anterior areas.

anteroventral area. Front part of ventral surface of either valve.

anteroventral depression. Gently hollowed area between velate structure and median lobe in some beyrichiids, located anteroventrally.

antimerodont. Type of merodont hingement in which crenulate terminal projections of one valve (usually smaller) are separated by a crenulate furrow, with reverse arrangement of elevations and depressions in opposite valve (e.g., *Clithrocytheridea*).

antiplenate. Direction away from plenate end of carapace or valves, or referring to features in half of valves opposite plenate end.

apical list. Part of prionodont hinge on proximal side of teeth or sockets.

area. Somewhat arbitrarily delimited portion of valve surface as generally defined in lateral view; descriptive terms for individual areas include anterior, anteromedian, anterodorsal, anteroventral, dorsal, dorsomedian, median, mid-anterior, mid-dorsal, mid-posterior, mid-ventral, posterior, posteromedian, posterodorsal, posteroventral, ventral, and ventromedian.

axis. Straight line in sagittal plane connecting most widely separated edges of carapace or valve, invariably equal to or exceeding length.

basal selvage. Ridge between distal extremities of teeth in prionodont type of hingement.

beak. Anteroventral projection of free border of carapace (e.g., *Cypridea*); not equivalent to rostrum.

beyrichiid dimorphism. Sexual dimorphism characterized by development of anteroventral, ventral, or slightly posteroventral pouch-like structure (crumina) in valves of heteromorphs (presumed females), incompletely separated from domicilium by partition (e.g., *Beyrichia, Hibbardia, Phlyctiscapha*).

bilamellar. Double-walled part of free margin of some podocopid valves formed by welding of duplicature to outer lamella at expense of vestibule.

bilobate. Valves characterized by presence of only two rounded elevations (lobes) separated by a linear depression (sulcus) (e.g., *Dilobella, Parabolbina*); also may refer to two-lobed hinge teeth.

bipartite interterminal furrow. Double linear depression, typically crenulate, between extremities of hinge (e.g., *Loxoconcha*).

bisulcate. Valves characterized by presence of two linear depressions (sulci) separating three rounded elevations (lobes) (e.g., *Beyrichia, Kloedenia*).

blood canals. Branched grooves on interior of some valves inferred to mark position of blood vessels (e.g., *Leperditia*).

border (or margin). Periphery of carapace or valve as seen in lateral view.

brood pouch. Gently to strongly swollen portion of heteromorphous (presumed female) carapace, diverse in origin and actually unknown in function, located posteriorly (e.g., *Kloedenella*) or posteroventrally (e.g., *Hibbardia*), ventrally (e.g., *Treposella*), or anteroventrally (e.g., *Beyrichia, Zygobolba*).

bulb. Very prominent spheroidal protuberance of valve, commonly in position of posterior lobe (L_3), that may extend above hinge line (e.g., *Falsipollex*).

calcareous layer. Relatively thick shell layer largely composed of calcium carbonate between outer and inner chitinous layers of carapace, commonly only part preserved in fossils.

capitulum. Wide prominence at anterior end of tignum in highly developed tooth-and-socket hingement.

CARAPACE. Protective covering of ostracode soft

parts, including appendages, forming two nearly symmetrical valves joined together by hinge along dorsal margin; mostly hard and calcareous but soft and uncalcified in most Archaeocopida and many Myodocopida.

cardinal angle. Junction between hinge line and anterior or posterior free margin.

cardinal corner. Area immediately adjacent to cardinal angle.

cardinal socket. Major hollow at or near one or both extremities of hinge, for reception of tooth borne by opposite valve; may have smooth, bifid, or crenulate floor and may occur in either valve.

cardinal tooth. Major prominence at or near one or both extremities of hinge, fitting into socket of opposite valve; may have smooth, bifid, stepped, or crenulate summit and may occur in either valve.

CARINA (pl., **CARINAE**). In Palaeocopida, ridge or frill parallel to velate structure on its dorsal side, or compressed ridge appearing as forward-directed structure on anterior part of carapace (compare histium); in Podocopida, any well-defined, somewhat strongly projecting ridge on outer surface, as in many cytherids and cytherellids.

carinal bend. Rather sharp angulation *(Umbiegungskante)* along distal edge of carina separating lateral and marginal surfaces of valve.

carinal crest. Attenuated projecting summit of carinal structure.

carinal ridge. Linear elevation with more or less rounded summit in adventral region connecting anterior lobe (L_1) and second posterior lobe (L_4) in some genera (e.g., *Tetradella, Tallinnella, Ogmoopsis*); compare histium.

CAUDAL PROCESS. Posterior projection of valve border generally above mid-height (e.g., *Loxoconcha*) or posteroventral and directed upward (e.g., *Cyprosina*).

caudal siphon. Posteroventral opening in valve borders (e.g., *Entomoconchus*) or produced as tubular structure (e.g., *Cyprosina*).

channel. Groove between valve margin and velate structure (e.g., *Ctenobolbina, Dicranella*), or depression of hinge line below dorsal margin.

chevron muscle scar. Inverted V-shaped muscle impression on interior of valves (e.g., *Leperditia*).

CHITINOUS LAYER. Thin waxy or transparent layer of chitin forming part of ostracode carapace, one covering outer side and another inner side of the calcareous shell layer.

closed velum. Ventrally incurved velate structures of right and left valves with distal edges meeting to inclose a false pouch (dolon) (e.g., *Uhakiella*).

compound socket. Depression in hinge line with bifid or crenulate floor (e.g., *Alatacythere*).

connecting lobe. Rounded linear elevation of valve surface confluent with 2 or more subvertically trending lobes (e.g., *Ceratopsis, Mastigobolbina, Tallinnella*).

contact margin. Edge part of valves exclusive of hinge, in contact when valves are closed, its distal limit comprising free edge.

corner. Area between front or rear part of dorsal border and anterior or posterior margin.

COSTA (pl., **COSTAE**). Rib on valve surface (e.g., *Glyptopleura, Cytherelloidea*).

crenulate. Notched by alternating small ridges and depressions, as in prionodont hingement.

crest. Small straight or curved ridge on valve surface (e.g., *Piretella*).

CRUMINA (pl., **CRUMINAE**). Saclike semiclosed space developed in ventral part of heteromorph domicilium (e.g., *Beyrichia kloedeni*).

cruminal dimorphism. Same as beyrichiid dimorphism.

DENTICLE. Small, delicate, spinelike projection differentiated according to location on carapace (e.g., dorsal, marginal); small toothlike projection of hinge element.

DENTICULATE. Bearing a series of small spinelike or toothlike projections.

depression. Broad gentle concavity on carapace surface without distinct limit.

DIMORPHISM. Development within a species of two shapes of adult carapaces, that of inferred females (heteromorphs) being moderately to strikingly differentiated from the form of adult inferred males (classed with juvenile stages of both sexes, which they resemble, as tecnomorphs).

DISTAL. Direction outward from mid-region of ostracode body.

distal line of adhesive strip. Margin on outer side of junction of duplicature and outer lamella.

distal zone of duplicature. Part of duplicature between line of concrescence and distal line of adhesive strip.

dolon. Cavity (false pouch) formed by distally incurved parts of velate structure.

DOMICILIAR DIMORPHISM. Type of dimorphism marked by slightly to strongly swollen posterior portion of carapace in females (e.g., *Kloedenella*).

DOMICILIUM. Part of carapace exclusive of projecting velate structures.

DORSAL. Upper part of ostracode, when in normal position, comprising region that contains hinge, eyes, antennules, antennae, and stomach.

dorsal area. Part of valve surface adjacent to dorsal border, divisible into anterodorsal, mid-dorsal, and posterodorsal areas.

dorsal denticle. Small solid spinose projection on dorsal margin, chiefly different from dorsal spine only in small size.

dorsal margin (or border). Part of valve outline adjacent to hinge line, somewhat above or coinciding with it.

dorsal plica. Linear elevation of valve surface adjacent and parallel to dorsal margin.

dorsal spine. More or less prominent, solid or hollow pointed projection of valve on dorsal margin (e.g., *Ctenonotella, Rakverella*).

DORSAL VIEW. Appearance of carapace or valve as seen from above with line of sight in sagittal plane and normal to hinge line.

dorsomedian area. Part of valve surface intermediate between median and mid-dorsal areas.

DORSUM. More or less flattened area of carapace surface adjacent to hinge line and set off from lateral surface of valves.

DUPLICATURE. Part of border in which calcareous peripheral portion of inner lamella is in contact with outer lamella or separated from it by vestibule, generally narrow but in some genera (e.g., *Cytheretta*) considerably extended.

EDGE. Distal limit of valve periphery touching opposite valve when carapace is closed.

entire (velate structure). Frill or velate ridge extending from anterior to posterior cardinal angles.

ENTOMODONT. Type of four-element hingement corresponding to antimerodont or hemimerodont except for development of short anterior segment of median ridge as crenulate toothlike projection, remainder of median ridge being smooth or finely crenulate, opposite valve with reverse arrangement of elevations and depressions (e.g., *Progonocythere*); see note under amphidont.

epicline. Dorsum that projects above hinge line.

extralobate area. Part of valve surface not involved in lobation and sulcation.

extralobate groove. Linear depression along inner margin of extralobate area adjoining ventral lobe (e.g., *Eobeyrichia*).

EYE TUBERCLE. Polished transparent rounded protuberance in anterodorsal region of valve forming lens of eye (e.g., Cytheracea) or marking inferred position of eye (e.g., Leperditicopida).

false pouch. Chamber formed by distally incurved frills of some heteromorphs (equivalent to dolon) (e.g., *Piretella, Eurychilina, Tallinnella*); differs from true pouch in being developed outside of domicilum.

false radial pore canal. Tubule through outer lamella from line of concrescence but not penetrating chitin of adhesive strip.

fissure. Narrow steep-sided groove on valve surface subparallel to free margin (e.g., *Beyrichia kloedeni*).

fissus. Same as fissure.

FLANGE. Ridge along valve margin of some podocopids formed by projection of outer lamella as narrow brim.

flange groove. Surface of duplicature between selvage and flange.

flange strip. Part of duplicature forming flange groove and in some genera the flange also.

flexure. Lateral offset of hinge line as seen in dorsal view.

FREE EDGE. Line of contact between closed valves except along hinge line; marks distal limit of contact margin and may lie inside free margin.

FREE MARGIN (OR BORDER). Anterior, ventral, and posterior parts of margin where valves are not held together by hingement.

FRILL. Wide velate structure that commonly extends beyond free edge of valves, mostly striate or septate radially (e.g., *Hollinella, Oepikium, Dibolbina, Chilobolbina*); commonly developed as double-walled outfold of shell.

frontal rounding. Curvature of carapace surface toward anterior valve margin as seen in longitudinal sections.

frontal section. Same as longitudinal section.

furrow. Shallow groove on valve surface.

geniculum. Abrupt lateral bend of sulcus.

girdle socket. Depression in hinge line with horseshoe-shaped outline defining a small tooth, generally anterior.

GONGYLODONT. Type of hingement characterized by presence in one valve of finely crenulate median ridge between anterior rounded tooth bounded by sockets and posterior pair of rounded teeth separated by deep socket, opposite valve with reverse arrangement of elevations and depressions (e.g., *Loxoconcha*).

granuloreticulate. Surface ornamentation of valves consisting of granules arranged in intersecting rows.

granulose. Surface ornamentation of valves consisting of more or less closely spaced minute protuberances, generally without distinct pattern, like grains of sandpaper.

HEIGHT. Maximum dimension of carapace or valve from dorsal to ventral margins measured perpendicularly to direction of length.

hemiamphidont. Type of amphidont hingement in which anterior toothlike projection of one valve is smooth or stepped (but not crenulate), whereas posterior projection is notched or crenulate, opposite valve with reverse arrangement of elevations and depressions (e.g., *Brachycythere, Alatacythere*); see note under amphidont.

hemimerodont. Type of merodont hingement in which one valve has crenulate ridges at extremities of hinge with smooth-floored furrow between, opposite valve with reverse arrangement of elevations and depressions; like antimerodont hingement except for smooth median furrow and opposing ridge (e.g., *Palaeocytheridea*).

HETERODONT. Hingement of valves effected by combination of tooth-and-socket and ridge-and-groove types, characterized by pointed or slightly crenulate teeth in one or both valves associated with ridge (hinge bar) in one valve and groove in other (e.g., Cytheridae).

heteromorph. Adult inferred female carapace in dimorphic genera (compare tecnomorph).

HINGE. Part of valves along or near dorsal margin serving for articulation.

hinge area. Surface involved in hingement of valves, commonly differentiated into anterior and posterior areas containing more complex elements and between these an interterminal area with simpler structures.

hinge bar. Smooth or finely crenulate ridge in interterminal hinge area of one valve, fitting into groove in opposite valve.

hinge flange. Structure of hinge corresponding to flange of contact margin and continuous with it in many species.

hinge flange groove. Furrow on distal side of hinge interterminal area corresponding to flange groove of contact margin and continuous with it in some species; termed accommodation groove in some Cytheracea because furrow serves to accommodate dorsal margin (hinge flange) of opposite valve when carapace is open.

HINGE LINE. Line along which valves articulate, seen when carapace is complete; it may coincide with dorsal margin or be depressed below it.

hinge list. Structure of hinge area corresponding to list of contact margin.

hinge margin. Part of dorsal border (or margin) of valves adjoining hinge.

hinge selvage. Structure of hinge area corresponding to selvage of contact margin and continuous with it in some species.

hinge selvage groove. Structure of hinge area corresponding to selvage groove of contact margin.

HINGEMENT. Collective term for structures comprising articulation of valves, classifiable in several types; same as hinge.

histial dimorphism. Type of dimorphism characterized by development of histium in heteromorphs.

histium. Adventral ridge or flange confluent with connecting lobe in some heteromorphs (e.g., *Sigmoopsis, Glossomorphites*), in quadrilobate valves forming ventrolaterally projected continuation of ventral end of connecting lobe and with same position in nonsulcate valves, commonly dimorphic.

holamphidont. Type of amphidont hingement in which both terminal toothlike projections of one valve are smooth or stepped (not crenulate) and separated by a long median furrow that may be smooth or finely crenulate, opposite valve with reversed arrangement of elevations and depressions (e.g., *Amphicytherura, Trachyleberis, Pterygocythereis*); see note under amphidont.

holomerodont. Type of merodont hingement in which crenulate toothlike ridges at extremities of hinge in one valve are separated by crenulate median ridge, opposite valve with depressions for reception of these elevations; like paleomerodont hingement except for finely crenulate nature of median ridge and furrow (e.g., *Haplocytheridea*).

horn. Dorsal part of lobe projecting more or less strongly above hinge line (e.g., *Beyrichia, Ceratopsis*).

horseshoe-shaped ridge. Prominent U-shaped elevation of valve surface bordering median sulcus

(S_2) on anterior, ventral, and posterior sides in some genera (e.g., *Bolbibollia, Zygobolbina*).

hump. Low but rather large dorsal inflation of valve surface projecting above hinge line.

hypocline. Dorsum inclined slightly downward-outward so that hinge line is not concealed in lateral view.

inflation. Large gently domed part of valve surface without distinct borders; sometimes used as synonym of width.

INNER CHITINOUS LAYER. Thin layer of transparent chitin secreted on inner side of calcareous shell layer.

INNER LAMELLA. Thin layer covering body in anterior, ventral, and posterior parts of carapace, chitinous except for calcified marginal parts forming duplicature; entirely distinct from inner chitinous layer on inside of outer lamella.

INNER MARGIN (OF DUPLICATURE). Proximal limit of duplicature.

inner selvage contact line. Proximal line between selvage and remainder of duplicature.

inner surface. Interior of carapace, in fossils comprising inner side of calcareous layer, which originally was covered by inner chitinous layer.

INSTAR. Ontogenetic stage comprising one of several successive forms assumed by animal between successive molts.

interterminal hinge area. Median part of hinge area generally bearing ridge-and-groove structures.

KIRKBYAN PIT. Central or subcentral steepwalled depression on valves of some palaeocopid genera, usually ovate and interrupting reticulate ornament (e.g., Kirkbyidae); inferred to mark location of adductor-muscle attachment.

kloedenellid dimorphism. Type of dimorphism characterized by inflated posteroventral part of domicilium in heteromorphs (e.g., *Kloedenella*); also termed domiciliar dimorphism.

knob. Prominent rounded protuberance of valve surface differentiated from surrounding area by distinct angulation (e.g., *Hollinella*).

knurling. Pointed projection of hinge line of one valve into that of opposite valve as seen in dorsal view.

L₁, L₂, L₃, L₄, Symbols respectively indicating lobes from front to rear parts of valve surface in many genera (e.g., *Tetradella, Ctenoloculina, Quadrijugator*).

LATERAL SURFACE. Flattened side of valve.

LATERAL VIEW. Appearance of carapace or valve as seen from side in direction normal to sagittal plane.

left plenate. Valve with plenate end at left as seen in lateral view.

LEFT VALVE. Half of carapace covering left side of ostracode (symbol, LV).

LENGTH. Maximum dimension of carapace or valve (a) in direction parallel to hinge line, according to customary procedure in measuring straight-backed ostracodes such as most palaeocopid forms, or (b) in direction along axis drawn between farthest anterior and posterior extremities, applicable to curve-backed ostracodes such as many podocopid forms.

LINE OF CONCRESCENCE. Proximal line of junction of duplicature with outer lamella, coinciding with inner border of chitinous adhesive strip.

lip. Inward projection of duplicature.

list. Ridge on proximal side of selvage on contact margin, absent in many carapaces.

list strip. Part of duplicature on proximal side of list extending to inner margin, commonly bearing septa.

lobate area. Part of valve surface bearing lobes and associated sulci.

lobate dimorphism. Type of dimorphism characterized by inflation of lobe in posteroventral or lateroventral part of heteromorph carapace (e.g., *Zygobolbina, Bonnemaia*), without evidence of partition separating brood pouch from domicilium.

lobate tooth. Two-lobed hinge tooth.

lobation. Pattern of elevated portions of valve surface defined as lobes.

LOBE. Rounded major protuberance of valve surface, generally best developed in dorsal part of carapace; also used for part of hinge tooth (e.g., posterior tooth of *Quadracythere*).

lobodont. Type of hingement resembling entomodont except for rounded lobate nature of anterior toothlike projection of median ridge of one valve and rounded socket in opposite valve at anterior extremity of median furrow (e.g., *Acanthocythere*).

loculus (pl., **loculi**). Deep pitlike depression in ventral or anteroventral surface of some heteromorphs (e.g., *Tetrasacculus, Ctenoloculina*), formed by transverse processes joining velum with marginal ridge.

longitudinal. In direction of length.

longitudinal rounding. Lateral rounding of carapace or valve as seen in longitudinal sections.

longitudinal section. Any section of carapace or valve parallel to direction of length and in plane normal to sagittal plane.

lophodont. Type of three-element hingement consisting of short toothlike ridges at extremities of median groove in one valve (usually smaller) and reverse arrangement of elevations and depressions in opposite valve, all hinge elements smooth (e.g., *Eucythere, Cushmanidea*).

lunule. Crescentric concave or convex area at edge of a socket.

MANDIBULAR MUSCLE SCAR. Attachment mark on valve interior for muscle leading to mandibular appendage, typically distinguished by position in front of adductor muscle scars, generally below and may be in front of antennal muscle scar.

MARGIN (OR BORDER). Periphery of carapace or valve as seen in lateral view.

marginal ridge (or rim). Linear elevation of valve adjoining free edge.

marginal structure. Feature developed near free edge of valve and parallel to it but independent; may include marginal ridge, denticles, tubercles, spines.

marginal surface. Flattened area adjacent to free edge of valve and set off from lateral surface.

median area. Part of valve surface located nearest middle, approximately equidistant from dorsal and ventral borders and likewise from anterior and posterior borders.

MEDIAN LOBE. Lobe (L_2) next behind anterior lobe (L_1) and in front of median sulcus (S_2), weakly developed generally and entirely absent in some lobate carapaces.

MEDIAN SULCUS. Sulcus (S_2) next behind anterior sulcus (S_1) in bisulcate and trisulcate carapaces or only sulcus present in unisulcate forms; generally most prominent sulcus.

MERODONT. Type of three-element hingement characterized by crenulate toothlike projections at hinge extremities of one valve and long median ridge or furrow between, opposite valve with reverse arrangement of elevations and depressions; variations accompanied by development of accommodation groove above median element of larger valve include paleomerodont, holomerodont, antimerodont, and hemimerodont.

mid-anterior area. Middle part of anterior area, intermediate between anterodorsal and anteroventral areas.

mid-dorsal area. Middle part of dorsal area, intermediate between anterodorsal and posterodorsal areas.

mid-posterior area. Middle part of posterior area, intermediate between posterodorsal and posteroventral areas.

mid-ventral area. Middle part of ventral area, intermediate between anteroventral and posteroventral areas.

MOLT. Carapace cast off in molting (ecdysis); act of casting off shell.

monolamellar. Single shell thickness along free margin, in podocopids where junction of duplicature and outer lamella is simple turn-over.

MUSCLE SCAR. Mark on shell interior for attachment of muscle, generally distinguishable by localized differences in texture of surface, elevation, depression, or delimiting narrow groove, also discernible in many specimens by coating carapace with oil or water and by converting calcareous shell substance to fluorite.

MUSCLE-SCAR PATTERN. Arrangement of all muscle scars on valve interior.

NODE. Protuberance of intermediate size on valve surface (larger than tubercle, smaller than knob), clearly distinct from lobes.

nonsulcate. Valve surface evenly elevated, unilobate, lacking sulcus.

NORMAL PORE CANAL. Tubule piercing approximately at right angle almost any part of valve, commonly with enlarged proximal part lined with chitin; in living ostracodes carries hair (seta) that projects from surface.

notch. Sharp indentation of valve margin, in ostracodes with beak consisting of anterodorsally directed indentation behind beak.

nuchal furrow. Median sulcus in some Myodocopida (e.g., *Cypridella*).

ocular sinus. Hollow in shell substance beneath eye tubercle, communicating with valve interior and accommodating soft parts of eye.

ocular tubercle. Same as eye tubercle.

OUTER CHITINOUS LAYER. Thin covering of chitin over exterior of calcareous shell layer, which comprises first-formed part of shell after molting.

OUTER LAMELLA. Relatively thick mineralized shell layer inclosed between thin chitinous layers which conceals and protects soft parts of body and appendages.

outer selvage contact line. Distal line between selvage and remainder of duplicature.

OUTLINE. Boundary of carapace or valve as seen from any direction, but generally referring to side view, extending around all projections.

overhang. Same as overreach.

OVERLAP. Closure of valves in such manner that contact margin or selvage of one valve extends over that of other valve.

overreach. Projection in lateral view of one valve beyond the other along dorsum.

paleomerodont. Type of merodont hingement in which extremities of elongate median ridge are elevated as crenulate toothlike projections, opposite valve with reverse arrangement of elevations and depressions (e.g., *Schuleridea*).

papilla (pl., **papillae**). Small steep-sided prominences of nipple-like form on valve surface.

papillose. Surface covered with small steep-sided protuberances termed papillae.

paramphidont. Type of amphidont hingement in which both terminal toothlike projections of one valve are notched or crenulate and elongate median element comprises smooth or finely crenulate furrow that ends forward in smooth-floored deep socket, opposite valve with reverse arrangement of elevations and depressions (e.g., *Cythereis*); see note under amphidont.

peripheral lock. Closure of valves in inequivalved carapaces with sharp edges of smaller valve fitting in furrow along all or nearly all of larger valve margin (e.g., *Cytherella, Cytherelloidea*).

PIT. Relatively large, more or less circular deep hollow in valve surface.

PITTED. Surface of carapace marked by medium to relatively large steep-sided depressions.

plenate end. Convexly prominent wide part of carapace having swing.

PORE. Minute orifice on outer surface of valve which is opening of pore canal.

PORE CANAL. Minute tubular passageway extending through shell.

postadductorial area. Part of carapace or valve behind median sulcus (S_2).

POSTERIOR. Portion of carapace or valve covering sex organs and anus, direction opposite to anterior, backward.

posterior area. Surface of either valve adjacent to rear border and extending forward not farther than center.

posterior cardinal angle. Angle between posterior margin and hinge line.

posterior corner. Area immediately adjoining posterior cardinal angle.

posterior frontal point. Posterior point in a frontal (i.e., longitudinal) section, where the two valves meet.

posterior hinge area. Part of hinge area adjacent to posterior corner, generally containing complex elements.

posterior horn. Projection of posterior lobe above hinge line, in some carapaces (e.g., Beyrichiidae) inner and outer posterior horns.

posterior lobe. Rounded elevation of valve surface behind median sulcus (S_2), commonly developed chiefly in posterodorsal region and indicated by symbol L_3; in trilobate carapaces probably corresponds to L_3+L_4 of quadrilobate forms.

posterior longitudinal point. Posterior point in longitudinal section of carapace where two valves meet.

posterior margin (or border). Rear part of outline of carapace of either valve as seen in lateral view; forms part of free margin.

posterior sulcus. Groovelike depression behind posterior lobe (L_3), present only in quadrilobate carapaces, indicated by symbol S_3.

POSTERIOR VIEW. Appearance of carapace or valve as seen from rear in direction parallel with hinge line or axis.

posterodorsal area. Surface of valve adjacent to and including posterior corner.

posteromedian area. Rear part of median surface of either valve, intermediate between tracts designated median and mid-posterior.

postplete. Greatest height behind mid-length of valve or carapace.

POUCH. More or less prominent swollen part of heteromorph carapace in ventral or anteroventral region (e.g., *Beyrichia, Apatobolbina*) or extending into posteroventral region, may be delimited by internal partitions in each valve into partly inclosed sacs (cruminae) (e.g., *Phlyctiscapha*).

preadductorial area. Part of domicilium in front of median sulcus (S_2).

preplete. Greatest height in front of mid-length of valve or carapace.

primitiid sulcus. So-called median sulcus in unisulcate forms, including juveniles of multisulcate genera.

PRIONODONT. Type of hingement resembling adont but distinguished by presence of crenulations along ridge and groove.

PROXIMAL. Direction inward toward middle part of ostracode body.

proximal zone of duplicature. Part of duplicature extending from inner margin to line of concrescence.

pseudovelum. Frill-like marginal or submarginal rim comprising single layer of shell, not compressed double-layered outfold of valve walls.

PUNCTUM (pl., **PUNCTA**). Small pitlike depression in valve surface.

PUNCTATE. Surface bearing many minute depressions resembling pin pricks.

PUSTULA (pl., **PUSTULAE**). Small protuberance on valve surface with pore at summit.

QUADRILOBATE. Valve distinguished by presence of four lobes.

rabbeted. Shell closure in which one valve bears recess for holding edge of opposite valve.

RADIAL PORE CANAL. Tubule extending through adhesive strip from inner to outer surface of duplicature.

restricted (velate structure). Velate ridge or frill confined to anterior and ventral parts of valve.

RETICULATE. Surface having a netlike pattern of small intersecting crests, striae, or rows of tubercles (e.g., *Amphissites, Hermanites*).

RIDGE. Elongate elevation of valve surface, commonly distinguished by location (e.g., ventral ridge), shape (e.g., horseshoe-shaped ridge, as in *Bolbibollia, Zygobolbina*; sickle-shaped ridge, as in *Drepanella*), or morphological significance (e.g., velate ridge, carina).

ridge-and-groove hingement. Articulation of valves characterized chiefly by ridge along hinge line of one valve fitting into groove of other (e.g., adont, prionodont).

right plenate. Valve with plenate end at right as seen in lateral view.

RIGHT VALVE. Half of carapace covering right side of ostracode, anterior margin at right as seen in lateral view (symbol, RV).

rim. Same as marginal ridge or rim.

ROSTRAL INCISURE. Gape below rostrum in front margin of valves for protrusion of antennae (e.g., *Cypridina*).

ROSTRAL NOTCH. Indentation of anterior margin of valves below rostrum, associated with opening termed rostral incisure.

rostral sinus. Same as rostral notch.

ROSTRUM. Anterior beaklike projection of valve margins overhanging an incisure or notch, generally at mid-height of valves or above (e.g., *Cypridina*).

S_1, S_2, S_3. Symbols respectively indicating sulci from front to rear parts of valve surface in many genera (e.g., *Tetradella, Quadrijugator*); S_1 and S_3 most commonly suppressed, S_2 being major sulcate structure and occurring alone in unisulcate genera (e.g., *Eurychilina*).

SAGITTAL PLANE. Plane bisecting ostracode longitudinally and dorsoventrally.

SAGITTAL SECTION. Any section of carapace or valve in or parallel to sagittal plane.

scalloped. With series of convex warps (e.g., frill of *Parabolbina*).

SCHIZODONT. Type of hingement resembling amphidont but having anterior tooth and socket of one valve both bifid, opposite valve with reverse arrangement of elevations and depressions (e.g., *Paijenborchella*).

scrobicula. Small groove at base of hinge tooth.

second posterior lobe. Rearmost lobe (L_4) in quadrilobate genera (e.g., *Tetradella*).

SELVAGE. Middle ridge of contact margin comprising principal ridge of duplicature and serving to seal valves when closed.

selvage apophysis. Thin lobelike overlapping projection of valve margin (mostly left) at position of podocopid incurvature of margin.

selvage discontinuity. Mid-ventral offset or gape in selvage at position of podocopid incurvature of margin (mostly in right valve) (e.g., many cytherids).

selvage fringe. Thin part of selvage reinforced structurally by slender ridges normal to selvage line.

selvage groove. Part of duplicature surface between selvage and list.

selvage line. Line formed by tapering edge of selvage.

selvage strip. Part of duplicature forming selvage and selvage groove.

semisulcus. Junction of lateral surface of valve with knob, bulb, node, or lobe, differing from sulcus in being bordered by a protuberance only on one side.

septum. Small ridge on list-strip part of duplicature.

SHELL. Calcareous and chitinous substance composing carapace.

sickle-shaped ridge. Narrow, strongly arcuate ridge on side of valve, shaped like sickle (e.g., *Drepanella*).

sieve-type pore canal. Wide normal pore canal partially closed by an internal apparently perforate plate.

SIMPLE HINGEMENT. Edge of one valve fitting against or under edge of other.

simple velum. Velate structure having simple flangelike form or forming ridge.

SOCKET. Well-defined hollow or pit in hinge area of one valve for reception of tooth in hinge of opposite valve.

SPINE. Solid or hollow, more or less elongate projection from valve surface, with rounded or sharply pointed distal extremity.

SPINOSE. Valve surface or margin characterized by presence of somewhat numerous spines; also may refer to distal edge of frill.

spur. Flattened spinelike projection comprising modification of velate structure in tecnomorphs of some dimorphic genera (e.g., *Hollinella, Falsipollex, Ctenoloculina*).

stria (pl., **striae**). Minute furrow on surface of valve.

striate. Surface characterized by many subparallel striae, generally spaced closely; on frill commonly arranged normal to edge.

submedian tubercle. Prominent node or humplike development of anteromedian portion of valves in some podocopids (e.g., trachyleberids) commonly forming hub of convergent surface ornament; expressed as muscle-scar pit on interior of valves.

subhistial field. Area between histium and free edge of valve.

sulcation. Pattern of linear depressed areas defined as sulci.

SULCUS. More or less prominent groove or trench on valve surface trending dorsoventrally and generally best developed in dorsal half of carapace; in some genera may be reduced to faint depression.

SURFACE. Exterior of valve or carapace, unless interior is specified.

SURFACE ORNAMENT. Relatively subordinate elevations, depressions, and varied sorts of markings on valve surface, mostly useful in taxonomy.

swing. Displacement of ventral part of valve or carapace in forward direction so as to produce a tapered appearance of posterior region associated with sloping, nearly straight posteroventral margin; direction reversed in some genera.

taxodont. Type of hingement; used by several authors for prionodont, merodont, antimerodont, etc.

tecnomorph. Specimen of dimorphic species other than adult inferred female (heteromorph); includes adult inferred males and juveniles of both sexes.

THICKNESS. Distance from outer to inner surface of valve (not same as width).

tignum. Long ridge or bar between large tooth and socket on hinge area of one valve.

TOOTH. Localized projection on hinge area fitting into socket of opposite valve, aiding articulation of valves.

tooth-and-socket hingement. Articulation effected by teeth on hinge area of one valve fitting sockets in opposite valve, considerably varied in form of teeth and sockets and to some extent in their placement.

toothlet. Minute tooth, generally in series.

transverse rounding. Curvature of valve or carapace in transverse section.

transverse section. Any section through carapace or valve in plane coinciding with line that represents width or parallel to this plane and normal also to sagittal plane; so-called vertical section intersecting sides approximately at right angles.

TRILOBATE. Valves having three lobes (L_1, L_2, L_3, or L_3+L_4) and two sulci.

TRISULCATE. Valves having three sulci (S_1 S_2 S_3) and four lobes.

TUBERCLE. Low rounded prominence of intermediate size (larger than granule, smaller than knob) on valve surface, common along free margin.

TUBERCULATE. Surface ornament characterized by many tubercles.

unilobate. Valves with evenly elevated surface, without sulci (e.g., *Apatobolbina*).

unisulcate. Valves having a single sulcus (S_2), bilobate (e.g., *Dilobella*).

vallum. Part of hinge between two adjacent sockets (e.g., prionodont hingement).

VALVE. One of the two halves of carapace, hinged at upper (dorsal) edge, classed as left valve and right valve.

velar dimorphism. Type of dimorphism characterized by velate features or structures developed in association with the velum (e.g., *Hollinella, Uhakiella, Abditoloculina*).

velate ridge. Low, generally rounded ridge in position of velum, commonly serrate (typically found in juveniles of hollinellids).

velate structure. Any elongate ridge- or frill-like projection of carapace in position subparallel to valve margins, provided that if more than one such projection occurs on each valve, the velate structure is most adventral (compare carina, histium); typically developed as double-walled outfold of shell (*see* frill).

velum. Wide sail-like velate structure.

ventral. Lower part of ostracode in normal position, comprising region containing mouth, maxillae, and thoracic legs; also part of carapace or either valve covering this region.

ventral area. Surface of valve adjacent to ventral border, divisible into anteroventral, mid-ventral, and posteroventral areas.

ventral lobe. Rounded elevation (lobe) extending subparallel to ventral border and located generally near it (e.g., *Hollinella*); same as connecting lobe.

ventral margin (or border). Part of free margin of valve along ventral side.

ventral ridge. Like ventral lobe but more linear and sharply defined, commonly coalescent with subvertically trending lobes (e.g., *Tetradella, Sigmoopsis*) (same as histium).

ventral selvage furrow. Elongate depression of external valve surface slightly distant from outer margin at position of podocopid incurvature of margin (e.g., *Campylocythere*).

ventral view. Appearance of carapace or valve seen from below in line of sagittal plane.

ventromedian area. Surface of valve intermediate between median and mid-ventral areas.

vestibule. Space between duplicature and outer lamella.

welded duplicature. Type of duplicature in valves having inner margin coincident with line of concrescence, forms without vestibule having wide

duplicature; adhesive strip continuous from its distal boundary to inner margin.

width. Greatest dimension of carapace or valve perpendicular to directions of length and height.

zygal ridge. Generally U-shaped ridge uniting adventral parts of median and posterior lobes in some beyrichiids.

ECOLOGY OF OSTRACODE ASSEMBLAGES

By Richard H. Benson

[University of Kansas]

CONTENTS

Ostracodes have become very successful inhabitants of every aquatic habitat. These microcrustaceans are found in sulphur springs, stagnant ponds, lakes, swamps, streams, brackish lagoons, estuaries, tide pools, salt marshes, epicontinental seas, and on the floor of ocean basins. As fossils they may occur in sediments deposited in all these environments, being particularly useful as paleoecological indicators of brackish- and fresh-water sediments. In most ostracode-bearing sediment, carapaces or separated valves are present in sufficient quantity to be treated statistically. Assemblages of species and genera endemic to particular habitats can be established, and growth stages and ecologic variants can be identified from fossils.

The living animal is benthonic or pelagic, although pelagic ostracodes are rare as fossils. Most benthonic ostracodes crawl on and burrow in the bottom sediments, or crawl

and swim among aquatic plants. Some forms are thought to swim for short distances just above the bottom, but all ostracodes are ultimately dependent directly or indirectly on the nature of the bottom. Its prevailing benthonic habit therefore makes the ostracode intimately associated with the environment of deposition of the bottom sediments.

FRESH-WATER OSTRACODES

Ostracodes are found living in almost every pond, lake, spring, stream, or river at least part of the year. They inhabit temporarily filled drainage ditches and stagnant ponds in poorly drained fields, as well as permanent lakes where wave action may be varyingly strong.

The earliest known fresh-water ostracodes inhabited shallow ponds, coal-forming swamps, and sluggish streams of Early

Pennsylvanian time. They are found as fossils associated with the remains of fresh-water pelecypods, worms, and scales of fishes. The genera thus identified include *Cypridopsis* and *Candona,* which are represented by species now living.

Fresh-water ostracode assemblages have also been identified from Permian, Jurassic, and Tertiary strata. Many fresh-water species now living were present during the Pliocene.

Almost all Recent fresh-water ostracodes are smooth Cyprids, except for *Limnocythere,* which is a cytherid. They differ from most marine ostracodes in the relatively unornamented nature of their carapace. Some fresh-water marls and limestones are almost entirely composed of the smooth valves of such ostracodes. Judging from the number of fossil species represented in a typical example of these deposits, a reconstructed living assemblage seems to denote many habitats. Actually, however, such a deposit commonly represents only a few habitats that occur in and on the bottom and among the grasses. One species may occupy a certain ecologic niche during only part of the year and after the laying of eggs by the females, the adults die. The same niche may then be occupied by a second and even a third species within a year, depending chiefly on mildness or harshness of the winter. Most fresh-water ostracodes produce several broods of young in a season but may be present in a given habitat only a few months. Large fossil populations are the result. As each individual molts some seven to nine times, a very large number of carapaces and separated valves can accumulate in the sediment in a short time.

Hoff (5) found that most fresh-water species live in (1) temporary still waters, including ponds and ditches; (2) temporary running waters, such as intermittent streams; (3) permanent still waters ranging from small ponds to large lakes; and (4) permanent running waters. Some species may be present in all four environments but are more successful in one than in the others.

Zonation and subdivision are present within any one of these habitats. The nature of the sediment encourages or excludes burrowers. Fine, organic-rich muds are more heavily populated by burrowers, such as *Candona* and *Chlamydotheca,* than are coarse sands. Gyttja and other organic oozes are devoid of ostracodes, presumably because of their lack of oxygen. The euphotic zone allows attached plants to grow, and with them a prolific fauna of swimming and crawling ostracodes develops. *Cypridopsis* is found in swarms around the roots of floating vegetation. Fossil ostracodes are seldom found in river sand deposits because of the poor opportunity for preservation and the instability of the sediment.

Changes in temperature with changing latitude are reflected in the presence or absence of warm- and cold-water ostracode species. Also some species are known to be syngamic in one part of their temperature range and parthenogenetic in another.

The stomach contents of various fresh-water ostracodes have been examined and found to contain traces of diatoms, protozoans, bacteria, and algae. Most ostracodes are scavengers; some may be predators; only one *(Entocythere)* is known to be parasitic.

Variations in acidity and alkalinity of the water in which ostracodes live are expressed more by the presence or absence of intolerant species than by thinning or thickening of the carapace. For example, *Candona* is intolerant of acid waters. Slightly alkaline stagnant ponds are likely to contain very prolific ostracode faunas. Ostracodal limestone may be found forming in this alkaline environment.

BRACKISH-WATER OSTRACODES

The ecologic flexibility of ostracodes is well demonstrated by the presence of large populations belonging to endemic assemblages in brackish-water estuaries and lagoons. Few invertebrates possess sufficient tolerance to withstand the wide variations in salinity that characterize this environment; fewer still are represented in the fossil record. Ostracodes are the most abundant microfossils present in brackish-water sediments, and they contribute significantly to the volume of sediments in some brackish lagoons. Accompanying a decline in the number and importance of marine Foraminifera with decreasing salinity, ostracodes show progressive increase. This relationship between two important microfossil groups provides an important method for the recognition of ancient shorelines.

The earliest indications of brackish-water ostracodes are discovered in Silurian rocks of the Central Appalachians. Because of the absence of modern forms until the Jurassic, brackish-water assemblages of older strata are identified mainly by association with sedimentary features and apparent paleogeographic position. The Bathonian strata of France contain the first brackish-water faunas with ostracode species belonging to presently living families. In Purbeckian and Wealdian sediments of Britain, a complete transition from fresh-water to brackish to marine assemblages is preserved. The Pascagoula clay (Miocene) of Mississippi contains a brackish-water assemblage that is generically similar to those living today. Species of the genera *Cyprideis, Cytheridea, Haplocytheridea, Cytherura,* and *Loxoconcha* typically are present in Recent assemblages.

Where a gradient in salinity exists from fresh-water to marine, the ostracode fauna is divisible into four intergrading biofacies. The limnic or fresh-water ostracodes are scarce in waters more saline than 2 o/oo. The brackish-water biofacies contains a population that is most abundant in waters having about 10 o/oo salinity. However, this biofacies is sometimes represented by two adjacent and transitional faunas, an oligohaline (0.2-2 o/oo) assemblage and a more saline brachyhaline (mesohaline, 2-17 o/oo) assemblage. Marine ostracodes seldom survive to reproduce in waters with less than 17 o/oo salinity.

Brackish-water ostracodes are found sometimes in hypersaline lagoons (Fig. 30). The tolerance of ostracodes usually associated with estuaries and brackish lagoons for great changes in salinity allows them to live in lagoons too saline for most normal marine ostracodes. In Florida Bay, where waters are periodically saline in excess of 55 o/oo, species of *Haplocytheridea, Loxoconcha,* and *Cyprideis* produce large populations. Most evaporite-bearing strata do not contain ostracodes, although an assemblage interpreted to represent a hypersaline environment has been found in the Devonian of Russia.

The genus *Cyprideis,* a typical brackish-water ostracode, inhabits waters of all degrees of salinity from fresh to marine. Species of this genus have both smooth and nodose forms of the same instar stages present in the same population in different individuals, but the number of nodose dimorphs increases in proportion to the smooth dimorphs as salinity increases. Ornate ostracodes increase in abundance throughout the entire ostracode population as the conditions of normal marine environment are approached. Extraction of salts for the development of a heavier and more complicated carapace may require less energy in waters with a higher concentration of calcium.

MARINE OSTRACODES

Geological evidence indicates that the marine environment was first to be inhabited by ostracodes and it is represented by the greatest diversity of forms throughout the fossil record. Most marine ostracodes possess complicated exoskeletons that in some way reflect the surrounding marine, benthonic habitat. Carboniferous species of *Cypridina* and a few other fossil pelagic ostracodes have been described, but most fossil marine ostracodes were crawlers, burrowers, and near-bottom swimmers.

The geographic and environmental distribution of Recent marine ostracodes is still very incompletely known. Information about their ecology is restricted to just a few well-studied areas. G. S. Brady (2, 3) described the ostracode faunas from seas surrounding the British Isles and those contained in collections acquired by the Challenger Expedition (1873-76). Most of the genera described in his reports no longer include the variety of species he assigned to them. G. O. Sars (11) described many species from the North Atlantic near Norway and Greenland, along with some data on their environment. The first exhaustive ecologic study of the ostracodes of a marine area was undertaken by G. W. Müller (9) in 1894 on the Bay of Naples fauna. In the 1930's, Remane (10) included ostracodes in his general ecologic studies of the North Sea and Baltic areas. Elofson (4) in 1941 published on the ecology of the Skagerak marine ostracode fauna with additional information concerning North Atlantic coastal forms. More recently KRUIT (7) has described the ostracode fauna of the Rhone Delta and surrounding bays. HORNIBROOK (6) has studied the Recent faunas of New Zealand and found some living palaeocopid species. BENSON (1) published the first ecologic study primarily devoted to marine

ostracodes from North America in reporting on the fauna of Todos Santos Bay, Baja California (Fig. 30).

Interpretation of the paleoecology of Paleozoic marine ostracodes is entirely the product of (1) comparison of their carapace structures with those of living forms whose

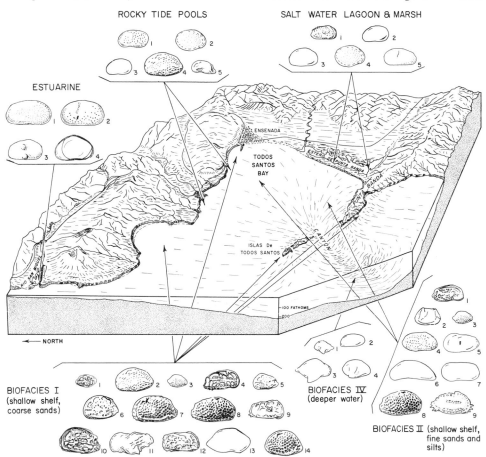

FIG. 30. Distribution of the various ostracode biofacies in the Todos Santos Bay Region, Baja California, Mexico, and their major constituents. Species found in the individual biofacies are listed below.
BIOFACIES I. (1) *Hemicytherura* sp. cf. *H. clathrata;* (2) *Brachycythere lincolnensis;* (3) *Cytherura bajacala;* (4) *Bradleya diegoensis;* (5) *Quadracythere regalia* (molt); (6) *Hemicythere jollaensis;* (7) *Quadracythere regaila* (adult); (8) *Hemicythere californiensis;* (9) *Bradleya pennata;* (10) *Bradleya aurita;* (11) *Paracytheridea granti;* (12) *Cythereis glauca;* (13) *Bairdia* sp. aff. *B. verdesensis;* (14) *Brachycythere driveri.*
BIOFACIES II. (1) *Brachycythere* sp.; (2) *Palmenella carida;* (3) *Cytherura bajacala;* (4) *Leguminocyt hereis corrugata;* (5) *Cytherura* sp. cf. *C. gibba;* (6) *Paracypris pacifica;* (7) *Cytherella banda;* (8) *Hemicythere californiensis;* (9) *Pterygocythereis semitranslucens.*
BIOFACIES IV. (1,3) *Cytheropteron pacificum;* (2) *Bythocypris actites;* (4) *Cytheropteron newportense.*
SALT-WATER LAGOON AND MARSH. (1) *Puriana pacifica;* (2,3) *Xestoleberis aurantia;* (4) *Loxoconcha lenticulata;* (5) *Cyprideis (Goerlichita) castus.*
ROCKY TIDE POOLS. (1) *Haplocytheridea maia;* (2) *Loxoconcha lenticulata;* (3) *Xestoleberis aurantia;* (4) *Brachycythere lincolnensis;* (5) *Caudites fragilis.*
ESTUARINE. (1) *Cyprideis (Goerlichia) miguelensis;* (2) *Cyprideis (Goerlichia) stewarti,* (3) *Cyprideis (Goerlichia)* sp.; (4) *Cypridopsis vidua.*

morphology is otherwise quite different, (2) speculation on the function of extinct structural features of the carapace, (3) examination of associations with a few long-range ostracodes like *Bairdia* which are living today, and lastly, (4) association with other marine fossil organisms. Some study has been devoted to the association of ostracodes with various lithologic types of sedimentary rocks.

One can see, therefore, that knowledge of the ecology of marine ostracodes is still in its early stages, partly because Recent forms living in deeper waters are relatively inaccessible and that paleoecology suffers from the lack of knowledge about the habits and environmental requirements of extinct species.

INFLUENCE OF PHYSICAL AND OR-GANIC ENVIRONMENTAL FACTORS ON THE DISTRIBUTION AND FORM OF MARINE OSTRACODES

Assuming that environment is the principal agent of natural selection and the catalyst for the evolution of the many varied shapes and structures found in ostracode exoskeletons, there should be correlation between some of these structures and the environmental factors to which they are exposed.

The shapes of the appendages of living ostracodes reflect their use for swimming, crawling, or burrowing. The swimmers have long natatory setae that form fanlike oars with the antennules. The antennae have less well-developed natatory setae for the same purpose. The burrowers have no natatory setae; instead, the antennae and antennules of these ostracodes are shaped for digging. The anterior appendages of the crawlers are very similar but are modified more for grasping.

The shape and ornament or lack of ornament of the carapace also reflects the environment. The shape, weight, thickness, and rigidity of the valves, and the hingement correspond to the type of substrate of the burrower and crawler or currents that affect the swimmer. The carapace of swimmers generally is smooth, high in proportion to length, with thin light-weight valves and simple hinges. These swimming forms need to move about freely, unimpeded by a massive exoskeleton.

Crawlers, which occasionally swim short distances, are not true swimmers. Some never leave the bottom and wander over the surface of various sediments. Most of these forms are highly ornamented and possess thick, strong valves wtih flat venters. Some venters are extended laterally into keels, alae, or vela for support on soft oozes. Species inhabiting areas with coarser sediments tend to be reticulate or spinose, the spines encasing setae that extend the sensory apparatus of the ostracode outward so as to feel interstices among the sand grains. Further development of the surface of the carapace into a reticulate system of ribs strengthens the valves against impact in an environment of shifting sands. Possibly this coarse shell also makes them a little less digestible by their natural enemies; however, small fish avoid eating even the thin-shelled ostracodes.

The carapaces of most burrowers, which live in soft sediments, are smooth like those of swimmers but they are much more elongate. Burrowers that inhabit the interstices of coarse-grained substrates, relatively few in number, are very small, short, and usually possess robust carapaces.

Many of the physical and organic environmental factors that affect an ostracode are not expressed in the morphology of individuals so much as in distribution of populations. In the light of present knowledge, it is difficult, for example, to tell what morphologic feature allows or compels *Echinocythereis dasyderma* BRADY to live in waters 12,000 feet deep and to explain why *Puriana rugipunctata* (HOWE) lives in or on shallow calcareous sediments. It is true that these ostracodes are very different in appearance but what selective factor has placed them in their respective habitats is not now evident from examination of their shapes or ornament.

The following environmental factors are important in determination of the ranges and locations of certain living marine ostracode species and assemblages.

DEPTH

Benthonic ostracodes live on the floor of modern seas and oceans at all depths, but they are most abundant in shallow seas of shelf areas. Bathymetric pressure seems to exert little or no effect on ostracodes, but other factors such as fading light, diminishing plant life, stability and change in composition of the bottom sediment do affect the benthonic ostracode. These factors affect the fresh-water ostracode living in deeper lakes, as well as marine ostracodes. *Echinocythereis* and *Cytheropteron* are two deep-water ostracode genera in present oceans, but these genera contain shallow-water species. *Hemicythere, Xestoleberis* and *Cythere* (s. s.) are typical tide-pool genera, but are occasionally found in shelf sediments. Intermediate between the deep-water genera and those adapted for life in tide pools are the normal marine shallow-shelf forms that are represented by many species in late Mesozoic and Cenozoic sediments. These genera are influenced considerably by the nature of the bottom sediment.

BOTTOM SEDIMENT OR SUBSTRATE

The texture and stability of the sediment composing the substrate exerts a strong influence on marine ostracodes, just as it does on fresh-water forms. Even the plant dwellers are indirectly affected by the plant's preference for a particular substrate. Smooth-shelled forms are predominant in fine-grained muds, the rougher more ornate ostracodes being found in coarser, or more calcareous sediments. Such terms as *endopelose* (silt and clay burrowers), *endopsammon* (sand burrowers), *epipelose* (silt and clay surface wanderers), and *epipsammon* (sand-surface crawlers) have been suggested by REMANE (10) and ELOFSON (4) for ostracode assemblages typical of certain bottom sediments, emphasizing the control of the substrate over the character of the associated assemblage.

LITHOLOGIC AND MICROPALEONTOLOGIC ASSOCIATION

Most workers on fossil ostracodes collect their specimens from calcareous shales and sands, particularly the shale partings in limestone sequences. Even Recent marine ostracodes are most abundant in sediments that will become limestones, calcareous shales, or calcareous sands. However, the ease with which sediments can be broken down by washing influences the reported productivity of a given stratigraphic unit as much as does the actual abundance of a fauna. Therefore, not much reliance can be placed on the reported lack or relative lack of ostracode faunas in limestones and siliceous shales.

Ostracodes are rarest in euxinic black shales and fine muds, evaporites, and well-sorted sands. Pyritized ostracodes have been reported from black shales, but they are more likely to be found in transition beds above and below such strata. Ostracode faunas have been described from both red and green shales. The influence of redox potential will be discussed below.

SALINITY

Except in lagoons and estuaries where the marine water becomes brackish because of continental run-off, and in confined shallow basins where evaporation is dominant, variations in salinity of open marine waters are not sufficiently large to influence marine ostracodes appreciably.

Few marine species and genera are successful inhabitants in waters with less than about 17 o/oo salinity, and then only for short periods during heavy rains in tide pools, salt-water lagoons, and marshes.

The ability of normal marine ostracodes to live in salinities greater than 55 o/oo has not been studied. The ostracodes of Florida Bay, which has salinities up to 55 o/oo, compose a special assemblage of tolerant marine species and euryhaline brackish-water species.

TEMPERATURE

Changes in the temperature of the sea northward and southward along the coasts of continents, along the edges of off-shore water masses, and during seasonal heating of shallow waters are reflected in the geographic distribution of vegetatively stenothermal species and time of reproduction during the year of the reproductively stenothermal species.

In warm tropical waters more species are present than in colder waters, but the total number of individuals in either area is dependent on the productivity of that area. Cold-water faunas are distinct from warm-water faunas. For example, the shallow neritic faunas of Norway and northern Scotland contain species different from those found in comparable faunas of southern Ireland and the Bay of Biscay. Possibly or probably as result of contemporary temperature differences, the Miocene ostracode faunas of Florida contain many species that are absent in the Miocene faunas of Virginia and Maryland.

Depth zonation, using the predominance of certain ostracode species along given isotherms, has not been attempted to the extent that it has with the foraminifers of the Pliocene and Recent of California. Some ostracodes are restricted to the 2°C. isotherm in the abyssal depths of the ocean, but this fauna is poorly known.

REDOX POTENTIAL AND HYDROGEN-ION CONCENTRATION

Minor variations in the acidity, alkalinity, and oxygen content that normally are found in marine environments have little or no determinable influence on the number or kind of ostracodes that may be present in a given area. The pH seldom falls into the acid range except below the sediment-water interface. It is not known whether burrowing ostracodes are restricted to nonacid sediments or whether they would not burrow in carbon-rich oozes devoid of oxygen.

Few ostracodes are found in euxinic environments or in black shales. Little is known about the minimum oxygen requirements of living marine ostracodes, and few places in present oceans are available for this kind of study. Fossil ostracodes are found in red and green shales and are sometimes associated with glauconitic sands.

PLANT DOMINANCE

Marine salt-marsh grasses such as the turtle grass *(Thalassia)* of Florida and the Gulf Coast and the eel grass *(Zostera)* of the Pacific Coast offer protection for ostracode populations. These ostracodes are absent in the intervening bare spots. A filigreed coralline alga growing in a tide pool can teem with species of *Xestoleberis* and *Cythere,* whereas a neighboring different type of alga may be associated with numerous individuals of species of *Loxoconcha* or *Hemicythere.* Ostracodes living on or near green algae commonly are different from those of brown algae.

FOOD AND NUTRIENTS

Most of the discrete particles found in the stomachs of marine ostracodes are too small to identify except by color.

The level of the phosphate and nitrate content of the water is reflected in the rise and fall of the ostracode population. Areas of upwelling, such as off the Pacific coast of Baja California, are luxuriant in all life and the general ostracode population is large. Warm waters, low in nutrient content, are correspondingly low in ostracode population except near the mouths of rivers draining mature limestone, and phosphate-rich inland areas. The faunas off the mouths of rivers in Florida are richer than surrounding areas of the sea floor.

ASSOCIATION WITH OTHER ANIMALS

Ostracodes live in association with most invertebrates that are now found preserved as fossils. They are found in the brachiopod-trilobite-bryozoan assemblages of the lower Paleozoic and are found in most of the normal marine benthonic faunas through and including the mollusk-foraminifer faunas of today. KUMMEROW (8) notes that ostracodes appear and disappear abruptly in the stratigraphic record as though occurring in pockets on the sea floor. They are associated with many "dwarfed" faunas, but are not themselves dwarfed. In sediments where conodonts are plentiful, ostracodes are rare, but they are found together in the transition zones between their respective preferred habitats. In sediments where calcareous foraminifers are abundant, ostracodes are not. Ostracodes are almost invariably absent in *Globigerina* ooze. The exchange in abundance between marine foraminifers and brackish ostracodes might be used to indicate marine transgression and regression. Brackish-water ostracodes and oysters lived in the same brackish-water habitat during the Cretaceous.

OSTRACODE BIOFACIES

Many assemblages of ostracode species have been described as characteristic of particular stratigraphic facies of Recent sedimentary environments. It is not possible to do more than list some of these assemblages or facies, as each worker has developed his own classification.

Fresh-water assemblages. Temporary still-water (pond), temporary running water, permanent still-water (lake), permanent running water (river), and prodelta.

Brackish-water assemblages and biofacies. Oligohaline, mesohaline, brachyhaline, polyhaline, marginal bay, midbay, lower bay, and estuarine.

Marine assemblages and biofacies. Saltwater lagoon, salt-tidal flats, salt-marsh channel, subtidal channel, bay mouth, open bay, open gulf, sandy bottom (epipsammon and endopsammon), mud-bottom (epipelos and endopelos), tide pool (supralittoral), abyssal, deep water, shallow water, etc.

OSTRACODE BIOGEOGRAPHY

Marine, fresh-water, and brackish Recent and Cenozoic ostracode faunas have been described from many parts of the world. A similarity exists between some of these that justifies their placement in some of the conventional biogeographic realms. Although the local migrations of some European and American faunas have been described or implied to explain the recurrence of forms at various stratigraphic levels or affinities with forms from other areas of equal age, little study has been devoted to establish the geographic extent of the various ostracode species. The following provinces or realms have known ostracode faunas, and on cursory examination appear to contain faunas more related to those in the same realm than to those of other realms.

Marine. Celtic (Britain and North Sea), Boreal (Norway, Greenland, and North Atlantic), Lusitanian (Mediterranean), Transatlantic (North America, east Coast), Caribbean (Trinidad, Gulf of Mexico, southern Florida), Californian, Japonic, Novo Zelandic.

The Indo-Pacific, West African, Patagonian, Peruvian, and Panamanian ostracode faunas are poorly known.

Fresh-water. Holarctic, Neotropical, Ethiopian and Australian. The Oriental ostracodes are largely undescribed.

REFERENCES

Benson, R. H.

(1) 1959, *Ecology of Recent ostracodes of the Todos Santos Bay region, Baja California, Mexico:* Univ. Kansas Paleont. Contr., Arthropoda, Art. I, p. 1-80 (Lawrence).

Brady, G. S.

(2) 1880, *Report on the Ostracoda dredged by the H.M.S. Challenger during the years 1873-1876:* Rept. Voyage Challenger, Zool. v. 1, pt. 3, p. 1-184.

————, & Norman, A. M.

(3) 1889, *A monograph of the marine and fresh-water Ostracoda of the North Atlantic and of Northwestern Europe. Section 1. Podocopa:* Roy. Dublin Soc. Sci. Trans., ser. 2, v. 4, p. 63-270.

Elofson, Olof

(4) 1941, *Zur Kenntnis der mariner Ostracoden Schwedens mit besonder Berüchsichtigung des Skageraks:* Zool. Bidr. f. Uppsala, v. 19, p. 215-534.

Hoff, C. C.

(5) 1942, *The ostracods of Illinois, their biology and taxonomy:* Illinois Biol. Mon., v. 19, nos. 1-2, p. 1-196 (Urbana).

Hornibrook, N. de B.

(6) 1952, *Tertiary and Recent marine Ostracoda of New Zealand; their origin, affinities, and distribution:* New Zealand Geol. Survey, Paleont. Bull., no. 118, 82 p. [reprinted 1953] (Wellington).

Kornicker, L. S.

(6a) 1958, *Ecology and taxonomy of Recent marine ostracodes in the Bimini area, Great Bahama Bank:* Inst. Marine Science, Univ. Texas, v. 5, p. 194-300 (Austin).

Kruit, C.

(7) 1955, *Sediments of the Rhone Delta. I. Grain size and microfauna:* Verh. geol.-mijnb. Genoot. Ned. Kolon., Geol. ser., v. 15, p. 357-514 ('s-Gravenhage).

Kummerow, E. H. E.

(8) 1949, *Fortschritte und Irrewege der Ostracodenkunde in Deutschland:* Neues Jahrb f. Min. Geol. Pal., Abt. B, p. 209-215 (Monatshefte).

Müller, G. W.

(9) 1894, *Die Ostracoden des Golfes von Neapel und der angrenzenden Meeres Abschnitte:* Fauna u. Flora Neapel, v. 21, 404 p., 40 pl. (Berlin).

Remane, A.

(10) 1933, *Verteilung und organization der benthonischen microfauna der Kieler Bucht:* Wiss. Meeresuntersuch., Abt. Kiel, N.F., Band 21, Heft 2, p. 163-221 (Kiel & Leipzig).

Sars, G. O.

(11) 1922-1928, *An Account of the Crustacea of Norway. Ostracoda:* Bergen Mus., v. 9, p. 1-277 (Oslo).

TECHNIQUES FOR PREPARATION AND STUDY OF FOSSIL OSTRACODES

By I. G. SOHN

[United States Geological Survey]

CONTENTS

This outline of methods for extracting microfossils from sediments is based on the techniques of preparing sedimentary rocks for the extraction of ostracodes in the micropaleontological laboratory of the U. S. Geological Survey, Washington, D. C. The same methods are applicable to preparation of other microfossils, and most have been described previously. The purpose of this outline is to give a step-by-step description of methods that can be used with a minimum of equipment. Additional information is available in TRIEBEL (1958), POKORNÝ (1958), WITWICKA, BIELECKA, STYK, & SZTEJN (1958), and JEKHOWSKY (1959).

The "preparation" of sediments for extraction of microfossils may be said to have three goals: (1) release of fossils from the enclosing matrix; (2) cleaning of fossils from adhering matrix so that morphologic characters can be observed; and (3) con-centration of the fossils by reducing the volume of material to be examined.

A—RELEASE OF FOSSILS FROM MATRIX

The first step in undertaking work on fossil ostracodes is to remove them from the sedimentary matrix in which they occur. Methods for accomplishing this vary considerably because of differences in the nature of the fossil-bearing sedimentary materials and in some samples because of the mode of preservation of the fossils. Generally the methods of disaggregation used in preparing ostracodes for study yield specimens in bulk, but occasionally effort must be directed to individual fossils found embedded in hard rock. Attention here will be given to methods of separating the ostracodes from enclosing sediment; this will be followed by description of procedures for cleaning and concentrating the fossils.

(1) BOILING IN WATER

Most soft sediments break down by boiling them in water to which one or two tablespoons of soda ash are added. Stir occasionally in order to prevent burning of the pot. Sediments that have interstitial water do not break down as readily as those that are dry; consequently it is good practice to air-dry collections prior to boiling. An empirical test for disaggregation by boiling is to bite off a small piece of sediment—if it disintegrates in the mouth, the sediment will break down more or less readily by boiling in water. Add a tablespoon of salt to bentonitic sediments in order to inhibit swelling.

(2) SPECIAL TREATMENT

Sediments that will not break down by boiling in water are treated by using hydrogen peroxide, gasoline or sodium acetate, freezing and thawing, and, as a last resort, by mechanical crushing.

(a) HYDROGEN PEROXIDE AND GASOLINE METHOD

Dry the sample, cover it in a saucepan with 15 per cent hydrogen peroxide, and let it soak from two to 24 hours. Most samples will break down. Add water and boil. Some calcareous fossils are corroded by the hydrogen peroxide, and consequently its use is limited.

The storing of hydrogen peroxide in concentrations of more than 15 per cent is a potential fire hazard. The recommendations of manufacturers and distributors as to storage and handling should be followed.

Dry the sample, cover it in a saucepan with commercial gasoline (which is stored in a metal container), and soak from 15 minutes to one hour. Filter gasoline back into container, add two tablespoons of soda ash, cover the sample with water and boil. Varsol, which is not as flammable as gasoline, and also less expensive, may be substituted for gasoline.

(b) SODIUM ACETATE METHOD

Rocks are comminuted in nature by freezing and thawing. The force exerted by the formation of ice crystals in pores breaks the rock. This is accomplished in the laboratory by use of sodium acetate. Break the dry specimens to about 1 cubic inch, place in a beaker or crucible, and cover with sodium acetate. Add 4 or 5 drops of water, cover the beaker, and place on low heat. The sodium acetate will melt and soak into the pores of the rock. Remove from heat and cool by placing in a pan of water. The sodium acetate will crystallize and rupture the specimens. Add a few drops of water and repeat the process. It will be noted that the first crystallization weakens the rock so that it will subsequently break down more readily. Continued melting and solidifying will not only disintegrate the rock, but also will break up the fossils; consequently, it is best to decant the melted solution of sodium acetate into a second beaker to which are transferred the larger pieces of rocks, and repeat the process.

The sodium acetate that remains with the broken sediment will be dissolved when the sediment is boiled and washed.

Occasionally the force of crystallization will break the glass beaker, but if the pan of cooling water is clean, the uncontaminated sediment can be recovered. Crystallization is accelerated by adding a few grains of sodium acetate to act as nuclei for the crystals.

(c) FREEZING AND THAWING

Results comparable to those yielded by the sodium-acetate method can be obtained by the slower method of freezing in a refrigerator and thawing. The broken sediment should be removed after each thawing to prevent breaking of the fossils, and the sediment then may be boiled.

(d) MECHANICAL CRUSHING

This process of disaggregation consists of applying a crushing rather than breaking force to the rock fragments, generally by means of an electrically driven laboratory crusher. The same results can be obtained with an iron mortar and pestle, or by placing small chunks of rock in a canvas bag and applying a crushing force by use of a wood mallet or the handle of a hammer. Boiling samples prepared in this way serves no purpose, and so the next step is washing, as described in B.

(e) SOLUTION OF MATRIX IN ACID

Silicified fossils can be removed from

limestone by dissolving the matrix with hydrochloric acid. Certain fossils are more successfully removed by use of acetic acid, or the more expensive but quicker-acting formic acid.

Wash the rock to remove all adhering mud, break into chunks, and place in glass vessel. Cover with water and add sufficient acid to start bubbling. Periodically decant, add water and acid. Neutralize the mud residue by adding soda ash and then proceed with washing.

Calcareous and chitinous fossils in a siliceous matrix can be removed with hydrofluoric acid. This method is usually avoided because of the corrosive nature of H_2F; procedures are described in handbooks on pollens and spores.

B—CLEANING

The most generally useful method of cleaning ostracodes after they have been freed from a sedimentary matrix is by washing in clear water and drying. This is adapted especially for preparation of bulk samples.

(1) WASHING

The broken sediment and fossils are washed through a battery of sieves. This operation reduces the volume two ways; unbroken pieces of rock and larger fossils are separated from the microfossiliferous fraction, and finer material is removed. The sieves used in washing tend to become clogged, however, because (a) fine sediment lodges in holes of the mesh, and (b) commonly too much material is put in the sieve.

For convenience in washing, two kinds of sieves (A,B) are used. Sieve A is 10 inches in diameter and 5 inches deep, with 200-mesh copper screen reinforced on the bottom by 16-mesh copper screen; a 1-inch sleeve below the screen protects the mesh and prevents contamination from the bottom. The edge of this sleeve is perforated by 0.25-inch semicircular holes 2.5 inches apart. These holes permit the water to flow out from beneath the screen. Sieve B is 9 inches in diameter and 5 inches deep, with a 16-mesh copper screen. This sieve fits into sieve A, serving to catch particles larger than most microfossils. Three copper angles are soldered on the outside of sieve B 1.5

inches below the top in order to keep the bottom of this sieve well above the screen of sieve A, and three legs 0.5 inch long on the bottom of the sieve protect the screen. The mesh is attached on the inside of the sieves by smooth solder for ease of cleaning.

Standard sieves of 16- and 100- or finer mesh can be used. For Foraminifera a 200-mesh screen is necessary. Tilt the standard sieves by propping one end with a spoon handle or piece of wood in order to permit the water to escape from beneath the screen.

The sample is washed with a hose attached to a double-feed swivel faucet. If the fine sediment clogs the screen, the end of the hose is pinched in order to increase the velocity of the water and held close to the screen; this invariably breaks the seal and the fine sediments can be washed out. Washing is continued until the water escaping from the screen is clear.

(2) DRYING

The washed sediment can be dried either in the sieve, if the sieve is not needed for a second sample, or transferred to paper towels or newspapers for drying. Spread several thicknesses of papers and turn the sieve over on the paper; most of the wet sediment will fall out. With the hose, wash the remaining sediment to the center of the screen and drop gently face down on the paper. The remaining sediment will bounce off the screen. Repeat if necessary. Fold the paper over several times, insert an identifying label, and set aside to dry. Remove any large fossils from the plus 16 fraction, which can be discarded, dried in the same manner as the minus 16 fraction and stored, or subjected to other methods of disintegration.

C—CONCENTRATION

Concentration of fossils is accomplished by dividing the sample into various fractions according to particle size, by use (where suitable) of specific gravity separations, and by hand-picking of specimens, generally under a low-power microscope.

(1) SIEVING

The dried sample is conveniently separated into size-determined parts by use of different screens. The minus 16 plus 200-mesh sediment is dry-sieved through a series of 40-, 60-, 80-, and 100-mesh screens, or 150-mesh if very small fossils are found.

In order to avoid contamination the sieves are dipped for a few minutes in an aqueous solution of methylene blue, after each use. Any specimens (except pyritized fossils) in the mesh are then colored blue, being then easily recognized as contaminants should they become incorporated in subsequently sieved samples (Beckmann, 1959). An alcohol solution of methylene blue dries faster than a water solution.

(2) HEAVY-LIQUID SEPARATION

Many types of microfossils can be concentrated by use of heavy liquids, and various techniques for accomplishing this are described. Ostracodes do not usually lend themselves to such treatment and must be hand-picked under a binocular microscope.

(3) PICKING

Samples obtained from the 40- and 60-mesh screens are conveniently picked under low power, whereas those from smaller screen sizes require medium power.

Various picking trays have been described. The simplest is a flat-bottomed black-painted tray with 0.15-inch or higher sides that is open on one end. Horizontal and vertical lines desirably are painted or grooved on the picking surface. The distance between the lines should correspond approximately to the field of view seen through the microscope. Scatter an evenly distributed layer of prepared sediment on the tray so that each particle can be seen. By moving the tray back and forth under the microscope every grain can be examined and the fossils picked out.

For lifting the fossils and transporting them to a hollowed slide or other receptacle, it is best to use a 00 sable or camel-hair brush. Dip the brush in water and pass it over the back of the other hand while the handle is rolled between the fingers. This removes excess water and makes a point on the tip of the brush. The wet brush will pick up particles when they are touched and drop them in a microslide when the brush with an adhering particle is swept over the bottom of the slide. Some workers wet the brush with saliva; from the standpoint of the worker, this is not advisable because the brush may become contaminated with harmful substances such as dyes and carbolic or other acid used to inhibit bacterial growth in the glue—not to mention that specimens adhering to the brush may be swallowed. Always examine the brush under the microscope before dipping in water in order to insure that specimens inadvertently picked up by the brush are not dropped into the water.

D—PREPARATION OF INDIVIDUAL SPECIMENS

Many ostracodes are so fragile that they are destroyed by boiling; other specimens have ornaments that break when crushed from hard rocks. These have to be handled on an individual basis. Slabs of rock may be broken along bedding planes and the surface examined with a microscope. Each specimen is stained with malachite green in alcohol. This serves two purposes; it helps locate the specimen on the slab and shows where the matrix is breaking. A groove is scraped around the specimen far enough from it to insure against accidental breaking. From the groove a cut below the specimen can be made so as to allow its removal. Use sharp needles, a vibratool or a dental drill.

Although the various methods of disaggregating sediments and subsequent washing commonly result in clean specimens, matrix sometimes adheres to the outside of specimens or fills the inside of valves. This extraneous material needs to be removed in order to observe shell structures. Such cleaning may be done manually, chemically, or by mechanical methods.

For manual cleaning the specimen on a slide is covered with a drop of water and the matrix removed with a sharp needle. The drop of water, in addition to possibly softening the matrix, prevents the specimen from popping out of the slide when touched with the needle. A chewed wood toothpick makes an excellent stiff-bristled microscopic broom to sweep out matrix from a valve. When sharpened, the toothpick can be used instead of a needle on fragile specimens.

For cleaning of specimens chemically, matrix can be removed from individual carapaces or valves by soaking in hydrogen peroxide in a concave glass slide, observing progress of reaction with a microscope. A toothpick may be used to sweep away the matrix. Hydrofluoric acid will remove siliceous matrix from calcareous specimens and at the same time render the specimens translucent (SOHN, 1956). The valves of hollow carapaces are sometimes dissociated by the gas pressure that is generated in this process.

Mechanical methods of cleaning also may be employed. Experiments with the use of ultrasonic vibrations have proved successful in cleaning individual ostracode specimens, but if incipient fractures are present, the specimens tend to break in this process. Ultrasonic treatment is based on high-frequency acoustical waves that are transmitted to water or other liquids by means of electrostrictive or magnetostrictive devices called transducers. The transducers are designed for various industrial uses, either as integral parts of stainless steel tanks or as units immersible in existing tanks. A generator transmits the ultrasonic energy to the transducer. This process is very good for disaggregating some groups of fossils such as diatoms and foraminifers, although not usable generally for recovery and cleaning of ostracodes because of tendency toward breakage.

For ease in ultrasonic treatment of individual specimens, the following procedure is recommended. The fossils are covered with about 0.25-inch of water in a beaker that is about 0.5-inch in diameter. This beaker is placed in a larger beaker to which water is added in such amount that the small beaker does not float and tip over. The larger beaker then is placed in the transducer tank in which the depth of water is controlled to prevent floating and tipping of the large beaker. About one minute of ultrasonic treatment is sufficient to clean most specimens.

E—MOUNTING SPECIMENS

It is customary to glue ostracode specimens to micropaleontological slides, and for this purpose all types of glue, including Duco cements, have been used. Most glues and Duco cement may in time contract upon hardening and thus rupture some specimens. Experience has shown that a dilute solution of gum tragacanth, to which a few drops of phenol or oil of cinnamon are added in order to prevent the develop-

ment of molds, is admirably suited for the purpose of mounting small fossils. The specimens are easily unglued by use of a wet brush. Should it be necessary to remove all traces of the gum tragacanth from a specimen, it may be immersed in alcohol; then the gum tragacanth forms a milky cohesive gel-like substance that can be teased away easily with a needle.

F—STUDY WITH REFLECTED LIGHT

Ostracodes usually are examined with reflected light. Finer details are better seen when the specimen is either coated or stained. For best results in photography specimens should be coated. TRIEBEL (1958) describes methods of preparing specimens for photography.

(1) AMMONIUM CHLORIDE COATING

BASSLER & KELLETT (1934, p. 9, fig. 2) describe an apparatus whereby a thin film of ammonium chloride sublimate is deposited on the specimen through the combination of fumes of concentrated hydrochloric acid and ammonia. This method is not satisfactory for microscopic specimens because of large grain-size of the sublimate commonly caused by high humidity. HESSLAND (1949, p. 115) describes a method which is an improvement on that reported by BRANSON & MEHL (1933, p. 17), and by COOPER (1935, p. 357), for obtaining a fine-grained deposit of ammonium chloride sublimate. A simplification of HESSLAND's method is to use glass tubing of 2- or 3-mm. inside diameter about 4 inches long. One end is drawn out to form a fine nozzle, and ammonium chloride powder is inserted through the other end. The wider end is then sealed with plastic wood or plaster of paris. When the tube is heated, a jet of ammonium chloride is released through the nozzle. The vapor can then be directed over the specimen and a fine-grained sublimate is deposited on the specimen. A vial prepared in this manner has been in use by me for several years.

(2) MAGNESIUM OXIDE COATING

A small piece of magnesium ribbon held by forceps when ignited will serve to white-coat a specimen that is passed over the magnesium oxide fumes. This method is advantageous in that the film remains on the specimen until it is washed off. The light emitted by burning magnesium can be injurious to eyesight and therefore care should be taken not to look directly at the light.

(3) STAINING

Any kind of stain causes fine detail to stand out on specimens and for this purpose almost all sorts of ink and food-coloring preparations can be used (Artusy & Artusy, 1956). Malachite green (HENBEST, 1931, p. 358) dissolved in alcohol has proved to be most suitable for staining ostracodes because of rapidity with which it dries. This stain can be removed by washing with alcohol.

(4) SILVER NITRATE COATING

LEVINSON (1951) has described a technique for depositing a metallic film of silver nitrate on ostracodes. For such treatment a clean specimen is heated for about three seconds, allowed to cool to room temperature, and then painted with 5 per cent silver nitrate solution. After 15 seconds the excess is drawn off with filter paper and the specimen is reheated over a bunsen flame for one minute. The resulting metallic film is permanent, but has the disadvantage of obscuring pore-canals and muscle scars.

G—STUDY WITH TRANSMITTED LIGHT

The muscle scars, pore canals, and duplicature structures of ostracodes are best observed in transmitted light, and methods of making the valves translucent have been devised. The same methods can be used to observe certain structures with reflected light.

(1) LIQUID IMMERSION TECHNIQUE

In some instances water is adequate to observe the structures with transmitted light but usually glycerin or an immersion oil is used. WAGNER (1957, p. 17) soaked ostracode specimens for several hours in castor oil in order to make them translucent.

(2) CANADA BALSAM TECHNIQUE

Specimens mounted in Canada balsam will show structures by transmitted light that are otherwise not seen.

(3) HYDROFLUORIC ACID TECHNIQUE

Calcareous specimens can be converted to fluorite by use of hydrofluoric acid (SOHN, 1956). Fluorite is more translucent in water or glycerin than calcite, and many specimens that do not show any shell structure in the calcitic state, will exhibit muscle scars and marginal structures when converted to fluorite.

H—STUDY WITH POLISHED AND THIN SECTIONS

Overlap, certain types of hingements, and duplicatures of ostracodes can be observed by use of polished surfaces and thin sections. The specimen is mounted on a glass slide in Canada balsam, bioplastic, or other suitable medium and ground in the same manner as thin sections of rocks. When the desired point on the specimen is reached, the specimen is turned over on the slide and the other side then is ground to make a thin section.

It is possible to reconstruct the structure of a complete ostracode carapace by means of a series of polished surfaces (KESLING & SOHN, 1958, p. 518) records of which can readily be made photographically and by aid of camera lucida. When Canada balsam is used, the specimen is oriented with a toothpick or a warm needle. It is easier to observe wall structures if the specimen is stained prior to mounting in the cement. The stain sometimes penetrates the shell material so that the inside border of the shell can be seen.

SYLVESTER-BRADLEY (1941, p. 6) has used the following method on large specimens with thin shells, obtaining excellent results. First he made a drawing of the specimen to be studied. The specimen was then broken with a needle, and the lines of fracture recorded on the drawing. Each fragment was then mounted on its edge with a gum tragacanth smear, and examined.

REFERENCES

Artusy, R. L., & Artusy, J. C.
(1) 1956, *The use of food coloring as a new technique for staining microfossils:* Jour. Paleont., v. 30, p. 969-970, 1 text fig. (Tulsa).

Bassler, R. S., & Kellett, Betty
(2) 1934, *Bibliographic Index of Paleozoic Ostracoda:* Geol. Soc. America, Spec. Paper 1, 500 p., 24 text fig. (New York).

Beckmann, von Heinz
(3) 1959, *Verunreiningung von Mikroproben beim Schlämmen:* Paläont. Zeitschr., v. 33, no. 1/2, p. 124 (Berlin).

Branson, E. B., & Mehl, M. G.
(4) 1933, *Conodont studies no. 1; Conodonts from Harding sandstone of Colorado; from the Bainbridge (Silurian) of Missouri; from the Jefferson City (Lower Ordovician) of Missouri:* Missouri Univ. Studies, v. 8, no. 1, 72 p., 4 pl., 1 text fig. (Columbia).

Cooper, C. L.
(5) 1935, *Ammonium chloride sublimate apparatus:* Jour. Paleont., v. 9, p. 357-359, 2 text fig. (Tulsa).

Henbest, L. G.
(6) 1931, *The use of selective stains in paleontology:* Jour. Paleont., v. 5, p. 355-364. (Tulsa).

Hessland, Ivar
(7) 1949, *Investigations of the Lower Ordovician of the Siljan District, Sweden: I. Lower Ordovician ostracods of the Siljan District, Sweden:* Geol. Inst. Univ. Uppsala, Bull., v. 33, p. 97-408, 26 pl.

Jekhowsky, B. de
(8) 1959, *Une technique standard de préparation des roches pour l'étude des microfossiles organiques:* Rev. Inst. Fr. Pétrole, v. 14, no. 3, p. 315-320 (Paris).

Kesling, R. V., & Sohn, I. G.
(9) 1958, *The Paleozoic ostracode genus Alanella Bouček, 1936:* Jour. Paleont., v. 32, p. 517-524, pl. 78, 3 text fig. (Tulsa).

Levinson, S. A.
(10) 1951, *The Triebel technique for staining ostracodes:* Micropaleontologist, v. 5, p. 27 (New York).

Pokorný, Vladimir
(11) 1958, *Grundzüge der Zoologischen Mikropaläontologie:* Deutsch. Verslag Wiss., Berlin, Band 1, Veb.

Sohn, I. G.
(12) 1956, *The transformation of opaque calcium carbonate to translucent calcium fluoride in fossil Ostracoda:* Jour. Paleont., v. 30, p. 113-114, pl. 25 (Tulsa).

Sylvester-Bradley, P. C.
(13) 1941, *The shell structure of the Ostracoda and its application to their paleontological investigation:* Ann. & Mag. Nat. History, ser. 11, v. 8, p. 1-33, 18 text fig. (London).

Triebel, Erich
(14) 1958, *Ostracoden:* in Freund, Hugo, Handbuch der Mikroskopie und der Technik, v. 2, pt. 3, p. 193-236, 8 pl. (Frankfurt a-M.).

Wagner, C. W.
(16) 1957, *Sur les ostracodes du Quaternaire Récent des Pays-bas et leur utilisation dans l'étude géologique des dépôts Holocènes:* Mouton & Co., 259 p., 50 pl. ('s-Gravenhage).

Witwicka, E., Bielecka, W., Styk, O., & Sztejn, J.
(15) 1958, *The methods of working out microfossils:* Poland, Inst. Geol. Bull. 134, 156 p., 1 pl., 69 text fig.

IDENTIFICATION OF FOSSIL OSTRACODES IN THIN SECTION

By S. A. Levinson
[Humble Oil & Refining Company]

CRITERIA USED FOR IDENTIFICATION

Research has revealed that various genera and species of ostracode fossils can be identified in thin section. This is possible because of variability in such characteristics as shell layering, shell thickness, overlapping of the edges of the carapace, the position and nature of ridges, frills and spines on the surface of the valves, and shell structures reflected by the deep vertical groove or trench in the valves, referred to as the sulcus.

PROCEDURES FOR PREPARATION OF THIN SECTIONS

Two methods are used in the sectioning of free specimens for the study of shell characteristics of known species. The method discussed first is used for most Paleozoic and Mesozoic species; the second method is used for fragile Paleozoic and Mesozoic forms and for all Cenozoic forms.

(1) In the first method, a small quantity of thermoplastic (such as Lakeside) is heated on a clean, oil-free frosted glass slide to just above the temperature at which it becomes fluid. As quickly as possible, the slide is placed under a binocular microscope (magnification about $30\times$) and the specimen is introduced into the thermoplastic. The thermoplastic will harden in approximately 30 seconds but will probably remain fluid for a sufficient length of time for the

specimen to be oriented in any desired position. However, a fine needle heated over a flame may be used to keep the thermoplastic fluid; the needle also is a satisfactory tool for orienting the specimen in the desired position. The slide is then ground by hand, using a figure eight motion, on carborundum paper (e.g., No. 400-A Tufbak, Behr-Manning) until the desired position on the specimen is reached. A smooth surface is obtained on the thermoplastic by adding a few drops of mineral oil on the carborundum paper. Suction cups (such as are obtainable from toy darts or arrows) of the same diameter as the slide afford an easy method for holding the slide against the carborundum paper. The specimen can be examined periodically from the reverse side of the slide to see progress of the grinding.

The slide is then washed in carbon tetrachloride to remove all adhering mineral oil. The thermoplastic is next melted in the vicinity of the specimen by using the heated needle (if the entire slide is heated the thermoplastic tends to run to edges of the slide). With the heated needle the specimen is oriented so that its flat side (that portion of the specimen previously ground down) is flush against the slide. Only slight pressure is needed to assure close contact between the specimen and the slide. The slide is again ground on No. 400-A carborundum paper and mineral oil until clear structures are obtained; then it is again washed with carbon tetrachloride.

The specimen may be stained by using a few drops of Heeger's solution (made by acidifying a solution of potassium ferricyanide with hydrochloric acid). The solution should be allowed to remain on the specimen for not more than 10 seconds; the slide then must be thoroughly washed with water. If the specimen is stained it must be covered the same day.

For covering the specimen the procedure is as follows. In a small evaporating dish, 5 drops of castolite hardener (catalyst) is mixed with 15 ml. of castolite (plastic). This is sufficient to cover approximately 10 slides. Two drops of this mixture are put directly on the specimen and a clean, oil-free cover glass is placed directly over the mixture. The cover glass must be firmly pressed against the slide to remove the air bubbles and excess mixture.

After all slides of a batch have been covered, the slides are placed approximately 7 inches from an infra-red heat lamp for 30 to 45 minutes and then allowed to set for 15 minutes. A single-edged razor blade is used to remove excess castolite from the slide. The slides are next washed in a mixture of 2 parts acetone to 1 part carbon tetrachloride. As a final step the slide is washed with a detergent soap. With a little practice approximately 30 specimens can be sectioned and covered during a normal eight-hour day.

(2) As noted above, the second method is used primarily for fragile Paleozoic and Mesozoic and all Cenozoic forms. In a large evaporating dish mix 12 drops or 1 ml. of castolite hardener with 25 ml. of castolite. Immediately pour the mixture into a plastic ice cube tray containing a dozen 1.5 cm. (approximately 0.5-inch) cells. If a 0.25-inch slit is made in the walls of each individual cell, it will facilitate the removal of the castolite cubes. Each cell should be filled approximately half full. When the tray is placed 7 inches from an infra-red heat lamp for 3 minutes at a temperature of 63°C., the mixture quickly hardens. Overheating causes the castolite to crack. The specimen is immediately placed on the top surface of the hardened castolite and oriented in the desired position. The castolite is sufficiently tacky to retain the specimen in a fixed position. Another mixture of castolite and castolite hardener is prepared and poured into the cells, filling them. If a double batch is originally made, the portion to be used at this time will have jelled and may be difficult to pour.

The tray is next placed back under the heat lamp for 3 minutes and then the material should set for 12 hours for complete hardening. After the heat-lamp treatment it is possible to pop the cubes from the tray to place them again under the lamp for an additional minute. This speeds the hardening process.

Using a diamond lap, the cube is slowly ground on one plane until the specimen is encountered. The cube is then ground by hand to the desired position, using No. 400-A carborundum paper and mineral oil, after which the slide is washed in carbon tetrachloride. A small amount of Lakeside thermoplastic is placed on a frosted slide and

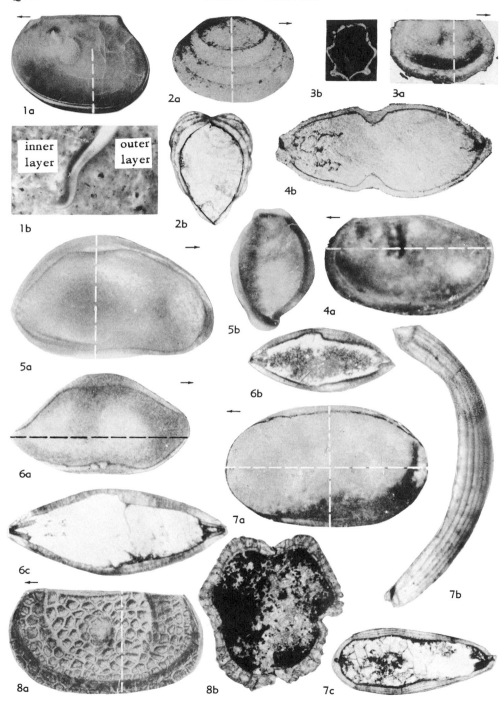

Fig. 31. Representative features shown by ostracode thin sections (lines on views of carapaces indicate location of sections illustrated).——*1. Isochilina; 1a,* LV lat., ×2; *1b,* part of section, ×13.——*2. Cryptophyllus; 2a,* R lat., ×38; *2b,* transv. sec., ×50.——*3. Eridoconcha; 3a,b,* R lat., transv. sec., ×45.——*4. Limnoprimitia; 4a,* L lat., ×35; *4b,* long. sec., ×47.——*5. Bairdiacypris; 5a,* R lat., ×21; *5b,* transv. sec., ×19.——*6. Bairdia; 6a,* R lat., ×36; *6b,c,* long. secs., ×33, ×46.——*7. Cavellina; 7a,* L lat., ×48; *7b,* transv. sec. LV, ×115; *7c,* long. sec., ×37.——*8. Amphissites; 8a,* L lat., ×50; *8b,* transv. sec., ×75 (Levinson, n).

melted and the ground side of the cube is pressed firmly against the slide. The cube is again ground on the diamond lap until the desired thickness is nearly approached. Final grinding is done on carborundum paper with mineral oil. The slide is washed with carbon tetrachloride, stained, and covered as described above.

A phase-contrast microscope with a polaroid filter has proved to be the most satisfactory instrument for examining ostracode thin sections.

EXAMPLES OF THIN SECTION IDENTIFICATION

Observations have shown that members of the family Leperditiidae (L.Ord.-M.Dev.) have a single primary shell layer comprising the external expression of the valve and a secondary layer internal to the primary layer, usually restricted to the dorsal and ventral portions of the carapace (Fig. 31,*1*). As some genera and species which are members of this family show minor variations in the basic shell structure, this type of shell layering can be used to recognize forms with a limited stratigraphic range.

The genera *Eridoconcha* (U.Ord.) and *Cryptophyllus* (M.Ord.-U.Jur.) and some species of *Amphissites* (M.Dev.-M.Perm.) possess a many-layered shell formed by the incomplete molting of the valves, with the newly formed shell cemented to one or more older shells (Fig. 31,*2*). Species of *Cryptophyllus* are considered to be important subsurface markers of the Bromide, Tulip Creek, and Oil Creek formations of the Simpson Group (M.Ord.) in Texas and the Mid-Continent area, and species of this genus can be readily identified in thin section.

The genus *Cavellina,* a smooth-shelled ostracode (?Sil., M.Dev.-Penn., ?Perm.) has been found to possess from 4 to 9 layers of shell material with the total thickness of the layers aproximately the same as the shell thickness of other ostracode genera (Fig. 31, 7). Preliminary studies suggest that the number of layers may have stratigraphic significance.

Species of six genera of the family Kirkbyidae have been sectioned and all show a two-layered shell structure. In these forms (Fig. 31,*8*) the inner layer contains the pore canals and is knoblike in cross section, with intermediate areas filled by the outer layer. In some specimens the inner layer is laminated and the outer layer is prismatic, as in the shell structure of Gastropoda and Pelecypoda. It is believed that in the ostracodes either of these layers may dominate or be developed to the exclusion of the other, which offers an additional criterion for identification of one-layered forms.

In the majority of ostracodes, one valve is larger than the other, edges of the larger valve overlapping the smaller. Some forms overlap only along the venter, whereas others may have the overlap restricted to one of the free margins or dorsum. Variations in the amount and nature of overlap are quite conspicuous in thin sections and can be used to identify many forms (Fig. 31,*5*).

As seen in thin sections, major ridges and frills appear as if the shell has been folded to form these features (Fig. 31,*3*). In addition, the shell is indented at the inner surface where folding occurs, and a darkened area bisecting this extension of the shell is commonly developed. These criteria and observation as to length and width of the extension, as seen in thin sections, permit the identification of such features.

In a number of genera, the position of the sulcus is reflected, in thin section, by a pronounced thickening of the shell where this feature occurs (Fig. 31,*4*). Thus, identification of this character is permitted, providing an important criterion for the separation of Paleozoic forms.

The genus *Bairdia* is a smooth-shelled ostracode frequently encountered in rocks from Middle Silurian to Recent age. Longitudinal sections of *Bairdia* are distinctive, for an inner extension of shell is observed at either the anterior or posterior margin or at both margins (Fig. 31,*6*); in different species the extensions range from very thin and elongate to short and stubby. Sections of some species of *Bairdia* show an abrupt thinning of the shell from mid-length to the posterior extremity. Such variations suggest that thin sections of species of this genus (abundant in Mississippian to Permian beds) may be readily used for age determinations.

CLASSIFICATION OF OSTRACODA

By H. W. Scott

[University of Illinois]

CONTENTS

INTRODUCTION

Taxonomy of the ostracodes has undergone major changes in recent years. Early authors classified the Palaeocopida primarily on the basis of sulcation, lobation, and ornamentation. Dimorphic features were mostly disregarded in distinguishing family-group taxa and only occasionally used at the generic level. While confusion existed in classification of the palaeocopid ostracodes, students of other groups, mostly Mesozoic and Cenozoic podocopids, were making progress in taxonomy by using muscle-scar patterns and hinge structures. In this manner they were able to relate many fossil genera to their living relatives.

EARLY WORK

The oldest published figures of an ostracode known to me are found in a paper by GODEHEU DE RIVILLE (1760); they represent a Recent form. A few years later, O. F. MÜLLER (1772) gave illustrations of other

living species, indicating that they differed from insects and stating that shortly he intended to describe them as representatives of a new genus. Such a paper appeared (MÜLLER, 1776) containing definition of the new genus *Cypris*, to which ten described species were referred. Later, he (MÜLLER, 1785) added details to observations on *Cypris* and introduced another new genus, *Cythere*. All this work was based on modern ostracodes.

The date when the first fossil ostracode was recognized is not known certainly, but among oldest publications concerned with such fossils is one by DESMAREST (1813). This is cited by SOWERBY (1825), who described and figured *Cypris faba* from Tertiary deposits of France on the basis of DESMAREST's work.

The first known observation of a Paleozoic ostracode seems to have been made by DALMAN (1826) when he described a fossil named *Battus*, classified by him with the

Trilobita. In 1834 KLÖDEN published description of a new species of ostracode as *Battus tuberculatus* (now assigned to the genus *Beyrichia*); *Battus* is not available for the fossils described by DALMAN and KLÖDEN because it had been used (1777) for lepidopteran insects. The oldest acceptable name for a Paleozoic ostracode is *Entomoconchus* M'COY, 1839. Publication of this genus was followed (M'COY, 1844) by the introduction of *Bairdia* and (M'COY, 1846) *Beyrichia*. The latter is the oldest generic name applied to a palaeocopid ostracode.

DEVELOPMENT OF OSTRACODE CLASSIFICATION

As somewhat incomplete background for discussion of the classification of fossil and living ostracodes adopted in this *Treatise*, it is desirable to take account at least briefly of selected main contributions by previous workers.

A monograph by G. O. SARS (1866) represents the earliest basic work devoted to major classification of the Ostracoda. In this publication he introduced four new groups designated as suborders (ostracodes as a whole being defined as an order) and two new families. SARS' classification, based almost entirely on Recent genera, is summarized as follows. No Paleozoic fossils were included by him.

Classification of Ostracoda by Sars, 1866

Suborder PODOCOPA, nov.
  Family Cypridae Baird, 1845 [*recte* Cyprididae]
  Family Cytheridae Baird, 1850
Suborder MYODOCOPA, nov.
  Family Cypridinadae Baird, 1850 [*recte* Cypridinidae]
  Family Conchoeciadae, nov [=Halocyprididae Dana, 1853]
Suborder CLADOCOPA, nov.
  Family Polycopidae, nov.
Suborder PLATYCOPA, nov.
  Family Cytherellidae, nov.

The first comprehensive effort to deal with classification of Paleozoic ostracodes is recorded in E. O. ULRICH's (1894) report on Ordovician Ostracoda from the upper Mississippi Valley region of the United States. No suborders were recognized, 13 families (of which three were new) being arranged simply as divisions of ostracodes as a whole. An outline of the classification given by ULRICH follows.

Classification of Lower Paleozoic Ostracoda by Ulrich, 1894

Order OSTRACODA
  Family Leperditiidae Jones, 1856
  Family Beyrichiidae Matthew, 1886
  Family Barychilinidae, nov.
  Family Entomidae Jones, 1873 [=Entomozoidae Pribyl, 1951]
  Family Cypridinidae Baird, 1850
  Family Entomoconchidae Brady, 1868
  Family Polycopidae Sars, 1866
  Family Cytherellidae Sars, 1866
  Family Cytheridae Baird, 1850
  Family Thlipsuridae, nov.
  Family Cypridae Baird, 1845 [*recte* Cyprididae]
  Family Beecherellidae, nov.
  Family Darwinulidae Brady & Norman, 1889

In a report by FREDERICK CHAPMAN (1901) on Silurian fossils from Gotland, a partial classification of ostracodes is given by T. R. JONES, as follows.

Classification of Silurian Ostracoda by Jones, 1901

Order OSTRACODA
  Section PODOCOPA Sars, 1866
    Family Leperditiidae, Jones, 1856
      Subfamily Aparchitinae, nov.
      Subfamily Beyrichiinae, nov. [*recte* Matthew, 1886]
    Family Cytheridae Baird, 1850
  Section CYPRIDIDA, nov.
    Family Cyprididae Baird, 1845
    Family Bairdiidae Sars, 1888
  Section PLATYCOPA Sars, 1866
    Family Cytherellidae Sars, 1866 [*recte* Jur.-Rec.]

A milestone in classificatory study of Paleozoic ostracodes is represented by a contribution to the Silurian volume of the Maryland Geological Survey by ULRICH & BASSLER (1923). The Silurian formations of the middle Appalachian Mountains region are rich in well-preserved ostracodes that are very useful for zonal subdivision of the strata. These fossils furnish the main basis for recognition of numerous new genera and several families and subfamilies. The classification given in this report is summarized as follows.

Classification of Paleozoic Ostracoda by Ulrich & Bassler, 1923

Order OSTRACODA
  Family Leperditiidae Jones, 1856
  Family Aparchitidae, nov. [*recte* Jones, 1901]
  Superfamily Beyrichiacea [Matthew, 1886]

Family Beyrichiidae Jones [*recte* Matthew, 1886]
Family Primitiidae, nov.
    Subfamily Eurychilininae, nov.
Family Zygobolbidae, nov.
    Subfamily Zygobolbinae, nov.
    Subfamily Drepanellinae, nov.
Family Kloedenellidae, nov.
Family Kirkbyidae Ulrich & Bassler, 1906
Superfamily Cypridacea, nov. [*recte* Baird, 1845]
    Family Thlipsuridae Jones [*recte* Ulrich, 1894]
    Family Beecherellidae Ulrich, 1894
    Family Bairdiidae, nov. [*recte* Sars, 1888]
    Family Cypridae Zenker [*recte* Baird, 1845]
    Family Cytherellidae Sars, 1866
    Family Entomidae Jones, 1873 [=Entomozoidae Pribyl, 1951]
    Family Cypridinidae Sars [*recte* Baird, 1850]
    Family Entomoconchidae Ulrich [*recte* Brady, 1868]
    Family Polycopidae Sars, 1866
    Family Darwinulidae Jones [*recte* Brady & Norman, 1889]
    Family Barychilinidae Ulrich, 1894
Superfamily Cytheracea, nov. [*recte* Baird, 1850]
    Family Cytheridae Zenker [*recte* Baird, 1850]

In 1934 a comprehensive catalogue of then-known Paleozoic ostracodes was undertaken by R. S. Bassler and Betty Kellett. This represents a considerable expansion of knowledge that is expressed in recognition of a notably enlarged number of families. Three superfamilies are distinguished but no subordinal category or categories. The arrangement of Paleozoic Ostracoda given in this report is as follows.

Classification of Paleozoic Ostracoda by Bassler & Kellett, 1934

Superfamily Leperditacea, nov. [*recte* Leperditiacea Jones, 1856]
    Family Leperditiidae Jones, 1856
    Family Leperditellidae Ulrich & Bassler, 1906
Superfamily Beyrichacea Ulrich & Bassler, 1923 [*recte* Matthew, 1886]
    Family Beyrichiidae Jones [*recte* Matthew, 1886]
    Family Primitiidae Ulrich & Bassler, 1923
        Subfamily Primitiinae, nov. [*recte* Ulrich & Bassler, 1923]
        Subfamily Eurychilininae Ulrich & Bassler, 1923
    Family Zygobolbidae Ulrich & Bassler, 1923
        Subfamily Zygobolbinae Ulrich & Bassler, 1923
        Subfamily Kloedeninae Ulrich & Bassler, 1923 [*recte* Kloedeniinae]
        Subfamily Drepanellinae Ulrich & Bassler, 1923
    Family Kloedenellidae Ulrich & Bassler, 1923
    Family Kirkbyidae Ulrich & Bassler, 1906
    Family Glyptopleuridae Girty, 1910
    Family Youngiellidae Kellett, 1933

Superfamily Cypridacea Ulrich & Bassler, 1923 [*recte* Baird, 1845]
    Family Cypridae Zenker [*recte* Cyprididae Baird, 1845]
    Family Thlipsuridae Jones [*recte* Ulrich, 1894]
    Family Beecherellidae Ulrich, 1894
    Family Bairdiidae Lienenklaus [*recte* Sars, 1888]
    Family Cytherellidae Sars, 1866
    Family Entomidae Jones, 1873 [=Entomozoidae Pribyl, 1951]
    Family Cypridinidae Sars, 1866
    Family Entomoconchidae Jones, Kirkby, & Brady [*recte* Jones, 1868]
    Family Barychilinidae Ulrich, 1894

For many years F. M. Swartz, of Pennsylvania State University, has been working intensively on stratigraphy and paleontology of middle Paleozoic formations of the Appalachian region. Independently, and associated with others, he has published several important papers concerned mainly with descriptions of ostracodes collected from Silurian and Devonian strata of this region. One of his papers (Swartz, 1936) contains an outline of ostracode classification that is worthy of notice. It is incomplete in that various groups are omitted from consideration.

Classification of Middle Paleozoic Ostracoda by Swartz, 1936

Superfamily Beyrichacea [*recte* Beyrichiacea Matthew, 1886]
    Division family Beyrichiidae
        Family Beyrichiidae, Ulrich, 1894 [*recte* Matthew, 1886]
        Family Zygobolbidae Ulrich & Bassler, 1923
    Division family Primitiidae
        Family Primitiidae, Ulrich, 1923
        Family Hollinidae, nov.
        Family Tetradellidae, nov.
        Family Drepanellidae, nov. [*recte* Ulrich & Bassler, 1923]
        Family Acronotellidae, nov.
        Family ?Primitiopsidae, nov.
        Family ?Aechminidae (Leperditacea?), nov.
    Division family Kloedenellidae
        Family Kloedenellidae Ulrich & Bassler, 1923
        Family ?Glyptopleuridae Girty, 1910
    Division family Kirkbyacea
        Family Kirkbyidae Ulrich & Bassler, 1906
        Family Youngiellidae Kellett, 1933
    [Division not indicated]
        Family Leperditellidae Ulrich & Bassler, 1906
        Family Beecherellidae Ulrich, 1894
        Family Bairdiidae Sars, 1888
        Family Cytherellidae Sars, 1866

Until the time of its publication, the most

comprehensive effort to treat the classification of Paleozoic ostracodes was reported in a paper by Gunnar Henningsmoen (1953) devoted to known straight-hinged forms. In the arrangement of family groups, both superfamilies and suborders were distinguished. A summary of Henningsmoen's classification is as follows.

Classification of Paleozoic Straight-hinged Ostracoda by Henningsmoen, 1953

Suborder Paleocopa, nov.

  Superfamily Beyrichiacea Ulrich & Bassler, 1923 [*recte* Matthew, 1886]

    Family Sigmoopsiidae, nov. [*recte* Sigmoopsidae]

      Subfamily Sigmoopsiinae, nov. [*recte* Sigmoopsinae]

      Subfamily Glossopsiinae, nov. [=Quadrijugatoridae Kesling & Hussey, 1953]

    Family Tetradellidae Swartz, 1936

      Subfamily Tetradellinae Swartz, 1936 (incl. Dilobellinae)

      ?Subfamily Piretellinae Öpik, 1937

      Subfamily Bassleratiinae Schmidt, 1941 (incl. Ctenentominae)

    Family Primitiidae Ulrich & Bassler, 1923

    Family Eurychilinidae Ulrich & Bassler, 1923

      Subfamily Eurychilininae Ulrich & Bassler, 1923

      ?Subfamily Euprimitiinae Hessland, 1949 [=Eurychilinidae Ulrich & Bassler, 1923]

      ?Subfamily Primitiopsiinae Swartz, 1936 [*recte* Primitiopsinae]

    Family Aparchitidae Jones, 1901

    Family Drepanellidae Ulrich & Bassler, 1923

      Subfamily Drepanellinae Ulrich & Bassler, 1923

      ?Subfamily Bolliinae Bouček, 1936 (incl. Ulrichiinae Schmidt, 1941)

      ?Subfamily Aechmininae Bouček, 1936

    ?Family Acronotellidae Swartz, 1936

    Family Beyrichiidae Jones, 1894 [*recte* Matthew, 1886]

      Subfamily Beyrichiinae Jones, 1894 (incl. Kloedeninae Ulrich & Bassler, 1923 [*recte* Matthew, 1886]

      Subfamily Zygobolbinae Ulrich & Bassler, 1923

    Family Hollinidae Swartz, 1936

    Family Kloedenellidae Ulrich & Bassler, 1908

      Subfamily Kloedenellinae Ulrich & Bassler, 1908

      Subfamily Beyrichiopsiinae, nov. [*recte* Beyrichiopsinae]

      Subfamily Glyptopleurinae Girty, 1910

    Family Kirkbyidae Ulrich & Bassler, 1906 (incl. Amphissitinae Cooper, 1941)

    ?Family Youngiellidae Kellett, 1933

    ?Family Miltonellidae Sohn, 1950

    ?Family Alanellidae Bouček, 1936

  Superfamily Leperditacea Bassler & Kellett, 1934 [*recte* Leperditiacea Jones, 1856]

    Family Leperditiidae Jones, 1856

      Subfamily Leperditiinae Jones, 1856

      Subfamily Isochilininae Swartz, 1949

    ?Family Leperditellidae Ulrich & Bassler, 1906

      Subfamily Leperditellinae Ulrich & Bassler, 1906

      ?Subfamily Conchoprimitiinae, nov.

      ?Subfamily Eridoconchinae, nov.

Suborder Podocopa Sars, 1866

    Family Quasillitidae Coryell & Malkin, 1936 (incl. Graphiodactylidae Kellett, 1936)

      Subfamily Quasillitinae Coryell & Malkin, 1936

      Subfamily Ropolonellinae Coryell & Malkin, 1936

The first all-inclusive effort to classify Recent and fossil ostracodes is contained in a textbook on micropaleontology prepared by Vladimir Pokorný (1958). This is especially noteworthy because of recognition given to the importance of muscle-scar patterns, hinge structures, characters of the duplicature, and dimorphic features as guides in classification. Notice of Pokorný's separation of the Leperditiida from Palaeocopida, inclusion of platycopines and podocopines in the Podocopida, and assignment of cladocopines with myodocopines in the Myodocopida is important. The arrangement of suprageneric taxa adopted by Pokorný differs in various ways from that accepted in the *Treatise* but approaches it in many ways. An outline of Pokorný's classification follows.

Classification of Recent and Fossil Ostracoda by Pokorný, 1958

Subclass Ostracoda Latreille, 1806

  Order Leperditiida Pokorný, 1953

    Family Leperditiidae Jones, 1856

      Subfamily Leperditiinae Jones, 1856

      Subfamily Isochilininae Swartz, 1949

  Order Beyrichiida Pokorný, 1953

    Family Beyrichiidae Matthew, 1886

      Subfamily Beyrichiinae Matthew, 1886

      Subfamily Zygobolbinae Ulrich & Bassler, 1923

      Subfamily Treposellinae Henningsmoen, 1954

    Family Tetradellidae Swartz, 1936

      Subfamily Tetradellinae Swartz, 1936

      Subfamily Sigmoopsidinae Henningsmoen, 1953 [*recte* Sigmoopsinae]

      Subfamily Quadrijugatorinae Kesling & Hussey, 1953

CLASSIFICATION ADOPTED IN TREATISE

GENERAL DISCUSSION

Up to the present a satisfactory basis for classification of all Ostracoda has not been found. No single morphological feature can be used to define orders, superfamilies, and families. Criteria used for separating the families of one superfamily may be entirely different from those used in another superfamily. Similarly, distinction of genera within a family is often schematic and inconsistent. On the other hand, some genera and families possess such striking shell characters that they are readily recognizable and traceable throughout long expanses of time. For example, the shape of the carapace of *Bairdia* is a very diagnostic feature that has persisted at least from late Paleozoic to Recent.

The most primitive orders—Archaeocopida, Leperditicopida, and Palaeocopida—

have no living representatives (possibly except in Punciidae) and therefore must be distinguished by differences in the shell. The thin carapace of the archaeocopids distinguishes them from the leperditicopids. The thick shells and large compound muscle scars serve to separate the leperditicopids from all other orders. The palaeocopids are differentiated from the two more primitive orders first by their small, fairly simple muscle scars, and secondly by the presence of one or more such features as lobation, sulcation, strong ornamentation, and dimorphic structures. None of the three primitive orders possess a duplicature.

The Podocopida are represented by many living species. This order was first defined by zoologists on the basis of the soft parts. Only in recent years have the shell features been carefully described, most of this work being done by paleontologists. In the podocopids the type of hingement, muscle-scar pattern, and outline of the carapace are used for classification. The Cypridacea are ovate to elongate-ovate in outline, with a convex dorsum, and an incurvature along the medial portion of the venter. They are readily separated from the Cytheracea by lack of a highly ornamented surface. Also, the Cytheracea, with few exceptions, have strongly developed hinge elements and distinct muscle scars.

Various criteria are used to divide ostracodes into superfamilies. In the palaeocopids, dimorphic structures are of primary importance. Secondarily, the presence or absence of sulci and lobes, the general outline of the valves, in conjunction with major ornamentation and the presence or absence of a velum, may be used. Of these, the kind of dimorphic structure present is the most important and is used for all dimorphic forms. The well-developed S_2 or pit, the strongly asymmetrical valves and kloedenellid dimorphism distinguish the Kloedenellacea; a smooth carapace, channeled hinge, and asymmetrical valves are characteristics of the Paraparchitacea; anteroventral or ventral cruminal dimorphism separates the Beyrichiacea from the nondimorphic Drepanellacea; a dimorphic velate structure sets representatives of the Hollinacea apart from the nondimorphic Drepanellacea; the carapace outline of Youngiellacea separates the group from all others; and the reticulate pattern, combined with presence of a median pit, distinguishes the Kirkbyacea.

Superfamilies among the myodocopids are determined for the most part by the presence and nature of the rostrum.

Among the podocopids, superfamilies are based mostly on differences in hingement, muscle-scar patterns, and to some extent on outline. The undifferentiated hinge and convex back of the cytherellids readily separate them from the cytherids, which have a complex hinge and convex or straight back. The large number of closely spaced muscle scars and the elongate form of the Darwinulacea are unique to this group. The simple hinge and carapace shape of the Bairdiacea combine to make these features of major importance in classification. Characterization of the Cypridacea is difficult, but this group may be separated from others by ovate outline, convex back, incurved midventer, simple hingement, lack of major sulci or lobes, and generally unadorned surface.

Family differentiation is based on various major features depending on the order within which the family falls. In the Leperditicopida, the two families are distinguished by the symmetry of the carapaces; the leperditiids have asymmetrical valves, the larger overlapping the smaller around the free margin, whereas the isochilinids have subequal valves.

Families within the palaeocopids are distinguished by differences in such morphological features as variation in dimorphic structures, differences in ornamentation (e.g., reticulate or costate surfaces), degree of sulcation and lobation, and variation in outline.

In the Kloedenellacea, Glyptopleuridae are costate, whereas the Sansabellidae are noncostate. The strongly obtuse anterior cardinal angle of the Kloedenellidae separates this family from other members of the superfamily. In some families adventral dimorphic structures are used for distinction. The velate structures of Hollinidae differentiate them from histial dimorphic features of the Sigmoopsidae. Other families may be separated on differences in outline of the carapace or hingement.

Among the podocopids, families commonly are defined to a considerable extent by variations in hingement and muscle-scar patterns. Major differences in outline and ornamentation may be helpful but are likely to be confusing. Hingement and muscle scars are by far the most important, and lack of knowledge of these factors often makes classification insecure.

Ostracode genera are interpreted rather narrowly. This practice has resulted in the creation of many monospecific genera and has given rise to a few monogeneric families. Such features as reversal of valve overlap around the free margin is still evaluated differently by various workers; some consider it to lack even specific value, whereas others assign it generic value. *Sansabella* may be found in Late Mississippian sediments, in a single sample, with specimens identical in all respects except for reversal of overlap. This is equally true for *Paraparchites*.

Distinction of genera is based, for the most part, on major shell features, but not commonly on ornamentation. Hinge characters, muscle-scar patterns, outline, lobation and sulcation, and adventral structures are most important. Though outline of the carapace in different genera is often found to be similar, other characters may be quite unlike. The outline is usually constant within a genus but may vary slightly, especially as modified by dimorphism or as represented by instar stages. Hinge characters are specially significant in delineating post-Paleozoic genera. The carapaces of many genera of the Cytheracea have similar outlines and are separated from one another primarily on the basis of hinge characters. Such forms as *Archicythereis, Cythereis,* and *Oligocythereis* look somewhat alike externally but hinge structures differ greatly. In post-Paleozoic ostracodes the valves often are found separated from each other, and therefore hinge details may be observed. This is seldom true in palaeocopids and thus hingement is not an important aid to classification in this group; more information is constantly being gathered and in time it is hoped that the hingement of all families will be known.

Muscle-scar patterns are very important for classification of ostracodes at the generic level. They have primary value among the platycopines, metacopines, and podocopines. Adductor and mandibular scars are diagnostic of many genera of these sub-orders. They are not well enough known among the palaeocopids to have significance in generic classification.

Lobation and sulcation aid generic classification but must be used with caution. Similar sulcation may be found in wholly unrelated genera and dissimilar sulcation may occur in very closely related forms. Sulci and lobes are more important in the Hollinacea, Beyrichiacea, and Kloedenellacea than in any other groups. The number of sulci and lobes usually remains constant in each genus but it may vary within a family.

Specific differentiation between ostracodes is based on differences in ornamentation, modification of outline, and the size and shape of various structural features such as alae, sulci, pore canals, and lobes. However, all of these characters may vary from instar to instar and between the sexes; therefore, caution must be used in evaluating the significance of any observed variation. KESLING (1954) has discussed in detail factors affecting speciation.

Observation of instars indicates that ornamentation becomes more complex with advancing age of the ostracode. Increase in complexity has been observed in many genera; *Glyptopleura, Amphissites, Beyrichia, Cythereis, Amphizona, Eridoconcha, Loxoconcha,* and others show this tendency. The number of costae, size of reticulations, number or size of spines, and numerous other ornamental features may vary in an ontogenetic series. That the younger instars are simpler is known to be true, but as usual, some exceptions have been observed and others may be discovered. In at least one species of *Healdia,* a pair of posterior spines is considerably reduced in the adult, whereas they are long and prominent features of the instars.

The question of degree of individual variation among ostracodes has not been fully explored. It has been reported in a few cases, but it is sometimes difficult or impossible to determine whether the observed variation is natural or due to the nature of preservation. In some species the thin outer layer may be reticulate and the next layer smooth or differently ornamented. In the Kirkbyidae it is not uncommon to find shells in which one portion of the valve differs from another in ornamentation, yet examination of many specimens may show that the observed variation is a matter of preservation rather than natural variation. Differences between individuals are usually observed in reticulate, granulose, papillose, or punctate species.

Dependence on molds and casts of the carapace interior for creation of new species has led to many unnecessary complications. Of course, exterior features of ornamentation cannot possibly be preserved in such fossils. Features of hingement also are lost; sulci and nodes may be more subdued, and external adventral features of dimorphism and overlap may be completely missing. Steinkerns may yield important supplemental information, but they are seldom by themselves an adequate basis for erection of new species.

Sexual dimorphism has been recognized in many ostracodes, both fossil and living. In some, the valves strikingly reflect dimorphism, for the inferred females may be posteriorly inflated *(Kloedenella)*, or have velate frills *(Hollinella)*, or develop large adventral pouches *(Beyrichia)*, or posterodorsal inflation *(Cypris)*, whereas the inferred males may possess none of these features and appear relatively simple and unornamented. Lack of recognition of dimorphism is a factor that always must be considered in ostracode studies. It can best be understood by studying populations from a single zone, and best of all by examination of a population from a single bedding plane.

ORDER ARCHAEOCOPIDA

The order Archaeocopida has been erected to include a group of Cambrian and ?Early Ordovician ostracodes. The zoological affinities of the group are not certain, but they appear to be most closely related to the Ostracoda, having many features in common with the palaeocopids.

The carapace of the archaeocopids is only slightly calcified. Its high chitin content makes it more or less flexible and therefore commonly strongly wrinkled. The hinge is long and straight. An eye tubercle is usually prominent in all families except the Indianidae. The four currently recognized families are separated on the basis of outline of carapace, presence or absence of puncta, folds, and eye tubercles.

Though archaeocopids have been reported from Early Ordovician strata, they are essentially Cambrian organisms. They have been described from North America and Europe and were probably widely distributed in Cambrian seas. Relations of the Archaeocopida to other ostracodes are indicated diagrammatically in Figure 32.

ORDER LEPERDITICOPIDA

The Leperditicopida were a very successful Ordovician-Devonian group. Many specimens have been found with well-preserved internal markings on the carapace. They are characteristically straight-backed and possess a compound muscle scar composed of many small units. The shell is usually thick and has one or more secondary layers both dorsally and ventrally. The large muscle-scar pattern and secondary shell layers are not observed in any other ostracodes. These features are so striking that their true relationship to other groups is unknown. The leperditiids appear in the Ordovician as highly differentiated ostracodes, and in so far as known, no other group developed from them. The order has a wide geographic range, being common in Europe and North America. Its stratigraphic range is limited to the Ordovician-Devonian part of the column, most representatives occurring in Ordovician and Silurian deposits. Inferred relations to other ostracode orders are illustrated diagrammatically in Figure 32.

ORDER PALAEOCOPIDA

The Palaeocopida are a group of Paleozoic ostracodes (other than Punciidae) possessing characters that clearly distinguish them from more recent genera. Chief among these are the nature of muscle scars, dimorphic structures, marginal extensions, and the dorsal surface.

A review of the diagnoses of palaeocopid families shows that they have not been differentiated on any single carapace feature. Some families are set apart from others on one or more of the following criteria: shape or outline of the carapace, costation, reticulation, sulcation, smoothness, and various types of dimorphism. Of these features dimorphism is by far the most important. Genera within families are separated for the most part on degree of lobation, sulcation, adventral structural developments, and major ornamental features. Hinge characters, when known, may be used to define genera and possibly families. Muscle-scar patterns are of value, but are seldom preserved in the palaeocopids. Some members of this assemblage show no evidence of dimorphism and may have reproduced parthenogenetically.

The inferred relations of palaeocopids to other ostracode orders are indicated in Figure 32 and those of superfamilies are shown in Figure 33; both diagrams indicate known stratigraphic distribution of these taxa.

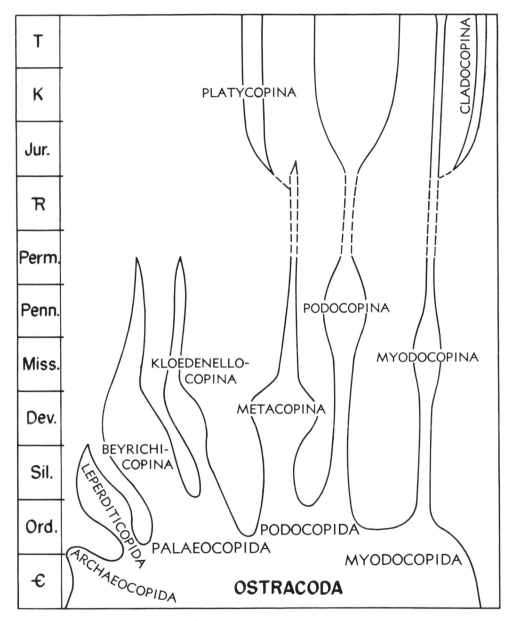

Fɪɢ. 32. Diagram representing stratigraphic distribution and inferred relationships of ostracode orders and suborders (Scott & Sylvester-Bradley, n).

SUBORDER BEYRICHICOPINA

The Beyrichicopina are one of the most abundant groups of ostracodes. With exception of the Punciidae, they are restricted to the Paleozoic and are especially prolific in the early Paleozoic (Figs. 32,33).

The straight back, subequal valves, convex free edge, and lack of inner calcareous lamellae are common to all. In addition, some form of velate structure is present in most, and lobes, sulci, and carinae are common structural features in many.

A well-developed velum is present in most of the Beyrichicopina. It is especially prominent in the Hollinacea, where it is modified in various ways as a dimorphic structure. In the Drepanellacea, the velum is represented by a pseudovelum which is not known to be related to dimorphism; in the Primitiopsacea the velum is modified into a dolon. In the Kirkbyacea a velum or pseudovelum is recognized, but it is not related to dimorphism.

One of the characteristic features of the Beyrichicopina is subequality in size of the valves. Both valves are beveled along the free margin, so that when closed the valves meet without apparent overreach. This feature alone serves to separate them from the Kloedenellocopina.

The Beyrichicopina are more strongly lobate and sulcate than any other group of ostracodes. Lobation and sulcation are exceptionally strong in the Beyrichiacea and Hollinacea.

POKORNÝ (1958) has pointed out the striking difference in outline between the Podocopida and the group here referred to as Beyrichicopina. In the former the carapace commonly has a convex back (some cytherids have straight backs) and a ventral edge that is concave medially. This is in sharp contrast to the straight back and convex free edge of the Beyrichicopina. In the latter group the basic shape or outline is seen in larval stages. Though the posterior end is more acuminate in the molts, the shape of the dorsal and ventral edges is established at an early stage. In the podocopids the larval stage is subtriangular, with

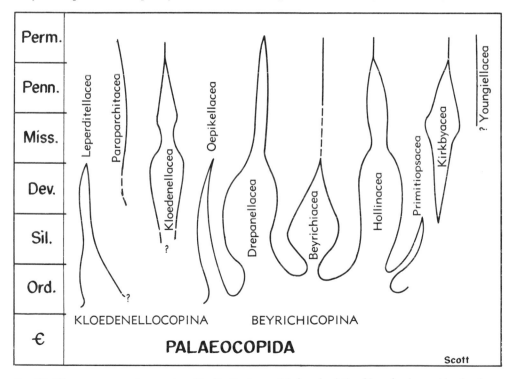

FIG. 33. Diagram representing stratigraphic distribution and inferred relationships of palaeocopid suborders and superfamilies (Scott, n).

a ventral margin that is straight to gently concave or convex. The development of the concave ventral edge in many adult podocopids appears to represent an advanced trait, the growth of which is recognizable in ontogeny of the individual.

Another aspect of shape is the "forward swing" of most Beyrichicopina. The ventral margin is convex, with the greatest degree of truncation posterior. This produces a carapace in which the anterior half is higher than the posterior half, resulting in a "forward ventral swing" which is in contrast to the "posterior swing" of the Leperditiidae.

The straight back and convex venter with forward swing are present in the Early Ordovician ostracodes, contemporaneous with the convex back and concave venter of the podocopids and the backward swing of the leperditiids.

The Beyrichiacea are among the most abundant and striking of early Paleozoic palaeocopids. The anteroventral to ventral dimorphic pouch, which is an enlargement of the carapace wall, is the distinguishing feature. This swelling is not formed by a frill of the eurychilinid type but is a development of the carapace wall. It is recognized only in the Beyrichiidae and Zygobolbidae. For this reason these two families are separated from the nondimorphic Drepanellacea. The family Beyrichiidae ranges from the nonsulcate *Apatobolbina* with its dimorphic lobe, through the bilobed *Bolbiprimitia,* to the strongly lobate and sulcate *Beyrichia.* Most of the Beyrichiacea are reticulate but a few are unornamented. Hingement, so far as known, is by means of primitive bar and groove. The superfamily ranges from Ordovician to Devonian but had its most striking development in the Silurian.

The Drepanellacea are separated from Hollinacea because they are nondimorphic, and as such, lack the marginal structures so characteristic of the latter group. The margin of most Drepanellacea is raised as a thick rim. This structure is represented by a relatively narrow, flat, marginal, smooth rim, sometimes referred to as a velum but more appropriately called a pseudovelum. It extends only slightly below the ventral surface and shows no evidence of being associated with dimorphism. No marginal rim or velate structure is present in the

Aechminidae or Richinidae. Rare reports of such occurrences need checking. Most Drepanellacea are ornamented with reticula or nodes, or both. The group seems to be related to the dimorphic Hollinacea. The most peculiar family is the Aechminidae, in which the large dorsal spine makes them appear wholly unrelated to other Drepanellacea. However, if the two nodes of *Ulrichia* were reduced to one, a form not greatly unlike *Aechmina* would be produced. Perhaps *Aechminaria,* with one large and one small node next to an intervening sulcus, is an intermediate stage between *Ulrichia* and *Aechmina.*

The Hollinacea are a major group of palaeocopids represented from Ordovician to Permian (Fig. 33). They are all ornamented in some fashion. Lobation, sulcation, and nodes are common structural features. The valves are subequal in size. Dimorphism in the form of a velate structure is present, the velum being modified in various ways as a dimorphic organ; in some, loculi are developed along the velum, either outside (e.g., *Tetradella*) or inside (e.g., *Ctenoloculina*). Dimorphic structures commonly result in carapaces that are strikingly dissimilar in the two sexes. Loculate and nonloculate discrete lobes or merged lobes may be merely marks of dimorphism.

The Kirkbyacea contain a group of very distinctive reticulate palaeocopids which are among the most highly ornamented of Paleozoic genera. Reticula occur in all and costae and nodes are common to many. The kirkbyan pit, which characterizes the entire assemblage, is represented by a break in the reticulate pattern, that probably defines the position of the adductor muscle scar. Hingement of the Kirkbyacea resembles that found in the Kloedenellacea, suggesting a possible relationship of these superfamilies, both of which may be related to the earlier Leperditellacea. The Kirkbyacea are abundant from Devonian to Permian, probably attaining a climax in Late Mississippian. Reported Silurian occurrences have not been confirmed.

The Oepikellacea are a small group of early Paleozoic ostracodes. Some doubt exists as to their true nature, because the type of *Aparchites* has not been restudied and its true characters are not fully known. Present information indicates that the

Aparchitidae and Pribylitidae are nondimorphic. However, the Oepikellidae have a well-developed velar dolon and perhaps should be placed with the Hollinacea. The tecnomorphic valves look very much like *Aparchites,* and the family is provisionally included in the Oepikellacea. The velum is a prominent feature in the heteromorphs of Oepikellidae, but in the Aparchitidae and Pribylitidae the velum is absent or weakly developed as a row of spines or a faint ridge parallel to the free edge. Solutions to problems of classification must await further study of *Aparchites* and related genera.

The Primitiopsacea represent a small assemblage of Middle Ordovician to Middle Devonian ostracodes. In the posterior part of the carapace they developed velar dimorphism characterized by open or closed dolonal flanges; if closed, an extradomiciliar chamber is produced. Two genera are known, one from Sweden and one from North America.

The Youngiellacea include a group of small subrectangular nondimorphic ostracodes with a prionodont hingement. The surface varies from smooth to reticulate or costate. The superfamily is not known to be closely related to any other palaeocopid. Minute size of the carapace, prionodont hingement, and elongate shape separate it from all other groups. *Moorites* has a general resemblance to *Cytherelloidea* in outline and ornamentation but hingement differs. It may be that affinities of the group are closer to the Platycopina than the Palaeocopida but the Youngiellacea are tentatively assigned to the latter. They are known only from Mississippian and Pennsylvanian formations.

SUBORDER KLOEDENELLOCOPINA

POKORNÝ (1958) recognized that the kloedenellids differ greatly from the beyrichiids and recorded them as *incertae ordinis.* He included in this listing *Paraparchites* and related genera. Also considered to be uncertain in ordinal assignment were the Leperditellidae and Conchoprimitiidae. Undoubtedly these groups differ greatly from other ostracodes and where to classify them has been a vexing problem, because all do not seem to have features in common that would allow placing them in a single group.

The Kloedenellacea are a large group of palaeocopids containing several important families and many genera. They are subrhomboidal to subrectangular in outline and have asymmetrical valves. One valve strongly overlaps around all or a portion of the free margin of the smaller valve. Most genera have a sharply defined S_2, though in a few species the sulcus may be weak or represented by a pit. In the genus *Dizygopleura* three sulci are present. The surface of the valves ranges from smooth to highly ornamented. Ornamentation is primarily of two types—reticulate, as in *Geisina,* and costate, as in *Glyptopleura.* Rarely are spines present. Hingement is fairly well known and consists of hinge tongue-and-groove in all genera and in some forms a connecting link between cardinal teeth and sockets. Dimorphism is represented by a swelling of the posterior portion of the carapace and is referred to as kloedenellid in type. The dimorphic swelling is not always readily recognized.

The Kloedenellacea are represented in the Silurian by several genera, among them *Kloedenella* and *Dizygopleura.* The origin of the group is not certainly known but possibly the lobation and sulcation of *Dizygopleura* are closely related to those of the zygobolbids. This may be more apparent than real, because the types of dimorphism are strikingly different. By Mississippian time the superfamily had developed several important branches—costate glyptopleurids, smooth sansabellids, reticulate miltonellids, and others.

The Kloedenellacea and Paraparchitacea have some features in common and are placed here in the new suborder Kloedenellocopina (Fig. 33). The straight back and unequal valves are common to all. The larger valve overreaches and overlaps all or a portion of the free margin of the smaller valve. The hinge is straight, producing for the most part well-defined cardinal corners (e.g., *Sansabella*), but some carapaces are rounded at one or both ends (e.g., *Paraparchites*). The ventral margin is usually convex but exceptions are found in *Kloedenella,* where the ventral edge may vary from distinctly concave to straight or gently convex.

Many kloedenellocopine forms are sulcate, ranging from unisulcate to trisulcate; some are smooth and nonsulcate. None are typically lobate or nodose, as is so characteristic of the Beyrichicopina. S_2 is usually represented as a prominent sulcus or by a pit. In *Sansabella* a faint sulcus or pit may be present, but in some species no evidence of either may be seen. In *Glyptopleura*, S_2 is often partially obliterated by longitudinal costae.

Dimorphism is recognizable in the Kloedenellacea by swelling of the posterior portion of the female carapace. Velate, histial, or lobate dimorphism such as occurs in the Beyrichicopina is lacking. Dimorphism in the Paraparchitacea has not been conclusively shown or disproved. Adult specimens show that the greatest width commonly is medial, but in some the greatest width is behind the mid-length. These may be dimorphs. If so, the type of dimorphism is close to that found in the Kloedenellacea.

The Leperditellacea constitute one of the most difficult of all ostracode groups to classify. It includes in part the old Primitiidae, which constituted a classificatory wastebasket into which many diverse forms were previously dumped. Recent examination of the type of *Leperditella* by LEVINSON shows that it is closely related to *Primitia,* the only difference being a poorly defined S_2 in *Leperditella*. This discovery has clarified many problems in the classification of the primitiids and leperditellids. The superfamily contains what often has been thought of as the true primitiids. They are nonvelate, straight-hinged, unisulcate ostracodes. The sulcus S_2 may be sharply or broadly outlined. The valves are unequal and dimorphism has been reported in only one genus. Little is known about the hinge, but it is believed to be simple (adont) without cardinal teeth or sockets. The surface is smooth, punctate, or reticulate. Spines are present in *Parahealdia*. The group appears early in the Ordovician and becomes important in Middle and Late Ordovician. Its numbers are reduced in the Silurian and Devonian, and only one genus, *Coryellina* (Penn.-Perm.), is recognized in the late Paleozoic. Inclusion of the Leperditellacea in the Kloedenellocopina is not wholly satisfactory. This superfamily, in common with the kloedenellids, has a straight hinge, unequal valves with overlap around the free margin, and a definite S_2. It differs in lacking kloedenellid dimorphism and in having a more convex ventral outline. The hinges may differ, but more study is needed on this subject.

Members of the Paraparchitacea are here considered to be related to the Kloedenellidae but distinct from them because of being nonsulcate, possessing a shorter hinge, and being unornamented, except for a few species that bear one or two spines (Fig. 33). The valves are asymmetrical, the larger strongly overlapping the smaller around the free margin, as in other Kloedenellacea. However, dimorphism has not been recognized. The generally smooth, nonsulcate surface, nonvelate margin, channeled hinge, and strong overlap distinguish the group. They developed in the Middle Devonian, possibly as a parthenogenetic offshoot from the dimorphic Kloedenellidae. They attained their greatest abundance in the Mississippian.

ORDER PODOCOPIDA

SARS (1866) erected the Podocopa on the basis of locomotor appendages to include families named Cypridae [*recte* Cyprididae] and Cytheridae. Later, SARS (1888) added the Bairdiidae and BRADY & NORMAN (1889) added the Darwinulidae and Paradoxostomatidae. SARS had observed that an antenna (second antenna of European usage) was modified as a walking structure, rather than a swimming organ. Therefore, all living podocopids are anatomically related in possessing antennae modified for use as ambulatory organs. Because these appendages are not adapted for fossilization, paleontologists must rely on preserved hard parts for classification. Fortunately, the Podocopida possess carapace features that distinguish them from other orders. All have calcified shells with well- to poorly-developed inner calcareous lamellae. Furthermore, they have a musclescar pattern consisting for the most part of a few secondary scars that usually are well preserved in fossil specimens. Inferred relationships of the podocopids to other ostracode orders are shown diagrammatically in Figure 32; subdivisions of the Podocopida and their stratigraphic occurrence are illustrated in Figure 34.

SUBORDER PODOCOPINA

The Podocopina are represented in early

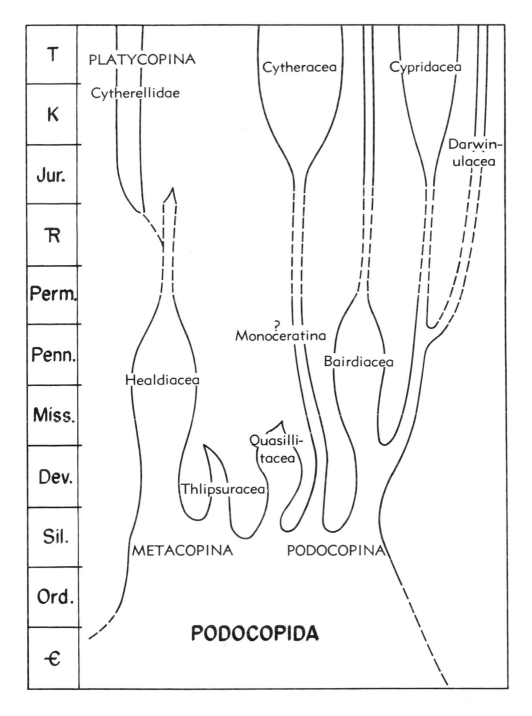

Fig. 34. Diagram representing stratigraphic distribution and inferred relationships of podocopid suborders and superfamilies (Scott & Sylvester-Bradley, n).

Paleozoic deposits by the Bairdiacea. The ancestral stock of the Bairdiacea is unknown, but the occurrence of a well-developed duplicature in forms as old as Silurian is interesting. Possibly sections of pre-Silurian fossils may show an earlier development of this important structural feature. The superfamily ranges from Ordovician to Recent. Throughout this time very little change in morphology of the group is observed. Shape, hingement, muscle-scar pattern, and wide duplicature remained more or less constant, but the number of muscle scars decreased slightly.

The Cypridacea probably developed from the bairdiid stock in early Paleozoic time. They became a very successful group and have invaded a great variety of habitats, including marine, brackish-, and fresh-water. They are a very difficult group for the paleontologist to classify because of their general lack of distinguishing external characteristics. Outline, minor features of the hinge and free margin, and muscle scars are the most usable criteria. The inner lamella is typically developed in cyprids from late Paleozoic to Recent. Whether or not this structure is present in forms described from the early and middle Paleozoic remains questionable; only further work can clarify this important point.

The Darwinulacea are a small, monotypical group without known close relatives. Their elongate-ovate shell with radially arranged muscle scars is typical of the assemblage. They are fresh-water ostracodes that may have been derived from some late Paleozoic cyprid stock.

The Cytheracea are one of the largest groups of ostracodes. Though most genera are represented by living species, classification of the superfamily has not been wholly satisfactory. The most important features of the carapace for purposes of classification are the hinge and muscle scars. Secondarily, outline and such features as ornamentation and alae are of some aid. The mid-ventral incurvature of the valve margin is one of the most characteristic features of the Podocopina. A similar incurvature occurs in the thlipsurids, quasillitids, and a few of the kloedenellids, however. The Cytheracea show great range in form and ornamentation. Classification of the group is based largely on differences in dentition, supplemented by muscle scars. These criteria admittedly are rather weak and no uniformity of opinion exists in evaluating them when applied at family and subfamily levels.

The origin of the Cytheracea is not clear. Seemingly, the group was derived in late Paleozoic time from *Monoceratina*-like forms or from quasillitids. The shape, incurved ventral margin, primitive inner lamella, and tripartite hinge of the quasillitids suggest possible relationship with the Cytheracea. The shape and stratigraphic occurrence of *Monoceratina,* however, suggest that this genus may be a connecting link between the Palaeocopida, on one hand, and cytherids, on the other (Fig. 34). The true relationship is unknown at present.

SUBORDER METACOPINA

One of the most important problems in classification of the ostracodes has concerned disposition of the cavellinids, healdiids, quasillitids, and thlipsurids. The morphology of these groups shows certain characters that are transitional between those typical of the Platycopina and Podocopina.

The Healdiidae differ from other Healdiacea in two important respects: (1) the hinge contact is posterodorsal when the long axis of the carapace is oriented horizontally, and (2) the smooth lateral surface is interrupted posteriorly by a marginal rim, or one or two spines, or both rim and spines. Dimorphism has been recognized by differences in outline of the dimorphs and greater posterior thickness of the inferred females, but criteria for separating the dimorphs needs further study. The stock is primarily of late Paleozoic age, one genus *(Hungarella)* has been reported from the Lower Jurassic.

The Quasillitidae possess many characteristics reminiscent of the Beyrichicopina and Podocopina. The muscle scar of *Euglyphella, Quasillites, Bufina,* and others of the group is represented by a circular boss. On some specimens numerous small secondary scars within the circular spot have been recognized. Though some spots commonly are bald, the absence of secondary scars is probably due to lack of preservation. The complex scars are comparable in general form to the scars found in healdiids; the simple round scars are not unlike those of the palaeocopids, but in no specimen are they cytherid. The terminal ribs and spines of *Quasillites, Bufina,* and *Parabufina* are very similar to those of the healdiids and the outline of the carapace in the latter two genera is comparable to the form of *Healdia.* In features of outline, ornamentation, and muscle scars the quasillitids may be compared with the healdiids.

Hingement is not known for all of the quasillitids. Described hinges vary from those represented by arcuate sockets at ends of the hinge in one valve (with corresponding teeth in the opposite valve) to those in which the anterior socket is divided into numerous minute depressions that received crenulations of the corresponding tooth; the sockets and teeth are connected by a tongue and groove. This type of hingement is not known to occur in the healdiids but is similar, though not identical, to some found in the Cytheracea. In outline and form *Euglyphella* is rather similar to some of the Trachyleberididae but hingement and muscle-scar patterns differ.

The evidence points strongly toward a close relationship between the Healdiidae and Quasillitidae, and they are here included in a new suborder named Metacopina. The relationship of the healdiids and quasillitids indicates that the two groups are near the main stem from which the Podocopina developed. The presence of a calcified primitive inner lamella in some forms and complex hinge in most is considered important enough to warrant placing the group in the Podocopida, but difference of the muscle-scar pattern from that seen in the Podocopina serves to distinguish the Metacopina.

Calcareous inner lamellae are absent or poorly developed in the Thlipsuridae. The outline of the carapace in this family is ovate, the valves are unequal (LV overlapping RV on free margin), and the ends are rounded. The thlipsurids seem to be closely related to the healdiids in outline, and the left valve overreaches the right. There is a very close relationship between the Healdiidae and Cavellinidae in muscle-scar patterns. It seems rather clear that the Thlipsuridae, Healdiidae, and Quasillitidae are closely related and they are here included in the Metacopina.

The Krausellidae and Pachydomellidae are small, poorly known Paleozoic families.

Shell characteristics indicate that they belong to the Metacopina.

SUBORDER PLATYCOPINA

SARS (1866) established the Platycopa to include the family Cytherellidae. Members of this family, as now defined, are restricted to Mesozoic-Cenozoic sediments. Many have been described from the Paleozoic, but they are considered to belong to the Cavellinidae.

The platycopines are composed of only one family, the Cytherellidae. The cytherellids are very closely related to the cavellinids and some workers would prefer to include the latter with the platycopines; however, there is a major difference in the muscle-scar patterns of the two groups. The muscle scar in the cavellinids is composed of many units, whereas the scar in the cytherellids is made up of only a few units. In outline and in shape the cytherellids and cavellinids are almost identical, and there is no doubt in the minds of most workers that the two groups are very closely related. They are so closely related that at one time *Cavellina* was thought to be a synonym of *Cytherella.* However, the muscle-scar pattern of *Cavellina* is similar to that found in *Healdia* and it seems more desirable to give a high priority to muscle-scar patterns in classification than it does to shape. By including the cavellinids in the metacopines we are bringing together a group of ostracodes with a common muscle-scar pattern and excluding the cytherellids because of a difference in this pattern.

Though the Cytherellidae are retained as the only family representing the Platycopina, it is believed that they developed directly from the Cavellinidae (Fig. 34). In the development of the cytherellids they retained the cavellinid shape for the most part, but were subjected to a reduction in the number of units and the shape of the muscle scar. As presently conceived, all Paleozoic forms previously referred to *Cytherella* belong to the Cavellinidae. The reduction in the number of muscle-scar units from the cavellinids to the cytherellids must have taken place for the most part in early Triassic time.

The cytherellids show dimorphism by inflation of the posterior portion of the female carapace.

ORDER MYODOCOPIDA

The presence of a well-developed rostral incisure and rostrum characterizes ostracodes assigned to the order Myodocopida. This assemblage is very unequally divided into an Ordovician-to-Recent suborder named Myodocopina, which contains numerous families, and a Recent suborder designated Cladocopina, which contains only a single family (Fig. 32).

SUBORDER MYODOCOPINA

The Myodocopina, which include many Paleozoic genera, range from small to large in size of the carapace, some attaining a length of 30 mm. One of the most characteristic features is the presence of a rostrum and rostral notch along the front margin in many, but not all of them. In some Pennsylvanian species of *Cypridinella* the inner surface of the carapace is modified by a complex design of ridges arranged in a diamond-shaped pattern. Externally, the valves are mostly smooth but may be marked by a swelling just above the center. Some genera developed a nuchal furrow, a feature most often found in genera having a poorly developed rostrum.

The Myodocopina can be divided into two groups, one in which the rostral incisure is well developed (superfamilies Cypridinacea and Halocypridacea) and another in which this incisure is not well developed (superfamilies Entomozoacea, Entomoconchacea, and Thaumatocypridacea). This grouping is rather artificial, because genera without a rostral incisure may be more closely related to some incisure-bearing forms than they are to each other. Some workers believe that the Entomozoacea are ancestral to the Thaumatocypridacea but evidence is inconclusive. The Entomozoacea and Entomoconchacea are not certainly classifiable as myodocopids.

The Entomozoacea are large myodocopids with a nuchal furrow but no rostrum. The nuchal furrow is used in orientation, anterior direction being considered to lie on

TABLE. *Characters of Ostracode Orders and Suborders*

| Order | ARCHAEO-COPIDA | LEPERDITI-COPIDA | PALAEOCOPIDA | | PODOCOPIDA | | | MYODOCOPIDA | |
Suborder			BEYRICHO-COPINA	KLOEDENEL-LOCOPINA	PODOCOPINA	METACOPINA	PLATYCOPINA	MYODO-COPINA	CLADOCOPINA
Valve Size	Subequal to unequal	Subequal to unequal	Equal to subequal	Unequal	Unequal to subequal	Unequal	Unequal	Equal to subequal	Equal to subequal
Duplicature	Absent	Absent	Absent	Absent	Present	Absent or poorly developed	Absent	Present	Present
Dimorphic Structures	None	None	None or various (a)	None or posterior swelling	None or posterior swelling	None or posterior swelling	None or posterior swelling	None or relative convexity	None or relative convexity
Dorsal Margin	Straight	Straight	Straight	Convex to straight	Convex to straight	Convex to straight or angled	Convex	Convex to straight or sinuate	Convex
Ventral Margin	Convex	Convex	Convex	Convex or rarely sinuate	Sinuate to convex	Convex to sinuate	Convex to straight or sinuate	Convex	Convex
Muscle Scars	Unknown	Very large, compound	Circular spots or unknown	Circular spots or unknown	Few discrete scars	Circular group (b)	Cluster of 8-14 biserial scars	Complex set of scars or unknown	Mostly median cluster of 3 scars
Major Lateral Features	Smooth, punctate or ridged	Mostly smooth	Sulcate, nodose, lobate, costate, reticulate	Sulcate to smooth, costate, reticulate	Smooth to reticulate, costate, alate, spinose, nodose, or sulcate	Smooth to reticulate, posterior spines or deep fissures	Smooth to costate	Costate to smooth	Reticulate punctate

(a) May include velate, pseudovelate, loculate, histial, lobate, structures associated with posteroventral, ventral, or antero-ventral swellings.

(b) Composed of several to many individual scars.

its concave side. The superfamily is restricted to the Paleozoic.

The Entomoconchacea are large forms similar to the Entomozoacea but with a posterior siphon. They range from Devonian to Carboniferous but are nowhere abundant.

The Thaumatocypridacea are represented by a single rare genus *(Thaumatocypris)* which contains several Jurassic species and one living form.

The Cypridinacea are the most abundant myodocopids. They are characterized by a rostrum overhanging an anterior incisure, and a caudal siphon or nuchal furrow may be present. In the Sarsiellidae the rostrum may be absent, being usually found in males but lacking in females. Extreme forms of dimorphism exist. Members of this superfamily have been reported from the Ordovician. They have been identified certainly from the Silurian, are most abundant in the Carboniferous, and many genera, especially those in the Cypridinidae, Sarsiellidae, and Cylindroleberididae, are found in modern seas.

SUBORDER CLADOCOPINA

The Cladocopina contain a small group of ostracodes with a subcircular outline and three closely spaced muscle scars. Only one family, Polycopidae, is included in the suborder. Of its three genera, *Polycope, Polycopsis,* and *Parapolycope,* the last two are known only from modern seas.

SUMMARY OF CHARACTERS

The more or less diagnostic morphological characters of the orders and suborders of ostracodes recognized in the *Treatise* are summarized in the table on p. Q90.

REFERENCES

Bassler, R. S., & Kellett, Betty
(1) 1934, *Bibliographic Index of Paleozoic Ostracoda:* Geol. Soc. America, Spec. Paper 1, 500 p., 24 fig. (New York).

Brady, G. S., & Norman, A. M.
(2) 1889, *A monograph of the marine and freshwater Ostracoda of the North Atlantic and of Northwestern Europe. Section I, Podocopa:* Roy. Dublin Soc., Sci. Trans., ser. 2, v. 4, p. 63-270, pl. 8-23.

Desmarest, A. G.
(3) 1813, Nouv. Bull. des Sciences.

Henningsmoen, Gunnar
(4) 1953, *Classification of Paleozoic straight-hinged ostracods:* Norsk Geol. Tidsskr. v. 31, p. 185-288 (Oslo).

Jones, T. R.
(5) 1901, *On some fossils of Wenlock age from Mulde, near Klinteberg, Gotland,* by Frederick Chapman; with notes by T. Jones: Ann. & Mag. Nat. History, ser. 7, v. 17, p. 141-160 (London).

Kesling, R. V.
(6) 1954, *Ostracods from the Middle Devonian Dundee limestone in northwestern Ohio:* Univ. Michigan Mus. Paleont. Contr., v. 11, no. 8, p. 167-186, 3 pl. (Ann Arbor).

Klöden, K. F.
(7) 1834, *Die Versteinerungen der Mark Brandenberg, Insonderheit Diejenigen, Welche Sich in den Rollsteinen und Blöcken der Sübbaltischen Ebene Finden:* 378 p. (Berlin).

McCoy, F.
(8) 1839, *On a new genus of Entomostraca:* Roy. Geol. Soc. Ireland, v. 2, p. 91-94 (Dublin).
(9) 1844, *A synopsis of the characters of the Carboniferous limestone fossils of Ireland:* Dublin Univ. Press, 207 p.
(10) 1846, *A synopsis of the Silurian fossils of Ireland, collected from the several districts by Richard Griffith:* (Dublin), 68 p.

Müller, O. F.
(11) 1772, *Observations of some bivalve insects found in common water:* Roy. Soc. Philos. Trans. for 1771 (1772), v. 61, p. 230-246, pl. 7 (London).
(12) 1776, *Zoologiae Danicae Prodromus, seu Animalium Daniae et Norvegiae Indigenarum Characters, Nomina, et Synonyma Imprimis Popularium:* (Havniae), p. 198-199.
(13) 1785, *Entomostraca seu Insecta Testacea, Quae in Aquis Daniae et Norvegiae Reperit, Descripsit et Iconibus Illustravit:* (Lipsiae et Havniae), p. 48-67.

Pokorný, Vladimir
(14) 1958, *Grundzüge der Zoologischen Mikropaläontologie:* v. 2, p. 66-453 (Berlin).

Riville, Godeheu de
(15) 1760, *Mémoire sur la mer lumineuse. Math. et Phys. Mém. por les savants étrangers:* Acad. Roy. Sci., v. 3, p. 269-276, pl.. 10, fig. 2 and 4 (Paris).

Sars, G. O.
(16) 1866, *Oversigt af Norges marine ostracoder:* Norske Vidensk. Akad. Förhandlingar, p. 1-130 (Oslo).
(17) 1888, *Nye Bidrag til Kundskaben om Middlehavets Invertebratfauna:* Archiv för Mathematik og Naturvidenskab, v. 12, p. 173-324, pl. 1-20 (Oslo).

Sowerby, J.

(18)   1825, *The Mineral Conchology of Britain:* v. 5, p. 136 (London).

Swartz, F. M.

(19)   1936, *Revision of the Primitiidae and Beyrichiidae with new Ostracoda from the Lower Devonian of Pennsylvania:* Jour. Paleont., v. 10, p. 541-586 (Tulsa).

Ulrich, E. O.

(20)   1894,(1897), *The Lower Silurian Ostracoda of Minnesota:* Minnesota Geol. Nat. History Survey, Rept., v. 3, pt. 2, p. 629-693 (Minneapolis).

————, **& Bassler, R. S.**

(21)   1923, *Paleozoic Ostracoda: Their morphology, classification and occurrence:* Maryland Geol. Survey, Silurian vol., p. 271-391 (Baltimore).

SUMMARY OF CLASSIFICATION AND STRATIGRAPHIC DISTRIBUTION

By R. C. Moore

[University of Kansas]

The tabular outline of classification that follows is accompanied by statement of the reported stratigraphic range of each taxon and by numbers indicating the count of recognized genera and subgenera in each. Where only a single number is given, this refers to genera, but if two numbers appear, the first indicates genera and the second subgenera (e.g., "5; 2" denotes 5 genera and 2 subgenera, the latter figure being exclusive of nominotypical subgenera). Also, the outline affords a useful means of explicit statement of the authorship of systematic descriptions or diagnoses, except that contributions of material on individual genera are recorded only in the text. The several authors are indicated by code letters listed as follows.

Authorship of Systematic Descriptions

Benson, R. H. ....BN
Berdan, J. M. ....BE
Bold, W. A. van den ....BO
Hanai, Tetsuro ....HA
Hessland, Ivar ....HE
Howe, H. V. ....HO
Kesling, R. V. ....KE
Levinson, S. A. ....LE
Moore, R. C. ....MO
Reyment, R. A. ....RE
Scott, H. W. ....SC
Shaver, R. H. ....SH
Sohn, I. G. ....SO
Stover, L. E. ....ST
Swain, F. M. ....SW
Sylvester-Bradley, P. C. ....SY

The stratigraphic distribution of orders, suborders, superfamilies, and families of Ostracoda recognized in the *Treatise* is indicated graphically in Figure 35. Taxa of differing rank are segregated and plotted in the order of their first known appearance in the geologic record. Numerals associated with names on the diagram are keyed to the following list of suprageneric divisions so that cross references are made easily. For example, the family Hollinidae, which appears on part 2 of Figure 35 is numbered 67 (at left of the name) and accompanied (at right) by the numeral 22, because this family has position 22 in the tabular summary of "Suprageneric Divisions of Ostracoda." Oppositely, working from the systematically arranged list of taxa, if one wishes to find the position of Hollinidae on the stratigraphic distribution diagram, the italic numeral 67 refers him to the proper place on the figure.

Genera are similarly plotted in other diagrams distributed through parts of the text devoted to systematic descriptions. These are identified conveniently by reference to the alphabetically arranged list of families given with the explanation of Figure 35. It is hoped that the compilation of data in these ways will be found useful for various purposes.

Suprageneric Divisions of Ostracoda

[The bracketed index numbers at the left margin of the tabular outline are for cross reference to and from the stratigraphic-distribution diagram (Fig. 35), numbers in roman type corresponding to those that follow names of taxa in the diagram and those in italic type corresponding to those that precede these names in the diagram.]

          Ostracoda *(subclass)* (896; 15). *L.Cam.-Rec.* (SY)
[1-1]   Archaeocopida *(order)* (12). *L.Cam.-M.Cam., ?U.Cam.-?L.Ord.* (SY)
[2-37]   Bradoriidae (2). *L.Cam.-M.Cam.* (SY)

[3-39]   Beyrichonidae (4). *L.Cam.-M.Cam.,*
         ?U.Cam.-?L.Ord. (SY)

[4-36]   Hipponicharionidae (3). *L.Cam.* (SY)

[5-38]   Indianidae (3). *L.Cam.-M.Cam.* (SY)

[6-2]   Leperditicopida *(order)* (15). *?U.*
         Cam., L.Ord.-U.Dev. (SC)

[7-40]   Leperditiidae (10). *?U.Cam., L.Ord.-*
         U.Dev. (SC)

[8-45]   Isochilinidae (5). *L.Ord.-M.Dev.* (SC)

[9-3]   Palaeocopida *(order)* (261; 2). *L.Ord.-*
         M.Perm., ?Rec. (SC)

[10-4]   Beyrichicopina *(suborder)* (191; 2).
         L.Ord.-M.Perm., ?Rec. (SC)

[11-21] Beyrichiacea *(superfamily)* (31; 2).
         M.Ord.-L.Perm. (LE)

[12-66] Beyrichiidae (25). *M.Ord.-U.Dev.,*
         ?L.Carb.-?L.Perm. (LE-BE-MO)

[13-63] Zygobolbidae (5;2). *?M.Ord.-?L.*
         Sil., M.Sil., ?Dev. (BE)

         Family Uncertain (1). *L.Sil.* (SC)

[14-22] Drepanellacea *(superfamily)* (35).
         M.Ord.-M.Perm. (SC)

[15-61] Drepanellidae (6). *M.Ord.-U.Dev.*
         (SC)

[16-85] Aechminellidae (7). *L.Dev.-M.*
         Perm. (SO)

[17-64] Aechminidae (6). *M.Ord.-M.Miss.*
         (LE)

[18-59] Bolliidae (9). *M.Ord.-M.Dev.* (SC)

[19-74] Kirkbyellidae (1). *M.Sil.-M.Penn.,*
         ?U.Penn. (SO)

[20-62] Richinidae (4). *M.Ord. - U.Dev.*
         (SC)

         Family Uncertain (2). *M.Ord.-U.*
         Ord. (SC)

[21-14] Hollinacea *(superfamily)* (80). *L.*
         Ord.-M.Perm. (SC-MO)

[22-67] Hollinidae (19). *M.Ord.-M.Perm.*
         (KE)

[23-54] Bassleratiidae (6). *?L.Ord., M.*
         Ord. (LE)

[24-58] Chilobolbinidae (2). *M.Ord.-M.Sil.,*
         ?U.Sil. (LE)

[25-43] ?Eurychilinidae (10). *L.Ord.-U.*
         Dev. (LE-MO)

[26-42] Piretellidae (8). *L.Ord. - U.Ord.*
         (HE)

[27-41] Quadrijugatoridae (13). *L.Ord.-U.*
         Ord. (KE)

[28-44] Sigmoopsidae (16). *L.Ord.-U.Ord.,*
         ?L.Sil.-?U.Sil. (KE-HE)

[29-57] Tetradellidae (3). *M.Ord.-L.Sil.*
         (SC-KE)

[30-56] Tvaerenellidae (2). *L.Ord.-U.Ord.*
         (HE)

         Family Uncertain (1). *M.Ord.* (SC)

[31-29] Kirkbyacea *(superfamily)* (22). *?L.*
         Dev., M.Dev.-M.Perm. (SO)

[32-105] Kirkbyidae (4). *L.Miss-M.Perm.*
         (SO)

[33-95] Amphissitidae (3). *M.Dev.-M.Perm.*
         (SO)

[34-93] Arcyzonidae (5). *M.Dev.* (KE)

[35-109] Cardiniferellidae (1). *U.Miss.* (SO)

[36-104] Kellettinidae (3). *?L.Miss., M.*
         Miss.-M.Perm. (SO)

[37-87] ?Placideidae (3). *L.Dev.-M.Perm.*
         (SO)

[38-98] ?Scrobiculidae (3). *?M.Dev., L.*
         Carb.(Miss.)-M.Perm. (SO)

[39-20] Oepikellacea *(superfamily)* (9). *L.*
         Ord.-M.Penn. (HE)

[40-55] Oepikellidae (1). *M.Ord.-U.Ord.*
         (HE)

[41-65] Aparchitidae (5). *L.Ord.-M.Penn.*
         (HE)

[42-76] Pribylitidae (3). *U.Sil. - M.Dev.*
         (HE)

[43-25] Primitiopsacea *(superfamily)* (7).
         M.Ord.-M.Dev. (HE)

[44-70] Primitiopsidae (7). *M.Ord.-M.Dev.*
         (HE)

         Primitiopsinae (5). *M.Ord.-M.Dev.*
         (HE)

         Leiocyaminae (2). *M.Sil.-U.Sil.*
         (HE)

[45-31] Youngiellacea *(superfamily)* (4). *?U.*
         Dev., L.Miss.-U.Penn. (SO)

[46-99] Youngiellidae (3). *L.Miss.-U.Penn.*
         (SO)

         Family Uncertain (1). *U.Dev.* (SO)

[47-35] Punciacea *(superfamily)* (2). *Rec.*
         (SY)

[48-142] Punciidae (2). *Rec.* (SY)

         Superfamily and Family Uncertain
         (1). *Ord.* (BE-SC)

[49-5]   Kloedenellocopina *(suborder)* (51). *L.*
         Ord.-U.Jur. (SC)

[50-24] Kloedenellacea *(superfamily)* (27).
         ?U.Ord., L.Sil.-M.Perm. (SO)

[51-69] Kloedenellidae (5). *?U.Ord., L.Sil.-*
         U. Penn., ?L. Perm. - ?M. Perm.
         (SO)

[52-96] Geisinidae (4). *M.Dev.-M.Perm.*
         (SO)

[53-97] Glyptopleuridae (3). *?M.Dev., M.*
         Miss.-M.Perm. (SC)

[54-100] Beyrichiopsidae (7). *U.Dev.-M.*
         Perm. (SO)

[55-101] Lichviniidae (4). *U.Dev.-M.Perm.*
         (SO)

[56-107] ?Miltonellidae (3). *?U.Miss., M.*
         Perm. (SO)

[57-108] Sansabellidae (1). *M.Miss-M.Penn.*
         (SO)

[58-16] Leperditellacea *(superfamily)* (19).
         L.Ord.-U.Jur. (LE-MO)

[59-50] Leperditellidae (19). *L.Ord.-U.Jur.*
         (LE-MO)

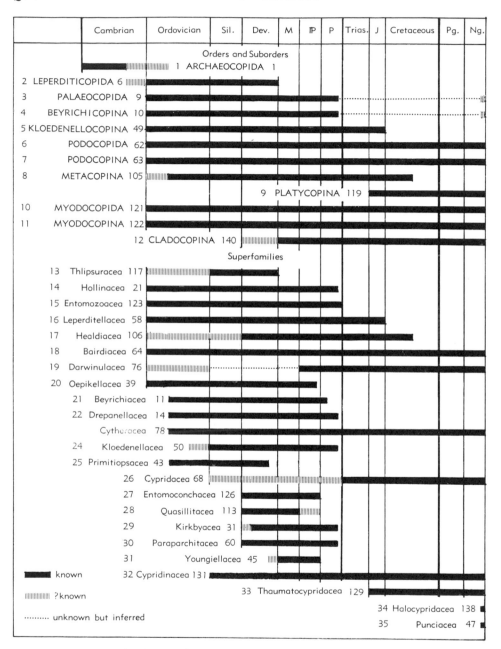

Fɪɢ. 35. Stratigraphic distribution of suprageneric ostracode taxa, geologic periods plotted according to relative time values (Moore, n). The numbers following the names of taxa indicate systematic placement as given in the preceding tabular outline of ostracode classification. Also, an alphabetical list of families is accompanied by index numbers referring to the serially arranged numbers that precede the names of taxa on the diagram; this facilitates location of any selected family as plotted with respect to stratigraphic occurrence.

Ostracode Families with Index Numbers

[Figures cited with families show stratigraphic distribution of component genera]

Acronotellidae—68 (Fig. 184)	Aechminidae—4 (Fig. 57)	Aparchitidae—65 (Fig. 105)
Aechminellidae—85 (Fig. 57)	Amphissitidae—95 (Fig. 94)	Arcyzonidae—93 (Fig. 94)

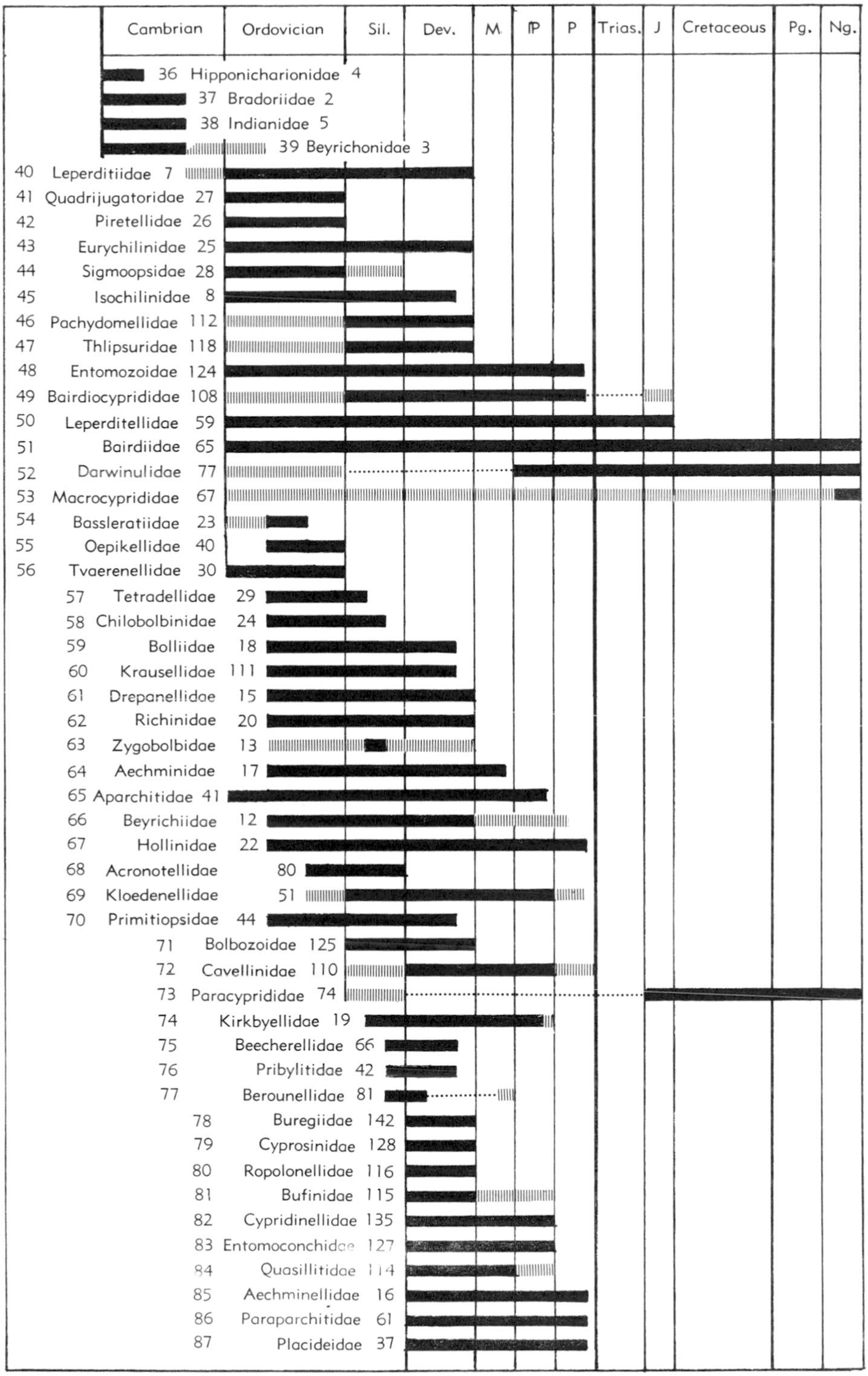

	Cambrian	Ordovician	Sil.	Dev.	M	ℙ	P	Trias.	J	Cretaceous	Pg.	Ng.

36 Hipponicharionidae 4
37 Bradoriidae 2
38 Indianidae 5
39 Beyrichonidae 3
40 Leperditiidae 7
41 Quadrijugatoridae 27
42 Piretellidae 26
43 Eurychilinidae 25
44 Sigmoopsidae 28
45 Isochilinidae 8
46 Pachydomellidae 112
47 Thlipsuridae 118
48 Entomozoidae 124
49 Bairdiocyprididae 108
50 Leperditellidae 59
51 Bairdiidae 65
52 Darwinulidae 77
53 Macrocyprididae 67
54 Bassleratiidae 23
55 Oepikellidae 40
56 Tvaerenellidae 30
57 Tetradellidae 29
58 Chilobolbinidae 24
59 Bolliidae 18
60 Krausellidae 111
61 Drepanellidae 15
62 Richinidae 20
63 Zygobolbidae 13
64 Aechminidae 17
65 Aparchitidae 41
66 Beyrichiidae 12
67 Hollinidae 22
68 Acronotellidae 80
69 Kloedenellidae 51
70 Primitiopsidae 44
71 Bolbozoidae 125
72 Cavellinidae 110
73 Paracyprididae 74
74 Kirkbyellidae 19
75 Beecherellidae 66
76 Pribylitidae 42
77 Berounellidae 81
78 Buregiidae 142
79 Cyprosinidae 128
80 Ropolonellidae 116
81 Bufinidae 115
82 Cypridinellidae 135
83 Entomoconchidae 127
84 Quasillitidae 114
85 Aechminellidae 16
86 Paraparchitidae 61
87 Placideidae 37

Fig. 35 (*Continued*).

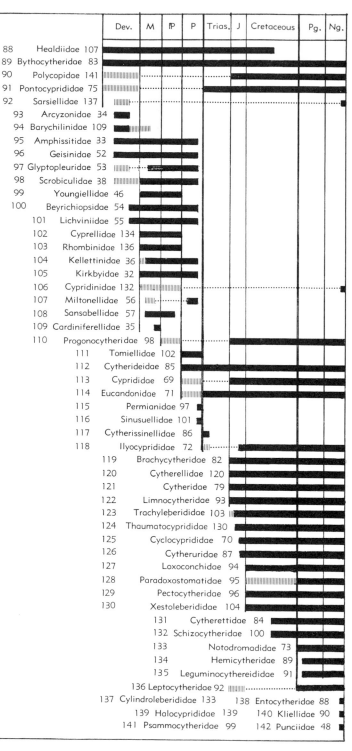

FIG. 35 *(Continued)*.

[60-*30*] Paraparchitacea *(superfamily)* (5). L.Dev.-M.Perm. (SC)

[61-*86*] Paraparchitidae (5). *L.Dev.-M. Perm.* (SC)

Suborder, Superfamily, and Family Uncertain (19). *M.Ord.-Penn.*

[62-6] Podocopida *(order)* (530;11). *L.Ord.-Rec.* (SY)

[63-7] Podocopina *(suborder)* (444;9). *L. Ord.-Rec.* (SY-SW-HO)

[64-*18*] Bairdiacea *(superfamily)* (21). *L. Ord.-Rec.* (SH)

[65-*51*] Bairdiidae (14). *L.Ord.-Rec.* (SH)

[66-*75*] Beecherellidae (3). *U.Sil.-M.Dev.* (BE)

[67-*53*] Macrocyprididae (2). *?L.Ord.-?Mio., Plio.-Rec.* (SY)

Family Uncertain (2). *?L.Dev., U. Miss., U.Cret.* (SH-SO-BE)

[68-26] Cypridacea *(superfamily)* (140;7). *?Sil.-?Perm. Trias.-Rec.* (SW)

[69-*113*] Cyprididae (74;1). *?Perm., L.Jur.-Rec.* (SW)

Cypridinae (58;1). *L.Jur.-Rec.* (SW)

Cypridopsinae (6). *?Perm., U. Cret.-Rec.* (SW)

Disopontocypridinae (4). *Oligo.-Rec.* (SW)

Candoninae (6). *Tert.-Rec.* (SW)

[70-*125*] Cyclocyprididae (5). *U.Jur.-Rec.* (SW)

[71-*114*] Eucandonidae (25). *?Perm., L. Trias.-Rec.* (SW)

[72-*118*] Ilyocyprididae (8;6). *Trias.-Rec.* (SW)

Ilyocypridinae (6). *?Trias., U. Jur.-Rec.* (SW)

Cyprideinae (2;6). *Trias.-L.Cret.* (SW)

[73-*133*] Notodromadidae (3). *Paleoc.-Rec.* (SW)

[74-*73*] Paracyprididae (6). *?Sil., Jur.-Rec.* (SW)

[75-*91*] Pontocyprididae (7). *?Dev., Trias.-Rec.* (SW)

Family Uncertain (12). *M.Ord.-U. Jur.* (SW)

[76-*19*] Darwinulacea *(superfamily)* (2). *?Ord., Penn.-Rec.* (SW)

[77-*52*] Darwinulidae (2). *?Ord., Penn.-Rec.* (SW)

[78-*23*] Cytheracea *(superfamily)* (281;2). *M.Ord.-Rec.* (HO)

[79-*121*] Cytheridae (6). *L.Jur.-Rec.* (HO)

[80-*68*] Acronotellidae (2). *U.Ord.-U.Sil.* (KE)

[81-*77*] Berounellidae (2). *U.Sil.-L.Dev., L.Carb* (SO-BE)

[82-*119*] Brachycytheridae (15). *L.Jur.-Rec.* (HO)

[83-*89*] Bythocytheridae (8). *L.Dev.-Rec.* (SY-KE)

[84-*131*] Cytherettidae (5). *U.Cret.-Rec.* (HO)

[85-*112*] Cytherideidae (42). *Perm.-Rec.* (HO)

Cytherideinae (23). *Perm-Rec.* (HO)

Cuneocytherinae (4). *L.Jur.-Mio.* (HO)

Eucytherinae (4). *Jur.-Rec.* (HO)

Krithinae (3). *U.Cret.-Rec.* (HO)

Neocytherideidinae (8). *L.Jur.-Rec.* (HO)

[86-*117*]   Cytherissinellidae (2). *L.Trias.* (BO-RE)

[87-*126*]   Cytheruridae (18). *U.Jur.-Rec.* (RE)

[88-*138*]   Entocytheridae (2;2). *Rec.* (HO)

[89-*134*]   Hemicytheridae (14). *Eoc.-Rec.* (HO)

[90-*140*]   Kliellidae (2). *Rec.* (HO)

[91-*135*]   Leguminocythereididae (5). *Eoc.-Rec.* (HO)

[92-*136*]   Leptocytheridae (4). *?Jur., Tert.-Rec.* (HO)

[93-*122*]   Limnocytheridae (12). *L.Jur.-Rec.* (HO)

[94-*127*]   Loxoconchidae (6). *Cret.-Rec.* (HO)

[95-*128*]   Paradoxostomatidae (11). *?Cret., Eoc.-Rec.* (HO-SY)

Paradoxostomatinae (7). *?Cret., Eoc.-Rec.* (HO-SY)

Microcytherinae (1). *Mio.-Rec.* (HO-SY)

Cytheromatinae (3). *Oligo-Rec.* (HO-SY)

[96-*129*]   Pectocytheridae (4). *L.Cret.-Rec.* (HA)

[97-*115*]   Permianidae (1). *U.Perm.* (BO-RE)

[98-*110*]   Progonocytheridae (20). *?Penn., L.Jur.-Rec.* (HO)

Protocytherinae (10). *Jur.-Cret.* (HO)

Protocytherinae (10). *?Penn., L.Jur.-Cret.* (HO)

[99-*141*]   Psammocytheridae (1). *Rec.* (HO)

[100-*132*]   Schizocytheridae (5). *U.Cret.-Rec.* (HO)

[101-*116*]   Sinusuellidae (1). *U.Perm.* (BO)

[102-*111*]   Tomiellidae (5). *Perm.* (BO-HO)

[103-*123*]   Trachyleberididae (33). *?L.Jur., M.Jur.-Rec.* (SY)

[104-*130*]   Xestoleberididae (4). *L.Cret.-Rec.* (HO)

Family Uncertain (51). *M.Ord.-Rec.* (HO-RE-BO)

[105-*8*]   Metacopina *(suborder)* (71). *?L.Ord., M.Ord.-L.Cret.* (SY)

[106-*17*]   Healdiacea *(superfamily)* (41). *?Ord.-?Sil., Dev.-L.Cret.* (SH)

[107-*88*]   Healdiidae (9). *L.Dev.-L.Cret.* (SH)

[109-*49*]   Bairdiocyprididae (12). *?Ord., Sil.-Perm., ?Jur.* (SH)

[109-*94*]   ?Barychilinidae (4). *M.Dev., ?L.Miss.* (KE)

[110-72]   ?Cavellinidae (10). *?L.Sil.-?U.Sil., L.Dev. - U.Penn., ?L. Perm. - ?U. Perm.* (BN)

[111-*60*]   Krausellidae (3). *M.Ord.-M.Dev.* (BE)

[112-*46*]   Pachydomellidae (3). *?L.Ord.-?U. Ord., L.Sil.-U.Dev.* (BE-SO)

[113-*28*]   Quasillitacea *(superfamily)* (15). *L. Dev.-U.Miss., ?L.Penn.-?U.Penn.* (SC)

[114-*84*]   Quasillitidae (7). *L.Dev.-U.Miss., ?L.Penn.-?U.Penn.* (SO-ST)

[115-*81*]   Bufinidae (4). *L.Dev.-U.Dev., ?L. Miss.-?U.Penn.* (SO-ST)

[116-*80*]   Ropolonellidae (3). *L.Dev.-U.Dev.* (SO)

Family Uncertain (1). *M.Dev.*

[117-*13*]   Thlipsuracea *(superfamily)* (15). *?L.Ord. - ?U.Ord., L.Sil. - U.Dev.* (SC)

[118-*47*]   Thlipsuridae (15). *?L.Ord.-?U. Ord., L.Sil.-U.Dev.* (KE)

[119-*9*]   Platycopina *(suborder)* (3;2). *L.Jur.-Rec.* (SC-SY)

[120-*112*]   Cytherellidae (3;2). *L.Jur.-Rec.* (RE)

Suborder and Order Uncertain (12). *M.Ord.-Rec.*

[121-*10*]   Myodocopida *(order)* (57;2). *L.Ord.-Rec.* (SY)

[122-*11*]   Myodocopina *(suborder)* (53;2). *L. Ord.-Rec.* (SY)

[123-*15*]   Entomozoacea *(superfamily)* (15). *L.Ord.-U.Perm.* (SY)

[124-*48*]   Entomozoidae (14). *Ord.-U.Perm.* (SY)

Entomozoinae (9). *Ord.-U.Perm.* (SY)

Bouciinae (1). *U.Sil.* (SY)

Richterininae (4). *L.Dev.-U.Dev.* (SY)

[125-*71*]   Bolbozoidae (1). *L.Sil. - U.Dev.* (SY)

[126-*27*]   Entomoconchacea *(superfamily)* (5). *L.Dev.-U.Carb.* (SY)

[127-*83*]   Entomoconchidae (4). *Dev.-Carb.* (SY)

Entomoconchinae (2). *Dev.-Carb.* (SY)

Oncotechmoninae (2). *M.Dev.* (SY)

[128-*79*]   Cyprosinidae (1). *L.Dev.-U.Dev.* (SY)

[129-*33*]   Thaumatocypridacea *(superfamily)* (1). *M.Jur.-Rec.* (SY)

[130-*124*]   Thaumatocyprididae (1). *M.Jur.-Rec.* (SY)

[131-*32*]   Cypridinacea *(superfamily)* (27;2). *Sil.-Rec.* (SY)

[132-*106*]   Cypridinidae (13;1). *?Carb., Rec.* (SY)

Cypridininae (8;1). *?L.Carb.-?U. Carb., Rec.* (SY)

	Philomedinae (5). *?L.Carb.-?U. Carb., Rec.* (SY)	[138-*34*]	Halocypridacea *(superfamily)* (5). *Rec.* (SY)
[*133-137*]	Cylindroleberididae (3;1). *Rec.* (SY)	[139-*139*]	Halocyprididae (5). *Rec.* (SY)
[*134-102*]	Cyprellidae (1). *L.Carb.-U.Carb.* (SY)	[140-*12*]	Cladocopina *(suborder)* (4). *?L.Dev.- ?U.Dev., L.Miss.-Rec.* (SY)
[*135-82*]	Cypridinellidae (4). *L.Dev.-U.Carb.* (SY)	[141-*90*]	Polycopidae (3). *?L.Dev.-?U.Dev., L.Jur.-Rec.* (SY)
[*136-103*]	Rhombinidae (2). *L.Carb.-U.Carb.* (SY)		Family Uncertain (1). *Miss.* Order and Suborder Uncertain (21). *Dev.*
[*137-92*]	Sarsiellidae (3). *?M.Dev., Rec.* (SY) Family Uncertain (1). *L.Sil.-U.Sil.*	[142-*78*]	Buregiidae (1). *Dev.* (SH) Family Uncertain (20). *L.Ord.-Mio.* Nomina Dubia (78)

SYSTEMATIC DESCRIPTIONS

By R. H. Benson, J. M. Berdan, W. A. van den Bold, Tetsuro Hanai, Ivar Hessland, H. V. Howe, R. V. Kesling, S. A. Levinson, R. A. Reyment, R. C. Moore, H. W. Scott, R. H. Shaver, I. G. Sohn, L E. Stover, F. M. Swain, and P. C. Sylvester-Bradley

---

CONTENTS

Subclass OSTRACODA Latreille, 1806

[=Ostrachoda Latreille, 1802; Ostrapoda Straus, 1821] [Type-genus designated Sylvester-Bradley, herein, *Cypris* Müller, 1776] [Diagnosis by P. C. Sylvester-Bradley, University of Leicester]

Laterally compressed Crustacea with bivalve carapace, more or less calcified, and hinged along dorsal margin, enclosing bisegmented body with head undifferentiated, bearing 4 pairs of cephalic appendages, 1 to 3 pairs of thoracic appendages and a pair of furcal rami, but no abdominal appendages. *L.Cam.-Rec.*

Order ARCHAEOCOPIDA Sylvester-Bradley, n. order

[=*Bradorina* Raymond, 1935] [Type-genus, designated Sylvester-Bradley, herein, *Bradoria* Matthew, 1899] [Diagnosis and discussion by P. C. Sylvester-Bradley, University of Leicester]

Hinge line long, straight or sinuous. Eye tubercles prominent in most families, absent in Indianidae. Shell only slightly calcified, more or less flexible. Surface finely punctate or wrinkled in most species, ornamented with strong folds in some, smooth in others. *L.Cam.-M.Cam., ?U.Cam.-?L.Ord.*

The range of the Ostracoda is commonly regarded as extending from earliest Ordovician (Canadian) to Recent, Cambrian genera described by Matthew (1886, 1896, 1899, 1902) and referred by him to the ostracodes being regarded by most workers as representatives of some other order of Crustacea. Ulrich & Bassler (1931), who monographed these forms, came to the conclusion that they could not be ostracodes (1) because the main muscle scar is situated close to the anterocardinal angle, just behind and beneath the eye tubercle; 2) because the shell is less calcareous than in typical ostracodes and in many species flexible; (3) because the valves are not completely separated along the dorsal margin; and (4) because the free margins do not close tightly, but show a narrow gape. Ulrich & Bassler classed the various Cambrian genera in 3 families, and placed them as members of the order Conchostraca in the Branchiopoda (which they ranked as a superorder). This was rather a surprising decision, for the forms in question differ far more radically from true Conchostraca than from Ostracoda. At the same time Ulrich & Bassler stated that they believed these forms to be probable ancestors of the Ostracoda.

RAYMOND (1935, 1946) removed the genera in question from the Conchostraca, creating for them a separate order, Bradorina, assigned to the subclass Archaeostraca. He also regarded them as ancestral to the Ostracoda. Other orders of the Archaeostraca, according to RAYMOND, included the Ceratocarina, Rhinocarina, and Discinocarina, but not the Ostracoda, which were placed in a distinct subclass. Once again it is concluded that the Bradorina differ from these other Archaeostraca more fundamentally than from the Ostracoda. Seemingly they are more closely related to the Ostracoda than to any other Crustacea and therefore taxonomic arrangement should express this inferred relationship. Either the Bradorina should be regarded as a superorder of equal standing with the Ostracoda, and included with Ostracoda in a subclass of their own, or they should be regarded as an order of the Ostracoda, equivalent in rank to other orders of the Ostracoda.

A review of the evidence used by ULRICH & BASSLER for removal of the group from the Ostracoda reveals weakness in some of their arguments. They state that "in all the forms studied by us, with the exception possibly of certain species placed in the emended genus *Indiana,* the main muscle spot is located close to the anterocardinal angle just behind and beneath the ocular tubercle, whereas in the Ostracoda what is regarded as the corresponding scar is located somewhere near the middle of the valves." Surprisingly, however, the only mention of a muscle scar in the descriptive part of their monograph refers to the scar developed in *Walcottella,* of the family Bradoriidae, and in this genus the muscle impression is developed in the median third of the length, in a position exactly analogous to that of the adductor in most Ostracoda. Moreover, the adductor muscle scar in many (perhaps all) ostracodes of the family Leperditiidae is developed "behind and beneath the ocular tubercle," more closely so in forms without chevron scars (e.g., *Eoleperditia*).

The thinner, flexible, less calcareous shell stands as a valid distinction of the Cambrian genera from Paleozoic ostracodes with thick calcareous shells. The ostracode carapace is not by any means invariably calci-fied, however, and in the Myodocopida most species of the family Halocyprididae possess uncalcified shells. The family includes the most abundant and widely distributed of all Recent ostracodes, but the lack of calcification has led to a poor fossil record. Certainly post-Cambrian ostracodes with calcified shells must have evolved from Cambrian and Precambrian ancestors with uncalcified valves. The distinction is therefore valid but not fundamental.

ULRICH & BASSLER's third criterion for separating the group from the Ostracoda was their contention that some reason exists "for believing that the valves were always tightly joined along the back, often perhaps by the fusion of the cardinal edges." This is not a valid distinction, for Ostracoda having a chitinous rather than calcareous carapace (e.g., Halocyprididae) are likewise joined along the dorsal margin, though ostracodes with a calcareous carapace have valves which are quite distinct from each other.

The last distinction listed by ULRICH & BASSLER is that the free margins of the valves in the Bradoriidae do not overlap each other, but are separated by a narrow gape. In fossil Ostracoda compression usually results in a tightly closed carapace, but in Recent ostracodes the valves often gape after death of the animals. Most specimens figured by ULRICH & BASSLER are of single valves or carapaces exposed on one side only. The few specimens that show a continuous gape possibly suggest that the valves could not be closed in life, but the evidence is inconclusive.

This review suggests that the Cambrian fossils under discussion differ from later Ostracoda, but the differences are not profound. The Cambrian forms almost certainly are ancestral to the post-Cambrian ostracodes, and are here regarded as true Ostracoda belonging to a primitive order for which the name Archaeocopida is introduced.

Characters of the Archaeocopida suggesting their relationship with the Leperditicopida are the similar general shape (especially the long, straight hinge line), and the possession of an eye tubercle in the same position and of the same nature as that developed in *Leperditia*. The genus

Cambria suggests relationship with the Beyrichiacea.

The stratigraphic occurrence of genera assigned to the Archaeocopida is indicated in Figure 36.

Family BRADORIIDAE Matthew, 1902

[Materials for this family prepared by P. C. SYLVESTER-BRADLEY, University of Leicester]

Surface finely punctate or wrinkled, otherwise unornamented except for development of either an eye tubercle at anterodorsal corner or an anteromedian tubercle corresponding to muscle scar. *L.Cam.-M.Cam.*

Bradoria MATTHEW, 1899 [**B. scrutator;* SD ULRICH & BASSLER, 1931] [*=Bradorona* MATTHEW, 1903]. Eye tubercle more or less conspicuous. Greatest length in upper third of carapace; consequent forward and backward projection of anterior and posterior margins rounded in most species but angular in some. *L.Cam.-M.Cam.*, N.Am.-Eu.——FIG. 37,*1b.* **B. robusta,* L.Cam., Can. (Nova Scotia); RV lat., ×15 (Ulrich & Bassler, 1931).——FIG. 37,*1a.* *B. scrutator* (MATTHEW), L.Cam., Can. (Nova Scotia); LV lat., ×8 (Ulrich & Bassler, 1931).

Walcottella ULRICH & BASSLER, 1931 [**W. apicalis*]. Surface finely punctate; prominent anteromedian tubercle; eye tubercle weak or absent. *M.Cam.*, N.Am.——FIG. 37,*2a.* **W. apicalis,* USA(Ariz.); LV lat., ×10 (Ulrich & Bassler, 1931).——FIG. 37,*2b,c.* *W. concentrica* ULRICH & BASSLER, USA (Ariz.); *2b,c,* ?LV lat., RV lat., ×10 (Ulrich & Bassler, 1931).

Family BEYRICHONIDAE
Ulrich & Bassler, 1931

[Materials for this family prepared by P. C. SYLVESTER-BRADLEY, University of Leicester]

Carapace subtriangular, corneous, smooth, or rarely punctate; compressed anterodorsally behind eye tubercle, the resulting postocular hollow bordered by variously arranged ridges or nodes. *L.Cam.-M.Cam.*, *?U.Cam.-?L.Ord.*

Beyrichona MATTHEW, 1886 [**B. papilio;* SD S. A. MILLER, 1889] [*=Escasona* MATTHEW, 1902]. Postocular hollow bordered by ill-defined ridges; carapace compressed also in posteroventral region to give a second dorsal hollow only slightly less marked than postocular hollow. *L.Cam.*, Eu.-N. Am.——FIG. 38,*1a.* **B. papilio,* Can.(N.B.); RV lat., ×15 (Ulrich & Bassler, 1931).——FIG. 38, *1b.* *B. tinea* MATTHEW, Can.(N.B.); carapace opened out, ×15 (Ulrich & Bassler, 1931).

Aluta MATTHEW, 1896 [**A. flexilis*]. Anterodorsal corner sharply angular, acute in most species. Postocular hollow bordered anteroventrally by prominent tubercle or ridge. *L.Cam.-M.Cam.*,

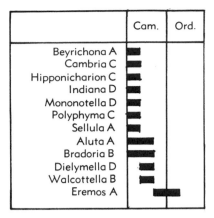

FIG. 36. Stratigraphic distribution of archaeocopid ostracodes (Moore, n). Classification of genera in families indicated by letter symbols (A.—Beyrichonidae, B.—Bradoriidae, C.—Hipponicharionidae, D—Indianidae.)

cosmop.——FIG. 38,*4a.* **A. flexilis,* M.Cam., Can. (N.B.); LV lat., ×15 (Ulrich & Bassler, 1931).——FIG. 38,*4b.* *A. troyensis* (FORD), L.Cam., USA (N.Y.); LV lat., ×6 (Ulrich & Bassler, 1931).

?Eremos MOBERG & SEGERBERG, 1906 [**E. bryograptorum*]. Carapace elongate, otherwise like *Aluta* in outline. Anterior tubercle in upper third (?ocular tubercle) joined ventrally to marginal ridge. *L.Ord.(Tremadoc.),* NW.Eu.——FIG. 38,*2.* **E. bryograptorum,* Swed.; LV lat. (Moberg & Segerberg, 1906).

Sellula WIMAN, 1902 [**S. fallax*]. Like *Beyrichona* but with an obtuse triangular projection of anterior border. *L.Cam.,* N.W.Eu.——FIG. 38,*3.* **S. fallax,* Swed.; LV lat., ×20 (Wiman, 1902).

Family HIPPONICHARIONIDAE
Sylvester-Bradley, n. fam.

[Materials for this family prepared by P. C. SYLVESTER-BRADLEY, University of Leicester, with additions by IVAR HESSLAND and R. A. REYMENT, University of Stockholm]

Carapace with narrow marginal rim; strongly developed ridges or lobes parallel border or cover whole surface. *L.Cam.*

Hipponicharion MATTHEW, 1886 [**H. eos;* SD S. A. MILLER, 1889]. Hinge line straight; carapace subtriangular in lateral view, with strongly developed lobes paralleling anterior and posterior margins. *L.Cam.,* N.Am.-Eu.——FIG. 39,*1a.* **H. eos,* Can. (N.B.); RV lat., ×10 (Matthew, 1886).——FIG. 39,*1b.* *H. minus* MATTHEW, Can.(N.B.); RV lat., ×10 (Ulrich & Bassler, 1931).

?Cambria NECKAJA & IVANOVA, 1956 [**C. sibirica*]. Carapace large. Hinge line sinuous, shorter than length of valve, terminating in acute projections, plenate end regarded as anterior; terminal lobes extending downward from dorsal margin to about

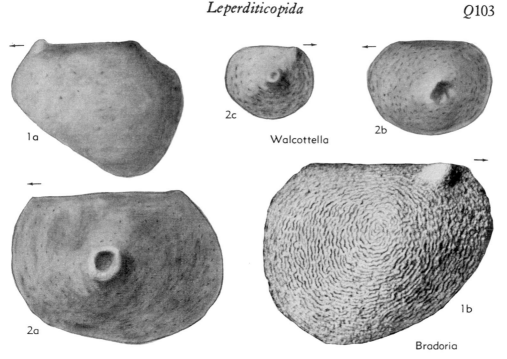

FIG. 37. Bradoriidae (p. Q102).

0.5 height of valve, anterior one sickle-shaped, convex toward anterior; lobes separated by sulcate depression; elongate lobe paralleling ventral margin, perhaps analogous to velum of Palaeocopida. Surface may be papillate. *L.Cam.*, Sib.——FIG. 40, *1*. **C. sibirica;* RV lat., ×10 (Neckaja & Ivanova, 1956).

Polyphyma GROOM, 1902 [**P. lapworthi*]. Hinge line straight; carapace elongate; anterior, ventral, and posterior margins evenly arcuate in lateral view, almost symmetrical, or highest point slightly posterior (when antiplenate end is regarded as anterior, as in other members of suborder but not as in *Cambria*). *L.Cam.*, Eu.——FIG. 39,2. **P. lapworthi*, Eng.(Malverns); RV lat., ×20 (Groom, 1902).

Family INDIANIDAE Ulrich & Bassler, 1931

[*=Indianitidae* ULRICH & BASSLER, 1931 (obj.)] [Materials for this family prepared by P. C. SYLVESTER-BRADLEY, University of Leicester]

Carapace smooth, unornamented, with no eye tubercle; surface polished, in some species finely punctate. *L.Cam.-M.Cam.*

Indiana MATTHEW, 1902 [**I. lippa;* ICZN pend.] [*=Indianites* ULRICH & BASSLER, 1931 (obj.)]. Carapace inequilateral, antiplenate end being taken as anterior (in conformity with most other members of order bearing eye tubercles). *L.Cam.*, N. Am.-Eu.——FIG. 39,3. **I. lippa*, Can.(Nova Scotia); *3a-c*, carapace R, dors., ant., ×10 (Ulrich & Bassler, 1931).

Dielymella ULRICH & BASSLER, 1931 [**D. recticardinalis*]. Like *Indiana* but anterodorsal corner produced to form small rostrum. *M.Cam.*, N.Am.-Eu. ——FIG. 39,5. **D. reticardinalis*, USA(Ariz.); RV lat., ×4 (Ulrich & Bassler, 1931).

?Mononotella ULRICH & BASSLER, 1931 [**Primitia ?fusiformis* MATTHEW, 1895]. Like *Indiana* but valves apparently fused along dorsal margin. *L. Cam.*, N.Am.——FIG. 39,4. **M. fusiformis* (MATTHEW), Can.(N.B.); *4a,b*, carapace R, dors., ×6 (Ulrich & Bassler, 1931).

Order LEPERDITICOPIDA
Scott, n. order

[*=Leperditiida* POKORNÝ, 1953] [Type genus designated SCOTT, herein, *Leperditia* ROUAULT, 1851] [Diagnosis and discussion by H. W. SCOTT, University of Illinois]

Hinge long, straight; shell thick, well calcified; surface usually smooth, some finely ornamented to nodose; valves strongly unequal to subequal; adductor muscle scar large, composed of numerous secondary elements. *?U.Cam., L.Ord.-U.Dev.*

The leperditiid ostracodes are universally recognized as a group distinct from all others. Their large size and strongly calcified shells are adequate to distinguish them readily. They are commonly some four or

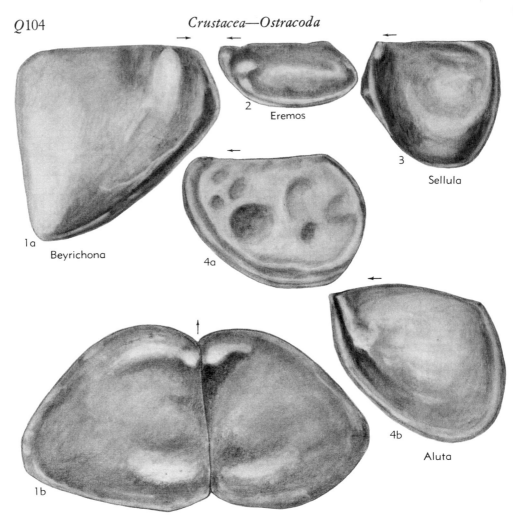

Fig. 38. Beyrichonidae (p. Q102).

five times larger than other ostracodes and the shell is usually proportionately thicker.

Internally the leperditiids show unique muscle-scar patterns. The adductor muscle scar is very large and is composed of as many as 200 small secondary scars. The diameter of the scar is so great that it is equal to approximately 30 percent of the height of the shell. The size of the adductor muscle and the number of fibers composing it are greater than found in any other ostracode. Other muscle scars, representing anterior appendages, as well as scars along the hinge, have been found. The interior surface commonly is marked by venose lines, especially in the ventral half of *Eoleperditia* and the anterior portion of *Isochilina*.

Abundant molts of leperditiids can usually be found with adults. The thick shells were favorable for fossilization and numerous very small molts can be observed in most populations. The number of instars is not definitely known, and no evidence of dimorphism has been reported. Strong muscles were necessary to handle the heavy, large shells so common in this group. Shell characters of the group indicate a benthonic habitat.

The leperditiids are believed to be true ostracodes. The presence of venose lines and large compound muscles probably represent primitive characters. They may be related to the thinly calcified archaeocopids, especially forms with a long straight dorsal

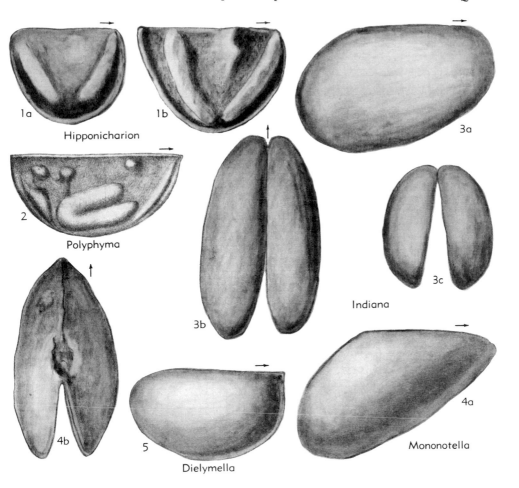

FIG. 39. Hipponicharionidae, Indianidae (p. Q102-Q103).

margin. They do not appear to be directly related to the smaller palaeocopids. No direct evidence pertaining to their ancestry has been found, nor is there any indication that they were the ancestors of other Paleozoic groups. They seem to represent an isolated assemblage of Cambrian to Devonian benthonic Ostracoda that reached a climax of development in Mohawkian (M.Ord.) times and remained abundant through the Middle Silurian.

Reported stratigraphic occurrences of leperditicopid genera are indicated graphically in Figure 41.

Family LEPERDITIIDAE Jones, 1856

[*nom. correct.* S. A. MILLER, 1889 (*pro* Leperditidae JONES, 1856)] [=Leperditiadae JONES, 1870] [Materials for this family prepared by H. W. SCOTT, University of Illinois, with additions by W. A. VAN DEN BOLD, Louisiana State University] [Includes Hermannininae ABUSHIK, 1960]

Large, thick-walled; cardinal angles well defined; greatest height medial to posterior; dorsal margin straight, valves strongly unequal, surface smooth to slightly ornamented; adductor muscle scar large, composed of numerous secondary scars (Figs. 43, 29); anterodorsal eye tubercle present in most species. *?U.Cam., L.Ord.-U.Dev.*

This family comprises one of the most important early Paleozoic assemblages of Ostracoda. It is represented by several genera, some of which are among the largest known ostracodes. They commonly occur in such great numbers as to cover rock surfaces along bedding planes. They differ from all other Ostracoda in large size of the adductor muscle scar. Internal features are unusually well preserved for Paleozoic specimens (Fig. 29). Numerous molts of various instars

Cambria

FIG. 40. Hipponicharionidae (p. Q102).

often are found associated with one another. Thus, they represent one of the best groups of Paleozoic Ostracoda for the study of instar stages.

The ostracode, described as *"Leperditia harrisi"* (FREDERICKSON, 1946), from Upper Cambrian rocks of Oklahoma is probably a leperditiid, although indeterminable generically. As such, it is the oldest member of the order so far recorded.

Leperditia ROUAULT, 1851 [*L. britannica*]. Nonsulcate, unornamented, smooth or punctate; posterodorsal swelling in LV but absent in RV; chevron-like muscle scar subjacent to eye tubercle and anterodorsal to larger adductor scar; hinge finely denticulate; inner margin shoulder of overlapped edge of LV serving as stop for RV. *L.Sil.-U.Dev.*, cosmop.——FIG. 45,5. *L. britannica,* Dev., Fr.; *5a,b,* carapace R, L, ✕5 (Oehlert, 1877).——FIG. 42,3. *L. scalaris* JONES, Sil., USA(Pa.); LV int. (muscle scar), ✕10 (Swartz, 1949).——FIG. 43. *L. sp.*, U.Sil., Swed.(Gotl.); LV lat., showing eye protuberance in anterodorsal region and impression of large muscle area slightly below and behind it, ✕3.8 (Triebel, 1941).

Anisochilina TEICHERT, 1937 [*A. punctulifera*]. Like *Eoleperditia,* but valves subequal, internal details unknown. *Ord.,* Greenl.——FIG. 42,5. *A. punctulifera;* LV ext., ✕10 (Teichert, 1937).

Briartina KEGEL, 1932 [*Leperditia quenstedti* GÜMBEL, 1874]. Like *Leperditia* but carapace more subrectangular, ends subequally rounded; dorsum subparallel to venter. *Dev.,* Eu.——FIG. 42,4. *B. quenstedti,* M.Dev., Ger.; LV lat., ✕8 (Kegel, 1932).

Eoleperditia SWARTZ, 1949 [*Cytherina fabulites* CONRAD, 1843]. Nonsulcate, unornamented, thick-shelled; RV overlapping free margin of LV, nondenticulate hinge, no marginal flattening; large adductor muscle scar composed of numerous secondary scars, other muscle scars and venose lines commonly present, lacking subocular chevron mark; prongs on inner margin of RV. *M.Ord.,*

cosmop.——FIG. 29, 42,2. *E. fabulites* (CONRAD); 42,2a,b, carapace (Ontario) R, L, ✕5 (Swartz, 1949); 42,2c, LV (Illinois) int., showing muscle scar, ✕10 (Scott, 1951); 29, RV (Illinois) int., showing eye spot, muscle scar, and tubercles along inner ventral margin, ✕10 (Scott, 1951).

Gibberella ABUSHIK in MANDELSTAM et al., 1958 [*Leperditia chmielewski* SCHMIDT, 1900]. Carapace elongate (up to 2 cm.), with large "eye" node and short deep triangular groove behind it, dorsal margin behind groove with boss that may project over it; marginal groove commonly present; "chevron" typically tail-shaped, not V-shaped. Surface smooth, coarsely pitted, or pustulose. Interior of valves with 30 to 40 oval, narrowly aligned muscle scars. *L.Sil.(Llandov.)-M.Sil.(Wenlock.),* N.Eu.(Baltic-Novaya Zemlya)-NE. Asia(E.Sib.).——FIG. 44,1a,b. G. lenaica ABUSHIK, L.Sil.(Llandov.), N.Zemlya; 1a,b, LV lat., dors., ✕1 (Mandelstam, 1958).——FIG. 44,1c. G. jejuna ABUSHIK, L.Sil.(Llandov.), W.Sib.; LV lat., ✕3 (Mandelstam, 1958). [BOLD.]

Hermannina KEGEL, 1933 [*pro* Hermannella PAECKELMANN, 1922, *non* CANU, 1891] [*Hermannella waldschmidti* PAECKELMANN, 1922] [=*Chevroleperditia* SWARTZ, 1949]. Like *Leperditia* but postdorsal swelling absent. *M.Sil.-M.Dev.* Eu.——FIG. 42,1a-e. *H. waldschmidti,* M.Dev., Ger.; 1a, RV lat., ✕10; 1b, post. hinge, ✕10; 1c-e, carapace L, vent., post., ✕7 (Kegel, 1933).——FIG. 42,1f. H. welleri SWARTZ, RV lat., muscle scars, ✕15 (Swartz, 1949).

Heterochilina POULSEN, 1937 [*H. obliqua*]. Like *Leperditia* but RV with curved ridge near free margin in mid-ventral or anteroventral position

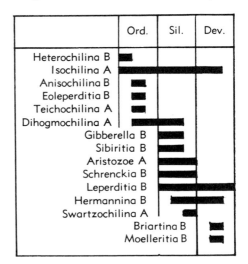

	Ord.	Sil.	Dev.
Heterochilina B	■		
Isochilina A		■■■■■	■■■■
Anisochilina B	■■		
Eoleperditia B	■		
Teichochilina A	■		
Dihogmochilina A	▌		
Gibberella B		■■■	
Sibiritia B		■■■	
Aristozoe A		■■■■	
Schrenckia B		■■	
Leperditia B		■■	■■■■■
Hermannina B		■■■	■■■
Swartzochilina A		▌	■■
Briartina B			■■
Moelleritia B			■■

FIG. 41. Stratigraphic distribution of leperditicopid ostracodes (Moore, n). Classification of genera in families is indicated by letter symbols (A.—Isochilinidae, B.—Leperditiidae).

and subparallel to it. *L.Ord.*, Greenl.——Fig. 45, *1.* *H. obliqua; 1a,b*, both RV lat., ×7 (Poulsen, 1937).

Moelleritia Abushik in Mandelstam *et al.*, 1958 [*Leperditia mölleri* Schmidt, 1883]. Carapace large (1.5 to 8 cm.), with large "eye" node, behind which dorsal margin shows a projecting boss; marginal rim well developed, broad, long, usually well separated from ventral margin; overlap of valves not strong, longitudinal axis oblique; in-terior with large chevron-shaped or triangular muscle area containing more than 200 scars, ventral rim of area marked by elongate, triangular, closely spaced scars and its inside by scattered smaller scars of varying shape, usually very small near anterior side. Surface smooth or punctate. *M.Dev.*, E.Eu.(Urals-Novaya Zemlya).——Fig. 44, *3.* *M. moelleri;* muscle-scar pattern, ×40 (Mandelstam, 1958). [Bold.]

Schrenckia Glebovskaia, 1949 [*Leperditia grandis*

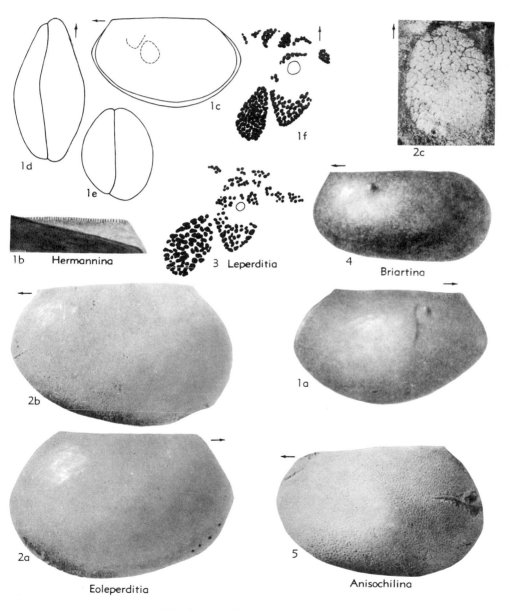

Fig. 42. Leperditiidae (p. *Q*106-*Q*107).

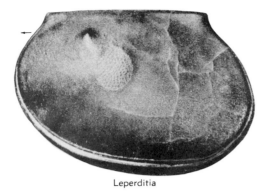

Leperditia

Fig. 43. Leperditiidae (p. *Q*108).

SCHRENCK, 1852]. Like *Leperditia* but differing in presence of 2 or more pitlike structures in ventral region of RV and in being most inflated in ventral half of carapace. *Sil.*, E.Eu.(USSR)-N. Asia(Sib.).——Fig. 44,*4. S. multa* ABUSHIK, U. Sil.(Ludlov.), E.Sib.; *4a-c,* 3 RV lat., ×2,×2,×3 (Mandelstam, 1958). [BOLD.]

Sibiritia ABUSHIK in MANDELSTAM *et al.,* 1958 [**Leperditia wiluiensis* SCHMIDT, 1873]. Carapace medium in size (0.7 to 1.7 cm.), valves very unequal, strongly overlapping, smooth; "eye"

spot distinct; marginal ridge narrow, commonly absent; longitudinal axis oblique, anterior and posterior portions of RV with 2 to 5 deep pits corresponding in interior to nodes against which margin of LV closes but they may join to a narrow groove; front of middle part of rim commonly with 1 to 4 rounded shallow pits which in interior form oblique nodes, apparently contributing to closure of valves. Chevron-shaped muscle area weakly developed, with 5 to 28 small muscle scars grouped in shape of triangle with concave base near "eye" node, irregular angular shape and irregularly arranged but rather regular in each triangle. Resembles *Eoleperditia* SWARTZ in development of the ventral margin, but differs in its larger size, characters of the muscle "chevron," closure by pits, better development of "eye" node, and presence of a narrow marginal rim. *L. Sil.(Llandov.)-M.Sil.(Wenlock.),* NE. Asia (E. Sib.).——Fig. 44,*2. S. ventriangularis* ABUSHIK, L.Sil.; *2a-d,* carapace R, L, ant., vent., ×3 (Mandelstam, 1958). [BOLD.]

Family ISOCHILINIDAE Swartz, 1949

[*nom. transl.* ABUSHIK, 1960 (*ex* Isochilininae SWARTZ, 1949)]
[Materials for this family prepared by H. W. SCOTT, University of Illinois]

Subequivalved leperditiids with flattened borders along free margin. *L.Ord.-M.Dev.*

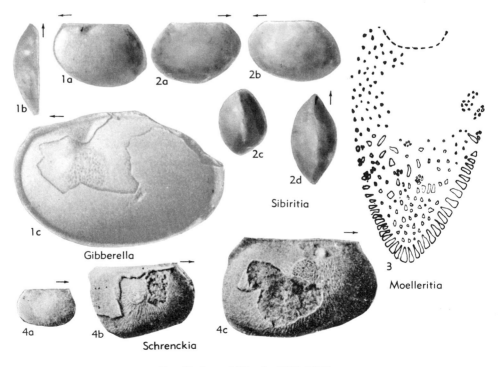

1a 2a 2b 1b 2c 2d

Sibiritia

1c

Gibberella

3

Moelleritia

4a 4b 4c

Schrenckia

Fig. 44. Leperditiidae (p. *Q*106-*Q*108).

Isochilina Jones, 1858 [*Leperditia (Isochilina) ottawa Jones, 1858; SD Bassler & Kellett, 1934] [= Hogmochilina, Holtedahlina, Paenaequina Solle, 1935; Holtedahlites Solle, 1936 (pro Holtedahlina Solle, 1935, non Foerste, 1909)]. Valves subequal, flattened border extending along free margin but weak or absent mid-ventrally in some shells; small pits in some species distributed along ventral flat border of RV, reflecting inner nodes; faint to strong eye tubercle anterodorsal, reflected by internal pit; faint to strong depression or sulcus (S_2) dorsal to adductor scar. *L.Ord.-M.Dev.*, N.Am.——Fig. 46,1. *I. ottawa*, M.Ord., N.Am.; *1a*, RV lat., ×7; *1b*, LV lat., ×10; *1c*, RV int. (muscle scars), ×20; *1d*, LV lat., (muscle scars), ×20; *1e*, RV lat. (ventral border, with pits) ×20; *1f*, RV int., ×10 (Swartz, 1949).——Fig. 45,3. *I. ampla* Ulrich, M.Ord., Tenn.; RV (holotype) int., ×1.3 (Swain, 1957).——Fig. 31,1. *I.* sp.; thin sec. showing 2 shell layers, ×13 (Levinson, n.).

Aristozoe Barrande, 1868 [*A. amica*]. Like *Isochilina* but with one or more nodes in anterodorsal area; cardinal corners well defined to rounded. [*A. amica* is probably an early molt of the adult that Barrande described as *A. memoranda*.] *Sil.*, Bohemia.

Dihogmochilina Teichert, 1937 [*Isochilina grandis* var. *latimarginata* Jones, 1891] [=Dihogmochilus Neave, 1950]. Like *Isochilina* but with forked sulcus. *M.Ord.-M.Sil.*, Arct.——Fig. 45,4. *D. latimarginata*, Sil.; LV lat., ×4 (Teichert, 1937).

Swartzochilina Scott, 1956 [*Dihogmochilina straitcreekensis* Swartz, 1949]. Like *Isochilina* but sulcus splits adductor scar in two, and area anterodorsal to eye spot marked by one or more strong rounded pits; flattened border along entire free margins. *U.Sil.*——Fig. 46,2. *S. straitcreekensis*, USA(Va.); *2a,b*, LV lat., RV lat., ×7 (Swartz, 1949).

Teichochilina Swartz, 1949 [*Isochilina jonesi* Wetherby, 1881]. Like *Isochilina* but RV overlapping LV, submarginal pits and corresponding inner nodes of RV lacking; flattened border along entire free margins. *M.Ord.*, N.Am.——Fig. 45,2. *T. jonesi* (Wetherby), USA(Ky.); *2a,b*, carapace L, post., ×2 (Wetherby, 1881).

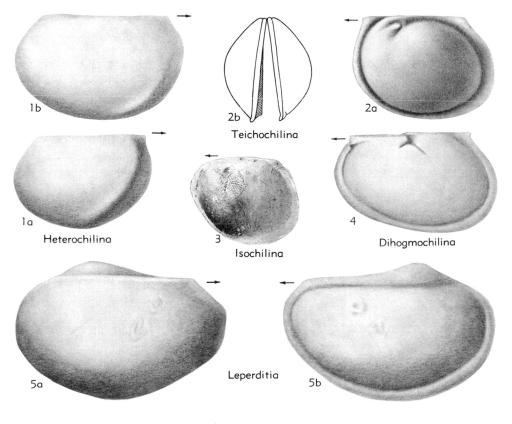

1b

2b
Teichochilina

2a

1a
Heterochilina

3
Isochilina

4
Dihogmochilina

5a

Leperditia

5b

Fig. 45. Leperditiidae, Isochilinidae (p. Q106-Q109).

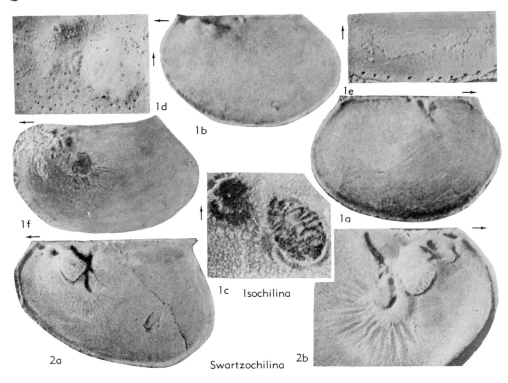

FIG. 46. Isochilinidae (p. *Q*109).

Order PALAEOCOPIDA
Henningsmoen, 1953

[*nom. correct.* SCOTT, herein (*pro* Paleocopa HENNINGSMOEN, 1953)] [=Order Beyrichiida POKORNÝ, 1953 *(partim)*] [Type Genus—*Beyrichia;* SD SCOTT, herein [Diagnosis and discussion prepared by H. W. SCOTT, University of Illinois]

Dorsal margin long and straight; surface smooth to ornamented; lobes, sulci, ventral and adventral structures common; calcareous inner lamella absent; dimorphic or nondimorphic; soft parts unknown. *L.Ord.-M. Perm., ?Rec.*

The Palaeocopida have been one of the most misunderstood of all ostracode groups, even though they have been studied for over 100 years. Much of the confusion in earliest work resulted from poor illustrations. Later, genera and species were added in large number without due regard to criteria of classification. In particular, students of the palaeocopids disregarded information available on dimorphism and ontogeny of modern forms. As a result, new genera, species, and even families often were based on dimorphs or instar stages. To add to the confusion, ornamental features such as number of spines borne by the carapace were used as a basis of generic classification. Disregard for such morphological features as hingement, muscle-scar pattern, and dimorphism, accompanied by emphasis on lobation and ornamental characters, kept classification in a constant state of disorder. One of the main attempts in this *Treatise* is to bring order to classification of the great group of palaeocopid ostracodes.

To arrive at the most satisfactory classification we have had the opinions and constant help of many workers but do not claim to have resolved all problems. Some monogeneric families have been recognized; the muscle scars and hinge structures are unknown in some genera, and evidence pertaining to dimorphism is incomplete or lacking in others. Future studies will certainly fill many of the gaps. We hope that future creation of new families and genera will be based on adequate material, fully illustrated and studied, so that questions of orientation, ontogeny, dimorphism, hingement, muscle scars, and marginal features will be known.

T. R. Jones and E. O. Ulrich were major early contributors to knowledge of Ordovician and Silurian ostracode classification. R. S. Bassler joined Ulrich (1923) in attempting to make known many of the early Paleozoic groups and in presenting the first extensive classification. Prior to their work, the Leperditiidae, Beyrichiinae, Aparchitinae, Leperditellidae, Kirkbyidae, Kloedenellinae, and Glyptopleuridae had been proposed. These family-group taxa were based on lobation, sulcation, and overlap as main criteria of classification. Proposal of them laid the foundation for further extensive work. Other workers accepted these groups for many years and concepts of them were still dominant with appearance of the Bassler & Kellett "Bibliography" in 1934. This bibliography gave students an available source of information and many new workers were attracted to ostracode studies.

The appearance of Swartz's (1936) classification was the first major step toward recognition of the importance of ventral and adventral dimorphic features. Swartz made many emendations and additions to the classification of forms now assigned to the palaeocopids. Hessland (1949) recognized the importance of dimorphism and ontogeny in classification and of the muscle scars for use in orientation. Henningsmoen (1953) first proposed that the straight-backed Paleozoic ostracodes, without frontal openings or calcareous inner lamella, be included in a new order, which he named Paleocopa. He included the Leperditiidae in this new order. Henningsmoen also recognized the importance of dimorphism in classification and thus agreed with Swartz's concepts. His recognition of velar and carinal dimorphic structures (partly including those now termed as histial) was an important step in delineating major taxa. Kesling, in many papers published in the 1950's, added much to knowledge of the structure and classification of the palaeocopids. His monograph (1951) on *Cypridopsis,* though based on a nonpalaeocopid modern species, was a major contribution to understanding of ontogeny and the molting process. The observations made could be applied in part to the Paleozoic genera.

New fields of research were opened and many students began to report on ostracode populations instead of single specimens. Pokorný (1953) separated the Leperditiida from the palaeocopids and postulated the existence of a heart in the leperditiids. His 1958 text was the first to give a comprehensive classification of all ostracodes.

Orientation of the palaeocopids has been a very troublesome problem for a long time, as discussed in a previous chapter. Bonnema (1909, 1913, 1930) called attention to the anterior position of the muscle scars in some forms and the posterior dimorphic swelling in others; Geis (1932) showed that posterior swelling is common in the females of many modern species, that the anterior end is the highest in many, and applied these observations to classification of Mississippian palaeocopids; Swartz (1936, 1949), Triebel (1941), Hessland (1949), Levinson (1950), Scott (1951), Kesling (1951), Henningsmoen (1953), and Jaanusson (1957) contributed to knowledge of muscle scars, hingement, ventral and adventral structures, lobation, sulcation, and other elements in relation to orientation. The classification here adopted for palaeocopids is based primarily on dimorphism, relationship of valves along the free margin, and muscle scars; secondarily it takes account of outline or form. Lobation, sulcation, and ornamentation play a minor part.

Suborder BEYRICHICOPINA Scott, n. suborder

[=Beyrichiida Pokorný, 1953] [Diagnosis prepared by H. W. Scott, University of Illinois]

Ostracodes with a straight back and convex free margin; lobes and sulci common; valves subequal; duplicature lacking; marginal structures commonly present as velum, pseudovelum, histium, or carina; dimorphic or nondimorphic. *L.Ord.-M.Perm., ?Rec.*

The Beyrichicopina represent a natural group of palaeocopids characterized by the presence of one or more of the following features: dimorphic structures, lobes, sulci, and subequal valves. A velum or pseudovelum is commonly present. The Beyrichiacea show dimorphism by the development of a ventral to anteroventral lobe or "pouch"; the Hollinacea are dimorphic, as indicated by locular, histial, or velate structures (except for the Quadrijugatoridae and Bassleratiidae, which are nondimorphic and may not belong to the superfamily); the Primitiopsacea are dimorphic, as shown by the development of a posteroventral dolon; the Oepikellacea are dimorphic in part. The Kirkbyacea are nondimorphic.

Superfamily BEYRICHIACEA Matthew, 1886

[*nom. transl.* ULRICH & BASSLER, 1923 (*ex* Beyrichiidae MATTHEW, 1886, *nom. transl. et correct.* ULRICH, 1894, *ex* Beyrichinae MATTHEW, 1886)] [Diagnosis by S. A. LEVINSON, Humble Oil & Refining Company]

Straight-hinged ostracodes with subequal ends or forward swing, mostly with well-developed lobes and sulci and showing tendency to have carinal, velate, and marginal structures; dimorphism well defined in most families. *M.Ord.-L.Perm.*

The stratigraphic distribution of genera assigned to the Beyrichiacea is indicated graphically in Figure 47.

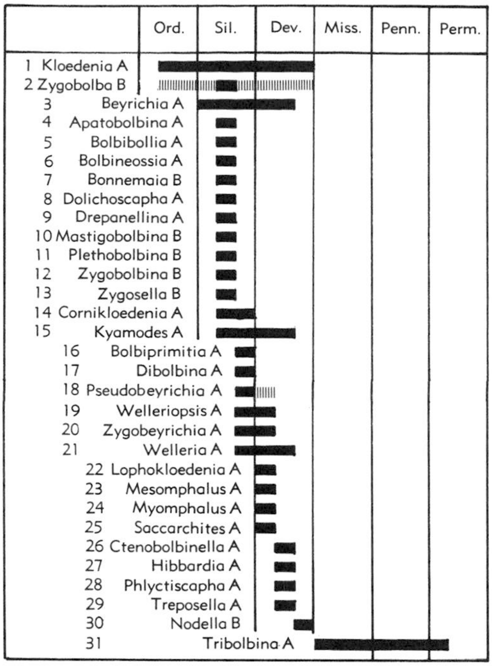

Family BEYRICHIIDAE Matthew, 1886

[*nom. transl. et correct.* ULRICH, 1894 (*ex* Beyrichinae MATTHEW, 1886)] [=Kloedeninae (*recte* Kloedeniinae) ULRICH & BASSLER, 1923; Treposellinae HENNINGSMOEN, 1954] [Materials for this family prepared by S. A. LEVINSON, Humble Oil & Refining Company, and R. C. MOORE, University of Kansas, with contributions on some genera by JEAN BERDAN, United States Geological Survey]

Straight-hinged, nonsulcate to trisulcate, marginal ridge or frill commonly present; dimorphism well marked, females invariably with a cruminal pouch. *M.Ord.-U.Dev., ?L.Carb.-?L.Perm.*

Beyrichia M'COY, 1846 [**B. klödeni* (ICZN pend.)] [*non* Beyrichia BOLL, 1847] [=*Eobeyrichia, Mitrobeyrichia, Neobeyrichia, Nodibeyrichia, Velibeyrichia* HENNINGSMOEN, 1954]. Trilobate, L2 smallest lobe, rarely projecting above hinge line; marginal ridge or frill developed in some species; surface commonly granulose or pitted; female with globular to subovate anteroventral pouch. *L.Sil.-M.Dev.*, Eu.-N.Am.-Austral.——FIG. 48,*1a-c.* B. *tuberculata* (KLÖDEN), U.Sil.(Ludlov.), Gotl.; *1a,* ♂ RV lat., ×14; *1b,c,* ♀ RV lat., vent., ×14 (Kesling, 1957).——FIG. 48,*1d-f.* B. *buchiana* JONES, U.Sil.(Ludlov.), Gotl. (type species of *Neobeyrichia; 1d,* ♂ RV lat., ×16; *1e,f,* ♀ RV lat., vent., ×14 (Kesling, 1957).——FIG. 48,*1g-i.* B. *salteriana* JONES, U.Sil.(Ludlov.), Gotl.; *1g,* ♂ RV lat., ×31; *1h,i,* ♀ RV lat., vent., ×26 (Kesling, 1957).——FIG. 48,*1j-l.* B. *fittsi* ROTH, L.Dev., USA (Okla.); *1j,* ♂ RV lat., ×24; *1k,l,* ♀ RV lat., vent., ×24 (Kesling, 1957).——FIG. 48,*1m.* B. *jonesii* BOLL, M.Sil.(Wenlock.), Gotl.; ♀ carapace transv. sec. showing cruminal pouches, ×24 (Kesling, 1957).——FIGS. 27C, 28. B. *moodeyi* ULRICH & BASSLER, M.Sil.(Clinton.), USA (N.Y.);

FIG. 47. Stratigraphic distribution of beyrichiacean ostracodes (Moore, n). Classification of genera in families is indicated by letter symbols: A—Beyrichiidae, B—Zygobolbidae. An alphabetical list of genera provides cross references to the serially numbered generic names on the diagram.

Generic Names with Index Numbers

Apatobolbina—4	*Mesomphalus*—23
Beyrichia—3	*Myomphalus*—24
Bolbibollia—5	*Nodella*—30
Bolbineossia—6	*Phlyctiscapha*—28
Bolbiprimitia—16	*Plethobolbina*—11
Bonnemaia—7	*Pseudobeyrichia*—18
Cornikloedenia—14	*Saccarchites*—25
Ctenobolbinella—26	*Treposella*—29
Dibolbina—17	*Tribolbina*—31
Dolichoscapha—8	*Welleria*—21
Drepanellina—9	*Welleriopsis*—19
Hibbardia—27	*Zygobeyrichia*—20
Kloedenia—1	*Zygobolba*—2
Kyamodes—15	*Zygobolbina*—12
Lophokloedenia—22	*Zygosella*—13
Mastigobolbina—10	

type species of *Velibeyrichia*, 27C, transv. sec. ♀ LV, enlarged (Wainwright, 1959); 28, serial transv. secs. (Wainwright, 1959). [In the opinion of BERDAN, the generic names published by HENNINGSMOEN (1954) are desirably used for subgenera of *Beyrichia*. Possibly *Velibeyrichia*, which includes most North American species assigned to *Beyrichia*, should have the status of an independent genus. —MOORE.]

Apatobolbina ULRICH & BASSLER, 1923 [**A. granifera*]. Nonsulcate, frilled; female with pouch on anteroventral portion of frill, pouch extending

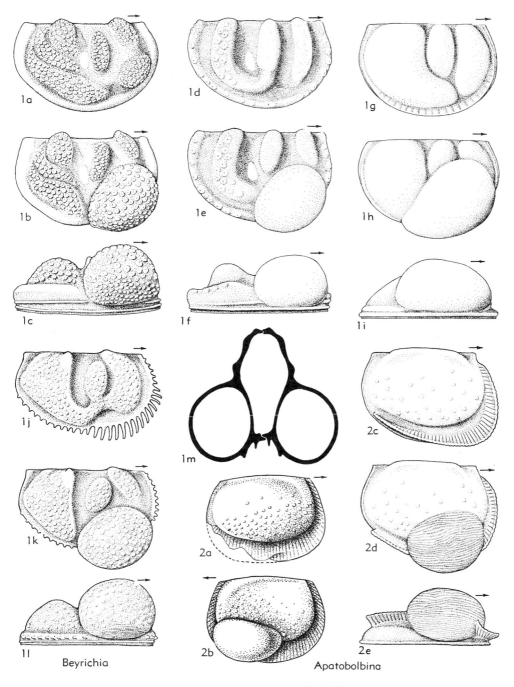

Beyrichia

Apatobolbina

Fig. 48. Beyrichiidae (p. Q112-Q114).

dorsally to about 0.3 height of valve. *M.Sil.*, N. Am.-Eu. (Ger.-Norway).——Fig. 48,*2a,b*. *A. granifera,* Clinton., USA (Pa.); *2a,b*, ♂ RV lat., ♀ LV lat., ✕16 (Kesling, 1951).——Fig. 48, *2c-e. A.* sp., Manistique F., USA (Mich.); *2c*, ♂ RV lat., ✕22; *2d,e*, ♀ RV lat., vent., ✕22 (Kesling, 1957).

Bolbibollia Ulrich & Bassler, 1923 [*B. labrosa*]. Resembles *Bollia* in having small horseshoe-shaped ridge around moderately deep S_2; female with anteroventral pouch, monotypic. [Trend of the sulcus and bordering horseshoe-shaped ridge is slightly variable but mostly about normal to the dorsal margin—not strongly oblique, as represented in originally published figures.] *M.Sil.*, Can. (Anticosti).——Fig. 49,*1*. *B. labrosa: 1a,b*, ♂ LV lat., ♀ RV lat. (syntypes), ✕40; *1c*, ♀ LV lat. (paratype), ✕40; *1d-f*, ♂ RV lat., ♀ RV lat., vent., ✕42, ✕37, ✕37 (*1a-c*, Sohn, n; *1d-f*, Kesling, 1957).

Bolbineossia Kesling, Heany, Kaufmann, & Oden, 1958 [*B. didictyosa*]. Sulcation represented only by central pit (S_2), valves otherwise evenly convex, subovate in side view, surface of known species reticulate. Females with wide frill extending onto protuberant anteroventral brood pouch. *M.Sil.*, N.Am.——Fig. 49,*4*. *B. didictyosa,* Manistique Gr., USA (Mich.); *4a*, ♂ RV lat.; *4b,c*, ♀ RV lat., vent.; all ✕25 (Kesling et al., 1958).

Bolbiprimitia Kay, 1940 [*Halliella fissurella* Ulrich & Bassler, 1923]. Resembling *Halliella* in form but with S_2 developed into thin, elongate deep sulcus, smooth ridge at free borders, surface reticulate; female with ventral-median pouch. *U. Sil.*, E.N.Am.——Fig. 49,*3*. *B. fissurella* (Ulrich & Bassler), Tonoloway Ls., USA (W.Va.); *3a*, ♂ RV lat., ✕40; *3b,c*, ♀ RV lat., vent., ✕33 (Kesling, 1957).

Cornikloedenia Henningsmoen, 1954 [*Drepanellina ventralis* Ulrich & Bassler, 1923]. Trilobate, with a posteroventral alate projection. *M.Sil.-U. Sil.*, NE.N.Am.——Fig. 50,*5*. *C. ventralis,* M.Sil., USA(Md.); ♂ RV (holotype) lat., ✕30 (Berdan, n). [Berdan].

Ctenobolbinella Kummerow, 1953 [*C. carinata*]. Large prominent knob surrounded by furrow in dorsal half of valve, slightly in front of middle; female with lens-shaped, elongate pouch in anteroventral area. *M.Dev.*, Eu.(Ger.).——Fig. 52,*3*. *C. carinata;* ♀ RV lat., ✕30 (Kummerow, 1953).

Dibolbina Ulrich & Bassler, 1923 [*D. cristata*]. Somewhat like *Eurychilina* with wide frill around free margins but strongly dimorphic; small central node with adjoining arcuate sulcus, generally most prominent behind it; narrow ridge may be developed from near posterodorsal margin to median-ventral area parallel to proximal margin of frill; females with well-developed anteroventral pouch. *U.Sil.*, E.N.Am.-NW.Eu.——Fig. 49,*2a,b*. *D. cristata,* Tonoloway Ls., USA (W.Va.); *2a,b*, ♂

LV lat., ♀ LV lat., ✕27 (Kesling, 1951).—— Fig. 49,*2c-e*. *D. steusloffi* (Krause), Ludlov., Gotl.; *2c-e*, ♂ RV lat., ♀ RV lat., vent., ✕30 (Kesling, 1957).

Dolichoscapha Kesling & Ehlers, 1958 [*D. escharota*]. Only female carapace known, hinge line long and straight; valves nonsulcate, velate ridge extending to each corner; surface reticulate except for central smooth spot. Brood pouch ventral, elongate. *M.Sil.*, N.Am.——Fig. 51,*3*. *D. escharota,* Schoolcraft Dol., USA (Mich.); ♀ LV (holotype) lat., ✕57 (Kesling, 1958).

Drepanellina Ulrich & Bassler, 1923 [*D. clarki*]. Carapace with submarginal, distinct ventral lobe parallel to free margins; L_2 and L_3 strongly developed into elongate lobes which may extend above dorsum; females with anteroventral portion of marginal ridge inflated and L_2 and L_3 tending to merge with marginal ridge. *M.Sil.*, E.N.Am.—— Figs. 26, 50,*4*, 52,*5*. *D. clarki,* Clinton., USA (Md.-Pa.); 26, serial transv. secs., ♀ LV; 50,*4a,b*, ♂ LV (syntype) lat., ♀ RV (syntype) lat., ✕15 (Berdan, n); 52, *5a,b*, ♀ RV lat., ♂ RV lat., ✕15 (Ulrich & Bassler, 1923). [Levinson].

?Halliella Ulrich, 1891 [*H. retifera;* SD Miller, 1892]. Straight-backed, with coarsely reticulate surface; S_2 prominent, posterior lobe larger than anterior; velate, dimorphic. [Characters poorly known; possibly the holotype is a male representative of *Beyrichia lyoni* Ulrich.] *M.Dev.*, USA(Ky.). [Scott.]

Hibbardia Kesling, 1953 [*Amphissites lacrimosus* Swartz & Oriel, 1948]. Bilobate, with short frill around free border and marginal ridge around free edge; subcentral pit at ventral end of shallow sulcus (S_2); surface reticulate. Female with cruminal pouch in posteroventral portion of each valve. *M.Dev.*, E.N.Am.——Fig. 27*A*, 51,*1*. *H. lacrimosa* (Swartz & Oriel), Ludlowville F., USA (N.Y.); 27*A*, transv. sec. ♀ carapace, enlarged; 51,*1a-c*, ♂ RV lat., ♀ RV lat., vent., ✕21; 51,*1d*, ♀ carapace transv. sec. showing pouches, ✕33 (Kesling, 1957).

Kloedenia Jones & Holl, 1886 [*Beyrichia wilckensiana* Jones, 1855]. Trilobate, S_1 shorter than or equal to S_2, both well developed but usually restricted to dorsal half of valves; L_1 and L_3 smoothly rounded and united ventrally, without transverse furrows; L_2 small, globular or subovate; free border with narrow ridge; surface pustulose, reticulate or smooth. Females with large, well-defined pouch in anteroventral area but comprising extension of ventral lobe not distinctly separated from body of valve. *M.Ord.-U.Dev.*, N.Am.-Eu.——Fig. 53,*1a-f*. *K. wilckensiana,* U.Sil. (from Pleist. drift), Ger.; *1a,b*, ♂ LV (hypotype) lat., slightly inclined vent., ✕20; *1c-f*, ♀ RV (hypotype) lat., dors., vent., post., ✕20 (Kesling & Wagner, 1956). ——Fig. 53,*1g-i. K. sussexensis* (Weller), U.Sil. (Decker Ls.), USA (N.J.); *1g*, ♂ LV lat.; *1h,i,*

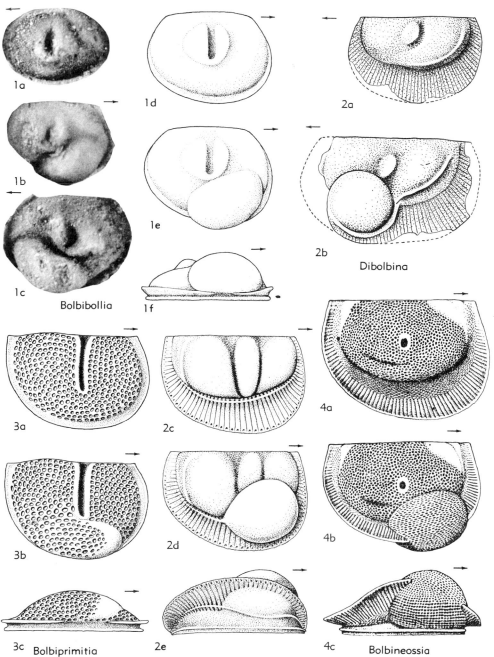

FIG. 49. Beyrichiidae (p. Q114).

♀ RV lat., post.; all ×17 (Swartz & Whitmore, 1956].——Fig. 52,*1*. *K. normalis* Ulrich & Bassler, Sil., USA(Md.); *1a-c,* ♀ LV lat., ♀ LV lat., ♂ RV lat., ×30 (Ulrich & Bassler, 1923). [Levinson-Berdan].

Kyamodes Jones, 1888 [***K. whidbornei*] [=*Kyammodes* Ulrich & Bassler, 1908]. Quadrilobate or trilobate, L_2 may be small; lobes tend to project above hinge line; sulci confined to dorsal half of carapace; LV conspicuously overreaching RV on ventral margin; surface smooth. Female un-

known in type species but in others assigned to genus large pouch observed in anteroventral area. *M.Sil.-M.Dev.,* Eu.-N.Am. —— Fig. 50,*2*. **K. whidbornei,* Dev., Eng.; *2a-e,* ? ♂ carapace R, L, dors., vent., post., ×20 (Jones, 1888).——Fig. 51, *4. K. tricornis* (Ulrich & Bassler), U.Sil. (McKenzie F.), USA (Md.); *4a,b,* ♂ LV lat., ♀ LV lat., ×19 (Kesling, 1951).——Fig. 52,*4. K. kiesowi* (Krause), Sil. (from drift), Ger.; ♀ LV lat., ×20 (Ulrich & Bassler, 1923). [Levinson-Berdan.]

Lophokloedenia Swartz & Whitmore, 1956 [**Bey-

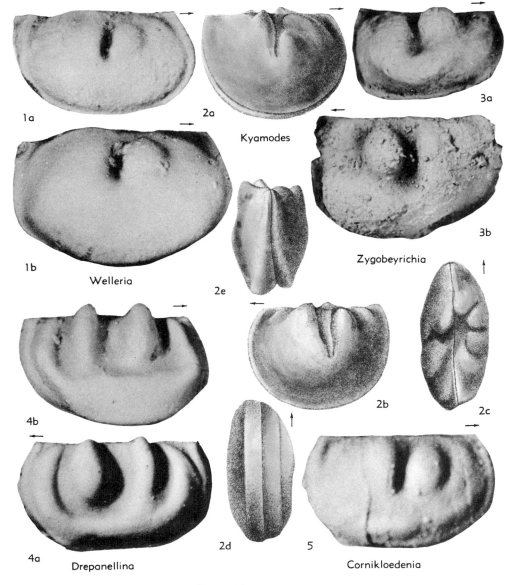

1a

2a

Kyamodes

3a

1b

Welleria

2e

Zygobeyrichia

3b

4b

2b

2c

4a

Drepanellina

2d

5

Cornikloedenia

Fig. 50. Beyrichiidae (p. Q114-Q122).

richia manliensis WELLER, 1903]. Like *Kloedenia* but L_1 extending backward along dorsal margin as indistinct crest that passes above L_2 and ends in low node at dorsal end of S_2; pouch well defined, anteroventral in position. Surface reticulate or pustulose. *L.Dev.,* N.Am.——FIG. 54,*2d-g. *L. manliensis,* Manlius Ls., USA (N.J.); *2d,e,* ♂

LV (holotype) lat., dors., ×17, ×22; *2f,* ♂ RV lat., ×17; *2g,* ♀ LV lat., ×17 (SWARTZ & WHITMORE, 1956).——FIG. 54,*2a-c. L. kummeli* (WELLER), Manlius Ls., USA (N.J.); *2a,* ♂ LV lat., ×17; *2b,c,* ♀ RV lat., post., ×15 (Swartz & Whitmore, 1956). [BERDAN.]

Mesomphalus ULRICH & BASSLER, 1913 [**M. hart-*

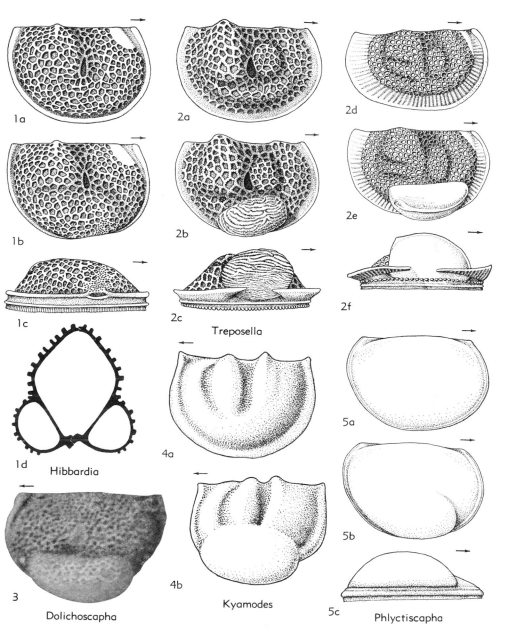

FIG. 51. Beyrichiidae (p. Q114-Q122).

leyi.] [=*Mezomphalus* ZASPELOVA, 1952]. Subquadrate elongate; L_2 poorly defined, S_1 suppressed and S_2 confined to dorsal half of valve, arcuate; fissus on L_3 extending on ventral side of S_2; pouch elongate, occupying most of ventral border; surface granulose or punctate or both. *L.Dev.,* E.N.Am.——FIG. 52,7. **M. hartleyi,* USA(Md.); 7a,b, ♀ LV lat., ♂ LV lat., ×20 (Ulrich & Bassler, 1923).

Myomphalus SWARTZ & WHITMORE, 1956 [**M. dorsinodosus* SWARTZ & WHITMORE, 1956]. Binodose, S_1 obsolete, L_2 a small node below dorsal

margin, S_2 weak and narrow, L_3 broad, with small node or spine developed near dorsal end of S_2; surface smooth or obscurely granulose. Females with anteroventral pouch which is poorly defined dorsally. *L.Dev.,* N.Am.——FIG. 53,3. **M. dorsinodosus,* Manlius Ls., USA (N.J.-N.Y.); 3a, ♂ LV (syntype) lat., ×25; 3b, ♀ RV (syntype) lat., ×25 (Swartz & Whitmore, 1956). [BERDAN.]

Phlyctiscapha KESLING, 1953 [**P. rockportensis*]. Large, smooth, tumid, dorsal-median portion of each valve extending above hinge line as a hump;

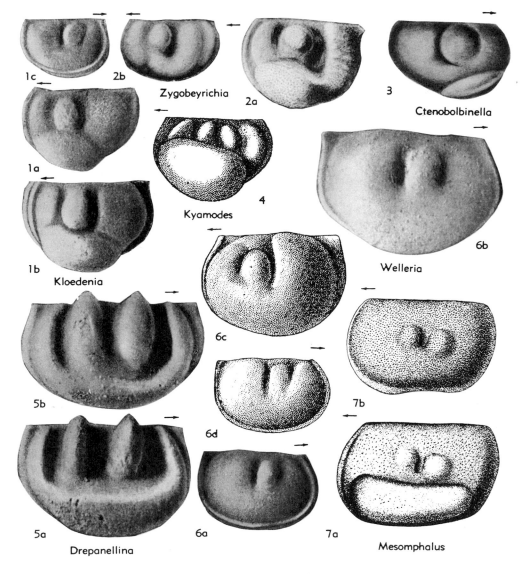

FIG. 52. Beyrichiidae (p. Q114-Q122).

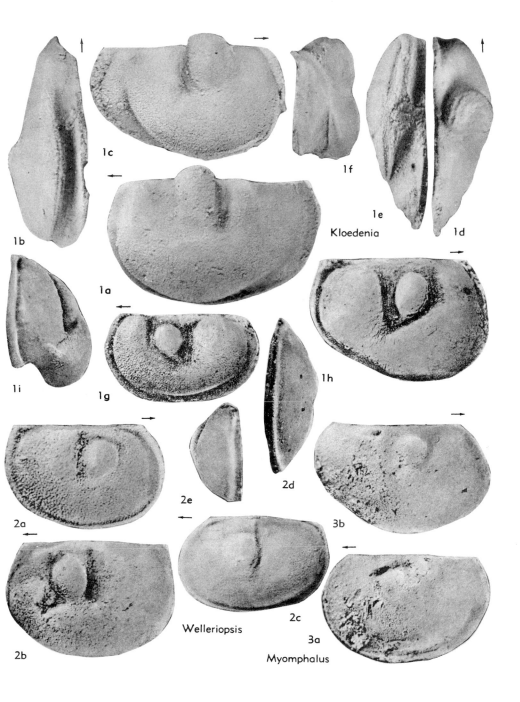

1c

1f

1e

Kloedenia

1d

1b

1a

1i

1g

1h

2a

2d

2e

3b

2b

Welleriopsis

2c

3a

Myomphalus

FIG. 53. Beyrichiidae (p. Q116-Q122).

direction of overlap variable, marginal ridge around free edge, small velate ridge parallel to marginal ridge. Female with pouch in ventral or posteroventral area, posterior part of pouch confluent with rest of valve. *M.Dev.,* N.Am.——Figs. 27B, 51,5. *P. rockportensis,* USA (Mich.); 27B,

♀ carapace, transv. sec., enlarged (Kesling, 1953); 51,5a-c, ♂ RV lat., ♀ RV lat., vent., ×22 (Kesling, 1957).

Pseudobeyrichia Swartz & Whitmore, 1956 [**P. perornata*]. Trilobate, with L_2 large, elevated subglobular, subcentral; L_1 and L_3 lower and nearly

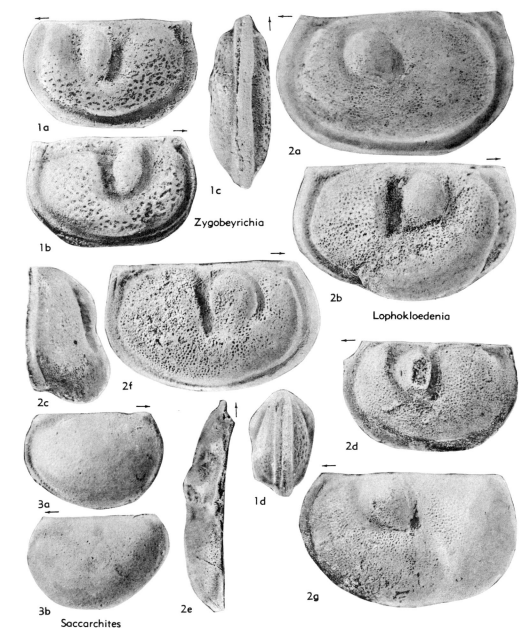

1a
1c
Zygobeyrichia
1b
2a
2b
Lophokloedenia
2f
2c
2d
3a
1d
2g
3b
Saccarchites
2e

Fig. 54. Beyrichiidae (p. Q118-Q122).

equal; S_1 shallow, S_2 well developed, fissus absent; pouch large, anteroventral, involving most of L_1 and not clearly separated from it; surface reticulate to pustulose. Hinge line intrenched, lateral surface separated from ventral region by angulation or elevated rim that is roughly parallel to free margin. *U.Sil., ?L.Dev.,* N.Am.——Fig. 56,2. **P. perornata,* ?L.Dev., USA (N.J.); *2a,b,* ♂ LV (holotype) lat., ♀ LV lat., ×45 (Swartz and Whitmore, 1956). [Berdan.]

?**Saccarchites** Swartz & Whitmore, 1956 [**S. saccularis*]. Lobes and sulci suppressed, surface smooth with scattered punctae. Females with poorly defined pouch in anteroventral position overhanging narrow brim. *L.Dev.,* N.Am.——Fig. 54,3. **S. saccularis,* Manlius Ls., USA (N.Y.); *3a,* ♂ RV (syntype) lat., *3b,* ♀ LV (syntype) lat., ×20 (Swartz & Whitmore, 1956). [Berdan.]

Treposella Ulrich & Bassler, 1908 [**Beyrichia lyoni* Ulrich, 1891]. Small rounded node in anteromedian area with elongate node in posteromedian area, nodes separated by distinct sulcus; low ventral ridge becoming obscure toward front and rear; with short frill on free margins; sur-

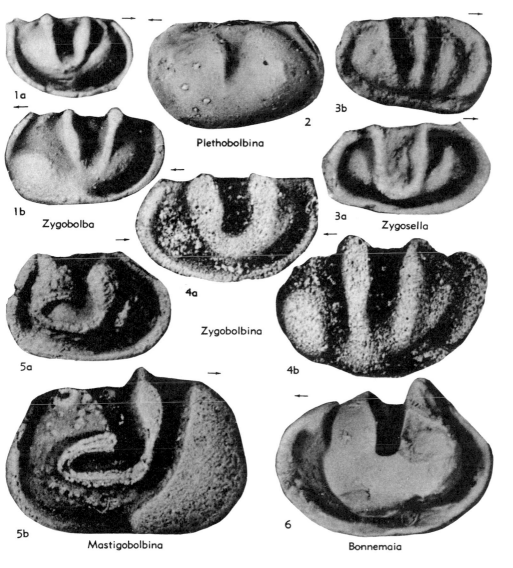

FIG. 55. Zygobolbidae (p. *Q122-Q123*).

face coarsely reticulate. Female with ovate pouch centrally located on ventral border. *M.Dev.*, N.Am. ——Fig. 51,2*a-c.* **T. lyoni* (Ulrich), Jeffersonville Ls., USA (Ky.-Ind.); *2a-c,* ♂ RV lat., ♀ RV lat., vent., ×21 (Kesling, 1957).——Fig. 51,2*d-f.* *T. stellata* Kesling, Centerfield Ls., USA (N.Y.); *2d-f,* ♂ RV lat., ♀ RV lat., vent., ×20 (Kesling, 1957). [According to Berdan, *Kozlowskiella* Přibyl (1953), is closely related to *Treposella* and appropriately is assignable to the Beyrichiidae. It is here included in Palaeocopida and family.—Moore.]

?**Tribolbina** Latham, 1932 [**T. carnegiei*]. Large with 2 long sulci, one extending from anterodorsal margin almost to ventromedian border, other from slightly behind dorsomedian area to join first at ventromedian area, sulci together forming distinct "V"; knoblike lobe may be developed between the sulci near the anterodorsal margin. Female with elongate pouch in anteroventral area. *L.Carb.-L.Perm.*, N.Am.-Eu.

Welleria Ulrich & Bassler, 1923 [**W. obliqua*]. Trilobate, S_1 and S_2 shallow, short; lobes low; surface smooth in type species. Females with ill-defined pouch that consists merely of ventral swelling. *U.Sil-M.Dev.*, NE. N.Am.-?Eu.——Figs. 50,*1,* 52,6. **W. obliqua,* U.Sil.(Tonoloway F.), USA (W.Va.); 50,*1a,b,* ♂ and ♀ RV (syntypes) lat., ×15 (Berdan, n); 52,6*a,b,* ♂ RV lat., ♀ RV lat., ×15; 52,6*c,d,* ♀ LV lat., ♂ RV lat., ×15 (Ulrich & Bassler, 1923). [Berdan.]

Welleriopsis Swartz & Whitmore, 1956 [**W. diplocystulis*]. Trilobate, S_1 and S_2 tending to unite on dorsal side of L_2, which does not reach dorsal margin; marginal rim narrow; surface smooth, or finely punctate-reticulose. Dimorphic pouch extending from anteroventral to posteroventral part of valve, fairly well defined dorsally and indented below L_2. *U.Sil.-L.Dev.*, N.Am.——Fig. 53,2*a,b.* **W. diplocystulis,* L.Dev. (Manlius Ls.), USA (N.J.); *2a,* ♂ RV (syntype) lat., ×35; *2b,* ♀ LV (syntype) lat., ×30 (Sohn, n).——Fig. 53, 2*c-e. W. jerseyensis* (Weller), U.Sil. (Decker Ls.), USA (N.J.); *2c-e,* LV (holotype) lat., vent., post., ×20 (Sohn, n). [Berdan.]

Zygobeyrichia Ulrich, 1916 [**Z. apicalis*]. Trilobate, L_1 separated from L_2 and L_3 by shallow ventral extension of S_1; surface reticulate. Pouch large, anteroventral. *U.Sil.-L.Dev.*, NE. N.Am.-Eu.—— Figs. 50,*3,* 52,2. **Z. apicalis,* L.Dev., USA (Me.); 50,*3a,* ♂ RV (topotype) lat., ×15; 50,*3b,* ♀ LV (topotype) lat., ×15 (Berdan, n); 52,2*a,b,* ♀ LV lat., ♂ LV lat., ×15 (Ulrich & Bassler, 1923).—— Fig. 54,*1. Z. barretti* (Weller), U.Sil. (Decker Ls.), USA (N.J.); *1a-d,* ♂ carapace (holotype) L, R, vent., post., ×25 (Swartz & Whitmore, 1956). [Berdan.]

Family ZYGOBOLBIDAE Ulrich & Bassler, 1923

[=Nodellinae Zaspelova, 1952] [Materials for this family prepared by Jean Berdan, United States Geological Survey]

Shape semielliptical, dorsum straight, cardinal angles distinct; median sulcus deep and conspicuous; L_1 suppressed or attenuate, L_2 and L_3 tending to unite ventrally, raised above other lobes so that general appearance of carapace is commonly bilobate, less commonly trilobate, or quadrilobate; free margins bordered by rim which is a ventrally directed flexure of shell. Dimorphism shown by anteroventral or ventral pouch. Hinge simple, muscle scar unknown. *?M.Ord.-?L. Sil., M.Sil., ?Dev.*

Zygobolba Ulrich & Bassler, 1923 [*Beyrichia decora Billings, 1866]. L_2 and L_3 narrow, forming a prominent U- or V-shaped ridge; surface smooth, finely punctate, or reticulate. *?Ord., M. Sil., ?Dev.*

Z. (Zygobolba). Pouch large, oval, situated anteroventrally. *?Ord., M.Sil., ?Dev.,* N. N.Am.-N. Eu.——Fig. 55,*1.* **Z. (Z.) decora* (Billings), M.Sil., Can.; *1a,* ♂ RV (topotype) lat., ×15; *1b,* ♀ LV lat., ×15 (Berdan, n).

Z. (Zygobolbina) Ulrich & Bassler, 1923 [**Zygobolbina conradi* Ulrich & Bassler, 1923]. Pouch bilobed, anteroventral in position, anterior part subglobular, divided by shallow sulcus from posterior part, which is a low swelling. *M.Sil.,* NE. N.Am.——Fig. 55,*4.* **Z. (Z.) conradi* (Ulrich & Bassler), USA (Md.); *4a,* ♂ LV (syntype) lat., ×15; *4b,* ♀ LV (syntype) lat., ×15 (Berdan, n).

Z. (Zygosella) Ulrich & Bassler, 1923 [**Zygosella vallata* Ulrich & Bassler, 1923]. Pouch narrow and ridgelike, confined to anterior part of carapace. *M.Sil.,* NE. N.Am.——Fig. 55,*3.* **Z. (Z.) vallata,* USA(Md.-W.Va.); *3a,* ♂ RV (holotype) lat., ×15; *3b,* ♀ RV lat., ×15 (Berdan, n).

Bonnemaia Ulrich & Bassler, 1923 [**B. celsa*]. L_2 and L_3 broad, sharply set off from surface of valve by angular contact, L_1 narrow (if present), S_1 very narrow (if present); L_3 in most species with sigmoidally curved angular crest on its posterior side; anteroventral pouch ovate, poorly defined dorsally; surface smooth where known, ornamentation of most species unknown. *M.Sil.,* NE. N.Am.——Fig. 55,6. **B. celsa,* USA (Md.); ♂ LV lat., ×15 (Berdan, n).

?**Glymmatobolbina** Harris, 1957 [**G. quadrata*]. Subquadrate, valves equal; S_2 prominent, S_1 weak; posterior portion of carapace broadly swollen, connected ventrally below S_2 to anterior lobe. Monospecific. [Resembles *Plethobolbina.*] *M.Ord.,* USA (Okla.). [Scott.]

Mastigobolbina Ulrich & Bassler, 1923 [**M. typa*]. L_1 narrow, L_2 pyriform, typically produced ven-

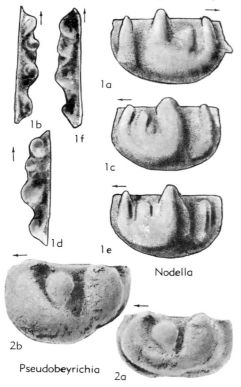

Nodella

Pseudobeyrichia

FIG. 56. Beyrichiidae, Drepanellidae (p. Q122-Q125).

trally and posteriorly as curved carina ("flagellum") that extends onto L_3, S_1 very narrow; pouch large, subovate, in anterior part of valve; surface smooth, reticulate, or pustulose. *M.Sil.*, NE. N.Am.——Figs. 27, 55,5. **M. typa*, USA (Md.); 55,5a, ♂ RV (syntype) lat., ×15; 55,5b, ♀ RV lat., ×15 (Berdan, n).

?**Plethobolbina** Ulrich & Bassler, 1923 [**P. typicalis*]. L_1 suppressed, L_2 and L_3 very large and full, S_2 only sulcus present, rim narrow and flattened; pouch inconspicuous; anteroventral swelling. Surface smooth or reticulate. *M.Sil.*, NE. N.Am.——Fig. 55,2. **P. typicalis*, USA (Pa.); ♂ LV (holotype) lat., ×15 (Berdan, n).

Family UNCERTAIN

?**Gillatia** Öpik, 1953 [**G. trinacria*]. Straight-backed, trilobate, with marginal rim, lobes not connected ventrally, but broad swelling of valves occurs on ventral side of sulci. *L.Sil.*, Austral.(Victoria). [Scott.]

Superfamily DREPANELLACEA Ulrich & Bassler, 1923

[*nom. transl.* Polenova and Zanina, 1960 (*ex* Drepanellinae Ulrich & Bassler, 1923)] [Diagnosis and discussion by H. W. Scott, University of Illinois]

Carapace subquadrate to subrectangular, with or without marginal rim; surface smooth, pitted, or reticulate; one or more nodes present, some species terminated by spines; nodes rarely lobate; S_2 poorly to well defined; hingement by tongue and groove; dimorphism unknown. *M.Ord.-M. Perm.*

The Drepanellacea include very diverse genera which upon first examination seem to be unrelated. The marginal rim which is so characteristic of *Bollia* and *Ulrichia* might appear to be of major importance, but when such forms as *Paraechmina* are studied, we note that the marginal rim is expressed in several ways. It may be complete and prominent (e.g., *P. spinosa* Hall), more or less lobate (e.g., *P. postica* Ulrich & Bassler, or completely absent (e.g., many *Aechmina* and all Richinidae). Gradations in development of the marginal rim are such that they cannot be used as a key to the superfamily. The number and position of nodes varies from the single dorsomedial spinose node of *Aechmina* to the multinodose form of *Cornigella*. Comparison of a smooth binodose, marginally rimmed *Ulrichia* with multinodose *Cornigella* might lead to the conclusion that these forms were entirely unrelated, whereas all gradations actually may be found between number and position of the nodes, degree of ornamentation, and strength of the marginal rim. These gradations imply relationship, which, in addition to absence of typical lobation or any form of recognizable dimorphism, warrants the inclusion of all indicated families in a single superfamily, the Drepanellacea.

The stratigraphic distribution of drepanellacean genera is indicated graphically in Figure 57.

Family DREPANELLIDAE Ulrich & Bassler, 1923

[*nom. transl.* Swartz, 1936 (*ex* Drepanellinae Ulrich & Bassler, 1923)] [=Neodrepanellinae, Nodellinae Zaspelova, 1952; Depranellidae Howe, 1955] [Materials for this family prepared by H. W. Scott, University of Illinois, with additions from Ivar Hessland, University of Stockholm]

Carapace subquadrate to subrectangular; hinge long, dorsum straight, cardinal angles subequal and well defined; lateral surface lobate or with nodose lobes, nodes commonly terminating as spines with one or more overreaching dorsum; surface smooth, granular, pitted, or finely reticulate; marginal ridge or carina subparallel to free margin, poorly developed posteriorly; valves subequal. Hingement by simple tongue and groove but unknown in some forms. Dimorphism unknown. *M.Ord.-U.Dev.*

Drepanella ULRICH, 1890 [*pro Depranella* ULRICH,

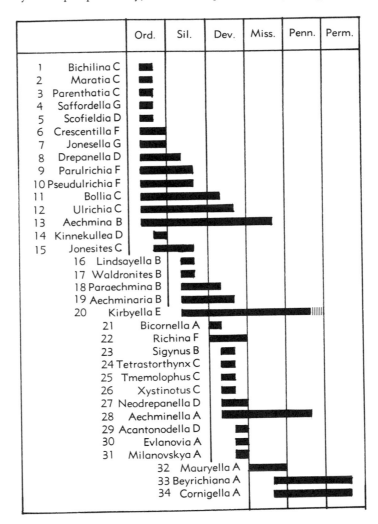

FIG. 57. Stratigraphic distribution of drepanellacean ostracode genera (Moore, n). Classification of the genera in families is indicated by letter symbols (A—Aechminellidae, B—Aechminidae, C—Bolliidae, D—Drepanellidae, E—Kirkbyellidae, F—Richinidae, G—Family Uncertain) and an accompanying alphabetical list furnishes a cross reference to the serially arranged numbers on the diagram.

Generic Names with Index Numbers

1890 (ICZN pend.)] [*Drepanella crassinoda* UL-RICH, 1890]. Subrectangular to subquadrate; dorsal border straight, cardinal angles subequal, well defined; carinate rib subparallel to anterior and ventral free margin, terminating anterodorsally as spine which overreaches dorsal margin, prominent spine at posterodorsal cardinal position; carinate ridge and posterodorsal spine representing external ornamental features not reflected internally; 2 prominent lobes in dorsomedial area separated by S_2, lobes smooth or marked by small secondary spines, which may terminate dorsally in a spine; surface smooth, granular, or pitted; valves subequal; no velum or subcarinate ridges known. Hingement in RV consisting of long straight groove, corresponding bar presumed to exist in LV; dimorphism unknown. *M.Ord.-L.Sil.,* N.Am.——FIG. 59,*1*. *D. crassinoda* (ULRICH), M.Ord.(Stones River F.), USA(Ky.); *1a,b,* RV lat., int., ×40 (Scott, n).

Acantonodella ZASPELOVA, 1952 [*A. terciocornuta*] [=*Acanthodella* HOWE, 1955]. Amplete or slightly postplete, with shallow sulcal depression between presulcal node and postsulcal oblique lengthened elevation, another dorsal elevation extending conformingly behind this one; adventral structure developed as ridge extending along entire free margin (anterior end may form knob dorsally). Dimorphism not reported. Surface pitted. Closely similar to *Drepanella, U.Dev.,* Eu. (USSR).——FIG. 58,*3*. *A. terciocornuta; 3a,b,* RV (holotype) lat., dors., ×50 (Zaspelova, 1952). [HESSLAND.]

?Kinnekullea HENNINGSMOEN, 1948 [*K. waerni*]. Straight dorsal margin; ends and ventral margin broadly rounded; arcuate ridge extending subparallel to anterior margin and in some specimens extending backward along ventral margin, ridge may project dorsally as spine or terminate in low node; rib tends to merge ventrally into smooth, reticulate, or pitted valve surface; no adventral structure; dimorphism unknown. [The sigmoid lobe is believed to be an external feature, not reflected interiorly. The genus cannot belong to the Bolliidae or Richinidae because in these families the lobes are reflected internally; also, it lacks the marginal rim of the bolliids. This form may represent an instar stage of some drepanellid.—SCOTT.] *U.Ord.,* Eu. (Swed.).——FIG. 58,*4*. *K. waerni;* LV lat., ×28 (Henningsmoen, 1948).

Neodrepanella ZASPELOVA, 1952 [*Drepanella tricornis* BATALINA, 1941] [=*Tetracornella* ZASPELOVA, 1952]. Anterodorsal quarter bearing 3 dorsally directed spines and rounded node; velate structure present, complete or incomplete posteriorly and merging or not merging with anterodorsal spine; no posteroventral spine. Surface reti-

culate. *M.Dev.-U.Dev.,* Eu.——FIG. 58,*2*. *N. tricornis* (BATALINA), U.Dev., Russia; *2a,b,* RV (neotype) lat., dors., ×50; *2c,d,* LV (paratype) lat., dors., ×50; *2e,f,* RV (paratype), lat., int., ×70 (Zaspelova, 1952).

?Nodella ZASPELOVA, 1952 [*N. svinordensis*]. Valves with 4 spinose dorsal lobes, anterior 3 being joined ventrally but posterior separated from others by sulcus; posteroventral spine may be present; velate structure present or absent. Muscle scar defined as small rounded node below median lobe. Surface smooth. *U.Dev.,* Eu.(Russia).——FIG. 56,*1*. *N. svinordensis; 1a,b,* RV (holotype), lat., dors., ×50; *1c,d,* LV (paratype) lat., dors.; *1e,f,* LV (paratype) lat., dors.; all ×50 (Zaspelova, 1952).

Scofieldia ULRICH & BASSLER, 1908 [*Drepanella bilateralis* ULRICH, 1894]. Like *Drepanella* but marginal ridge broader, dorsomedial node (L_2) smaller, and S_2 and S_3 merge at base of L_2. *M. Ord.,* N.Am.——FIG. 58,*1*. *S. bilateralis* (ULRICH), Decorah Sh., USA (Minn.); *1a,b,* LV lat., int., ×20 (J. R. Cornell, n).

Family AECHMINELLIDAE Sohn, n. fam.

[Materials for this family prepared by I. G. SOHN, United States Geological Survey]

Small, straight-backed, lobate or nodose, with some or all lobes terminating in spines; overlap slight; surface reticulate, pitted, or smooth; free margins smooth. Hinge ridge-and-groove type. *L.Dev.-M.Perm.*

Aechminella HARLTON, 1933 [*A. trispinosa*] [=*Balantoides* MOREY, 1935; *Boursella* TURNER, 1939]. Quadrilobed or trilobed with posterior spine, sulci vertical. *M.Dev.-L.Penn.,* N.Am.——FIG. 60,*1*. *A. trispinosa,* Penn., USA (Okla.); *1a,b,* LV (holotype) dors., lat., ×60 (Sohn, n).

Beyrichiana KELLETT, 1933 [*B. permiana*] [=?*Mammoides* BRADFIELD, 1935]. Bilobed in dorsal half, with posterolateral spine or node on each valve; differs from *Aechminella* in lack of distinct sulcus. *U.Miss.-L.Perm.,* N.Am.——FIG. 60,*5*. *B. ? mammilata,* Penn., Okla. (type species of *Mammoides*); RV (holotype) lat., ×30 (Sohn, n).

Bicornella CORYELL & CUSKLEY, 1934 [*B. tricornis*]. Differs from *Beyrichiana* in having 2 backwardly pointed spined lobes in dorsal half. *L.Dev.,* N.Am.——FIG. 60,*2*. *B. tricornis,* USA (Okla.); *2a,b,* LV lat., dors., ×120 (Sohn, n).

Cornigella WARTHIN, 1930 [*C. minuta* (=*Beyrichia tuberculospinosa* JONES & KIRKBY, 1886)]. Subhemicircular valves with single upward pointing large spine near dorsal margin, surface with scattered small nodes. *U.Miss.-M.Perm.,* Eu., USA.

Evlanovia EGOROV, 1950 [*E. tichonovitchi*]. Differs from *Beyrichiana* in absence of posterior spine and in having 2 nodes near ventral margin, one subcentral, other anteroventral. *U.Dev.*, Eu. (Russia).

——FIG. 60,3. *E. tichonovitchi; 3a,* ♂ RV lat.; *3b,c,* ♀ RV (holotype) lat., dors.; all ×45 (Egorov, 1950).

Mauryella ULRICH & BASSLER, 1923 [*M. mam-

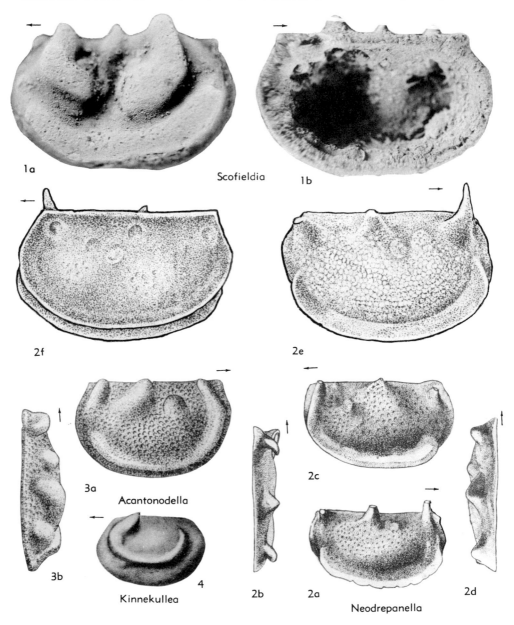

FIG. 58. Drepanellidae (p. Q125).

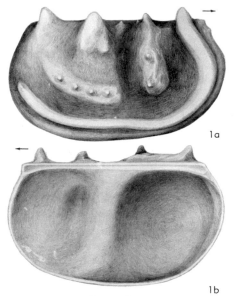

Drepanella

FIG. 59. Drepanellidae (p. *Q*125).

milata] [=*Verrucosella* CRONEIS & GALE, 1938; *Verrucolsella* NEAVE, 1950]. Rounded nodes superposed on reticulated valve. *Miss.*, N.Am.

?**Milanovskya** EGOROV, 1950 [**M. bicornis*]. Backward-pointing spines below dorsal margin near each end. *U.Dev.*, Eu. (Russia).——FIG. 60,4. **M. bicornis; 4a,b,* ♀ LV (holotype) lat., dors.; *4c,d,* ♂ LV lat., dors.; all ×45 (Egorov, 1950).

Family AECHMINIDAE Bouček, 1936

[*nom. transl.* SWARTZ, 1936 (Oct.) (*ex* Aechmininae BOUČEK, 1936) (July)] [Materials for this family prepared by S. A. LEVINSON, Humble Oil & Refining Company]

Straight-hinged with distinct spine; base of spine in dorsal-median area. Commonly with rounded pit variously located adjacent to base of spine; marginal ridge or marginal spines or papillae present in some species. *M.Ord.-M.Miss.*

Aechmina JONES & HOLL, 1869 [**A. cuspidata*]. Simple, with distinct dorsal-median spine, commonly with marginal spines or papillae. *M.Ord.-M.Miss.*, N.Am.-NW.Eu.-Austral.—— FIG. 61,*1.* **A. cuspidata,* Sil.(Wenlock.), Eng.; LV lat., ×25 (Jones & Holl).——FIG. 61,*4. A.* sp. cf. *A. cuspidata,* Dev.(Helderberg.), USA (Md.); RV lat., ×60 (Bouček, 1936).

Aechminaria CORYELL & WILLIAMSON, 1936 [**A. nodosa*]. Like *Aechmina* but with distinct rounded

pit at anteroventral side of base of spine. *M.Sil.-M.Dev.*, N.Am.——FIG. 61,*5. A. robusta,* M.Sil. (Waldron), USA (Ind.); RV lat., ×25 (Coryell & Williamson, 1936).

Lindsayella CORYELL & WILLIAMSON, 1936 [**L. rugosa*]. Like *Waldronites* but without submarginal ridge. *M.Sil.*, N.Am.——FIG. 61,*2.* **L. rugosa,* Waldron Sh., USA (Ind.); RV lat., ×40 (Coryell & Williamson, 1936).

Paraechmina ULRICH & BASSLER, 1923 [**Cytherina spinosa* HALL, 1852 (*non* REUSS, 1846; ICZN pend.)]. Like *Aechminaria* but with well-defined marginal ridge. *M.Sil.-L.Dev.*, N.Am.——FIG. 61, *7.* **P. spinosa* (HALL), M.Sil., USA (N.Y.); LV lat., ×45 (Hall, 1852).

Sigynus KESLING, 1953 [**S. dictyotus*]. Small pit ventral to base of dorsomedian spine, surface reticulate. *M.Dev.*, N.Am.——FIG. 61,*6.* **S. dictyotus,* Arkona, Can. (Ont.); RV lat., ×60 (Kesling, 1953).

Waldronites CORYELL & WILLIAMSON, 1942 [*pro Cornulina* CORYELL & WILLIAMSON, 1936 (*non* CONRAD, 1853)] [**Cornulina bispinosa* CORYELL & WILLIAMSON, 1936]. Like *Aechmina* but with submarginal ridge from mid-posterior to anterior cardinal angle where it terminates in small spine; irregular crescent-shaped ridge connecting major spine with anterodorsal spine. *M.Sil.*, N.Am.—— FIG. 61,*3.* **W. bispinosa* (CORYELL & WILLIAMSON), Waldron Sh., USA (Ind.); RV lat., ×25 (Coryell & Williamson, 1936).

Family BOLLIIDAE Bouček, 1936

[*nom. transl.* SCOTT & WAINWRIGHT, herein (*ex* Bolliinae BOUČEK, 1936)] [=Ulrichiinae SCHMIDT, 1941] [Materials for this family prepared by H. W. SCOTT, University of Illinois, and JOHN WAINWRIGHT, Shell Oil Co., with some additions by IVAR HESSLAND, University of Stockholm, as recorded]

Carapace subquadrate to subrectangular; hinge long, cardinal angles subequal and well defined; lateral surface distinguished by presence of 2 dorsomedial nodes which are separate or joined ventrally to form U-shaped lobe, area between nodes representing S_2; distinct marginal rim (pseudovelum) extending around all or a portion of free margin; lateral surface of valves (other than lobes and marginal rim) flat to inflated and smooth to coarsely reticulate, pitted or granulose; valves subequal. Hingement by tongue and groove. Dimorphism unknown (except in *Bichilina*). [Distinguished from the Richinidae mainly by presence of a marginal rim.] *M.Ord.-M.Dev.*

Bollia Jones & Holl, 1886 [**B. uniflexa;* SD S.A. Miller, 1892 (*fide* Warthin, 1948)]. Dorsal margin straight, long; cardinal angles distinct, sub-ᴄ｡..ꞁ˙ ᵐꞌᵃᵍinal rim well defined, extending around entire free margin, ...˙ incᴏnspicuous in posterodorsal area; 2 dorsomedial lobes connected ventrally to form a U-shaped lobe surrounding S_2, surface within marginal ridge and exclusive of lobes nearly flat or gently convex and ornamented with pits, puncta, or reticula. *M.Ord.-L.Dev.,* Eu.-N.Am.——Fɪɢ. *62,1. *B. uniflexa,* M.Sil., Eng.; *1a,b,* carapace R, dors., ×15 (Jones & Holl, 1886).——Fɪɢ. 62,2. *B. ungula* Jones, L.Dev. (Onondaga F.), USA (Pa.); LV lat., ×25 (Swartz & Swain, 1941).——Fɪɢ. 62,3. *B. subaequata* Ulrich, M.Ord. (Decorah Sh.), USA

(Minn.); *3a,b,* LV lat., RV lat., ×45 (J. R. Cornell, n.).

?Bichilina Sarv, 1959 [recorded by Sarv as *Bichilina* Neckaja, in coll.] [**B. prima* Sarv, 1959 (described as *B. prima* Neckaja, in coll.)]. Amplete ᴏⲓ ⲣⲓⲉⲡⲓⲉ.ᵉ ⲉⲇⱳivalved, monosulcate; sulcus deep and subcircular in outline, surrounded ᵇᵧ ᵇⲟⲅˢᵉᵇoe-shaped ridge; dorsal plica present, surface smooth or tuberculate. Dimorphic, adventral structure developed as more or less wide flange along entire free margin, of equal width and slightly concave or wider and somewhat convex in anteroventral part. *M.Ord.,* NW.Eu.——Fɪɢ. 62,9. **B. prima,* Est.; *9a,b,* RV (holotype) (tecnomorph) lat., and RV (heteromorph) int., ×33 (Sarv, 1959). [Hessland] [Classed by Scott in Bolliidae but in opin-

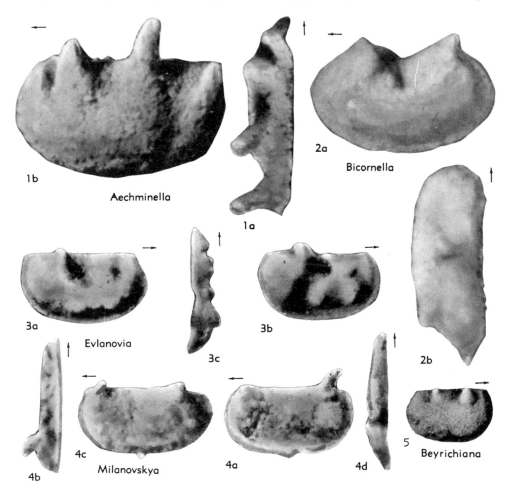

1b — Aechminella

1a

2a — Bicornella

2b

3a — Evlanovia

3b

3c

4b

4c — Milanovskya

4a

4d

5 — Beyrichiana

Fɪɢ. 60. Aechminellidae (p. Q126-Q127).

ion of HESSLAND probably belongs in Tvaerenellidae.]

?**Jonesites** CORYELL, 1930 [*pro Placentula* JONES &
HOLL, 1886 (*non* LAMARCK, 1822)] [**Primitia
excavata* JONES & HOLL, 1869] [=?*Placentella*
WILSON, 1935]. Subovate, cardinal angles obtuse;
dorsal margin straight or in some specimens convex, marginal ridge parallel to free margin and
continuing around dorsal border; thickened dorsal
rim may rise above hinge line; S_2 defined by a
raised U-shaped ridge that is continuous with
marginal rim; anterior arm of ridge may be thickened to form a node but in some species ventral
part of this ridge may be missing. Surface within
marginal ridge (except ridge bounding S_2) depressed, reticulate. *U.Ord.-M.Sil., Eu.-N.Am.*——
FIG. 62,*14*. **Jonesites primitia excavata* (JONES &

HOLL), M.Sil. (Woolhope Ls.), Eng.; LV, ×44
(H. N. Coryell, 1930).

Maratia KAY, 1940 [**M. mara*]. Small, subovate,
valves subequal; dorsal margin straight, ventral
broadly convex; S_1, S_2, S_3 present, all confined to
dorsal half of valves, L_1, L_2, L_3, and L_4 distinguished, L_1 and L_4 tending to fade ventrally
into general shell surface, L_2 smaller than L_3,
adventral bend *(Umbiegungskante)* sharp, forming broad venter; surface coarsely pitted to punctate; S_2, L_2, and L_3 reflected internally more
strongly than other sulci and lobes. [Like *Ulrichia*
but lacks sharp marginal rim and sulci, and lobes
more numerous.] *M.Ord., N.Am.*——FIG. 62,*7*.
*M. mara, M.Ord. (Decorah Sh.), USA (Minn.);
7a-c, RV lat., LV lat., int., ×50 (J. R. Cornell, n).

?**Parenthatia** KAY, 1940 [**Moorea punctata* ULRICH

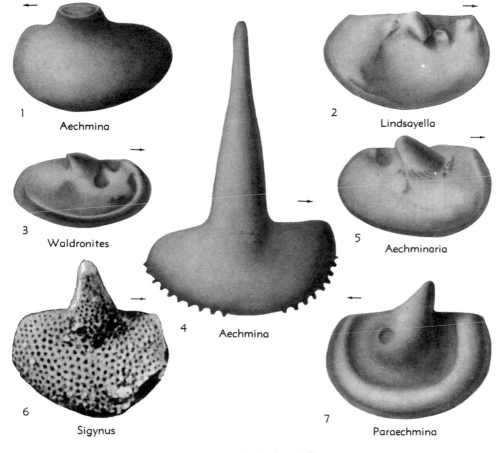

1 Aechmina

2 Lindsayella

3 Waldronites

4 Aechmina

5 Aechminaria

6 Sigynus

7 Paraechmina

FIG. 61. Aechminidae (p. Q127).

Crustacea—Ostracoda

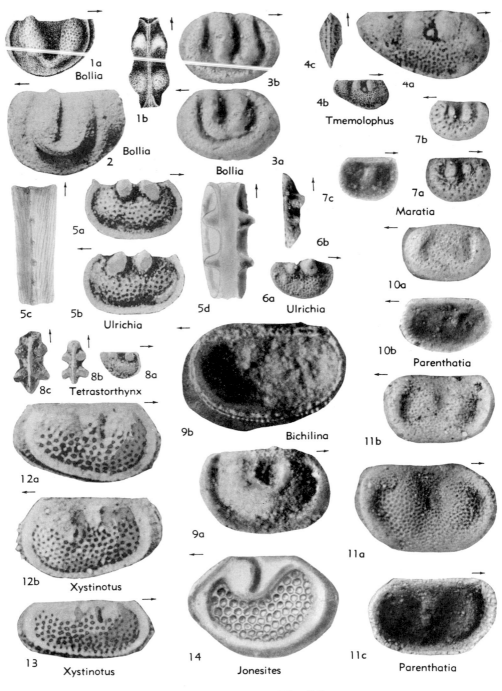

Fɪɢ. 62. Bolliidae (p. *Q*128-*Q*131).

1894]. Small, valves subequal, hinge long and straight; cardinal angles poorly defined, anterior commonly more obtuse; marginal rim sharply defined along anterior and posterior margins, less well developed at mid-venter; S_2 indistinct or well defined, when present dividing valve into 2 gently swollen lobes which merge ventrally into general surface of shell, lobes joined to form broad "U" poorly defined at base; lobes separated from anterior and posterior marginal rims by shallow depressions; surface finely reticulate. [*P. punctata* probably represents a young instar, S_2 and the "U"-lobe being undeveloped. Specimens of *P. camerata* found associated with *P. punctata* seem to represent adults. *Parenthatia* may belong to the *Leperditellidae*, but the presence of a marginal rim and the subequality of the valves are closer to the Bolliidae. In ventral view the marginal rim produces the typical wide, flat to gently concave, surface common to the bolliids.] *M.Ord.,* N.Am.——Fig. 62,*10.* *P. punctata* (Ulrich), M. Ord. (Decorah Sh.), USA (Minn.); *10a,b,* LV lat., RV int., ×50 (J. R. Cornell, n).——Fig. 62, *11. P. camerata* Kay, M.Ord. (Decorah Sh.), USA (Minn.); *11a-c,* RV lat., LV lat., int., ×50 (J. R. Cornell, n.).

Tetrastorthynx Kesling, 1953 [**T. diabolicus*]. Like *Tmemolophus* but dorsomedial nodes widely spaced and considerably more distinct; surface reported to be highly granulose; original surface may have been finely reticulate. Dimorphism unknown, *M.Dev.,* N.Am.——Fig. 62,*8.* *T. diabolicus,* Arkona Sh., Can. (Ont.); *8a-c,* RV lat., carapace dors., carapace (holotype) dors., ×42 (Kesling, 1953).

Tmemolophus Kesling, 1953 [**T. margarotus*]. Like *Ulrichia* but dorsal nodes more lobelike, with distinct intervening sulcus; velate structure paralleling ventral free margin only, marginal costa sometimes present in some species; postero-ventral margin more truncate; like *Bollia* but lacks ventral connection to dorsal nodes. [The subequal size of the nodes and their dorsomedial position exclude this genus from the Hollinidae.] *M.Dev.,* N.Am.——Fig. 62,*4.* *T. margarotus,* Arkona Sh., Can. (Ont.); *4a,* carapace (holotype) R, ×68; *4b,c,* carapace R, vent., ×30 (Kesling, 1953).

Ulrichia Jones, 1890 [**U. conradi*] [=?*Warthinia* Spivey, 1939; *?Limbatula* Zaspelova, 1952]. Small, surface reticulate; with 2 prominent dorsomedial nodes; marginal rim distinct. [Like *Bollia* except that dorsomedial lobes are not united ventrally; they may extend slightly above the dorsal margin.] *M.Ord.-M.Dev.,* cosmop.——Fig. 62,*5.* **U. conradi,* M.Dev.(Bell Sh.), USA (Mich.); *5a,b,* RV lat., LV lat., ×40 (Kesling, 1952); M. Dev.(Hamilton Gr.), USA (N.Y.), *5c,d,* vent. dors., ×40 (Scott, n.).——Fig. 62,*6. U. nodosa*

(Ulrich), (type species of *Warthinia*), U.Ord. (Maquoketa F.), USA (Iowa); *6a,b,* RV lat., dors., ×35 (J. H. Burr, Jr., n).

Xystinotus Kesling, 1953 [**X. wrightorum*]. Like *Ulrichia* but without pronounced nodes; lateral surface reticulate, dorsal part of each valve smooth; with marginal costa or row of minute tubercles. *M.Dev.,* N.Am.——Fig. 62,*12.* **X. wrightorum,* Arkona Sh., Can. (Ont.); *12a,b,* carapace (holotype) R, carapace (paratype) L, ×68 (Kesling, 1953).——Fig. 62,*13. X. subnodatus* (Turner), Arkona Sh., Can. (Ont.); carapace R, ×68 (Kesling, 1953).

Family KIRKBYELLIDAE Sohn, n. fam.

[Materials for this family prepared by I. G. Sohn, United States Geological Survey]

Subquadrate, small essentially equivalved, reticulated, sulcate, with distinct to subdued subventral horizontal lobe that terminates in posterior spine, and with or without one marginal rim. Narrow ridge-and-groove hingement with minute terminal teeth. *M. Sil.-M.Penn., ?U.Penn.*

Kirkbyella Coryell & Booth, 1933 [**K. typa*]. Horizontal lobe with backward-pointing spine below sulcus. *M.Sil.-M.Penn., ?U.Penn.,* N.Am.—— Fig. 63,*1.* **K. typa, ?*Penn., USA (Tex.); *1a-c,* RV (holotype) lat., int., dors., ×60 (Sohn, n).

Family RICHINIDAE Scott, n. fam.

[Materials for this family prepared by H. W. Scott, University of Illinois, with additions by Ivar Hessland, University of Stockholm]

Subovate, with 2 well-defined to somewhat obscure dorsomedial nodes, S_2 separating nodes; hinge shorter than greatest length, cardinal angles indistinct; surface smooth to reticulate; marginal rim lacking; dimorphism unknown. *M.Ord.-U.Dev.*

The Richinidae represent a rare and poorly known group of Ordovician to Devonian ostracodes. They differ from other families of the Drepanellacea in lack of a marginal rim and the ill-defined nature of the cardinal corners. The surface of most described species is recorded as smooth, but this may be a matter of preservation. Faint sculpturing of a reticulate nature has been reported on *Richina* and others. Study of more material may show that this group definitely belongs to the Bolliidae but until the presence of a marginal rim can be demonstrated they should be placed in a separate family. It is not certain that all de-

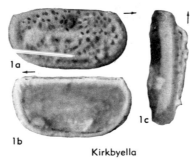

Kirkbyella

FIG. 63. Kirkbyellidae (p. Q131).

scribed forms are represented by original shell material; some may be steinkerns and need restudy. Several species have been described from Bohemia, Canada, and the United States.

Richina CORYELL & MALKIN, 1936 [**R. truncata*]. Subovate, ends rounded, cardinal angles indistinct; 2 dorsomedial nodes separated by S_2; sides slightly convex rather than nearly flat, as in *Ulrichia;* marginal rim lacking; surface smooth to finely reticulate. [This description is based on examination of topotype material with original shell preserved. SWARTZ & WHITMORE (1956) believe that *Richina* is more closely related to *Bollia* than to *Ulrichia,* as indicated by the yoke-ridge of *R. zygalis,* Silurian. *Richina* is believed to be closely related to the Bollidae, differing primarily in lack of a marginal rim.] *L.Dev.-U.Dev.,* N.Am.——FIG. 64,1. **R. truncata,* M.Dev.(Hamilton), USA (N.Y.); LV lat., ×40 (Coryell & Malkin, 1936).

Crescentilla BARRANDE, 1872 [**C. pugnax*]. Small; dorsal border straight, hinge short, ends broadly rounded; with 2 conical and somewhat rounded protuberances overreaching dorsal border near each cardinal extremity; S_2 with broad flat floor. [This genus differs from *Parulrichia* in the dorsal extension of the 2 protuberances beyond the dorsal margin.] *M.Ord.-U.Ord.,* Eu. (Czech.-Eng.). ——FIG. 64,5. **C. pugnax,* RV lat., ×30 (Schmidt, 1941).

Parulrichia SCHMIDT, 1941 [**Primitia diversa* JONES & HOLL, 1886]. Small, valves equal, dorsal margin straight; 2 conical nodes on lateral surface near dorsal margin, may be terminated by spines; nodes separated by wide S_2; surface smooth or weakly sculptured. [This genus differs from *Pseudulrichia* in wider spacing of the nodes and broader sulcus; otherwise, they are very similar.] *M.Ord.-M.Sil.,* Eu. (Czech.-Eng.). —— FIG. 64, 4a-c. **P. diversa* (JONES & HOLL), M.Sil.(Wen-

lock.), Eng.; *4a,* RV (lectotype, SCHMIDT, 1941) lat., ×30; *4b,c,* RV (another specimen), vent., dors., ×30 (Jones & Holl, 1866).——FIG. 64,*4d,e.* *P. diversa antiqua* SCHMIDT, M.Ord.(Caradoc.), Boh.; *4d,e,* carapace L, dors., ×50 (Schmidt, 1941). [SCOTT-HESSLAND.]

Pseudulrichia SCHMIDT, 1941 [**Leperditia bivertex* ULRICH, 1879]. Subovate, ends broadly rounded; hinge shorter than greatest length; valves equal; marginal rim lacking; with 2 low and indistinct to well-defined dorsomedial nodes, separated by S_2; surface smooth to slightly sculptured. [Differs from *Richina* in slightly sharper cardinal angles and nodes that merge more gently into shell.] *M.Ord.-M.Sil.,* Eu.-N.Am.——FIG. 64,*3a-d.* **P. bivertex* (ULRICH), M.Ord.(Trenton.), USA(Ky.); *3a,b,* LV (lectotype) lat., dors., ×14 (*3a,* Schmidt, 1941; *3b,* Ulrich, 1879); *3c,d,* carapace, L, dors., ×50 (Schmidt, 1941).——FIG. 64,*3e,f.* *P. simplex* (KAY), M.Ord.(Decorah Sh.), USA(Minn.); *3e,f,* LV lat., int., ×50, ×45 (J. R. Cornell, n). [SCOTT-HESSLAND.]

Family UNCERTAIN

Jonesella ULRICH, 1890 [**Leperditia crepidiformis* ULRICH, 1879] [=*Vogdesella* BAKER, 1924 (*pro Melanella* WADE, 1911, *non* MOREY, 1924)]. Subovate, with U-shaped lobe in anterior half and prominent S_2 inside this lobe; valves subequal, without marginal structures or rim; interior marked by U-channel reflecting exterior lobe; hingement by bar and groove. Dimorphism unknown. [This genus lacks the marginal rim of either Bollidae or Drepanellidae. The U-lobe is reflected interiorly by a corresponding channel which is situated much farther anteriorly than the same feature in *Bollia*. It differs from features of the Richinidae in the ventral connection of the 2 dorsal lobes to form the U-lobe. Very similar to specimens of *Dilobella (Ctenobolbina) fulcrata* (ULRICH) here interpreted to be ?males. *M.Ord.-U.Ord.,* N.Am.-Eu.——FIG. 64,2. **J. crepidiformis* (ULRICH), U.Ord.(Eden.), USA(Ky.); LV lat., ×18 (Ulrich & Bassler, 1923).

Saffordellina BASSLER & KELLETT, 1934 [**Saffordellina muralis* (ULRICH & BASSLER, 1923)] [*pro Saffordella* ULRICH & BASSLER, 1923 (*non* DUNBAR, 1920)]. Large, straight-backed ostracodes; lateral surface marked with several nodes in medial and anterodorsal quarter, prominent sigmoid ridge subparallel to ventral and anterior margins; venose-like lines radiating from nodes. Proper disposal of this genus cannot be made until the type is restudied. Hinge, marginal, and interior structures are unknown. It may belong to the Drepanellidae. *M.Ord.,* N.Am.——FIG. 64,6. **S. muralis* (ULRICH & BASSLER), USA(Tenn.); RV lat., ×? (Ulrich & Bassler, 1923).

Superfamily HOLLINACEA Swartz, 1936

[*nom. transl.* KESLING, herein (*ex* Hollinidae SWARTZ, 1936)] [Diagram and discussion by H. W. SCOTT, University of Illinois, and R. C. MOORE, University of Kansas]

Carapace straight-backed, nearly equi-valved, sulcate, lobate, velate; velar di-morphism except in Quadrijugatoridae and Bassleratiidae, which nondimorphic fam-ilies may not belong to the superfamily. *L.Ord.-M.Perm.*

The Hollinacea are a large assemblage of dimorphic straight-hinged palaeocopids that chiefly characterize lower Paleozoic marine strata, especially those of Ordovician age. Only a few forms occur in post-Devonian rocks, among which the long-ranging *Hol-linella* extends into Middle Permian forma-tions. The stratigraphic distribution of hollinacean ostracode genera is summarized graphically in Figure 65.

Family HOLLINIDAE Swartz, 1936

[=Ctenoloculininae JAANUSSON & MARTINSSON, 1956] [Mate-rials for this family prepared by R. V. KESLING, University of Michigan, with contribution from JEAN BERDAN, United States Geological Survey]

Carapace slightly inequivalved, with overlap of LV on RV, mostly with strongly developed lobation, including bi-, tri-, and quadrilobate types. L_3 large and bulbous in many genera; velar structures more or less prominent, restricted to anterior and ventral parts of free border. Hinge line straight, hingement consisting of subtri-angular pits or furrows at ends of LV hinge for reception of corners of RV. Dimorphism distinct, shown primarily by form of velar structures. Surface commonly papillate (KESLING, 1952; SWARTZ, 1936). *M.Ord.-M. Perm.*

Ctenobolbina, included in the family with question, is the only Ordovician genus other than *Grammolomatella* and, possibly, *Eohol-*

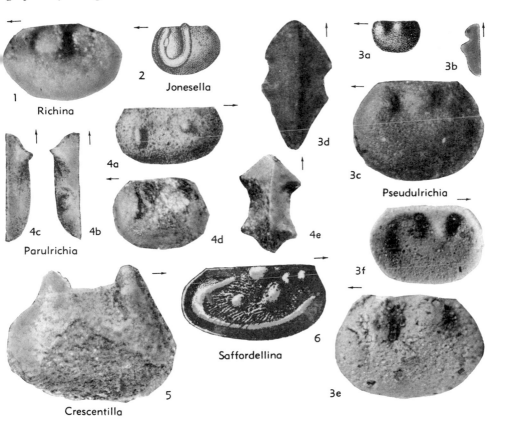

FIG. 64. Richinidae (p. Q132).
(Note: FIG. 64 also includes 2 genera of "Family Uncertain.")

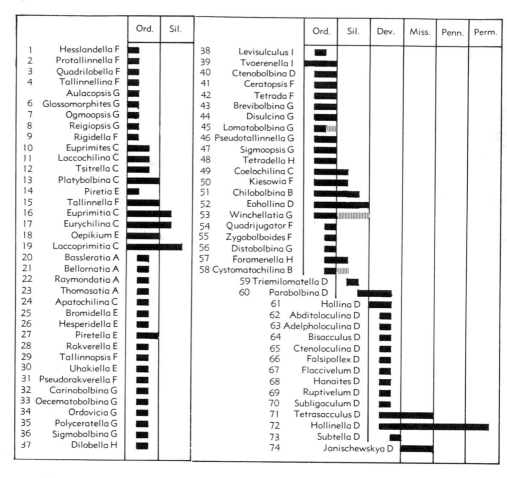

FIG. 65. Stratigraphic distribution of hollinacean ostracode genera (Moore, n). Classification of the genera in families is indicated by letter symbols (A—Bassleratiidae, B—Chilobolbinidae, C—Eurychilinidae, D—Hollinidae, E—Piretellidae, F—Quadrijugatoridae, G—Sigmoopsidae, H—Tetradellidae, I—Tvaerenellidae) and an accompanying alphabetical list furnishes a cross reference to the serially arranged numbers on the diagram.

Generic Names with Index Numbers

lina. It differs from typical hollinid genera in 3 significant ways: in lacking discernible dimorphism in the type species, in showing L_3 no more emphasized than other lobes, and in having S_3 developed as a distinct sulcus. In the type species, *C. ciliata*, however, the restricted frill, configuration of S_2, papillate surface, and general outline strongly suggest that it may have been ancestral to the typical hollinids.

Eohollina appears to have lobation and dimorphism similar to that of *Hollina*. There remains some question of its position, however, because the incurved frill of the female also resembles that in *Bromidella*. On the basis of carapace lobation, *Eohollina* is tentatively included in the Hollinidae.

Other new Ordovician genera created by HARRIS (1957) and assigned to Hollinidae lack dimorphism, and their relationships are even more doubtful. They are *Acanthobolbina*, *Ballardina*, and *Haplobolbina*. Possibly, as their author suggests, they are related to *Ctenobolbina*, but it is equally possible that they are sigmoopsids, in which the existence of dimorphism has not yet been established. They are placed in Palaeocopida of uncertain position.

Hollina ULRICH & BASSLER, 1908 [*Ctenobolbina insolens* ULRICH, 1908]. Distinguished by lobation, L_1 being a distinct lobe, L_2 a node, L_3 a large bulb, and L_4 commonly indistinct (but in some species comparable to L_1), with ventral lobes present below L_1 and L_3; female with large strongly incurved frill in anterior and ventral parts of each valve; male with 2 hollow, blunt ventral projections of ventral lobes in each valve (Kesling, 1952; Ulrich, 1908). *L.Dev.-M.Dev.*, N.Am.——FIG. 66,*1*. *H. insolens* (ULRICH), M. Dev., USA(Ind., Falls of Ohio); *1a,b*, ♂ RV lat., int,. ×30; *1c,d*, ♀ LV lat., int., ×30 (Kesling, 1952).

Abditoloculina KESLING, 1952 [*A. insolita*]. Lobation as in *Hollinella*, with large spurlike lateral projections in each sex of known species; female with frill and several loculi, male with narrow interrupted frill or velar ridge in each valve. M.Dev., N.Am.——FIG. 66,*2*. *A. insolita*, M.Dev., USA(Ind., Falls of Ohio); *2a-c*, ♂ RV lat., vent., int., ×30; *2d-f*, ♀ LV lat., vent., int., ×30 (Kesling, 1952).

Adelphobolbina STOVER, 1956 [*Ctenobolbina papillosa* ULRICH, 1891]. Valves with large inflated lobes for L_1 and L_3, L_2 completely fused with L_1, L_4 low and forming posterior 0.2 of valve, separated from L_3 by a long curved semisulcus; S_2 deep, geniculate, long; in some species L_1 and L_3

connected by a ventral lobe, in others separated by a ventral extension of S_2; surface papillose. Dimorphism not pronounced, primarily in form of the frill, which in females is broader and incurved in anteroventral part, but in males is narrow and flared out; in some species, females have a wide carapace and papillae in channels between the frills and submarginal or marginal ridges, whereas males have narrow carapace and lack ornamentation in the channels. M.Dev., N.Am.——FIG. 66, *3e,f*. *A. papillosa* (ULRICH), USA(Ind., Falls of Ohio); *3e,f*, ♂ RV lat., vent., ×30 (Kesling & Peterson, 1958).——FIG. 66,*3a-d*. *A. megalia* (KESLING & TABOR), Hamilton, USA(Mich.); *3a,b*, ♂ carapace R, vent., ×35; *3c,d*, ♀ carapace R, vent., ×25 (Kesling & Tabor, 1953).

Bisacculus STEWART & HENDRIX, 1945 [*B. bilobus*]. Valves bilobed, with long curved S_2 in each valve extending from dorsal border entirely or nearly to ventral border; female with 2 loculi in each valve, male with very little velar development. M.Dev., N.Am.——FIG. 66,*5*. *B. bilobus*, Hamilton, USA(Ohio); *5a,b*, ♀ carapace L, vent., ×50; *5c*, ♂ carapace L, ×50 (Stewart & Hendrix, 1945).

?**Ctenobolbina** ULRICH, 1890 [*Beyrichia ciliata* EMMONS, 1855]. Valves distinctly trilobate, with long S_2 and S_3 concave toward front, nearly reaching ventral border; L_3 not swollen, not extending above hinge line; long (but restricted) velar ridge or narrow frill. Dimorphism obscure or absent. [Genus formerly included Devonian species now assigned to *Adelphobolbina*.] M.Ord.-U.Ord., N. Am.——FIG. 66,*4d*. *C. ciliata* (EMMONS), U.Ord., USA(Ohio); RV lat., ×18 (Ulrich, 1890).——FIG. 66,*4a-c*. *C. alata* ULRICH, U.Ord., USA(Ohio); *4a-c*, carapace R, dors., vent., ×22 (Kesling, 1951).—— FIG. 67,*1*. *C. obliqua* ULRICH, M.Ord.(Edinburg F.), USA(Va.); *1a-d*, ♂ RV lat., dors., vent., int., ×20; *1e,f*, ♀ LV lat., RV int., ×20 (J.C.Kraft, n).

Ctenoloculina BASSLER, 1941 [*Tetradella cicatricosa* WARTHIN, 1934]. Carapace with L_1, L_2, and L_3 appearing as vertically elongate flat-topped ridges, L_4 being shield-shaped and elevated less than other lobes, in known species with lobes highly ornamented in contrast to smooth remainder of valves; female with frill and loculi, male with spurs at ventral ends of L_1, L_2, and L_3, those of L_2 and L_3 being large and recurved in most species. M.Dev., N.Am.-Eu.——FIG. 68,*1*. *C. cicatricosa* (WARTHIN), Hamilton, USA(Mich.); *1a,b*, ♀ LV lat., vent., ×40; *1c*, ♂ LV lat., ×40 (Kesling); *1d,e*, ♀ carapace L, vent., ×37 (Kesling, 1951).

?**Eohollina** HARRIS, 1957 [*Beyrichia irregularis* SPIVEY, 1939]. Trilobate, L_1 lobate and continuous with ventral lobe in some species, L_2 appearing as a small node adjacent to the deep S_2, and L_3 large and bulbous in some species but forming only a small knob in others. Dimorphism in form

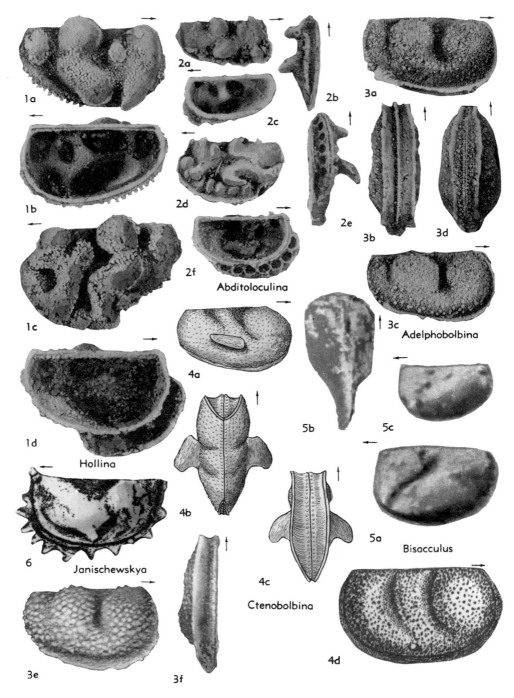

FIG. 66. Hollinidae (p. Q135-Q137).

of velar structure, which is a ridge in males and an incurved frill in females. [Generic boundaries not distinctly drawn, and certain species assigned here seem closely allied to *Bromidella*.] *M.Ord.-U.Sil.*, N.Am.——Fig. 68,2a. **E. irregularis* (Spivey), U.Ord., USA(Iowa); LV lat., ×25 (Spivey, 1939).——Fig. 68,2b-d. *E. depressa* (Kay), M.Ord., USA(N.Car.); *2b*, ♂ RV lat.; *2c,d*, ♀ LV lat., int.; all ×30 (Kay, 1940).

Falsipollex Kesling & McMillan, 1951 [**F. altituberculatus*]. Lobation as in *Hollinella*, L_1 and L_4 vertically elongate, low, and inconspicuous in many species, L_2 a node set below hinge line, L_3 large and bulbous; S_1 distinct but shallow, S_2 deep, well defined; female with frill, male with 2 velar spurs in each valve (Kesling, 1951). *M.Dev.*, N.Am.——Fig. 68,3n,o. **F. altituberculatus*, Hamilton, USA(Mich.); *3n*, ♀ RV lat.; *3o*, ♂ LV lat.; ×50 (Kesling & McMillan, 1951).——Fig. 68,3a-e. **F. valgus* Kesling, Hamilton, USA (Mich.); *3a,b*, ♀ LV lat., int.; *3c-e*, ♂ carapace L, dors., vent.; all ×30 (Kesling, 1952).——Fig. 68,3f,g. *F. equipapillatus* Kesling & Weiss, Hamilton, USA(Mich.); *3f,g*, ♀ RV lat, ♂ RV lat, ×30 (Kesling & Weiss, 1953).——Fig. 68,3h-m. *F. laxivelatus* Kesling, USA(Mich.); *3h-j*, ♀ carapace R, vent., ant., ×30; *3k-m*, ♂ carapace L, vent., ant.; ×30 (Kesling, 1951).

Flaccivelum Kesling & Peterson, 1958 [**Winchellatia teleutaea* Kesling & Tabor, 1952]. Each valve with distinctive lobation, L_2 inconspicuous and partly fused with L_1, L_3 swollen but not bulbous, and posterior region low and convex; S_2, the only prominent sulcus, long and sinuous, deeper in dorsal part but reaching to velar structure in some species; female with a broad incurved frill continuous with lateral surface, in no way demarked at its proximal junction with rest of valve; male with small velar ridge, scarcely more than a crest, lying along bend between lateral and marginal surfaces. Surface smooth to finely granulose in known species. *M.Dev.*, N.Am.——Fig. 69,1. **F. teleutaea* (Kesling & Tabor), USA (Mich.); *1a,b*, ♂ RV lat., vent.; *1c,d*, ♀ RV lat., vent.; all ×30 (Kesling & Tabor, 1952).——Fig. 69,1e-h. *F. informis* (Ulrich), USA(Ind., Falls of Ohio); *1e,f*, ♂ RV lat., ant., ×30; *1g,h*, ♀ LV lat., ant., ×30 (Kesling & Peterson, 1958).

Grammolomatella Jaanusson, 1957 [**Biflabellum vestrogothicum* Henningsmoen, 1948]. Unisulcate, S_2 long, geniculate, sigmoidal, extending to frill in female valves; L_3 not delineated; L_1 and L_2 fused to form an anterior lobe; female with restricted, radially striate frill; male with 2 spurs in each valve, which in some species are pointed and connected by a narrow flange, but in others terminate bluntly. *M.Ord.-Sil.*, Eu.(Scand.-Eng.-Aus.).——Fig. 69,2. **G. vestrogothicum* (Henningsmoen), M.Ord., Swed.; *2a,b*, ♂ and ♀ LV lat., ×30 (Jaanusson, 1957).

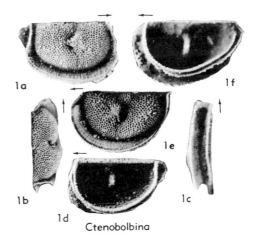

Fig. 67. Hollinidae (p. Q135).

Hanaites Pokorný, 1950 [**Halliella (Hanaites) givetiana*] [=*Proplectrum* Kesling & McMillan, 1951]. Valves showing L_1 as distinct lobe, L_2 obscure or lacking, and L_3 as prominent small node or knob; S_2 broad, with slight swelling in posterior part; velate ridge long and curved; anterior or anteroventral part of each valve bearing palmate or spurlike projection; known species with reticulate surface. Dimorphism in shape of velar ridge (Stover, 1956). *M.Dev.*, N.Am.-Eu.——Fig. 69, 3. *H. platus* (Kesling & McMillan), Hamilton, USA(Mich.) (type species of *Proplectrum*); *3a,b*, ♀ carapace L, vent., *3c*, ♂ carapace L; all ×25 (Stover, 1956); *3d*, RV lat. (immature), ×50 (Kesling).

Hollinella Coryell, 1928 [**H. dentata*] [=*Basslerina* Moore, 1929; *Hollites* Coryell & Sample, 1932]. Lobes consisting of low gently arched L_1 confluent in most species with ventral lobe, L_2 distinctly nodelike (in some forms partly confluent with L_1) and set below dorsal border, L_3 large and bulbous, L_4 ill defined, ventral lobe (prominent in many species) connecting L_1 and L_4 and located between frill and ventral end of wide S_2; female with somewhat incurved long frill extending from anterior corner of valves to posteroventral part; male with outward-flaring frill that in most forms is narrower than frill of female. Dimorphism distinct to very indistinct. *M. Dev.-M.Perm.*, N.Am.-Eu.——Fig. 69,4a,b. **H. dentata*, Penn., N.Am.; *4a,b*, ♂ RV lat., ♀ LV lat., ×30 (Kesling, 1957).——Fig. 69,4c-g. *H. oklahomaensis* (Harlton), Penn., USA(Okla.); *4c-e*, ♂ carapace R, vent., post.; *4f,g*, ♀ carapace R, post.; all ×30 (Kesling, 1957).

?Janischewskya Batalina, 1924 [**J. digitata*]. Poorly known; type species with long S_2 and spinose frill. *L.Carb.*, Eu.——Fig. 66,6. **J. digitata*, USSR.; LV lat., ×? (Swartz, 1936).

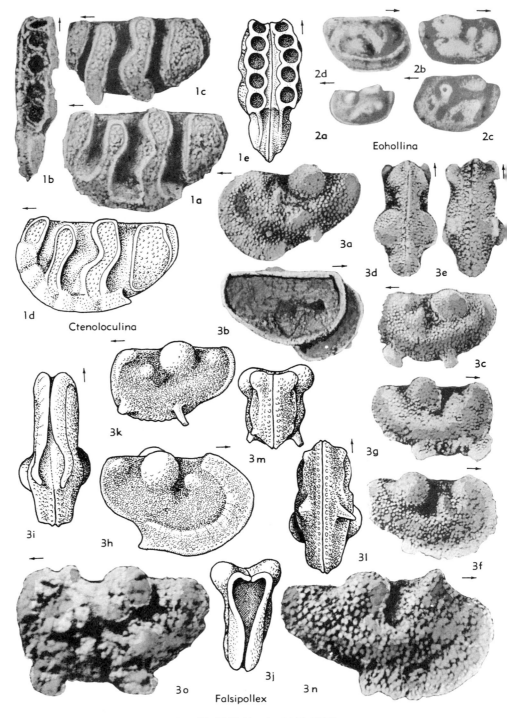

FIG. 68. Hollinidae (p. Q135-Q137).

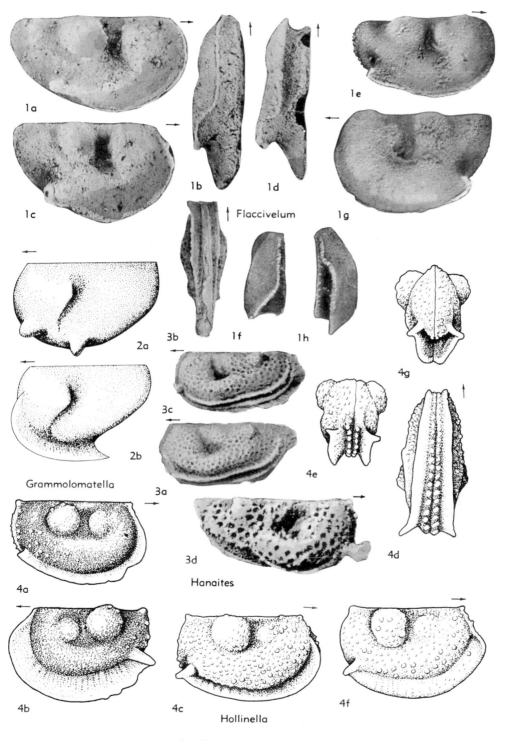

1a 1b 1d 1e 1g 1c

↑ Flaccivelum

3b 1f 1h

2a

2b

Grammolomatella 3a

3c

3d

Hanaites

4a

4b 4c 4f

Hollinella

4e 4g 4d

Fig. 69. Hollinidae (p. Q137).

Parabolbina SWARTZ, 1936 [**Ctenobolbina granosa* ULRICH, 1900]. Only distinguishable sulcus consisting of prominent S_2, which extends from hinge about halfway to ventral border; female with incurved, scalloped frill; male with 2 velar spurs on each valve. *M.Sil.-M.Dev.*, N.Am.-Eu.——FIG. 70, *1a*. **P. granosa* (ULRICH), L.Dev., USA(N.Y.); *1a*, ♀ RV lat., ×34 (Ulrich, 1900).—— FIG. 70,*1b,c*. *P. limbata* SWARTZ, L.Dev., USA (Pa.); *1b,c*, ♀ RV lat., ♂ LV lat., ×40 (Swartz, 1936).

Ruptivelum KESLING & WEISS, 1953 [**R. bacculatum*]. Lobation as in *Hollinella;* female with uninterrupted frill; male with frill divided into 2 segments (KESLING, 1953). *M.Dev.*, N.Am.—— FIG. 70,*3*. **R. bacculatum*, Hamilton, USA(Mich.); *3a,b*, ♀ LV lat., vent., ×30; *3c,d*, ♂ LV lat., vent., ×30 (Kesling & Weiss, 1953).

Subligaculum KESLING & MCMILLAN, 1951 [**S. scrobiculatum*]. Lobation resembling that of *Parabolbina* but with sulcate extensions from ventral end of S_2; female with incurved, scalloped frill; male with short anteroventral frill and posteroventral spur (Kesling, 1951). *M.Dev.*, N.Am.—— FIG. 70,*2*. **S. scrobiculatum*, Hamilton, USA (Mich.); *2a,b*, ♀ RV lat., int., ×50; *2c,d*, ♂ LV lat., int., ×50 (Kesling & McMillan, 1951).

?Subtella ZASPELOVA, 1952 [**S. prima*]. Valves with 2 dorsal tubercles or nodes; posterior third of shell flattened; velate structure prominent ventrally but reduced toward cardinal angles. Surface smooth. *U.Dev.*, Eu.(Russia).——FIG. 71,*1a,b*. **S. prima; 1a,b*, LV (holotype), lat., dors., ×50 (Zaspelova, 1952).——FIG. 71,*1c,d*. *S. latimarginata* ZASPELOVA; *1c,d*, RV (holotype) lat., dors., ×50 (Zaspelova, 1952). [BERDAN.]

Tetrasacculus STEWART, 1936 [**T. bilobus*] [=*Pterocodella, Workmanella* CRONEIS & GALE, 1938]. Valves bilobed, with long curved S_2; female with posterior loculus incomplete in some Devonian species; male with small projection on ventral end of front lobe or bearing small velar ridge. *M.Dev.-U.Miss.*, N.Am., *L.Carb.*, USSR. FIG. 70,*4a-c*. **T. bilobus*, M.Dev.(Hamilton), USA (Mich.-Ohio); *4a,b*, ♀ carapace R, vent., ×50; *4c*, ♂ carapace R (Kesling & McMillan, 1951).—— FIG. 70,*4d-h*. *T. mirabilis* (CRONEIS & GALE), Miss., Ill.; *4d,e*, ♂ carapace L, vent., ×50; *4f-h*, ♀ carapace L, vent., ant., ×50 (Kesling, 1951).

Triemilomatella JAANUSSON & MARTINSSON, 1956 [**T. prisca*]. Unisulcate, S_2 deep and extending at least down to midheight; L_1 and L_2 confluent, L_3 not delimited from L_4, ventral lobe joining anterior with posterior lobes; surface papillose. Frill of each female valve terminating posteroventrally in a spine, separated from marginal or submarginal structure by a broad channel marked by several shallow depressions resembling loculi; frill of male like that of female but separated into 2 parts, a short anteroventral and long ventral, without

a spine, separated from marginal or submarginal loculi-like pits on marginal surface. *M.Sil.*, NW.Eu. (Gotl.-Eng.).——FIG. 70,*5*. **T. prisca*, M.Sil., Gotl.; *5a-c*, ♂ carapace L, vent., ant.; *5d-f*, ♀ carapace L, vent., ant.; all ×45 (Jaanusson & Martinsson, 1956).

Family BASSLERATIIDAE E.A. Schmidt, 1941

[*nom. transl.* LEVINSON, herein (*ex* Bassleratinae SCHMIDT, 1941] [Materials for this family prepared mainly by S. A. LEVINSON, Humble Oil & Refining Company, and R. C. MOORE, University of Kansas]

Small, subovate to subrectangular with one or more marginal ridges and anterocentral node; other ridges and nodes commonly within marginal ridges. *?L.Ord.*, *M.Ord.*

Bassleratia KAY, 1934 [**B. typa*] [=?*Laddella* SPIVEY, 1939]. Prominent 2 marginal ridges and anterocentral node, outer ridge parallel to free border, inner ridge more elevated; central vertical ridge commonly present. *M.Ord.*, N.Am.——FIG. 72,*1*. **B. typa*, Can.(Ont.); *1a-d*, RV (holotype) lat., dors., vent., ant., ×45 (Bradfield, 1935). [*Hilseweckella* HARRIS, 1957 (**H. rugulosa*) is added as ?junior synonym of *Bassleratia* on basis of study by H. W. SCOTT (1960) of type specimen. ——FIG. 134A,*2*. *?B. rugulosa* (HARRIS), M.Ord., USA(Okla.); *2a,b*, LV lat., ant., ×35 (161).— MOORE.]

Bellornatia KAY, 1934 [**B. tricollis*]. Single marginal ridge parallel to dorsal and free borders, inner oval ridge located in posterodorsal part of valve, 3 nodes inclosed by inner ridge. *M.Ord.*, N.Am.——FIG. 72,*2*. **B. tricollis*, USA(Iowa-Minn.); *2a*, RV (holotype) lat., ×60 (Bradfield, 1935); *2b-d*, LV lat., RV lat., vent., ×50 (J. M. Cornell, n) (*2a*, Iowa; *2b-d*, Minn.).

Lennukella JAANUSSON, 1957 [**Drepanella europaea* ÖPIK, 1937]. Straight-backed, equivalved, unisulcate, with large preadductorial knob behind S_1; dorsal ridge strong, continued forward as ornamental ridge in front of knob; prominent ventral and anteroventral carinal ridge; velar structure developed as narrow ridge that in side view conceals subvelar field. *M.Ord.(Llandeil.-Caradoc.)*, NW. Eu.(Est.-Swed.).——FIG. 72A,*5*. **L. europaea* (ÖPIK), Swed.; *5a,b*, RV lat., carapace vent., ×30 (36).

Raymondatia KAY, 1934 [**R. goniglypta*]. Small, with single marginal ridge parallel to dorsal and free borders; prominent inner posterior ridge extending vertically from middle third of posterodorsal border; at mid-height of valve turning abruptly backward at a right or slightly obtuse angle; anterocentral node commonly inflated; reaching dorsal border. *M.Ord.*, N.Am.——FIG. 72,*3*. **R. goniglypta*, USA(Iowa-Minn.); *3a,b*, RV (holotype) lat., dors., ×60; *3c-e*, LV lat.,

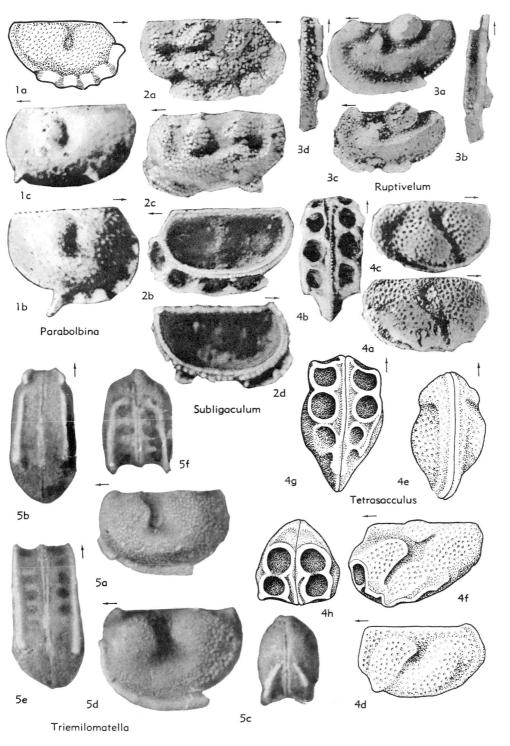

Ruptivelum

Parabolbina

Subligaculum

Tetrasacculus

Triemilomatella

FIG. 70. Hollinidae (p. Q140).

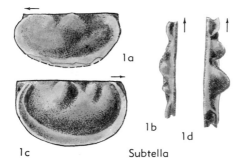

FIG. 71. Hollinidae (p. Q140).

RV lat., int., ×50 (J.R. Cornell, n) (*3a,b*, Iowa; *3c-e*, Decorah Sh., S.Minn.).

?**Steusloffia** ULRICH & BASSLER, 1908 [**Strepula linnarssoni* KRAUSE, 1889; SD ULRICH & BASSLER, 1923]. Straight-backed, subovate in outline, with single long, broad sulcus (S_2) located behind large knob (L_2) in anterodorsal region and with distinct posteroventral lobe (L_3); adults bearing narrow crests on and in front of presulcal lobe, and on and behind posteroventral lobe; velar structure moderately broad, becoming ridgelike or obsolete before reaching posterior cardinal corner; dimorphism known. *L.Ord.-M.Ord.*, NW.Eu.——FIG. 72A, *1.* **S. linnarssoni* (KRAUSE), ?M.Ord.(boulder in glacial drift), N.Ger.; LV lat.(reconstr.), ×35(36).——FIG. 72A,*2. S. costata* (LINNARSSON), M.Ord., Swed.; LV lat.(reconstr.), ×35(36).——FIG. 72A, *3. S. multimarginata* ÖPIK, M.Ord., Est.; LV lat. (reconstr.), ×35(36).——FIG. 72A,*4. S. rigida* ÖPIK, M.Ord., Est.; LV lat., ×20(58).

Thomasatia KAY, 1934 [**T. falcicosta*]. Like *Bassleratia* in having 2 marginal ridges and antero-central node but with outer ridge distinctly developed only at anterior and ventral borders, obscure at rear; inner ridge parallel to anterior border and commonly bifurcating in ventrocentral area, may curve dorsally in anterocentral area with development of arcuate ridge from posterior cardinal angle ending abruptly near outer marginal ridge at ventral border. *M.Ord.*, N.Am.——FIG. 72,4a-f. **T. falcicosta*, Can.(Ont.) USA(Minn.); *4a-d*, RV (holotype) lat., dors., vent., ant., ×40; *4e-f*, LV lat., int., ×50 (J. R. Cornell, n) (*4a-d*, Ont.; *4e,f*, Decorah Sh., S. Minn.).——FIG. 72, *4g-l. T.* sp., Edinburg F., USA(Va.); *4g-k*, RV lat., LV lat., int., dors., vent., ×30; *4l*, LV hinge, ×60 (J.C. Kraft, n).

Family CHILOBOLBINIDAE Jaanusson, 1957

[*nom. transl.* LEVINSON, herein (*ex* Chilobolbininae JAANUSSON, 1957] [Materials for this family prepared by S. A. LEVINSON, Humble Oil & Refining Company, and R. C. MOORE, University of Kansas, with additions by IVAR HESSLAND, University of Stockholm]

Carapace straight-hinged, moderately and rather evenly convex, with centrally located pit or short sulcus; velate frill well developed; dimorphism distinct, females with prominent ventral pouch. *M.Ord.-M.Sil.*

Chilobolbina ULRICH & BASSLER, 1923 [**Primitia dentifera* BONNEMA, 1909] [=*Chilobolba* BONNEMA, 1938]. Wide striate frill; very short median furrow or pit; females with long ovate brood pouch located mid-ventrally or anteroventrally. *M.Ord.-M.Sil.*, N.Am.-Eu.——FIG. 73,2. *C. hartfordensis* ULRICH & BASSLER, M.Sil., USA(Md.); LV lat. ×? (Levinson, 1951).——FIGS. 73,1, 74. *C.* sp., M.Ord.(Edinburg F.), USA(Va.); 73,1a-d, LV lat., int., dors., vent., ×20; 73,1e, RV int., ×20; 74, LV transv. sec., ×40 (Kraft, n).——FIG. 75,1. *C. dentifera* (BONNEMA), ?U.-M.Sil., USA (Md.); *1a*, ♂ RV lat.; *1b-d*, ♀ RV lat., int., vent.; all ×20 (Kesling, *et al.*, 1958).

Cystomatochilina JAANUSSON, 1957 [**Primitia (Ulrichia?) umbonata* KRAUSE, 1892]. Nonsulcate or with small shallow sulcal pit at or below mid-height, predominant knoblike presulcal node; dimorphic, with very wide frill extending along entire free margin, concave anteriorly and ventrally or convex anteroventrally and ventrally. *U. Ord., ?L.Sil.*, Eu.(Baltoscandia).——FIG. 76,1. **C. umbonata* (KRAUSE), U.Ord.(Ashgill.) erratics from S. Bothnian area, Swed.; *1a*, LV lat., (heteromorph with partly convex frill, reconstr.), ×30; *1b*, LV lat. (tecnomorph, reconstr.), ×35 (36). [HESSLAND.]

?Family EURYCHILINIDAE Ulrich & Bassler, 1923

[*nom. transl.* HENNINGSMOEN, 1953 (*ex* Eurychilininae ULRICH & BASSLER, 1923)] [=Euprimitiinae HESSLAND, 1949] [Materials for this family prepared by S. A. LEVINSON, Humble Oil & Refining Company, and R. C. MOORE, University of Kansas, with additions by others as recorded]

Straight-hinged, unisulcate or with pit, marginal frill or velate structure, commonly showing dimorphic variations but females lacking domiciliar pouch. *L.Ord.-U. Dev.*

Eurychilina ULRICH, 1889 [**E. reticulata*] [=*Actinochilina* JAANUSSON, 1957]. Hinge long; S_2 wide and deep, anterior edge raised to form a node; frill wide, curved, and radiate, in females of some species sausage-shaped in section. *L.Ord.-L.Sil.*, N.Am.-Eu.——FIG. 77,3. **E. reticulata*, M.Ord. (Trenton.), USA(Minn.); *3a*, LV lat., ×18 (Jones & Brady, 1874); *3b,c*, ♀ LV lat., ♀ LV int., ×17, ×20 (J.R. Cornell, n).——FIG. 77,2. *E. sp.* M.Ord.(Edinburg F.), USA(Va.); *2a,b*, ♀ LV lat., ♀ RV int., ×20 (J.C. Kraft, n).——FIG. 77, *1. E.* sp., M.Ord. (Edinburg F.), USA(Va.); *1a,b*, ♀ LV lat., ♂LV lat., ×20; *1c*, ♀ LV transv. sec. through sulcus showing flange on inner side of frill, ×45 (J.C. Kraft, n).

Apatochilina ULRICH & BASSLER, 1923 [**Eurychilina*

obesa ULRICH, 1890]. Like *Coelochilina,* with narrow radiate frill, S_2 faint near dorsal border. *M.Ord.,* E.N.Am.-Eu.——FIG. 77,4. *A. obesa* (ULRICH), M.Ord., USA(Ky.); ?LV lat., ×18 (Kesling, 1953).——FIG. 78,1. *A.* sp., M.Ord.

(Edinburg F.), USA(Va.); *1a-d,* RV lat., RV lat., RV int., LV lat., ×20; *1e,* RV transv. sec. through midpoint, ×45 (J.C. Kraft, n).

Bicornellina ZASPELOVA, 1952 [*B. bolchovitinovae*]. Valves with prominent upwardly directed spine

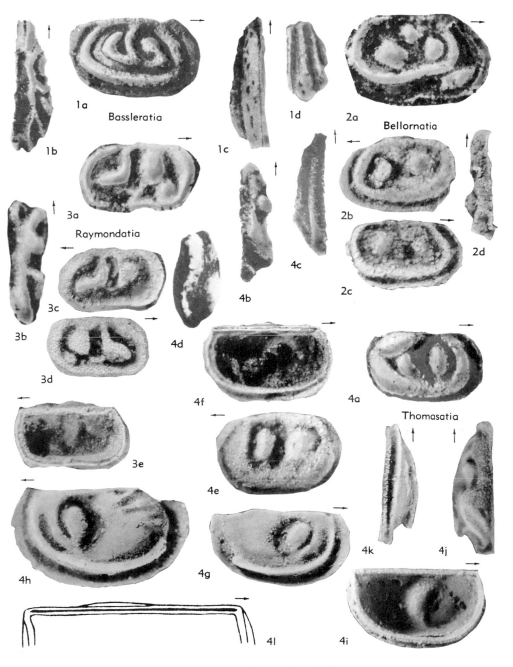

FIG. 72. Bassleratiidae (p. Q140-Q142).

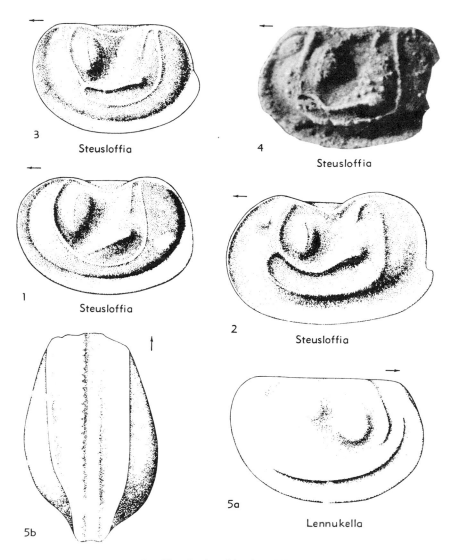

FIG. 72A. Bassleratiidae (p. Q140-Q142).

in antero- and posterodorsal regions, united to frill or disconnected; frill complete, striated; no anterodorsal node. Muscle scar obscure on exterior but forming distinct node on interior of valve. Surface smooth. *U.Dev.*, Eu.(Russia).——FIG. 77, 5. *B. bolchovitinovae; 5a,b,* LV (holotype) lat., dors.; *5c,* RV (paratype) lat., all ×70 (ZASPELOVA, 1952). [BERDAN.]

Coelochilina ULRICH & BASSLER, 1923 [*Eurychilina aequalis* ULRICH, 1890]. S_2 narrow and shallow, node absent, frill sausage-shaped in section. *M. Ord.-L.Sil.*, N.Am.-Eu.——FIG. 78,3. *C. aequalis* (ULRICH), M.Ord., USA(Ky.); ?LV lat., ×18 (Kesling, 1953).

Euprimites HESSLAND, 1949 [*E. recticulogranu-lata*]. Like *Euprimitia* but with horseshoe-shaped ridge enclosing ventral part of sulcus. *L.Ord.-M. Ord.*, Eu.——FIG. 78,2. *E. recticulogranulata,* L. Ord., Swed.; *2a-c,* RV (holotype) lat., vent., ant., ×30 (Hessland, 1949).

Euprimitia ULRICH & BASSLER, 1923 [*Primitia sanctipauli* ULRICH, 1894]. S_2 straight, narrow, deepest at ventral end; with presulcate node and narrow frill or velate ridge; surface reticulate. *L. Ord.-L.Sil.*, N.Am.-Eu.——FIG. 78,4. *E. sancti-pauli* (ULRICH), M.Ord. (Decorah sh.), USA (Minn.); *4a,b,* RV lat., RV lat., ×20, ×50 (*4a,* Ulrich, 1894; *4b,* J.R. Cornell, n).——FIG. 78,9. *E. labiosa* ULRICH, M.Ord.(Edinburg F.), USA(Va.); *9a-c,* LV lat., RV lat., RV int., ×40; *9d-e,* RV

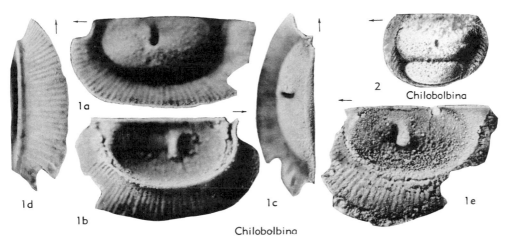

FIG. 73. Chilobolbinidae (p. Q142).

transv. sec. through sulcus, RV hinge, ×60 (J.C. Kraft, n).

Laccochilina HESSLAND, 1949 [*Eurychilina estonula* ÖPIK, 1935] [=*Eobromidella* HARRIS, 1957; *Prochilina* JAANUSSON, 1957]. Like *Eurychilina* but S_2 a pit and presulcate node prominent; frill sausage-shaped. *L.Ord.-M.Ord.,* Eu.-E.N.Am.——FIG. 78, *11*. *L. dorsoplicata* HESSLAND, L.Ord., Swed.; RV (holotype) lat., ×30 (Hessland, 1949).

Laccoprimitia ULRICH & BASSLER, 1923 [*Primitia centralis* ULRICH, 1890]. S_2 round and pitlike, usually located centrally, narrow depressed zone occurring along free margin or with narrow flange; presulcate node may be present. *L.Ord.-M.Sil.,* E. N.Am.-Eu.——FIG. 78,*8*. *L. centralis* (ULRICH), U.Ord., USA(Ky.); LV lat., ×14 (Hessland, 1949).——FIG. 78,*7*. *L. ventroturgida* HESSLAND, L.Ord., Swed.; *7a-c*, RV (holotype) lat., dors., ant., ×30 (Hessland, 1949).

Platybolbina HENNINGSMOEN, 1953 [*nom. subst.* pro *Platychilina* THORSLAND, 1940 (*non* KOKEN, 1892)] [*Primitia distans* KRAUSE, 1889]. Like *Chilobolbina* except S_2 very faint to absent, with prominent subcentral muscle spot. *L.Ord.-U.Ord.,* Eu.-E.N.Am.——FIG. 78,*10*. *P. tiara* HESSLAND, U.Ord., Norway; ♀ LV lat., ×24 (ÖPIK, 1937). ——FIG. 78,*6*. *P. umbonata* (KRAUSE), U.Ord., Norway; LV lat. (presulcal node broken off), ×22 (Henningsmoen, 1953).

?**Tsitrella** SARV, 1959 [*T. lamina*]. Preplete, lengthened, ventral margin mainly parallel to dorsal, flattened, equivalved; nonlobate, with very short dorsal sulcus; low adventral ridge developed in some shells; surface reticulate. Dimorphism not reported. *L.Ord.-M.Ord.,* NW.Eu.——FIG. 78, *5*. *T. lamina*, L.Ord., Est.; RV (holotype) lat., ×37 (Sarv, 1959). [HESSLAND.]

Family PIRETELLIDAE Öpik, 1937

[=Oepikiumidae JAANUSSON, 1957] [Materials for this family prepared by IVAR HESSLAND, University of Stockholm]

Amplete or preplete, cardinal angles well defined, S_2 permanently distinct (other sulci may be indicated), L_2 in most genera developed as prominent presulcate knob (other lobes generally not present), dorsal plica and lobal crests present in most genera, generally C_1 and C_3, which may be united ventrally and, dorsally, with dorsal plica; as a rule scattered tubercles on surface, in some species seemingly forming indistinct linear patterns; dimorphic, velar ridge or flange in

Chilobolbina

FIG. 74. Chilobolbinidae (p. Q142).

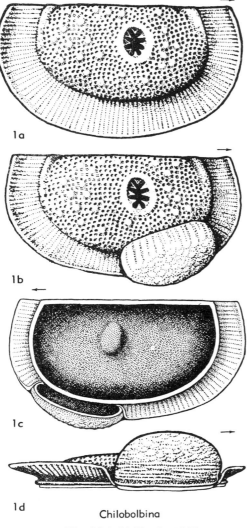

1a

1b

1c

1d   Chilobolbina

Fig. 75. Chilobolbinidae (p. Q142).

tecnomorphs and incurved frill forming anteroventral closed or almost closed chamber in heteromorphs (frill may be continued posteriorly by isolated spines). *L.Ord.-U.Ord.*

Piretella Öpik, 1937 [*P. acmaea] [=*Duhmbergia* SCHMIDT, 1941 (holotype seemingly late instar of *Piretella*)]. Unisulcate, L_2 forming a large semiglobal swelling; C_1-C_3 form approximately a semicircle, ends united dorsally in adult and late larval specimens by crest or row of tubercles generally continuing backward (=dorsal plica), free end declining adventrally; velar structure wide,

entire or extending along anterior border and mainly anterior half of ventral border, incurved type composed of large spines with rounded ends, distinctly separated from each other, continued posteriorly by similar spines at greater distance. *M.Ord.-U.Ord.,* NW.Eu.(Baltoscandia).——FIG. 79,1. *P. acmaea,* U.Ord., Est.; *1a,* tecnomorph RV (holotype) lat., ×20; *1b-e,* heteromorph carapace, R, dors., vent., ant., ×20 (58).——FIG. 81,1. *P. margaritata* Öpik, U.Ord., Est.; *1a,b,* tecnomorph and heteromorph LV lat., ×25 (Kesling, 1951).

Bolbina HENNINGSMOEN, 1953 [*Bollia ornata* KRAUSE, 1896]. Straight-backed, subelliptical, ends rather evenly rounded; S_2 well marked, medium in length, L_2 forming swollen area joined ventrally to posteroventral lobe, which may terminate in conical spurlike process (possibly dimorphic); velate structure short, ridgelike. *Ord.* (from glacial drift), Ger. [HESSLAND refers to *Piretellidae*.]

Bromidella HARRIS, 1931 [*B. reticulata*]. Sinuous sulcus in the dorsal part of each valve, passing around the upper ends of L_2 and L_3, separating an attenuated M-shaped dorsal ridge from the rest of the valve. Surface spinose in type species. Female with frill incurved to form a false pouch. *M.Ord.,* N.Am.——FIG. 80,*1g,h.* *B. reticulata,* M.Ord., Okla.; *1g,h,* ♀ RV lat., int., ×25 (Swartz, 1936).——FIG. 80,*1a,b.* *B. rhomboides*

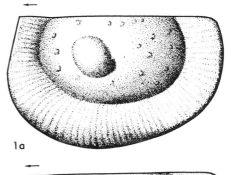

1a

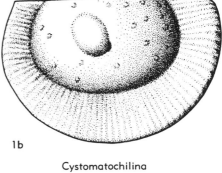

1b

Cystomatochilina

Fig. 76. Chilobolbinidae (p. Q142).

KAY, M.Ord., USA(Minn.); *1a,b*, ♂ LV lat., ♀ RV int., ×50 (J.R. Cornell).——FIG. 80,*1c-f*. *B. depressa* KAY, M.Ord., USA(Minn.); *1c,d*, ♂

LV lat., RV lat., ×50; *1e,f*, ♀ RV lat., int., ×50 (J.R. Cornell).

?**Hesperidella** ÖPIK, 1937 [**P. esthonica* BONNEMA,

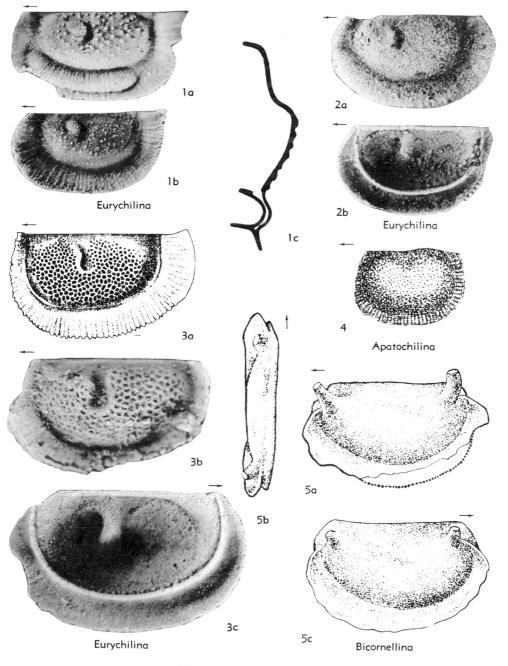

1a

1b

Eurychilina

1c

2a

2b

Eurychilina

3a

4

Apatochilina

3b

5b

5a

3c

5c

Eurychilina

Bicornellina

FIG. 77. Eurychilinidae (p. Q142-Q144).

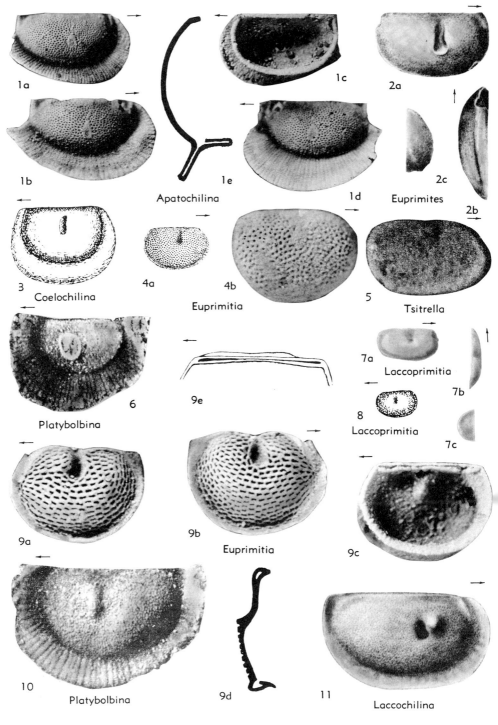

1a

1b

Apatochilina

1c

1e

1d

2a

2c

Euprimites

2b

3

Coelochilina

4a

4b

Euprimitia

5

Tsitrella

6

Platybolbina

9e

7a

Laccoprimitia

8

Laccoprimitia

7b

7c

9a

9b

Euprimitia

9c

10

Platybolbina

9d

11

Laccochilina

Fig. 78. Eurychilinidae (p. Q143-Q145).

1909]. Unisulcate, sulcus deep and crescent-shaped; L_1 developed as large rounded node; sulcus and node surrounded by crest $(?C_1-C_3)$, with part of dorsal crest (plica) forming approximate circle, which may be incomplete if dorsal section of $?C_1$ is lacking; scattered tubercles tending locally toward linear arrangement; velar structure confluent with dorsal plica; dimorphic, velar flange straight and moderately wide in tecnomorphs, widened anteroventrally in heteromorphs, slightly convex, undulating radially; subvelar field slightly concave. *M.Ord.* NW.Eu.(Baltoscandia).——Fig. 79,2. **H. esthonica* (BONNEMA), Est.; *2a,* RV tecnomorph, lat., ×40 (Öpik, 1937); *2b,* RV heteromorph, lat. (Swed.), ×36 (Jaanusson, 1957).

Oepikium AGNEW, 1942 [**Biflabellum tenerum* ÖPIK, 1935] [=*Biflabellum* ÖPIK, 1935 (*non* DÖDERLEIN, 1913); *Öpikium* AGNEW, 1942; *Öpikum* HENNINGSMOEN, 1953 (*nom. null.*)]. S_2 prominent but not deep, extending from the dorsal border to the frill. L_2 nodelike, not clearly separated from L_1. Female with frill incurved. Male with broad flat or flaring frill; frill in type species broadest among known ostracodes. *L.Ord.-U.Ord.,* Eu.——Fig. 81,3. **O. tenerum,* Est.; *3a,b,* ♀ and ♂ LV lat. (reconstr.), ×20 (Kesling, n).

Piretia JAANUSSON, 1957 [**P. geniculata*]. Distinctly unisulcate, with poorly developed presulcal node and no ornamental crests or dorsal plica; dimorphic; velar structure not reaching postero-

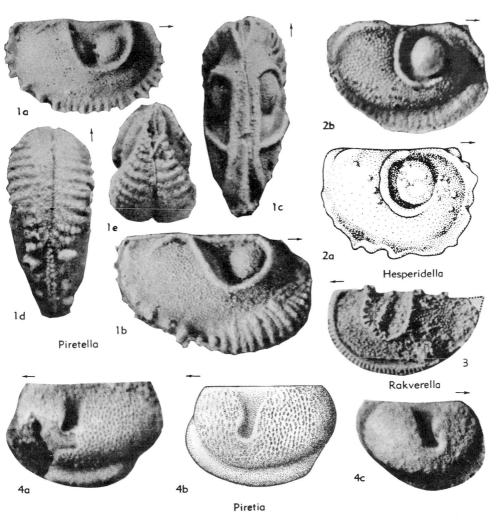

1a

1e

1c

2b

2a

Hesperidella

1d

1b

Piretella

3

Rakverella

4a

4b

4c

Piretia

Fig. 79. Piretellidae (p. *Q*146-*Q*150).

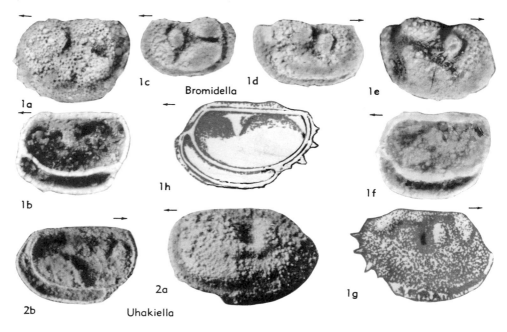

FIG. 80. Piretellidae (p. Q146-Q150).

dorsal corner, forming ridge to slightly concave flange *or* flange that is strongly convex ventrally and anteroventrally. *L.Ord.*, NW.Eu.(Baltoscandia).——FIG. 79,4. *P. geniculata, Swed.; 4a, LV heteromorph (holotype) lat., 4b, same (reconstr.); 4c, RV (tecnomorph) lat.; all ×35 (Jaanusson, 1957).

Rakverella ÖPIK, 1937 [*R. spinosa*]. Preplete, posterodorsal angle acute; dorsal ends of C_1-C_3 generally extending beyond hinge line, dorsal extensions corresponding to C_2 and C_4 may occur, crest sides may be spiny; velar structure dimorphic, extending along anterior margin and generally entire ventral margin, narrow in tecnomorphic specimens, wide and convex in heteromorphs, ends of velar spines may be free and protruding so as to give velar structure a crenulate or spiny appearance; subvelar area not well known. *M.Ord.*, NW.Eu.(Baltoscandia).——FIGS. 79,3; 81,2b. *R. spinosa, Est.; 79,3, LV (holotype) lat., ×20 (Öpik, 1937); 81,2b, same (reconstr.), ×33 (Öpik; Kesling, n).——FIG. 81,2a. R. bonnemai ÖPIK, Est.; RV (holotype) lat. (reconstr.), ×33 (Kesling, n).

Uhakiella ÖPIK, 1937 [*U. coelodesma*]. L_1 a prominent node; L_2 a distinct node ventrally connected to a low ridge leading to L_3; L_3 a prominent lobe, ventrally nearly confluent with a long ventral ridge. Frill of female incurved to form a false pouch. This genus is closely related to *Bromidella* of the same age in N.Am. *M.Ord.*, Eu.——FIG. 80,2a. *U. coelodesma, Est.; ♀ LV lat., ×20 (Öpik, 1937).——FIG. 80,2b. U. kohtlensis ÖPIK, Est.; ♀ LV int., ×20 (Öpik, 1937).

Family QUADRIJUGATORIDAE
Kesling & Hussey, 1953

[*nom. correct.* JAANUSSON, 1957 (*pro* Quadrijugatidae KESLING & HUSSEY, 1953)] [=Ceratopsinae NECKAJA, 1958] [Materials for this family prepared by R. V. KESLING, University of Michigan, with contributions by IVAR HESSLAND, University of Stockholm, and others as recorded]

Carapace nearly equivalved, subquadrate, subelliptical, or subovate in side view, more or less subquadrate in ventral and end views, dorsal border long; each valve quadrilobate, lobes in some genera composed of 4 equal distinct elongate ridges, in others with L_1 partly confluent with L_2 and L_3 with L_4; S_2 invariably present, long; S_1 and S_3 restricted ventrally by confluence of lobes in some genera but otherwise appearing as long sulci; marginal and velate structures invariably present, but variously developed. [In most genera, lobes joined ventrally to ridge in position of histium as in Sigmoopsidae, but without ventral edge to form a histial structure; in the aberrant *Kiesowia*, however, this ridge, like the lobes, dissected and discontinuous, scarcely recognizable in some species. Most genera nondimorphic but a few (e.g., *Tallinnella*) with velate structure of females wider anteroventrally than corresponding part of males. As additional species are studied, it is possible that the dimorphic genera may be segregated in another family, but at present dimorphic features of several genera are not well understood. The family is closely related to the Sigmoopsidae, differing mainly in lack of histium and associated dimorphism.] *L. Ord.-U.Ord.*

Quadrijugator KESLING & HUSSEY, 1953 [*Bollia

permarginata FOERSTE, 1917]. Quadrilobate, with 4 nearly equal strongly elevated ridges, some or all joined to a ventral ridge, valves with low velate and marginal ridges close together. RV with hinge consisting of ridge highest at ends and indistinct at center; LV hinge with long groove, deepest at ends and shallow medially, with long shallow groove in central part. *U.Ord.*, N.Am.—— FIG. 82,*1.* **Q. permarginatus,* USA(Mich.); *1a,b,* carapace R, dors., ×60; *1c,d,* RV lat., vent., ×60 (Kesling & Hussey, 1953).

Ceratopsis ULRICH, 1894 [*Beyrichia chambersi*

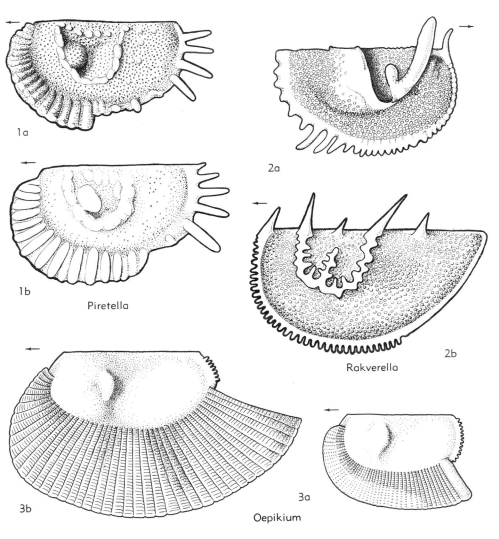

1a

1b
Piretella

2a

2b
Rakverella

3a

3b
Oepikium

FIG. 81. Piretellidae (p. Q146-Q150).

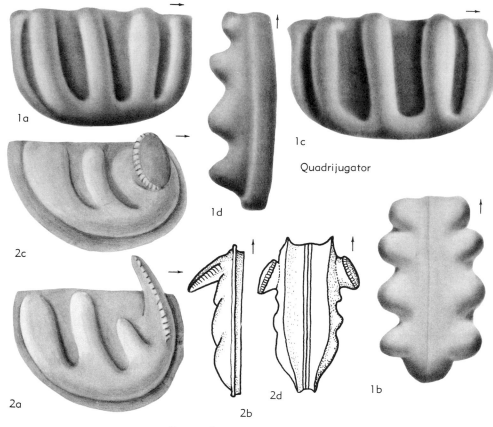

FIG. 82. Quadrijugatoridae (p. Q151-Q152).

MILLER, 1875] [=*Ceratella* ULRICH, 1890 (*non* GRAY, 1869) (*nom. nud.*)]. Valves quadrilobate but lobes unequal, L_1 being large and ornate (in different species extending well above hinge line, produced as long spine, or developed as mushroom-shaped process), L_2 short, not reaching hinge line, L_3 with form of long ridge, and L_4 an elongate lobe or ridge; in most species S_1 narrower than S_2 or S_3, ventral lobe prominent and ridgelike in some species, nearly indistinguishable in others. *M.Ord.-U.Ord.,* N.Am.-Eu.——FIG. 82, *2a,b.* **C. chambersi* (MILLER), M.Ord., USA (Minn.); *2a,* RV lat., ×30; *2b,* vent., ×27.5 (Ulrich, 1894).——FIG. 82,*2c,d.* *C. oculifera* (HALL), U.Ord., USA(Ohio); *2c,d,* carapace R, dors., ×25, ×20 (Kesling, 1951).

Hesslandella HENNINGSMOEN, 1953 [**Ctenentoma macroreticulata* HESSLAND, 1949]. Unisulcate, with long sulcus which may be geniculated, traces of S_1 and S_3 may be seen; presulcal node distinct; adventral structure developed as velar ridge, entire; subvelar area channeled; dimorphism not observed. *L.Ord.,* Eu.——FIG. 85,*1.* **H. macroreticulata,* Swed.; *1a,* LV lat.; *1b,* RV (holotype) ant., both ×30 (Hessland, 1949). [HESSLAND.]

Kiesowia ULRICH & BASSLER, 1908 [**Beyrichia dissecta* KRAUSE, 1892]. Quadrilobate but with each lobe reduced to a node or to 2 or more nodes in a group, boundaries of sulci poorly defined; velar structure present as a ridge resembling that of *Quadrijugator* or a broad frill as in some *Tetradella* species; ventral region dissected into nodelike areas rather than forming a connecting ridge as in other genera. *M.Ord.-L.Sil.,* Eu.—— FIG. 83,*1a.* **K. dissecta* (KRAUSE), M.Ord., Ger. (drift); RV lat., ×15 (Kesling, 1951).——FIG. 83,*1b.* *K. radians* (KRAUSE), M.Ord., Ger. (drift); RV lat., ×17 (Kesling, 1951).

Protallinnella JAANUSSON, 1957 [**Beyrichia grewingki* BOCK, 1867]. Seemingly equivalved; quadri-

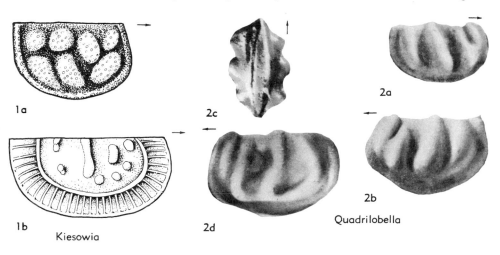

1a
2c
2a
1b   Kiesowia   2d   2b   Quadrilobella

FIG. 83. Quadrijugatoridae (p. Q152-Q153).

lobate, with lobes extending to or beyond dorsal margin, except L_2; velar structure entire, in some specimens slightly convex in anterior and anteroventral parts and developed as wide, thin, fragile flange; dimorphism likely but not proved; subvelar field broad, except posteriorly and anterodorsally. *L.Ord.*, Eu.(Baltoscandia).——FIG. 84,3. **P. grewingki* (BOCK), Swed.; *3a*, RV (matrix incompletely removed), lat., ×45; *3b*, RV (slightly broken), lat., ×45 (Hessland, 1949). [HESSLAND.]

Pseudorakverella SARV, 1959 [**P. optata*]. Amplete or slightly preplete, distinctly lobate, with large L_1, L_3 and L_4 covering much of lateral surface, sharp-ridged and with wide bases, L_2 small and developed as dorsoventrally elongate node, L_1 and L_4 united by connecting lobe; adventral structure a narrow ridge running along entire free margin. Dimorphism not observed. Surface smooth. *M.Ord.*, NW.Eu.——FIG. 84,2. **P. optata*, Est.; *2a-c*, RV (holotype) lat., int. dors., ×20 (Sarv, 1959). [HESSLAND.]

Quadrilobella IVANOVA, 1955 [**Q. recta*]. Very similar to *Quadrijugator*, possibly synonymous, differing only in having velar ridge much farther from marginal ridge and in having L_2 closely associated with L_1, so that S_1 is much narrower than other sulci and in some species very shallow ventrally (in this respect resembling the arrangement of L_1 and L_2 in *Ceratopsis*). *L.Ord.*, USSR.——FIG. 83,*2a-c*. **Q. recta* IVANOVA, W.Sib.; *2a,b*, RV lat., LV lat.; *2c*, carapace vent.; all ×15 (Ivanova).——FIG. 83,*2d*. *Q. arpilobata* IVANOVA, W. Sib.; LV lat., ×15 (Ivanova). [=*Tetradellina* HARRIS, 1957 (*fide* R. V. KESLING).]

Rigidella ÖPIK, 1937 [**Steusloffia mitis* ÖPIK, 1935]. Quadrilobate, lobes crested, united ventrally by connecting crest, L_2 shorter than other

lobes but prominent; S_1 and S_3 generally developed as semisulci; dorsal plica may be present; probably dimorphic, velate flange entire, moderately wide, widest anteroventrally, concave in some specimens but in others incurved anteriorly and anteroventrally; subvelate field comparatively high. *L.Ord.* N.Eu.(Baltoscandia).——FIG. 85,2. **R. mitis*, Est.; *2a,b*, ♂ LV lat., RV lat., ×43 (*2a*, Hessland; *2b*, Jaanusson, 1957). [HESSLAND.]

Tallinnella ÖPIK, 1937 [**T. dimorpha*]. Quadrilobate, lobe distinct, extending to dorsal border or above, except L_2 which is shorter and united with connecting lobe or developed as isolated knob; L_1 swollen ventrally, L_1, L_2, and L_3 ridgelike, extending to dorsal border or above; adventral structure forming thick velate ridge or velum that is widest anteroventrally; dimorphism observed in type species only, indicated by broader anteroventral part of velum in heteromorphs; marginal structure tuberculate. *L.Ord.-U.Ord.*, Eu.——FIG. 85,3. **T. dimorpha*, M.Ord., Est.; *3a,b*, tecnomorph LV lat., int., ×20 (317); *3c,d*, heteromorph LV lat., int., ×20 (58); *3e*, RV vent. (Swed.), ×15 (36). [HESSLAND.]

Tallinnellina JAANUSSON, 1957 [*Tetradella teres HESSLAND, 1949]. Quadrilobate, with lobes extending to or even somewhat above dorsal margin, except L_2, which is comparatively long but not reaching dorsal margin; dorsal plica may be developed; velar structure wide and frill-like, running along entire free margin, concave or convex anteriorly and anteroventrally; probably dimorphic; entire subvelate field broad. *L.Ord.*, Eu.(Baltoscandia).——FIG. 84,*1c*. **T. teres*, Llanvirn., Swed.; RV (holotype), lat., ×50 (30).——FIG. 84,*1a,b*. *T. lanceolata* (HESSLAND), Swed.; *1a,b*, LV (holotype) lat., ×48, ×35 (*1a*, 30; *1b*, 36). [HESSLAND.]

Tallinnopsis Sarv, 1959 [*Tetradella calkeri Bonnema, 1909]. Distinctly quadrilobate, lobes joined
ventrally by connecting lobe, sulci open dorsally;
adventral structure developed as low narrow ridge
separted from connecting lobe by narrow furrow;
dimorphism not observed. *M.Ord.*, N.Eu.——Fig.
86,2. *T. calkeri* (Bonnema), Est.; *2a,b*, RV lat.,
LV lat., ×17, ×32 (*1a*, Bonnema, 1909; *1b*, Sarv,
1959). [Hessland.]

Tetrada Neckaja, 1958 [*Tetradella memorabilis
Neckaja, 1953]. Distinctly quadrilobate and
deeply sulcate, with sulci as a rule closed dorsally; adventral structure narrow, in some shells
indistinct; dimorphism not observed; surface pitted.
M.Ord.-U.Ord., Eu.(NW.Russia).——Fig. 86,3.
T. memorabilis (Neckaja); *3a,b*, LV (holotype)
lat., vent., ×46 (Neckaja, 1953). [Hessland.]

Zygobolboides Spivey, 1939 [*Z. grafensis*]. Dorsal margin straight, ventral broadly rounded; valves
subequal; cardinal angles obtuse, posterior more
rounded than anterior; greatest height slightly
behind mid-length; trilobate, L_2 confluent with L_1,
L_3, and L_4 distinct; terminal lobes (L_1, L_4) subparallel to free margin and usually connected
ventrally by narrow ridge, medial lobe (L_3)
gently arcuate and connected to ventral ridge;
lobes approximately equal to height of valve;
S_2 and S_3 prominent, open dorsally, usually closed
ventrally by ventral ridge, which internally is reflected poorly or not at all but otherwise interior
shows 3 prominent sulci and 2 ridges that reflect
external features; dimorphism probably shown by
inflation of posterior lobes, but not now established.
U.Ord., N.Am.——Fig. 86,1. *Z. grafensis*, U.Ord.,

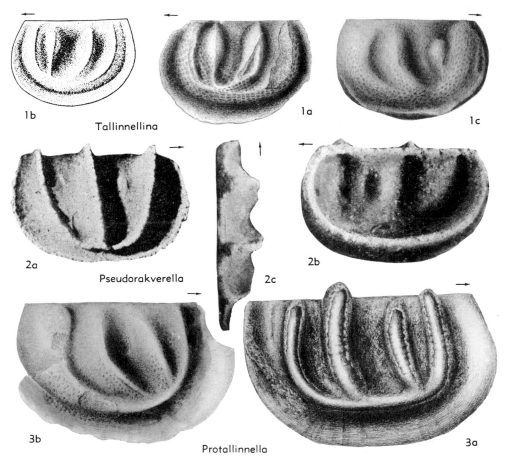

1b   Tallinnellina   1a   1c

2a   Pseudorakverella   2c   2b

3b   Protallinnella   3a

Fig. 84. Quadrijugatoridae (p. Q152-Q154).

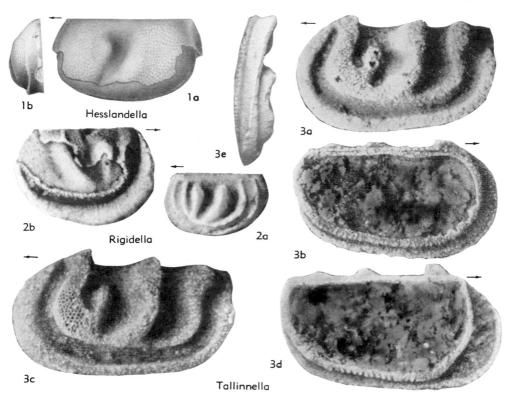

FIG. 85. Quadrijugatoridae (p. Q152-Q153).

USA(Iowa); *1a,b,* LV lat., RV lat., ×35, ×25; *1c,d,* LV int., RV int., ×50; *1e,* LV dors., ×50 (J.H. Burr, Jr., n).

Family SIGMOOPSIDAE
Henningsmoen, 1953

[*nom. correct.* JAANUSSON, 1957 (*pro* Sigmoopsiidae HEN-NINGSMOEN, 1953)] [=Glossopsiinae HESSLAND, 1949; Glosso-morphitinae HESSLAND, 1954; Sigmoopsidinae POKORNÝ, 1958] [Materials for this family prepared by R. V. KESLING, University of Michigan, with contributions from IVAR HESSLAND, University of Stockholm, and others as recorded]

Carapace nearly equivalved, with straight hinge line; valves subquadrate to subelliptical in outline and bi-, tri-, or quadrilobate, with gently convex posterior region in most genera forming extralobate area, lobes joined to ventral ridge that by its position corresponds to histium in most genera; S_2 uniformly present, generally long and sinuous, S_3 present in trilobate genera; marginal and one or two parallel structures developed in most species as low unornamented ridges. Dimorphism observed in all genera, expressed in form of outermost ridge or flangelike structure (termed "carina" by HENNINGSMOEN, 1953, and "histium" by JAANUSSON, 1957), here called histium; ridge between marginal and histial structures present in some genera, here called velar, but may not be homologous to velar structures in Hollinidae. [This family seems closely related to the contemporary Quadrijugatoridae.] *L.Ord.-U.Ord., ?L.Sil.-?U.Sil.*

Sigmoopsis HENNINGSMOEN, 1953 [**Ceratopsis platyceras* ÖPIK, 1937]. Valves subelliptical, in most species with distinct forward swing, tri- or quadrilobate, lobes being unequal, L_1 rather large, in some species with backward-directed spine, also in some partly or wholly confluent with small L_2, which does not reach dorsal border, L_3 long, curved, and rather strongly convex, L_4 broad, bordered posteriorly by narrow extralobate area; S_1 short and narrow (if present), S_2 long and sinuous, S_3 long, curved, narrow; histium developed as low rounded or sharply crested ridge, dimorphic; velar ridge present. *M.Ord.-U.Ord.*, Eu.——FIG. 88,1. **S. platyceras* (ÖPIK), M.Ord., Est.; *1a,b,* ♀ RV lat., ♂ LV lat., ×20 (Kesling, 1951).

Aulacopsis HESSLAND, 1949 [*A. bifissurata*]. Valves subtriangular in outline, bi-, tri-, or quadrilobate, lobes being unequal, L_1 dorsally confluent with L_2 and L_3 with L_4; S_2 long, extending from hinge line to ventral part of valve, S_1 and S_3 (if present) narrow fissures confined to ventral region (some species having both S_1 and S_3, others only S_3, and at least one species having none). *L.Ord., Eu.*——Fig. 87,4. **A. bifissurata*, Swed.; RV lat., ×60 (Kesling, 1951).

?Brevibolbina SARV, 1959 [*B. dimorpha*]. Domicilium approximately amplete, equivalved, unisulcate (sulcus shallow, short, somewhat sinuous, situated mainly at about mid-height of dorsal area), presulcal node low, circular base, posteroventral broad spine or knob directed backward. Dimorphic; tecnomorphs with narrow ridge along part of anteroventral and ventral margins (may be lacking); heteromorphs with wide convex flange along same section of free margin. *M.Ord.-U.Ord. NW.Eu.*——Fig. 88,5. **B. dimorpha*, U.Ord., Est.; *5a,b*, heteromorph LV (holotype) lat. (reconstr.), int.; *5c,d*, tecnomorph RV ext., int.; all ×35 (Sarv, 1959). [HESSLAND.]

Carinobolbina HENNINGSMOEN, 1953 [*Ctenobolbina*

estona ÖPIK, 1937]. Closely similar to *Sigmoopsis* but with much shorter histial structure that is confined to anteroventral and ventral areas. *M. Ord., Eu.*——Fig. 88,4. **C. estona* (ÖPIK), M.Ord., Est.; *4a,b*, ♀ and ♂ RV lat., ×25 (Henningsmoen, 1953).

Distobolbina SARV, 1959 [*D. nabalaensis*]. Preplete, generally considerably higher in anterior part, many strongly convex, equivalved; unisulcate, with slightly arcuate sulcus extending from dorsal area almost to ventral margin, indistinct dorsally but fairly deep in central and ventral parts, rather conspicuous presulcal node with circular base, second node occurring dorsal to sulcus and swollen short lobe developed behind ventral part of sulcus extending mainly parallel to corresponding part of ventral margin. Dimorphic, with adventral structure forming narrow ridge near margin in tecnomorphs and strongly convex anteroventral flange in heteromorphs. Surface tuberculate and spinose. *U.Ord., NW.Eu.*(Baltoscandia).——Fig. 88,9. **D. nabalaensis*, Est.; *9a*, heteromorph LV lat. (holotype), *9b*, tecnomorph LV lat., both ×35 (Sarv, 1959). [HESSLAND.]

?Disulcina SARV, 1959 [*Ctenobolbina perita* SARV,

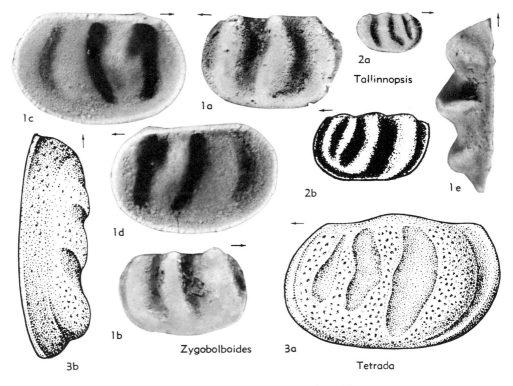

2a
Tallinnopsis

1c

1a

1d

2b

1e

1b

Zygobolboides   3a
Tetrada

3b

FIG. 86. Quadrijugatoridae (p. Q154-Q155).

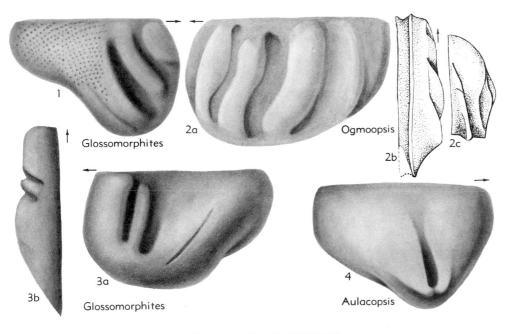

FIG. 87. Sigmoopsidae (p. Q156-Q159).

1956]. Dorsal margin long, cardinal angles distinct, carapace generally somewhat higher in posterior part and slightly preplete, equivalved, bisulcate; sulci arcuate, anterior situated mainly in centrodorsal area, posterior mostly in central area and in some shells extending more toward ventral margin, adventral structure developed as narrow ridge along anterior and ventral margins. Dimorphism not observed. Surface smooth. *M.Ord.-U.Ord.*, NW.Eu.——FIG. 88,2. *D. perita* (Sarv), U.Ord., Est.; *2a*, RV lat. (holotype); *2b*, LV lat. (reconstr.); *2c*, RV int.,; all ×32 (Sarv, 1959). [HESSLAND.]

Glossomorphites HESSLAND, 1954 [*pro Glossopsis* HESSLAND, 1949 (*non* BUSH, 1904)] [**Glossopsis lingua* HESSLAND, 1949]. Valves quadrilobate but lobes unequal, L_1 being elongate (linguiform in some species), L_2 ridgelike, L_3 and L_4 confluent dorsally though separated by a fissure ventrally; sulci unequal, S_1 elongate and rather narrow, S_2 deeper and wider than other sulci and reaching from hinge line to ventral part of valve, S_3 a fissure confined to middle and ventral parts of valve; ventral ridge or histium developed only along anterior and ventral sections of free border. *L.Ord.*, Eu.——FIG. 87,3. **G. lingua* (HESSLAND), Swed.; *3a,b*, LV lat., dors., ×50 (Kesling).——FIG. 87,1. *G. alatus* (HESSLAND), Swed.; RV lat., ×60 (Kesling, 1951).

Lomatobolbina JAANUSSON, 1957 [**Ctenobolbina*

mammillata THORSLAND, 1940]. Unisulcate, S_2 long and sigmoidal; L_2 a small vertically elongate node in front of geniculum, more or less fused with L_1; ventral part of rear half of valve inflated, forming posteroventral lobe, in some species with node or spine at its top, in this respect differing from closely similar *Sigmobolbina*; characteristic marginal flange, in some radially striate, broadest posteroventrally. Dimorphism in histium; in females moderately broad, flangelike, but in male a wedgelike ridge, in ventral view bowed farthest outward in anteroventral region; velar structure, if present, posteriorly confluent with histial structure, apparently broader in females than in males. *M.Ord., ?U.Ord.,* Eu.——FIG. 88,7. **L. mammillata* (THORSLUND), M.Ord., Swed.; *7a,b*, ♂ and ♀ RV lat., ×25.——FIG. 88,8. *L. craspedota* JAANUSSON, M.Ord., Swed.; *8a,b*, ♂ and ♀ LV lat., ×37.5 (Jaanusson, 1957).

Oecematobolbina JAANUSSON, 1957 [**O. nitens*]. Unisulcate, S_2 broad dorsally, narrowing ventrally, geniculate; no velar structure; a broad, radially striate frill-like marginal structure in some species. Histium in male valve a ridge, in most species bearing 2 ridgelike thickenings and, in some, 2 rows of oblong pits or depressions; in female valve broader, flangelike, internally hollow and partitioned into chambers by radial septa. *M.Ord.,* NW.Eu.(Est.-Swed.-Ire.).——FIG. 88,6. **O. nitens,* M.Ord., Swed.; *6a,b*, 2 ♀ LV lat.; *6c,d*, ♂ LV lat.,

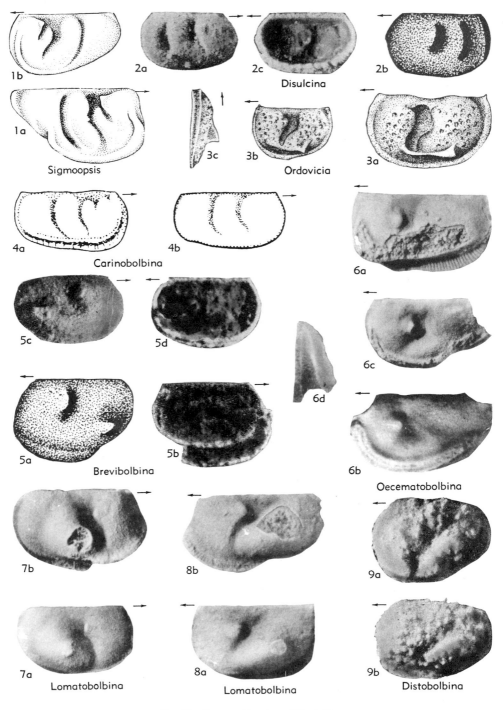

FIG. 88. Sigmoopsidae (p. Q155-Q160).

ant., all ×37.5 (Jaanusson, 1957).——Fig. 89,6. *O. ctenolopha* (Öpik), M.Ord., Est.; ? ♂ LV lat., ×33 (Kesling, 1951).

Ogmoopsis Hessland, 1949 [**O. nodulifera*]. Valves quadrilobate, with subequal elongate ridgelike lobes but L_2 slightly shorter than others; all sulci long, extending from hinge line to ventral ridge or histium. [Genus seems closely related to *Quadrijugator*, from which it differs in having lobes of varied form, with distal surfaces gently convex instead of round, and in having greater space between marginal and velar ridges.] *L.Ord.*, Eu.——Fig. 87,2. **O. nodulifera*, Swed.; *2a*, LV lat., ×50; *2b,c*, LV vent., ant., ×40 (Kesling, 1951).

?Ordovicia Neckaja, 1956 [**O. porchowiensis*]. Amplete or somewhat preplete, monosulcate, with deep, long, slightly sinuous sulcus, LV larger than RV, cardinal corners distinct (in many

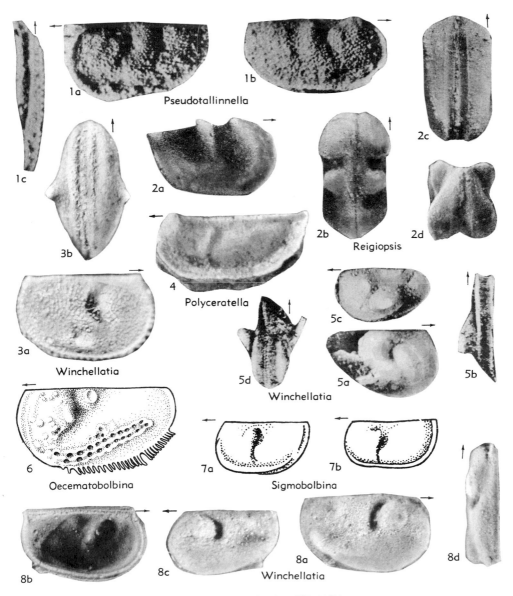

FIG. 89. Sigmoopsidae (p. Q159-Q161).

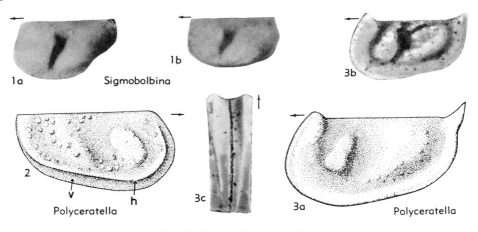

FIG. 90. Sigmoopsidae (p. *Q*160).

shells extended into short spines), presulcal node small and indistinctly set off, dorsal broad swellings on each side of sulcus in adult specimens rising above hinge, ridge on ventral side of sulcus and extending backward from it approximately parallel to ventral margin and generally ending in spine, adventral structure developed as ridge running along entire free margin, broadest anteroventrally, narrowing anteriorly and posteriorly, marginal ridge along free margin. Dimorphism not reported. Surface tuberculate. [Apparently closely related to or possibly congeneric with *Hesslandella*.] *M.Ord.*, NW.Eu.——FIG. 88,3. *O. porchowiensis*, Russia(Pskow area); *3a*, LV (holotype) lat.; *3b,c*, LV (juv.) lat., vent.; all ×27 (Neckaja, 1958). [HESSLAND.]

Polyceratella ÖPIK, 1937 [*Ulrichia kuckersiana* BONNEMA, 1909]. Quadrilobate, L_1 and L_4 more or less parallel to anterior and posterior borders and connected by a ventral lobe; L_3 broad dorsally but constricted ventrally, either confluent with the ventral lobe or separated from it by a furrow; L_2 a knob, separated from ventral lobe by a furrow or almost obsolete; S_2 long, wide dorsally and narrow ventrally; ridge- or flangelike velar structure in all species. Histium formed as projecting edge of ventral lobe, in females broad and frill- or flangelike, widest anteroventrally, in males a narrow ridge. *M.Ord.*, Swed., Est.——FIG. 90,3. *P. kuckersiana* (BONNEMA), Swed., Est.; *3a*, ♀ LV lat., ×35 (Jaanusson); *3b,c*, ♀ carapace (holotype) L, vent., ×20 (58).——FIG. 90,2. *P. bonnemai* (THORSLUND), Swed.; ♂ RV lat., ×35 (*h*, histial ridge; *v*, velar flange) (36).——FIG. 89, 4. *P. tetraceras* (ÖPIK), Est.; ♂ LV lat., ×20 (58).

Pseudotallinnella SARV, 1959 [*P. scopulosa*]. Preplete, equivalved, quadrilobate, with comparatively low lobes that may be partitioned into separate knobs, sulci shallow; adventral structures consist-

ing of inner narrow ridge running along anterior and ventral margins and outer ridge different in length but conforming to this. Dimorphism may be indicated by distinctly channeled area between adventral structures in heteromorphs. Surface coarsely tuberculate. *M.Ord.-U.Ord.*, NW.Eu.—— FIG. 89,1. *P. scopulosa*, U.Ord., Est.; *1a*, heteromorph LV (holotype) lat.; *1b,c*, RV lat., vent.; all ×20 (Sarv, 1959). [HESSLAND.]

?Reigiopsis SARV, 1959 [*R. oepiki*]. Preplete, equivalved, posterior margin forming an acute angle with dorsal margin which is long and straight; dorsal area with 2 large backwardly directed cones separated by sulcal depression; adventral structures comprising outer bend and inner low ridge (apparently velar). Dimorphism not observed. *L.Ord.*, NW.Eu.(Baltoscandia).——FIG. 89,2. *R. oepiki*, Est.; *2a-d*, carapace (holotype) R lat., dors., vent., post., ×18 (Sarv, 1959). [HESSLAND.]

Sigmobolbina HENNINGSMOEN, 1953 [*Entomis oblonga* STEUSLOFF var. *kuckersiana* BONNEMA, 1909]. Generally unisulcate but some species with traces of S_1 and S_3, S_2 being sigmoidal as in *Sigmoopsis;* histium of females may be flangelike and of males ridgelike, also in some species terminating posteriorly in spine; velar ridge narrow but distinct. *M.Ord.*, Eu.——FIG. 89,7. *S. kuckersiana* (BONNEMA), M.Ord., Est.; *7a,b*, ♀ and ♂ LV lat., ×21 (Pokorný, 1958).——FIG. 90,1. *S. sigmoidea* JAANUSSON, M.Ord., Swed.; *1a,b*, ♀ and ♂ LV lat., ×25 (Jaanusson, 1957).

Winchellatia KAY, 1940 [*W. longispina*]. Lobation resembling that of the Devonian hollinid *Flaccivelum* but L_3 smaller and not inflated dorsally; prominent posteroventral projection; female with wide frill-like histium separated from free edge by smooth channel, male with histial ridge. [Genus differs from *Lomatobolbina* in having its marginal structure poorly developed and in lacking any form of velar structure.] *M.Ord.-U.Ord.*, N.Am.-Eu.——Fig. 89,5. **W. longispina,* M.Ord. (Trenton.), USA(Iowa); *5a,b,* ♀ RV lat., vent.; *5c,d,* ♂ carapace L, vent., all × 30 (Kay, 1940). ——Fig. 89,3. *W. minnesotensis* Kay, M. Ord., USA(Minn.-Iowa); *3a,b,* ♂ carapace R, vent., ×50 (J. R. Cornell, n.).——Fig. 89,8. *N. lansingensis,* M.Ord.(Decorah), USA(Minn.-Iowa), *8a,b,* ♀ RV lat., LV int.; *8c,d,* ♂ LV lat., dors.; all ×50 (J. R. Cornell, n).

Family TETRADELLIDAE Swartz, 1936

[Materials for this family prepared by H. W. Scott, University of Illinois, and R. V. Kesling, University of Michigan]

Straight-hinged ostracodes with subequal valves; bilobate to quadrilobate; dimorphism by anterior and anteroventral loculi occupying position distal to carinate ridge, in some between velate structure and carinate ridge. *M.Ord.-L.Sil.*

Tetradella Ulrich, 1890 [**Beyrichia quadrilirata* Hall & Whitfield, 1875]. Subquadrate to subovate, with straight dorsal margin and broadly convex ventral margin; valves subequal, quadrilobate; lobes simple or divided, L_1 and L_4 merging ventrally to form continuous lobe subparallel to free margin, L_2 and L_3 joining ventrally with

ventral ridge; loculi along anterior and anteroventral margin situated between L_1 and ventral ridge and velum (abvelate); dimorphism by abvelate loculi. *M.Ord.-U.Ord.,* cosmop.——Fig. 91, *1.* **T. quadrilirata* (Hall & Whitfield), M.Ord., USA(N.Y.); *1a-c,* ♀ LV lat., dors., ant., ×27 (Kesling, 1951).——Fig. 92,1. *T.* sp. cf. **T. quadrilirata* (Hall & Whitfield), U.Ord. (Richmond.), USA(Ohio); *1a-d,* ♀ RV lat., dors., ♀ LV lat., dors.; *1e,f,* ♂ RV lat., vent.; *1g,h,* ♀ RV lat., vent.; all ×30 (Kesling & Hussey, 1953). ——Fig. 91,5. *T. perornata* Öpik, M.Ord., Est.; *? ♂* LV lat., ×40 (Kesling, 1951).——Fig. 91,6. *T. marchica* (Krause), M.Ord., USA(Pa.); *? ♂* LV lat., ×18 (Kesling, 1951).——Fig. 91,2. *T. lunatifera* (Ulrich), M.Ord., USA(Minn.); *2a-c, ? ♂* RV lat., vent., ant., ×27 (Kesling, 1951).—— Fig. 91,4. *T. ellipsellina* Kay, M.Ord.(Decorah Sh.), USA(Minn.); *4a-c,* ♀ LV lat., vent., ♀ RV lat., ×30 (J. R. Cornell, n).

Dilobella Ulrich, 1890 [**D. typa*]. Small, subovate, valves subequal; L_1 and L_2 fused to form large anterior lobe with dorsal end slightly above hinge line, in some species ventrally connected to L_3 forming U-shaped lobe; L_3 large, extending above hinge line in some species, S_2 long and deep, nearly vertical but curved slightly forward adventrally; both dimorphs with velate ridge near

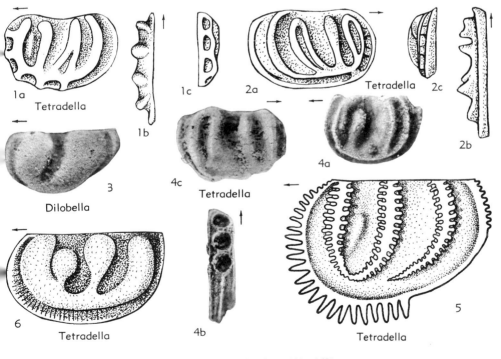

FIG. 91. Tetradellidae (p. Q161-Q162).

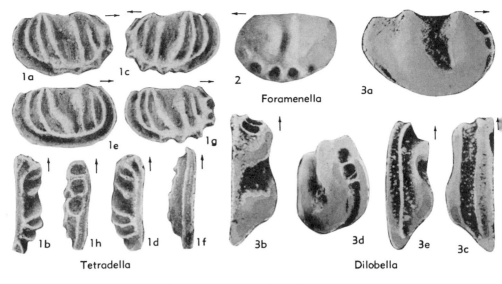

FIG. 92. Tetradellidae (p. Q161-Q162).

marginal ridge, males with bend along U-shaped lobe but without true carina, females with anterior and anteroventral abvelate loculi between velar ridge and carinate edge of U-shaped lobe, carina nearly or completely hiding loculi in lateral view, continuous with lateral surface of U-shaped lobe and hence nearly same as histium. *M. Ord.*, N.Am. —— FIG. 92,3. **D. typa*, Ord. (Trenton.), USA(Minn.); *3a-d*, ♀ RV lat., dors., vent., ant.; *3e*, ♂ LV vent., all ×30 (Kay, 1940). ——FIG. 91,3. *D. simplex* KAY, M.Ord.(Decorah

Sh.), USA(Minn.); ♂ LV lat., ×33 (J. R. Cornell, n).

Foramenella STUMBUR, 1956 [**Euprimitia parkis* NECKAJA, 1952]. Small, elongate oval, convex, nearly equivalved; with conspicuous S_2 slightly in front of mid-length giving each valve bilobate appearance; females with 5 loculi in each valve, lateral velate ridge, edges of loculi with slight development of rims but without bordering carinal structure. *U.Ord.-L.Sil.*, Est.——FIG. 92,2. **F. parkis*, ♀ LV lat., ×33 (Sarv, 1959).

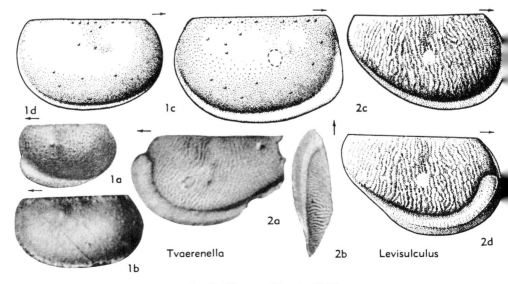

FIG. 93. Tvaerenellidae (p. Q163).

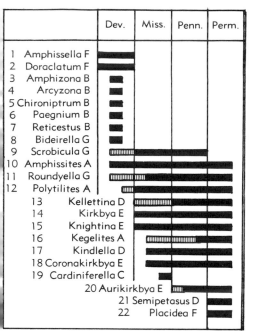

	Dev.	Miss.	Penn.	Perm.
1 Amphissella F				
2 Doraclatum F				
3 Amphizona B				
4 Arcyzona B				
5 Chironiptrum B				
6 Paegnium B				
7 Reticestus B				
8 Bideirella G				
9 Scrobicula G				
10 Amphissites A				
11 Roundyella G				
12 Polytilites A				
13 Kellettina D				
14 Kirkbya E				
15 Knightina E				
16 Kegelites A				
17 Kindlella D				
18 Coronakirkbya E				
19 Cardiniferella C				
20 Aurikirkbya E				
21 Semipetasus D				
22 Placidea F				

FIG. 94. Stratigraphic distribution of kirkbyacean ostracode genera, with indicated family assignments (A—Amphissitidae, B—Arcyzonidae, C—Cardiniferellidae, D—Kellettinidae, E—Kirkbyidae, F—Placideidae, G—Scrobiculidae) (Moore, n). The accompanying alphabetical list of generic names furnishes a cross reference to the serially arranged numbers on the diagram.

Generic Names with Index Numbers

Amphisella—1	*Kellettina*—13
Amphissites—10	*Kindlella*—17
Amphizona—13	*Kirkbya*—14
Arcyzona—4	**Knightina**—15
Aurikirkbya—20	*Paegnium*—6
Bideirella—8	*Placidea*—22
Cardiniferella—19	*Polytilites*—12
Chironiptrum—5	*Reticestus*—7
Coronakirkbya—18	*Roundyella*—11
Doraclatum—2	*Scrobicula*—9
Kegelites—16	*Semipetasus*—21

Family TVAERENELLIDAE Jaanusson, 1957

[*nom. transl.* HESSLAND, herein (*ex* Tvaerenellinae JAANUSSON, 1957)] [Materials for this family prepared by IVAR HESSLAND, University of Stockholm]

Unisulcate to nonsulcate or with sulcal depression, presulcal node or knob or spine flattened and indistinct; no surface ornamentation or indications of crests; with velar dimorphism, flange in heteromorphs being slightly to moderately convex and edges not in contact in closed carapaces. *L.Ord.-U.Ord.*

Tvaerenella JAANUSSON, 1957 [**Primitiella carinata* THORSLUND, 1940]. Nonsulcate or provided with sulcal depression, presulcal node small and indistinct; dimorphic, with velar ridge *or* slightly convex and moderately wide flange that is widest anteroventrally and generally rather long. *L. Ord.-U.Ord.*, NW.Eu.——FIG. 93,*1*. **T. carinata* (THORSLUND), M.Ord., Swed.; *1a*, LV (holotype), heteromorph with anteroventrally curved frill, lat., ×22; *1b*, LV tecnomorph with velar ridge, ×25; *1c*, RV, heteromorph (reconstr.), lat., ×35; *1d*, RV, tecnomorph (reconstr.), lat., ×35 (*1a*, Thorslund, 1940; *1b-d*, Jaanusson, 1957).

Levisulculus JAANUSSON, 1957 [**L. lineatus*]. Preplete, dorsal margin long, shallowly unisulcate (narrow sulcal or semisulcal depression), presulcal node small; dimorphic, velar structure not reaching posterodorsal corner, straight and narrow to wide *or* wide and distinctly convex, widest anteroventrally. *M.Ord.*, NW.Eu.(Baltoscandia). ——FIG. 93,*2*. **L. lineatus*, M.Ord., Swed.; *2a,b*, LV heteromorph (holotype), lat., vent., ×45, ×37; *2c*, RV tecnomorph with straight velum (reconstr.), lat., ×35; *2d*, RV heteromorph with convex velum (reconstr.), lat., ×35 (Jaanusson, 1957).

Family UNCERTAIN

Echinoprimitia HARRIS, 1957 [**E. imputata*]. Small, subrectangular, straight-backed; valves subequal, S_2 present; marginal spines present along anterior and part of ventral margins; spine anterodorsal to S_2. [The smallness of the specimen on which the genus was based and the row of short marginal spines indicates that this represents an instar stage of a genus resembling those included in the Hollinidae or Eurychilinidae. Possibly it is a molt of *Eurychilina papillata* HARRIS. Both were found in the same zone at the same locality.] *M.Ord.*, USA (Okla.). [SCOTT.]

Superfamily KIRKBYACEA Ulrich & Bassler, 1906

[*nom. transl.* SOHN, herein (*ex* Kirkbyidae ULRICH & BASSLER, 1906)] [Diagnosis by I. G. SOHN, United States Geological Survey]

Reticulate, straight-backed, with or without lobes, nodes and carinae; ridge-and-groove hingement, with or without terminal dentition; valves subequal, overlap slight; free margin of one valve rabbeted to receive opposing valve; one or more marginal rims, dimorphism unknown. *?L.Dev., M.Dev.-M.Perm.*

The stratigraphic distribution of kirkbyacean genera is shown graphically in Figure 94.

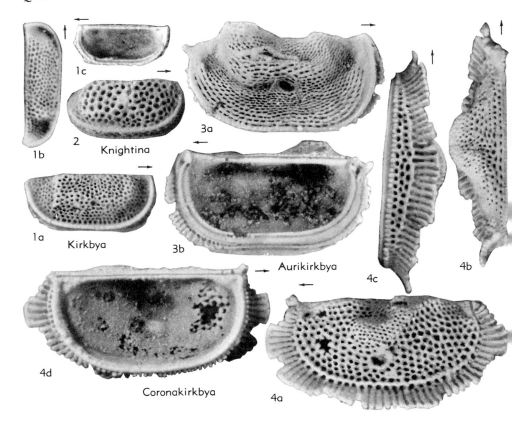

FIG. 95. Kirkbyidae (p. Q164).

Family KIRKBYIDAE Ulrich & Bassler, 1906

[=Kirkbijidae SPIZHARSKY, 1939] [Materials for this family prepared by I. G. SOHN, United States Geological Survey]

Reticulate, lobed or unlobed, with well-developed kirkbyan pit, and without nodes or carinae. *L.Miss.-M.Perm.*

Kirkbya JONES, 1859 [*Dithyrocaris permiana* JONES, 1850]. Elongate, greatest length at or near dorsal margin; posterior cardinal angle acute; lateral surface evenly convex or with posterior shoulder; 2 marginal rims. *L.Miss.-M.Perm.*, N.Am.-Eu.-Asia.——FIG. 95,1. *K.* sp., Perm., USA(Tex.); *1a,b,* RV lat., dors., ×40; *1c,* RV int., ×30 (73).——FIG. 96,4. *K. canyonensis* HARLTON, Penn., USA(Tex.); ant. sec. through LV (*ir,* inner ridge; *or,* outer ridge), ×90 (73).

Aurikirkbya SOHN, 1950 [*Kirkbya wordensis* HAMILTON, 1942]. Two dorsal lobes connected ventrally by ridge, terminal teeth well developed, marginal rim thicker than shell wall. ?*L.Penn.*, *M.Penn.-M.Perm.*, N.Am.——FIGS. 95,3, 96,2. *A. wordensis* (HAMILTON), USA(Tex.); 95,*3a,b,* RV

lat., int., ×30 (333); 96,2, post. section through LV at kirkbyan pit, ×55 (73).

Coronakirkbya SOHN, 1954 [*C. fimbriata*]. Large, centrally lobed, 2 marginal rims; reticulation of lobe smaller than those of valve. ?*L.Miss.-?M. Penn.*, *U.Penn.-M.Perm.*, N.Am.——FIG. 95,4. *C. fimbriata*, Perm., USA(Tex.); *4a-d,* LV (holotype) lat., dors., vent., int., ×30 (73).

Knightina KELLETT, 1933 [*Amphissites allorismoides* KNIGHT, 1928] [=*Tenebrion* ZANINA, 1956]. Cardinal angles obtuse, with well-developed posterior shoulder, 1 or 2 marginal rims, no terminal dentition. *L.Miss.-M.Perm.*, N.Am.-Eu.-Asia. 95,2, 96,1. *K. allorismoides* (KNIGHT), Penn., USA(Mo.); 95,2, RV lat. (topotype), ×57; 96,1, ant. sec. through RV, ×90 (73).

Family AMPHISSITIDAE Knight, 1928

[Materials for this family prepared by I. G. SOHN, United States Geological Survey]

Carapace with one or more nodes and well-developed kirkbyan pit, usually carved into or near central node. *M.Dev.-M.Perm.*

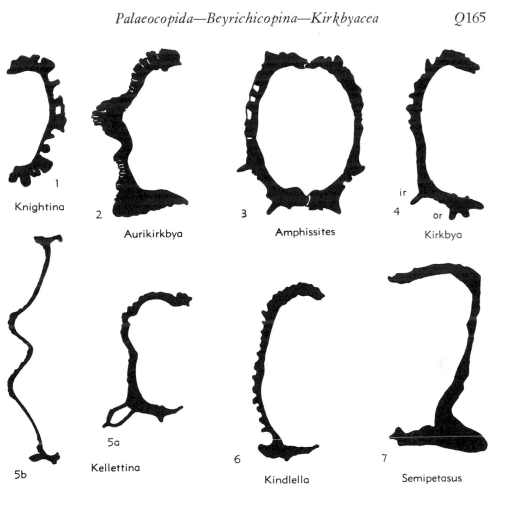

FIG. 96. Sections of kirkbyacean genera (Sohn, n).

Amphissites GIRTY, 1910 [*A. rugosus] [=*Albanella* HARRIS & LALICKER, 1932; *Binodella* BRADFIELD, 1935; *?Ectodemites* COOPER, 1941]. Median node flanked laterally by carinae which may or may not superpose elongate nodes, and which are connected by dorsal carina subparallel to and joining hinge near cardinal angles; with terminal dentition and 2 marginal rims. *M.Dev.-M.Perm.,* N.Am.-Eu.-Asia.——FIG. 98,3. *A. rugosus,* Miss., Ark.; *3a-d,* carapace (lectotype ROUNDY, 1926), L lat., dors., vent., post., ×40 (Sohn, n).——FIG. 96,3. *A. centronotus* (ULRICH & BASSLER), Penn., USA(Tex.); cross section of carapace in front of central node, ×90 (73).——FIG. 97,1. *A. primus* (COOPER), U.Miss., USA(Ill.), (type species of *Ectodemites*); *1a-c,* carapace L, dors., vent., ×30 (Sohn, n). [*Brillius* BRAYER, 1952, is judged to be based on a male corresponding to female *Ectodemites* and accordingly classed doubtfully as synonym of *Amphissites*—SCOTT.]

Kegelites CORYELL & BOOTH, 1933 [*pro Girtyites* CORYELL & BOOTH, 1933 (*non* WEDEKIND, 1914)] [*Girtyites spinosus* CORYELL & BOOTH, 1933] [=*Kirkbyites* JOHNSON, 1936]. Like *Amphissites,* without dorsal carina or anterior node; posterior node trending toward cardinal angle, projecting above hinge line. *M.Miss.-M.Perm.,* N.Am.——FIG. 98,1. *K. spinosus* (CORYELL & BOOTH), Penn., USA(Tex.); *1a-c,* carapace (holotype), L lat., R lat., vent., ×80 (Sohn, n).

Polytylites COOPER, 1941 [*P. geniculatus]. Like *Amphissites,* without carinae but with terminal nodes well developed. *?U.Dev., L.Miss.-M.Perm.,* N.Am.-Eu.-Asia.——FIG. 98,2a-d. *P. geniculatus,* Miss., USA(Ill.), *2a-d,* carapace (holotype) L. lat., R lat., dors., vent., ×40 (Sohn, n).——FIG. 98, *2e,f. P. digitatus* SOHN, M.Perm., USA(Tex.); *2e,f,* LV (holotype), lat., dors., ×40 (73).

FIG. 97. Amphissitidae (p. Q165).

Family ARCYZONIDAE Kesling, n. fam.

[Materials for this family prepared by R. V. KESLING, University of Michigan]

Carapace with subequal valves, subquadrate to subelliptical or subovate in outline, with marginal and velar structures in all genera but carina only in some; side of valves with large subcentral pit but no node, interior of valves marked by node with clustered adductor muscle scars in position corresponding to external pit; RV hinge consisting of groove that fits edge of LV, hinge line straight. Surface reticulate. *M. Dev.*

Ostracodes of this family previously have been included in the Kirkbyidae on account of their straight hinge line, reticulate surface, and central pit; they differ in the much larger size of the pit and in lacking a central node. The family may contain the ancestors of Kirkbyidae.

Arcyzona KESLING, 1952 [*Amphissites diadematus* VAN PELT, 1933]. Carapace with velar structure in all species consisting of narrow frill, simple ridge, or broadly rounded ridge covered by reticulation of narrow crests; carina, if present, developed as frill-like flange, ridge, or reticulate elevation. Surface coarsely reticulate. *M.Dev.,* N.Am.-Eu.——FIG. 99,*1a.* *A. diademata* (VAN PELT), USA(Mich.); RV lat., ×40 (201).——FIG. 99,*1b. A. rhabdota* KESLING, USA(Mich.); RV lat., ×40 (201).——FIG. 99,*1c. A. aperticarinata* KESLING & WEISS, USA(Mich.); RV lat., ×40 (213).

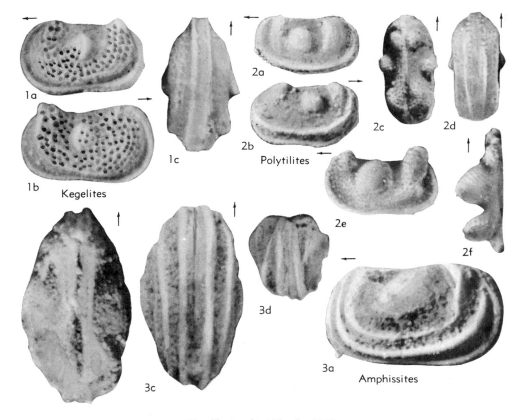

FIG. 98. Amphissitidae (p. Q165).

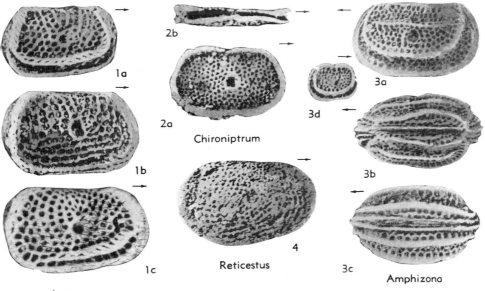

2b

Chironiptrum

2a

1a

1b

1c

Arcyzona

3a

3d

3b

4

Reticestus

3c

Amphizona

Fig. 99. Arcyzonidae (p. Q166-Q167).

Amphizona Kesling & Copeland, 1954 [*A. asceta*]. Carapace with dorsal ridge, central horizontal ridge, frill, and well-developed carina; low node in front of central pit; very young instars lack central ridge and node. Surface reticulate. *M.Dev.*, N.Am.——Fig. 99,3. *A. asceta*, USA (N.Y.); 3a-c, adult carapace L lat., dors., vent., ×35; 3d, young instar R lat., ×35 (207).

Chironiptrum Kesling, 1952 [*C. oiostathmicum*]. Carapace with frill confluent with flangelike dorsal ridge, no carinae. Surface finely reticulate. *M. Dev.*, N.Am.——Fig. 99,2. *C. oiostathmicum* USA(Mich.); 2a,b, RV lat., dors., ×40 (201).

Paegnium Kesling, 1957 [*P. tanaum*]. Carapace elongate, RV overlapping LV; hinge line very long; narrow frill or velar ridge around free border, continuous with a narrower dorsal ridge, in this feature resembling *Chironiptrum;* no carina; low marginal ridge. Surface reticulate; central pit smaller than in other genera of family, scarcely larger than meshes of reticulation. *M.Dev.*, N.Am.——Fig. 100,1. *P. tanaum*, Hamilton, USA (N.Y.); 1a-d, carapace (holotype) L, R, dors., vent., ×85 (204).

Reticestus Kesling & Weiss, 1953 [*R. acclivitatus*]. Carapace lacking distinct frill, velar ridge, and carina, but lateral surface separated from marginal surface by distinct smooth rounded bend that apparently represents velar structure. Surface reticulate. *M.Dev.*, N.Am.——Fig. 99,4. *R. acclivitatus*, USA(Mich.); RV lat., ×30 (213).

Family CARDINIFERELLIDAE Sohn, 1953

Differs from Kirkbyidae in absence of marginal ridges and in presence of a primitive merodont hinge. *U.Miss.*

Cardiniferella Sohn, 1953 [*C. bowsheri*]. Subovate; reticulate except for smooth marginal area; hinge incised; overlap slight. *U.Miss.*, N.Am.——Fig. 101,1. *C. bowsheri*, USA(Tex.); 1a, RV (holotype), lat.; 1b, carapace (paratype), dors., ×40 (334).

Family KELLETTINIDAE Sohn, 1954

[*nom. transl.* Sohn, herein (*ex* Kellettininae Sohn, 1954)]
[Materials for this family prepared by I. G. Sohn, United States Geological Survey]

Carapace without a well-defined kirkbyan pit. *?L.Miss.(?L.Carb.), M.Miss.-M.Perm.*

Kellettina Swartz, 1936 [*Ulrichia robusta* Kellett, 1933]. Two unequal large nodes on each side of approximate mid-length, not extending below mid-height; well-developed marginal rim. *?L.Carb.*, Eu.(Russ.); *L.Penn.-M.Perm.*, N.Am.-Eu.-Asia.——Fig. 102,3. *K. robusta* (Kellett), Perm., USA(Kans.); LV (holotype), ×40 (Sohn, n).——Fig. 96,5. *K. vidriensis* Hamilton, Perm., USA(Tex.); 5a,b, ant. transv. and long. secs. through LV, ×45 (73).

Kindlella Sohn, 1954 [*K. fissiloba*]. Shallow marginal rim; lobes extending below mid-height. *M.*

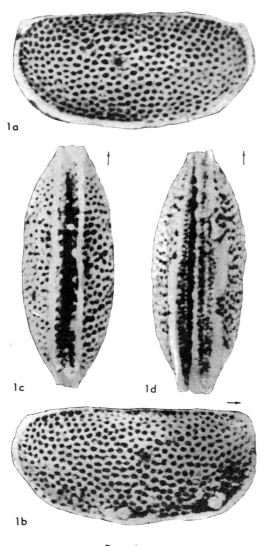

1a

1c

1d

1b

Paegnium

FIG. 100. Arcyzonidae (p. Q167).

Miss.-M.Perm., N.Am.-Eu.——FIGS. 96,6, 102,*1.* *K. fissiloba*, M.Perm., USA(Tex.); 102,*1a,b*, carapace (holotype) R, L, ×40 (Sohn, n); 96,6, ant. transv. sec. through LV, ×95 (73).

Semipetasus SOHN, 1954 [*S. signatus*]. Large, with confluent elongate lobes, minute terminal dentition. *L.Perm.-M.Perm.*, N.Am.——FIGS. 96,7, 102,2. *S. signatus*, M.Perm., USA(Tex.); 102,*2a,b*, LV (holotype), lat., dors., ×20 (Sohn, n); 96,7, RV ant., transv. sec. through RV, ×48 (73).

?Family PLACIDEIDAE Schneider, 1956

[Materials for this family prepared by I. G. SOHN, United States Geological Survey] [Proper assignment of the Placideidae and Scrobiculidae is unknown at this time. SOHN believes that they are not Kirkbyacea because they lack the kirkyban pit. He would assign them questionably to the Podocopida. A reported muscle-scar pattern with as many as 40 units suggests a possible relationship with the metacopines. However, outline, hingement, free margin, and a strongly reticulated surface do not conform with other metacopines. These families display characters somewhat transitional between the Kirkbyacea and the metacopines and perhaps future studies may show that this is their true position.]

Straight-backed, subquadrate reticulated, with or without dorsocentral node, and with marginal rim; adductor muscle scar an irregularly rounded rosette with up to 40 spots, tongue-and-groove hingement; straight marginal pore canals. *L.Dev.-M. Perm.*

Placidea SCHNEIDER, 1956 [*Amphissites lutkevichi* SPIZHARSKY, 1939]. No dorsocentral node, ventral margin slightly concave, more than 1 mm. long. *L.Perm.-M.Perm.*——FIG. 103,*3a.* *P. lutkevichi* (SPIZHARSKY), Russian Platform; LV ×40 (50). ——FIG. 103,*3b. P. trituberculata* SCHNEIDER, Russian Platform, carapace L, ×40 (239).

?Amphissella STOVER, 1956 [*A. papillosa*]. Differs from *Placidea* in straight to slightly convex ventral margin and small size (less than 0.75 mm. in length). *Dev.*——FIG. 103,*1.* *A. papillosa*; M. Dev. (Hamilton), N.Y.; carapace L (holotype), ×40 (348).

?Doraclatum STOVER, 1956 [*D. compandium*]. Differs from *Amphisella* in possessing a dorsocentral node. *Dev.*——FIG. 103,*2.* *D. compandium*, M.Dev. (Hamilton), N.Y.; carapace L (holotype), ×40 (348).

?Family SCROBICULIDAE Posner, 1951

[See note under Placideidae.] [Materials for this family prepared by I. G. SOHN, United States Geological Survey, with addition by R. H. SHAVER, Indiana University and Indiana Geological Survey]

1a

Cardiniferella

1b

FIG. 101. Cardiniferellidae (p. Q167).

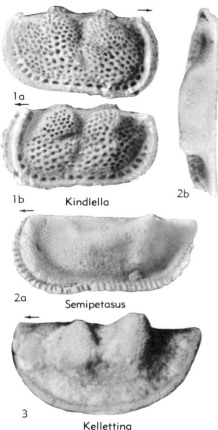

1a

1b Kindlella

2a Semipetasus

2b

3 Kellettina

Fig. 102. Kellettinidae (p. Q167-Q168).

Subquadrate or suboval, small, straight-hinged, inequivalved, reticulated, with slightly impressed tongue-and-groove hingement, subcentral roseate muscle scar, no marginal rims. *?M.Dev., L.Carb.(Miss.)-M.Perm.*

Scrobicula POSNER, 1951 [*Cytherella? scrobiculata* JONES, KIRKBY & BRADY, 1884]. Dorsal margin of larger valve curved, overreaching impressed hinge. *?M.Dev.-?U.Dev., L.Carb.,* Eu.——FIG. 104,2. *S. scrobiculata* (figured specimen), L.Carb., Russia; *2a-c,* LV lat., vent., dors., ×65 (281).

?Bideirella STOVER, 1956 [*B. reticulata*]. Resembling *Roundyella* in small size, subequal valves, subrectangular outline in lateral view, reticulated surface, smooth central spot, and smooth marginal band, but differing in presence of 2 low vertical ridges near ends of each valve. Morphology of hinge, contact margin, and adductor muscle scar unknown. *M.Dev.,* N.Am.——FIG. 104,1. *B. reticulata; 1a,* carapace (holotype), lat.; *1b-d,* cara-

paces (paratypes), lat., lat., dors.; all ×30 (348). [SHAVER.]

Roundyella BRADFIELD, 1935 [*Amphissites simplicissimus* KNIGHT, 1928] [*=Scaberina* BRADFIELD, 1935]. Dorsum straight, surface with or without scattered papillae and small spines. *?M.Dev.-?L. Miss., M.Miss.-M.Perm.,* N.Am.——FIG. 104,3. *R. simplicissima,* Penn., USA(Mo.), *3a,b,* ?RV lat., int., ×30 (Sohn, n).

Superfamily OEPIKELLACEA Jaanusson, 1957

[*nom. transl.* HESSLAND, herein (*ex* Oepikellidae JAANUSSON, 1957)]

Median sulcus absent, surface smooth, pitted, reticulate, or nodular; velar structure present. *L.Ord.-M.Penn.*

The Aparchitidae are included provisionally in this superfamily. The genus *Aparchites* is so poorly known that it is not clear at this time whether or not this is a valid family.

The stratigraphic distribution of genera assigned to the Oepikellacea is shown graphically in Figure 105.

Family OEPIKELLIDAE Jaanusson, 1957

[Materials for this family prepared by IVAR HESSLAND, University of Stockholm]

Carapace nonsulcate, dimorphic; tecnomorphic specimens generally amplete, ad-

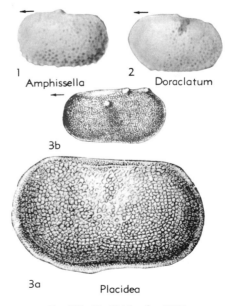

1 Amphissella

2 Doraclatum

3b

3a Placidea

Fig. 103. Placideidae (p. Q168).

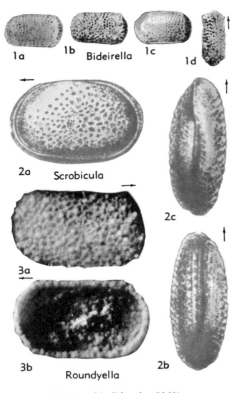

1a 1b Bideirella 1c
1d

2a Scrobicula

2c

3a

3b 2b
Roundyella

Fig. 104. Scrobiculidae (p. Q169).

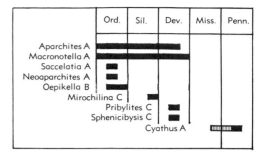

	Ord.	Sil.	Dev.	Miss.	Penn.
Aparchites A					
Macronotella A					
Saccelatia A					
Neoaparchites A					
Oepikella B					
Mirochilina C					
Pribylites C					
Sphenicibysis C					
Cyathus A					

Fig. 105. Stratigraphic distribution of oepikellacean ostracode genera, with indicated family assignments (A—Aparchitidae, B—Oepikellidae, C—Pribylitidae) (Moore, n).

1d

1c

1a Oepikella

1b

2a Mirochilina 2b

Fig. 106. Oepikellidae, Pribylitidae (p. Q170, Q173).

ventral structure in heteromorphs developed as velar dolon in ventral or anteroventral position, in tecnomorphs forming narrow ridge or seemingly not developed. *M.Ord.-U.Ord.*

Oepikella Thorslund, 1940 (as *Öpikella*) [*Öpikella tvaerensis* Thorslund, 1940]. Inequivalved, LV overlapping RV along entire free margin; adductor muscle scar large (in some species surrounded by vascular marks); heteromorphs with moderate wide anteroventral incurved frill. *M.Ord.-U.Ord.,* Eu.-N.Am.——Fig. 106,*1*. ?*O. frequens* (Steusloff), M.Ord. (Edinburg F.), USA(Va.); *1a,b,* LV (heteromorph) lat.; LV (heteromorph) int.; *1c,d,* RV (tecnomorph lat., LV (tecnomorph) lat.; all ×20 (J. C. Kraft, n).——Fig. 110,*1*. *O. tvaerensis* (Thorslund), M.Ord., Swed.; *1a,b,* RV (holotype), lat., post., ×15; *1c,* RV heteromorph, lat. [=*O. asklundi* (Thorslund)], ×15 (369).

?Family APARCHITIDAE Jones, 1901

[Materials for this family prepared by Ivar Hessland, University of Stockholm]

Carapace nonsulcate, inequivalved, largely amplete; adductor muscle scar not visible

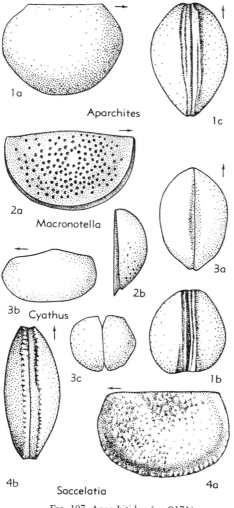

Fig. 107. Aparchitidae (p. *Q171*).

exteriorly or indistinct but may also be well defined; velar structure developed as low ridge which may be tuberculate, extending along ventral border and in some along part of end borders or entire free border; subvelar area channeled; dimorphism not observed. *L.Ord.-M.Penn.*

Aparchites JONES, 1889 [**A. whiteavesi;* SD S. A. MILLER, 1889]. Carapace generally swollen; hinge line of varying length, distinctly shorter than carapace; cardinal angles as a rule distinctly obtuse; adductor muscle scar not visible exteriorly; velar ridge smooth or tuberculate. *L.Ord.-M.Dev., Eu.-N. Am.-Austral.-Asia.* —— FIG. 107,*1.* **A. whiteavesi,* M.Ord., Can.(Man.); *1a-c,* carapace (holotype) R lat., end, vent., ×10 (186).——FIG. 108,*2a.* ?*A. kauffmanensis* SWAIN, M.Ord.(Chazy.), USA(Pa.); RV lat., ×20 (356).——FIG. 108,*2b-e.* ?*A. crossotus* KESLING, M.Dev., USA(Mich.); *2b-e,* RV (holotype) lat., lat. showing muscle scar, int., vent., ×40 (201).——FIG. 108,*2f-h.* A. *fimbriatus* (ULRICH), M.Ord., USA(Minn.); *2f,* ?heteromorph carapace L lat., ×22; *2g,h,* ?heteromorph carapace L, vent., ×22 (J. R. Cornell, n).——FIG. 109,*1.* A. *fimbriatus* (ULRICH), M.Ord. (Edinburg F.), USA (Va.); *1a,b,* LV lat., LV int., ×10, ×9; *1c,* carapace vent., ×13; *1d,e,* LV lat., LV lat. (juv. instars), ×13; *1f,* muscle scar, ×22 (J. C. Kraft, n). [In checking proofs, HESSLAND notes that Figures 108 and 109, illustrating species added by the editor, are questionably assignable to *Aparchites;* moreover, he judges that *"A. fimbriatus"* actually belongs with oepikellid forms.—Ed.]

?**Cyathus** ROTH & SKINNER, 1930 [**C. ulrichi*]. Carapace very small (max. length about 0.5 mm., possibly only larval specimens known), elongated, tumid, with poorly defined cardinal angles; dorsal margin somewhat shorter than carapace, depressed between largely equal umbones; adductor muscle scar not visible exteriorly; low ridge along ventral border interpreted as velar structure. *M. Penn., N.Am.*——FIG. 107,*3.* **C. ulrichi,* M.Penn., USA(Colo.); *3a-c,* carapace (holotype) dors., L lat., end view ×48 (298).

Macronotella ULRICH, 1894 [**M. scofieldi*] [=*Baltonotella* SARV, 1959]. Carapace mainly regularly arched, greatest thickness in central part; dorsal margin same length as carapace or somewhat shorter; cardinal angles generally well defined but may also be rounded; velate ridge smooth; adductor muscle scar indicated exteriorly by distinct rounded spot which is smooth like peripheral part of valves; remaining part of surface distinctly punctate or reticulate. *L.Ord.-Dev., N.Am.-Eu.* ——FIGS. 107,*2,* 108,*3.* **M. scofieldi,* M.Ord.; 107, *2a,b,* RV (syntype, Ky.) lat., end view, ×15 (83); 108,*3,* LV (Minn.) lat., ×50 (J. R. Cornell, n).

?**Neoaparchites** BOUČEK, 1936 [**Primitia obsoleta* JONES & HOLL, 1865]. Like *Aparchites* but not indisputably known whether adventral structure is developed, possibly a velar ridge; cardinal angles rounded. ?*Sil.* (Pleist. erratic), Eu.

?**Saccelatia** KAY, 1940 [**Aparchites arrectus* UL-RICH, 1894] [=*Saccaletia* KAY, 1940 (invalid original spelling)]. Carapace small (length less than 1 mm.), with straight and long dorsal margin; cardinal angles well defined, obtuse; carapace thickest centroventrally, outline lentiform in ventral view; adductor muscle scar not visible exteriorly; adventral row of tubercles or spines or ridge interpreted as velar structure. *M.Ord., N. Am.-Eu.*(Balt.).——FIGS. 107,*4,* 108,*1a-c.* **S. arrecta* (ULRICH), M. Ord., USA(Minn.); 107,*4a,b,* carapace (holotype) L, vent., ×40 (194); 108,*1a,*

LV lat., ×50; 108,*1b*, RV int., ×50; 108,*1c*, cara-
pace dors., ×50 (108,*1a-c*, J. R. Cornell, n).——
FIG. 108,*1d-f*. *S. bullata* KAY, M.Ord., USA
(Minn.); *1d-f*, carapace L, dors., vent., ×45 (J. R.

Cornell, n).——FIG. 108,*1g,h*. *S. arcamuralis* KAY,
M.Ord., USA(Minn.); *1g,h*, RV lat., LV lat., ×45
(J. R. Cornell, n).

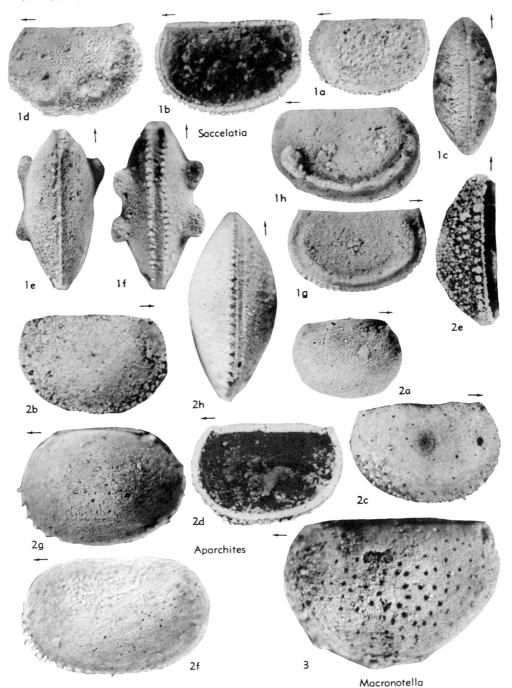

FIG. 108. Aparchitidae (p. Q171).

Aparchites

Samarella

Fig. 109. Aparchitidae, Paraparchitidae
(p. Q171, Q194).

Family PRIBYLITIDAE Pokorný, 1958

[Materials for this family prepared by IVAR HESSLAND, University of Stockholm, with addition by JEAN BERDAN, United States Geological Survey]

Carapace small, nonsulcate, or with very shallow S_2 with long straight hinge line, and well-defined cardinal angles, generally amplete or slight preplete or postplete; thickest in ventral half; adventral structure interpreted as velar variously indicated by sharp bend or ridge, with or without tubercles. Dimorphism not observed. *U.Sil.-M. Dev.*

Pribylites POKORNÝ, 1950 [*P. moravicus*] [=*Parapribylites* POKORNÝ, 1950] Carapace very small (length less than 0.6 mm., possibly only representing larval specimens) with bulbous anterior cardinal areas, amplete, preplete, or slightly postplete, trapezoidal in ventral view, RV overlapping

LV ventrally; adductor muscle scar not visible exteriorly, sharp bend or short ridge (in some reduced to ventral spine) may indicate velar structure, marginal structure may be denticulate; subvelar area approximately straight. *M.Dev., Eu.*——FIG. 110, 2. *P. moravicus*, Czech.; 2a-c, carapace (holotype), L lat., vent., dors., ×60 (275).

?**Mirochilina** BOUČEK, 1936 [*M. jarovensis*]. Small, straight-hinged, with obtusely angled cardinal extremities, subtriangular in transverse section, lateral and ventral surfaces of valves disposed almost perpendicularly to each other; with weak median sulcus that ends in shallow median pit; narrow striated frill on anteroventral, ventral, and posterior contact margins; surface smooth. *U.Sil. (Ludlov.), Eu.*——FIG. 106,2. *M. jarovensis*, Czech.; 2a,b, carapace (holotype) L, dors., ×40 (10). [BERDAN.]

Sphenicibysis KESLING, 1952 [*S. hypoderota*] [=*Sphenicibys* POKORNÝ, 1958 (errore)]. Carapace subtriangular in end views, amplete or gently preplete; RV distinctly more swollen in ventral half than LV; adductor muscle scar not visible exteriorly; sharp adventral bend with small tubercles interpreted as velar structure; marginal structure indicated by low ridge; subvelar area straight. *M.Dev., N.Am.*——FIG. 110,3. *S. hypoderota*, USA(Mich.); 3a-c, carapace (holotype) L lat., R lat., vent., ×50 (201).

Superfamily PRIMITIOPSACEA Swartz, 1936

[*nom. transl.* HESSLAND, herein (*ex* Primitiopsidae SWARTZ, 1936)] [Diagnosis by IVAR HESSLAND, University of Stockholm]

Unisulcate (sulcus or sulcal pit corresponding to S_2) or nonsulcate; hingement may consist of median groove and corresponding ridge with lateral pits and prominences. Dimorphic, with more or less closed posterior velar chamber in heteromorphs but adventral structures may or may not be developed in tecnomorphs. *M.Ord.-M.Dev.*

The stratigraphic distribution of genera assigned to the Primitiopsacea is indicated in Figure 111.

Family PRIMITIOPSIDAE Swartz, 1936

[=Primitiopsididae POKORNÝ, 1958] [Materials for this family prepared by IVAR HESSLAND, University of Stockholm]

Characters of superfamily. *M.Ord.-M.Dev.*

Subfamily PRIMITIOPSINAE Swartz, 1936

[*nom. transl.* HESSLAND, 1949 (*ex* Primitiopsidae SWARTZ, 1936)] [=Primitiopsiinae HENNINGSMOEN, 1953]

Carapace with distinct sulcus or sulcal pit, indistinct presulcal node, velar structure

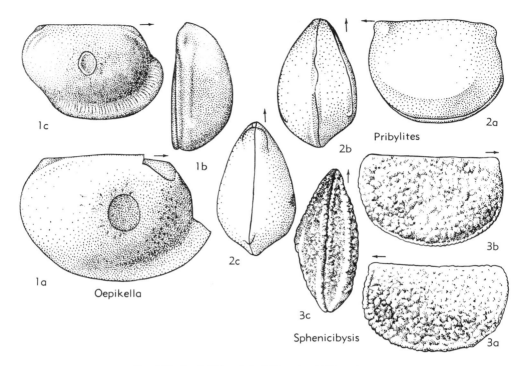

Fig. 110. Oepikellidae, Pribylitidae (p. Q170, Q173).

forming bend or ridge in both sexes except along posterior part in heteromorphs where closed or open pouch is developed; surface generally reticulate. *M.Ord.-M.Dev.*

Primitiopsis Jones, 1887 [**P. planifrons*]. Carapace rather elongate in heteromorph specimens, RV overlapping LV; hinge depressed between dorsal ridges, dorsum narrow; sulcal structure developed as S_2 depression with adductorial pit in ventral part. Dimorphic; velar structure forming ventral bend posteriorly, in heteromorphs constituting pouch by meeting of velar flanges, surface of flanges smooth, marginal structure tuberculate. *M.Ord. - M.Sil.,* N.Am.(Okla.) - Eu.(Scand.).——Fig. 112,3. **P. planifrons,* M.Sil.(Wenlock.), Swed.; *3a-c,* heteromorph carapace (neotype, Martinsson, 1955) L, vent., dors., ×25 (52).

Clavofabella Martinsson, 1955 [**C. incurvata*]. Inequivalved, RV overlapping LV; hinge consisting of median groove with corresponding ridge and lateral elongate pits and sockets; unisulcate (sulcal pit). Dimorphic, showing velar structure developed as ridge that in heteromorphs widens to form posterior flanges which do not meet; surface pitted or reticulate, marginal structure tuberculate. *M.Sil.,* Eu.(Scand.-Eng.).——Fig. 112,2. **C. incurvata,* Wenlock., Swed.; *2a-d,* heteromorph carapaces with velar frill, L lat., vent. oblique, dors., vent.,

×25; *2e-g,* tecnomorph carapace lacking velar frill, L lat., dors., vent., ×25 (52).

Limbinaria Swartz, 1956 [**L. multipunctata*]. Nonlobate, with deep sulcal pit, domicilium largely amplete, with distinct cardinal angles, RV overlapping LV along entire free margin. Dimorphic, adventral structure developed as narrow ridge in tecnomorphs and wide flat flange along posterior and adjacent parts of ventral margin in hetero-

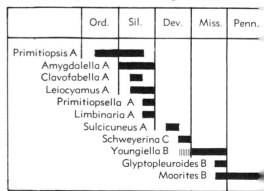

	Ord.	Sil.	Dev.	Miss.	Penn.
Primitiopsis A	▬▬▬				
Amygdalella A		▬▬			
Clavofabella A		▬			
Leiocyamus A		▬			
Primitiopsella A		▬			
Limbinaria A		▬			
Sulcicuneus A		▬▬▬			
Schweyerina C			▬		
Youngiella B			▦▬▬▬▬		
Glyptopleuroides B				▬▬	
Moorites B				▬▬▬	

Fig. 111. Stratigraphic distribution of (A) primitiopsacean and (B) youngiellacean ostracode genera, including (C) *Schweyerina,* of uncertain family (Moore, n).

FIG. 112. Primitiopsidae (p. Q174-Q175).

morphs, distinct dorsal plica along entire dorsal margin. Surface coarsely pitted or reticulate. *U. Sil.*, N.Am.——FIG. 114,2. **L. multipunctata,* Tonoloway Ls., USA(Va.); *2a,* heteromorph LV (holotype) lat.; *2b,* tecnomorph RV lat.; both ×45 (359).

Primitiopsella POLENOVA, 1960 [*pro Leperditellina* POLENOVA, 1955 (*non* NECKAJA, 1955)] [**L. miranda* POLENOVA, 1955]. Dorsal margin almost straight to slightly convex; tecnomorph carapace amplete, almost equivalved, nonlobate, RV slightly overlapping LV; shallow sulcal depression; dimorphic, some shells with closed broad swelling parallel to posterior margin (or additionally along posterior part of ventral margin) may be connected to short adventral ridge paralleling ventral margin (which in tecnomorphs seems to continue along posterior margin). Surface smooth or finely pitted. *U.Sil.(Ludlov.),* USSR(Urals)-Est.——FIG. 113,1. **P. miranda* (POLENOVA), Urals; *1a-c,* carapace (heteromorph) L lat., dors., vent.; *1d,* carapace (outline) R lat. (tecnomorph); ×41 (279). [HESSLAND-REYMENT.]

Sulcicuneus KESLING, 1951 [**S. porrectinatium*]. Carapace subtriangular in end view, LV larger than RV; unisulcate, median sulcus *(S₂)* well de-

veloped, extending to about mid-height, presulcal node small; adductor muscle spot composed of scars (6 in holotype of type species) arranged in rosette around central scar. Dimorphic, velar structure forming ridge that may be tuberculate, in heteromorphs forming posterior flanges which may meet or form a gap; velum marginally tuberculate. *M.Dev.,* N.Am.——FIG. 112,1. **S. porrectinatium,* Traverse Gr., USA(Mich.); *1a-c,* carapace (holotype) R lat. (heteromorph) dors..

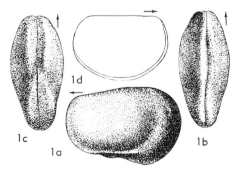

FIG. 113. *Primitiopsella miranda* (POLENOVA) (Primitiopsidae, (p. Q175).

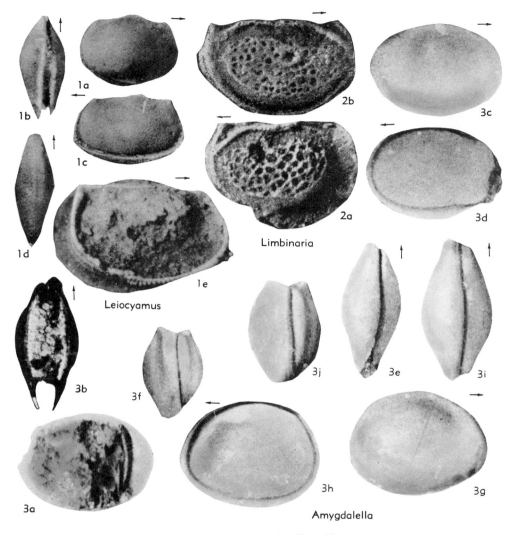

FIG. 114. Primitiopsidae (p.*Q174-Q177*).

vent.; *1d,e,* carapace L (heteromorph); *1f,* RV int.; *1g,* RV int. (tecnomorph); all ×40 (200).

Subfamily LEIOCYAMINAE Martinsson, 1956

Velar structure in one sex only consisting of flange, widest posteriorly and there forming a chamber; without sulcus or sulcal pit, no preadductorial node. Surface smooth or pitted. *M.Sil.-U.Sil.*

Leiocyamus MARTINSSON, 1956 [**L. apicatus*]. Inequivalved, RV larger than LV, nonsulcate; hinge depressed between dorsal ridges or umbones (larger on LV) forming simple groove in one valve and corresponding ridge in other. Dimorphic,

[HESSLAND has reported (too late for incorporation in *Treatise* systematic text) that recent studies by MARTINSSON (The primitiopsid ostracodes from the Ordovician of Oklahoma and the systematics of the family Primitiopsidae: Uppsala Univ. Geol. Inst., Bull. v. 38, p. 139-154, 1960) serve to establish the primitiopsid nature of species from Middle Ordovician strata of Oklahoma desribed by HARRIS (Ostracoda of the Simpson Group; Oklahoma Geol. Survey, Bull. 75, 333 p., 10 pl., 1957). Accordingly, the stated stratigraphic range of the family is revised to include the Middle Ordovician occurrences. MARTINSSON'S revised arrangement of primitiopsid genera is indicated in the following outline: Subfamily PRIMITIOPSINAE Swartz, 1936 (*Primitiopsis, Clavofabella, Limbinaria, Primitiopsella*); subfamily ANISOCYAMINAE Martinsson, nov. (*Anisocyamus,* Martinsson, nov., based on *Primitiopsis elegans* Harris, 1957, as type species); Subfamily SULCICUNEINAE Martinsson, nov. (*Sulcicuneus*); Subfamily POLENOVULINAE Martinsson, nov. (*Polenovula* Martinsson, nov., based on *Leperditellina? crassa* Polenova, 1955, as type species, and *Viazoviella* Martinsson, nov., based on *Leperditellina miranda* Polenova, 1955, as type species).—ED.]

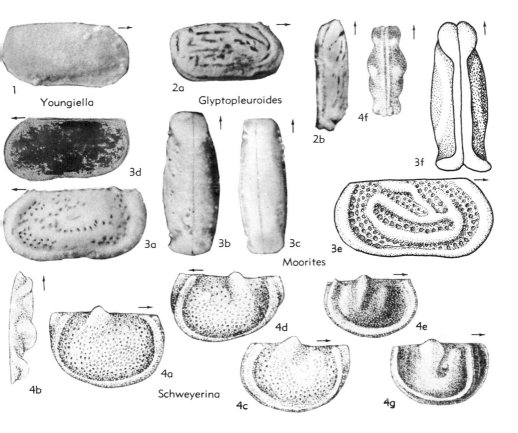

FIG. 115. Youngiellidae and Family Uncertain (p. *Q*177-*Q*178).

velar structure in heteromorphs developed as flange which is broad and convex posteriorly and continued forward along ventral side of valve, not forming a closed chamber. *M.Sil.-U.Sil.*, NW.Eu. ——FIG. 114,*1.* *L. apicatus,* Wenlock., Swed.; *1a-b,* carapace of posteriorly chambered specimen (holotype), R lat., vent., ×25; *1c,d,* carapace of unchambered specimen, L lat., vent., ×25; *1e,* LV of chambered specimen, int., ×55 (243).

Amygdalella MARTINSSON, 1956 [*A. subclusa*] [=*Amygalella* LEVINSON, 1957 (errore)]. Inequivalved, RV larger than LV, hinge depressed between dorsal ridges; nonsulcate; dimorphic, with velate structure in females consisting of posterior pouch which can be almost closed (no extension along ventral margin). *Sil.,* NW.Eu.(Baltoscandia). ——FIG. 114,*3.* *A. subclusa,* Baltoscandia (drift boulder); *3a,* RV (heteromorph) (holotype), int. showing post. chamber; *3b,* chambered carapace, long. sec.; *3c-f,* chambered carapace R, L lat., vent., ant.; *3g-j,* unchambered carapace R, L lat., vent., ant.; all ×20 (106).

?Superfamily YOUNGIELLACEA Kellett, 1933

[*nom. transl.* SOHN, herein (*ex* Youngiellidae KELLETT, 1933)] [Diagnosis by I. G. SOHN, United States Geological Survey]

Carapace minute, suboblong, with straight taxodont hinge bearing vertical teeth and sockets. Surface smooth, reticulate or ridged. *?U.Dev., L.Miss.(L.Carb.)-U.Penn.*

The stratigraphic distribution of youngiellacean ostracode genera is plotted in Figure 111.

Family YOUNGIELLIDAE Kellett, 1933

Characters of superfamily. *L.Miss.(L. Carb.)-U.Penn.*

Youngiella JONES & KIRKBY, 1895 [*pro Youngia* JONES & KIRKBY, 1886 (*non* LINDSTRÖM, 1885)] [*Youngia rectidorsalis* JONES & KIRKBY, 1886]. Smooth or faintly reticulate, without marginal rim. *L.Carb.(Miss.),* Eu.-N.Am.——FIG. 115,*1.*

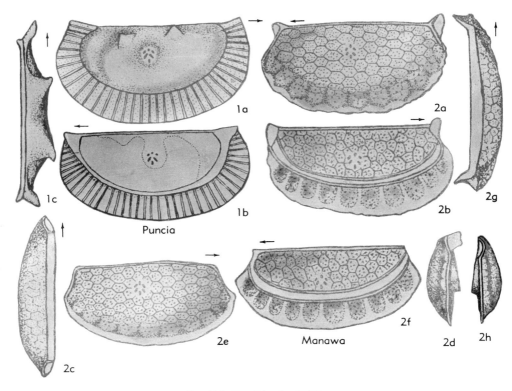

1a, 1c, 1b Puncia
2a, 2g, 2b, 2e, Manawa, 2f, 2d, 2h, 2c

FIG. 116. Punciidae (p. Q179).

*Y. *rectidorsalis* (Jones & Kirkby), L.Carb. (Scremerston Gr.), Eng.; carapace R, ×80 (Sohn, n).

?**Glyptopleuroides** Croneis & Gale, 1938 [*G. *insculptus*]. Unpitted, with marginal rim and bifurcating elongate ridges on surface of valves. U. Miss., N.Am.——Fig. 115,2. *G. *insculptus*, Ill.; 2a,b, RV lat., dors., ×40 (Sohn, n).

Moorites Coryell & Billings, 1932 [*M. *hewetti* (=*Glyptopleurina? *minuta* Warthin, 1930)] [=?Hardinia Coryell & Rozanski, 1942]. Rims along end margins, surface with coarse pits in elongate undefined depressions between rims. Miss.-U.Penn., N.Am.——Fig. 115,3. *M. *minutus* (Warthin), U.Penn., Tex.; 3a-c, carapace L, dors., vent., ×80; 3d, RV converted to fluorite, int. viewed with transmitted light to show hinge, ×80 (Sohn, n); 3e,f, carapace R, vent., ×60 (200).

Family UNCERTAIN

Schweyerina Zaspelova, 1952 [*S. *ovata*]. Bilobed, with marginal ribs and reticulate surface. U.Dev., Eu.——Fig. 115,4a-d. *S. *ovata*, USSR; 4a,b, carapace (holotype) R, dors.; 4c,d, RV (paratype) lat., LV lat.; all ×70 (406).——Fig. 115,4e-g. *S. *normalis*; 4e,f, carapace (holotype) R, dors.; 4g, RV lat.; all ×50 (406).

?Superfamily PUNCIACEA Hornibrook, 1949

[*nom. transl.* Sylvester-Bradley, herein (*ex* Punciidae Hornibrook, 1949)] [Diagnosis by P. C. Sylvester-Bradley, University of Leicester]

Carapace small, delicate, and differing from most other Palaeocopida by discrete muscle-scar pattern. *Rec.*

Family PUNCIIDAE Hornibrook, 1949

[Materials for this family prepared by P. C. Sylvester-Bradley, University of Leicester]

Carapace elongate, semi-elliptical with long straight hinge-line, with or without terminal articulating brackets. Contact margins extended by wide, double-walled, septate frill; muscle-scar pattern in ring of 6 scars. Soft parts unknown. [Family based on 2 monotypical genera discovered in the South Pacific from 2 dredgings. The carapaces bear a remarkable resemblance to the Ordovician-Devonian Eurychilinidae and seem to comprise the only surviving members of the Palaeocopida. This is agreed to by Betty Kellett Nadeau after studying type specimens. H. W. Scott (1960) thinks that

Puncia and *Manawa* should be classified in the Eurychilinidae, treating Punciidae as a junior synonym of Eurychilinidae. This would call for reporting the range of Eurychilinidae and Hollinacea as L.Ord.-Rec. —MOORE.] *Rec.*

Puncia HORNIBROOK, 1949 [**P. novozealandica*]. Frill wide, divided into about 40 compartments by radial septa. Shallow median sulcus contains muscle-scar pattern. *Rec.*, S.Pac.——FIG. 116,*1*. **P. novozealandica; 1a-c*, RV lat., int., dors., ×90 (174).

Manawa HORNIBROOK, 1949 [**M. tryphena*]. Frill wide, divided into about 11 compartments and bearing a translucent rim; no median sulcus. RV

hinge terminating at both ends in brackets which overlap bevels at ends of LV hinge. *Rec.*, S.Pac. ——FIG. 116,*2*. **M. tryphena; 2a-d*, LV lat., int., dors., post., ×90; *2e-h*, RV lat., int., dors., post. ×90 (174).

Superfamily and Family UNCERTAIN

Craspedobolbina KUMMEROW, 1923 (1924) [**C. dietrichi*]. S_1 nearly obsolete, S_2 deep, L_2 not well defined anteriorly. Velate structure developed as frill in pouchless valves, and as rim in pouched valves, or may be rim in both. Surface granulose, reticulate, or smooth. *Ord.*(boulder in glacial drift), N.Ger. [BERDAN-SCOTT.]

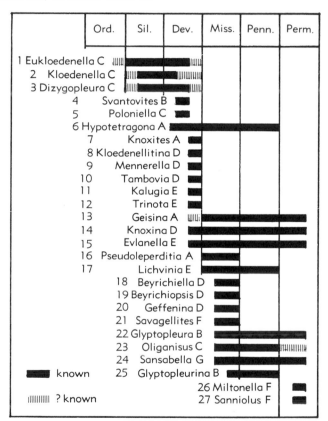

FIG. 117. Stratigraphic distribution of kloedenellacean ostracode genera (Moore, n). Classification of the genera in families is indicated by letter symbols (A—Geisinidae, B—Glyptopleuridae, C—Kloedenellidae, D—Knoxinidae, E—Lichviniidae, F—Miltonellidae, G—Sansabellidae) and the following alphabetical list furnishes a cross reference to the serially arranged numbers on the diagram. [For Knoxinidae, read Beyrichiopsidae.]

Generic Names with Index Numbers

Beyrichiella—18	Glyptopleura—22	Knoxites—7	Sanniolus—27
Beyrichiopsis—19	Glyptopleurina—25	Lichvinia—17	Sansabella—24
Dizygopleura—3	Hypotetragona—6	Mennerella—9	Savagellites—21
Eukloedenella—1	Kalugia—11	Miltonella—26	Svantovites—4
Evlanella—15	Kloedenella—2	Oliganisus—23	Tambovia—10
Geffenina—20	Kloedenellitina—8	Poloniella—5	Trinota—12
Geisina—13	Knoxina—14	Pseudoleperditia—16	

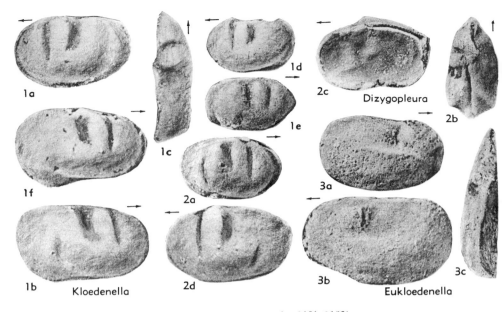

FIG. 118. Kloedenellidae (p. Q181-Q182).

Suborder KLOEDENELLOCOPINA Scott, n. suborder

[Diagnosis prepared by H. W. Scott, University of Illinois]

Strongly calcified carapaces with smooth or ribbed, sulcate or nonsulcate surface; dorsal margin straight, ventral margin convex or somewhat rarely concave; valves unequal, larger overreaching smaller along all or part of free edge; dimorphism by posterior inflation or nondimorphic. *L.Ord.-U.Jur.*

The Kloedenellocopina are not a wholly natural group. They include both dimorphic and nondimorphic forms. The most conspicuous feature common to all is strong overlap of the smaller valve by the larger around all or a portion of the free margin. The Kloedenellacea show dimorphism by swelling of the posterior part of the female carapace. Dimorphism in the Paraparchitacea may be represented by a slight enlargement of the posterior half of the carapace. Some specimens of an adult population show the greatest width to be medial, whereas other specimens of the same population show that the greatest width is behind the mid-length. These two types are probably dimorphs but further work on a large number of specimens is needed.

Superfamily KLOEDENELLACEA Ulrich & Bassler, 1908

[*nom. transl.* Scott, herein (*ex* Kloedenellinae ULRICH & BASSLER, 1908)] [Diagnosis by H. W. Scott, University of Illinois]

Carapace with strongly unequal valve larger overlapping smaller around all o part of free margin. Cardinal angles round ed, hinge line impressed, hinge straigh of tongue-and-groove type, with anterio toothlike overlap or with overlapping valv extending to hinge. Surface smooth, o reticulated, with pit, sulci and with o without costae. Dimorphic in width o posterior part, nonvelate, except Knoxinidac [A dimorphic inner partition has been re ported in the Lichviniidae. A row of spine along the free margins of both valves i some specimens of *Sansabella* has been in terpreted as a velate structure or pseudc velum; it is believed to be merely orna mental and not to represent an outfoldin of the inner chitin layer, thus being unr lated to dimorphism.] *?U.Ord., L.Sil.-M Perm.*

The stratigraphic distribution of Klo denellacean ostracode genera is shown Figure 117.

Family KLOEDENELLIDAE Ulrich & Bassler, 1908

[*nom. transl.* ULRICH & BASSLER, 1923 (*ex* Kloedenellinae ULRICH & BASSLER, 1908)] [=Dizygopleurinae EGOROV, 1950] [Materials for this family prepared by I. G. SOHN, United States Geological Survey]

Straight-backed, subrhomboidal, sulcate, small; posterior 0.7 of hinge incised; one valve overlapping other along free margins and with anterior toothlike process that fits over exterior of opposing valve; rest of hinge ridge-and-groove. Dimorphic in width of posterior. *?U.Ord., L.Sil.-U.Penn., ?L. Perm.-?M.Perm.*

Kloedenella ULRICH & BASSLER, 1908 [**Kloedenia pennsylvanica* JONES, 1889]. LV overlapping RV,

1

Dizygopleura

2c

2a

2d

Kloedenella

2b

3a

3c

3b

3d

Eukloedenella

FIG. 119. Kloedenellidae (p. *Q*181-*Q*182).

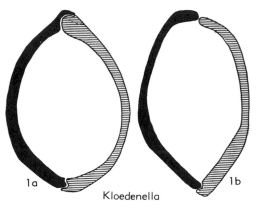

Fig. 120. Kloedenellidae (p. *Q*181).

2 sulci in anterodorsal half. *Sil.-Dev., ?Miss.,* N. Am.-Eu.-Asia.——Figs. 118,*1,* 119,*2,* 120,*1. K. nitida* Ulrich & Bassler, Sil., USA(Md.); 119, *2a-d,* carapace (holotype) R lat., dors., L lat., vent., ×40 (Sohn, n); 120,*1a,b,* transv. secs., ant. 3rd, post. 3rd, ×67 (Sohn, n).——Fig. 118,*1d-f. K. bipustulata* Swartz & Whitmore, L.Dev. (Manlius Ls.), N.Y.; *1d,e,* ♂ LV lat. RV lat. (syntypes), ×30; *1f,* ♀ RV lat., ×30 (78).—— Fig. 118,*1a-c. K. parvisulcata* Swartz & Whitmore, U.Sil. (Decker Ls.), N.J.; *1a,* ♂ LV (holotype) lat., ×30; *1b,c,* ♀ RV lat., dors., ×30 (78).

Dizygopleura Ulrich & Bassler, 1923 [**D. swartzi*]. LV overlapping RV, 3 sulci, at least one of which reaches ventral 0.3 of valve. *Sil.-Dev.,* N.Am.-Eu.——Figs. 24, 119,*1. *D. swartzi,* Sil., USA(Md.); 119,*1,* lectotype (herein), ×40 (Sohn, n); 24, serial transv. secs., ×20 (399). ——Fig. 118,*2. D. angustisulcata* Swartz & Whitmore, U.Sil. (Decker Ls.), USA(N.J.); *2a,b,* ♂ carapace (holotype) R, dors.; *2c,* ♂ RV int.; *2d,* ♀ LV lat.; all ×30 (78).

?Eukloedenella Ulrich & Bassler, 1923 [**E. umbilicata*]. LV overlapping RV, one sulcus in anterodorsal half of valve. [Eight syntypes (USNM 63622) comprise 8 specimens that lack most if not all shell material.] *?U.Ord., Sil.-Dev.,* N.Am.-?Eu. ——Fig. 119,*3. *E. umbilicata,* Sil., USA(Md.); *3a-d,* carapace (syntype) R lat., dors., L lat., vent., ×40 (Sohn, n).——Fig. 24. *E. sinuata* Ulrich & Bassler, Sil., USA(Pa.); serial transv. secs., ×20 (399).——Fig. 118,*3. E. cicatrix* Swartz & Whitmore, U.Sil. (Decker Ls.), USA(N.J.); *3a,* ♂ RV (syntype) lat., ×30; *3b,c,* ♀ LV (syntype) lat., dors., ×30 (78). [=*Punctoprimitia* Stewart, 1945 (*fide* H. W. Scott.)]

Oliganisus Geis, 1932 [**O. sulcatus*] [=*?Ellipsella* Coryell & Rogatz, 1932; *?Neokloedenella* Croneis & Funkhouser, 1939; *Oliganiscus* Neave,

1940]. RV overlapping LV, with or without shallow anterodorsal sulcus. *M.Miss.-U.Penn., ?L. Perm.-?M.Perm.,* N.Am.——Fig. 121,*1a,b. *O. sulcatus,* U.Miss., USA(Ind.); *1a,b,* carapace (holotype) L, dors., ×30 (Sohn, n).——Fig. 121,*1c,d. O. punctatus* Geis, U.Miss., USA(Ind.); *1c,d,* carapace (holotype) L, dors., ×30 (Sohn, n).

?Poloniella Gürich, 1896 [**P. devonica*]. Differs from *Dizygopleura* in ventral union of terminal sulci; seemingly lacks anterior toothlike process. [Juvenile *Dizygopleura* have joined terminal sulci, but have anterior toothlike process.] *Dev.,* Poland.

Family GEISINIDAE Sohn, n. fam.

[?=Perprimitiinae Egorov, 1950] [Materials for this family prepared by I. G. Sohn, United States Geological Survey]

Straight-backed; smooth, punctate, or pitted, subquadrate to subelliptical, small; anterodorsal half sulcate; overlap slight; probably ridge-and-groove hinge; inner lamella narrow, of even width along free margins, feathering towards cardinal angles; dimorphic in width of posterior part. [Sohn suggests that this family belongs to the Podocopida because of a duplicature in *Geisina.* It is being included in the Kloedenellacea until further evidence is obtained.] *M.Dev.-M.Perm.*

Geisina Johnson, 1936 [**Beyrichiella gregaria* Ulrich & Bassler, 1906] [*?Nuferella* Bradfield, 1935; *?Perprimitia* Croneis & Gale, 1939; *Hastifaba* Cooper, 1946; *?Neobeyrichiopsis* Tasch, 1953; *?Perijonesina* Hou, 1955; *?Prosopeionum* Stover, 1956]. Bisulcate, with anterior sulcus shallow or missing; RV overlapping LV to cardinal angles; hinge incised behind sulcus; postero-

Fig. 121. Kloedenellidae (p. *Q*182).

dorsal spine present. *?Dev., Miss.-Perm.,* N.Am.-Eu.-Asia.——Fɪɢ. 122,*3.* **G. gregaria* (Uʟʀɪᴄʜ & Bᴀssʟᴇʀ), U.Penn., USA(Mo.); RV (syntype) lat., ×40 (Sohn, n).

Hypotetragona Mᴏʀᴇʏ, 1935 [**H. impolita*] [=*?Janetina* Cᴏʀʏᴇʟʟ & Mᴀʟᴋɪɴ, 1936; *Kloedenellina, Gillina* Cᴏʀʏᴇʟʟ & Jᴏʜɴsᴏɴ, 1939; *?Limnoprimitia* Kᴜᴍᴍᴇʀᴏᴡ, 1949; *Knoxiella, ?Kules-*

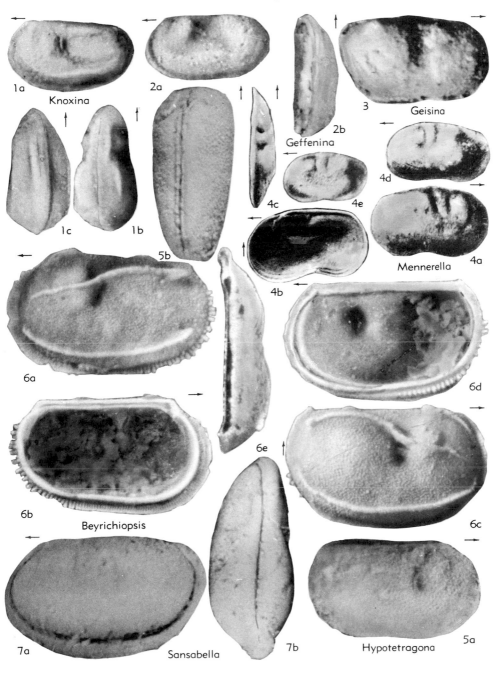

Fɪɢ. 122. Geisiriidae, Beyrichiopsidae, Sansabellidae (p. Q182-Q183, Q185-Q187).

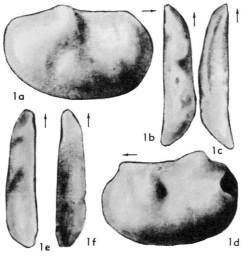

Knoxites

Fig. 123. Geisinidae (p. Q184).

chovkia EGOROV, 1950; *Plavskella* SAMOILOVA, 1951]. Differs from *Geisina* in absence of posterodorsal spine. *M.Dev.-Penn.,* N.Am.-Eu.-Asia. ——FIG. 122,5. **H. impolita,* Miss., USA(Mo.); *5a,b,* carapace (holotype) R lat., dors., ×40 (Sohn, n).

Knoxites EGOROV, 1950 [**K. menneri*]. Differs from *Geisina* in having a posterodorsal tubercle. *U.Dev.,* Eu.——FIG. 123,*1.* **K. menneri,* Russia; *1a-c,* ♀ RV (holotype) lat., dors., vent.; *1d-f,* ♀ LV lat., dors., vent.; all ×40 (144).

?Pseudoleperditia SCHNEIDER, 1956 [**P. tuberculifera*]. Like *Geisina* but differs in having anterodorsal spine and lacking an incised hinge. *L.Carb.,* USSR.——FIG. 124,*1.* **P. tuberculifera; 1a,b,* carapace L, dors., ×30 (238).

Family GLYPTOPLEURIDAE Girty, 1910

[Materials for this family prepared by H. W. SCOTT, University of Illinois]

Carapace costate, straight-backed, cardinal angles obtuse; RV larger than LV overlapping it around free margin; S_2 present as short sulcus or pit; hinge channel terminated by anterior and posterior cardinal teeth; dimorphic. *?M.Dev., M. Miss.-M. Perm.*

Glyptopleura GIRTY, 1910 [**G. inopinata*] [=*Ceratopleurina, Glyptopleurites* CORYELL & JOHNSON, 1939; *Mesoglypha* COOPER, 1941]. Subquadrate to subovate; dorsum straight in lateral view, venter broadly convex, cardinal angles obtuse; with 2 or more transverse simple or bifurcated costae; S_2 as short sulcus or pit; hinge channel terminated by strong anterior and posterior cardinal teeth, teeth of RV fit into sockets on LV; dimorphic.

M.Miss.-M.Perm., N.Am.-Eu.(Eng.-Russia).——FIG. 125,*1a-c.* **G. inopinata,* U.Miss., USA(Ill.); *1a-c,* carapace L, dors., post., ×27 (Scott, n).——FIG. 125,*1d.* *G. adunca* CRONEIS & THURMAN, U. Miss., USA(Ill.); ♀ carapace dors., ×40 (Scott, n).——FIG. 125,*1e,f.* *G. varicostata* CRONEIS & THURMAN, U.Miss., USA(Ill.); *1e-f,* RV int. showing cardinal teeth, LV int. showing hinge sockets, ×60 (Scott, n).

Glyptopleurina CORYELL, 1928 [**G. montifera*] [=*Idiomorpha* CRONEIS & GALE, 1938 (*non* FÖRSTER, 1869); *Idiomorphina* CRONEIS & GALE, 1939 (*pro Idiomorpha*)]. Carapace ornamented with inosculating costae that may terminate in either or both anterodorsal and posterodorsal lobes or nodes; faint frill rarely present; otherwise like *Glyptopleura. M.Miss.-Penn.,* N.Am.——FIG. 125, *2.* **G. montifera,* M.Miss., USA(Ill.); *2a-c,* ♀ carapace L, dors., post., ×40 (Scott, n).

?Svantovites POKORNÝ, 1950 [**S. primus*]. Carapace subquadrangular, dorsum straight to gently arched, venter straight to broadly convex; greatest height in anterior portion, greatest width in posterior half; distinctly inequivalve, larger valve (left) overlapping smaller around free margin; with costae (interconnected in some by fine ribs) oblique to dorsal surface; hingement by cardinal teeth (RV) and sockets (LV) and connecting hinge bar and groove. Similar to *Glyptopleura* but lacking pit or S_2 and with more oblique costae. *M.Dev.,* Eu.(Czech.).——FIG. 125,*3.* **S. primus; 3a-c,* carapace R, dors., vent.; *3d,* LV int., ×77 (Scott, n).

Pseudoleperditia

Fig. 124. Geisinidae (p. Q184).

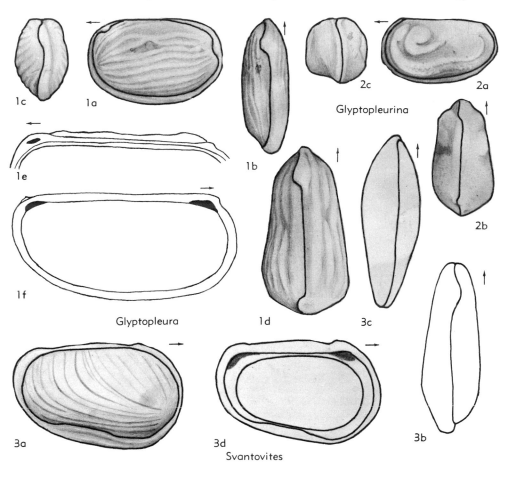

FIG. 125. Glyptopleuridae (p. *Q*184).

Family BEYRICHIOPSIDAE
Henningsmoen, 1953

[=Knoxinae EGOROV, 1950 (invalid because this family-group taxon contains no nominate genus, such as *Knoxa, Knoxus,* or *Knoxum,* from which the subfamily name Knoxinae could be derived); likewise, Knoxidae EGOROV, 1950 (*nom. transl.* POLENOVA, 1960, in USSR Treatise) is unavailable. Knoxinae and Knoxidae are not derivable from *Knoxina* and regardless of EGOROV's intentions as to type species of Knoxinae, this name is not emendable to Knoxininae.—Editor.] [Materials for this family prepared by I. G. SOHN, United States Geological Survey] [Includes ?Mennerellidae POLENOVA, 1960]

Straight-backed, small, bilobate or trilobate; velate, with or without crests on valve surface; ridge-and-groove hinge; overlap slight, even, restricted to free margins; dimorphism exhibited in width of posterior part. *U.Dev.-M.Perm.*

Beyrichiopsis JONES & KIRKBY, 1886 [**B. fimbriata;* SD ULRICH & BASSLER, 1908] [=*Deloia* CRONEIS & THURMAN, 1939; *Leightonella* CRONEIS & GALE, 1939; *Denisonella* CRONEIS & BRISTOL, 1942 (*pro Denisonia* CRONEIS & BRISTOL, 1939; *non* KREFT, 1869); *?Lokius* CORYELL & JOHNSON, 1939]. Bilobate or trilobate, with one or more elongate crests. *L.Carb.(Miss.),* ?Eu.-N.Am.——FIG. 122,6. *B. fortis* JONES & KIRKBY, L.Carb., Scot.; *6a,b,* LV lat., int., ×40; *6c-e,* RV lat., int., dors., ×40 (Sohn, n).

Beyrichiella JONES & KIRKBY, 1886 [**B. cristata*] [=*Kirkbyia* COSSMAN, 1899 (*pro Synaphe* JONES & KIRKBY, 1896; *non* HUEBNER, 1825); *?Kirkbyina* ULRICH & BASSLER, 1908]. Bilobate, crest parallel and adjacent to dorsal margin; sulcus submedian, curves backward at ventral 0.3 of valve. *L.Carb.,* Eu.(G.Brit.).

?Geffenina CORYELL & SOHN, 1938 [**G. marmerae*]. Differs from *Knoxina* in absence of crests. *Miss.,* N.Am.——FIG. 122,2. **G. marmerae,* U.Miss., USA(W.Va.); *2a,b,* LV (topotype), lat., vent., ×40 (Sohn, n).

Kloedenellitina EGOROV, 1950 [**Beyrichia? sygmae-*

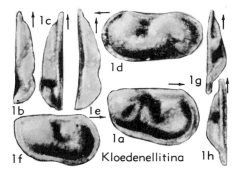

FIG. 126. Beyrichiopsidae (p. Q185).

formis BATALINA, 1941]. Differs from *Geffenina* in presence on posterior lobe of vertical sulcus that opens ventrally. *U.Dev.*, Eu.——FIG. 126,*1.* **K. sygmaeformis*, Russia; *1a-c,* ♀ RV (neotype) lat., dors., vent.; *1d,e,* ♀ LV lat., vent.; *1f-h,* ♂ RV lat., dors., vent.; all ✕25 (144).

Knoxina CORYELL & ROGATZ, 1932 [**K. lecta*] [=?*Coryella* HARRIS & LALICKER, 1932; ?*Chesterella* CRONEIS & GUTKE, 1939; ?*Mennerites* EGOROV, 1950]. Trilobate, small, with or without subdued horizontal crests, velum usually not preserved but reflected as inconspicuous flange. *U.Dev₂-M.Perm.,* N.Am.-?Eu.——FIG. 122,*1.* **K. lecta,* M.Perm., USA(Tex.); *1a-c,* carapace (holotype) L, dors., vent., ✕40 (Sohn, n).

?**Mennerella** EGOROV, 1950 [**M. tuberosa*]. Differs from *Knoxina* in ventrolateral inflation instead of crest, and in tuberculate lobes. *U.Dev.,* Eu.—— FIG. 122,*4.* **M. tuberosa,* Russia; *4a-c,* ♂ RV lat., int., dors.; *4d,e,* ♀ LV lat., immature specimens; all ✕22.5 (144).

?**Tambovia** SAMOILOVA, 1951 [**T. prima*] [=?*Marginia* POLENOVA, 1952]. Bilobate, with one or more crests parallel to each of end margins. *U.Dev.,* Eu.——FIG. 127,*1.* *T. sculpta* (POLENOVA), Russia; *1a,b,* carapace (holotype) L, dors., ✕40; *1c,d,* carapace (paratype) L, dors., ✕40 (277). [**T. prima* probably is based on juvenile instars.]

Family LICHVINIIDAE Posner in Egorov, 1950

[*nom. transl.* SOHN, herein (ex Lichvininae POSNER in EGOROV, 1950)] [=Lichwininae POSNER in POLENOVA, 1952] [Materials for this family prepared by I. G. SOHN, United States Geological Survey]

Straight-backed, subelliptical, punctate, or reticulate; small; with ocular protuberance and one or more subcentral sulci or pits; hinge of RV with groove in anterior third and posterior bar, that of LV with anterior bar and posterior groove; marginal rim along all margins, disconnected along dorsal margin where posterior part may continue as loop around concavity. Dimorph-

ism indicated by internal ?partition in posterior 0.3 of female carapace. *U.Dev.-M. Perm.*

Lichvinia POSNER in EGOROV, 1950 [**L. lichvinensis*]. Marginal rim extending as loop around ?single concavity. *Carb.,* Eu.——FIG. 128,*2. L. malevkensis* POSNER in EGOROV, Carb., Russia; *2a,* ♂ LV lat.; *2b-d,* ♀ carapace L, dors., vent.; all ✕40 (144).

Evlanella EGOROV, 1950 [**E. ljaschenkoi*]. Marginal rim not extending around single concavity, which has elongate ventral rib below it. *U.Dev.-M.Perm.,* Eu.——FIG. 128,*1.* **E. ljaschenkoi,* U.Dev., Russia; *1a,b,* ♂ RV (holotype) lat., vent.; *1c,d,* ♀ LV (paratype) lat., vent.; *1e,* ♀ RV lat.; all ✕ 40 (144).

Kalugia EGOROV, 1950 [**K. ivanovi*]. With 2 subcentral pits. *U.Dev.,* Russ.——FIG. 128,*3.* **K. ivanovi; 3a,b,* ♀ LV (holotype) lat., int.; *3c,d,* ♀ LV lat., dors.; *3e,f,* ♂ RV lat., vent.; all ✕40 (144).

?**Trinota** HOU, 1955 [**T. costata*]. Ventral ridge and 3 nodes, with central node above subcentral pit. *U.Dev.,* E.Asia.——FIG. 129,*1.* **T. costata,* China(Hupeh); *1a,b,* RV (holotype) lat., LV (paratype) lat., ✕? (176).

?Family MILTONELLIDAE Sohn, 1950

[Materials for this family prepared by I. G. SOHN, United States Geological Survey]

Straight-backed, reticulated, sulcate, with narrow, shallow groove that extends backward from anterior cardinal angle and curves around and below center of valve subconcentric with free margin; marginal frill narrow or dentate. *?U.Miss., M. Perm.*

Miltonella SOHN, 1950 [**M. shupei*]. Groove outlining 0.7 of lateral surface. *M.Perm.,* N.Am.—— FIG. 130,*3.* **M. shupei,* M.Perm. (Leonard or Word), USA(Tex.); *3a,* carapace (holotype) dors.

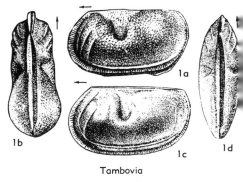

Tambovia

FIG. 127. Beyrichiopsidae (p. Q186).

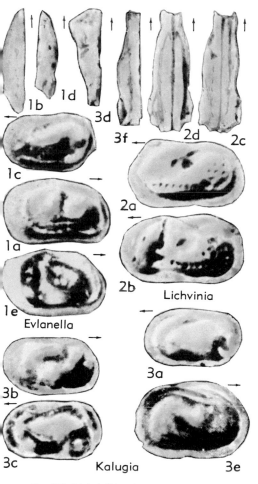

FIG. 128. Lichviniidae (p. Q186).

(333); *3b,* LV (paratype), int. (333); *3c,* LV lat. (73); all ×20.

Sanniolus SOHN, 1954 [**S. sigmoides*]. Sine-like lobe within area outlined by miltonellid groove. Perm., N.Am.——FIG. 130,2. **S. sigmoides,* Perm. (Leonard or Word), USA(Tex.); *2a-c,* RV (holotype), lat., int., dors., ×54 (73).

?Savagellites PŘIBYL, 1953 [*pro Savagella* GEIS, 1932 (*non* FOERSTE, 1920)] [**Kirkbya lindahli* ULRICH, 1891]. Differs from *Miltonella* in nearness of lateral groove to free margins with termination at dorsoposterior angle. *U.Miss.,* N.Am. ——FIG. 130,1. **S. lindahli,* USA(?Ill.); *1a-c,* carapace L, dors., vent., ×? (385).

Family SANSABELLIDAE Sohn, n. fam.

[Materials for this family prepared by I. G. SOHN, United States Geological Survey]

Straight-backed, unlobed, ?velate, small, with or without subcentral pit; larger valve overlapping opposite valve along free margin; ridge-and-groove hinge; surface texture unknown. Dimorphism indicated by width of posterior part. *M.Miss.-M.Penn.*

Sansabella ROUNDY, 1926 [**S. amplectens*] [=*Persansabella* CORYELL & SOHN, 1938; *?Lamarella, Carboprimitia* CRONEIS & FUNKHOUSER, 1939; *Reversabella* CORYELL & JOHNSON, 1939; *?Lochriella* SCOTT, 1942]. Dorsum incised along entire length of hinge; pseudovelum of some specimens preserved as spines on free margins of both valves; reversal of overlap usual. *M.Miss.-M.Penn.,* N. Am.-Eu.-Asia.——FIG. 122,7. **S. amplectens,* Miss., USA(Tex.); *7a,b,* carapace (lectotype, herein) ?L. lat., dors., ×60 (Sohn, n).

?Superfamily LEPERDITELLACEA Ulrich & Bassler, 1906

[*nom. transl.* LEVINSON, herein (*ex* Leperditellidae ULRICH & BASSLER, 1906)] [Diagnosis by H. W. SCOTT, University of Illinois]

Unisulcate, smooth or ornamented, straight-backed ostracodes. Includes many forms previously classified as primitiids. *L. Ord.-U.Jur.*

Except for a single genus *(Coryellina)* all known representatives of the Leperditellacea are restricted to pre-Carboniferous strata; they are most numerous and varied in Ordovician deposits. The stratigraphic dis-

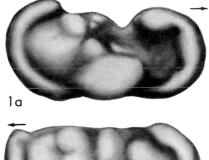

Trinota

FIG. 129. Lichviniidae (p. Q186).

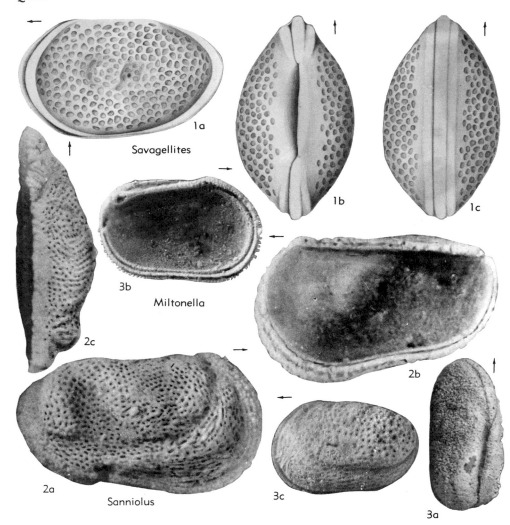

FIG. 130. Miltonellidae (p. *Q*186-*Q*187).

tribution of leperditellacean ostracode genera is shown graphically in Figure 131.

Family LEPERDITELLIDAE Ulrich & Bassler, 1906

[=Primitiidae ULRICH & BASSLER, 1923; Conchoprimitiinae, Eridoconchinae HENNINGSMOEN, 1953] [Materials for this family prepared by S. A. LEVINSON, Humble Oil & Refining Company, and R. C. MOORE, University of Kansas, with contributions from others as recorded, with additions by IVAR HESSLAND, University of Stockholm, and H. W. SCOTT, University of Illinois]

Straight-hinged, unisulcate, without marginal frill, velate structures, or pronounced flanges at extremities of hinge line. *L.Ord.-U.Jur.*

Leperditella ULRICH, 1894 [**L. rex* CORYELL & SCHENCK, 1941 (*nom. subst. pro Leperditia inflata* ULRICH, 1892, *non* MURCHISON, 1839; *nec* MUNSTER, 1830)]. Elongate oval, with barely perceptible, shallow S_2; anterior and posterior ends evenly rounded. *M.Ord.-U.Ord.*, N.Am.-Eu.——FIG. 132, *1*. **L. rex* CORYELL & SCHENCK, *M.Ord.*, USA (Ky.); *1a,b*, LV (holotype) lat., ×10, ×15 (*1a*, 386; *1b*, Levinson, n).

?Aparchitella IVANOVA, 1955 [**A. major*]. Large, straight-backed ostracodes with convex free margin; surface partially or wholly reticulate; valves unequal, large left valve overlaps right along margin; tubercle in anterodorsal area of left valve, reflected internally by deep pit; shallow furrow

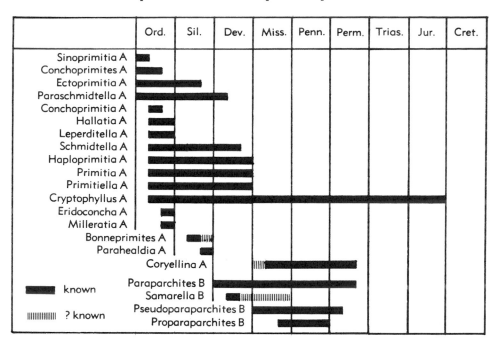

FIG. 131. Stratigraphic distribution of (A) leperditellacean and (B) paraparchitacean ostracode genera (Moore, n).

(S₂?) behind tubercle; large spine on right valve in anteroventral position directed anterolaterally, spine absent on left valve. Muscle-scar pattern unknown. *Ord.*, Russia.——FIG. 134*A,1*. **A. major*; *1a,b*, RV (young stage), LV (int. mold) ×15 (179).

Bonneprimites SWARTZ & WHITMORE, 1956 [**Primitia bonnemai* SWARTZ, 1936] [=*Eoprimitia* HARRIS, 1957]. Like *Leperditella* but with S₂ prominent, slightly curved. *M.Sil., ?U.Sil.*, Eu.-?N.Am.——FIG. 134,5. **B. bonnemai* (SWARTZ), M.Sil., Gotl.; LV lat., ×33 (74).

Conchoprimites HESSLAND, 1949 [**C. reticulifera*]. Somewhat preplete, LV larger than RV; unisulcate, short S₂ being generally rather deep, with anterior margin steeper than posterior; low presulcal node; groove (inferred retention line of larval stage), in some shells combined with steplike difference in level, extending along anterior or posterior margin or both, or along entire free margin; adventral structures and dimorphism not observed. *L.Ord.-M.Ord.*, Eu.(Baltoscandia).——FIG. 134,7*a*. **C. reticulifer*, L.Ord.(Llanvirn.), Swed.; LV (holotype) lat., ×30 (30).——FIG. 134,7*b*. *C. tolli* BONNEMA, M.Ord., Est.; LV lat., ×13 (58). [HESSLAND.]

Conchoprimitia ÖPIK, 1935 [**C. gammae*] [=*Conchoides* HESSLAND, 1949]. Like *Conchoprimites* but without sulcus; concentric grooves

parallel to free margin due to retained molts, as in *Conchoprimites*, *L.Ord.-M.Ord.*, Eu.-N.Am.——FIG. 132,9*a,b*. **C. gammae*, L.Ord.(Arenig.), Est.; *9a,b*, carapace (holotype) L, vent., ×20.——FIG. 134,9. *C. symmetrica* (ULRICH), M.Ord. (Decorah Sh.), USA(Minn.); *9a-c*, LV lat., int., RV int., ×20 (J. R. Cornell, n.).——FIG. 132,9*c-f*. *C. sp.*, M.Ord. (Edinburg F.), USA(Va.); *9c-f*, carapace L, R, dors., LV int., ×20 (J. C. Kraft, n). [*Hyperchilarina* HARRIS, 1957, classed as synonym of *Conchoides* and accordingly of *Conchoprimitia*.——SCOTT.] [According to HESSLAND, differences in gibbosity of carapaces suggest that *Conchoprimitia* is dimorphic. In his opinion, forms illustrated in Figs. 132,9*c-f*, and 134,9 are doubtfully assignable to this genus.——ED.]

Coryellina BRADFIELD, 1935 [**C. capax*]. Unisulcate, S₂ prominent; posterior border angularly truncated with straight posterodorsal half disposed at right angle with dorsal margin and straight posteroventral half that meets upper portion at 120-degree angle; posterior spine usually present near mid-height; hinge line with small cardinal flanges. *?L.Miss., U.Miss.-M.Perm.*, N.Am.-Eu.——FIG. 132,7. **C. capax*, M.Penn.(Deese F.), Okla.; *7a,b*, carapace R, dors., ×50 (11).

Cryptophyllus LEVINSON, 1951 [**Eridoconcha oboloides* ULRICH & BASSLER, 1923]. Umbonate, each valve with 2 to 6 wide concentric ridges parallel to free margins of carapace, separated from one

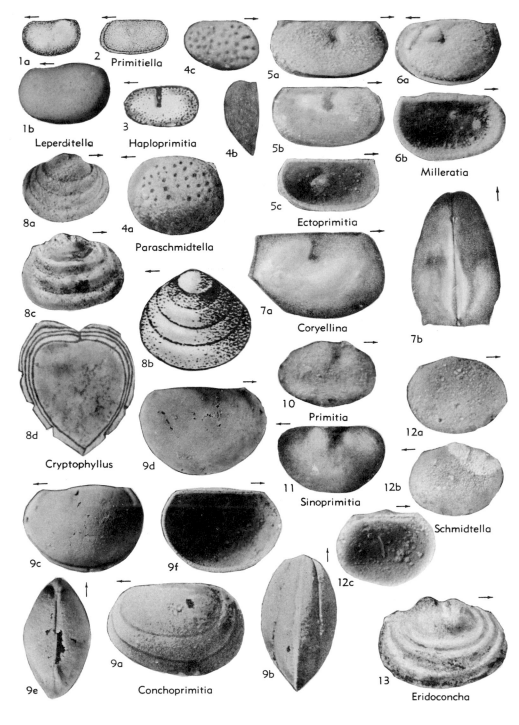

1a
2 Primitiella
4c
5a
6a
1b
3
Leperditella   Haploprimitia
4b
5b
6b
Milleratia
8a
4a
Paraschmidtella
5c
Ectoprimitia
7a
Coryellina
7b
8c
8b
10
Primitia
12a
8d
9d
11
Sinoprimitia
12b
Cryptophyllus
Schmidtella
9c
9f
12c
9e
9a
Conchoprimitia
9b
13
Eridoconcha

Fig. 132. Leperditellidae (p. Q188-Q193).

another by narrow V-shaped troughs; S_2 short and usually seen only in dorsal view; L_2 and L_3 developed in some specimens into small lobes, L_3 invariably larger than L_2. Similar to *Eridoconcha* except that ridges are separated from one another by narrow V-shaped troughs; differs from *Conchoprimitia* and *Conchoprimites* in shortness of hinges. [Transverse sections of carapaces demonstrate that the parallel grooves and ridges represent distal parts of retained molts.] *M.Ord.-U.Jur.*, Eu.-N. Am.——Fig. 132,*8a,b*. **C. oboloides* (ULRICH & BASSLER), M.Ord. (Decorah Sh.), USA(Minn.); *8a*, RV lat., ×50 (J. R. Cornell, n); *8b*, ?LV lat., ×40 (86).——Fig. 132,*8c,d*. *C. sulcatus* LEVINSON, U. Ord. (Richmond.), USA(Ind.); *8c*, ?RV lat., ×40; *8d*, carapace transv. sec., showing shell layers of retained molts; ×80 (49).

Ectoprimitia BOUČEK, 1936 [**Primitia corrugata* KRAUSE, 1892]. Like *Primitiella* except for deeper sulcus, that is more extended dorsally and with steeper anterior and posterior margins; cardinal angles sharp, unequal. *Ord.*, Eu.-N.Am.——Fig. 132,*5. E.* sp., M.Ord. (Edinburg F.), USA(Va.); *5a-c*, RV lat., RV lat., LV int., ×20 (J. S. Kraft, n).

Eridoconcha ULRICH & BASSLER, 1923 [**E. rugosa*]. Commonly umbonate with 1 to 5 narrow concentric ridges separated from one another by U-shaped troughs. S_2 short to elongate, L_2 and in some species L_3 developed into lobes, when developed L_3 always larger. Similar to *Cryptophyllus* except that ridges are separated by U-shaped troughs; ridges marking edges of individual molts not completely shed. *U.Ord.*, N.Am.——Fig. 132, *13. E. multiannulata* LEVINSON, U.Ord. (Richmond.), USA(Ohio); RV lat., ×50 (49).——Fig. 134,*3. E. marginata* (ULRICH), U.Ord. (Maquoketa F.), USA(Iowa); *3a-d*, LV lat., RV int., dors., vent., ×33 (118).

Hallatia KAY, 1934 [**H. healeyensis*]. Small, valves subequal, elongate oval, hinge line straight, ventral margin broadly convex, moderately deep short sulcus dividing valves into 2 broad, relatively flat lobes; cardinal angles obtuse; valves extended posteroventrally into histial-like structure; surface smooth. Hinge of tongue-and-groove type; interior of valves divided by ridge corresponding to S_2 sulcus of exterior, broad chambers thus formed uniting ventrally. [Dimorphism doubtfully recognizable in some species (e.g., *H. convexa*), which, if established, would indicate propriety of assigning genus to Sigmoopsidae.] *M.Ord.-U.Ord.*, N. Am.——Fig. 133,*1a-d. H.* sp., M.Ord. (Edinburg F.), USA(Va.); *1a-d*, RV lat., dors., vent., int., ×24 (J. C. Kraft, n).——Fig. 133,*1e-g. H. convexa* KAY, M.Ord. (Decorah Sh.), USA(Minn.); *1e-g*, RV lat., LV lat., LV int., ×36 (J. R. Cornell, n). [MOORE-SCOTT.]

Haploprimitia ULRICH & BASSLER, 1923 [**Primitia minutissima* ULRICH, 1894]. Like *Primitia* except

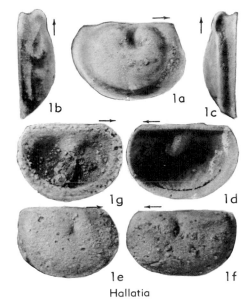

1a, 1b, 1c, 1g, 1d, 1e, 1f

Hallatia

FIG. 133. Leperditellidae (p. *Q*191).

that S_2 is very thin and straight; carapace usually elongate similar to *Primitiella*. *L.Ord.-U.Dev.*, Eu.-N.Am.——Fig. 132,*3. *H. minutissima* (ULRICH), M.Ord., USA(Minn.); LV lat., ×30 (83).

?Kayina HARRIS, 1957 [**K. hybosa*]. Straight-backed ostracodes with the left valve slightly larger than the right and overlapping it around free margin; greatest height and width posterior; posterodorsal knob on left valve; faint internal partition suggests primitive sulcus similar to that found in *Leperditella*. *M.Ord.*, N.Am.——Fig. 134*A,3. *K. hybosa*, M.Ord. (Tulip Creek Ls.), USA(Okla.); *3a-d*, LV, RV, post., dors., ×17 (161).

Milleratia SWARTZ, 1936 [**Beyrichia cincinnatiensis* MILLER, 1875]. Hinge long, straight; S_2 extending from dorsal margin to mid-height, commonly terminating ventrally in a pit; L_2 and L_3 in dorsal half of valve usually strongly inflated, L_3 larger than L_2; faint marginal ridge may be present near free borders. Males more elongate than females. *M.Ord.-U.Ord.*, N.Am.(Ohio-Ind.-Mich.).——Fig. 134,*4. *M. cincinnatiensis* (MILLER); *4a,b*, ♀ RV lat., ♂ RV lat., ×40 (49); *4c-h*, LV lat., int., RV lat., LV int., dors., carapace vent., ×33 (J. H. Burr, Jr., n); (*4a,b*, Richmond., Ohio; *4c-h*, Maquoketa F., Iowa).——Fig. 132,*6. M.* sp., M.Ord. (Edinburg F.), USA(Va.); *6a,b*, LV lat., int., ×30 (J. C. Kraft, n).

Parahealdia CORYELL & CUSKLEY, 1934 [**P. pecorella*]. Unisulcate, with 2 posterior spines; surface

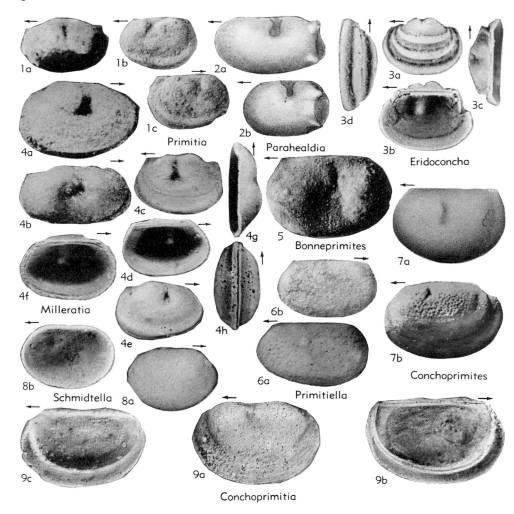

FIG. 134. Leperditellidae (p. Q189-Q193).

granulose or smooth. *L.Dev.*, N.Am.——FIG. 134, 2. **P. pecorella*, Haragan Sh., USA(Okla.); *2a,b*, ♂ LV lat., ♀ LV lat., ×13 (125).

Paraschmidtella SWARTZ, 1936 [**P. dorsopunctata*]. Amplete, cardinal angles obtuse, umbones broad and protruding above hinge line; nonsulcate; surface deeply pitted; dimorphism not observed. *Ord.-Dev.*, N.Am.-Eu.-E.Asia.——FIG. 132,*4a,b*. **P. dorsopunctata*, L.Dev., USA(Pa.); *4a,b*, LV (holotype, internal mold), lat., post., ×30 (74).——FIG. 132,*4c*. *P. ovata* (KAY), M.Ord. (Decorah Sh.), USA(Minn.); RV lat., ×50 (J. R. Cornell, n). [HESSLAND.]

Primitia JONES & HOLL, 1865 [**Beyrichia mundula* JONES, 1885 (ICZN pend.)]. Unisulcate, S₂ prominent and slightly curved. *L.Ord.-U.Dev.*, Eu.-N.

Am.——FIG. 134,*1a*. **P. mundula* (JONES), Sil., Ger.; *1a*, LV lat., ×33 (387).——FIG. 134,*1b,c*. *P. tumidula* ULRICH; U.Ord. (Maquoketa Sh.), USA(Iowa); *1b,c*, LV lat., RV lat., ×23 (J. R. Cornell, n).——FIG. 132,*10*. *P. mammata* ULRICH, M.Ord. (Decorah Sh.), USA(Minn.); RV lat., ×50 (J. R. Cornell, n).

Primitiella ULRICH, 1894 [**P. constricta* ULRICH]. Like *Leperditella* except more elongate. *L.Ord.-U.Dev.*, Eu.-N.Am.——FIG. 132,*2*. **P. constricta* ULRICH, M.Ord., USA(Minn.); LV lat., ×30 (83).——FIG. 134,*6a*. *P. unicornis* (ULRICH), M.Ord. (Decorah Sh.), USA(Minn.); *6a*, LV lat., ×33 (J. R. Cornell, n).——FIG. 134,*6b*. *P. plattevillensis* KAY, M.Ord. (Decorah Sh.), USA(Minn.); *6b*, RV lat., ×30 (J. R. Cornell, n).

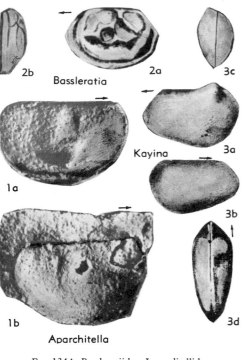

2b

Bassleratia

2a

3c

Kayina

3a

1a

3b

1b

Aparchitella

3d

Fig. 134A. Bassleratiidae, Leperditellidae (p. Q140, Q188, Q191).

Schmidtella Ulrich, 1892 [**S. crassimarginata*]. Subovate, moderately convex, most inflated in dorsal region, which tends to be umbonate; RV slightly larger than LV, overlapping it ventrally; no sulcus present but faint central pit and elevation in some species; carapace may retain immature molts, as in *Cryptophyllus* and *Conchoprimitia*. *M. Ord.-M.Dev.,* N.Am.-Eu.——Fig. 134,8. *S. incompta* Ulrich, U.Ord. (Maquoketa F.), USA (Iowa); *8a,b,* RV lat., int., ×33 (J. R. Cornell, n).——Fig. 132,12. *S. umbonata* Ulrich, M.Ord. (Decorah Sh.), USA(Minn.); *12a-c,* RV lat., LV lat., int., ×50 (J. R. Cornell, n).

Sinoprimitia Hou, 1953 [**S. sinensis*]. Unisulcate, with median oblique sulcus extending to dorsal margin, small node at each side of sulcus; surface smooth; similar to *Milleratia* but nodes smaller and valves not as inflated, *L.Ord.,* E.Asia(China). ——Fig. 132,11. **S. sinensis;* LV lat., ×? (175).

Superfamily PARAPARCHITACEA Scott, 1959

[Diagnosis by H. W. Scott, University of Illinois]

Nonsulcate, nonlobate, nonvelate palaeocopid ostracodes with unequal valves, the larger overlapping the smaller around all or most of the free margin. *L.Dev.-M.Perm.*

The Paraparchitacea are a relatively small assemblage of mostly robust late Paleozoic ostracodes. Their stratigraphic distribution is shown in Figure 131.

Family PARAPARCHITIDAE Scott, 1959

[Materials for this family prepared by H. W. Scott, University of Illinois]

Nonsulcate, nonlobate, nonvelate, and smooth to punctate, some with posterodorsal spine; dorsum straight to gently convex;

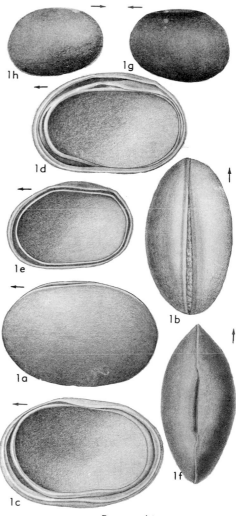

1h

1g

1d

1e

1b

1a

1c

1f

Paraparchites

Fig. 135. Paraparchitidae (p. Q194).

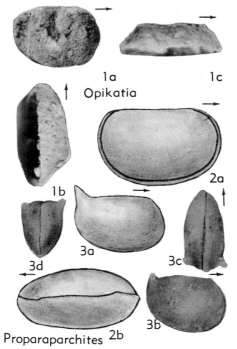

1a
Opikatia
1c

1b

2a

3a

3c

3d

3b

Proparaparchites 2b

Pseudoparaparchites

Fig. 136. Paraparchitidae, Palaeocopida, Suborder and Family Uncertain (p. Q194-Q195).

valves unequal, larger overlapping smaller along free margin, subovate to elongate-ovate, ends broadly rounded; hinge channel straight or interrupted at ends by faint to moderately strong posterior and anterior cardinal indentations where overlap begins; one valve may slightly overreach other dorsally, but dorsal shoulders usually of equal height. L.Dev.-M.Perm.

Paraparchites ULRICH & BASSLER, 1906 [*P. humerosus] [=Antiparaparchites CORYELL & ROGATZ, 1932; Ardmorea BRADFIELD, 1935; Coelonella STEWART, 1936; Microcoelonella CORYELL & SOHN, 1938]. Subovate to elongate-ovate, smooth except for posterodorsal spine in a few species; LV usually overlapping RV along free margin; hinge channeled; cardinal teeth poorly to well defined on exterior at cardinal points; greatest height medial or forward, greatest width medial in males, posterior in females, forward swing pronounced in some species. Dimorphism generally but not invariably observable. Valves may appear double- or triple-layered, owing to retention of molts. L. Dev.-M.Perm., N.Am.——FIG. 135,1a-f. *P. humerosus, L.Perm. (Admire), USA(Kans.); 1a,b,

carapace (lectotype) L (slightly tilted to show hinge channel), vent.; 1c-e, RV int., 3 specimens showing instar valves retained inside adult; 1f, carapace (paratype) dors.; all ×21 (324).——FIG. 135,1g,h. P. reversus (CORYELL & ROGATZ), L. Perm., USA(Tex.) (type species of Antiparaparchites); 1g,h, carapace L, R, ×30 (128).

Paraparchitella COOPER, 1946 [*P. ovata]. Like Paraparchites except that hinge channel is shallow and narrow; greatest width in posterior half; ventral overlap strong. L.Penn.(Seville Ls.), USA(Ill.). [SCOTT.]

Proparaparchites COOPER, 1941 [*P. ovatus]. Like Paraparchites but carapace subrectangular, without pronounced swing, ends similarly rounded. U. Miss.-Penn., N.Am.——FIG. 136,2. *P. ovatus, U.Miss., USA(Ill.); 2a,b, carapace R, dors., ×64 (Scott, n).

Pseudoparaparchites KELLETT, 1933 [*P. kansensis] [=Microparaparchites CRONEIS & GALE, 1938]. Like Paraparchites but strongly swollen in posterodorsal area and bearing spines on posterodorsal corners of each valve; smooth or punctate; greatest width medial in males, posterior in females. Miss.-Penn., N.Am.——FIG. 136,3a. *P. kansensis, Penn., USA(Kans.); carapace R, ×40 (Scott, n).——FIG. 136,3b-d. P. brazoensis (CORYELL & ROGATZ), Penn., USA(Tex.); 3b-d, carapace R, dors., post., ×32 (21).

?Samarella POLENOVA, 1952 [*S. crassa]. Dorsal margin straight to gently convex, ventral margin convex; LV overlapping RV strongly along ventral margin and slightly along anterior and posterior ends, RV overlapping and overreaching LV along dorsal margin; hinge straight in dorsal view without channel or tooth indentations, greatest width medial; surface smooth to slightly granular. [This genus is related to Carboniferous forms previously classified as Paraparchites.] M.Dev., ?L.Carb., Eu.——FIG. 109,2. *S. crassa, USSR; 2a-c, carapace R, L, dors., ×50 (277).

PALAEOCOPIDA, Suborder and Family UNCERTAIN

[Materials for this section prepared by authors as severally recorded at end of generic descriptions]

Acanthobolbina HARRIS, 1957 [*A. loeblichi]. Elongate, hinge line straight, with prominent S_2 extending from dorsal border to slightly below mid-height, curved around weakly-developed small node (?L_2), very large spine projecting outward from ventral region; velar ridge present. M.Ord. Okla.——FIG. 136A,11. *A. loeblichi; 11a,b, carapace R, dors., ×30 (161) [KESLING.]

Ctenonotella ÖPIK, 1937 [*C. elongata]. Elongate with straight, very long hinge line bearing row of spines pointing backward; anterior cardinal angle obtuse, posterior angle acute, in some specimen prolonged into spine; velar structure wide, restricted, surrounding anterior, ventral, and part of posterior margins; L_2 and L_3 elongate and narrow, L_1 and L_4 lacking or indistinct, only S_2 well developed; subvelar area unknown. Dimorphism not observed. *M.Ord.*, Eu.——FIG. 136A,*14*. *C. elongata*, Est.; LV (holotype), lat., ×24 (200, from 58). [Probably belongs in Quadrijugatoridae.] [HESSLAND.]

Dicranella ULRICH, 1894 [*D. bicornis*]. Straight-hinged, with pronounced swing, L_1 a node or spine, in some species extending above hinge line; L_2, if present, a small node on base of L_1, in some species inconspicuous; L_3 a prominent spine; S_2 a sharply defined sulcus or pit, L_1 and L_3 being joined in some specimens to form low ridge below S_2; frill from anterior corner to posteroventral part of free border. [This genus may be close to *Piretella*, as SCHMIDT (69) suggested. According to KAY (194), dimorphism is indicated by a broad frill in inferred females and a narrow frill or velate ridge in inferred males.] *M.Ord.*, N.Am.——FIG. 136A,*16a-d*. *D. bicornis*, USA(Minn.); *16a,b*, LV lat., carapace vent., ×33 (J. R. Cornell, n); *16c,d*, LV lat., ant., ×20 (200).——FIG. 16e-g. *D. marginata* KAY, Decorah F., USA(Minn.); *16e*, RV lat., ×33 (J. R. Cornell, n); *16f,g*, LV lat., ant., ×20 (200) [KESLING.]

Editia BRAYER, 1952 [*E. elegantis*]. Like *Amphissites* but differs in absence of nodes and kirkbyan pit, and in presence of eye tubercle, which excludes genus from Kirkbyidae. *M.Miss.*, N.Am.-Eu.——FIG. 136A,*13*. *E. elegantis*, USA(Mo.); *13a,b*, carapace (topotype) R, dors., ×60 (Sohn, n)·

Haplobolbina HARRIS, 1957 [*H. arcuata*]. Hinge line straight, L_3 a node tangent to dorsal border, slightly anterior; low node at anterior corner (?L_1); sulci very weakly developed; valves wide ventrally, with shallow, broad channel on marginal surface; velar ridge complete. *M.Ord.*, N.Am.——FIG. 136A,*3*. *H. arcuata*, USA(Okla.); *3a-c*, RV lat., dors., ant., ×25 (161). [KESLING.]

?Karlsteinella BOUČEK, 1936 [*K. reticulata*]. Large, equivalved, with straight hinge; S_2 well developed, anterior and posterior lobes broad; anterior and posterior ends somewhat extended; surface reticulate. [Genus based on a single specimen; proper allocation to family impossible.] *U.Sil.(Ludlov.)*, Czech. [SCOTT.]

Kayatia ÖPIK, 1953 [*K. prima*]. Unisulcate (sulcus mainly in dorsal half); with dorsoventrally elongate nodes (one on each side of sulcus); inequivalved (LV larger, rabbeted to receive edge of RV); with 2 to 4 ridges that may be partly developed as frills concentric to free margin (possibly velar, histial and carinal structures); dimorphism not observed. *Sil.*, Austral.——FIG. 136A,*6*. *K. prima*; *6a,b*, LV (holotype) lat., oblique ventral view of silicified specimen showing adventral structures, ×13; *6c*,

LV dors., ×26; *6d*, carapace transv. sec., ×26 (271). [HESSLAND.]

Kozlowskiella PŘIBYL, 1953 [*non* BOUCOT, 1957] [*Ulrichia (Kozlowskiella) kozlowskii*]. Approximately amplete, nonlobate, with distinct rounded sulcal pit, single dorsal bulb on each side of sulcal pit (posterior larger); adventral structure consisting of ridge along entire free margin; surface pitted; dimorphism not reported. Hinge with ridge in LV and corresponding furrow in RV. *M.Dev.*, Poland.——FIG. 136A,*1*. *K. kozlowskii*, Givet.; *1a,b*, carapace R, dors., ×13 (160). [Probably belongs in Bassleratiidae.] [HESSLAND.]

Nanopsis HENNINGSMOEN, 1954 [*Beyrichia nanella* MOBERG & SEGERBERG, 1906]. Preplete, unisulcate or bisulcate (if sulcus-like depression parting 2 anterodorsal nodes corresponds to S_1); no adventral structures; dimorphism not observed. *L.Ord.*, NW.Eu.(Scand.).——FIG. 136A,*2*. *N. nanella* (MOBERG & SEGERBERG), M.Ord.(Swed.); *2a*, LV (lectotype, 167), lat., ×33; *2b*, RV (Norway), lat., ×17 (*2a*, 252; *2b*, 167). [Probably belongs in Leperditellacea.] [HESSLAND.]

Opikatia KAY, 1940 [*O. emaciata*] [=*Oepikatia* HENNINGSMOEN, 1953 (as *Öpikatia*); seemingly KAY's original spelling must be preserved, since no indication is given that the generic name is derived from ÖPIK]. Truncate oval, valves subequal, medium-sized; dorsal margin slightly sinuous, rather long, free margins well rounded; dimorphic, tecnomorphs with median sulcus or pit bordered on posterior side of distinct lobe, heteromorphs with 3 rounded nodes adjacent to posterodorsal, mid-posterior, and posteroventral margins and corresponding internal pouches; surface smooth or slightly pitted. *M.Ord.*, N.Am.——FIG. 136,*1*. *O. emaciata*, U.Decorah Sh., USA(Minn.); *1a-c*, RV lat., carapace dors., LV vent., ×50 (J. R. Cornell, n). [MOORE.]

Pinnatulites HESSLAND, 1949 [*Primitiella procera* KUMMEROW, 1924]. Nonsulcate; inequivalved (RV larger); posterior part of swollen ventral area projecting backward into small process that is somewhat depressed laterally; no adventral structure, unless developed as angled or rounded bend (*Umbiegungskante*); dimorphism not observed. [May belong among Leperditellidae.] *Ord.*, NW. Eu. (Baltoscandia)-E.Asia.——FIG. 136A,*4*. *P. procera* (KUMMEROW), L.Ord.(Arenig.), Swed.; *4a,b*, LV lat., ant., ×20; *4c*, RV lat., with shell partly removed to show internal mold, ×20 (30). [HESSLAND.]

Piretopsis HENNINGSMOEN, 1953 [*P. donsi*]. Quadrilobate, with straight, long hinge line, approximately amplete; L_1 with hornlike projection directed laterally backward; L_2 tiny, nodelike; L_3 well defined; L_4 low, poorly defined; crests on all lobes except L_2, C_4 apparently not confluent with C_1-C_3; velate structure wide and incurved in the only specimen found, extending along anterior and ventral borders. *M.Ord.*, Eu.——Fig. 136A,*15.* *P. *donsi*, Norway; RV ext. mold (holotype) lat., ×20 (166). [Hessland.]

Polyzygia Gürich, 1896 [*P. *symmetrica*]. Quadrilobate, with straight and long dorsal margin, practically amplete, LV larger than RV; lobes dis-

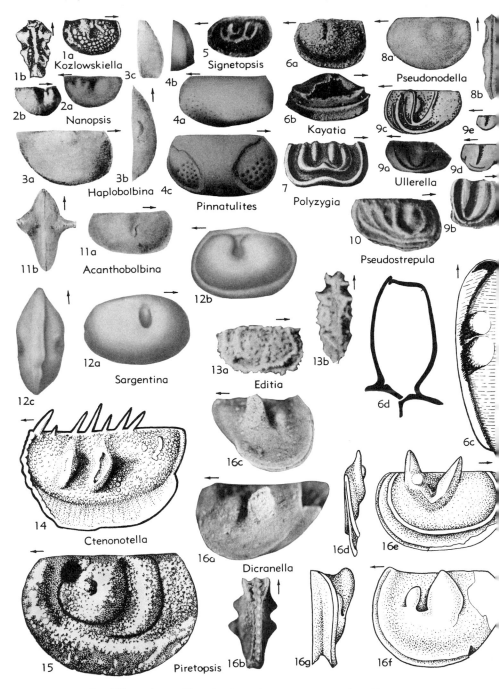

Fig. 136A. Palaeocopida, Suborder and Family Uncertain (p. Q194–Q197).

tinct, extending to dorsal margin, L_1 and L_4 joined by connecting structure; adventral structure probably velar, entire, indicated as *"Randwulst"*; subvelar area channeled; dimorphic, males being more elongate; surface may be spinulose. [Probably belongs to Bassleratiidae.] *M.Dev., Eu.(Pol., Russia).*——FIG. 136A,7. **P. symmetrica,* Pol.; RV (holotype) lat., ×20 (156). [HESSLAND.]

Pseudonodella ZASPELOVA, 1952 [**P. plana*]. Small, approximately amplete, flattened, shallow sulcal depression with low swellings on each side (anterior may have been formed by fusion of 3 lobes); adventral structure developed in many shells as low ridge; dimorphism not reported; surface smooth. *U.Dev.,* USSR.——FIG. 136A,8. **P. plana; 8a,b,* carapace L lat., dors., ×47 (Orientation according to Zaspelova) (407). [HESSLAND.]

Pseudostrepula ÖPIK, 1937 [**Strepula kuckersiana* BONNEMA, 1909]. Small, LV larger than RV, dorsal margin straight, hinge depressed, more or less preplete; L_2 knoblike, S_2 crescent-shaped, other lobes and sulci not clearly developed; crests bifurcated dorsally, 2 ventrally connected ridges may correspond to C_1-C_3, other crests questionable; velar structure entire, ridgelike, in some apparently wider, possibly indicating dimorphism; subvelar area gently channeled; marginal structure developed as ridge. *Ord., Eu.*——FIG. 136A, 10. **P. kuckersiana* (BONNEMA), M.Ord., Est.; RV lat., ×27 (58). [Probably belongs among Quadrijugatoridae.] [HESSLAND.]

Sargentina CORYELL & JOHNSON, 1939 [**S. allani*]. [=?*Semilukiella,* ?*Uchtovia* EGOROV, 1950; *Perimarginia* HOU, 1955]. Like *Sansabella* but differs in possessing subcentral sulcus and in curved dorsal margin of larger valve; spinose velum and reversal of overlap unknown. *?U.Dev., Miss.-Penn.,* N.Am.-Eu.-Asia. —— FIG. 136A,12. **S. allani,* Miss., USA(Ill.); *12a-c,* carapace (holotype) R, L, dors., ×30 (Sohn, n). [SOHN.]

Signetopsis HENNINGSMOEN, 1954 [**S. quadrilobata*]. Dorsal margins straight or slightly convex, outline almost amplete to postplete; quadrilobate to bilobate with comma-like ridge adherent to L_2 posteriorly ventral to S_2; adventral structure entire, developed as flange, widest posteriorly, where it is apparently dimorphic, confluent with entire dorsal ridge; subvelar area unknown. *M.Sil.-U.Sil.,* NW. Eu.——FIG. 136A,5. **S. quadrilobata,* Norway; tecnomorph LV (holotype) lat., ×24 (167). [May belong in Bassleratiidae.] [HESSLAND.]

Strepula JONES & HOLL, 1886 [**S. concentrica*]. Differs from *Amphissites* in absence of nodes and kirkbyan pit. [Referred to Tetradellidae by HENNINGSMOEN, 29.] *Sil., Eng.*

Ullerella HENNINGSMOEN, 1950 [*nom. subst. pro Ullia* HENNINGSMOEN, 1949 (*non* ROEWER, 1943)] [**Ullia ulli* DONS, 1949]. Preplete, S_2 distinct surrounded by 2 or 3 (rarely 4) U-shaped ridges

that generally reach dorsal margin, except posterior branch of outer ridge(s) which may be developed in ventral part only; inner arc may correspond to L_2 and L_3 being connected ventrally and an outer arc to L_1 and L_4 united by connecting lobe; velar structure comprising ridge or moderately wide frill; dimorphism not stated. [May belong in Sigmoopsidae.] *Ord.,* NW.Eu. ——FIG. 136A,9a,b. **U. ulli* (DONS), M.Ord. Norway; *9a,* LV (holotype) lat., ×7; *9b,* LV (cast from internal mold) showing velate structure, ×7 (141).——FIG. 136A,9c-e. *U. ventroplicata* HENNINGSMOEN, M.Ord., Norway; *9c,* LV lat., ×10; *9d,e,* LV (larval instars) lat., ×10 (166). [HESSLAND.]

Order PODOCOPIDA, Müller, 1894

[*nom. correct.* POKORNÝ, 1953 (*pro* Podocopa MÜLLER, 1894 =Podocopa SARS+Platycopa SARS)] [Diagnosis by P. C. SYLVESTER-BRADLEY, University of Leicester]

Ostracoda with dorsal margin curved, or, if straight, shorter than total length. Duplicature narrow or wide. Adductor muscle-scar pattern a circular aggregate of many scars in more primitive members of order (Metacopina); number of individual scars reduced in others, either aggregate and biserial (Platycopina), or discrete and variously arranged (Podocopina). Hinge margin undifferentiated in Platycopina and some Metacopina, commonly differentiated into three or more elements in Podocopina. *L.Ord.-Rec.*

The Podocopida were presumably derived from the Palaeocopida in Ordovician times, but may be polyphyletic.

Suborder PODOCOPINA Sars, 1866

[*nom. correct.* SWAIN, herein (*pro* Podocopa SARS, 1866)] [Diagnosis by F. M. SWAIN, University of Minnesota, and P. C. SYLVESTER-BRADLEY, University of Leicester; description and discussion by F. M. SWAIN and H. V. HOWE, Louisiana State University]

Ostracoda with four pairs of postoral appendages, usually with well-developed eyes, exopodite of antennae poorly developed, and having barlike furcal ramus; carapace without permanent aperture anteriorly, and with mid-ventral incurvature of outer valve margin. Muscle-scar pattern of discrete scars, in which adductor scars are commonly developed in a group distinct from scars of muscles operating appendages. Duplicature commonly wide, with or without vestibule; free-margin bearing a selvage, which overlaps that of opposite valve, thus sealing carapace when valves are closed. Hinge commonly differentiated into three or four elements, any or all of which may be denticulate. [Marine and fresh-water, rarely terrestrial; world-wide.] *L.Ord.-Rec.*

The antennules (first antennae of some authors) of Podocopina are uniramous (Fig. 137,*1*) and exopodites are lacking. The basal portion (protopodite) consists of a single divided podomere or of two podomeres; the

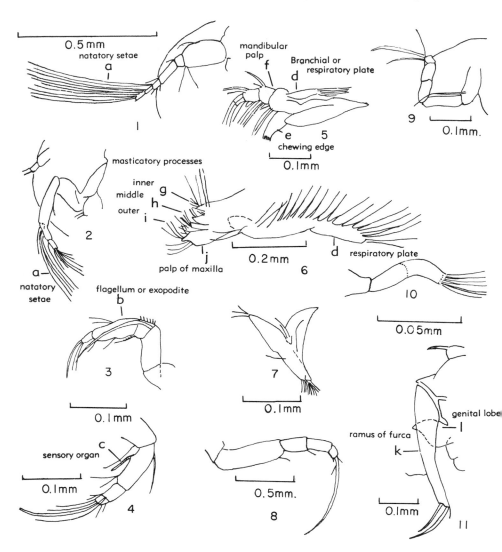

FIG. 137. Appendages of podocopine ostracodes (from 31).——*1. Cypricercus reticulatus* (ZADDACH), left antennule (*a*, natatory setae).——*2. Cypricercus reticulatus* (ZADDACH), left antenna (*a*, natatory setae).—— *3. Limnocythere verrucosa* HOFF, left antenna of female (*b*, flagellum or exopodite).——*4. Candona fluviatilis* HOFF, right antenna (*c*, sensory organ).——*5. Candona indigena* HOFF, left mandible (*d*, branchial or respiratory plate; *e*, chewing edge; *f*, mandibular palp).——*6. Candona acuta* HOFF, left maxilla (*d*, respiratory plate; *g*, inner masticatory process; *h*, middle masticatory process; *i*, outer masticatory process; *j*, palp of maxilla).——*7. Candona fluviatilis* HOFF, first thoracic leg of right side.——*8. Candona biangulata* HOFF, second thoracic leg.——*9. Candona suburbana* HOFF, third thoracic leg.—— *10. Limnocythere verrucosa* HOFF, brush-form sensory organ of male.——*11. Candona acuta* HOFF, furca of female (*k*, ramus of furca; *l*, genital lobe characteristic of many Candonidae).

endopodite in Cyprididae typically has five podomeres, reduced in fresh-water Cytheridae of North America to three or four podomeres (31, p. 43). The antennules generally bear short, stiff clawlike bristles for digging and climbing (in Cytheridae and Darwinulidae), or long-feathered swimming setae (Fig. 137,*1a*) (in most Cyprididae).

The antennae (Fig. 137,*2*) (second antennae of some authors) consist of a protopodite with one or two podomeres, and an endopodite with three or four podomeres. The exopodite comprises a scale having three setae in Cyprididae, a long bent seta (Fig. 137,*3b*) that contains an adhesive-secreting duct in the Cytheridae, and is completely vestigial in the Darwinulidae. The first podomere of the endopodite in living Cypridacea (except Eucandonidae) commonly bears a group of five natatory setae (Fig. 137,*2a*) located near the distal end; these setae are greatly reduced in the Darwinulidae and most living Cytheracea and are absent in the Eucandonidae. This same podomere bears a sensory organ (Fig. 137, *4c*) on the posterior margin. The ultimate podomere of the endopodite is claw-bearing. The antennae function in locomotion and feeding as sensory structures, and in males for grasping the female during copulation.

Mandibles are composed of two podomeres, a palp representing the endopodite and a modified exopodite that may form a branchial plate (Fig. 137,*5d*) or be reduced to a few setae. Its basal portion is highly chitinized (Fig. 137,*5e*) for attachment of mandibular muscles. The truncate distal end bears strongly chitinized teeth. The mandibular palp (Fig. 137,*5f*) is composed of the second podomere of the base of the mandible and three podomeres of the endopodite; it bears claws and setae.

Maxillae are characterized by a basal portion having three narrow distal masticatory processes (Fig. 137, *6g,h,i*) and a palp (Fig. 137,*6j*) of two or three podomeres. The exopodite is well developed, forming a branchial plate (Fig. 137,*6d*). The number and shape of the setae on the outer masticatory process, lying next to the palp, are taxonomically important; in some living Cytheracea the palp and processes may be reduced.

The first thoracic leg in living Cytheracea consists (Fig. 137,*7*) of a forward-directed protopodite with a pediform endopodite of three or four podomeres, and exopodite or branchial plate being absent. In living Cypridacea the first leg commonly is much modified, resembling the maxilla, the basal podomere ending in a setaceous masticatory process and the endopodite being modified as a palp that in many forms consists of a single podomere; the exopodite typically is a branchial plate but it may be reduced to one or a few setae, as in Eucandonidae and Cypridopsinae (Cyprididae); the palp in males may be modified as a prehensile claw of one or two podomeres for use in copulation.

The second thoracic legs are uniramous, consisting of a protopodite and a backward-directed pediform endopodite (Fig. 137,*8*) which terminates in a curved claw. The endopodite typically is composed of three or four podomeres, but an additional one may result from division of the next-to-last podomere. An exopodite is lacking.

The third thoracic legs typically have five podomeres (Fig. 137,*9*), resembling the second legs in living Cytheracea and in Darwinulacea, but in living Cypridacea the third legs are represented by basal podomeres and endopodites consisting of three podomeres, one of which may be divided. This leg in living Cypridacea is bent dorsally and modified for use as a cleaning foot.

Living cytheracean males may display structures interpretable as a fourth pair of thoracic appendages; these are the so-called "brushlike organs" (Fig. 137,*10*), which have a tuft of setae and are provided with nerves; they may be sexual sensory organs.

At the posterior end of the body in living Cypridacea are paired appendages termed furcal rami, which are believed to be remains of the abdomen. Each typically consists of a basal ramus (Fig. 137,*11k*) articulated with the body and ending distally in two setae and two claws. In the Cypridopsinae the furcal rami are reduced to small bases with a dorsal seta and terminal whiplike flagellum. Furcal rami in living Cytheracea are much reduced and in the Darwinulacea they are absent.

The nervous, respiratory, circulatory, excretory, and digestive systems (and processes) of Podocopa are described in an introductory section.

Paired ovaries in living cypridacean females lie in a space between lamellae of each valve; in living Cytheracea these lie lateral to the mid-gut in the body of the animal. The testes in living male Cypridacea lie in a cavity of the valves, and consist of about four branches that unite to form a vas deferens connected (after making a series of loops) to an ejaculatory duct. Testes in living Cytheracea adjoin the intestine in the body proper and no ejaculatory duct is present (68, 31).

Sars (65, p. 10) erected the suborder Podocopa [Podocopina] to include the families Cypridae [*recte* Cyprididae] (*Cypris, Cypria, Paracypris, Notodromas, Candona, Pontocypris, Argilloecia, Bairdia*) and Cytheridae (*Cythere, Cythereis, Cyprideis, Cytheridea, Cytheropsis, Ilyobates, Loxoconcha, Xestoleberis, Cytherura, Cytheropteron, Bythocythere, Pseudocythere, Sclerochilus, Paradoxostoma*). This was based on anatomical grounds, since the cited families and genera all possess an antenna (second

antenna of various European authors) modified essentially for walking rather than swimming, as observed in other known living ostracodes. The name is derived from Greek words meaning "foot-oar," the antennae being conceived to serve the ostracodes as oars.

Sars (312, p. 44-45) considered the mid-ventral carapace structure to be important in classification, stating that ". . . ventral edges of the valves in the oral region [are] conspicuously bent inwards and somewhat bowed, so as to overlap each other. . .," noting also that ". . . the peculiar closure of the valves in the oral region [of Podocopina] is very characteristic, no trace of such a closure being found in any of the forms belonging to the three preceding suborders [Platycopina, Myodocopina, Cladocopina], whereas in all known Podocopa [Podocopina] its existence may be demonstrated."

In 1888 (p. 288) Sars erected the family Bairdiidae for the genus *Bairdia* and in

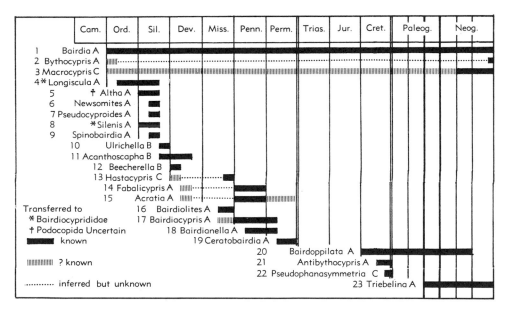

Fig. 138. Stratigraphic distribution of bairdiacean ostracode genera (Moore, n). Classification of the genera in families is indicated by letter symbols (A—Bairdiidae, B—Beecherellidae, C—Macrocpyrididae) and the following alphabetical list furnishes a cross reference to the serially arranged numbers on the diagram.

Generic Names with Index Numbers

Acanthoscapha—11	Bairdianella—18	Fabalicypris—14	Pseudophanasymmetria—22
Acratia—15	Bairdiolites—16	Hastacypris—13	Silenis—8
Altha—5	Bairdoppilata—20	Longiscula—4	Spinobairdia—9
Antibythocypris—21	Beecherella—12	Macrocypris—3	Triebelina—23
Bairdia—1	Bythocypris—2	Newsomites—6	Ulrichella—10
Bairdiacypris—17	Ceratobairdia—19	Pseudocyproides—7	

1923 added *Bythocypris* to this family. The family Darwinulidae was introduced by BRADY & NORMAN in 1889 (p. 121).

The Cyprididae, Darwinulidae, Cytheridae, and Bairdiidae of early authors are here considered as superfamilies, but regardless of the taxonomic rank assigned to them, it is evident that they are related anatomically in possessing antennae modified in such a manner as to permit their use as ambulatory organs. All have calcified shells with distinctive characteristics that may be used by paleontologists. Most cytherid genera have well-developed hinge teeth and muscle scars arranged in a vertical or nearly vertical row of four to six closing muscle scars, with additional mandibular scars in front. Cypridid genera have an irregular grouping of closing scars, with mandibular scars below and in front of the closing scars. Bairdiid and darwinulid genera exhibit muscle scars arranged in a more or less oval pattern, the darwinulid scars showing a somewhat radial pattern about a common center.

The Darwinulacea are sufficiently different from the Cypridacea in anatomical and shell features to warrant classification as separate superfamilies.

The Quasillitacea of CORYELL & MALKIN (22) resemble the Podocopina in over-all shape of their carapaces and possession of a tripartite hinge and calcareous inner lamella, but they show a muscle-scar arrangement suggestive of the Platycopina, which G. W. MÜLLER (53) considered merely a division of the Podocopa [Podocopids) because the platycopina antennae could be used for walking. The antennae of the Platycopina, however, differ distinctly from those of Podocopina in their strongly biramous flattened character, and typically in the Cytherellidae, the carapace also is different in shape and in the manner of articulation, having a peripheral lock, the furrow being in the larger right valve. The Quasillitacea here are placed in the Metacopina.

The Thlipsuracea, although classified in the present *Treatise* with the Metacopina, have several podocopine characters, including terminally differentiated hinge features (*Strepulites*), calcified inner lamellae (*Stibus*), a bairdiid type of muscle-scar pattern (*Thlipsura*), and, most importantly, mid-

ventral incurvature of the valve margins. Accordingly, this superfamily might be arranged with the Podocopina. Worthy of notice, however, are the lack of demonstrated occurrence in all genera of a calcified inner lamella, bairdiid muscle-scar, and ventral incurvature.

Superfamily BAIRDIACEA Sars, 1888

[*nom. transl.* SYLVESTER-BRADLEY, 1948 (*ex* Bairdiidae SARS, 1888)] [Diagnosis and discussion by P. C. SYLVESTER-BRADLEY, University of Leicester]

Convex-backed Podocopina with wide duplicature, wide vestibule at anterior end, narrow or wide vestibule at posterior end. Muscle-scar pattern consisting of discrete scars, more or less radially arranged. Hinge various, some genera having structures not developed in other superfamilies (e.g., *Macrocypris*); denticles develop in some forms along contact margins, and are not confined to hinge margin (e.g., *Bairdia*, *Bairdoppilata*). [Marine.] *L.Ord.-Rec.*

The main development of the superfamily was in Late Paleozoic time. *Bairdia* is the longest-ranging ostracode genus known (Ord.-Rec.), and is probably the only Paleozoic member of the Bairdiidae to survive into the Mesozoic (Fig. 138). *Bairdia* has often been regarded as a close relative of the fresh-water genus *Cypris* (Pleist.-Rec.), and united with that genus in the family Cyprididae. However, the geological history of the two genera has been distinct from at least Carboniferous time onward, and in this work each is regarded as the type of a distinct superfamily. The Bairdiacea underwent a major radiation in Paleozoic time. The Cypridacea are well represented but inadequately known in the late Paleozoic and Mesozoic; they underwent an extensive radiation in early Cenozoic time, when many of the genera known to occur in Recent sediments were evolved.

Family BAIRDIIDAE Sars, 1888

[=Nesideidae G. W. MÜLLER, 1912] [Materials for this family by R. H. SHAVER, Indiana University and Indiana Geological Survey, with additions by IVAR HESSLAND, University of Stockholm]

Carapace convex-backed, mostly with asymmetrical, angulated, convex and concave, rounded and acuminate so-called "bairdian" shape in lateral view; lateral outlines mostly symmetrically convex and terminally acuminate in dorsal view; LV larger than RV, both overreaching and overlapping it; with short, ridge-and-groove hingement and prominent duplicature and vestibule; muscle-scar pattern composed of several discrete spots. *L.Ord.-Rec.*

The family is especially characterized by a wide duplicature, which is well developed terminally, less so ventrally, with conspicuous vestibule, wide zone of fusion with the outer lamella, and long radial pore canals, but along the hinge margin the duplicature is either absent or so tightly compressed against the outer lamella that many contact marginal structures are absent (Figs. 139,*1,2*). The marginal ridges and grooves are differentially prominent or weak around the edges of either valve and are conspicuously modified dorsally into a short, ridge-and-groove hingement (Fig. 140). The adductor muscle-scar pattern forms a rosette or other shape, composed of discrete spots that commonly range in number from 7 to 12 (Fig. 139,*3-7*). The Bairdiidae have an unusually long range, being reported from early Paleozoic to the present, but basic

morphological characters vary little; slight modifications of the marginal elements and sculpturing are mostly employed for generic differentiation.

The present classification is more restricted than that of some authors in excluding healdiids, cypridids, and others previously assigned to this family, mostly on the basis of a "convex-back." The Bairdiidae are easily differentiated from the Healdiidae by their wide duplicatures and associated vestibules and by muscle scars containing fewer spots. They are less easily separated from the Cyprididae by their commonly more complex hingement, contact margins, and muscle scars. Bairdian shape, overworked for taxonomic purposes at the species level, becomes significant as a family character when emphasis is placed on the nature of marginal structures and muscle-scar pattern.

Bairdia McCoy, 1844 [**B. curtus*] [=*Nesidea* Costa, 1849; *Morrisitina* Gibson, 1955 (*pro Morrisites* Gibson, 1955, *non* Buckman, 1921); *Acratinella* Schneider, 1956]. Carapace mostly elongate

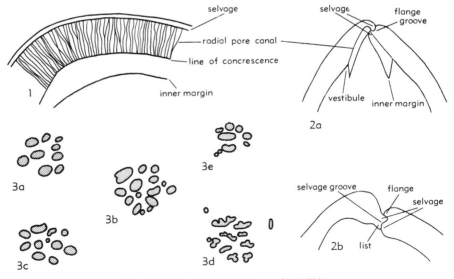

Fig. 139. Morphological features of Bairdiidae.

1. Bairdia minor Müller, Rec., Medit., showing radial pore canals at anterior extremity of valve interior, ×145 (53).

2. Bairdia oklahomensis Harlton, Penn., USA (Ill.); *2a*, long. sec. through anterior margin of carapace, ×65; *2b*, transv. sec. through hinge, ×120. (Shaver, n).

3. Bairdiid muscle-scar patterns. *3a, Bairdoppilata*

sp., U.Cret., USA(Tex.); from RV ext., ×100 (Shaver, n). *3b, Bairdia golcondensis* Croneis & Gale, U.Miss., USA(Ill.); from RV ext., ×100 (Shaver, n). *3c, B. menardensis* Harlton, Penn., USA(Ill.); from RV ext., ×75 (Shaver, n). *3d, B. formosa* Brady, Rec., Medit.; from LV int.; ×70 (Sylvester-Bradley). *3e, Bythocypris bosquetiana* (Brady), Rec., Medit.; from LV ext., ×60 (Shaver, n).

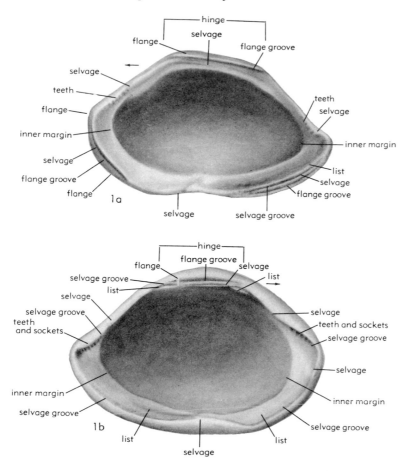

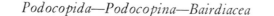

Fig. 140. Morphological features of typical bairdiid, *Bairdoppilata* sp., Upper Cretaceous of Texas; *1a,b*, right valve interior, left valve interior, ×57 (Shaver, n).

fusiform in lateral view, with broadly arched dorsum that becomes concave terminally, especially at the rear; venter centrally straight but curved upward terminally so that extremities are nearly at mid-height, anterior end generally higher and better rounded than the posterior, which generally is acuminate; in dorsal view, lateral outlines symmetrically convex and extremities acuminate; surface of valves smooth, punctate, or rarely with protuberances; LV larger than RV, mostly overreaching it around entire margin and overlapping it, especially ventrally, around contact margin. Short ridge-and-groove hingement commonly marked by prominent cardinal angles, especially in RV; contact margins complex, with wide duplicature, vestibule, and associated structures; adductor-muscle scar pattern of several

discrete spots (21, 43, 53). *Ord.-Rec.,* cosmop.——FIG. 141,*2a,b. B. oklahomensis* HARLTON, Penn., USA(Ill.); *2a,b,* carapace R, dors., ×35 (Shaver, n).——FIG. 141,*2c,d. B. formosa* BRADY, Rec., Medit.; *2c,d,* LV int., RV int., ×50 (363).

?**Acratia** DELO, 1930 [**A. typica*] [=*Acratina* EGOROV, 1953]. Carapace terminally acuminate in lateral view, especially posteriorly, with broadly arched dorsal border making acute angles with venter so that extremities are low and pointed; venter with anterior upswept concavity; internal morphology unknown. *?M.Dev., Penn., ?Perm.,* N.Am.-Eu.——FIG. 142,*1. *A. typica,* Penn., USA(Tex.); *1a,b,* carapace R, dors., ×36 (138).

Antibythocypris JENNINGS, 1936 [**A. gooberi*]. Carapace subreniform, resembling *Bythocypris* in lateral view; with short, simple, posterodorsal, ridge-and-groove hingement near that of *Bairdia;* surface coarsely punctate and ridged (181). [Although original orientation and reported RV over LV overlap by the author is incorrect, the genus is considered distinct from *Bythocypris* because of its sculpture, muscle scar, hinge, and contact-marginal structures.] *U.Cret.,* N.Am.——Fig. 144, 5. **A. gooberi,* USA(N.J.); *5a,b,* carapace R, dors., ×75 (Shaver, n).

Bairdiacypris Bradfield, 1935 [*non Bairdiocypris*

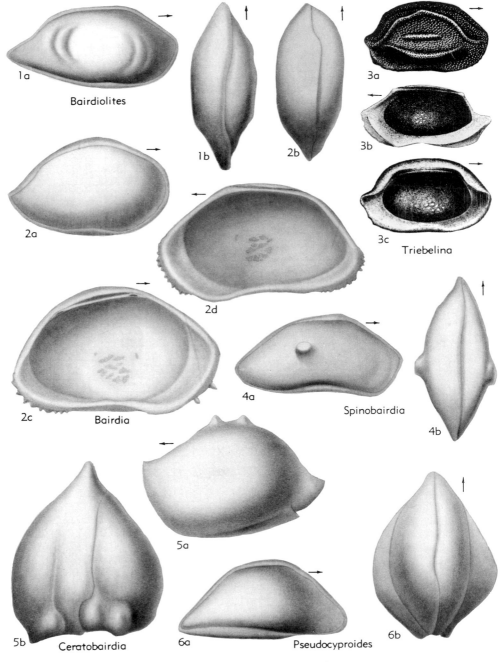

Fig. 141. Bairdiidae (p. *Q202-Q207*).

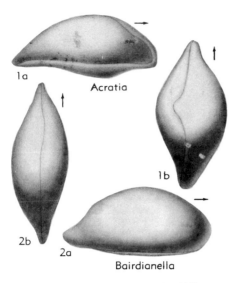

FIG. 142. Bairdiidae (p. Q203-Q205).

KEGEL, 1932 (*Bairdiocyprididae*)] [*B. deloi*] [=*Actuaria* SCHNEIDER, 1956]. Elongate, sub-reniform, resembling *Argilloecia* (Pontocyprididae) in lateral view but with slight angulation and nearly straight dorsal slopes suggestive of *Bairdia;* hinge, duplicature, and associated structures like those of *Argilloecia, Bairdia,* and *Fabalicypris* (11, 21). ?*U.Miss., Penn.-L.Perm.,* N.Am.-Eu.——FIG. 144,*1a,b.* *B. deloi,* Penn., USA(Okla.); *1a,b,* carapace R, dors., ×40 (Shaver, n).——FIG. 144, *1c,d.* B. *haydenbranchensis* (PAYNE), Penn., USA (Ill.); *1c,* long. sec. through ant. margin, ×65 (Shaver, n); *1d,* transv. sec. through hinge, ×65 (Shaver, n).

Bairdianella HARLTON, 1929 [*B. elegans*]. Type species differs from *Bairdia* in its lack of all but ventral overlap and by its low posterior area, both suggestive of molts of *Bairdia;* imperfectly known. *M.Penn.-L.Perm.,* N.Am.——FIG. 142,2. *B. elegans,* Penn., USA(Tex.); *2a,b,* carapace R, dors., ×50 (Shaver, n).

Bairdiolites CRONEIS & GALE, 1939 [*B. crescentis*]. Like *Bairdia* but with 2 crescentic ridges located antero- and posterocentrally on each valve (20). *U. Miss.,* N. Am. (Ill., Okla.)-Eu.(Eng.). —— FIG. 141,*1.* *B. crescentis; 1a,b,* carapace R, dors., ×60 (132).

Bairdoppilata CORYELL, SAMPLE & JENNINGS, 1935 [*B. martyni*]. Like *Bairdia* but each valve with short series of transverse teeth and sockets in antero- and posterodorsal positions, in selvage of RV and selvage groove of LV. *L.Cret.-Tert.,* cosmop., USA-Venez.-Alg.——FIGS. 139,*3,* 140,*1.* B. sp., U.Cret., USA(Tex.); 139,*3,* muscle-scar pattern from RV ext., ×100; 140,*1a,b,* RV int., LV int., ×57 (Shaver, n).

Bythocypris BRADY, 1880 [*B. reniformis* (=*Bairdia bosquetiana* BRADY, 1866)]. Carapace reniform in lateral view, with straighter venter and more rounded dorsum and extremities; mostly lacking angulation and asymmetry of *Bairdia,* but with similar overlap and overreach, ridge-and-groove hinge, duplicature, vestibule, and associated structures, including muscle scar (53). [Many modern species are defined partly on the basis of soft parts, whereas numerous Paleozoic species do not exhibit the muscle scar and marginal structure of Recent *Bairdia* and other Bairdiidae. Altogether, the genus is in an unsatisfactory condition, nearly all Paleozoic species being doubtfully assigned to it.] ?*L.Ord., Rec.,* cosmop.——FIG. 144,*4.* *B. reniformis,* Rec.; *4a-c,* carapace L, dors., RV lat., N.Atl., ×50 (13); *4d,* carapace vent., ×40 (13); *4e,* LV, from ext., ×40 (53).

Ceratobairdia SOHN, 1954 [*C. dorsospinosa*]. Like *Bairdia* but with ventral alae; one or more centrodorsal spines or knobs on LV. *M.Perm.-U.Perm.,* N.Am.——FIG. 141,*5.* *C. dorsospinosa,* USA (Tex.); *5a,b,* carapace L, post., ×30 (73).

Fabalicypris COOPER, 1946 [*F. wileyensis*]. Carapace tumid, elongate-elliptical in lateral view, mostly lacking suggestion of bairdian shape but apparently one end of a complete, *Bairdia-Bairdiacypris-Fabalicypris* gradational series; LV over RV overlap abruptly offset anteroventrally; hinge and contact margins near those of *Bairdia* (21). ?*M.*

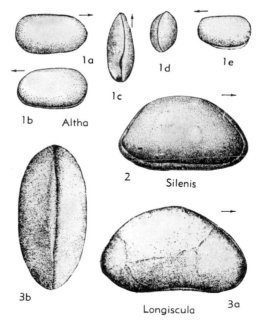

FIG. 143. Bairdiocyprididae (doubtful) and Podocopida (classification uncertain) (p. Q366-Q367, Q383-Q384).

Dev., Penn., N.Am.-?Eu.——Fɪɢ. 144,*2a,b.* **F. wileyensis,* Penn., USA(Ind.); *2a,b,* carapace R, vent., ×40 (21).——Fɪɢ. 144,*2c-e. F.* sp., Penn., USA(Ill.); *2c,* long. sec. through ant. margin, ×65 (Shaver, n); *2d,e,* transv. secs. through hinge and venter, ×65 (Shaver, n).

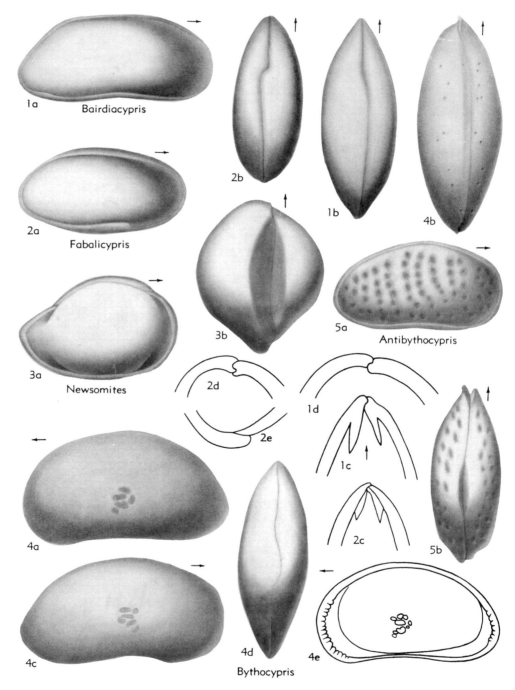

Fɪɢ. 144. Bairdiidae (p. Q203-Q207).

Newsomites Morris & Hill, 1952 [**N. pertumidus*]. Nearly bairdian in lateral view but carapace very tumid, nearly as thick as long, with lateral-dorsal inflation causing strong depression of hinge; reversal of normal LV over RV overlap and overreach common; imperfectly known. *M.Sil.*, N.Am.——Fig. 144,3. **N. pertumidus*, USA(Tenn.); *3a,b*, carapace R, dors., ×45 (254).

Pseudocyproides Morris & Hill, 1952 [**P. alatus*]. Bairdian in outline excepting abnormally low extremities, strong alate extensions jointly from lateral and ventral surfaces producing triangular outline in end view, imperfectly known. *M.Sil.*, N.Am.——Fig. 141,6. **P. alatus*, USA(Tenn.); *6a,b*, carapace R, dors., ×80 (Shaver, n).

Spinobairdia Morris & Hill, 1952 [**S. kellettae*]. Bairdian in outline but with large projecting spine subcentrally on each valve; imperfectly known. *M.Sil.*, N.Am.——Fig. 141,4. **S. kellettae*, USA (Tenn.); *4a,b*, carapace R dors., ×45 (Shaver, n).

Triebelina van den Bold, 1946 [**T. indopacifica*] [=*Glyptobairdia* Stephenson, 1946]. Differs from *Bairdia* in conspicuously pitted and ridged surface and presence of simple, weak tooth at each end of RV hinge-bar and corresponding sockets in LV. *Oligo.-Rec.*, cosmop.——Fig. 141,3. *T. coronata* (Brady), Rec., Cuba; *3a,b*, RV lat., int., ×40; *3c*, LV int., ×40 (343).

Family BEECHERELLIDAE Ulrich, 1894

[=Alanellidae Bouček, 1936] [Materials for this family prepared by Jean Berdan, U. S. Geological Survey]

Long, smooth, slightly asymmetrical ostracodes with adont hinge, and spines on posterior end and also in some carapaces anterior end of one or both valves. Muscle scar unknown. *U.Sil.-M.Dev.*

Beecherella Ulrich, 1891 [**B. carinata*]. Venter flat, carinate at outer margins. Carinae produced as spines anteriorly and posteriorly on each valve, posterior spine being larger. Cross-section subtriangular. Duplicature well developed. *L.Dev.*, NE.N.Am.——Fig. 145,3. **B. carinata*, N.Y.; *3a*, RV (lectotype) int. showing duplicature, ×23; *3b,c*, RV (topotype) int., lat., ×20; *3d*, LV lat., ×20 (Berdan, n).

Acanthoscapha Ulrich & Bassler, 1923 [**Beecherella navicula* Ulrich, 1891] [=*Alanella* Bouček, 1936]. Spindle-shaped, with ends of both valves acuminate, on LV both ends produced as spines which wrap around pointed ends of RV; contact margins of both valves a flattened flange that projects as a keel anteroventrally and posteroventrally. *U.Sil.-M.Dev.*, Eu.-N.Am.——Fig. 145,1a-c. **A. navicula*, L.Dev., N.Y.; *1a,b*, LV (lectotype), int., lat., ×23; *1c*, RV (topotype) lat., ×23 (Berdan, n).——Fig. 145,1d,e. *A. bohemica* (Bouček), U. Sil. (Budňany), Czech.; *1d*, RV (holotype) lat., ×40; *1e*, RV (topotype), ×40 (10).

?**Ulrichella** Bouček, 1936 [**U. remesi*]. Long, cylindrical ostracodes with anterior end of each valve spindle-shaped and posterior produced as a sharp spike; two slight constrictions perpendicular to hinge line divide shell into three nearly equal parts; hinge not known. *U.Sil.*, Eu.——Fig. 145,2. **U. remesi*, U.Sil., Czech.; *2a*, LV (holotype) lat., ×40; *2b*, RV (topotype) lat., ×40 (10).

Family MACROCYPRIDIDAE Müller, 1912

[*nom. correct.* Sylvester-Bradley herein, *pro* Macrocypridae *nom. transl.* Sylvester-Bradley, 1948 (*ex* Macrocyprinae Müller, 1912)] [Materials for this family by P. C. Sylvester-Bradley, University of Leicester]

Carapace elongate, with arcuate dorsal border, adductor muscle scars forming rosette pattern, anterior and posterior duplicature with wide vestibules. [All Recent forms are marine bottom-dwellers.] *?L.Ord.-?Mio., Plio.-Rec.*

Macrocypris Brady, 1867 [**Cythere minna* Baird, 1850]. Carapace smooth, compressed, elongate, dorsal margin arched, anterior margin rounded, ventral margin straight or concave, posterior acuminate; RV larger than LV, overreaching it on all margins except anterior. Selvage of both valves projecting prominently in center of ventral margin, that of RV overlapping LV when carapace is closed; duplicature with wide anterior and posterior vestibules; radial pore canals straight, crowded at anterior and posterior ends; normal pore canals small, few, scattered; hinge of 5 elements, in LV terminal elements (both anterior and posterior) being long, denticulate ridges that project, median element a smooth groove, locellate grooves (shorter than the other elements) between terminal elements and median element, RV with reverse arrangement; muscle-scar pattern consisting of central rosette of about 9 scars, with 3 others close above it and 2 small separated scars in front of and above main group. [Many fossils ranging from Ordovician upward have been assigned to this genus, but no pre-Tertiary fossils have been demonstrated to possess the muscle-scar pattern, duplicature and hinge characteristics of type species.] *?Ord.-?Mio., Plio.-Rec.*; cosmop.——Fig. 146,1; 244,8a,b. **M. minna* (Baird) Rec., Norway; *146,1a*, carapace L; *1b,c*, LV int., RV int.; *1d,e*, LV, RV vent.; *1f*, carapace vent.; all ×30 (Sylvester-Bradley, 1948); *244,8a,b*, LV dors., RV dors., ×125 (Sylvester-Bradley, 1948).

Macrocypria Sars, 1923 [**Macrocypris sarsi* Müller, 1912 (*pro *Bairdia angusta* Sars, 1866, *non Cythere (Bairdia) angusta* (Münster) Jones 1850)]. Like *Macrocypris*, but anterior acuminate. *Rec.*, Norway.——Fig. 146,2. **M. sarsi* (Müller) Rec., Norway; ♀ carapace L, ×40 (312).

BAIRDIACEA, Family UNCERTAIN

[Materials prepared by authors as recorded at end of generic descriptions]

Hastacypris CRONEIS & GUTKE, 1939 [*H. bradyi*]. Resembling *Macrocypris* in shape and overlap of RV over LV but described as lacking ventral overlap and concavity of *Macrocypris;* poor preservation of holotype of type species makes relationship to Paleozoic and Recent species of *Macrocypris* uncertain. ?*L.Dev., U.Miss.,* ?Eu.(Czech.)-N.Am.(Ill.).——FIG. 146A,*1.* *H. bradyi; 1a,b,* carapace L, dors., ×30 (133). [SHAVER.]

Pseudophanasymmetria SOHN & BERDAN, 1952 [*Phanassymetria foveata* VAN VEEN, 1936]. Small, fat, markedly asymmetrical; hinge straight, simple; smooth or punctate, and with a shallow posterodorsal depression. *U.Cret.,* Eu.——FIG. 146A,*2.* *P. foveata* (VAN VEEN), Maastr., Holl.; *2a,b,* carapace (paratype) R, ×113, dors., ×70;*2c,* LV int., ×70 (334a). [SOHN-BERDAN.]

Superfamily CYPRIDACEA Baird, 1845

[*nom. transl. et correct.* DANA, 1849 (*ex* Cypridae BAIRD, 1845)] [Diagnosis and discussion by F. M. SWAIN, University of Minnesota]

Carapace variable in size, shape, and surface ornamentation; calcareous or corneous; mid-ventral margin curved inward. Hinge of ridge-and-groove type or rabbeted; muscle scars a median group of spots, typically not radially or linearly arranged, with additional more anteroventral spots in most genera; inner lamellae present; line of concrescence and inner margin typically separated. [Fresh-water and marine.] ?*Sil.-?Perm., Trias.-Rec.*

The Cypridacea include most genera of living fresh-water ostracodes as well as a few marine forms. Fossil cyprids are recorded from fresh-water deposits as old as

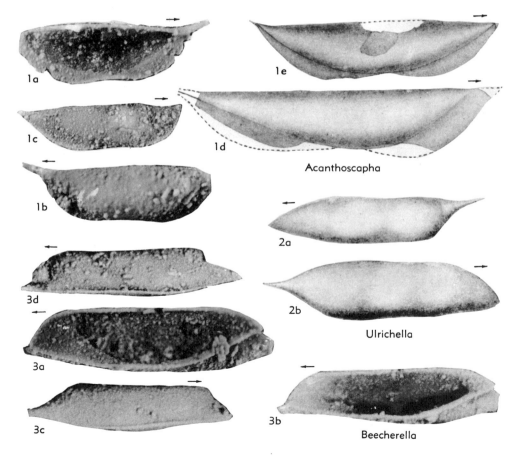

FIG. 145. Beecherellidae (p. Q207).

Carboniferous and from marine strata as far back as Ordovician.

Zoological characters. Definition of the superfamily by zoologists has been based mostly on characters of the soft parts. Although these are not preserved in fossils, the paleontologist cannot ignore features of soft-parts anatomy because these may be reflected in structures and form of the shell, as pointed out by TRIEBEL (82). The following diagnosis by HOFF (31) concisely states significant zoological characters.

Surface of the shell usually smooth, dorsal margin without interlocking teeth. Eyes developed

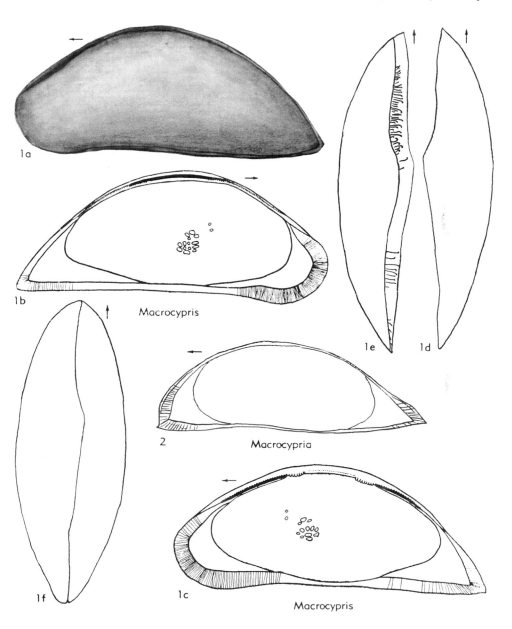

1a

1b   Macrocypris

1e   1d

1f   1c

2   Macrocypria

Macrocypris

FIG. 146. Macrocyprididae (p. Q207).

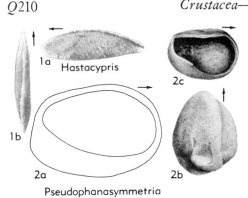

Pseudophanasymmetria

Fig. 146A. Bairdiacea, Family Uncertain (p. Q208).

to varying degrees; either separated or fused into a single median eye. The antennules with a basal portion of two or three podomeres and an endopodite of four or five podomeres, with swimming setae well developed. The antennae with a basal part of two podomeres and an endopodite of three or four podomeres. The expodite is reduced to a small scale-like appendage bearing at the most three setae. First thoracic leg not pediform but modified as a mouth part, with the anterior margin of the base adapted for feeding. The endopodite of the first leg forms a small palp in the female, but is enlarged to form prehensile organs in the male. The second thoracic leg has an endopodite of three or four podomeres and a strong distal claw. The third leg is bent dorsally and is probably used in cleaning the respiratory surfaces and other parts of the body. The third leg usually has three distal setae but the distal end may be modified for grasping. The furca is typically well developed and rod-shaped but may be reduced to a "flagellum" or whiplike structure (Cypridopsinae). The gonads are located within the valves of the shell. In the male, a portion of the vas deferens is modified to form an ejaculatory duct.

Reproduction. Probably in most freshwater ostracodes, including cyprids, reproduction is syngamic, but in some species it is partly or entirely parthenogenetic. Hoff (31) recognized four groups of Illinois fresh-water ostracodes, based on type of reproduction. With citation of representatives belonging to the Cyprididae, these are as follows:

Types of Reproduction in Fresh-water Ostracodes

1. Males unknown, reproduction parthenogenetic (several species of *Eucandona, Cypricercus, Cyprinotus* and *Cypridopsis*).

2. Males found in some localities, reproduction usually parthenogenetic (*Eucandona, Ilyocypris, Cyprinotus*).

3. Males found in small numbers, reproduction both syngamic and parthenogenetic (*Cypricercus, Potamocypris*).

4. Males invariably present, reproduction syngamic (*Eucandona, Cyclocypris, Cypria, Physocypria, Cypricercus, Notodromas, Cyprois*).

Some species (e.g., *Cyprinotus incongruens*) reproduce syngamically in one locality and parthenogenetically in another. According to Hoff, all Illinois cypridids studied by him are oviparous. On the other hand, a species of the Darwinulidae (*Darwinula stevensoni*) retains eggs in a posterior shell cavity during their development. Egg cases of many oviparous species can withstand freezing and desiccation, in fact, the eggs of some species require desiccation before development can begin (24). The egg shell of *Cypridopsis vidua* is about 0.1 mm. in diameter; it is double-walled and composed of chitin impregnated with calcite (45).

Ontogeny. Living fresh-water ostracodes are believed to pass through nine molt stages, of which the last comprises the sexually mature animal (31). The earlier instars possess fewer and simpler pairs of appendages than the later ones. The instar valves of many, though not all cypridids differ in shape from adult valves. Immature molts commonly are relatively shorter and higher and are more triangular in outline than the adults (325, 352).

Food. The food of fresh-water ostracodes consists of diatoms, bacteria, organic detritus, and among larger Cyprididae, bodies of dead animals.

Habitats. Although most fresh-water ostracodes belong in the Cypridacea, a few genera of the Cytheracea (including *Limnocythere, Cytherissa,* and *Entocythere*) and Darwinulacea occur in this general habitat. In central North America (Illinois) fresh-water ostracodes are recorded (31) in the following environments: (1) temporary ponds and ditches; (2) permanent lakes and swamps; (3) temporary streams and pools left in stream beds after flow has ceased; (4) permanent streams of all sizes; and (5) underground waters. In general, fewer species of ostracodes are present in running waters than in still waters. Restriction to one or another of these habitats is mostly on the species level. *Cyclocypris, Notodromas (Cypridacea), Limnocythere* (Cytheracea), and *Darwinula* (Darwinulacea) generally occur only in permanent still waters; however, nonmarine ostracodes probably are not found commonly in planktonic communities.

Regarding physical conditions affecting ostracode distribution, HOFF provides the following information which probably applies fairly generally. (1) Temperature of water has little effect on distribution. (2) Detailed knowledge is lacking as to influence of bottom conditions on the distribution of ostracode species, but present information suggests that relatively few species are affected by type of bottom; species of *Eucandona* prefer a mud bottom, probably owing to their crawling locomotion, and *Cypricercus reticulatus* occurs only on a mud bottom. (3) Presence or absence of currents is of major importance in distribution of many ostracodes, certain species of *Eucandona, Cypria, Cypricercus, Notodromas,* and *Darwinula* being confined to still waters, occurring mostly in plant zones where wave action is not pronounced. Several species of *Eucandona, Ilyocypris* and *Cyprinotus* typically occur in streams; in general, large, rectangular, compressed, ornamented species are likely to be found in running water, whereas more tumid species are more closely restricted to still water. (4) Strongly acid waters are not likely to contain ostracodes, particularly forms having large, heavily calcified shells; some species of *Cypria, Cypridopsis,* and *Physocypria* tolerate acid waters, probably because their small shells bear a thick organic covering.

Seasonal distribution. Many fresh-water ostracode species are seasonally distributed, some being found only in the spring, others in spring and summer, and a few (including species of *Eucandona* and *Ilyocypris*) throughout the year.

Classification and distribution. Genera of the Cypridacea are arranged here in seven families (Pontocyprididae, Paracyprididae, Cyprididae, Cyclocyprididae, Eucandonidae, Ilyocyprididae, and Notodromadidae). In addition some subfamilies are recognized, and several unassigned genera are grouped in a family-uncertain category.

Except for the subfamily Cyprideinae,

which became extinct in the Tertiary Period, all families and subfamilies of the Cypridacea are extant. The species and even genera of living cypridids are based mostly upon characteristics of the appendages and other decomposable structures, rather than upon carapace features, thus limiting attempts by paleontologists to compare fossil and living species. Nevertheless, a good deal of progress is being made in the comparative study of Tertiary and Quaternary members of the family; this has brought to notice slight but important carapace differences which permit use of these ostracodes in stratigraphy. When fossil fresh-water ostracodes have been studied more thoroughly, almost surely it will be necessary to design classification more suited to the needs of the paleontologist, but, in view of present limited knowledge of the fresh-water faunas, no attempt at reorganization of the families of Cypridacea is feasible.

Distribution. The doubtful record of a single cypridacean genus *(Pontocypris)* in Ordovician rocks and the likewise unconfirmed record of *Cypridopsis* and *Eucandona* in Permian strata are the only indications of extant genera of this superfamily in pre-Mesozoic deposits. Several genera are confined to Jurassic and Cretaceous rocks. A majority of cypridacean genera are exclusively Cenozoic, so far as known (Fig. 147).

Literature. Among more important published works on the Cyprididae, attention may be directed to papers by SARS (65, 68), G. W. MÜLLER (54), BRADY and co-workers (12, 13, 14, 15, 16), BRONSTEIN (18), HOFF (31), and KAUFMANN (40). The internal carapace structures are very important to an understanding of the fossil Ostracoda. Discussion of hinge, muscle spots, and other internal features of the Cyprididae are found in papers by HOWE and others (33, 35, 178), KESLING (44), MARTIN (51), SCOTT (323), SWAIN (349, 350), SYLVESTER-BRADLEY (79), TRIEBEL (82), and ZALANYI (405).

Family CYPRIDIDAE Baird, 1845

[*nom. correct.* BAIRD, 1850 (*pro* Cypridae BAIRD, 1845)] [Materials for this family prepared by F. M. SWAIN, University of Minnesota; diagnosis of a few genera by W. A. VAN DEN BOLD, Louisiana State University, and R. A. REYMENT, University of Stockholm]

Carapace variable in shape and size (to several mm. in length), typically subovoid to subtriangular, with greatest height median

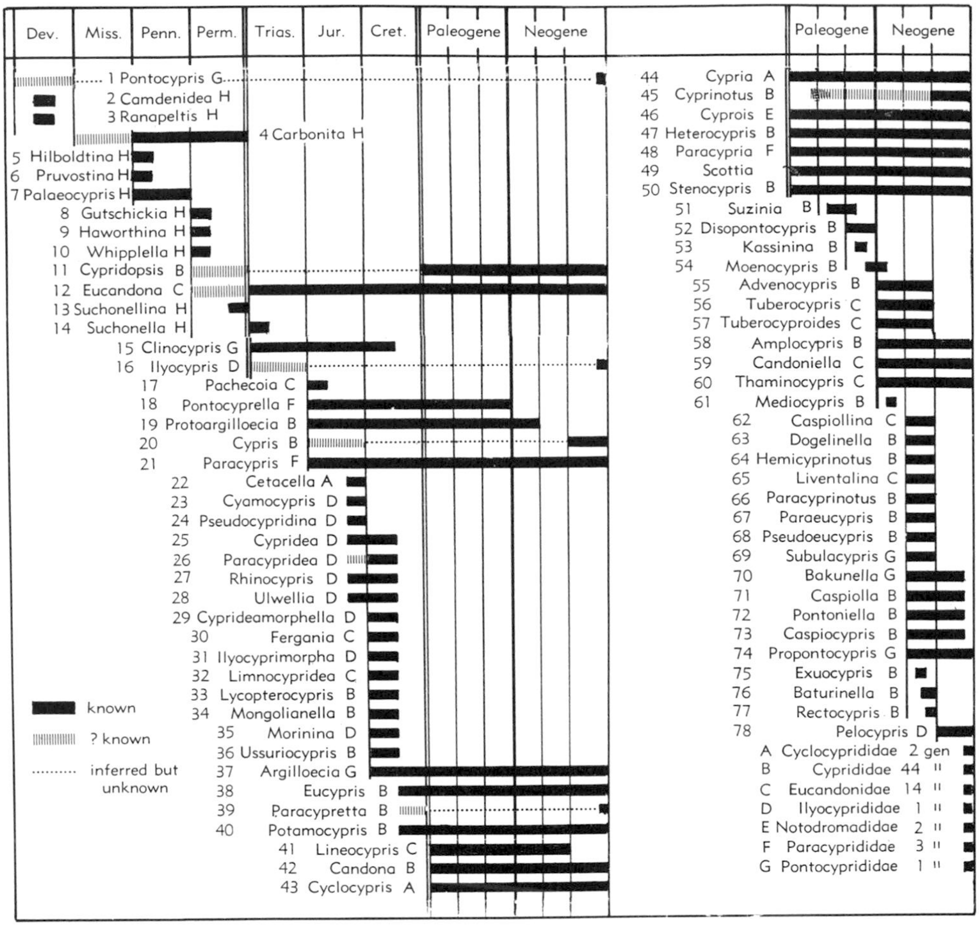

Chart column headers: Dev. | Miss. | Penn. | Perm. | Trias. | Jur. | Cret. | Paleogene | Neogene | (right block) Paleogene | Neogene

Left list:
1 Pontocypris G
2 Camdenidea H
3 Ranapeltis H
4 Carbonita H
5 Hilboldtina H
6 Pruvostina H
7 Palaeocypris H
8 Gutschickia H
9 Haworthina H
10 Whipplella H
11 Cypridopsis B
12 Eucandona C
13 Suchonellina H
14 Suchonella H
15 Clinocypris G
16 Ilyocypris D
17 Pachecoia C
18 Pontocyprella F
19 Protoargilloecia B
20 Cypris B
21 Paracypris F
22 Cetacella A
23 Cyamocypris D
24 Pseudocypridina D
25 Cypridea D
26 Paracypridea D
27 Rhinocypris D
28 Ulwellia D
29 Cyprideamorphella D
30 Fergania C
31 Ilyocyprimorpha D
32 Limnocypridea C
33 Lycopterocypris B
34 Mongolianella B
35 Morinina D
36 Ussuriocypris B
37 Argilloecia G
38 Eucypris B
39 Paracypretta B
40 Potamocypris B
41 Lineocypris C
42 Candona B
43 Cyclocypris A

Legend:
known
? known
inferred but unknown

Right list:
44 Cypria A
45 Cyprinotus B
46 Cyprois E
47 Heterocypris B
48 Paracypria F
49 Scottia
50 Stenocypris B
51 Suzinia B
52 Disopontocypris B
53 Kassinina B
54 Moenocypris B
55 Advenocypris B
56 Tuberocypris C
57 Tuberocyproides C
58 Amplocypris B
59 Candoniella C
60 Thaminocypris C
61 Mediocypris B
62 Caspiollina C
63 Dogelinella B
64 Hemicyprinotus B
65 Liventalina C
66 Paracyprinotus B
67 Paraeucypris B
68 Pseudoeucypris B
69 Subulacypris G
70 Bakunella G
71 Caspiolla B
72 Pontoniella B
73 Caspiocypris B
74 Propontocypris G
75 Exuocypris B
76 Baturinella B
77 Rectocypris B
78 Pelocypris D

A Cyclocyprididae 2 gen
B Cyprididae 44 "
C Eucandonidae 14 "
D Ilyocyprididae 1 "
E Notodromadidae 2 "
F Paracyprididae 3 "
G Pontocyprididae 1 "

Fig. 147. Stratigraphic distribution of cypridacean ostracode genera (Moore, n). Classification of the genera in families is indicated by letter symbols according to the following tabulation and for the purpose of locating any wanted genus an alphabetical list furnishes cross reference to the serially arranged numbers on the diagram. Index letters for families of Cypridacea: A—Cyclocyprididae; B—Cyprididae; C—Eucandonidae; D—Ilyocyprididae; E—Notodromadidae; F—Paracyprididae; G—Pontocyprididae; H—Family Uncertain.

Generic Names with Index Numbers

Advenocypris—55
Amplocypris—58
Argilloecia—37
Bakunella—70
Baturinella—76
Camdenidea—2
Candona—42
Candoniella—59
Carbonita—4
Caspiocypris—73
Caspiolla—71
Caspiollina—62
Cetacella—22
Clinocypris—15
Cultella[1]
Cyamocypris—23
Cyclocypris—43
Cypria—44
Cypridea—24
Cyprideamorphella—29

Cypridopsis—11
Cyprinotus—45
Cypris—20
Cyprois—46
Disopontocypris—52
Dogelinella—63
Eucandona—12
Eucypris—38
Exuocypris—75
Fergania—30
Gutschickia—8
Haworthina—9
Hemicyprinotus—64
Heterocypris—47
Hilboldtina—5
Ilyocyprimorpha—31
Ilyocypris—16
Kassinina—53
Limnocypridea—32
Lineocypris—41

Liventalina—65
Lycopterocypris—33
Mediocypris—61
Moenocypris—54
Mongolianella—34
Morinina—35
Pachecoia—17
Palaeocypris—7
Paracypretta—39
Paracypria—48
Paracypridea—26
Paracyprinotus—66
Paracypris—21
Pareucypris—67
Pelocypris—78
Pontocyprella—18
Pontocypris—1
Pontonella—72
Potamocypris—40
Propontocypris—74

Protoargilloecia—19
Pruvostina—6
Pseudocypridina—24
Pseudoeucypris—68
Ranapeltis—3
Rectocypris—77
Rhinocypris—27
Scottia—49
Stenocypris—50
Subulacypris—69
Suchonella—14
Suchonellina—13
Suzinia—51
Thaminocypris—60
Tuberocypris—56
Tuberocyproides—57
Ulwellia—28
Ussuriocypris—36
Whipplella—10

[1] *Cultella*, Trias., USSR, added in press.

to anteromedian, posterior end more pointed than anterior; LV generally larger than RV; surface commonly with tiny pits, pustules, or reticulations. Hinge line curved or rarely long and straight, consisting of groove in larger valve for reception of simple edge of smaller valve; muscle scars comprising rather compact median to anteromedian groups of 4 or 5 spots and 1 or 2 additional anteroventral spots. [Fresh-water and marine. Almost certainly a polyphyletic assemblage.] *?Perm., L.Jur.-Rec.*

Subfamily CYPRIDINAE Baird, 1845

[*nom. transl.* KAUFMANN, 1900 (*ex* Cyprididae BAIRD, 1850, *nom. correct. pro* Cypridae BAIRD, 1845)] [=Ctenocyprina (*partim*), Synopsida (*partim*), Zygopsida (*partim*) DADAY, 1900; Eucypridae ALM, 1915; Eucyprides (*partim*) SARS, 1925; Cypriformes SKOGSBERG, 1920, Cyprinotini+Eucyprini+Hungarocyprini+Scottini BRONSTEIN, 1947, Cypridina POKORNÝ, 1958] [Includes Mediocyprinae (*recte* Mediocypridinae), Baturinellinae, Advenocyprinae (*recte* Advenocypridinae SCHNEIDER, 1960)]

Carapace subtriangular, subelliptical, ovoid or rarely subquadrangular, with variable surface ornamentation; typically with anteroventral flangelike process, and posteroventral flanges, spines or caudal processes; marginal structures variable. Soft parts as described for family; furcal rami typically well developed, not reduced as in Cypridopsinae. *L.Jur.-Rec.*

Cypris O. F. MÜLLER, 1776 [**C. pubera;* SD BAIRD, 1846] [=*Eurycypris* G. W. MÜLLER, 1898]. Elongate subtriangular, dorsum strongly convex, angulate medially; venter flattened, slightly concave, terminal margins spinose, posterior end bluntly pointed, extended below; valves subequal; surface smooth or pitted. Inner lamellae moderate in width, broadest in front (68). [Fresh-water.] *?Jur., Pleist.-Rec.,* cosmop.——FIG. 148,*1.* **C. pubera,* Pleist., Eng.; *1a,b,* LV lat., RV lat.; *1c,d,* LV int., RV int.; *1e,* carapace vent.; all ×20 (Sylvester-Bradley, n).

Acocypris VÁVRA, 1895 [**A. capillata*]. Elongate acuminate, compressed; hinge margin nearly straight or slightly convex, venter slightly sinuous, ends narrowly rounded, extended below; LV with posteroventral blunt spinose projection, larger than RV; surface smooth [Fresh-water.] *Rec.,* Afr.——FIG. 148,*2a,b.* **A. capillata; 2a,b,* carapace L, dors., ×15 (393).——FIG. 148,*2c-e.* *A. hyalina* LOWNDES; *2c,d,* carapace L, dors.; *2e,* RV int.; all ×15 (217).

Advenocypris SCHNEIDER, 1956 [**A. alpherovi*]. Shell subtriangular, convex, valves dissimilar in shape, LV strongly overreaching RV; anterior end typically more steeply slanted and higher; posterior end of LV curved and strongly extended below, truncate to slightly concave above, posterior end

of RV less curved. Pore-canal zone narrow, with straight pore canals that commonly terminate in ampule-like expansions, inner lamellae broad terminally; contact line of valves not depressed, hinge of RV consisting of knife-edge ridge which fits over a rabbet groove on left. Surface smooth or ornamented. [Fresh-water.] *Mio.-Plio.,* SE.Eu.——
——FIG. 149,*2.* **A. alpherovi,* U.Mio.(Sarmat.) N.Caucasus; *2a,b,* carapace L, R, ×33; *2c,* LV int., ×64 (50).

Afrocypris SARS, 1924 [**A. barnardi*]. Very large, elongate, subquadrate, highest anteromedially, convexity moderate; dorsum nearly straight, venter slightly concave, ends rounded, truncated on dorsal slopes, posterior slightly narrowed to pointed below; LV larger than RV; surface smooth or with compressed terminal rims and surface nodes. Inner lamellae narrow. [Fresh-water.] *Rec.,* Afr.——FIG. 148,*3.* **A. barnardi; 3a,b,* carapace L, dors., × 10 (313).——FIG. 149,*1.* *A. biconica* KLIE; *1a,b,* LV lat., dors., ×10; *1c,* RV int., ×15 (217a).

Amplocypris ZALANYI, 1944 [**A. sinuosa;* SD SWAIN, herein]. Large, elongate subelliptical to subtrapezoidal; dorsal margin straightened medially with broadly obtuse, poorly defined anterior cardinal angle and less obtuse posterior cardinal angle; ventral margin nearly straight to slightly concave; anterior margin broadly rounded, truncated above; posterior margin sharply rounded and extended below, truncate above; convexity of valves low to moderate; surface mostly smooth. Inner lamellae relatively broad terminally and in some species ventrally; pore canals short and widely spaced. *Neog.* C.Eu.(Hung.).——FIG. 149, *5.* **A. sinuosa; 5a,b,* carapace R, dors., ×40 (406).

Astenocypris G. W. MÜLLER, 1912 [*pro Leptocypris* SARS, 1903 (*non* BOULENGER, 1900)] [**Leptocypris papyracea* SARS, 1903]. Elongate, highest medially, much compressed; dorsum moderately arched; venter sinuous, posterior margin truncate; surface longitudinally striated. Inner lamellae very narrow (309). *Rec.,* E.Asia.——FIG. 149,*3.* **A. papyracea* (SARS); *3a,b,* carapace L, dors., ×20 (309).

Baturinella SCHNEIDER, 1956 [**B. kubanica*]. Shell large, subreniform, elongate, length twice height, moderately convex, with greatest convexity median and on dorsal hump of RV; anterior margin rounded; posterior end rounded below, beveled above; dorsal margin straight, without cardinal angles; ventral margin concave anteromedially; surface smooth. Hinge of RV a median groovelike depression, at both ends of which are platelike teeth; hinge of LV a corresponding median knife-edge ridge with elongated depressions at each end; porecanal zone wide, canals closely spaced and numerous, some ampule-like at terminations without reaching margin of valve; inner lamellae of same width as canal zone, well developed terminally.

[Fresh-water.] *U.Plio.*, SE.Eu.——Fig. 149,4. **B. kubanica*, Caucasus (Kuban); *4a*, RV lat., ×27; *4b-d*, LV lat., dors., vent., ×27; *4e,f*, LV hinge, RV hinge, ×43 (50).

Bradycypris Sars, 1924 [**Cypris intumescens* Brady, 1907]. Short, tumid, highest in front of middle; LV larger than RV, overlapping it anteriorly; venter concave, ends nearly equally rounded; front end of RV with striated marginal zone; surface smooth. [Fresh-water.] *Rec.*, S.Afr.——Fig. 150,

2. **B. intumescens* (Brady); *2a,b*, carapace R, dors., ×30 (313).

Centrocypris Vávra, 1895 [**C. horrida*]. Strongly tumid, subquadrate; valves subequal, with anteromedian weak sulcus; surface spinose. [Fresh-water.] *Rec.*, Afr.——Fig. 149,6. **C. horrida; 6a,b*, carapace L, dors., ×30 (393).

Chlamydotheca Saussure, 1858 [**Cypris (Chlamydotheca) azteca*]. Subtriangular, relatively short and high, compressed; dorsum strongly arched,

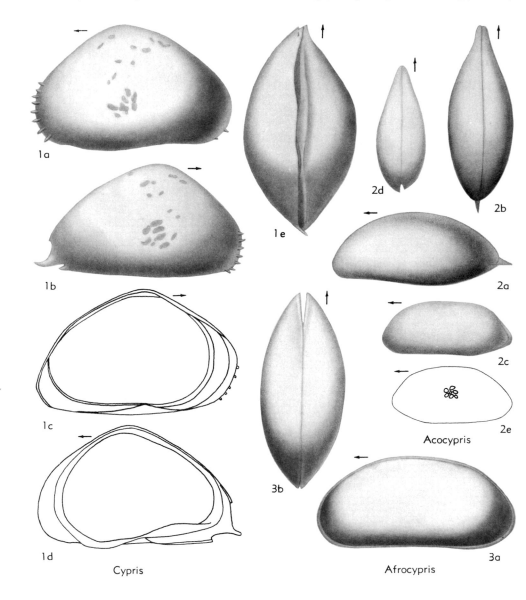

1a — 1b — 1c — 1d   Cypris

1e — 2d — 2b — 2a — 2c — 2e   Acocypris

3b — 3a   Afrocypris

Fig. 148. Cyprididae (Cypridinae) (p. Q213).

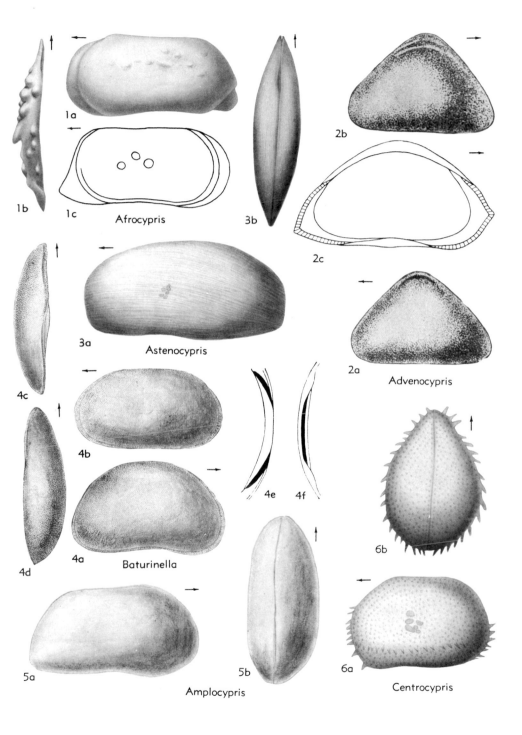

1a, 1b, 1c Afrocypris
2a Advenocypris, 2b, 2c
3a Astenocypris, 3b
4a, 4b, 4c, 4d, 4e, 4f Baturinella
5a, 5b Amplocypris
6a, 6b Centrocypris

Fig. 149. Cyprididae (Cypridinae) (p. *Q*213-*Q*214).

venter slightly sinuous, with ventroterminal flanges at one or both ends, posterior extremity more pointed; RV larger than LV; surface smooth or pitted (374). [Fresh-water.] *Rec.,* N.Am.-S.Am.——Fig. 150,*1a-e. C. rudolphi* Triebel, Brazil; *1a,b,* carapace R, dors., ×10; *1c,* LV int., ×10,

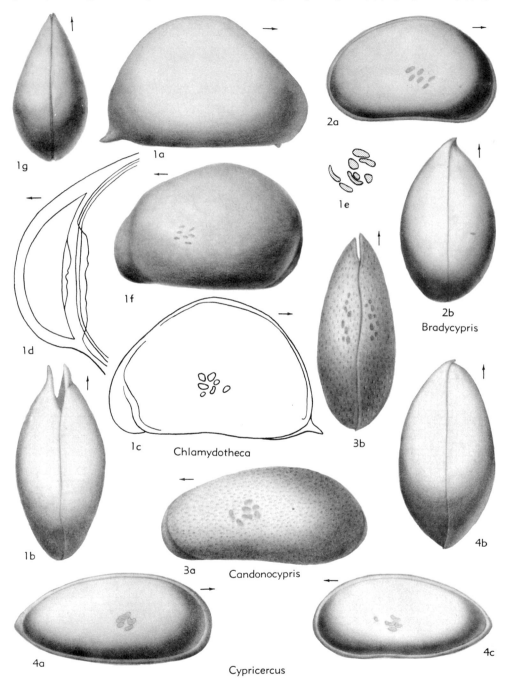

Fig. 150. Cyprididae (Cypridinae, Candoninae) (p. *Q*214-*Q*217, *Q*233).

1d, RV int., ant. part., ×20; *1e*, muscle scar, ×20 (374).——Fig. 150,*1f,g*. **C. azteca* (Saussure), Mexico; *1f,g*, carapace L, dors., ×15 (330).

Chrissia Hartmann, 1957 [**C. levetzovi*]. Carapace procumbent, greatest height less than half of length; shell wall thin and fragile, not distinctly punctate as in *Stenocypris*; LV without inner list along ventral and posterior margins, vestibule present, contact zone equally broad along anterior, ventral and posterior margins; radial pore canals scarcely visible, very small; false marginal pore canals each bearing a bristle distributed over most of surface, canals unbranched; fused secondary pore canals such as characteristically present along sinuous line of concrescence in *Stenocypris* lacking; lamellae not united by transverse septa. Males unknown. [Fresh-water.] *Rec.*, S.Afr.(Transvaal).——Fig. 151,*1*. **C. levetzovi*; carapace R, ×25 (164). [Reyment-Swain.]

Cypricercus Sars, 1895 [**C. cuneatus*]. Elongate, moderately convex; highest medially, dorsum not strongly arched, venter nearly straight, rear margin more narrowly rounded than front; LV larger than RV, overlapping it anteriorly and ventrally but RV may extend backward beyond LV. Anterior inner lamellae broad (313). [Fresh-water.] *Rec.*, Eu.-Afr.-Austral.——Fig. 150,*4*. **C. cuneatus*, S.Afr.; *4a,b*, ♀ carapace R lat., dors., ×15; *4c*, ♂ carapace L lat., ×30 (313).——Fig. 152,*1*. *C. fuscatus* (Jurine), Norway; RV lat. ×30 (314).

Cypriconcha Sars, 1926 [**Cypris barbata* Forbes, 1893]. Elongate, subelliptical, compressed; dorsum nearly straight, venter slightly sinuous, ends broadly rounded, slightly truncate above; valves subequal; surface smooth. Inner lamellae fairly broad, particularly toward rear. [Fresh-water.] *Rec.*, N. Am.——Fig. 152,*3*. **C. barbata* (Forbes); *3a,b*, RV lat., LV int., ×15 (315).

Cyprinotus Brady, 1886 [**C. cingalensis*] [=*Cypridonotus* Claus, 1892]. Subtriangular, valves compressed; dorsum strongly convex, subangulate medially, venter slightly concave, posterior margin angulate below midheight, truncate above and below; free margins of RV finely spinose; terminal margins of LV with hyaline border; LV overlapping RV terminally but RV may extend beyond LV dorsally; surface smooth. Inner lamellae rather narrow, broadest anteriorly (308). [Fresh-water.] *?Tert., Pleist.*(incl. *Rec.*), Eu.-N.Am.-Asia.
C. (Cyprinotus). Distinguished by soft parts. [Fresh-water.] *?Tert., Pleist.*(incl. *Rec.*), Eu.-Asia.——Fig. 152,*2a-c*. **C. cingalensis*, Rec., Ceylon; *2a-c*, carapace L, dors., ant., ×30 (108).——Fig. 152,*2d,e*. *C. salinus* (Brady), Rec., Eu.; *2d,e*, carapace L, dors., ×40 (314).
C. (Cyprinotoides) Masi, 1928 [**Cyprinotus (Cyprinotoides) somalicus*]. Distinguished by soft parts. [Fresh-water.] *Rec.*, E.Afr.(Somaliland).

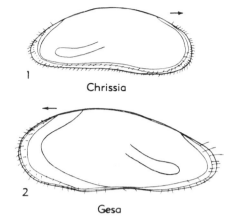

Chrissia

Gesa

Fig. 151. Cyprididae (Cypridinae) (p. Q217-Q221).

Dogelinella Schneider in Mandelstam *et al.*, 1957 [**D. taeniata*]. Carapace subequivalved, elongate, swollen, ends compressed, with straight dorsal margin. Surface smooth. *Plio.*, SW.Asia(Fergana). ——Fig. 153,*1*. **D. taeniata*; *1a,b*, carapace R, dors., ×40 (238a). [Bold.]

Dolerocypria Tressler, 1937 [**D. taalensis*]. Differs from *Dolerocypris* in soft parts. *Rec.*, SW.Pac. (Philippines).

Dolerocypris Kaufmann, 1900 [**Cypris fasciata* O.F.Müller, 1785]. Sublanceolate, highest medially, compressed; posterior margin narrower; RV larger than LV, overlapping it terminally; surface smooth. Inner lamellae very broad (68). [Fresh-water.] *Rec.*, Eu.-Asia-E.Indies. ——Fig. 154,*1*. **D. fasciata* (O.F.Müller), Eu.; *1a,b*, ♀ LV lat., int., ×30; *1c*, carapace dors., ×30; *1d,e*, RV lat. and int. ant. margin, ×30, ×100; *1f*, LV int. vent. margin, ×100; *1g*, carapace dors., ×30; (*1a-c*, 314; *1d-g*, 56).

Drieschia Brehm, 1923 [**D. mammillata*]. Shell tuberculate and generally as in *Centrocypris* but more elongate and more compressed. [Fresh-water.] *Rec.*, China.

Eucypris Vávra, 1891 [**Monoculus virens* Jurine, 1820; SD Sars, 1928] [=*Amphicypris* Sars, 1902]. Elongate-subovate, moderately convex; dorsum moderately arched, truncated behind and in front of greatest height, venter slightly concave to markedly sinuous, not appreciably flattened, ends rounded, not spinose, posterior margin narrow and extended ventral to mid-height; valves subequal, LV slightly larger than RV; surface smooth. Inner lamellae broad, concentrically striated (68). [Fresh-water.] *U.Cret.-Rec.*, cosmop. ——Fig. 154,*2a-c*. **E. virens* (Jurine), Rec., Eu.; *2a,b*, ♀ carapace L, dors.; *2c*, juv. carapace L, ×20 (314).——Fig. 154,*2d-f*. *E. crassa* O.F.

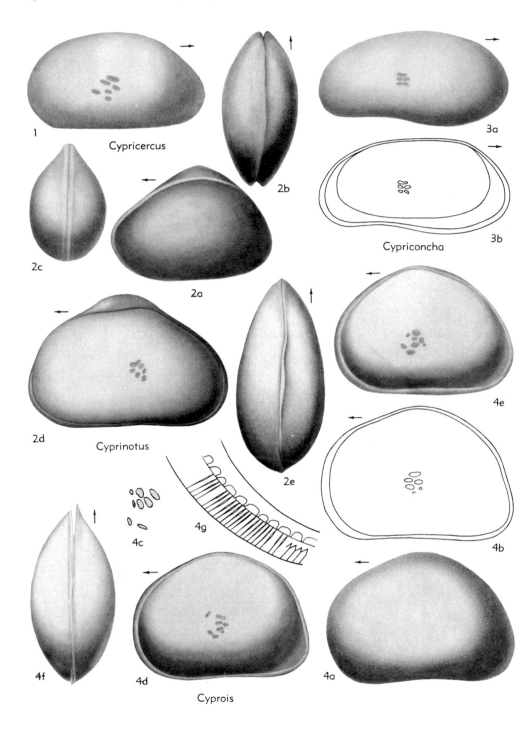

FIG. 152. Cyprididae (Cypridinae), Notodromadidae (p. Q217, Q245).

MÜLLER, Rec., N.Eu.; *2d*, ♀ LV int., *2e,f,* carapace L, dors.; all ×30 (314).

Exuocypris MANDELSTAM, 1956 [**E. extorris*]. Carapace kidney-like, elongate oval, rather large (1.25 by 0.75 mm.), thin-walled; anterior end truncated; exterior margin overhanging interior margin; surface smooth. Pore-canal zone occupying both exterior overhang of valve and part recessed within

outer margin, canals being long and widening toward outer ends; wide tonguelike, extended inner lamellae adjacent to pore canals; muscle scar cyprid-like. [Fresh-water, complete carapace unknown.] *M.Plio.,* SW.Asia(Turkmen).——FIG. 154,3. **E. extorris; 3a,b,* LV int., ant. and marginal areas showing pore canals, ×64 (50).

Gesa HARTMANN, 1957 [**G. dubia*]. Procumbent,

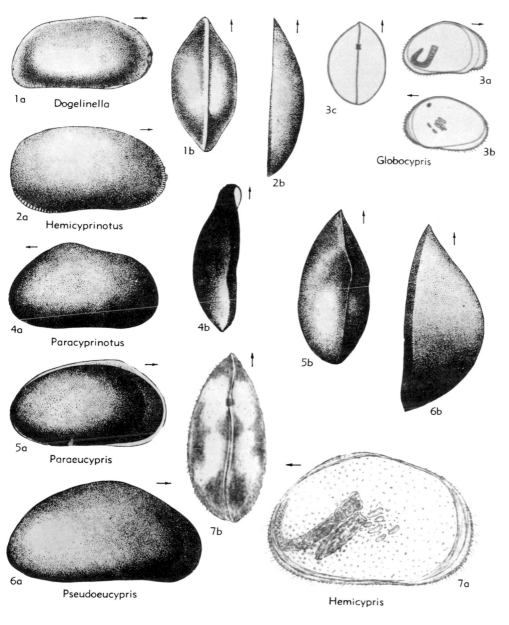

FIG. 153. Cyprididae (Cyprinidae) (p. Q217-Q227).

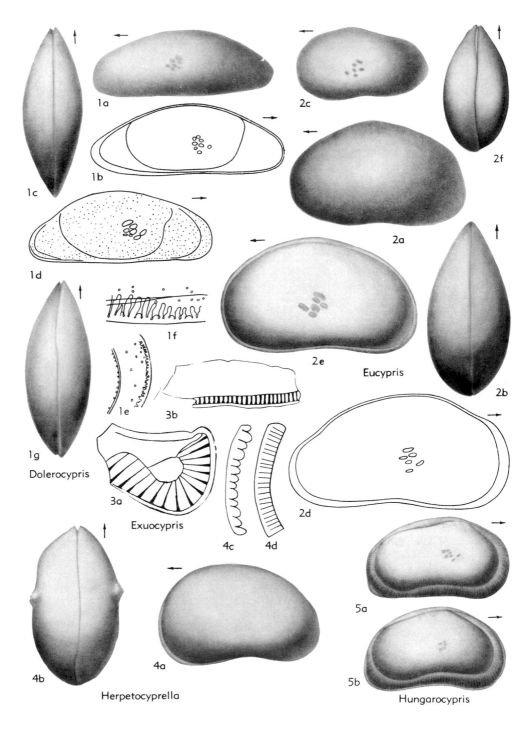

FIG. 154. Cyprididae (Cypridinae, Candoninae) (p. Q217-Q221, Q233).

slender, similar in form to various species of *Stenocypris;* carapace resembles *Chrissia levetzovi* in outline; surface features unknown; greatest height less than half of length, in front of middle; posterior sharply rounded, anterior regularly and broadly rounded; ventral margin with median conconcavity; inner margin and line of concrescence distinctly separated from margin anteriorly but posteriorly the two margins close together and parallel. Marginal pore canals not proven, few normal canals. Males unknown. [Fresh-water.] *Rec.,* S.Afr. (Transvaal).——FIG. 151,2. **G. dubia;* carapace L, ×55 (164). [REYMENT-SWAIN.]

Globocypris KLIE, 1939 [**G. trisetosa*]. Subovate, strongly tumid; venter sinuous; RV larger than LV, with extended anterior lip; surface pitted. Inner lamellae rather narrow. [Fresh-water.] *Rec.,* Afr.——FIG. 153,3. **G. trisetosa,* Kenya; *3a,b,* RV lat., LV lat.; *3c,* carapace dors., all ×15 (221).

Hemicyprinotus SCHNEIDER in MANDELSTAM *et al.,* 1957 [**H. valvaetumidus*]. Carapace elongate, moderately swollen; dorsal margin arched, posterior end vertically truncate or broadly rounded. Surface smooth. [Fresh-water.] *Plio.,* China (Sinkiang).——FIG. 153,2. **H. valvaetumidus; 2a,b,* RV lat., dors., ×30 (238a). [BOLD.]

Hemicypris SARS, 1903 [**Cyprinotus pyxidata* MONIEZ, 1892]. Like *Cyprinotus* but RV overlapping LV. [Fresh-water.] *Rec.,* E.Indies.——FIG. 153,7. **H. pyxidata* (MONIEZ); *7a,b,* LV lat., carapace dors., ×15 (309).

Heterocypris CLAUS, 1893 [**Cypris incongruens* RAMDOHR, 1808]. Subovate-elongate, medium-sized, moderately convex, thickest postmedially; dorsum moderately arched, subangulated postmedially, venter nearly straight to slightly concave, anterior margin narrowed; LV overlapping RV anteriorly and ventrally; general surface smooth except for scattered pits, edge of LV tuberculate anteriorly and posteroventrally (68). [Freshwater.] *Paleoc.-Rec.,* cosmop.——FIG. 155,5. **H. incongruens* (RAMDOHR), Rec., Eu.; *5a,b,* carapace R, dors.; *5c,* RV int.; *5d,e,* ♀ carapace dors., ♀ LV int.; all ×30 (316).

Homocypris SARS, 1924 [**H. conoidea*]. Elongate, highest behind middle, moderately convex, flattened medially; dorsum moderately arched, venter concave, terminal margins about equally rounded; front end compressed; valves perfectly equal; surface smooth. Inner lamellae very broad anteriorly. [Fresh-water.] *Rec.,* S.Afr.——FIG. 155,1. **H. conoidea; 1a-c,* carapace R, L, dors., ×20 (313).

Hungarocypris VÁVRA, 1906 [**Notodromas madaraszi* ORLEY, 1886]. Differs from *Candonocypris* in soft parts. *Rec.,* S.Eu.-Asia-N.Afr.——FIG. 154, 5. **H. madaraszi* (ORLEY); *5a,b,* ♀ LV int., ♂ LV int., ×15 (18).

Isocypris G.W.MÜLLER, 1909 [**I. priomena;* SD SWAIN, herein] [=*Hyalocypris* BRADY, 1913]. Elliptical, compressed; ends equally rounded; distinct marginal anterodorsal notch; surface smooth (313). [Fresh-water.] *Rec.,* S.Afr.——FIG. 155,4. **I. priomena; 4a,b,* carapace L, dors., ×30 (313).

Kassinina MANDELSTAM, 1960 [*pro Kassinia* MANDELSTAM, 1956 (*non* KHABAKOV, 1937)] [**Kassinia kassini* MANDELSTAM, 1956]. Carapace elongated-oval, kidney-like; LV overlapping RV; anterior end lower and more inclined than posterior; dorsal margin convex, ventral margin straight to slightly concave; inner lamellae moderately developed at ends of valves; surface densely tuberculate, presenting a shagreen appearance; LV hinge a groove open at ends for reception of sharpened edge of RV. *M.Oligo.,* SW.Asia(Kazakhstan).——FIG. 155, 6. **K. kassini* (MANDELSTAM); *6a,b,* LV lat., int., ×30 (50).

Liocypris SARS, 1924 [**L. grandis*]. Like *Homocypris* but much larger (4.5 mm.), equivalved, more compressed and more pointed posteriorly. *Rec.,* S.Afr.——FIG. 155,7. **L. grandis; 7a,b,* carapace L, dors., ×10 (313).

Lycopterocypris MANDELSTAM, 1956 [**Cypris faba* EGGER, 1910]. Carapace thin-walled, kidney-like, with maximum height in anterior third (about 0.85 by 0.45 mm.); LV larger than RV; anterior end more broadly rounded than posterior; dorsal margin strongly curved, ventral margin nearly straight; RV hinge formed of narrowed valve edge which fits over steplike groove in LV. *L. Cret.,* NE.Asia(Transbaikalia-Sib.-Mongolia).—— FIG. 155,2. *L. eggeri* MANDELSTAM, Transbaikalia; RV lat., ×27 (50).

Mediocypris SCHNEIDER, 1956 [**M. brodi*]. Carapace elongate, thick-walled, large, with length 3 times height; anterior margin rounded, posterior rounded below, truncate above; dorsal margin slightly convex, merging gradually with ends; ventral margin concave medially; surface sculptured by 4- and 5-sided pits; venter of each valve bearing narrow submarginal ridge or rim. Porecanal zone narrow, with straight pore canals; inner lamellae approximately of same width as pore-canal zone; RV hinge consisting of extended platelike valve edge corresponding to steplike depression in LV, which overlaps RV ventrally. [Fresh-water.] *M.Mio.,* SE.Eu.(Caucasus).——FIG. 156,1. **M.brodi,* N.Ossetia; *1a,b,* RV lat., int., *1c,d,* LV int., dors.; all ×27 (50).

Megalocypris SARS, 1898 [**M. princeps*]. Very large, elongate subelliptical, moderately convex; dorsum straight, with very obtuse cardinal angles; venter slightly concave, ends compressed, margins rounded, posterior end slightly pointed; valves subequal; surface smooth. Inner lamellae broadest anteriorly. [Fresh-water.] *Rec.,* S.Afr.——FIG. 156,2. **M. princeps; 2a,b,* carapace L, dors., ×7.5 (307).

Mesocypris DADAY, 1908 [*M. pubescens*]. Subreniform, moderately thick at each end but flattened medially; dorsum strongly arched, venter concave, posterior margin more pointed than anterior and extended ventrally; RV larger than LV; surface finely granulose. Inner lamellae broad, with median finger-like projection at each end (137). [Fresh-water.] *Rec.,* E.Afr.——FIG. 156,3. *M. pubescens; 3a,b,* LV lat., RV lat.; *3c,* carapace dors., ×50 (137).

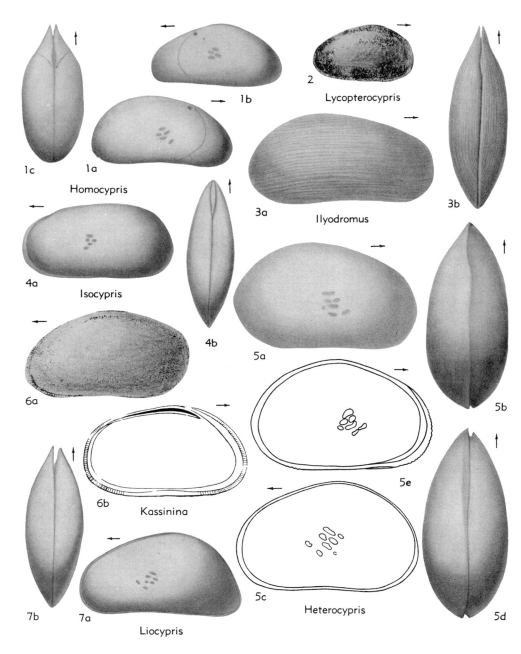

FIG. 155. Cyprididae (Cyprininae, Candoninae) (p. *Q221, Q233*).

Moenocypris TRIEBEL, 1959 [**M. francofurtana*]. Carapace kidney-shaped in lateral view, greatest height about in middle, anterior and posterior margins rounded, passing into lateral margins without angularity; LV larger than RV, with ventral overlap; surface smooth. Ventral margin of LV between line of concrescence and inner margin with isolated round spot in which both shell lamellae coalesce; selvage marginal in LV, near-marginal at ends of RV; zone of concrescence narrow; anterior and posterior pore canals short, anterior ones mostly forked, ventral margin with longer marginal canals and shorter submarginal canals, entire free margin also bearing finely branching secondary pore canals; central muscle field with 4 large adductor spots, 2 large mandibular spots, 2 small antennal spots; LV hinge a narrow furrow, margin of right fitting into this.

Females larger than males. [Fresh-water to brackish.] *U.Oligo.-L.Mio.,* Eu.(Ger.). —— FIG. 157,*1.* **M. francofurtana,* L.Mio., Mainz Basin; *1a,b,* ♀ RV lat. (showing muscle scars and traces of ovaries), ♂ RV lat. (coiled impressions of testicles), ×33; *1c,* ♀ RV int., central muscle field, ×125; *1d,* ♀ LV int., vent. margin, ×200 (381). [REYMENT.]

Mongolianella MANDELSTAM, 1956 [**M. palmosa*]. Medium large (length 1.2 mm.), thick-walled, elongate; LV larger than RV; anterior end higher than posterior, broadly rounded and slightly truncated above; posterior end pointed and extended below; dorsal margin convex, straightened medially; ventral margin concave; surface smooth; line of contact of valves not coinciding with interior margin of shell. RV hinge an anterior platelike tooth passing backward into ridge, LV hinge with

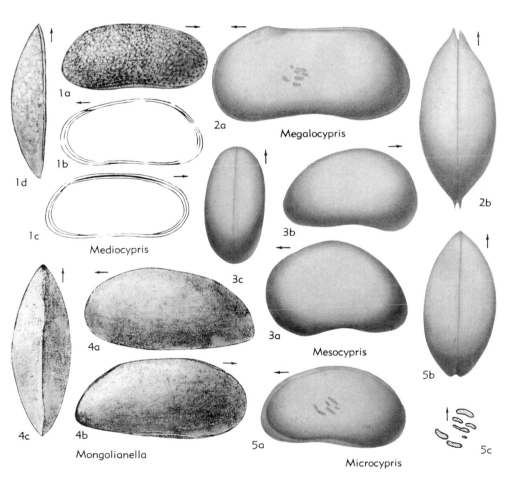

FIG. 156. Cyprididae (Cypridinae, Candoninae) (p. *Q221-Q223, Q234*).

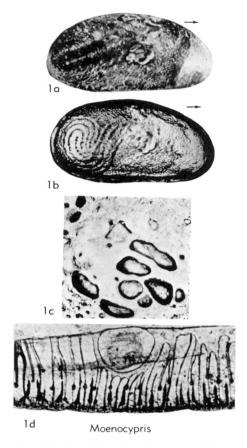

1a

1b

1c

1d   Moenocypris

Fig. 157. Cyprididae (Cypridinae) (p. Q223).

corresponding groove and anterior elongate socket; inner lamellae broadest at anterior end; adductor scar comprising anteromedian group of about 6 spots and 2 additional anteroventral spots. [Fresh-water.] *L.Cret.,* E.C.Asia(Transbaikalia-Mongolia).——Fig. 156,4, 158,1. *M. palmosa,* E. Transbaikalia; 156,4a-c, carapace L, R, vent.; 158,1a,b, LV int., RV int.; all ×27 (50).

Neocypridopsis KLIE, 1940 [*N. debilis*]. Reniform-subtriangular, not elongated, moderately convex; dorsum strongly and evenly arched, venter concave, posterior margin narrower, both ends extended below; LV larger than RV; surface pitted. Inner lamellae of moderate width. [Fresh-water.] *Rec.,* Eu.——Fig. 165,1. *N. debilis; 1a,b,* RV int., LV int.; *1c,* carapace dors., all ×100 (222).

Neozonocypris KLIE, 1944 [*N. congensis*]. Differs from *Zonocypris* mainly in soft parts. *Rec.,* Afr.

Oncocypria DADAY, 1908 [*O. mülleri*]. Differs from *Oncocypris* in presence of anteromedian sulcus and weaker surface ornamentation. Dimorphic,

males more pointed posteriorly and more elongate than females (137). *Rec.,* E.Afr.——Fig. 159,3. *O. muelleri; 3a-c,* carapace L, dors., RV int., ×50 (137).

Oncocypris G.W.MÜLLER, 1898 [*O. voeltzkowi*]. Small subovate to subquadrate, strongly inflated, thickest postmedially; dorsum moderately arched, venter slightly convex, sinuous; front portion compressed, margin with keel, anterior edges of valves bent toward right; LV larger than RV; surface rugose. Inner lamellae narrow, line of concrescence scalloped. [Fresh-water.] *Rec.,* Afr. ——Fig. 159,2a,b. *O. omercooperi* LOWNDES; *2a,b,* LV int., carapace dors., ×60 (233).——Fig. 159, 2c-e. *O. voeltzkowi; 2c-e,* carapace L, dors., LV int., ×75 (256).

Pachycypris CLAUS, 1893 [*P. incisa;* SD SWAIN, herein]. More elongate and inflated than *Chlamydotheca;* flange at anterior end; surface pitted or pustulose. *Rec.,* S.Am.——Fig. 160,2. *P. incisa; 2a,b,* carapace L, dors., ×25; *2c,* ant. margin, ×45 (19).

Paracypretta SARS, 1924 [*P. ampullacea*]. Subovate-reniform, very strongly inflated; dorsum arched, venter slightly concave, markedly flattened, ends rounded, posterior narrower than anterior; LV decidedly larger than RV, projecting beyond it forward; surface longitudinally striate. Inner lamellae broadest anteriorly; RV with thick, chitinized front marginal transverse septa. Parthenogenetic. [Fresh-water.] *?U.Cret., Rec.,* S.Afr.——Fig. 160, 3. *P. ampullacea,* Rec.; *3a-c,* carapace R, dors., ant., ×40; *3d,* RV int., ×40 (313).

Paracyprinotus SCHNEIDER in MANDELSTAM *et al.,* 1957 [*P. similis*]. Carapace elongate reniform, swollen, dorsal margin straight, forming obtuse angle with anterior margin, ends rounded, anterior end higher than posterior; radial pore canals straight, hinge simple, RV fitting into groove on

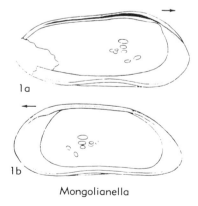

1a

1b

Mongolianella

Fig. 158. Cyprididae (Cypridinae) (p. Q224).

LV. Surface smooth. *Plio.,* C.Asia(Kazakhstan). ——Fig. 153,4. **P. similis; 4a,b,* LV lat., dors., ×30 (238a). [Bold.]

Paraeucypris Schneider in Mandelstam *et al.,* 1957 [**P. tota*]. Carapace elongate, ovate, smooth, ends rounded; dorsal margin arched, forming obtuse angle with anterior margin; hinge simple, RV fitting into groove on LV. *Plio.,* SW.Asia(Kirghizia).——Fig. 153,5. **P. tota; 5a,b,* carapace R, dors., ×30 (238a).

Platycypris Herbst, 1957 [**P. baueri*]. Elongate subelliptical, venter slightly sinuous, dorsum gently convex; ends rounded; valves compressed; inner lamellae very narrow except anteriorly; resembles *Scottia* but probably differs in structure of ap-

pendages. *Rec.,* Austral.——Fig. 161,1. **P. baueri; 1a,* ♂ carapace R; *1b,c,* ♀ carapace R, dors., ×25 (168).

Protoargilloecia Mandelstam, 1956 [**Bairdia silicula minor* Jones & Hinde, 1890]. Carapace small (length 0.45 mm.), thin-walled, smooth, podlike in form; RV larger than LV; anterior margin rounded, extended above, posterior margin extended and pointed below; dorsal margin convex, ventral margin convex in RV, concave in LV. RV hinge a minute groove for reception of sharpened edge of LV; inner lamellae weakly developed at anterior end; pore-canal zone narrow, perforated by thin, straight pore canals; adductor muscle scar like that of other cyprids. *Jur.-Mio.,* W.Eu.-SW.

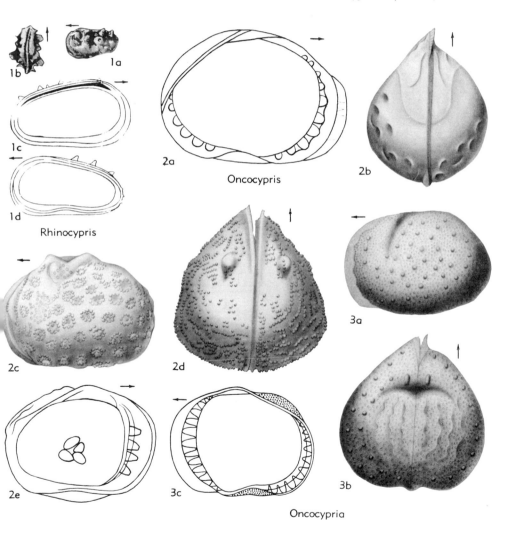

Rhinocypris

Oncocypris

Oncocypria

Fig. 159. Cyprididae (Cyprininae), Ilyocyprididae (p. Q224, Q241).

Asia(Kazakhstan).——Fig. 160,*1*. **P. minor*
(Jones & Hinde), Cret., Eng.; *1a,b,* LV lat., RV
lat., ×25 (50).

Pseudocypretta Klie, 1932 [**P. maculata*]. Subtri-

angular, tumid; dorsum strongly convex, umbonate
medially, venter nearly straight, ends nearly
equally rounded; RV overlapping LV dorsally;
surface smooth. *Rec.,* Afr.——Fig. 160,*5*. **P.*

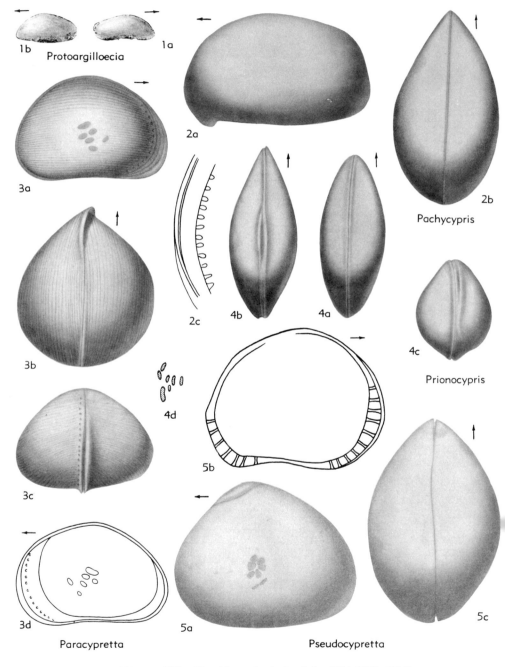

Fig. 160. Cyprididae (Cypridinae, Candoninae) (p. Q224-Q227, Q234).

maculata; 5a-c, LV lat., int., carapace dors., ×100 (217).

Pseudocypris DADAY, 1908 [**P. bouvieri*]. Subtriangular to ovoid; dorsum arched, broadly angulate medially, venter nearly straight, front margin rounded, rear narrower and acuminate ventrally; valves subequal; longitudinal flange or false keel parallel to ventral margin; surface smooth. Inner lamellae moderate in width (137). [Fresh-water.] *Rec.,* Afr.——FIG. 162,3. **P. bouvieri; 3a-c,* carapace R, L, dors., ×30 (137).

Pseudoeucypris SCHNEIDER in MANDELSTAM *et al.,* 1957 [**P. eboris*]. Carapace elongate, kidneyshaped, swollen; dorsal margin arched, posterior end usually truncate; hinge with sharp edge in RV which fits into groove in LV. Typical are outline and swollen ventral portion. Surface smooth. [Fresh-water.] *Plio.,* SW.Asia(Kazakhstan).——FIG. 153,6. **P. eboris; 6a,b,* RV lat., dors., ×20 (238a). [BOLD.]

Rectocypris SCHNEIDER in MANDELSTAM *et al.,* 1958 [**R. schwejeri* SWAIN, herein, *pro Bythocypris reniformis* SCHWEJER, 1949 (*non* BRADY, 1880)]. Carapace elongate-ovate, swollen, anterior end higher than posterior, which is usually truncate in upper part, narrowly rounded in lower; dorsal margin arcuate; inner lamella 2 or 3 times size of zone of concrescence, pore canals straight; hinge consisting of sharp margin of RV with projection in anterior part, fitting into groove in LV. Surface smooth. *U.Plio.,* SE.Eu.(Volga-Caucasus). [BOLD.]

Sclerocypris SARS, 1924 [**S. clavularis*]. Subquadrate, compressed; hinge margin straight and 0.5 to 0.75 of carapace length; anterior margin broad, truncate above, extended ventrally to provide a marginal anteroventral notch; posterior margin narrow, extended below mid-height; valves subequal; surface granular. [Fresh-water.] *Rec.,* Afr.——FIG. 162,2a-c. **S. clavularis; 2a-c,* carapace R lat., dors., RV int., ×15 (313).——FIG. 162,2d,e. *S. jenkinae* KLIE; *2d,e,* RV int., LV dors., ×10 (217).

Scottia BRADY & NORMAN, 1889 [**Cypris browniana* JONES, 1850] [*non Scottia* BOLIVAR, 1912] [=*Scottiana* CARUS, 1890 (obj.)]. Small, subovate, inflated; dorsum convex, venter nearly straight, posterior margin narrower; LV overlapping RV along hinge; surface with scattered pits. [Freshwater.] *Tert.-Rec.,* Eu.-?N.Am.; *Pleist.,* Eu.(Brit. Is.).——FIG. 162,4. **S. browniana* (JONES); *4a,b,* carapace R, dors., ×60 (15).

Stenocypria G.W.MÜLLER, 1901 [*Cypris fischeri* LILLJEBORG, 1883]. Differs from *Eucandonia* in soft parts and marginal shell structure. *Rec.,* Eu.-N. Am.——FIG. 162,1. **S. fischeri* (LILLJEBORG), Eu.; *1a,b,* carapace L, dors.; *1c,* LV lat.; *1d,* RV lat. post. margin, ×20 (*1a,* 18; *1b-d,* 258).

Stenocypris SARS, 1889 [**Cypris malcolmsonii* BRADY, 1886]. Very elongate, subelliptical, much compressed; dorsum nearly straight with broadly

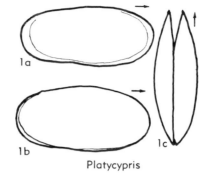

FIG. 161. Cyprididae (Cypridinae) (p. Q225).

obtuse cardinal angles, ventral margin slightly concave, ends about equally rounded, truncated above; LV slightly larger than RV; surface smooth. Inner lamellae very broad anteriorly, line of concrescence S-shaped. [Fresh-water.] *Tert.-Rec.,* Afr.-Austral.-E.Indies.——FIG. 163,1. **S. malcolmsonii* (BRADY); Rec., India; *1a,b,* carapace R, dors., ×30 (305a).

Strandesia STUHLMANN, 1888 [**Cypris (Strandesia) mercatorum* VÁVRA, 1895; SD VÁVRA, 1895)] [=*Acanthocypris* CLAUS, 1893; *Neocypris* SARS, 1901; *Spirocypris* SHARPE, 1903]. Elongate, subelliptical, moderately convex; hinge margin nearly straight, venter slightly sinuate, ends rounded, posterior slightly narrower and commonly with spinose extension; dorsum of RV with large winglike process; margin of LV tuberculate; valves subequal; surface smooth. Inner lamellae broad (393). [Fresh-water.] *Rec.,* S.Hemis.——FIG. 162,5. **S. mercatorum* (VÁVRA), Zanzibar; *5a,* carapace L, showing winglike process on RV, ×20; *5b,* carapace dors., ×20; *5c,* muscle scar, enlarged (393). ——FIG. 163,4a-c. *S. centrura* KLIE, Brazil; *4a-c,* carapace L, R, dors., ×30 (KLIE, 1940).——FIG. 163, *4d-f. S. bicuspis* (CLAUS) (type species of *Acanthocypris*), S.Am.; *4d,e,* carapace L, dors., ×40; *4f,* LV ant. end, ×100 (19).

Suzinia SCHNEIDER, 1956 [*pro Ilowaiskya* SUZIN, 1956 (*non* VIALOV, 1940)] [**Ilowaiskya transcaucasica* SUZIN, 1956]. Carapace smooth, elongate, with rounded ends; LV larger than RV, overlapping it ventroterminally; hinge of LV bearing groove, closed at its ends, for reception of knife-edge of RV. Inner lamellae broad anteriorly and ventrally, becoming narrower posteriorly; inner margin with deep re-entrant anteriorly where it nearly joins line of concrescence; broad porecanal zone thickly supplied with narrow straight canals. [Marine.] *M.Eoc.-L.Oligo.,* Caucasus-Armenia.——FIG. 163,2. **S. transcaucasica* (SUZIN), L.Oligo, Armenia; *2a,b,* carapace R lat., dors.; *2c,* RV int.; all ×30 (50).

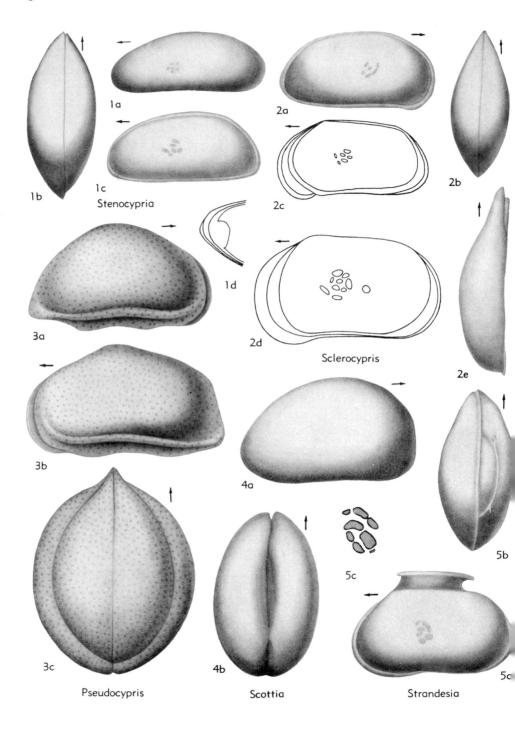

FIG. 162. Cyprididae (Cypridinae) (p. Q227).

Ussuriocypris MANDELSTAM, 1956 [*U. ussurica*]. Kidney-shaped, about 1.45 by 0.55 mm.; LV larger than RV, anterior end more uniformly rounded than posterior, which has truncate margin above; dorsal margin convex but straightened medially in most specimens, ventral margin nearly straight to slightly concave; surface smooth. Hinge consisting of benchlike depression in LV and sharpened edge in RV. *L.Cret.*, NE.Asia(E.Sib.).

——FIG. 163,3. *U. ussurica; 3a,b,* RV lat., LV lat., ×13 (50).

Zonocypris G.W.MÜLLER, 1898 [*Z. madagascarensis*]. Similar to *Paracypretta* but valves only slightly unequal, LV larger than RV; surface concentrically striated; sexes separate. [Fresh-water.] *Rec.*, Afr.——FIG. 163,5. *Z. madagascarensis; 5a,b,* carapace L, dors., ×80; *5c,* RV int., ×80 (256).

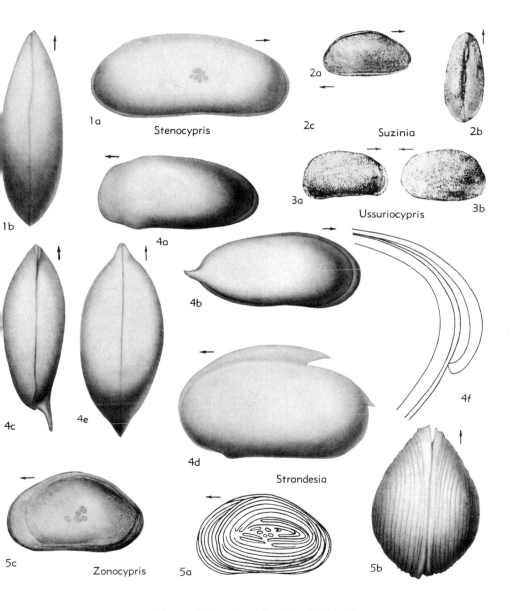

FIG. 163. Cyprididae (Cypridinae) (p. Q227-Q229).

Subfamily CYPRIDOPSINAE Kaufmann, 1900

[=Cypridopsides SARS, 1925; Cypridopsini BRONSTEIN, 1947]

Carapace small to medium-sized (generally less than 1 mm.), reniform to subtriangular; dorsum moderately to strongly arched, venter typically concave, ends rounded, extended below mid-height; valves more or less unequal; surface smooth, densely pitted, pustulose or spinose. Inner lamellae typically very narrow; furcal ramus reduced to a base which ends distally in a seta. [Fresh-water.] *?Perm., U.Cret.-Rec.*

Cypridopsis BRADY, 1867 [*Cypris vidua* O.F.MÜLLER, 1776; SD BRADY & NORMAN, 1889] [=Pionocypris BRADY & NORMAN, 1896 (obj.); Proteocypris BRADY, 1905]. Reniform, subtriangular, not elongated, moderately convex, thickest postmedially; dorsum strongly arched, angulated medially, truncated on either side of position of greatest height, venter concave, anterior margin narrower; RV slightly larger than LV; surface pitted, pustulose, spinose. In part parthenogenetic (44). [Fresh-water, brackish-water.] *?Perm., U.Cret.-Rec.*, cosmop.——FIG. 164,1. *C. vidua* (O.F.MÜLLER), Rec., Eu.-N.Am.; *1a,b,* carapace L, dors., ×70; *1c,d,* LV int., vent., ×70 (Sars, 1925); *1e,* ♀ carapace dors., ×70 (55).

Cypretta VÁVRA, 1895 [*Cypridopsis (Cypretta) tenuicauda*] [=Cypridella VÁVRA, 1895 (non DE KONINCK, 1841)]. Possibly more tumid than *Cypridopsis,* otherwise differing from that genus in soft parts. [Fresh-water.] *Rec.,* S.Afr.——FIG. 164,2a,b. *C. tenuicauda* (VÁVRA); *2a,b,* carapace R, dors., ×60 (393).——FIG. 164,2c-e. *C. dubiosa* (DADAY); *2c,* carapace dors.; *2d,e,* LV int., RV int., ×50 (18).

Cypridopsella KAUFMANN, 1900 [*Monoculus villosus* JURINE, 1820; SD SARS, 1928] [=Candonella CLAUS, 1891]. Differs from *Potamocypris* in stronger dorsal and lesser ventral overlap of LV by RV, more posterior position of greatest height and narrower inner lamellae. In part parthenogenetic (68). [Fresh-water.] *Rec.,* Eu.-S.Am.-?Asia.——FIG. 164,3. *C. villosa* (JURINE); *3a,b,* carapace L, dors.; *3c,* LV int.; all ×70 (314).

Cyprilla SARS, 1924 [*C. arcuata*]. Subreniform, compressed, thickest medially; dorsum strongly convex, venter slightly concave, terminal margins extended below, posterior truncate, bluntly pointed below; greatest length near venter; RV larger than LV, strongly overlapping dorsally but overlapped by LV ventrally and posteriorly; surface pitted. [Fresh-water.] *Rec.,* S.Afr.——FIG. 165,3. *C. arcuata; 3a,b,* carapace L, dors.; *3c,d,* LV int., RV int.; all ×100 (313).

Paracypridopsis KAUFMANN, 1900 [*P. zschokkei*] [=Poracypridopsis KAUFMANN, 1900]. Elongate, subquadrate, compressed; dorsum not strongly arched, angulated anteromedially, truncate in front and behind greatest height, venter concave, terminal margins extended below, posterior truncate above; RV overlapping LV dorsally but LV overlapping RV terminally; surface pitted or pustulose. Inner lamellae narrow, broadest anteriorly. [Fresh-water.] *Rec., Eu.*——FIG. 165,2. *P. zschokkei; 2a,* carapace L, ×50; *2b,* muscle scar, ×110 (40).

Potamocypris BRADY, 1870 [*Bairdia fulva* BRADY, 1868]. Elongate-reniform, compressed, highest medially; dorsum moderately arched, venter concave, terminal margins extended below, posterior slightly narrower; RV strongly overlapping LV dorsally and ventrally; surface densely pitted. Inner lamellae moderately broad at both ends (14). [Fresh-water.] *U.Cret.-Rec.,* cosmop.——FIG. 165, 4. *P. fulva* (BRADY), *Rec.,* Brit.Is.; *4a-c,* carapace L, R, dors., ×80; *4d,* RV int., ×80 (53).

Subfamily DISOPONTOCYPRIDINAE Mandelstam, 1956

[*nom. correct.* SWAIN, herein (*pro* Disopontocyprinae MANDELSTAM, 1956)]

Form variable, LV much larger than RV and commonly with thickened dorsal margin; anterior end higher than posterior; surface smooth or rarely perforated or reticulate. Pore-canal zone comparatively wide, with numerous straight canals; inner lamellae well developed anteriorly. *Oligo.-Rec.*

Disopontocypris MANDELSTAM, 1956 [*Pontocypris oligocaenica* ZALANYI, 1929]. Kidney-shaped, elongate oval, or subtrapezoidal; medium-sized (±0.8 mm.); anterior end higher than curved posterior end, which in RV is narrowly rounded and strongly extended below; dorsal margin straight, with distinct step at anterior end of RV; ventral margin nearly straight, but with mid-ventral incurvature of margin; surface smooth or rarely with honeycomb ornamentation and median pits; pore-canal zone narrow, with dense straight pore canals; inner lamellae broader anteriorly than elsewhere; hinge of LV groove, curved at ends, for reception of RV edge. *Oligo.,* SE.Eu.(Caucasus-Hung.).——FIG. 166,3. *P. oligocaenica* (ZALANYI), Caucasus; *3a,b,* LV lat., RV lat.; *3c,d,* LV int., RV int.; all ×30 (50).

Caspiocypris MANDELSTAM, 1956 [*Bairdia candida* LIVENTAL, 1929]. Reniform, elongate oval, about 1.2 mm. long; LV larger than RV; anterior end broadly rounded and usually higher than narrowly rounded posterior end, which is extended below, beveled above; dorsal margin straight to slightly concave; anterior cardinal angle more obtuse and less distinct than posterior; surface smooth or with pits or tubercles in middle part. Inner lamellae best developed anteriorly; pore-canal zone narrow, with widely spaced canals anterior steplike terminus; LV hinge a groove hinge formed of knifelike valve edge which has anterior steplike terminus; LV hinge a groove open at both ends. *Plio.-Rec.,* SE.Eu.(Yugosl.-Caucasus-Caspian Region-Roumania-Albania-living

in Caspian Sea).——FIG. 166,2. *C. candida* (LIVENTAL) *Rec.,* Caspian; *2a,b,* RV lat., LV lat.; *2c,d,* RV int., LV int., ×27 (50).

Caspiolla MANDELSTAM, 1960 [*pro Caspiella* MANDELSTAM, 1956 (*non* THIELE, 1928)] [**Bairdia*

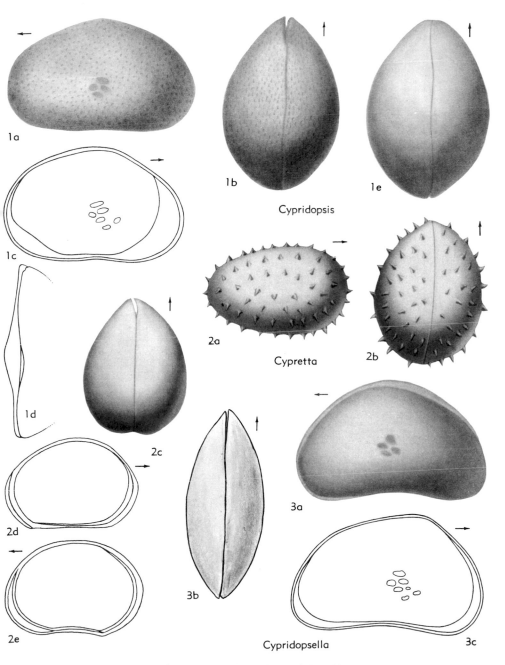

Cypridopsis

Cypretta

Cypridopsella

FIG. 164. Cyprididae (Cypridopsinae) (p. Q230).

acronasuta LIVENTAL, 1929]. Large, irregular reniform; LV larger than RV; anterior end high and broadly curved, posterior end pointed and extended below, truncated above; dorsal margin convex or straight; ventral margin straight to sinuous, margins curved inward mid-ventrally; surface smooth, rarely with angular tubercles; hinge of RV consisting of knifelike valve margin with small step

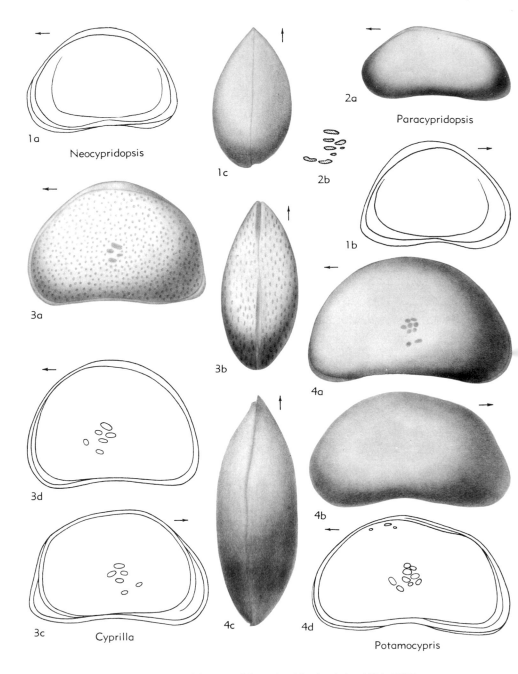

FIG. 165. Cyprididae (Cypridinae, Cypridopsinae) (p. Q224, Q230).

at anterior end; LV hinge a groove, open at both ends. Inner lamellae broadest anteriorly; pore-canal zone narrow, perforated by dense, narrow pore canals located ventrally; adductor muscle scar comprising anteromedian group of 6 spots and 2 additional more anteroventral spots. *Plio.-Pleist.,* SE.Eu.(Yugosl. - Caucasus).——— FIG. 166,4. **C. acronasuta* (LIVENTAL), Plio., Caucasus; *4a,b,* LV lat., RV lat., ×40; *4c,d,* LV int., RV int., ×30 (50).

Pontoniella MANDELSTAM, 1960 [*pro Pontonella* MANDELSTAM, 1956] [**Paracypria acuminata* ZALANYI, 1929]. Elongate, podlike, small (length about 1 mm.); LV larger than RV; anterior end high and broadly rounded; posterior pointed and strongly extended below, slightly concave above; dorsal margin slightly convex to straight; ventral margin concave; surface smooth or pitted; hinge of RV formed of knifelike margin with anterior step; LV with longitudinal groove, open at both ends, and overlapping outward extension of valve edge, weakly developed anteriorly. Inner lamellae best developed anteriorly; pore-canal zone narrow; canals straight, concentrated at front end; adductor muscle scar a group of 6 spots, median to slightly posteromedian in location. *Plio.-Pleist.,* SE.Eu.(Yugosl.-S.Russ.).——FIG. 166,1. **P. acuminata* (ZALANYI), U.Mio., Yugosl.; *1a,b,* LV lat., RV lat., ×25; *1c,d,* LV int., RV int., ×20 (50).

Subfamily CANDONINAE Daday, 1900

[=Herpetocypridinae KAUFMANN, 1900; Ctenocyprina *(partim)*, Euopsida *(partim)*, Synopsida *(partim)* DADAY, 1900; Candonides *(partim)* SARS, 1923; Eucyprides *(partim)* SARS, 1925; Herpetocyprini+Herpetocyprellini BRONSTEIN, 1947; Erpetocypridina POKORNÝ, 1958] [Includes Herpetocyprinae *(recte* Herpetocypridinae) SCHNEIDER, 1960]

Distinguished from other Cyprididae by rudimentary development of natatory setae on (posterior) antennae, resulting in lack of swimming power, but differing from Eucandonidae, in which swimming setae of the antennae are entirely absent. *Tert.-Rec.*

Candona BAIRD, 1845 [**Cypris reptans* BAIRD, 1835; SD BAIRD, 1846] [=*Erpetocypris* BRADY & NORMAN, 1889 (obj.); *Herpetocypris* SARS, 1890 (obj.)] [see *Eucandona*]. Large (2.5 mm.), elongate, subelliptical, compressed; dorsum nearly straight; venter concave; ends rounded, posterior broader; LV larger than RV; surface smooth; inner lamellae broadest anteriorly; complex radial canals; parthenogenetic. [Fresh-water.] *Tert.-Rec.,* cosmop.——FIG. 167,2. **Candona reptans* (BAIRD), Rec., Brit.Is.; *2a,b,* carapace L, dors., ×20; *2c,* RV int., ×20; *2d,* int. and vent. margin enlarged showing pore canals (314).

Candonocypris SARS, 1894 [**Cypris candonoides* KING, 1855]. Elongate suboblong, compressed; dorsum nearly straight to gently convex, venter slightly concave, ends rounded, anterior narrowed, greatest height postmedian; RV larger than LV,

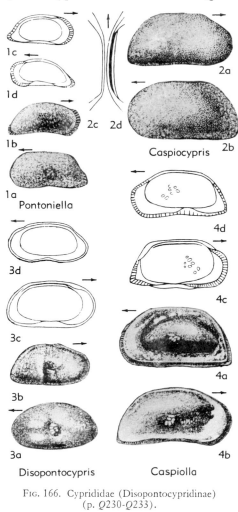

FIG. 166. Cyprididae (Disopontocypridinae) (p. Q230-Q233).

overlapping it strongly in front; surface smooth. Inner lamellae broad anteriorly. Parthenogenetic. [Fresh-water.] *Rec.,* ?Austral.-N.Z.-S.Afr.——FIG. 150,3. **C. candonoides* (KING), S.Afr.; *3a,b,* carapace L, dors., ×30 (67).

Herpetocyprella DADAY, 1909 [**H. mongolica*]. Differs from *Erpetocypris* in soft parts. *Rec.,* Asia. ——FIG. 154,4. **H. mongolica,* S.Asia, *4a,b,* carapace L, dors., ×30; *4c,d,* terminal pore canals, enlarged (18).

Ilyodromus SARS, 1895 [**Candona stanleyana* KING, 1855]. Elongate subelliptical, compressed; dorsum straight, with broadly obtuse cardinal angles, venter concave, ends rounded, posterior narrower; dorsal slopes truncated; LV slightly larger than RV; surface typically striated longitudinally. Inner lamellae very broad and shelflike. Parthenogenetic. [Fresh-water.] *Rec.,* N.Z.——FIG. 153,3. **I.*

stanleyana (KING); *3a,b,* carapace R, dors., ×30 (306).

Microcypris KAUFMANN, 1900 [**M. reptans*]. Elongate-subreniform, moderately convex; dorsum moderately and evenly arched; venter concave; anterior end narrower, LV slightly larger than RV, surface with scattered pits. [Fresh-water.] *Rec.,* Eu.——FIG. 156,5. **M. reptans; 5a,b,* carapace L, dors., ×40; *5c,* muscle scars, ×50 (40).

Prionocypris BRADY & NORMAN, 1896 [**Candona serrata* NORMAN, 1862]. Elongate, subreniform, moderately convex; dorsum gently arched, venter slightly concave, ends about equally rounded and typically serrate; valves subequal; surface smooth. Inner lamellae broadest anteriorly. [Fresh-water.] *Rec., Eu.-Asia.*——FIG. 160,4. **P. serrata,* W.Eu.; *4a,b,* carapaces dors., ×30; *4c,* carapace ant., ×30; *4d,* muscle scar, enlarged (107).

Family CYCLOCYPRIDIDAE Kaufmann, 1900

[*nom. transl.* SWAIN, herein (*ex* Cyclocypridinae KAUFMANN, 1900] [=Synopsida (*partim*) DADAY, 1900; Candocyprinae (*partim*) ALM, 1915; Cyclocyprides SARS, 1925; Cyclocyprinae HOFF, 1942; Cyclocyprini BRONSTEIN, 1947] [Materials for this family prepared by F. M. SWAIN, University of Minnesota]

Carapace small (less than 1 mm.), subovate to subtriangular, highest medially; dorsum strongly convex, venter nearly straight, ends rounded, anterior generally narrower; greatest length below mid-height; valves compressed to strongly convex, thickest in posterior half; valves subequal to strongly unequal; surface smooth, pitted, reticulated, or lined. Inner lamellae narrow. [Fresh-water.] *U.Jur.-Rec.*

Cyclocypris BRADY & NORMAN, 1889 [**Cypris globosa* SARS, 1863]. Subovate, tumid, highest medially; dorsum strongly convex, venter straight, anterior margin narrow, truncate above; RV slightly larger than LV; surface smooth, weakly pitted or reticulated (68). *Tert.-Rec., Eu.-Asia.-N.Am.*——FIG. 168,4. **C. globosa* (SARS), Rec., Eu.; *4a,b,* carapace L, dors., ×50; *4c-e,* ♀ carapace L, R, dors., ×60 (*4a,b,* 65; *4c-e,* 257).

?Cetacella MARTIN, 1958 [**C. inermis*]. Suboval in side view, LV larger than RV; characterized by regular finely ribbed surface ornament. Hinge composed of ridges and furrows; muscle field with 6 adductor spots, 2 smaller mandibular spots obliquely anteroventral from central field, and 1 or 2 antennal spots anterodorsal therefrom. Sexual dimorphism moderately strong, reflected in lengths and heights of carapaces. [Fresh-water to brackish environment.] *U.Jur.(Kimm.),* Eu.——FIG. 169,1. **C. inermis,* Ger.; *1a,b,* LV lat., RV lat., ×60; *1c,* carapace, dors., ×60 (242).

Cyclocypria DOBBIN, 1941 [**C. kincaidia*]. Differs from Cyclocypris in nature of soft parts. *Rec.,* W. N.Am.

Cypria ZENKER, 1854 [**Cypris exculpta* FISCHER, 1854 (=*Cypria punctata* var. *striata* ZENKER, 1854); SD BRADY & NORMAN, 1889]. Subovate, compressed, highest medially to postmedially; dorsum strongly arched, venter straight to slightly concave; anterior margin narrower and more extended; LV slightly larger than RV; surface smooth or punctate (68). *Tert.-Rec.,* cosmop.——FIG. 168, 2. **C. exculpta* (FISCHER), Rec., Eu.; *2a,b,* ♀ carapace L, dors.; *2c,* ♂ carapace R; *2d,e,* ♀ carapace L, dors.; all ×60 (314).

Physocypria VÁVRA, 1898 [**P. bullata*]. Subovate to subtriangular, moderately convex, highest postmedially; dorsum strongly arched to umbonate, venter slightly convex, anterior margin narrower; valves strongly unequal with either LV or RV larger, overlapping strongly dorsally and ventrally; free margins of LV or RV serrate or spinose (54). *Rec.,* cosmop.——FIG. 168,1a,b. P. pustulosa (SHARPE), Eu.; *1a,b,* carapace L, vent. (22).—— FIG. 168,1c-e. P. globula FURTOS, Ohio; *1c,d,* carapace L, dors., ×60; *1e,* RV int., ×60 (22).

Family EUCANDONIDAE Swain, n. fam.

[=Candoninae KAUFMANN, 1900; =Candoninae (*partim*) DADAY, 1900; =Synopsida (*partim*) DADAY, 1900; =Typhlopsida DADAY, 1900; =Candocyprinae (*partim*) ALM, 1915; =Candonini BRONSTEIN, 1947] [Materials for this family by F. M. SWAIN, University of Minnesota, with assistance on some genera by R. A. REYMENT, University of Stockholm, W. A. VAN DEN BOLD, Louisiana State University, and R. H. SHAVER, Indiana University and Indiana Geological Survey] [Includes Lineocyprinae (*recte* Lineocypridinae), Paracandoninae SCHNEIDER, 1960]

Shell medium to large, more or less elongate, subreniform, compressed to moderately convex; typically higher behind middle; dorsum arched, venter more or less concave, anterior margin typically more narrowly rounded than posterior; valves equal or unequal; surface smooth or pitted. Inner lamellae typically broad. *?Perm., Trias.-Rec.*

Eucandona DADAY, 1900 [**Candona balatonica* DADAY, 1894; SD SWAIN, herein] [=*Candona* BRADY & NORMAN, 1889 (*non* BAIRD, 1845, obj.); ?*Cyprida* GOLDENBERG, 1870]. [BRADY'S (1846) designation of *Cypris reptans* as type species of *Candona* takes precedence over selection of this same species by BRADY & NORMAN (15) as the type species of *Erpetocypris,* and likewise of their designation of *Cypris candida* O.F.MÜLLER, 1776; as type species of *Candona;* accordingly *Erpetocypris* and SARS' *Herpetocypris,* which is an invalid emendation of *Erpetocypris,* must be classed as objective junior synonyms of *Candona.*] Elongate, medium-sized, subreniform, moderately convex, highest posterior to middle; dorsum strongly arched to nearly straight, venter concave; ends rounded, anterior margin generally narrower than posterior, LV slightly larger but RV may overlap along venter; surface smooth or finely punctate; shell substance cloudy white in Recent forms.

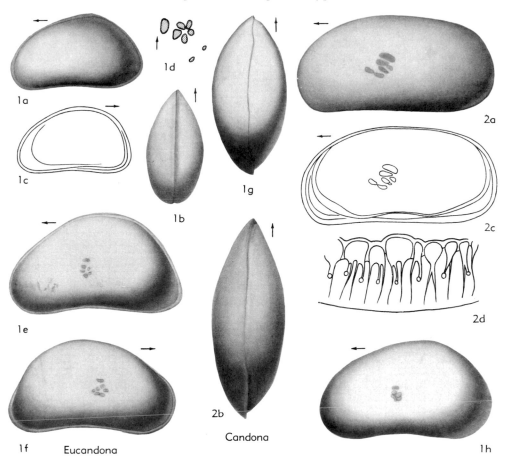

FIG. 167. Cyprididae (Candoninae), Eucandonidae (p. Q233-Q235).

Inner lamellae broadest anteriorly; radial canals few, simple, widely spaced. Males slightly larger than females, but nearly of same proportions; some species parthenogenetic (40, 68). [Freshwater.] *?Perm., Trias.-Rec.*, cosmop.——FIG. 170, *1*. **E. balatonica* (DADAY) Rec., S.Eu.-W.Asia; *1a*, ♀ carapace L; *1b*, ♂ carapace L; both ×43 (18).——FIG. 167,*1*. *E. candida* (O.F.MÜLLER), Rec., Eu.; *1a,b*, ♀ carapace L, dors., ×40; *1c*, LV int., ×40; *1d*, muscle-scar pattern, ×65 (*1a-c*, 55, 1776; *1d*, 40, 1900); *1e-g*, ♀ carapace L, R, dors., ×40; *1h*, ♂ carapace L, ×40 (55).

Arunella BRADY, 1913 [**A. subsalsa*]. Differs from *Eucandona* in soft parts; appendages resemble those of *Bairdia*. Rec., Eng.

Candocypria FURTOS, 1933 [**C. osburni*]. Differs from *Eucandona* in greater width of posterior inner lamellae. Rec., N.Am.——FIG. 171,*1*. **C. osburni*, Ohio; *1a,b*, ♀ carapace L, dors.; *1c*, ♂ RV int. showing marginal area and muscle spots; all ×50 (24).

?Candoniella SCHNEIDER, 1956 [**C. suzini*]. [Status questionable, may =*Paracandona;* not to be confused with *Candonella* CLAUS, 1891 (=*Cypridopsis*)]. Carapace thin-walled, elongate oval in lateral view, height small in relation to length; LV larger than RV, anterior and posterior margin well rounded; dorsal margin straight, ventral margin convex. Surface smooth or ornamented with tiny punctae. Zone of concrescence narrow, numerous straight pore canals, vestibule in anterior part; RV hinge with sharp, straight edge; muscle-scar patterns typical of the family Cyprididae. [Fresh-water.] *Neog.*, Eu.——FIG. 172,*3*. **C. suzini*, USSR; *3a,b*, RV lat., int.; *3c*, LV int.; ×46 (50). [REYMENT.]

Candonopsis VÁVRA, 1891 [**Candona kingsleyi* BRADY & ROBERTSON, 1870] [*non Candonopsis* ALMEIDA, 1950]. More elongate and more compressed than *Eucandona;* valves subequal; inner lamellae very broad anteriorly; females relatively shorter and higher than males (68). Rec., cosmop. ——FIG. 171,*2*. **C. kingsleyi* (BRADY & ROBERTSON); *2a,b*, RV int., carapace dors., ×60, ×50;

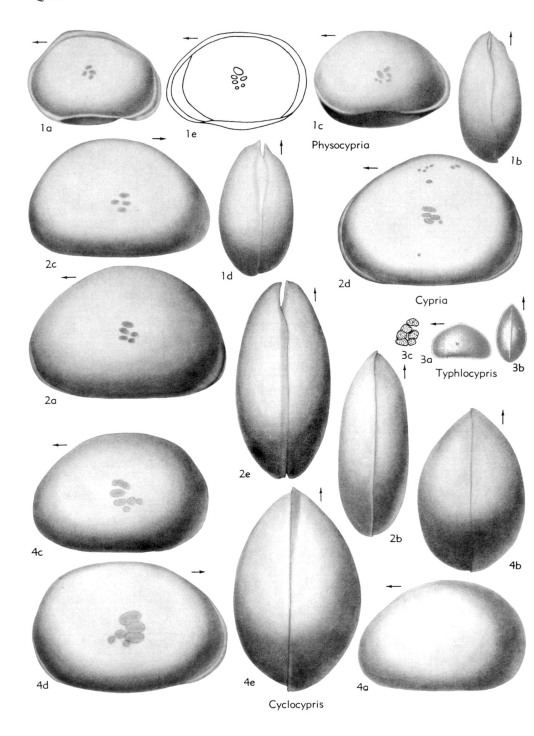

1a
1e
1c
Physocypria
1b
2c
1d
2d
Cypria
3c   3a
Typhlocypris   3b
2a
2e
4c
2b
4b
4d
4e
4a
Cyclocypris

Fig. 168. Cyclocyprididae, Eucandonidae (p. Q234-Q239).

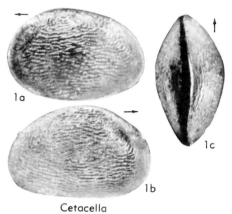

Cetacella

Fig. 169. Cyclocyprididae (p. Q234).

2c, muscle scar, ×110; *2d*, ♂ LV int., ×50; *2e*, ♂ carapace dors., ×50. (*2a-c*, 40; *2d,e*, 257).

Caspiollina Mandelstam, 1957 [*C. uschakensis*]. Carapace kidney-shaped, ends rounded, anterior lower than posterior, dorsal margin arched, ventral concave; LV overlapping along anterodorsal margin; inner lamella slightly broader than zone of concrescence, pore canals straight. *Plio.*, SW.Asia (W. Turkmenia).——Fig. 172,4. *C. uschakensis; 4a,b*, LV lat., RV lat., ×43 (238a).[Bold.]

Cryptocandona Kaufmann, 1900 [*C. vavrai*]. Small, subelliptical to subtriangular, highest medially, compressed; venter nearly straight; ends about equally rounded. Inner lamellae broad at both ends. *Rec.*, Eu.——Fig. 173,4. *C. vavrai; 4a,b*, RV int., carapace dors., ×40; *4c*, muscle scar, ×65 (40).

Fergania Mandelstam, in Mandelstam et al., 1957 [*F. ferganensis*]. Carapace trapezoid, posterior end obliquely truncate, both ends finely denticulate; hinge in LV with groove into which sharp dorsal edge RV fits. [Fresh-water.] *L.Cret.* (Alb.), SW. Asia(Fergana).——Fig. 172,6. *F. ferganensis;* LV lat., ×64 (238a). [Bold.]

Limnocypridea Lyubimova, 1956 [*L. abscondida*]. Shell large, (1.2 to 1.8 mm.), thick-walled, irregularly oval or trapezoidal, with rounded cardinal angles, posterior more obtuse, maximum curvature or margin at posteroventral side of valves; LV larger than RV, overlapping it around entire periphery but chiefly along dorsal and ventral margins, dorsal overlap very strong in some species; ends nearly equal in height, both truncated above, rounded and slightly extended below; dorsal margin straight; ventral margin concave medially; RV hinge an elevated rounded ridge that widens terminally and extends a short distance down dorsal slopes; LV with corresponding groove. Surface reticulate and tuberculate on ventral part or smooth. Pore-canal zone wide, well

developed terminally and ventrally; canals straight or slightly curved; inner lamellae broadest terminally. [Fresh-water.] *L.Cret.*, NE.Asia.——Fig. 174,1. *L. abscondida*, SE.Mongolia; *1a-c*, carapace L, R, dors., ×13; *1d,e*, LV int., RV int., ×17 (231).

Lineocypris Zalanyi, 1929 [*L. trapezoidea*]. Trapezoidal; straight dorsum with well-defined cardinal angles, venter slightly concave, anterior margin rounded, extended below and truncate above, posterior margin bluntly pointed and extended near venter, strongly truncate above; valves subequal; surface nearly smooth. Inner lamellae moderately wide anteriorly. [Fresh-water.] *Tert.*, Eu.

Liventalina Schneider in Mandelstam et al., 1958 [*Herpetocypris dagadjikensis* Markova, 1956]. Carapace elongate-ovate, narrow; dorsal margin arched, posterior end obliquely truncate and elongate; valves reticulate, especially in center; radial pore canals straight, vestibule present. *Plio.*, Aktchakyl stage; Turkmenia, Caucasus. [Bold.]

Nannocandona Ekman, 1950 [*N. faba*]. Like *Eucandona* but differing in soft parts. *Rec.*, Eu. (Swed.)

Pachecoia Almeida, 1950 [*P. rodriguesi*]. Carapace subtriangular in lateral view, with strongly arched dorsum, somewhat resembling *Bairdia*, differing from many cypridids in having beaklike posterior termination, lateral borders nearly symmetrical in dorsal view; LV larger than RV; contact-marginal structures include narrow calcified inner lamella, posterodorsal hinge short, simple, with slight ridge in LV articulating in groove of RV; adductor muscle scar consisting of about 6 loosely-grouped elliptical spots. [Fresh-water.] [Simpler than Bairdiidae and reminiscent of the

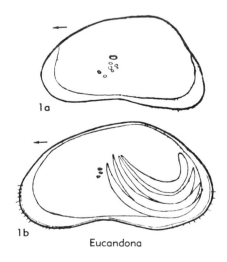

Eucandona

Fig. 170. Eucandonidae (p. Q235).

cypridid genus *Eucandona*.]. L.Jur., S.Am.(Braz.). ——Fig. 172,5. *P. rodriguesi;* outline of RV int., ×20 (90). [SHAVER.]

Paracandona HARTWIG, 1899 [*Candona euplectella* BRADY & NORMAN, 1889]. Elliptical, bean-shaped, tumid; dorsum nearly straight, venter slightly concave, ends broadly rounded; surface reticulate and tuberculate (24). *Rec.,* N.Am.-Eu.——Fig. 175,3. *P. euplectella* (BRADY & NORMAN), Scot.; *3a,b,* carapace L, dors., ×75 (257).

Parapontoparta HARTMANN, 1955 [*P. arcuata*]. Differs from *Pontoparta* in structure of appendages. [Marine.] *Rec.,* Brazil coast.

Pontoparta VÁVRA, 1901 [*P. rara*]. Resembles elliptical species of *Cryptocandona* but has greater convexity and much narrower inner lamellae. *Rec.,* Bismarck Arch.——Fig. 175,2. *P. rara; 2a,b,* ♀ carapace L, dors., ×70 (394).

Pseudocandona KAUFMANN, 1900 [*Candona pubescens* KOCH, 1837] [=*Metacandona* BRONSTEIN, 1930; *?Archicandona* BRONSTEIN, 1939]. Like *Eucandona* but smaller, with straighter hinge margin, narrower inner lamellae and pitted, reticulate,

or tuberculose surface (18). *Rec.,* Eu.-Asia.—— Fig. 175,1a-d. *P. tuberculata* BRONSTEIN, USSR; *1a-d,* carapace L, dors., RV int., LV int., ×30 (18).——Fig. 175,1e,f. *P. pubescens,* USSR.; *1e,f,* carapace L, dors., ×20 (18).——Fig. 175, *1g-i. P. bispinosa* BRONSTEIN (type species of *Metacandona*), USSR.; *1g,h,* ♂ carapace L, dors., ×40; *1i,* carapace L, ×40 (18).

?Riocypris KLIE, 1936 [*R. uruguayensis*]. Resembles triangular species of *Cryptocandona* but has greater convexity, more compressed anterior end, larger RV, and prominent selvage ridge in RV. *Rec.,* Uruguay.——Fig. 173,1. *R. uruguayensis; 1a-c,* RV int., LV int., carapace dors., ×30 (218).

Siphlocandona BRADY, 1910 [*Candona similis* BAIRD, 1845]. More elongate than *Eucandona;* dorsal margin straight or slightly arched, venter slightly concave, ends rounded, anterior slightly broader; valves equal. *Rec.,* Br.I.

Thalassocypria HARTMANN, 1957 [*T. aestuarina*]. [Marine.] *Rec.,* El Salvador.

Thalassocypris HARTMANN, 1955 [*T. elongata*].

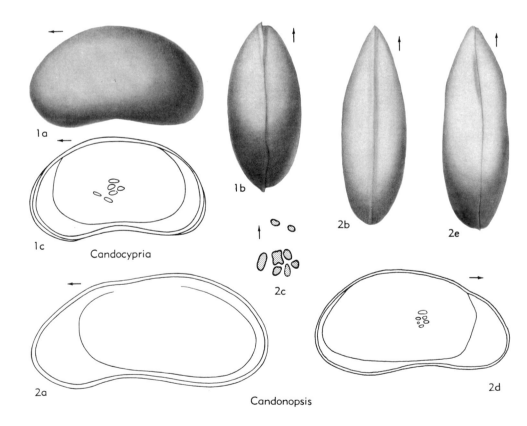

FIG. 171. Eucandonidae (p. Q235-Q237).

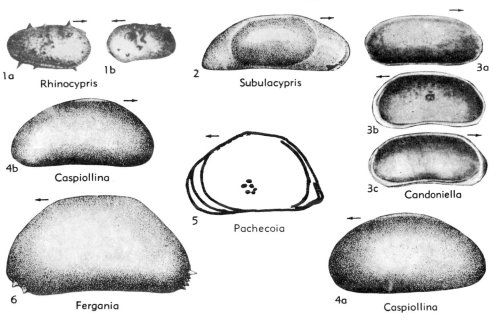

1a, 1b Rhinocypris
2 Subulacypris
3a, 3b, 3c Candoniella
4b Caspiollina
5 Pachecoia
6 Fergania
4a Caspiollina

Fig. 172. Eucandonidae, Ilyocyprididae (Ilyocypridinae), Pontocyprididae (p. Q235-Q237, Q241, Q247).

Similar to *Dolerocypris* and *Dolerocypria,* but with appendages relating it to Candonidae and Eucandonidae. [Marine.] *Rec.,* Brazil coast.

Thaminocypris ZALANYI, 1944 [*T. declinata*]. *Eucandona*-like shell with straight hinge line, subacuminate ventrally, produced anterior margin; broader and bluntly angulated posterior margin, smooth surface; narrow inner lamellae; dysodont hinge. *Neog.,* Eu.——Fig. 173,2. *T. declinata; 2a-d,* carapace R, L, dors., vent., ×30 (406).

Tuberocypris SWAIN, 1947 [*T. acuminatus*]. Like *Eucandona* but with prominent median swollen area on each valve. *U.Tert.,* USA(Utah).——Fig. 173,3. *T. acuminatus; 3a-d,* RV lat., int., vent.; RV vent., all ×40 (351).

Tuberocyproides SWAIN, 1947 [*T. dipleura*]. Elongate subquadrate, convexity moderate; dorsum flattened, with hinge surface nearly straight, venter concave, anterior margin broadly rounded, truncate above, posterior margin acuminate, strongly extended ventrally; LV larger, overlapping RV along free margins, ventral half with strongly elevated ventrally projecting lobes; general surface pustulose. Inner lamellae broadest anteriorly. *U. Tert.,* N.Am.——Fig. 173,5. *T. dipleura,* USA (Utah); *5a,b,* LV lat., int., ×40 (351).

Typhlocypris VEJDOVSKY, 1882 [*Cypris eremita* VEJDOVSKY, 1880]. Similar to *Eucandona* but with antennae adapted for swimming; shell subtriangular, dorsum strongly arched, RV umbonate and extending beyond LV dorsally. Possibly close

to *Pachecoia* ALMEIDA and *Bairdiocypris* KEGEL. *Rec.,* Eu.——Fig. 168,3. *T. eremita* (VEJDOVSKY); *3a,b,* carapace L, dors., ×17; *3c,* muscle scar, ×85 (391).

Family ILYOCYPRIDIDAE Kaufmann, 1900

[*nom. transl.* SWAIN, herein (*ex* Ilyocypridinae KAUFMANN, 1900] [Materials for this family prepared by F. M. SWAIN, University of Minnesota]

Shell subquadrate, compressed, thickest posteriorly, with one or more dorsomedian sulci; dorsum straight, venter straight to slightly concave, ends rounded, anterior broader; LV larger than RV; surface pitted, tuberculate or spinose. Inner lamellae rather narrow. [Fresh-water.] *Trias.-Rec. Cret.-Rec.*

Subfamily ILYOCYPRIDINAE Kaufmann, 1900

[=Iliocyprinae G.W.MÜLLER, 1900; Synopsida (*partim*) DADAY, 1900; Ilyocyprinae MASI, 1906; Eucypridae (*partim*) ALM, 1915; Ilyocyprides SARS, 1925]

Shell without anteroventral marginal notch and with soft parts as prescribed (KAUFMANN, 1900) for Ilyocypridinae. *?Trias., U.Jur.-Rec.*

Ilyocypris BRADY & NORMAN, 1889 [*Cypris gibba* RAMDOHR, 1808] [=Iliocypris MÜLLER, 1900; Iliocypris DADAY, 1900; Ilyocyprois MASI, 1906]. Elongate, subquadrate, compressed, bisulcate; with marginal compressed rim; surface pitted, pustulose or tuberculate; dimorphic posterior thickening in females. *?Trias., Rec.,* cosmop.——Fig. 176,*1.* *I. gibba* (Ramdohr), *Rec.,* N.Eu.-N.Am.; *1a,b,* carapace L, dors., ×60; *1c,* LV int., ×60 (82).

Cyprideamorphella Mandelstam, 1956 [**C. tarbagataiensis*]. Kidney-shaped, thick-walled, oval with maximum height in anterior third, length about 1.3 mm.; RV larger than LV; both ends broadly rounded, anterior end higher; dorsal margin straight, ventral slightly concave; surface pitted and with one or more submarginal ridges; adductor muscle scars large; RV hinge an elongate groove, enlarged at anterior end; that of LV a corresponding bar with anterior expanded part. [Fresh-water.] *L.Cret.,* NE.Asia.——Fig. 176,*2.* **C. tarbagataiensis,* E.Transbaikalia; *2a,* carapace L; *2b,c,* hinge of RV and LV; all ×27 (50).

Iliocyprella Daday, 1900 [**Ilyocypris bradyi* var.

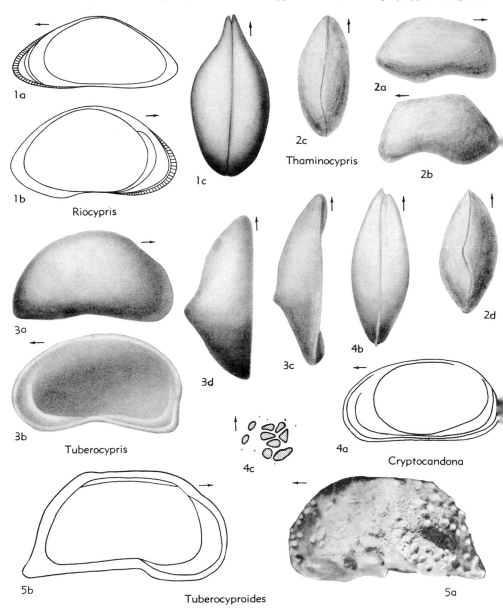

Fig. 173. Eucandonidae (p. Q237-Q239).

repons Vávra, 1891] [=*Ilyocyprella* Sars, 1925]. Differs from *Ilyocypris* in soft parts and lack of swimming power. *Rec.,* Eu.-Asia M.

Ilyocyprimorpha Mandelstam, 1956 [**I. palustris*]. Thick-walled, about 0.9 mm. in length, each valve with single obscure, wide, transverse median concavity or sulcus; RV larger than LV; anterior end higher than posterior; ends rounded; dorsal margin fairly straight; ventral margin slightly concave medially; surface with massive spines and weak reticulations; RV hinge consisting of anterior elongate broad groove at site of cardinal angle followed by median and posterior narrow groove; LV hinge an anterior flangelike tooth passing backward into narrow bar thickened at posterior end. Inner lamellae present terminally; pore canals few, narrow and straight; muscle scar a medial group of 4 or more spots plus one or more additional anteroventral spots. [Fresh-water.] *L.Cret.,* NE.Asia(S.Transbaikalia-Mongolia).——Fig. 176, *3*. **I. palustris,* Transbaikalia; *3a,b,* LV lat., LV int.; *3c,d,* hinge of LV and RV; all ×43 (50).

Pelocypris Klie, 1939 [**P. lenzi*]. Subquadrate; dorsum sinuous, with anterior high shoulder and medium dorsal depression, venter nearly straight; terminal margins spinose; dimorphic posterior thickening in females. *Pleist.-Rec.,* C.Am.——Fig. 176,*4a-d.* **P. lenzi; 4a,b,* ♀ carapace L, dors.; *4c,* ♂ RV int.; *4d,* ♂ LV dors.; all ×30 (215a). ——Fig. 176,*4e-i. P. zilchi* Triebel, Pleist., C.Am. (El Salvador); *4e,* ♂ LV lat.; *4f-i,* ♀ RV lat., int., vent., dors.; all ×40 (82).

Rhinocypris Anderson, 1940 [**Ilyocypris jurassica* Martin, 1940; SD Sylvester-Bradley, herein] [=*Origoilyocypris* Mandelstam, 1956]. Carapace very small, ovoid; shell thin; LV slightly larger than RV; internally a narrow shelf of uniform width all round except dorsally; shallow groove extending from middle of dorsal margin almost to center of each valve with smaller groove anterior to this, carapace widest in front. Hinge simple, knurled anteriorly and flexed to right posteriorly; slight overlap of LV over RV on all margins except dorsally, greatest ventrally; ventral margin slightly concave; surface covered with minute closely arranged blunt spines or pustules of uniform size. RV hinge on elevated ?smooth ridge. [Fresh-water.] *U.Jur.-L.Cret.,* W.Eu.-SW. Asia(Caspian).——Fig. 159,*1. R. cirrita* Mandelstam, L.Cret.; *1a,b,* carapace L, dors., ×20; *1c,d,* LV int., RV int., ×65 (50).——Fig. 172,*1.* **R. jurassica,* U.Jur. or L.Cret., Eng.; *1a,b,* carapace R, L, ×55 (50).

Subfamily CYPRIDEINAE Martin, 1940

[=Rostrocyprinae Anderson, 1939 (invalid, lacking nomino-typical genus)]

Shell subquadrate, compressed to moderately convex; dorsum nearly straight, venter slightly sinuate, converging posteriorly

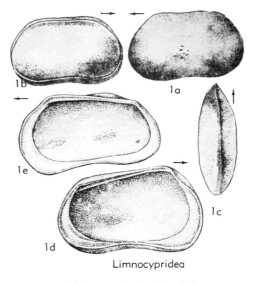

Limnocypridea

Fig. 174. Eucandonidae (p. Q237).

with dorsum; valves more or less unequal, each bearing anteroventral marginal notch and in typical forms ventral projection known as beak; surface smooth, pitted, pustulose, or weakly sulcate; hinge margin of larger valve grooved, with socket-like terminal, depressions commonly crenulate. [Fresh-water.] *Trias.-L.Cret.*

Cypridea Bosquet, 1852 [**Cypris granulosa* Sowerby, 1836; SD Sylvester-Bradley, 1947] [=*Dsunbaina* Galeeva, 1955]. *M.Jur.-L.Cret.,* Eu.-N.Am.-S.Am.-Afr.-Japan.

C. (Cypridea). LV larger than RV; surface punctate or reticulate, with or without granules, tubercles, or spines. *U.Jur.-L.Cret.,* Eu.-N.Am.-S.Am. ——Fig. 177,*1a-d.* **C. (C.) granulosa* (Sowerby), Jur., Eng.; *1a,b,* carapace L, R; *1c,d,* LV dors., RV dors.; all ×40 (79).——Fig. 177, *1e,f. C. (C.) propuncta* Sylvester-Bradley, Jur., Eng.; *1e,f,* LV int., RV int., ×50 (79).——Fig. 178,*1. C. (C.) wyomingensis* Jones, L.Cret. (Draney F.), USA(Idaho); *1a,* LV lat.; *1b-d,* carapace R, dors., vent., all ×35 (200).

C. (Cyamocypris) Anderson, 1939 [**Cypris valdensis* Fitton, 1836]. Large thin shell with exaggerated selvage. *U.Jur.,* Eu.——Fig. 177,*3a-c. C. (C.) tumescens* (Anderson), L.Cret., Eng.; *3a,b,* carapace R, L; *3c,* LV int.; all ×40 (*3a,b,* 91; *3c,* 79).——Fig. 177,*3d-f.* **C. (C.) valdensis* (Fitton), Eng.; *3d-f,* carapace L, R, vent., ×40 (51).

C. (Morinina) Anderson, 1939 [**M. dorsispinata*]. Distinguished by dorsal sulcus, large scattered normal canals, and smooth surface.*L.Cret.,*

Eu.——Fig. 179,3. *C. (M.) dorsispinata* (Anderson), L.Cret., Eng.; *3a-d,* carapace R, L, dors., vent., ×60 (91).

C. (Paracypridea) Swain, 1946 [**C. (P.) obovata*]. Like *C. (Cypridea)* but greatest height in postmedian position and RV overlapping LV. U.Jur. or L.Cret., Brazil.——Fig. 177,2. **C. (P.) obvata; 2a,b,* LV (paratype) lat., LV (holotype) int., ×30; *2c,* LV lat., showing muscle-scar pattern, ×30 (350).

C. (Pseudocypridina) Roth, 1933 [**P. piedmonti*] [=*Langtonia* Anderson, 1939]. LV larger than RV; beak much reduced and notch very slight or absent; ornamentation weak or absent (79). U.Jur., Eu.-N.Am.-S.Am.——Fig. 179,1. **C. (P.) piedmonti,* Brazil; *1a-c,* carapace L, dors., vent., ×30 (350).

C. (Ulwellia) Anderson, 1939 [**U. menevensis*]. Distinguished by reversal of valve overlap. U.Jur.-L.Cret., Eu.——Fig. 179,2. **C. (U.) menevensis*

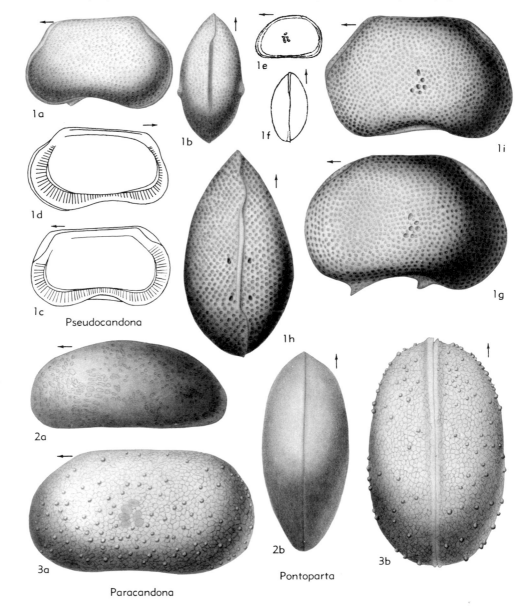

Pseudocandona

Paracandona

Pontoparta

Fig. 175. Eucandonidae (p. Q238).

Jur., Eng.; *2a-d,* carapace R, L, dors., vent., ×50 (91).

C. (Yumenia) Hou, 1958 [*Y. *oriformis*]. Like *C. (Cypridea)* in shape, overlap, and anteroventral marginal sinuosity, but lacks anteroventral beak

and notch. *L.Cret.,* Yumen Dist., Kansu. [Hanai.]

Cultella Lyubimova, 1959 [*C. *daedala*]. Medium-sized (0.85 mm.) thin-walled, irregularly oval, uniformly convex or with a small median to dorso-median swelling; LV larger than RV, slightly

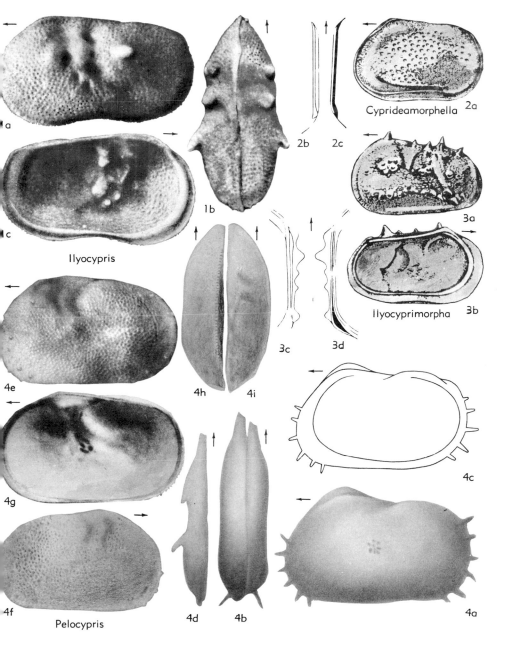

Fig. 176. Ilyocyprididae (Ilyocypridinae) (p. *Q*239-*Q*241).

overreaching RV along dorsal and ventral edges; terminal margins not of same height; anterior high, strongly truncated above, evenly rounded below; posterior end considerably lower and evenly rounded; dorsal margin straight and inclined toward posterior end; ventral margin straight or

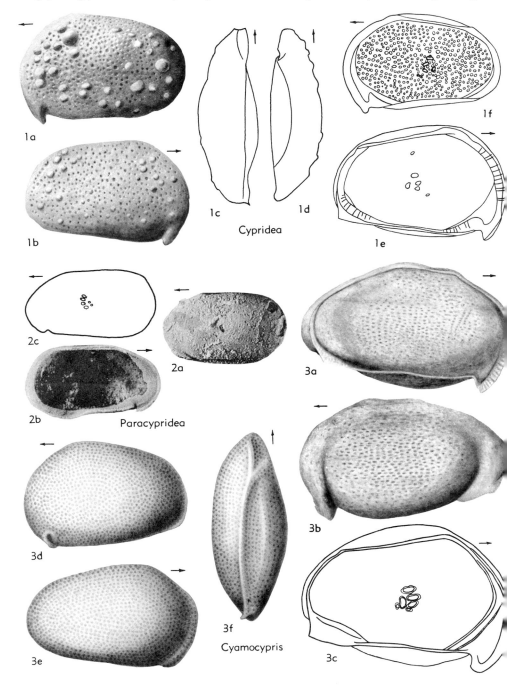

FIG. 177. Ilyocyprididae (Cyprideinae) (p. Q241-Q242).

slightly concave medially, with ends joined smoothly, valves pitted. Hinge consisting of groove in LV and corresponding projecting edge of RV; hinge groove of LV without terminal broadened portions of *Limnocypridea* and simpler than in *Cypridea*. *Trias.,* USSR(Tschelabinski Basin, Zatetschenski region).——Fig. 179A,*1.* **C. daedala;* RV lat., ×50 (231a).

Family NOTODROMADIDAE Kaufmann, 1900

[*nom. transl.* Swain, herein (*ex* Notodromadinae Kaufmann, 1900) [=Zygopsida *(partim)* Daday, 1900; Notodromides Sars, 1925; Notodrominae Hoff, 1942; Notodromini Bronstein, 1947] [Materials for this family prepared by F. M. Swain, University of Minnesota]

Carapace small to medium in size, subovate, inflated, with arched dorsum, flattened carinate venter and rounded ends; valves subequal; surface smooth or pustulose; prominent anterodorsal eye spots. Inner lamellae broadest anteriorly. [Fresh-water.] *Paleoc.-Rec.*

Notodromas Liljeborg, 1853 [*Cypris monacha O.F.Müller, 1776]. Similar to *Newnhamia* but with surface smooth except for ventral keels; female smaller than male, with flattened venter and typically with posteroventral short spine; male with venter angulated behind mid-length and lacking posteroventral spine (68). *Rec.,* Eu.-Asia-E.Indies-N.Am.——Fig. 180,*1.* **N. monachus* (O.F.Müller), Eu.; *1a-c.* ♀ carapace L, dors., vent., ×40; *1d,* ♂ carapace R, ×40; *1e,f,* ♀ carapace L, vent., ×40; *1g,h,* ♂ carapace L, dors., ×40; *1i,* LV int. ant. part, ×80 (*1a-d,* 314; *1e-i,* 55).

Cyprois Zenker, 1854 [*Cypris marginata Strauss, 1821]. Subovate, moderately convex; dorsum strongly arched, flattened on anterior slope, venter slightly concave, front margin broadly rounded but rear bluntly pointed, extended below; anterior and ventral marginal areas compressed; LV slightly larger than RV; surface smooth. Inner lamellae broad anteriorly; radial canals numerous toward front, fewer at rear (68). [Fresh-water.] *Paleoc.-Rec.,* Eu.-N.Am.——Fig. 152,*4.* **C. marginata* (Strauss), Rec., Eu.; *4a,b,* LV lat., RV int., ×30 (Sars, 1925); *4c,* muscle-scar pattern, ×50 (40); *4d-e,* ♀ and ♂ carapace L, ×30; *4f,* ♂ carapace dors.; ×30; *4g,* ♀ LV int. ant. margin, ×60 (256).

Newnhamia King, 1855 [**N. fenestrata*]. Small, subovate, strongly convex, thickest posteriorly; venter flattened and somewhat depressed, with boatshaped keel; free edges with narrow compressed rim; surface coarsely pustulose. Male smaller than female. *Rec.,* S.Hemis.——Fig. 180,*2.* **N. fenestrata,* Austral.; *2a-c,* ♀ carapace L, dors., vent., ×70 (392).

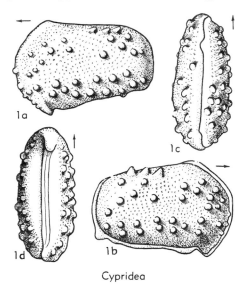

Cypridea

Fig. 178. Ilyocyprididae (Cyprideinae) (p. Q241).

Family PARACYPRIDIDAE Sars, 1923

[*nom. transl.* Swain, herein (*ex* Paracyprides Sars, 1923)] [Materials for this family prepared by F. M. Swain, University of Minnesota] [Includes Aglaiocyprinae (*recte* Aglaiocypridinae) Schneider, 1960]

Smooth, elongate, with wide duplicature, large anterior and posterior vestibules, and radial pore canals commonly branched. [Marine and fresh-water.] *?Sil., Jur.-Rec.*

Paracypris Sars, 1866 [**P. polita*]. Elongated, wedge-shaped, tapering to pointed-posterior; LV larger than RV, very broad inner lamellae, bifurcating radial pore canals. Shape resembles that of *Macrocypris* but carapace is smaller, dorsum less convex, and muscle-scar spots fewer. [Marine.] *?Sil., Jur.-Rec.,* cosmop.——Fig. 181,*1.* **P. polita,* Rec., Norway; *1a,b,* ♀ carapace L, dors., ×40; *1c,* ♀ LV int., ×40; *1d,* muscle scar, ×120 (314).

Aglaiella Daday, 1910 [**A. stagnalis*]. Differs from *Aglaiocypris* in soft parts. Radial pore canals branched. [Brackish-water.] *Rec.,* Egypt.——Fig. 181,*2.* **A. stagnalis; 2a,b,* ♀ carapace L, dors., ×50; *2c,d,* RV int., anteroventral margin, LV ext., posterior margin, ×200 (137).

Aglaiocypris Sylvester-Bradley, 1946 [1947] [*Aglaia pulchella Brady, 1868] [=*Aglaia* Brady, 1868 (*non* Renier,. 1804)]. Shell subelliptical in lateral view, highest medially; dorsal margin slightly convex, ventral margin straight to gently concave, ends broadly and nearly equally rounded, slightly extended below mid-height; valves subequal, with low convexity, greatest thickness median; surface smooth. Muscle scar a rosette group of several spots; inner lamellae moderate in width, broadest anteriorly (15). [Marine, shallow-water.] *Rec.*, N.Atl.——Fig. 181,3a,b. **A. pul-chella;* 3a,b, carapace L, dors., ×30 (12).——Fig. 181,3c-e. *A. complanata* (Brady & Robertson); 3c-e, RV int., carapace dors., LV and RV dors., ×80 (53).

Paracypria Sars, 1910 [**Paracypris tenuis* Sars, 1905]. Carapace like *Paracypris*, but with terminal LV overreach. [Brackish- and fresh-water.] *Tert.-Rec.*, Eu.-Afr.-Chatham I.——Fig. 181,4. **P. tenuis*, Rec., Chatham I.; 4a-c, carapace L, dors., RV int., ×75 (310).

Phlyctenophora Brady, 1880 [**P. zealandica*]. Dif-fers from *Paracypris* in soft parts. [Marine.] *Rec.*, S.Pac.

Pontocyprella Lyubimova, 1955 [**Bairdia harrisiana* Jones, 1849]. Small kidney-shaped, with thickened valves, LV overlapping RV; length about 1 mm.; anterior end rounded, extended and angulated above; presumed posterior end narrower, some-what extended below; dorsal margin convex; ven-tral margin straight to slightly concave; LV hinge a closed groove with pronounced overhang above; RV hinge formed of knifelike valve edge. Inner lamellae scarcely developed anteriorly, absent else-where; pore-canal zone rather broad, with straight, closely spaced canals. *Jur.-Paleog.*, C.Asia-W.Eu.-E.Eu.——Fig. 181,5a,b. **P. harrisiana* (Jones), Cret., Eng.; 5a,b, carapace L, R, ×40 (37).——Fig. 181,5c,d. *P. aktagensis* Mandelstam, L.Cret., C.Asia; 5c,d, carapace R lat., dors., ×30 (50).

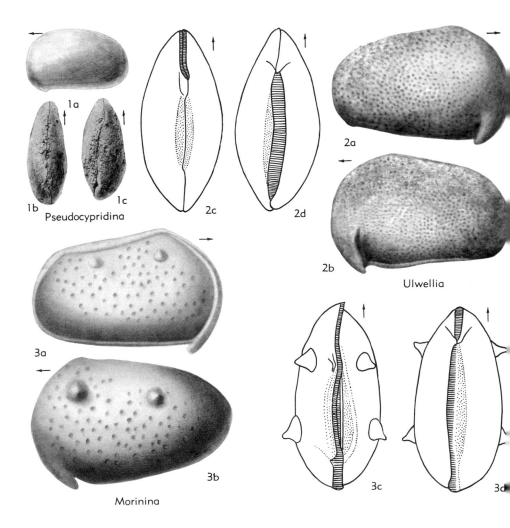

Fig. 179. Ilyocaprididae (Cyprideinae) (p. Q241-Q243).

Family PONTOCYPRIDIDAE
G. W. Müller, 1894

[*nom. transl.* SWAIN, herein (*ex* Pontocypridinae, *nom. correct* KAUFMANN, 1900, *pro* Pontocyprinae G.W.MÜLLER, 1894)] [=Pontocypridae ALM, 1915] [Materials for this family prepared by F. M. SWAIN, University of Minnesota, with assistance on some genera by W. A. VAN DEN BOLD, Louisiana State University] [Includes Argilloeciinae, Clinocyprinae (*recte* Clinocypridinae) MANDELSTAM, 1960]

Shell elongate; dorsum arched; ventral border nearly straight; posterior end generally more pointed than anterior; valves nearly equal; typically smooth. *?Dev., Trias.-Rec.*

Pontocypris SARS, 1866 [*Cypris serrulata* SARS, 1863; SD BRADY & NORMAN, 1889 (=*Cythere (Bairdia) mytiloides* NORMAN, 1862)] [=*?Cytheropsis* McCOY, 1849 (*non* SARS, 1866); *Erythrocypris* G.W.MÜLLER, 1894]. Medium size; elongate-acuminate, compressed, thickest anteromedially; dorsum arched, angulated in front of middle; posterior end strongly extended ventral to mid-height, acuminately pointed; valves subequal; surface smooth; posteroventral margin of RV serrate. Inner lamellae broad; muscle scar with few spots, forming a compact group. *?Dev., Rec.,* cosmop.——FIG. 182,2. *P. mytiloides, Rec., N. Atl., 2a,b,* ♀ carapace L, dors.; *2c,d,* ♂ LV lat., int.; all ×50 (*2a,b,* 65; *2c,d,* 67).

Argilloecia SARS, 1866 [*A. cylindrica*] [=*Argillaecia* BRADY, 1870]. Subelliptical, nearly cylindrical in cross section; mid-portion of shell somewhat flattened; posterior margin truncated; RV slightly larger than LV; inner lamellae typically very broad. Sexual dimorphism marked, males much smaller and more elongated than females. *Cret.-Rec.,* cosmop.——FIG. 182,1. *A. cylindrica; Rec.,* Norway; *1a,b,* ♀ LV lat., int.; *1c,* ♀ carapace dors.; *1d,* ♂ LV lat.; all ×70 (65).

Bakunella SCHNEIDER in MANDELSTAM *et al.,* 1958 [*Pontocypris dorsoarcuata* ZALANYI, 1929]. Carapace irregularly pyriform, swollen, with posterior end pointed ventrally, truncate in upper part, dorsal margin arched, ventral margin concave, anterior end rounded. Inner lamella 4 times broader than zone of concrescence. *Plio.-Pleist.,* SE.Eu., SW.Asia. [BOLD.]

Clinocypris MANDELSTAM, 1956 [*C. scolia*]. Shell elongate, subreniform or lanceolate, length about 0.9 mm., height 0.4 mm.; LV larger than RV; anterior end high and broadly rounded, posterior pointed and strongly extended below; dorsal margin curved, most convex in anterior third, ventral nearly straight, slightly concave; surface smooth or pitted; RV hinge consisting of narrowed valve edge that fits over benchlike margin of LV. Inner lamellae equally wide at both ends; pore-canal zone narrow, with numerous canals. [Freshwater.] *Trias.-L.Cret.,* SW.Asia(Caspian)-NE.Asia (Dzungaria-Mongolia).——FIG. 182,3. *C. scolia,* L.Cret., Dzungaria; RV lat., ×27 (50).

Cultella

FIG. 179A. Ilyocyprididae (Cyprideinae) (p. Q243-Q245).

Pontocypria G.W.MÜLLER, 1894 [*P. spinosa*]. Like *Pontocypris,* but more equal-ended and with more convex ventral border; surface pustulose. Muscle scar a circlet of spots around central spot, as in *Bairdia. Rec.,* Medit.——FIG. 182,4. *P. spinosa; 4a,b,* carapace L, dors., ×120 (53).

Propontocypris SYLVESTER-BRADLEY, 1947 [*Pontocypris trigonella* SARS, 1866]. Shell like *Pontocypris,* but less elongate, and less acuminate posteriorly, and less angulated dorsally; posteroventral margin of RV not serrate. *Pleist.-Rec.,* Atl.-Medit.——FIG. 182,5. *P. trigonella, Rec.,* N.Atl.; *5a,b,* ♀ carapace L, dors.; *5c,* ♂ RV lat.; all ×100 (65).

Subulacypris SCHNEIDER in MANDELSTAM *et al.,* 1957 [*S. subtilis*]. Carapace elongate, not high, irregularly triangular; dorsal margin arched, especially at 0.3 from anterior end, posterior drawn out ventrally, zone of concrescence narrow, with straight pore canals, inner lamella 3 or 4 times broader; hinge with ridge in RV, groove in LV. Surface smooth. Close to *Clinocypris* but with broader inner lamella in both ends. *Plio.,* E.Asia (N.China).——FIG. 172,2. *S. subtilis;* RV lat., ×63 (238a). [BOLD.]

CYPRIDACEA, Family UNCERTAIN

[Materials for this section prepared by F. M. SWAIN, University of Minnesota]

Camdenidea SWAIN, 1953 [*C. camdenensis*]. Subtriangular, valves inflated; LV larger than RV; dorsal margin strongly convex, venter concave medially, ends narrowly rounded, extended below, posterior extremity pointed; surface smooth. Hinge straight, short, consisting of ridge and groove; shell with terminal compressed marginal zones; inner lamellae broadest ventrally; mid-ventral margin turned inward; muscle scar large, medially located, consisting of sinuous small spots. [With *Ranapeltis* perhaps represents an early link between Cyprididae and Bairdiidae. Marine.] *Dev.,* N.Am.——FIG. 182A,2. *C. camdenensis,* USA (Tenn.); *2a,b,* R lat., L int., ×30 (Swain, n). [SWAIN.]

Carbonita STRAND, 1928 [*pro Carbonia* JONES, 1870 (*non* ROBINEAU-DESVOIDY, 1863)] [*Carbonia* *agnes* JONES, 1870]. Small, elongate, subquadrate, moderately convex, thickest posteriorly; hinge margin straight to slightly convex; venter slightly sinuous, ends rounded, anterior margin slightly narrower; small raised ridge near anteroventral margin; RV overlapping LV along free margins

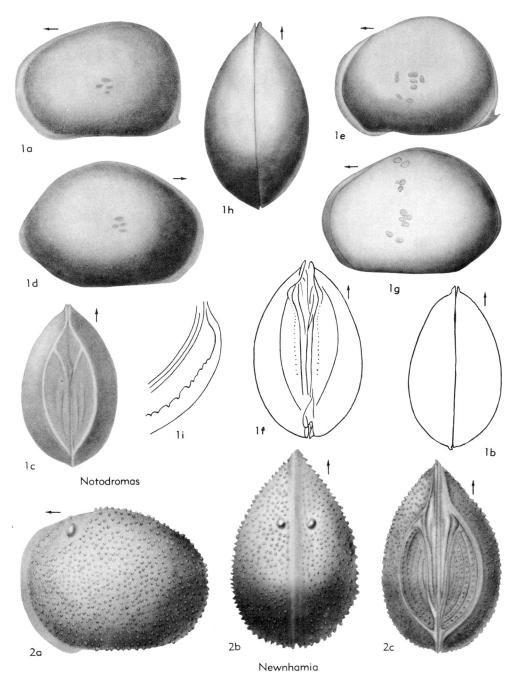

1a

1h

1e

1d

1g

1c

1i

1f

1b

Notodromas

2a

2b

2c

Newnhamia

FIG. 180. Notodromadidae (p. Q245).

but LV extending slightly beyond RV dorsally; surface smooth or pitted. [Fresh-water, ?marine.] ?*L.Carb., Penn.(U.Carb.)-Perm.,* Eu.-N.Am.——

FIG. 182A,*3.* **C. agnes* (JONES), U.Carb., S.Wales; *3a,b,* LV lat., dors., ×40 (183). [SWAIN.]

Gutschickia SCOTT, 1944 [**Whipplella ninevehensis*

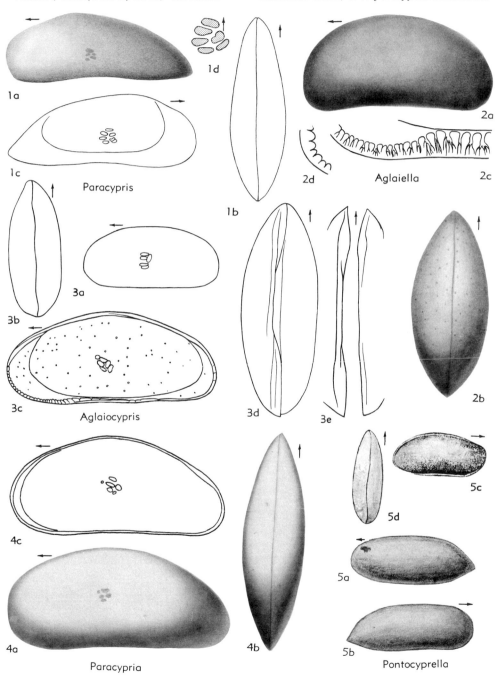

FIG. 181. Paracyprididae (p. Q245-Q246).

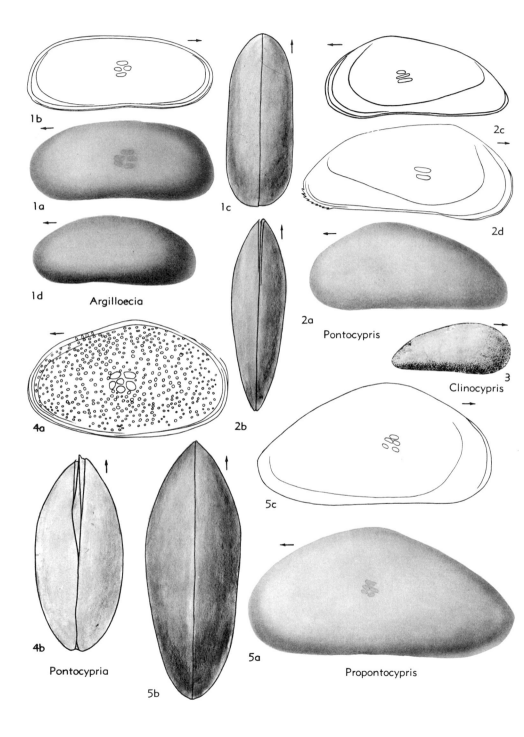

Fig. 182. Pontocyprididae (p. Q247).

HOLLAND, 1934]. Ovate-subtriangular, moderately tumid, ends compressed; dorsum strongly arched; venter nearly straight; terminal margins about equally rounded, extended below; LV overlapping RV dorsally but RV overlapping LV ventrally; surface smooth, pitted or granulose. [Freshwater.] *U.Penn.-L.Perm.*, N.Am.——FIG. 182A,4.

G. ninevehensis (HOLLAND), L.Perm., USA(Pa.); *4a-d,* carapace L, R, dors., vent., ×30 (323). [SWAIN.]

Haworthina KELLETT, 1935. [*Bairdia bulleta* HARRIS & LALICKER, 1932]. Resembling *Pontocypris* but its relationship uncertain to Paleozoic species doubtfully referred to *Pontocypris*. L.Perm., N.Am.

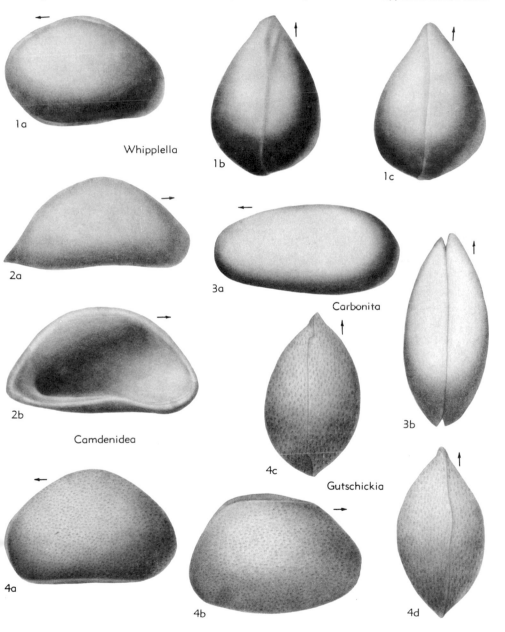

FIG. 182A. Cypridacea, Family Uncertain (p. Q247-Q253).

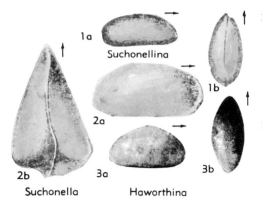

FIG. 182B. Cypridacea, Family Uncertain
(p. Q251-Q253).

(Tex.-Kans.).——FIG. 182B,*3*. *H. bulleta* (HAR-
RIS & LALICKER), USA(Kans.); *3a,b*, carapace R,
dors., ×30 (198) [SHAVER.]

Hilboldtina SCOTT & SUMMERSON, 1943 [**H. multiplicata*]. Elongate-ovate, compressed; dorsum
nearly straight, venter slightly convex, ends
rounded, posterior extremity narrower and more
pointed; RV overlapping LV; surface with longitudinal striae. [Fresh-water.] *L.Penn.(U.Carb.),
E.N.Am.-Eu.*——FIG. 182C,*3*. **H. multiplicata*,
L.Penn., USA(Tenn.); *3a-d*, carapace L, R, dors.,
vent., ×50 (326). [SWAIN.]

Palaeocypris BRONGNIART, 1876 [**P. edwardsii*].
Elongate, subquadrate to subreniform, highest anteromedially; dorsum convex in larger LV, straight
in RV, rear end narrower; LV with strong dorsal
extension beyond hinge line; surface pitted. Inner
lamellae of moderate width anteriorly; soft parts
known. [Fresh-water.] *Carb., Eu.(Fr.).* [SWAIN.]

Pruvostina SCOTT & SUMMERSON, 1943 [**P. wanlessi*]. Like *Whipplella* but hinge longer and with
hinge margin strongly impressed in prominent
channel [Fresh-water.] *L.Penn., N.Am.*——FIG.
182C,*2*. **P. wanlessi*, USA(Tenn.); *2a-d*, carapace
L, R, dors., vent., ×40 (326). [SWAIN.]

Ranapeltis BASSLER, 1941 [**R. typicalis*]. Subreni-

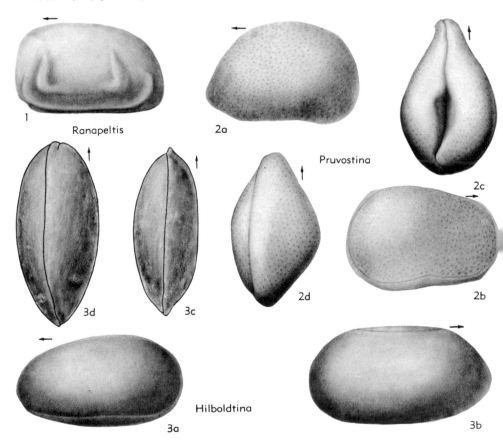

FIG. 182C. Cypridacea, Family Uncertain (p. Q252-Q253).

form-acuminate, strongly convex; dorsum arched but hinge surface nearly straight, venter slightly concave, front end broadly rounded, truncate above, posterior margin narrower, extended and acuminate below, truncate above; LV larger, overlapping RV along free margins; ventral surface with strongly elevated longitudinal ridges. Inner lamellae narrow; muscle scar with median group of radiating spots. [With *Camdenidea* perhaps represents early link between Cyprididae and Bairdiidae. Marine.] *M.Dev.,* N.Am.——Fig. 182C,*1,* 310B,*2.* **R. typicalis,* USA(Tenn.); 182C, *1,* LV lat., ×30 (Kesling); 310B,*2a,b,* LV lat., int., ×30 (Swain, n). [Swain.]

Scabriculocypris Anderson, 1940 [**S. trapezoides*]. Elongate subquadrate, anterior end more rounded than posterior; LV strongly overlapping RV, especially along venter; surface reticulate to spinose. Hinge resembling that of *Cythere.* U.Jur., Eng. [Topotype specimens examined by Sylvester-Bradley have muscle scars suggestive of Cyprida-cea.]

Suchonella Spizharsky, 1937 [**S. typica;* SD Swain, herein]. Dorsum strongly arched, ventral margin straight, somewhat concave in front; widely wedge-shaped in dorsal view; posteroventral lateral surface with broad protuberance or smaller elevated horn; LV larger than RV, overlapping it on ventral and posterior margins. Hinge of RV with hemicylindrical groove corresponding to bar on LV; middle of valve marked by thickening, which separates front half from rear where eggs probably were accommodated; muscle imprint located in front half of valve defined as round spot in which weak ridges and grooves radiate from center. [Fresh-water.] *L.Trias.*(topmost *Tatar.*), Kuznetsk Basin, Sib. [337, 338; the latter apparently was original description, but publication was delayed 2 years after first appearance of name in print.].——Fig. 182B,*2.* **S. typica,* Suchonai-N. Dvina Basin, USSR; *2a,b,* carapace R, vent., ×35 (338) [Swain.]

Suchonellina Spizharsky, 1937 [**Cythere (Cytherella?) inornata* M'Coy & Jones, 1850 (*non* M'Coy, 1844); SD Swain, herein]. Elongate oval, anterior end blunt, posterior end sharply produced below; LV larger than RV, overlapping it on ventral and posterior margins; dorsal margin straight or slightly convex; ventral margin straight or slightly concave; valves very strongly inflated, with greatest convexity closer to posterior than anterior edge; in dorsal view rear edge blunter than front edge. Muscle imprint in front half of valves, from interior appearing as small rounded convexity, from center of which radiate little ridges and depressions; hinge a cylindrical groove on RV and bar on LV; hinge edge of RV however, loosely overlapping LV. Length 0.5 to 0.7 mm., height 0.25 to 0.3 mm. [Fresh-water.] [337, 338; the former apparently was intended to be original

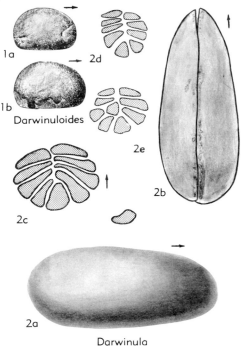

Fig. 183. Darwinulidae (p. Q254).

record but was delayed in publication.]. *U.Perm.,* Suchonai River Valley, USSR-Eng.——Fig. 182B, *1. S. yanichevski* Spizharsky, Tom River; *1a,b,* carapace R, dors., ×20 (337). [Swain.]

Whipplella Holland, 1934 [**W. cuneiformis*]. Subovate, strongly inflated in posterior half, compressed anteriorly; dorsum strongly arched, venter slightly concave, anterior margin narrower; valves unequal, overlap variable; surface pitted. [Freshwater.] *L.Perm.,* N.Am.——Fig. 182A,*1.* **W. cuneiformis,* USA(Pa.); *1a-c,* carapace L, vent., dors., ×50 (323). [Swain.]

Superfamily DARWINULACEA Brady & Norman, 1889

[*nom. transl.* Swain, herein (*ex* Darwinulidae Brady & Norman, 1889) (=Darwinellidae Brady, Crosskey, & Robertson, 1874)] [Diagnosis by F. M. Swain, University of Minnesota]

Shell elongate-ovate; more narrowly rounded and less convex in front; surface typically smooth; RV larger than LV. Hinge simple, undifferentiated; muscle scar comprising numerous radially arranged spots. [Fresh- or brackish-water.] *?Ord., U.Carb. (Penn.)-Rec.*

Family DARWINULIDAE Brady & Norman, 1889

With characters of superfamily. *?Ord., U. Carb.(Penn.)-Rec.*

The Darwinulidae is a small long-ranging family, apparently monotypical, that seems always to have been restricted to fresh or brackish waters. Species have been found with reasonable certainty as far back as Late Carboniferous and may range to the Ordovician. The family was defined on anatomical characteristics by BRADY & NORMAN as follows: "Antennae destitute of swimming setae and of poison gland and duct. Mandible-palp 3-jointed; the basal joint large and densely setiferous. Two pairs of jaws, the first bearing a large branchial plate, the second a smaller branchial plate and a pediform palp. Two pairs of feet external to the valves. Post-abdominal lobes sub-conical, small." Though in general it is not desirable to erect a superfamily on the basis of only 2 genera, the genera are unique in many ways and the most satisfactory way of handling them seems to be to place them in a separate family and superfamily.

Darwinula BRADY & ROBERTSON, 1885 [*pro Darwinella* BRADY & ROBERTSON, 1872 (*non* MÜLLER, 1865); *pro Polycheles* BRADY, 1870 (*non* HELLER, 1862)] [**Polycheles stevensoni* BRADY & ROBERTSON, 1870] [=*Cyprione* JONES, 1885 (an example of nested valves)]. Elongate, oblong or ovate; RV much larger than LV. Hinge formed by simple RV overlap; muscle scar comprising about a dozen elongate, radially arranged spots anterior to middle; calcified inner lamellae lacking (Brady & Norman, 1889). [fresh- or brackish-water.] *?Ord., U.Carb. (Penn.)-Rec.*, cosmop.——FIG. 183,*2a-c*. **D. stevensoni* (BRADY & ROBERTSON), Rec., Eu.; *2a,b*, ♀ carapace R, dors., ×80; *2c*, muscle scar (RV lat.), ×1,000 (*2a-c*, Triebel; *2b*, Sars).——FIG. 183,*2d,e*. *D.* sp., Rec., Eng.; *2d,e*, muscle-scar patterns RV, LV, ×200 (Sylvester-Bradley, n).

Darwinuloides MANDELSTAM, 1956 [**Darwinula oviformis* MANDELSTAM, 1947]. Shell ellipsoidal, ovoid, highly convex; LV larger than RV; anterior end more broadly rounded than posterior; dorsal margin convex or straight; ventral margin generally convex, but may be slightly concave medially; surface smooth; pore-canal zone not well developed; adductor muscle scar stated to be as in *Darwinula;* hinge unknown; dorsally, valve edges depressed to form shallow groove along line of contact. [Fresh-water.] *Perm.-Trias.*, Sib.(Kuznetzk Basin).——FIG. 183,*1*. **D. oviformis* (MANDELSTAM); *1a,b*, RV lat., RV lat., ×45 (50).

Superfamily CYTHERACEA Baird, 1850

[*nom. transl.* ULRICH & BASSLER, 1923 (*ex* Cytheridae BAIRD, 1850)] [Diagnosis and discussion prepared by H. V. HOWE, Louisiana State University, and P. C. SYLVESTER-BRADLEY, University of Leicester]

Anatomically, as outlined by SARS (1866), here belong those members of the Podocopina possessing 3 similar pairs of legs adapted for locomotion; antennules generally subpediform; exopodite of the antennae developed as a long, curved, rodlike flagellum containing ducts of glands lying on each side of body. Adductor muscle scars detectable on most carapaces, usually arranged in nearly vertical row of 4 elements, with 1 or 2 antennal scars and usually 3 mandibular scars in front. A few adopt families lack hinge teeth but most have a compound hinge characteristically divided into 3 or 4 elements, any or all of which may be dentate. *M.Ord.-Rec.*

The majority of cytheracean families are marine and many have highly ornamented carapaces. In our present state of knowledge, the dentition appears to be the firmest, though frankly an artificial basis for classification. It is supplemented by general shape and by characteristic subdivision of certain elements of the adductor muscle scars or of the antennal and mandibular scars. Likewise the character of the duplicature, the presence or absence of a vestibule, the number of radial pore canals and their nature, whether straight and simple, curved, bifurcating, or with bulbous enlargments, and the presence or absence of an eye tubercle are of family significance. The superfamily had its beginning early in the Paleozoic, but only in the Mesozoic and Cenozoic Eras does it become the dominant element in ostracode faunas. The stratigraphic distribution of cytheracean genera is indicated graphically in Figure 184.

Family CYTHERIDAE Baird, 1850

[Materials for this family prepared by H. V. HOWE, Louisiana State University] Includes Camptocytherides MANDELSTAM, 1960]

Ovate, subreniform to subquadrangular carapaces with subequal valves, RV tending to overlap LV dorsally and LV to overlap RV ventrally; smooth or reticulate with large, widely spaced, usually sievelike normal pore canals. Hinge typically antimerodont, though some but not all elements may be smooth; muscle scars in near-vertical row of 4 adductors, 1 or 2 antennal scars in front of upper part of row and usually a rounded mandibular scar below; marginal areas fairly broad, with only a few straight or wavy

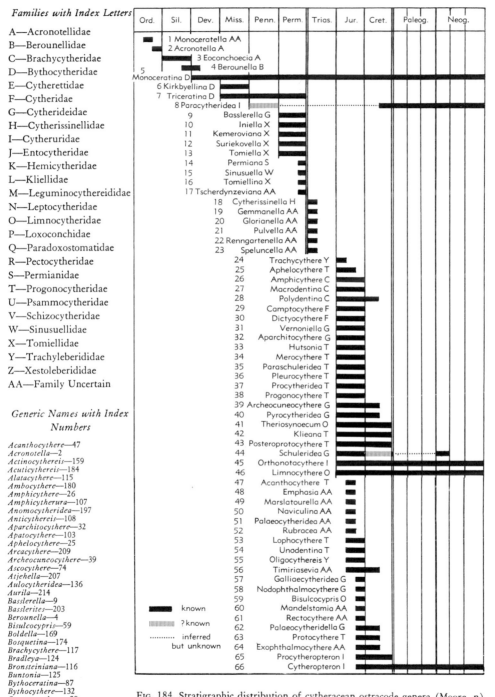

FIG. 184. Stratigraphic distribution of cytheracean ostracode genera (Moore, n). Classification of the genera in families is indicated by letter symbols according to the following tabulation and for the purpose of locating any wanted genus an alphabetical list furnishes a cross reference to the serially arranged numbers on the diagrams.

Clithrocytheridea—119
Cnestocythere—182
Costa—206
Cuneocythere—148
Cushmanidea—154
Cyamocytheridea—172
Cyprideis—198
Cytheralison—165
Cythereis—81
Cytheretta—152
Cytheridea—177
Cytherissa—178
Cytherissinella—18
Cytheroma—179
Cytheromorpha—133
Cytheropteron—66
Cytherura—89
Dictyocythere—30
Digmocythere—145
Diogmopteron—100
Dolocythere—96
Dolocytheridea—75
Dordoniella—99
Echinocythereis—126
Emphasia—48
Eoconchoecia—3
Eocytheropteron—84
Eucythere—121
Eucytherura—90
Euryitycythere—67
Exophthalmocythere—64
Falunia—191
Flexus—171
Galliaecytheridea—57
Gemmanella—19
Glorianella—20
Gubkiniella—68
Habrocythere—78
Haplocytheridea—120
Hemicythere—155
Hemicytheridea—193
Hemicytherura—216
Henryhowella—188
Hermanites—150
Hirsutocythere—151
Howeina—210
Hulingsina—200
Hutsonia—33
Idiocythere—142
Iniella—10
Isocythereis—95
Jonesia—196
Juvenix—69
Kalyptovalva—104
Kangarina—183
Kemeroviana—11
Kikliocythere—127
Kingmaina—83
Kirkbyellina—6
Klieana—42
Kobayashiina—211
Krithe—122
Leguminocythereis—149
Leniocythere—143
Limnocythere—46
Loculicytheretta—215
Looneyella—77
Lophocythere—53
Loxoconcha—92
Loxocythere—176
Macrodentina—27
Mandelstamia—60
Marslatourella—49
Mediocytherideis—194
Mehesella—128
Merocythere—34
Microcythere—204
Microcytherura—212
Microxestoleberis—163
Miracythere—170
Monoceratella—1
Monoceratina—5
Monsmirabilia—137
Munseyella—129
Murrayina—189
Mutilus—201
Naviculina—50
Neocyprideis—147
Neocythere—80
Neocytherideis—202
Neomonoceratina—205

Nephokirkos—138
Netrocytheridea—105
Nodophthalmocythere—58
Occultocythereis—144
Oligocythereis—55
Orionina—161
Orthonotacythere—45
Ovocytheridea—106
Paenula—102
Paijenborchella—123
Paijenborchellina—85
Palaeocytheridea—51
Palaeocytheridella—62
Palaeomonsmirabilia—135
Palmenella—217
Paracyprideis—91
Paracytheretta—118
Paracytheridea—8
Paracytheropteron—218
Paradoxostoma—93
Parakrithe—156
Paraschuleridea—35
Parataxodonta—97
Parexophthalmocythere—73
Pavloviella—70
Pellucistoma—186
Perissocytheridea—199
Permiana—14
Phacorhabdotus—109
Phlyctocythere—141
Platycythereis—82
Pleurocythere—36
Polydentina—28
Posteroprotocythere—43
Procytherettina—101
Procytheridea—37
Procytheropteron—65
Progonocythere—38
Protobuntonia—113
Protocythere—63
Protocytheretta—175
Pseudobythocythere—79
Pseudocythere—88
Pseudocytheridea—181
Pseudokrithe—131
Pterygocythere—114
Pterygocythereis—195
Pulvella—21
Puriana—190
Pyrocytheridea—40
Quadracythere—162
Rectocythere—61
Renngartenella—22
Rotundracythere—157
Rubracea—52
Ruttenella—168
Saida—167
Schizocythere—173
Schuleridea—44
Sclerochilus—158
Segmina—86
Semicytheridea—98
Semicytherura—219
Sinusuella—15
Speluncella—23
Sphenocytheridea—166
Spongicythere—146
Stillina—111
Suriekovella—12
Tanella—213
Taxodiella—71
Thalmannia—192
Theriosynoecum—41
Timiriasevia—56
Tomiella—13
Tomiellina—16
Trachycythere—24
Trachyleberidea—130
Trachyleberis—134
Triceratina—7
Triginglymus—140
Tscherdynzeviana—17
Unodentina—54
Urocythereis—208
Uroleberis—164
Veenia—110
Velarocythere—112
Vernoniella—31
Vetustocythere—139
Vicinia—72
Xestoleberis—94

#	Taxon	Jur.	Cret.	Paleog.	Neog.
67	Euryitycythere G	■			
68	Gubkiniella T	■			
69	Juvenix AA	■			
70	Pavloviella AA	■			
71	Taxodiella AA	■			
72	Vicinia AA	■			
73	Parexophthalmocythere Y	■			
74	Asciocythere G	▬			
75	Dolocytheridea G	▬			
76	Centrocythere T	▬			
77	Looneyella T	▬			
78	Habrocythere AA	▬			
79	Pseudobythocythere D	▬▬▬			
80	Neocythere T	▬▬			
81	Cythereis Y	▬			
82	Platycythereis Y	▬			
83	Kingmaina C		▬▬▬		
84	Eocytheropteron I		▬▬▬▬		
85	Paijenborchellina I		▬▬▬▬		
86	Segmina AA		▬▬▬		
87	Bythoceratina D		▬▬▬▬▬▬▬		
88	Pseudocythere D		▬▬▬▬▬▬▬		
89	Cytherura I		▬▬▬▬▬▬▬		
90	Eucytherura I		▬▬▬▬▬▬▬		
91	Paracyprideis G		▬▬▬▬▬▬▬		
92	Loxoconcha P		▬▬▬▬▬▬▬		
93	Paradoxostoma Q		▬▬▬▬▬▬▬		
94	Xestoleberis Z		▬▬▬▬▬▬▬		
95	Isocythereis Y		■		
96	Dolocythere R		▬		
97	Parataxodonta T		▬		
98	Semicytheridea AA		▬		
99	Dordoniella C		■		
100	Diogmopteron C		■		
101	Procytherettina T		■		
102	Paenula AA		■		
103	Apatocythere C		▬▬		
104	Kalyptovalva G		▬▬		
105	Netrocytheridea G		▬▬		
106	Ovocytheridea G		▬▬		
107	Amphicytherura V		▬▬▬▬		
108	Anticythereis Y		▬▬▬		
109	Phacorhabdotus Y		▬▬▬		
110	Veenia Y		▬▬▬		
111	Stillina AA		▬▬		
112	Velarocythere AA		▬▬		
113	Protobuntonia Y		▬▬▬		
114	Pterygocythere C		▬▬▬▬		
115	Alatacythere C		▬▬▬▬		
116	Bronsteiniana AA		▬▬▬▬		
117	Brachycythere C		▬▬▬▬		
118	Paracytheretta E		▬▬▬▬▬		
119	Clithrocytheridea G		▬▬▬▬▬	⋯	‖
120	Haplocytheridea G		▬▬▬▬▬	⋯	
121	Eucythere G		▬▬▬▬▬		
122	Krithe G		▬▬▬▬▬		
123	Paijenborchella Y		▬▬▬▬▬		
124	Bradleya Y		▬▬▬▬▬		
125	Buntonia Y		▬▬▬▬▬		
126	Echinocythereis Y		▬▬▬▬▬		
127	Kikliocythere C		▬▬▬▬		
128	Mehesella I		▬▬▬		
129	Munseyella R			▬▬▬	
130	Trachyleberidea Y		▬▬▬▬▬▬		
131	Pseudokrithe AA		▬▬▬▬▬▬		
132	Bythocythere D		▬▬▬▬▬▬		
133	Cytheromorpha P		▬▬▬▬▬▬		
134	Trachyleberis Y		▬▬▬▬▬▬		

FIG. 184 *(Continued).*

radial canals; a small vestibule may be present at the ends. *Jur.-Rec.*

Cythere O. F. Müller, 1785 [**C. lutea;* SD Brady & Norman, 1889] [=*Cytherina* Lamarck, 1818; ?*Cyclas* Eichwald, 1857 (*non* Lamarck, 1798)]. Carapace reniform in side view, elongate ovate dorsally; surface smooth, but finely pitted and bearing faint radial ribbing; RV overlapping LV dorsally, but LV overlapping RV ventrally; eye tubercle weak or lacking. Hinge antimerodont, with accommodation groove over slightly arched middle element of LV; adductor scars in slightly curved row of 4, with crescent-shaped antennal scar in front of uppermost adductor and oval mandibular scar in front of lower adductor; other scars weakly developed above adductors; marginal areas broad, with only trace of vestibule; radial canals few and wavy; normal canals widely spaced and funnel-shaped (15, 56, 98). *Rec., N.Atl.*——Fig. 185,*1*. **C. lutea; 1a,b,* ♀ carapace L, dors.; *1c,d,* ♂ LV int., RV int.; all ×60 (*1a,b,* 314; *1c,d,* 362).

Camptocythere Triebel, 1950 [**C. praecox*]. Carapace ovate, more broadly rounded in front; RV overlapping LV dorsally but LV overlapping RV ventrally; surface smooth except for large openings of widely spaced normal canals; adductor scars in somewhat curved row of 4, with 2 unequal antennal scars and oval mandibular scar in front; hinge similar to that of *Cythere* except front and median elements, which are essentially smooth. No eye. *Jur.*, Eu.——Fig. 186,*3*. **C. praecox,* Ger.; *3a,b,* RV lat., int., ×60; *3c,* LV lat., ×60; *3d,e,* carapace dors., ant., ×60; *3f,* LV ant. margin with radial canals, ×170; *3g,h,* LV and RV hinge margin, ×155 (all 376).

Cnestocythere Triebel, 1950 [**C. lamellicosta*]. Carapace of medium size, subrhomboidal in side view, about equivalved but RV overlapping LV along mid-hinge and LV with raised knob which reaches over RV at back of hinge, also overlapping mid-ventrally; ornamentation reticulate, with strong raised longitudinal ribbing and distinct eye spots. Hinge antimerodont, with strong terminal elements; normal canals widely spaced, sievelike; marginal areas broad, with very few radial canals; adductor scars in nearly vertical row of 4, with large vertically elongate antennal scar in front. *Mio.*, Eu.——Fig. 185,*5*. **C. lamellicosta,* Aus. (Vienna Basin); *5a,* LV lat., ×60; *5b,* carapace dors., ×60; *5c,* RV hinge, ×110 (all 376).

Loxocythere Hornibrook, 1952 [**L. crassa*] [=*Tetracytherura* Ruggieri, 1952]. Subquadrate to subtriangular in outline; RV overlapping LV

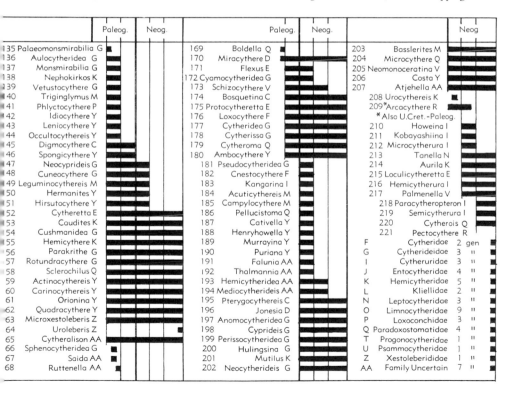

Fig. 184 (*Continued*).

dorsally; surface reticulate. Adductor scars in row of 4, with at least one in front; normal canals widely scattered; hinge similar to that of *Cythere* but median element only faintly crenulate; marginal areas fairly regular, with small vestibules at ends; radial canals few and straight. Eye tubercle

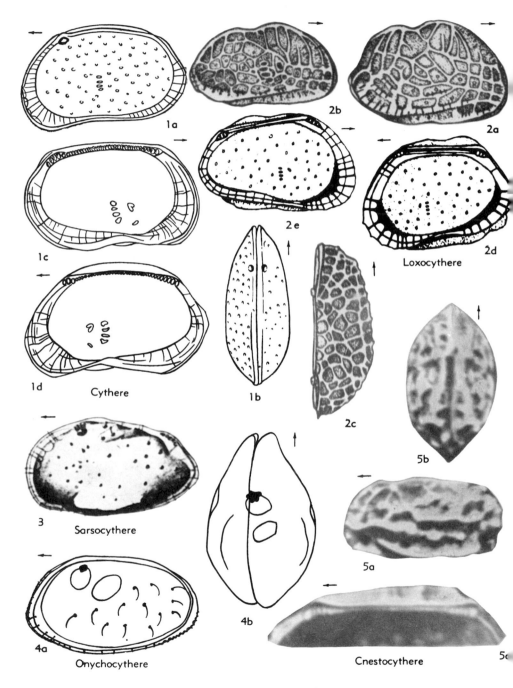

FIG. 185. Cytheridae (p. Q257-Q259).

lacking. Sexes distinct. *Oligo.-Rec.*, N.Z.——Fig. 185,2. **L. crassa*, Rec.; *2a,b*, ♀ RV lat.; ♂ RV lat.; *2c*, ♀ RV dors.; *2d,e*, ♀ RV int., LV int.; all ×75 (32).

Onychocythere Tressler, 1939 [**O. alosa*]. Carapace ovate, more tumid than *Cythere*, RV larger than LV; unornamented. Marginal areas relatively narrow, with few radial canals; muscle scars in oblique row of 4; chiefly distinguished by unequal development of thoracic legs, 2nd being much longer than others. *Rec.*, Fla.——Fig. 185,4. **O. alosa; 4a,b*, carapace L, dors., ×45 (371).

Sarsocythere Tressler & Smith, 1948 [**S. patuxiensis*]. Carapace stout, with prominent alae near venter, outline broadly rounded in front and somewhat pointed in rear; surface nearly smooth. Hinge well developed, with terminal crenulations; radial canals few, grouped. *Rec.*, E.N.Am.——Fig. 185,3. **S. patuxiensis*, Md.; LV lat., ×50 (372).

?Family ACRONOTELLIDAE Swartz, 1936

[Materials for this family prepared by R. V. Kesling, University of Michigan]

Carapace with subequal valves, straight hinge line, sulcus (S_2), and strong ventral projection that terminates sharply toward rear. *U.Ord.-U.Sil.*

Genera with various kinds of lobation have been assigned to this family, their chief common characters being a straight hinge line and prominent ventral projection. Teichert (368) and Triebel (82) think that the ventral projection, which in many species ends in a spine, is a recurrent homeomorphic feature instead of a mark of taxonomic close relationship. Nonsulcate genera (such as *Boucekites, Gravia, Monoceratella, Mooreina, Pinnatulites, Pullvillites, Russia,* and *Tricornina*) assigned to this family by various authors here are rejected from inclusion in it, being treated as *incertae sedis.* Also, *Vltavina* is excluded because it lacks a well-defined ventral projection.

Acronotella Ulrich & Bassler, 1923 [**A. shideleri*] [=*Acrotonella* Peneau, 1927 *(errore)*]. Elongate, with projecting anterior corner; S_2 long, sloping, reaching nearly to anteroventral border; L_2 nodose, low, indistinct; central ventral area bearing large spine that projects outward and slightly backward. *U.Ord.*, N.Am.——Fig. 186,1. **A. shideleri*, Ind.; *1a,b*, carapace L, dors., ×30 (86).

?Eoconchoecia Moberg, 1895 [**E. mucronata*]. Valves with long curved S_2, bearing blunt ventral and anterior projecting spines. *Sil.*, Eu.——Fig. 186,2. **E. mucronata*, Swed.; carapace R, ×? (251).

Acronotella

Eoconchoecia

Camptocythere

Fig. 186. Cytheridae, Acronotellidae (p. Q257-Q259).

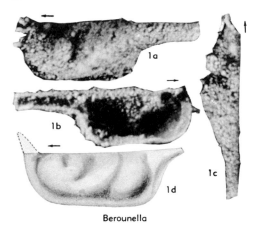

FIG. 187. Berounellidae (p. Q260).

?Family BEROUNELLIDAE Sohn & Berdan, 1960

[Materials for this family prepared by I. G. SOHN and JEAN BERDAN, U.S. Geological Survey]

Small, subquadrate, lobed and spinose, essentially symmetrical, with posterior extension of dorsal margin as a straight, narrow tube that may equal half or more of greatest length of main portion of carapace; no velate structure. Hingement simple, overlap inconspicuous; duplicature present. Surface smooth, possibly reticulate. *U.Sil.-L. Dev., ?L.Carb.*

Berounella BOUČEK, 1936 [*B. rostrata*]. Trilobate (?quadrilobate), spinose anterior margin, with or without additional spines on surface of valves. *U.Sil.-L.Dev.,* Eu.-N.Am.——FIG. 187,*1d.* *B. rostrata,* U.Sil., Czech.; LV lat., ×35 (10).—— FIG. 187,*1a-c.* *B.* sp., L.Dev., N.Y.; *1a-c,* LV lat., int., dors., ×50 (Sohn, n).

?Kirkbyellina KUMMEROW, 1939 [*K. styliolata*]. Subcentral sulcus surrounded by wide lobe which trends toward posterior extension. *L.Carb.* (Viséan), Ger.

Family BRACHYCYTHERIDAE Puri, 1954

[*nom. transl.* HOWE, herein (*ex* Brachycytherinae PURI, 1954] [=Pterygocytheridae PURI, 1957] [Materials for this family prepared by H. V. HOWE, Louisiana State University, with aid of R. A. REYMENT, University of Stockholm] [Includes Pterygocytherides MANDELSTAM, 1960]

Carapace fairly large, plump, especially ventrally, commonly with ventral ridge or ala; broadly and usually obliquely rounded anteriorly, narrower posteriorly, tending to angulation at or below middle; in end view subcircular, subtriangular, or even strongly triangular; eye tubercle usually present; surface smooth to reticulate. Hinge an amphidont development from merodont hingement of Progonocytheridae, ranging from paramphidont to hemiamphidont or holamphidont, commonly with accommodation groove; adductor muscle scars may be in vertical row of 4 but modified by subdivision of upper 2 scars and by fusion of lower 2 scars (differences may occur within same genus), antennal scar usually crescentor heart-shaped or occurring as 2 closely placed spots, mandibular scars oval, additional scars high in carapace; marginal areas fairly broad and regular, without vestibule; radial canals numerous and tending to be bulbous in mid-section; normal canals rather widely spaced. *Jur.-Rec.*

Brachycythere ALEXANDER, 1933 [*Cythere sphenoides* REUSS, 1854]. Carapace subtriangular to subovate in lateral view, plump ventrally, usually with small carina separating flattened ventral face from lateral face; ventral surface usually striated longitudinally, lateral surface smooth to weakly reticulate; eye spot distinct; anterior end broadly and obliquely rounded, posterior end subangulate at or below middle. Hinge hemiamphidont; adductor scars (in ALEXANDER's figure of supposed type species) in somewhat vertical row with uppermost scar L-shaped (possibly due to fusion of 2 scars), followed below by 2 oval scars obliquely side by side and below them a large oval scar, V-shaped antennal scar above oval mandibular scar in front of paired adductors (Paleocene species of this genus, however, with upper very oblique pair of adductors above 2nd less oblique pair, which in turn overlie 2 nearly fused elongate ovate scars); marginal areas regular, without vestibule; radial canals numerous, tending to be bulbous in mid-section; normal canals small, widely spaced [Characters of this genus cannot be defined adequately until topotype specimens from vicinity of Salzburg, Austria, are studied.] *U.Cret.-Rec.,* Eu.-N.Am.——FIG. 188,*1.* ?*B. sphenoides* (REUSS), U.Cret., Texas; *1a,b,* carapace R, RV int., ×45; *1c,* LV int., ×45; *1d-f,* int. marginal areas, ant., post., and muscle scars, ×90 (89).——FIG. 188,*4. B. ventricosa* (BOSQUET), Eoc., Fr.; *4a,b,* LV lat., dors. (U.Ypres.), ×60; *4c,* RV lat. (Lutet.), ×60; *4d,e,* LV int., RV hinge (U. Ypres.), ×75 (42).——FIG. 189,*1. B. plena* ALEXANDER, U.Cret., La.; *1a-c,* RV lat., dors., int., ×75 (200).

Alatacythere MURRAY & HUSSEY, 1942 [*Cythereis (Pterygocythereis?) alexanderi* HOWE & LAW, 1936 (*non* MORROW, 1934) (=*Alatacythere ivani* HOWE, 1951]. Identical in all essential features except hinge to *Pterygocythereis.* Hinge hemiamphi-

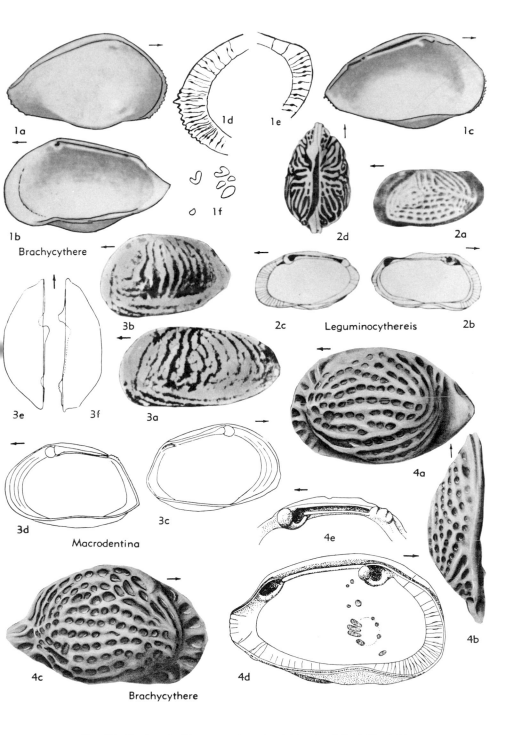

FIG. 188. Brachycytheridae, Leguminocythereididae (p.Q260-Q267, Q307).

dont, with very narrow accommodation groove
over median bar of LV in type species. [Might well
be considered a subgenus of *Pterygocythereis* but
differs in range.] *U.Cret.-Oligo.*, N.Am.-Eu.——
FIG. 189,2. *A. lemnicata* (ALEXANDER), Eoc., USA
(Tex.); *2a-c*, RV lat., dors., int., ×27 (200).——
FIG. 190,4. *A. ivani* HOWE, Oligo., Miss.; *4a,b*, LV
lat., dors., ×37.5 (35); *4c,d*, LV int., RV int.,
×37.5 (170).

Amphicythere TRIEBEL, 1954 [*A. semisulcata*].
Carapace ovate, ventrally inflated, narrowly compressed at ends; surface pitted and bearing nearly
vertical sulcus in region of muscle attachment;
eyenode distinct. Hinge strongly paramphidont;
closing muscles in row of 4, antennal and mandibular scars unknown; marginal areas broad, with
simple, straight radial canals. *Jur.*, Eu.——FIG.
192,1. *A. semiculcata*, Ger.; *1a-c*, carapace L,
R, vent., ×60; *1d,e*, LV & RV hinge, ×100
(379).

?Apatocythere TRIEBEL, 1940 [*A. simulans*]. Externally like *Cytheridea* but hingement of amphi-
dont-type (except that anterior socket of RV and
tooth of LV are lacking); marginal areas broad,
with numerous radial canals on anterior part that
curves above middle; muscle scars in vertical row
of 4, with a single scar in front. *U.Cret.*, Eu.——
FIG. 190,1. *A. simulans; 1a,b*, ♀ RV and LV
lat.; *1c,d*, ♂ carapace dors., ♀ LV dors.; *1e,f*,
♀ LV and RV int., ×45 (81).

Bosquetina KEIJ, 1857 [*Cythere pectinata* BOSQUET,
1852]. Carapace externally shaped like *Brachycythere* but lacks distinct eye tubercle. Hinge
weakly holamphidont; adductor scars in vertical
row of 4, with lower mandibular scar, 2 upper
antennal scars, and at least one scar higher in
carapace; marginal areas regular, with numerous
bulbous radial canals; normal canals widely spaced,
small. Dimorphous. *Oligo.-Rec.*, Eu.——FIG. 191,
5. *B. pectinata* (BOSQUET), Plio., Fr.(Perpignan);
5a-c, LV lat., int., dors., ×60; *5d*, RV hinge,
×60; (all 42).

Dictyocythere SYLVESTER-BRADLEY, 1956 [*Cythere
retirugata* JONES, 1885]. Carapace shaped like that
of *Macrodentina* but strongly reticulate. Hinge
holamphidont, with stepped anterior tooth in RV;
marginal areas regular, with few straight canals;
normal canals widely spaced. *Jur.*, Eu.——FIG. 193,
2. *D. retirugata*, Eng.; *2a,b*, carapace R, dors.,
×40 (367).——FIG. 193,3. *D.* sp., diagram of RV
hinge, ant. part, enlarged (235).

Digmocythere MANDELSTAM, 1958 [*Brachycythere
russelli* HOWE & LEA in HOWE & LAW, 1936].
Externally shaped like *Brachycythere*, with weak
eye spot. Hinge weakly paramphidont, anterior
tooth of RV usually showing crenulations; adductor scars comprising a small upper pair, below
them a larger oblique pair, then 2 single scars,
in front of adductors a lower rounded mandibular
scar, an upper crescent-shaped antennal scar, and

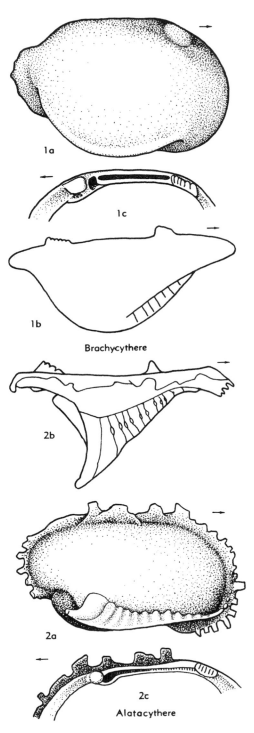

FIG. 189. Brachycytheridae (p. Q260-Q262).

at least 2 other scars high in carapace; marginal area regular, with numerous wavy radial canals; normal canals widely spaced, small. *Eoc.-Oligo.,* N.Am.

Diogmopteron HILL, 1954 [**Brachycythere lünensis* TRIEBEL, 1941]. Resembles *Alatacythere* but hingement partially reversed, accommodation groove being in RV instead of LV. *U.Cret.,* Eu.

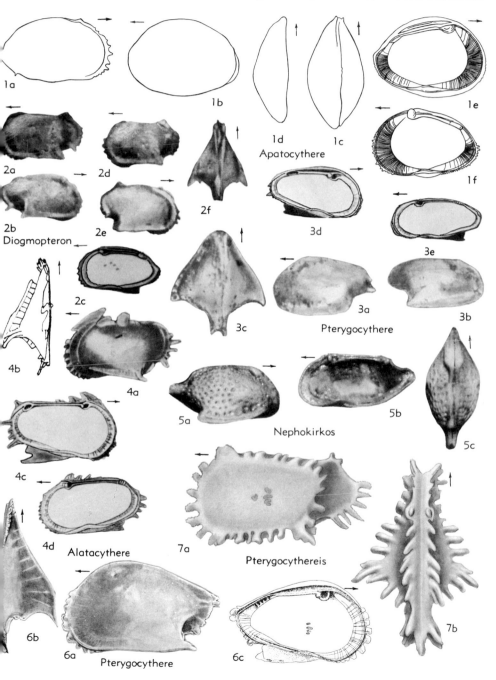

FIG. 190. Brachycytheridae, Hemicytheridae (p. *Q262-Q267, Q305*).

——FIG. 190,2. **D. luenensis*, Ger.; *2a,b*, ♂ LV
and RV lat.; *2c*, ♂ RV int.; *2d,e*, ♀ LV and RV
lat.; *2f*, ♀ carapace dors.; all ×35 (82, 169).

?**Dordoniella** APOSTOLESCU, 1955 [**D. strangulata*].
Resembles *Apatocythere* in general external outlines, including possession of external sulcus in
muscle area; LV overlapping RV along entire
periphery. Hinge similar to that of *Apatocythere*
but with crenulate posterior tooth; marginal areas
broad, with small vestibule; radial canals not
known; muscle scars in oblique row of 4, with

antennal scar in front of middle of row and
mandibular scar lower and farther forward. *U.
Cret.(Cenom.)*, Eu.——FIG. 193,1. **D. strangulata*, Fr.; *1a-c*, carapace R, L, dors.; *1d*, LV int.;
1e,f, RV int., dors.; all ×40 (92). [REYMENT.]

Kikliocythere HOWE & LAURENCICH, 1958 [**Cypridina favrodiana* BOSQUET, 1847]. Carapace large, very
inflated, superficially resembling *Brachycythere* but
sides rounded so that end view is nearly circular instead of triangular and larger LV overlaps RV at
ends of hinge line. Hinge holamphidont; adductor

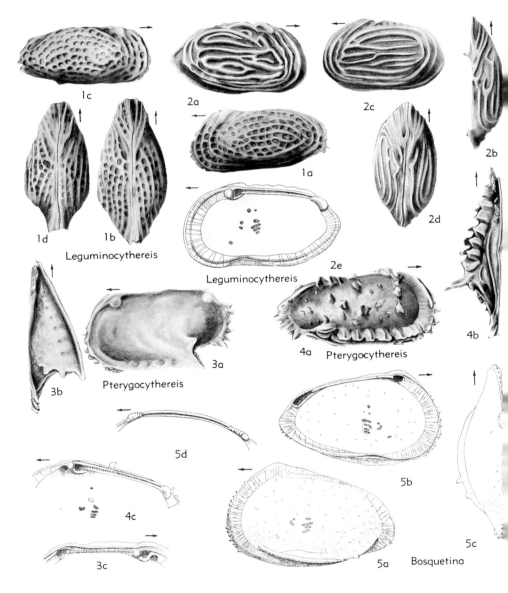

1c

1d   1b
Leguminocythereis

2a

1a

2c

2e
Leguminocythereis

2d

2b

3b

3a
Pterygocythereis

4a   Pterygocythereis

4b

5d

5b

4c

3c

5a   Bosquetina

5c

FIG. 191. Brachytheridae, Leguminocythereididae (p. Q262-Q267, Q307).

scars indistinct, but apparently in row of 4, with V-shaped antennal scar in front, other scars unknown; marginal areas regular, without vestibule; radial canals and normal canals unknown. *U.Cret.*

(Maastricht.), Eu.——Fig. 193,4. **K. favrodiana* (Bosquet), Holl.; *4a-c,* carapace L, dors., ant.; *4d,e,* RV int., LV int.; all ×30 (396).

Kingmaina Keij, 1957 [**Cythere forbesiana* Bos-

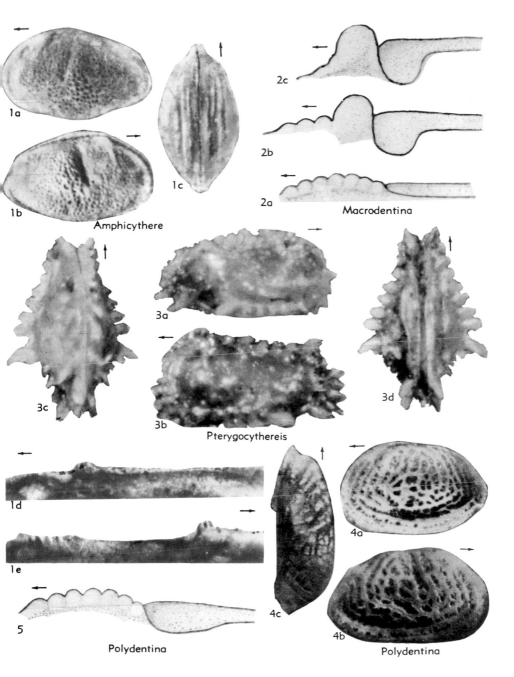

Fig. 192. Brachycytheridae (p. Q262-Q267).

QUET, 1852]. Carapace very inflated ventrally, with strong carina between rounded lateral and flattened ventral surfaces; obliquely rounded anteriorly, compressed posteriorly and subangulate below middle; surface ornamented with ridges or reticulate pits which have nearly vertical alignment; eye spot distinct; end view triangular. Hinge holamphidont but (like *Triginglymus*) with thickened protuberance of shell wall below and behind anterior hingement of each valve; adductors

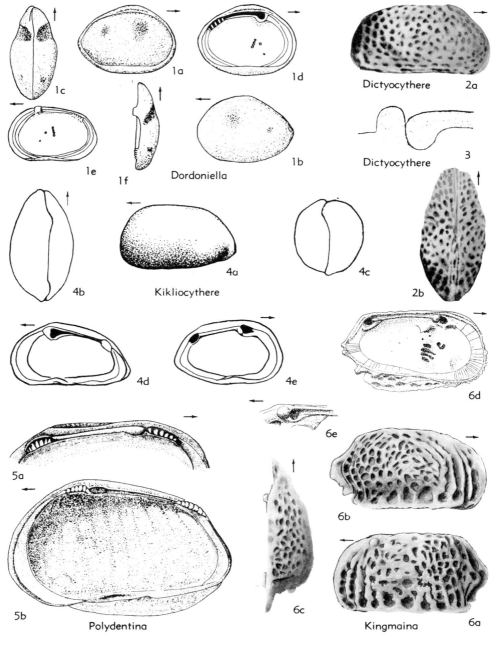

FIG. 193. Brachycytheridae (p. Q262-Q267).

in vertical row of 4 elongate scars in Tertiary type species (but in Cretaceous forms some of these scars are divided), mandibular scar subcircular, antennal scar U-shaped in type species but divided in Cretaceous species; marginal areas regular with numerous radial canals in anteroventral region, sparse elsewhere. *Cret.-Eoc.,* Eu.——Fig. 193,6. **K. forbesiana* (BOSQUET), Eoc., Fr.-Belg.; *6a-c,* LV lat., RV lat., RV dors. (U.Ypres., Paris Basin), ×60; *6d,e,* LV int., RV hinge ant. part (Barton, Belg.), ×60, ×93 (42).

Macrodentina MARTIN, 1940 [**M. lineata*] [*?=Rhysocythere* SYLVESTER-BRADLEY, 1956]. Dimorphous, with strong elongate-oval subequal valves, broadly rounded in front but less so posteriorly, dorsum arched, venter flattened and longitudinally ribbed, lateral surface finely to strongly pitted, reticulate or vertically ribbed; eye tubercle lacking. Hinge paramphidont, front tooth of RV having several stepped crenulations; muscle scars in vertical row of 4, mandibular scar oval, antennal scar crescent-shaped; on marginal areas regular, with rather widely spaced radial canals; normal canals widely spaced. *Jur.,* Eu.——Fig. 188,3. **M. lineata; 3a,* ♂ LV lat.; *3b,* ♀ LV lat.; *3c,d,* LV int., RV int.; *3e,f,* LV dors., RV dors.; all ×45 (51).——Fig. 192,2. *M.* sp., schematic diagram of ant. RV hinge; *2a,* young molt, *2b,* later instar, *2c,* adult, enlarged (235).

Polydentina MALZ, 1958 [**Clithrocytheridea? steghausi* KLINGLER, 1955]. Described as subgenus of *Macrodentina* with similar appearance but differing in less well-developed paramphidont hingement, anterior tooth of RV being a crenulate ridge and anterior end of median element of LV being a raised and thickened bar which fits in shallow socket behind anterior tooth of RV, hinge details variable. Sexual dimorphism strong. [Marine, brackish.] *Jur.-L.Cret.,* Eu.——Fig. 192,4; 193,5. **P. steghausi* (KLINGLER), U.Jur.(Kimm.), Ger.; *192,4a-c,* ♀ LV lat.; ♀ RV lat., ♀ RV dors., ×65; *193,5a,b,* ♀ LV and RV int., ×83 (235).——Fig. 192,5. *P.* sp., schematic diagram of RV hinge, ant. part (235).

Pterygocythere HILL, 1954 [**Cypridina alata* BOSQUET, 1847]. Identical in all features to *Alatacythere,* except for thicker shell material and hence broader accommodation groove in LV. *U.Cret.-Eoc.,* Eu.-N.Am.——Fig. 190,3. **P. alata* (BOSQUET), U.Cret.(Maastricht.), Holl.(Maastricht.); *3a,b,* LV lat., RV lat.; *3c,* carapace dors.; *3d,e,* LV int., RV int.; all ×37.5 (after 169).——Fig. 190,6. *P. hilli* KEIJ, Eoc. (Led.), Belg.; *6a,b,* ♀ LV lat., vent.; *6c,* LV int.; all ×60 (42).

Pterygocythereis BLAKE, 1933 [**Cythereis jonesii* BAIRD, 1850] [*=Fimbria* NEVIANI, 1928 (*non* BOHADSCH, 1761; *nec* MEGERLE, 1811; *nec* RISSO, 1826; *nec* COBB, 1894; *nec* BELON, 1896); *Pterigocythereis* VAN DEN BOLD, 1946 (*errore*)]. Carapace rather inflated ventrally, with strong pointed

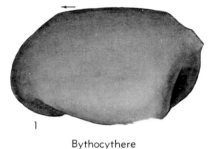

Bythocythere

Fig. 194. Bythocytheridae (p. Q268).

alae, which in type species are broken into a row of spines, each spine surrounding a canal; lateral surface generally smooth but in some species bearing tubercles or spines and in most species bearing a bladelike dorsal ridge; eye spot distinct; anterior and posterior margins spiny; ventral surface flattened and usually bearing longitudinal ribs and in some species nodes. Hinge typically holamphidont but in some European species middle elements may be crenulate; adductor scars in vertical row of 4, with top and bottom scars smallest, antennal scar V-shaped, mandibular scar oval, at least 3 additional scars high in carapace; marginal areas regular, with 8 to 10 groups of 2 to 4 radial canals, each group leading to a marginal spine; normal canals widely spaced (99, 266). *L.Mio.-Rec.,* N.Am., *Eoc.-Rec.,* Eu.——Figs. 190,7, 192,3. **P. jonesi* (BAIRD), Rec., N.Atl. (Scapa Flow); *190,7a,b,* carapace L, dors., ×50 (Sars); *192,3a-d,* RV lat., LV lat., carapace dors., carapace vent., ×50 (Howe, n).——Fig. 191,3. *P. cornuta* (ROEMER), Eoc. (Barton.), Belg.; *3a,b,* ♀ LV lat., vent., ×60; *3c,* LV hinge, ×75 (all 42).——Fig. 191,4. *P. fimbriata* (MÜNSTER), Oligo. (Rupel.), Belg.; *4a,b,* ♂ RV lat., vent., ×60; *4c,* RV hinge, ×75 (all 42).

Family BYTHOCYTHERIDAE Sars, 1926

[*nom. transl.* SYLVESTER-BRADLEY, herein (*ex* Bythocytherinae SARS, 1926)] [Materials for this family prepared by P. C. SYLVESTER-BRADLEY, University of Leicester, and R. V. KESLING, University of Michigan] [Includes Pseudocytherinae SCHNEIDER, 1960]

Hinge lophodont, with long median element, and various modifications. Short caudal process developed in most genera, and median dorsal sulcus in many. Adductor-muscle scars commonly in arcuate group of 5 or more scars, arranged within median sulcus if one is present. *Dev.-Rec.*

The Bythocytheridae comprise the longest-ranging family of the Cytheracea, *Monoceratina* being recorded continuously from Devonian to Recent. Doubt about the validity of identifying Paleozoic with post-Paleozoic species of *Monoceratina* has now been resolved, since SOHN reports that fluoritized Paleozoic specimens show duplicature and 5 adductor-muscle scars, as in post-Paleozoic forms.

Bythocythere SARS, 1866 [**B. turgida;* SD BRADY & NORMAN, 1889] [=*Bathocythere* NORMAN, 1867 (errore)]. Short, upturned caudal process; no median sulcus; venter inflated or alate. Hinge lophodont, with no accommodation groove; moderately wide anterior vestibule; radial pore canals sparse, straight; muscle-scar pattern with 5 adductor scars forming an arcuate group, concave toward front, and with 6th scar slightly more separated, directly above adductors; 2 antennal scars in front of adductor scars. *Tert.-Rec.,* cosmop.——FIGS. 194,*1,* 195,*4.* **B. turgida,* Rec., Eire; 194,*1,* LV lat.; 195,*4a,b,* RV int., LV int.; all ×60 (Sylvester-Bradley).

Bythoceratina HORNIBROOK, 1952 [**B. mestayerae*] [=*Bythocytherina* SOHN, 1957 (errore)]. Like *Monoceratina* but hinge differs in having long denticulate median element to otherwise lophodont hinge. Some species with caudal process below hinge line as in *Bythocythere* but others have this process in line with hinge (as in *Monoceratina*). *Cret.-Rec.,* N.Z.-Antarct.——FIG. 196,*3a-c.* **B. mestayerae,* Rec., N.Z.; *3a,* LV lat.; *3b,c,* RV int., dors.; all ×75 (32).——FIG. 196,*3d. B. maoria* HORNIBROOK, Mio., N.Z.; RV int., ×75 (32).——FIG. 196,*3e. B. tuberculata* HORNIBROOK, Rec., N.Z.; muscle scar, ×200 (32).

Jonesia BRADY, 1866 [**Cythere simplex* NORMAN, 1865 *(non Cythere auriculata simplex* CORNUEL, 1848) (=**Bythocythere acuminata* SARS, 1866)] [=*Macrocythere* SARS, 1926 (obj.); *Macrocytherina,* *Luvula* CORYELL and FIELDS, 1937]. Carapace more or less elongate, dorsal margin straight or convex, anterior end evenly rounded, ventral margin sinuous (concave in front of middle), posterior end sharply acuminate; no median sulcus. Hinge modified lophodont, with terminal elements reduced or missing, and median ridge in LV projecting farther at ends than in middle; muscle-scar pattern linear, with 5 scars sloping downward anteroventrally; wide anterior and posterior vestibules. *Mio.-Rec.,* cosmop.——FIG. 196,*1.* **J. acuminata* (SARS), Rec., Eng.-Norway; *1a-c,* carapace R, dors., vent., ×40; *1d,* LV lat., ×40; *1e,* muscle scar, ×85 (*1a-c,e,*107; *1d,* 315).

Miracythere HORNIBROOK, 1952 [**M. novaspecta*]. Median sulcus behind prominent anterodorsal swelling. Hinge modified lophodont, with median ridge of LV terminating backward in bladelike tooth which fits under posterior tooth of RV; adductor-muscle scars in closely-spaced row of 5. *U.Eoc.-Rec.,* N.Z.——FIG. 196,*2.* **M. novaspecta,* Rec.; *2a,* LV lat.; *2b,* carapace dors.; *2c,* LV (juv.)

dors.; *2d,* RV int.; *2e,* muscle scar; *2a-d,* ×75, *2e,* ×150 (32).

Monoceratina ROTH, 1928 [**M. ventrale (sic)*] [=*Bythocytheremorpha* MANDELSTAM, 1958]. Typically elongate, with long, straight dorsal margin terminating in caudal process; median sulcus extending from dorsal margin to near center of valve, commonly surrounded by crescentic lobe which may bear one or more thornlike spines. Hinge modified lophodont, dominated by long median ridge in LV which in some species has a swollen posterior extremity; anterior element of RV reduced or absent; posterior element of RV a short swollen ridge; adductor-muscle scars of Paleozoic forms in arcuate group of 5, with 2 antennal scars in front (some post-Paleozoic species assigned to genus have only 4 adductor scars). *Dev.-Rec.,* cosmop.——FIG. 195,*3a-c.* **M. ventralis* ROTH, Penn., Okla.; *3a,b,* ? ♂ carapace L; ? ♀ carapace L; *3c,* carapace vent.; all ×50 (297).——FIG. 195,*3d-f. Monoceratina* sp., Oligo., Tex.; *3d,e,* LV lat., dors., ×50; *3f,* muscle scar, ×150 (82).——FIG. 195,*3g. M. herburyensis* SYLVESTER-BRADLEY, M.Jur., Eng.; RV int., ×80 (Sylvester-Bradley).

?Pseudobythocythere MERTENS, 1956 [**P. goerlichi*]. Posterior end subtriangular; no caudal process. Hinge antimerodont; adductor impression of 4 scars in front of median sulcus; duplicature rather wide; radial pore canals sparse. *Cret.,* Ger.——FIG. 195,*2.* **P. goerlichi,* L.Alb.; *2a-c,* LV lat., int., carapace dors., ×60; *2d,* RV int., ×60 (250).

Pseudocythere SARS, 1866 [**P. caudata*]. Carapace compressed, without lateral expansions; caudal process continuous with dorsal margin; no median sulcus. Duplicature broad; radial pore canals sparse; anterior vestibule extending to half width of marginal zone; 5 adductor-muscle scars. *Cret.-Rec.,* cosmop.——FIG. 195,*5.* **P. caudata,* Rec., Medit.; *5a,* RV int.; *5b,* carapace dors., ×100 (53).

Triceratina UPSON, 1933 [**T. wrefordensis*]. Posterodorsal end attenuated; S_2 long, shallow; 1 or 2 nodes in anteroventral region and long backward-directed spine in ventral; surface punctate. *Miss.-Perm.,* N.Am.——FIG. 195,*1a,b.* **T. wrefordensis,* L.Perm., USA(Neb.); *1a,b,* carapace L, vent., ×75 (390).——FIG. 195,*1c-f. T. inconsueta* COOPER, U.Miss., Ill.; *1c-f,* carapace R, vent., ant., post., ×75 (20).

Family CYTHERETTIDAE Triebel, 1952

[Materials for this family prepared by H. V. HOWE, Louisiana State University]

Carapace moderately large, heavy, with obliquely rounded ends, posterior end narrower; dorsal margin of LV tending to be raised over terminal teeth of RV. Surface smooth, longitudinally pitted, or reticulate, in some genera bearing 3 longitudinal ribs;

without appreciable muscle node but a pit may be present on interior valve surface; no eye socket or tubercle. Hinge very strongly holamphidont, with terminal teeth of RV pointing away from each other, anterior tooth of LV pointing downward behind anterior socket, which is only weakly inclosed ventrally, median element of hinge smooth or finely crenulate; marginal area parallel to outer margin only in upper half of back end, elsewhere of variable width owing to irregular embayments, but net effect being great breadth; commonly with small anterior vestibule, from which long sinuous

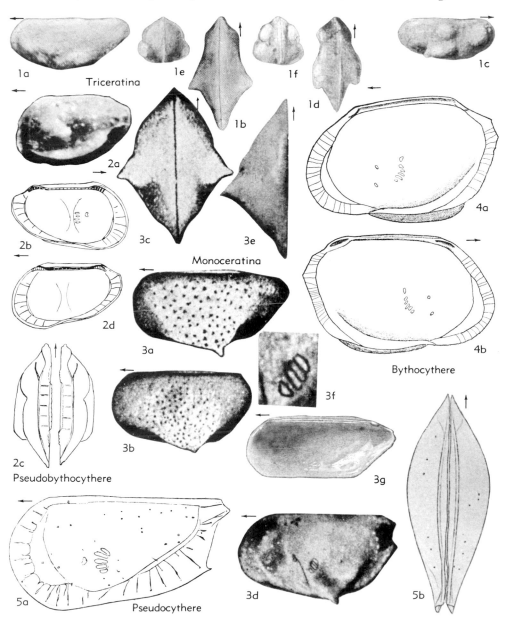

FIG. 195. Bythocytheridae (p. Q268).

to bulbous radial canals extend; muscle scars usually difficult to make out in their entirety owing to depth and thickness of shell, but consisting of vertical row of 4, in front of which antennal scars appear fused in V with mandibular scars much lower, near lobe of inner margin (several additional scars usually seen higher in carapace); nor-

mal canals obscure, if present. *U.Cret.-Rec.*

Cytheretta G.W.Müller, 1894 [**C. rubra* (=**Cytherina subradiosa* Roemer, 1838)] [=*Pseudocytheretta* Cushman, 1906; *Cylindrus* Neviani, 1928 (*non* Deshayes, 1824; *nec* Fitzinger, 1833; *nec* Herrmannsen, 1852); *Prionocytheretta* Mehes, 1941]. Carapace ovate, strong, LV larger and less elongate than RV and differently constructed; sur-

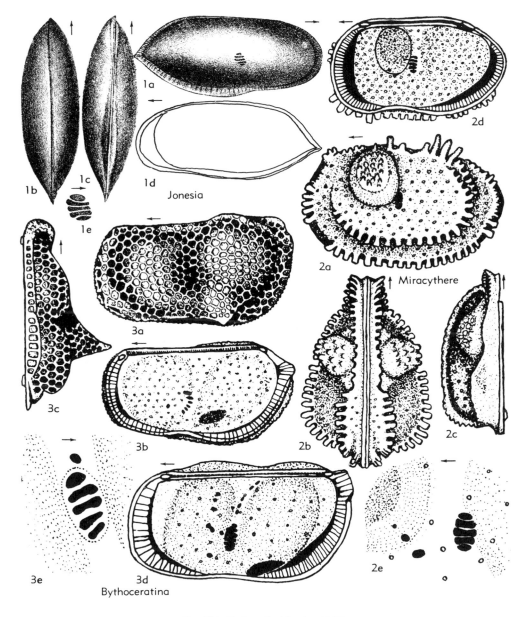

Fig. 196. Bythocytheridae (p. Q268).

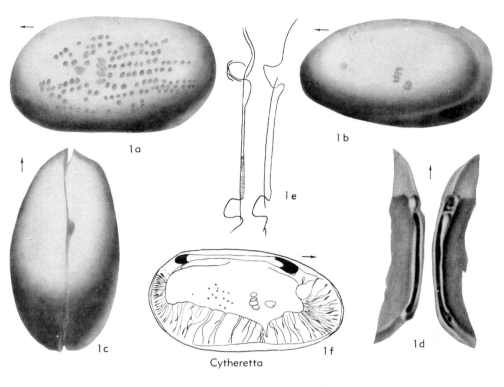

1a

1b

1e

1c

1d

1f

Cytheretta

FIG. 197. Cytherettidae (p. *Q270-Q271*).

face smooth, irregularly pitted, or with pittings arranged in longitudinal reticulations. Hinge heavy holamphidont; marginal areas irregularly broad, with long curved and branching radial canals; adductor scars in vertical row of 4, with V-shaped adductor scar in front and mandibular scars near ventral inner margin (53, 134, 266, 300). *Eoc.-Rec.,* Eu.-N.Am.——FIG. 197,1. *C. subradiosa (ROEMER), Rec., Medit.; 1a,b, LV lat., ×70; 1c, carapace dors., ×70; 1d,e, hinge, LV and RV, ×70, ×90; 1f, LV int., ×70 (all 53).——FIG. 198,2. C. bambruggensis KEIJ, Eoc.(Led.), Belg.; 2a,b, LV lat., RV lat., 2c, carapace dors.; 2d, LV int., all ×90 (all 42).

Flexus NEVIANI, 1928 [*Cythere plicata MÜNSTER, 1830] [=Eucytheretta PURI, 1958]. Carapace more elongate and tapering than *Paracytheretta,* smooth or reticulate, bearing 3 very elongate ridges; narrow posterior end of RV denticulate. Internal features like *Cytheretta. Oligo.-Mio.,* Eu. ——FIG. 199,2. *F. plicatus (MÜNSTER), Mio., Ger.; 2a,b, LV lat., RV lat., ×40 (378).

Loculicytheretta RUGGIERI, 1954 [*Cythere pavonia BRADY, 1866]. Carapace rather small for this family, elongate ovate but with a high posterior angulation; females with 3 circular pits or loculi on posteroventral margin of each valve similar to those of certain Paleozoic ostracodes, but males

without loculi, resembling *Cytheretta.* Surface bearing 3 longitudinal ridges which tend to merge near posterior angulation; pitted between ridges. Hinge holamphidont, in RV with sharp, high anterior tooth, nearly circular deep socket that tends to merge into a gradually narrowing furrow along hinge line, and blunt ovate posterior tooth, neither socket completely closed to interior in LV; marginal area broad and nearly regular, with numerous radial canals; muscle scars as in *Cytheretta. ?Eoc., Plio.-Rec.,* Eu.-Sp.-Italy). ——FIG. 200,1. *L. pavonia (BRADY), Plio., Italy; 1a-c, ♀ LV lat., dors., vent., ×50; 1d, ♀ RV dors., ×50 (303); 1e, Rec., Sp., ♂ carapace vent., ×53; 1f, ♀ RV int., post. part showing loculi and canals, ×130 (Reyment, n).

Paracytheretta TRIEBEL, 1941 [*P. reticosa]. Carapace externally somewhat resembling *Cythereis,* reticulate, with 3 longitudinal ribs and LV overlapping RV at anterior hingement but internally like *Cytheretta. U.Cret.-Rec.,* Eu.-N.Am.-S.Pac. ——FIG. 198,1. *P. reticosa, Paleoc., Ger.; 1a, LV lat., ×45; 1b,c, RV lat., int., ×45; (all 82).

Protocytheretta PURI, 1958 [*Cythere daniana BRADY, 1869]. Carapace elongate-ovate, plump, with rounded ends, posterior extremity usually showing slight angulation but bearing spines; surface with 3 longitudinal ribs, faint muscle node, and longitudinally reticulate pattern; LV overlapping RV strongly at anterior end of hinge and less so posteriorly. Hinge, marginal area, and muscle scars as in *Cytheretta. Oligo.-Rec.,* N.Am.——FIG. 199,*1.* **P. daniana* (BRADY), Rec., Gulf Mexico(Fla.); *1a-c,* carapace L, R, dors., ×33 (291).

Family CYTHERIDEIDAE Sars, 1925

[*nom. transl.* SYLVESTER-BRADLEY & HARDING, 1953 (*ex* Cytherideinae SARS, 1925)] [Materials for this family prepared by H. V. HOWE, Louisiana State University, with contributions on some genera by W. A. VAN DEN BOLD, Louisiana State University, and R. A. REYMENT, University of Stockholm]

Carapace ovoid to pear-shaped; surface smooth, pitted, reticulate, or in some forms longitudinally ribbed, with outline generally resembling that of *Cytheridea.* Adductor muscle scars in vertical or inclined row of 4, usually with 2 or more scars in front, but in some specimens with single heart-shaped scar; hinge adont to merodont, smooth to crenulate. Subfamilies differentiated mainly on basis of hingement, marginal areas, and radial canals. *Perm.-Rec.*

Subfamily CYTHERIDEINAE Sars, 1925

[Includes Dolocytherideinae, Clithrocytherideinae MANDELSTAM, 1960]

Surface smooth, pitted or reticulate. Hinge antimerodont or holomerodont, some genera (e.g., *Haplocytheridea*) containing species with reversed valves, normal LV hinge appearing in RV and vice versa; marginal areas widest anteriorly where inner margin in most genera departs from line of concrescence (with similar but smaller departure common posteriorly); radial canals variable in number but in most genera rather abundant; muscle scars in vertical row

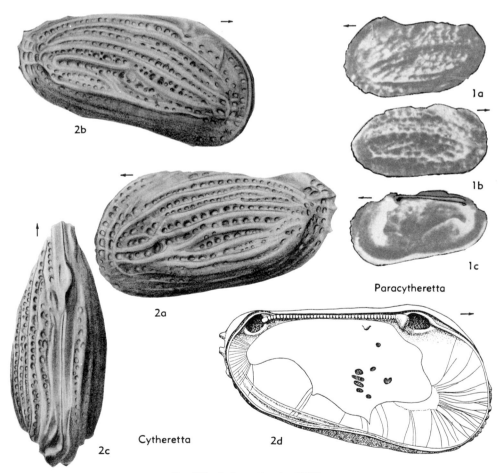

2b

1a

1b

1c

Paracytheretta

2a

2c

Cytheretta   2d

FIG. 198. Cytherettidae (p. Q271).

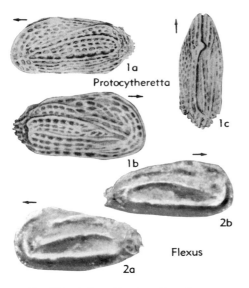

FIG. 199. Cytherettidae (p. Q271-Q272).

of 4, usually with 2 equally spaced ovate scars in front and commonly additional scars above and below. *Perm.-Rec.*

Cytheridea BOSQUET, 1852 [*Cythere mülleri* MÜNSTER, 1830; SD BRADY & NORMAN, 1889] [=*Eucytheridea* BRONSTEIN, 1930]. Carapace thick-shelled, subquadrangular in dorsal view, ovate-triangular in side view, anterior obliquely rounded, posterior oblique, bluntly pointed near venter, both ends tending to be denticulate. Hinge merodont, in LV with terminal crenulate sockets and crenulate median element set somewhat obliquely so that anterior portion projects as short bar, back portion being depressed as furrow; marginal area widest at ends where inner margin and line of concrescence may be separated; radial canals numerous, slightly thickened medially, occurring in groups on anterior; adductor muscle scars in vertical row of 4, with large distinct and smaller less distinct antennal scar in front of upper part usually fused together, and large mandibular scar in front of lower part with much smaller scar obliquely below and behind it, other scars higher in carapace (260, 341). *Oligo.-Rec.*, Eu.-?N.Am. ——FIG. 201,*1*. *C. muelleri* (MÜNSTER), U.Oligo., N.W.Ger.; *1a*, ♂ carapace (neotype) L, ×36; *1b,c*, ♂ and ♀ carapaces dors., ×36; *1d,e*, ♂ LV dors., ♂ RV dors., ×56; *1f*, ♂ LV int. ant., ×56; *1g*, muscle scars, ×100 (152).——FIG. 202, *4. C. pernota* OERTLI & KEIJ, Oligo.(Tongr.-Rupel.), Belg.; *4a,b*, ♀ RV lat., ♀ LV lat., ×60; *4c,d*, ♂ RV lat., LV lat., ×60; *4e*, ♀ RV int., ×75 (all 42).

Anomocytheridea STEPHENSON, 1938 [*Cytheridea*

floridana HOWE & HOUGH in HOWE *et al.*, 1935] [=*Amonocythere* SOHN, 1951]. Like *Cyprideis* in all characters except anterior portion of median hinge element which forms short blunt smooth bar instead of crenulate ridge, and antennal scars distinct instead of forming a "V." [Might be considered subgenus of *Cyprideis*.] *Mio.-Rec.*, N.Am. ——FIG. 203. *A. floridana* (HOWE & HOUGH), Mio., Fla.; *1a*, ♀ LV lat., ×45; *1b*, both valves dors., ×45; *1c*, muscle scars, enlarged; *1d,e*, RV and LV hinge, enlarged; (all 342).

Asciocythere SWAIN, 1952 [*Bythocypris rotundus* VANDERPOOL, 1928]. Carapace plump, subovate, LV with crenulate terminal sockets separated by smooth median bar, above which is well-defined accommodation groove. Marginal areas narrow, with numerous radial canals. [*Palaeocytheridea* resembles *Asciocythere* in all described characters except that type species has straighter dorsal outline. *L.Cret.*, N.Am.-?Eu.——FIG. 202,*1*. *A. rotunda* (VANDERPOOL), ?Trinity, N.Car. (subsurface); *1a,b*, ?♀ and ?♂ carapace R; *1c*, ?♀ carapace dors.; *1d*, LV int.; all ×40 (354).

Basslerella KELLETT, 1935 [*non* HOWE, 1935 (=*Basslerites* HOWE in CORYELL & FIELDS, 1937), *nec* BOUČEK, 1936 (=*Boucia* AGNEW, 1942)] [*B. crassa*]. Externally similar to *Cytheridea* but hinge of LV with crenulate groove at and in front of

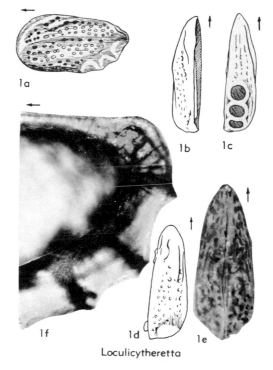

Loculicytheretta

FIG. 200. Cytherettidae (p. Q271).

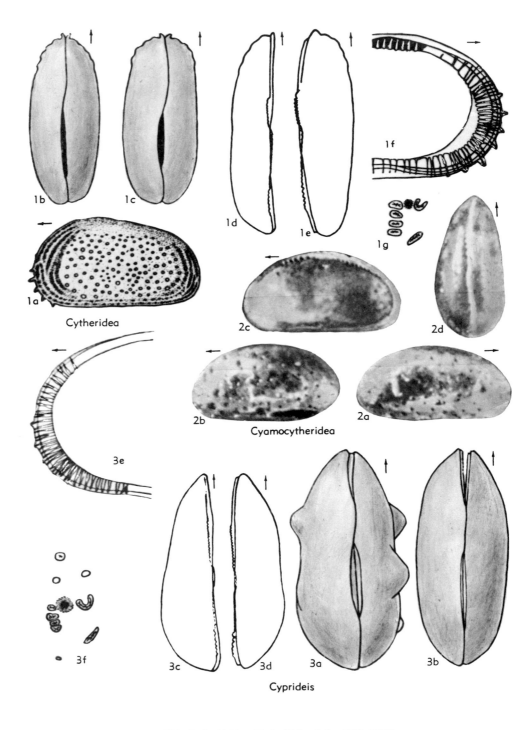

FIG. 201. Cytherideidae (Cytherideinae) (p. Q273-Q276).

center and marginal area of even medium width, inner margin not quite coinciding with line of concrescence. *Perm.*, USA.——Fig. 202,*3*. *B. crassa,* L.Perm., Kans.; *3a,b,* RV lat., LV int., ×60; *3c,* carapace dors., ×60 (198).

Clithrocytheridea STEPHENSON, 1936 [*Cytheridea garretti* HOWE & CHAMBERS, 1935] [=*Clithocy-*

theridea VAN DEN BOLD, 1946]. Like *Cytheridea* in outline but commonly ornamented with reticulations and ridges. Hinge antimerodont, LV with terminal crenulate sockets separated by raised crenulate bar (type species with flattened dorsal margin above median bar but some others referred to genus having distinct accommodation groove);

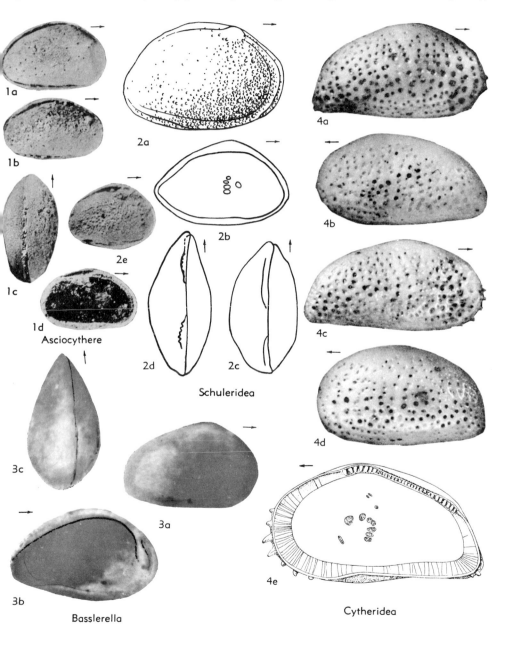

Asciocythere

Schuleridea

Basslerella

Cytheridea

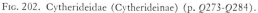

FIG. 202. Cytherideidae (Cytherideinae) (p. Q273-Q284).

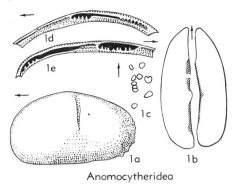

Anomocytheridea

Fɪɢ. 203. Cytherideidae (Cytherideinae) (p. *Q*273).

marginal area fairly broad, with numerous radial canals; muscle scars in vertical row of 4, with V-shaped antennal scar in front and 2 oblique mandibular scars below. *U.Cret.-Oligo., ?Rec.,* N. Am.-Eu.——Fɪɢ. 204,*1a-d.* **C. garretti* (Howᴇ & Cʜᴀᴍʙᴇʀs), Eoc., Ala.; *1a,* carapace R, ×45; *1b,c,* LV int., RV int., ×45; *1d,* LV lat., ×45; (all 33).——Fɪɢ. 204,*1e-g. C. lerichei* Kᴇɪᴊ, Eoc. (Lutet.), Fr.(Paris Basin); *1e,f,* ♂ LV lat., ♀ LV lat., ×60; *1g,* LV int., ×110 (all 42).

Cyamocytheridea Oᴇʀᴛʟɪ, 1956 [**Bairdia punctatella* Bosǫᴜᴇᴛ, 1852]. Resembling *Haplocytheridea* in most characters, but more oval, with both ends rounded instead of being acuminate ventrally, and in having a distinct anterior vestibule. Females plumper than males posteriorly. *Oligo.-Mio.,* Eu. (Paris Basin).——Fɪɢ. 201,*2. *C. punctatella* (Bosǫᴜᴇᴛ), Oligo. (Chatt.), Switz.; *2a,b,* ♂ RV lat.; *2c,* ♀ RV int.; *2d,* ♀ carapace dors.; all ×58 (57).

Cyprideis Jᴏɴᴇs, 1857 [**Candona torosa* Jᴏɴᴇs, 1850]. Similar to *Cytheridea* in shape but more ovate; type species with scattered tubercles on surface. Hinge an entomodont development similar to that in Progonocytherinae, consisting in LV of long crenulate anterior socket with about 15 pits, postjacent short high crenulate ridge merging into shallow furrow, and short posterior crenulate socket with about 6 pits; marginal areas regular, with fairly numerous straight radial canals; muscle scars in vertical row of 4 adductors, above which are 3 small scars, V-shaped antennal scar in front of top of row and mandibular scar in front of bottom of row, with 2nd mandibular scar near ventral margin. Sexual dimorphism pronounced; unlike *Cytheridea,* ripe ova received in shell cavity for development. [Habitat typically brackish-water.] (38, 68). *Mio.-Rec.,* Eu.——Fɪɢ. 201,*3; 207,1. *C. torosa* (Jᴏɴᴇs), Pleist.(NW. Ger.)-Rec.(NE.Atl.); *3a,b,* ♀ and ♂ carapaces dors., ×58; *3c,d,* ♀ LV and RV dors., ×55; *3e,* ♀ RV int. ant., ×90; *3f,* muscle scars, ×115

(201,*3a-g,* Pleist., NW.Ger., 152); 207,*1a,b,* ♂ RV lat., ♀ LV lat., ×40; *1c,* ♀ LV int., ×83; *1d,* ♀ carapace long. sec., ×47 (207,*1a-d,* Rec., NE.Atl., 88, by permission of Mouton & Co., The Hague).

Dolocytheridea Tʀɪᴇʙᴇʟ, 1938 [**Cytherina hilseana* F. A. Rᴏᴇᴍᴇʀ, 1841]. Carapace egg-shaped to subcylindrical; surface smooth. Hinge of LV with 2 crenulate sockets connected by very short furrow; middle of RV hinge occupied by smooth ridge. *L.Cret.,* Eu.——Fɪɢ. 205,*2. *D. hilseana* (Rᴏᴇᴍᴇʀ), Ger.; *2a,b,* carapace L, dors., ×45; *2c,* LV int., ×45 (373).

Galliaecytheridea Oᴇʀᴛʟɪ, 1957 [*G. dissimilis*]. Similar to *Asciocythere* but LV hinge has crenulate bar instead of smooth bar, and radial canals are rather few (10 to 15 in front); adductor scars in nearly vertical row of 4 with single antennal scar in front and 2 mandibular scars below it near ventral margin. *U.Jur.,* Eu.(Fr.).——Fɪɢ. 208,*1. *G. dissimilis,* L.Kimm.; *1a,b,* ♀ RV lat., ♀ LV lat.; *1c,d,* ♂ LV lat., ♂ RV lat.; *1e,f,* ♀ and *?* ♂ carapaces dors.; all ×50 (269).

Haplocytheridea Sᴛᴇᴘʜᴇɴsᴏɴ, 1936 [**Cytheridea montgomeryensis* Howᴇ & Cʜᴀᴍʙᴇʀs, 1935] [=*Leptocytheridea* Sᴛᴇᴘʜᴇɴsᴏɴ, 1937; *Phractocytheridea* Sᴜᴛᴛᴏɴ & Wɪʟʟɪᴀᴍs, 1939]. Shaped like *Cytheridea* in lateral view but ovate with pointed ends in dorsal view. Hinge holomerodont, with terminal crenulate sockets connected by crenulate furrow; marginal area moderately broad; with slight if any vestibules but with numerous radial canals; muscle scars essentially as in *Cytheridea. U.Cret.-Rec.,* N.Am.-Eu.——Fɪɢ. 204,*2a-c. *H. montgomeryensis* (Howᴇ & Cʜᴀᴍʙᴇʀs), U.Eoc., La.; *2a,* carapace R, ×80; *2b,c,* hinge LV and RV, ×80 (33).——Fɪɢ. 204,*2d,e. H. heizelensis* Kᴇɪᴊ, Eoc. (Barton.), Belg.; *2d,e,* ♀ LV lat., ♂ LV lat., ×90 (42).

Heterocyprideis Eʟᴏғsᴏɴ, 1941 [**Cythere (Cytheridea) sorbyana* Jᴏɴᴇs, 1857]. Carapace strong, plump, anterior end bearing spines and posteroventral corner of RV with single spine; surface somewhat rugose, lines and punctae tending to parallel margins both in molts and adults. Hinge antimerodont, with terminal crenulate sockets in LV separated by crenulate projecting bar (23, 38, 68). *Rec.,* North Sea.——Fɪɢ. 205,*3. *H. sorbyana* (Jᴏɴᴇs); *3a,* LV lat., ×60; *3b,* carapace dors., ×60; *3c,* ant. margin with radial canals, enlarged (*3a,b,* 314; *3c,* 23).

Kalyptovalva Howᴇ & Lᴀᴜʀᴇɴᴄɪᴄʜ, 1958 [**Cytheridea ovata* Bosǫᴜᴇᴛ, 1854]. Carapace solid, thickwalled, inequivalved, egg-shaped, LV much larger than RV, overlapping it on all sides. Hingement consisting of continuous furrow around periphery of LV, lower edge of which is thickened along hinge margin and upper edge crenulated at anterior end of hinge line; hinge margin in RV thickened into bar that bears crenulations on its anterior end; marginal area widest at ends but shell too thick for observation of radial canals; muscle scars in vertical row of 4 adductors with 2 scars in front. *U.Cret.*, Eu.(Belg.-Holl.).——Fig. 208,3. **K. ovata* (Bosquet), Maastricht., Holl.; *3a-c,* ♀ carapace R, dors., ant., ×35; *3d,e,* ♂ RV int., ♀ LV int., ×45 (34).

Neocyprideis Apostolescu, 1956 [*non* Hanai, 1959] [**N. durocortoriensis*] [=*Goerlichia* Keij, 1957]. Resembles *Cyprideis* in form (might be considered as subgenus) but median hinge element of LV is prominent undifferentiated crenulate bar.

Eoc.-Mio., Eu.——Fig. 205,1. *N. williamsoniana* (Bosquet), (type species of *Goerlichia*), Oligo., Belg.; *1a,* ♂ LV lat., ×60; *1b,c,* ♂ and ♀ RV lat., ×60; *1d,* ♀ RV int., ×75; *1e,f,* LV and RV dors. showing hinge, ×75 (all 42).——Fig. 208,4. **N. durcortiensis*, L.Eoc., Fr.; *4a,* carapace dors.; *4b,c,* LV dors., RV dors.; *4d,e,* LV int., RV int.; all ×50 (93).

Netrocytheridea Howe & Laurencich, 1958 [**Cypridina fusiformis* Bosquet, 1847]. Sexually dimorphous, inequivalve, fusiform, with rounded

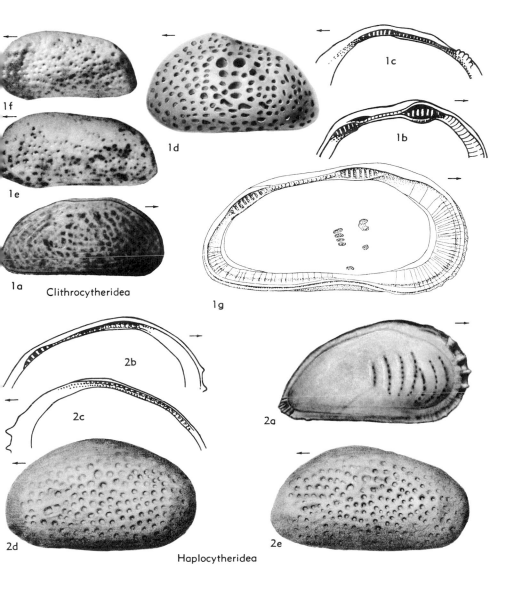

1f
1e
1a
Clithrocytheridea
1c
1b
1g
1d

2b
2c
2d
2a
2e
Haplocytheridea

Fig. 204. Cytherideidae (Cytherideinae) (p. Q275-Q276).

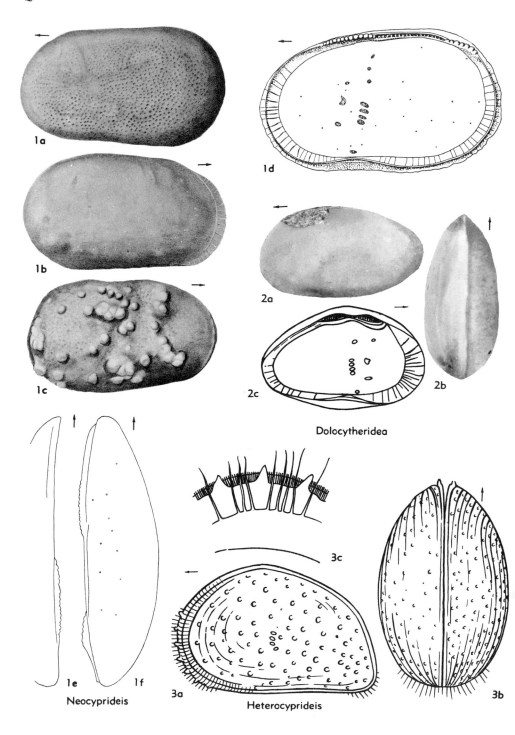

Dolocytheridea

Neocyprideis

Heterocyprideis

FIG. 205. Cytherideidae (Cytherideinae) (p. Q276-Q277).

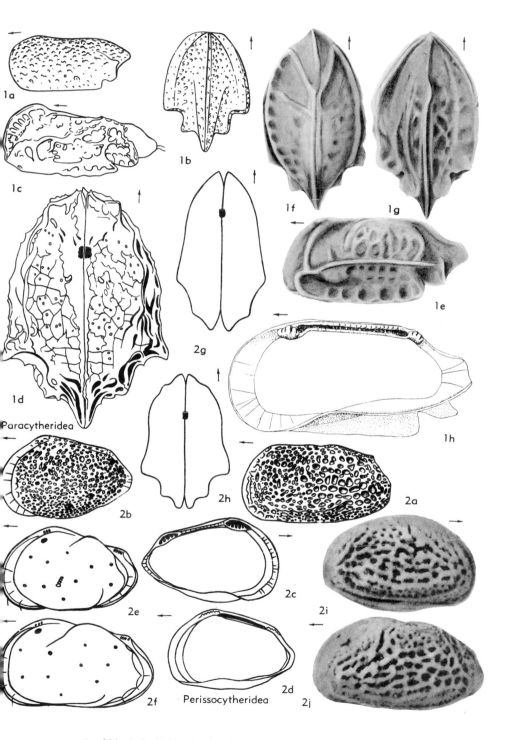

FIG. 206. Cytherideidae (Cytherideinae), Cytheruridae (p. *Q281, Q299*).

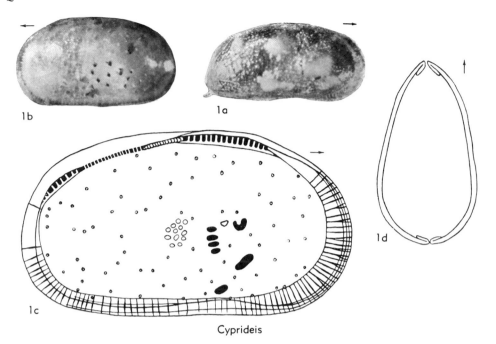

Cyprideis

Fig. 207. Cytherideidae (Cytherideinae) (p. Q276).

front and somewhat pointed rear; dorsal and ventral margins subparallel; carapace thickest in middle. Hinge antimerodont, in RV with high terminal crenulate cusps separated by shallow crenulate furrow. *U.Cret.*, Eu.——Fig. 208,2. *N. fusiformis* (Bosquet), Maastricht., Holl.; *2a,b,* carapace R, dors.; *2c,* RV int.; all ×50 (34).

Nodophthalmocythere Malz, 1958 [*N. vallata*]. Sexually dimorphous, in side view females triangular, oval, males elongate-oval. Similar in most characters to *Schuleridea* except surface bears bumps or longitudinal swellings and eye tubercles are much more prominent. Inner margin and line of concrescence coincident, zone of concrescence broad; marginal pore canals straight, simple, close; normal pore canals with external pits, numerous; adductor muscle scars in vertical row of 4, with 2 scars in front. [Marine.] *U.Jur.,* Ger.-Eng.).——Fig. 209,1. *N. vallata*, Fr.; *1a,b,* ♀ LV lat., ♀ RV lat., ×115, ×100; *1c,* ♀ LV hinge, ×135 (235). [Howe-Reyment.]

Ovocytheridea Grekoff, 1951 [*O. nuda*]. Shape similar to that of *Cytheridea,* but thickest behind middle, some species with posterior extension, smooth; hinge of RV with strongly crenulated terminal teeth and uniting crenulated furrow, also accommodation groove; strongly notched terminal sockets and uniting crenulated bar; muscle scars in row of 4, with another above in front; marginal areas with straight canals; sexual dimorphism pronounced. *U.Cret.(Coniac.-Santon.),* Afr.——

Fig. 209,2. *O. nuda*, Campan., Cameroons; *2a,b,* ♂ carapace R, dors., ×60; *2c-e,* ♀ carapace R, dors., vent., ×60 (293).——Fig. 209,3. *O. symmetrica* Reyment, Coniac., N.Nigeria; *3a-c,* carapace R, dors., vent., ×40 (293).——Fig. 209, 4. *O. apiformis* Reyment, Coniac., N.Nigeria; RV int., ×60 (293). [Howe-Reyment.]

Palaeocytheridella Mandelstam, 1958 [*Eucythere observata* Sharapova, 1937]. Carapace elongate-ovate, LV overlapping, with small spines or single spine at posterior end of ventral margin, anterior and posterior end commonly equal in height, dorsal margin straight; duplicature with numerous straight pore canals; hinge in LV with anterior elongate socket divided into 8 parts, median bar, and posterior socket similar to anterior. [This genus apparently was erected to take species that Lyubimova had assigned to *Palaeocytheridea* (with *Eucythere denticulata* as type). As such it is thought by Howe & Laurencich (34) to be congeneric with *Asciocythere* Swain. However, the shape of Swain's genus differs in being more triangular and accordingly, *Palaeocytheridella* may well be valid.] *U.Jur.-L.Cret.,* Eu.-Asia(Kazakhstan-Siberia).——Fig. 210,1. *P. observata* (Sharapova), L.Cret.(Neocom.), USSR; *1a,* RV lat.; *1b,c,* RV int., LV int.; all ×60 (329). [Bold.]

Perissocytheridea Stephenson, 1938 [*Cytheridea matsoni* Stephenson, 1935] [?=*Ilyocythere* Klie, 1939; *Iliocythere* Hartmann, 1953]. Carapace small, subpyriform, with nearly straight dorsum, convex venter, and broadly rounded anterior; surface with pits and ridges (smooth in *Ilyocythere* variant); sexual dimorphism strong. Hinge antimerodont; marginal areas broad with narrow terminal vestibules and rather widely spaced radial canals (342, 221). Brackish-water. *Mio.-Rec.,* N. Am.-S.Am. —— Fɪɢ. 206,*2a-d.* *P. *matsoni* (Stephenson), Mio., La.; *2a,* ♂ LV lat., ×75; *2b,c,* ♀ LV lat., int., ×75; *2d,* ♀ RV int., ×75 (all 340).——Fɪɢ. 206,*2e-h. P. gibba* (Kʟɪᴇ), Rec.,

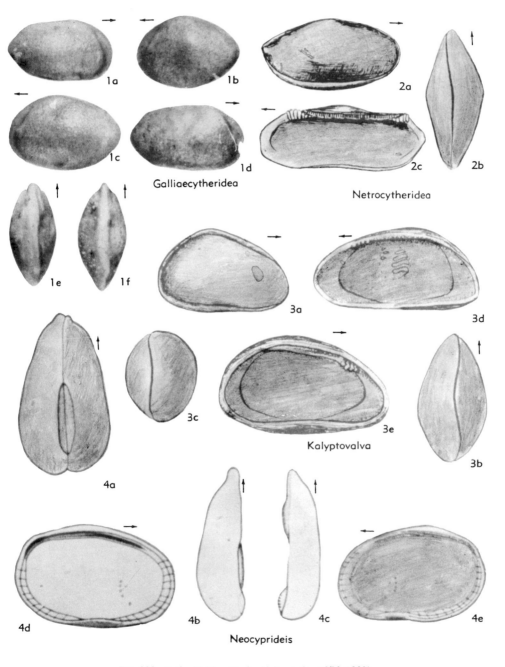

Galliaecytheridea

Netrocytheridea

Kalyptovalva

Neocyprideis

Fɪɢ. 208. Cytherideidae (Cytherideinae) (p. Q276-Q280).

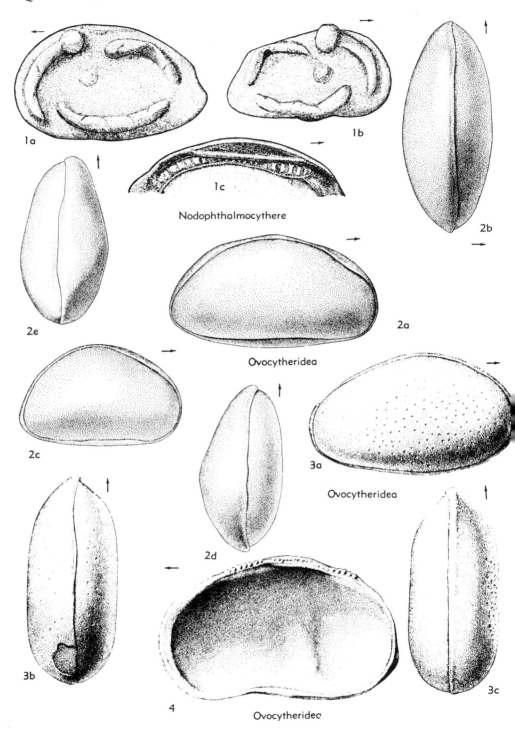

1a

1b

1c

Nodophthalmocythere

2b

2e

2a

Ovocytheridea

2c

3a

Ovocytheridea

2d

3b

4

3c

Ovocytheridea

Fig. 209. Cytherideidae (Cytherideinae) (p. *Q*280).

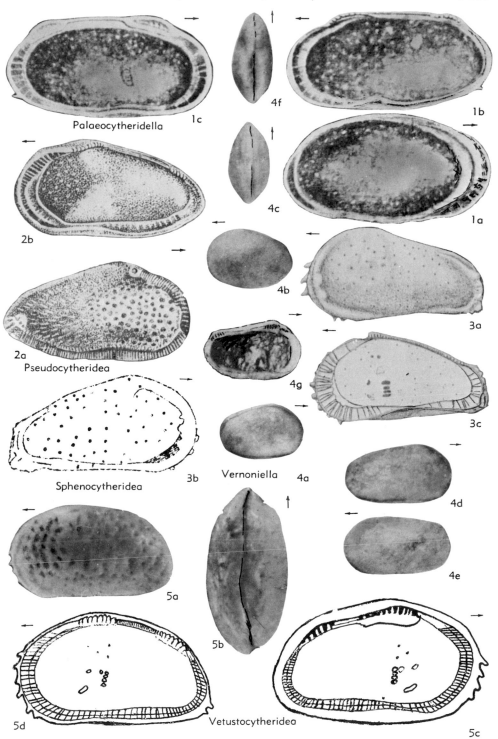

Fig. 210. Cytherideidae (Cytherideinae) (p. Q280-Q284).

Brazil (type species of *Ilyocythere*); 2e,f, ♂ LV lat., ♀ LV lat., ×75; 2g,h, ♂ and ♀ carapace dors., ×75 (all 221).——FIG. 206,2i,j. *P. rugata* SWAIN, Rec., Tex.; 2i, ♂ carapace R; 2j, ♀ LV lat., both ×75 (355).

Pseudocytheridea SCHNEIDER, 1949 [**Cytheridea? zalanyi* SCHNEIDER, 1939]. Carapace elongate, *Cytheridea*-like, with distinct eye spot; hinge with terminal elongate crenulate teeth in RV and median smooth groove. [Differs from *Schuleridea* in having hinge bar in LV, not RV; differs from *Clithrocytheridea* in having median hinge element not crenulate and in presence of eye spot.]. *U. Oligo.-Mio.*, Hung.-Caucasus-Crimea.——FIG. 210, 2. **P. zalanyi* (SCHNEIDER), Mio., W. Caucasus; 2a,b, RV lat., int., ×50 (321). [BOLD.]

Schuleridea SWARTZ & SWAIN, 1946 [**S. acuminata*] [=*Aequacytheridea* MANDELSTAM, 1947]. Shape resembling *Cytheridea*, LV larger than RV, overlapping it on all sides, being dorsally somewhat umbonate; distinct eye-tubercle in RV but less distinct in LV. Surface pitted to smooth. LV hinge originally described with crenulate terminal sockets and unknown median element but (as figured for another species by TRIEBEL, 379) shows straight smooth median furrow, above which is an accommodation groove; muscle scars in row of 4, with one in front of middle. *Jur.*, ?*L.Cret.*, *Mio.*, N.Am.-Eu.——FIG. 202,2a-d. **S. acuminata*, Jur., La.; 2a, carapace R, ×60; 2b,c,d, LV int., carapace dors., ×60 (77).——FIG. 202,2e. *S. hatterasensis* SWAIN, L.Cret. or Jur., N.Car. (subsurface); carapace R, ×35 (354).

Sphenocytheridea KEIJ, 1958 [**S. gracilis*]. Similar to *Haplocytheridea* in lateral outline and hinge but differs in having raised rim around anterior (fringed with small spines), ventral and posterior margins, and in having a distinct anterior vestibule where radial canals are clustered in groups of 3 or 4. Vertical row of 4 adductor scars and 1 antennal scar. Sexual dimorphism. *Eoc.(Lutet.)*, Eu. (Fr.).——FIG. 210,3. **S. gracilis*; 3a, LV lat.; 3b, RV lat. showing pattern of normal pore canals; 3c, RV int.; all ×100 (197). [HOWE-REYMENT.]

Vernoniella OERTLI, 1957 [**V. sequana*]. Carapace triangular to elliptical, smooth or pitted; sexual dimorphism strong; margins regular with few radial canals, RV hinge with terminal cusps divided 5 or 6 times, between which lies a smooth to finely crenulated furrow beneath a strong dorsal wall, LV with terminal notched sockets separated by strong median bar. *Jur.*, Eu.(Fr.). ——FIG. 210,4. **V. sequana*, Fr.; 4a-c, ♀ carapace R, L, dors.; 4d-f; ♂ carapace R, L, dors.; 4g, ♂ LV int.; all ×50 (269).

Vetustocytheridea APOSTOLESCU, 1956 [**Cytheridea (Vetustocytheridea) guitrancourtensis*]. Shaped and ornamented like *Haplocytheridea* but LV hinge with pronounced thickening of shell wall below anterior socket. *Eoc.*, Eu.(Fr.).——FIG. 210,

5. **V. guitrancourtensis* (APOSTOLESCU), Fr.; 5a,b, carapace L, dors., ×40; 5c,d, LV int., RV int., ×50 (93).

Subfamily CUNEOCYTHERINAE Mandelstam, 1960

Carapace ovate to subtrapezoidal in side view and ovate, subquadrate to subrhomboidal in dorsal view, rather sturdily constructed; surface smooth, pitted, or reticulate; of medium size; LV larger than RV, overlapping it except in front. LV hinge with furrow below dorsal margin, lower edge of which is distinctly thickened and may form knobs at ends, particularly in front; RV hinge developed as bar, usually strongest at anterior end; marginal areas broad, particularly anteriorly, with numerous long radial canals which tend to curve toward upper and lower margins; muscle scars in vertical row of 4, with crescent-shaped antennal scar in front of middle, usually 2 mandibular scars located near ventral margin, and in some species as many as 3 additional scars high in carapace. *Jur.-Mio.*

Cuneocythere LIENENKLAUS, 1894 [**C. truncata* (=**Bairdia marginata* BOSQUET, 1852)]. Ovate, thick-shelled, in dorsal view almost wedge-shaped; surface pitted to reticulate. Marginal areas broad, with shallow vestibule and numerous long curved radial canals; hinge a smooth groove in LV, with distinct barlike development below the groove, bar developing a small knob at its anterior end; muscle scars in vertical row of 4, with a crescent-shaped antennal scar in front and 2 mandibular scars near ventral margin (42, 228). *Eoc.-Mio.*, Eu.——FIG. 211,4. **C. marginata* (BOSQUET), Oligo., Belg.; 4a, ♀ LV (lectotype) lat.; 4b, ♀ RV lat.; 4c, ♂ carapace dors.; 4d, ♀ LV int.; all ×60 (42).

Archeocuneocythere MANDELSTAM, 1947 [**A. reniformis*]. Similar to *Cuneocythere* but LV hinge groove opening widely toward front and RV hinge bar widening from knife-edge at back to club-shaped in front. Anterior marginal canals straight. [Marine.] *Jur.-L.Cret.*, SW.Asia.——FIG. 211,2. **A. reniformis*, M.Jur., USSR(Kazakhstan); 2a, RV lat., ×35; 2b,c, RV and LV hinge, ×64 (237). [HOWE-REYMENT.]

Monsmirabilia APOSTOLESCU, 1955 [**Bairdia perforata* BOSQUET, 1852 (*non* ROEMER, 1838) (=**Monsmirabilia subovata* APOSTOLESCU, 1955, *nom. nov.*)]. Somewhat more ovate than *Cuneocythere*, with smooth to pitted surface; anterior tooth of LV much more strongly developed, and with well-developed accommodation groove above LV hingement (92, 42). *Eoc.(Ypres.-Barton.)*,

Eu.(Fr.-Belg.-Holl.).——Fɪɢ. 211,*1*. *M. subovata* (Aᴘᴏsᴛᴏʟᴇsᴄᴜ), Led., Belg.; *1a,b*, ♀ carapace R, dors.; *1c*, ♀ LV int.; *1d*, ♂ carapace R; all ×60 (42).

Paleomonsmirabilia Aᴘᴏsᴛᴏʟᴇsᴄᴜ, 1956 [**P. paupera*]. Differs from *Monsmirabilia* in having obliquely rounded posterior end in side view and in possession of distinct eye spots. *L.Eoc.*, Eu. (Paris Basin).——Fɪɢ. 211,*3*. **P. paupera; 3a,b*, carapace R, L, ×28; *3c,d*, carapace dors., LV int., ×32 (93).

Subfamily EUCYTHERINAE Puri, 1954

Carapace shaped like that of Cytherideinae but internal characters quite different. Hingement lophodont or antimerodont but some carapaces almost toothless; marginal areas very broad in front, with vestibule and relatively few radial canals; muscle scars usually in oblique row of 4 or more adductors, in front of which lies a V-shaped ?antennal scar (287). *Jur.-Rec.*

Eucythere Bʀᴀᴅʏ, 1868 [*pro Cytheropsis* Sᴀʀs, 1866 (*non* M'Cᴏʏ, 1849)] [**Cythere declivis* Nᴏʀᴍᴀɴ, 1865; SD Bʀᴀᴅʏ & Nᴏʀᴍᴀɴ, 1889]. Carapace shaped like that of *Cytheridea* except somewhat more triangular, males more elongate than females. Hinge lophodont, formed of interlocking flanges; marginal areas broad anteriorly where line of concrescence departs from inner margin, radial canals few; muscle scars in oblique row of 4 with large heart-shaped or V-shaped scar in front (divided in some Cretaceous species) (12, 267). *U.Cret.-Rec.*, Eu.-N.Atl.——Fɪɢ. 212,*2*. **E. declivis* (Nᴏʀᴍᴀɴ), Rec., N.Atl.; *2a,b*, carapace L, dors., ×150 (15).——Fɪɢ. 213,*5*. *E. triordinis* Sᴄʜᴍɪᴅᴛ, Eoc. (Lutet.), Belg.; LV int., ×90 (42).

Euryitycythere Oᴇʀᴛʟɪ, 1959 [**E. subtilis* Bᴀʀᴛᴇɴsᴛᴇɪɴ & Bʀᴀɴᴅ, 1959]. Carapace mainly triangular to trapezoidal in side view, with broad anterior margin, sides slightly inflated (apart from weak ventral wing), with narrow peripheral edge; eye tubercles present; weak, vertical impressed zone at about mid-length running from dorsal margin to wing; surface smooth, or with pitlets. Inner margin and line of concrescence coincident; marginal zone broad, with numerous threadlike pore canals (about 50 anterior), partly grouped in 3's; selvage well developed; LV hinge slightly arched with angled and broadened furrow, accommodation groove above anterior hinge section and RV with corresponding hinge bar. *L.Cret.(Valangin.-Hauteriv.)*, Eu.(Ger.-Fr.).——Fɪɢ. 214, *1a-c*. **E.*

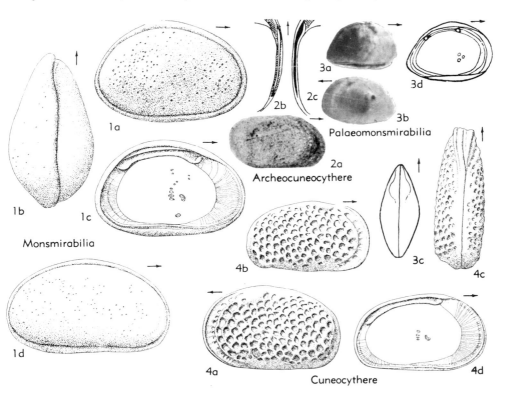

FɪG. 211. Cytherideidae (Cuneocytherinae) (p. Q284-Q285).

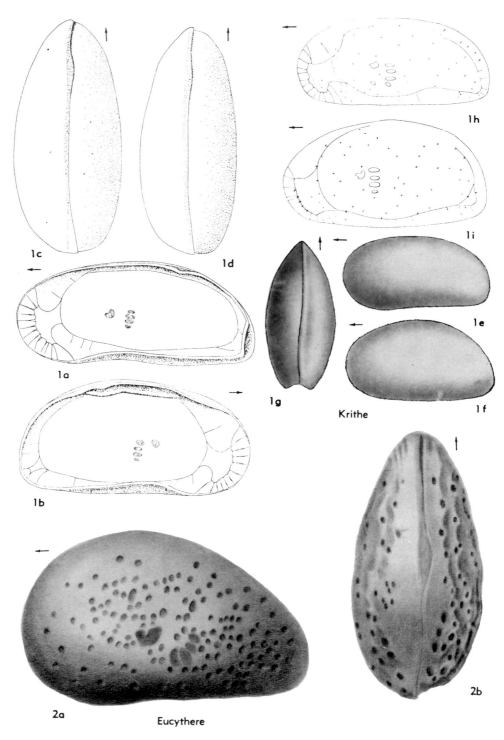

FIG. 212. Cytherideidae (Eucytherinae, Krithinae) (p. Q285-Q289).

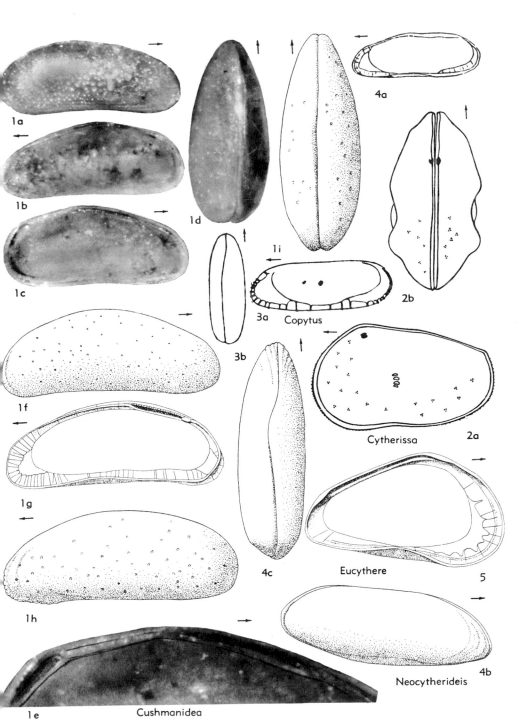

FIG. 213. Cytherideidae (Eucytherinae, Neocytherideidinae) (p. Q285-Q290).

subtilis Bartenstein & Brand, NW.Ger.; *1a,b,* LV lat., LV int.; *1c,* carapace dors.; all ×50 (96). ——Fig. 214,*1d-f.* E. *parisiorum* Oertli, Fr. (Paris Basin); *1d-f,* carapace R, L, dors., ×58 (270). [Reyment.]

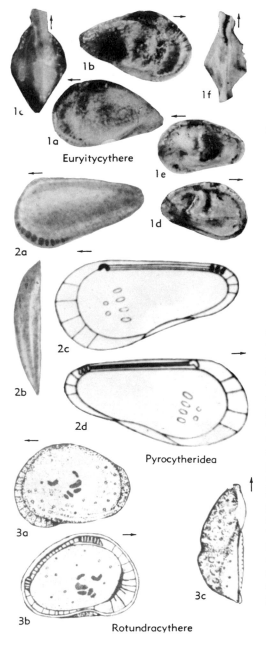

Euryitycythere

Pyrocytheridea

Rotundracythere

Fig. 214. Cytherideidae (Eucytherinae) (p. Q285-Q288).

Pyrocytheridea Lyubimova, 1955 [*P. pergraphica*]. Elongate pear-shaped in side view, front broadly rounded, rear narrower, with long margins converging; surface smooth, LV hinge with small anterior curved tooth, straight median furrow, and small crenulate posterior tooth; RV hinge with crescent-shaped anterior socket, long straight median bar, and posterior pitted socket; marginal areas widest in front, with very few unbranched radial canals, narrow anterior vestibule sometimes; muscle scars in vertical row of 4 adductors, with 2 small scars in front. *Jur.-L.Cret.*, USSR.——Fig. 214,*2. *P. pergraphica,* M.Jur.(Callov.), Samarskaia Luka; *2a,b,* LV lat., dors., ×50; *2c,d,* RV int., LV int. (diagrammatic), ×70 (230). [Bold-Reyment.]

Rotundracythere Mandelstam, 1958 [**"Eucythere rotundata* Hornibrook, 1952"]. Carapace short, LV overlapping RV, with high, obliquely rounded anterior end, posterior end narrowly rounded ventrally, dorsal margin slightly convex; radial pore canals few in anterior end, rather widely spaced (although author reports pore canals to be numerous), vestibule crescent-shaped; LV hinge with terminal, crenulate sockets and median crenulate bar. *Eoc.-Rec.*, N.Z.——Fig. 214,*3. *R. rotunda* (Hornibrook), Rec.; *3a-c,* LV (holotype) lat., int., dors., ×75 (32).

Subfamily KRITHINAE Mandelstam in Bubikan, 1958

Carapace elongate, dorsal margin nearly straight or weakly convex, ventral margin usually slightly concave, anterior end rounded, posterior usually obliquely truncate and commonly inturned so as to form a "V" when viewed from above; greatest thickness at or behind middle; surface smooth. Marginal areas very broad anteriorly, usually with large anterior vestibule from which only a moderate number of radial canals extend; small vestibule commonly present in posteroventral region; hingement essentially adont, with smooth or partially crenulate longitudinal furrow along dorsal margin of larger valve to receive sharp edge of opposite valve, reversal of hingement and valve size not unusual; muscle scars in vertical row of 4 adductors, in front of which usually lies a V-or U-shaped antennal scar and below which 1 or 2 mandibular scars may occur. *U.Cret.-Rec.*

Krithe Brady, Crosskey & Robertson, 1874 [**Cythere (Cytherideis) bartonensis* Jones, 1857; SD Brady & Norman, 1889] [=*Ilyobates* Sars, 1866 (*non* Kraatz, 1858)]. Carapace oblong, with greatest height at or behind mid-length, anterior end rounded, posterior obliquely truncate and usually inturned, though not in type species; moderately to strongly tumid in dorsal view; surface smooth, but bearing widely spaced large normal canals. LV hinge with longitudinal furrow for reception of sharp dorsal edge of RV (hinge reversed in original description, based partially on another species); muscle scars in vertical row of 4 adductors, with usually a U-shaped antennal scar in front but this may be divided into 2 or more scars; mandibular scars usually 2 small spots in front of lower part of row. Dimorphism pronounced, females being shorter and plumper than males (68, 38, 14). *U.Cret.-Rec.,* Eu.-N.Am.-Asia.——FIG. 212,*1a-d. K. papillosa* (BOSQUET), Mio., Fr.(Aquitaine); *1a,b,* ♂ RV int., ♀ LV int., ×75; *1c,d,* ♂ and ♀ carapace dors., ×75 (42).——FIG. 212,*1e-g.* *K. bartonensis* (JONES), *1e,f,* ♂ and ♀ LV lat.; *1g,* carapace dors. (Quat., Br. I.), ×60 (14).——FIG. 212,*1h,i.* *K. bartonensis* (JONES), ♂ and ♀ LV lat. (Belg., Barton.), ×75 (42).

Parakrithe VAN DEN BOLD, 1958 [*Cytheridea (Dolocytheridea) vermunti* VAN DEN BOLD, 1946]. Similar to *Krithe* in external form; sexually dimorphous; reversal of overlap, valve size, and hingement occurs. Hinge similar to that of *Krithe* but posterior portion of hinge furrow in larger valve may show faint crenulation; differs from *Krithe* in having shallow anterior vestibule and much longer radial canals as a result. *Eoc.-Rec.,* W.Indies-Eu.-Ind.O.——FIG. 215,*1.* *P. vermunti* (VAN DEN BOLD), L.Oligo., Trinidad; *1a,b,* ♀ carapace R, dors.; *1c,* ♂ carapace dors.; *1d-f,* ♂ RV int., ♀ RV int., ♀ LV int.; all ×60 (102). [BOLD.]

Parakrithella HANAI, *nom. subst.* herein [*pro Neocyprideis* HANAI, 1959 (*non* APOSTOLESCU, 1956)] [*Neocyprideis pseudadonta* HANAI, 1959]. Carapace like that of *Krithe* but lacks posterior incision in dorsal view. Marginal area broad; line of concrescence close to anterior margin except at anteroventral corner where it is suddenly close to inner margin; radial pore canals bifurcated; LV hinge with longitudinal bar that fits groove in RV and has a faint crenulation along posterior one-fourth. *Rec.,* E.Asia.——FIG. 215,*2.* *P. pseudadonta,* Rec., Japan; *2a-c,* RV lat., RV int., LV int., ×90 (28).

Subfamily NEOCYTHERIDEIDINAE Puri, 1957

[*nom. subst.* PURI, 1957 (*pro* Cytherideisinae PURI, 1952, =Cytherideidae, *nom. correct.* SYLVESTER-BRADLEY & HARDING, 1953, invalid because *Cytherideis* JONES, 1856, is junior synonym of *Cypridea* BOSQUET, 1852] [Includes Pontocytherinae MANDELSTAM, 1960]

Carapace externally shaped like that of Cytherideinae but in many genera more elongate, smooth, pitted, weakly to strongly reticulate, or even noded. Hinge generally lophodont, with terminal smooth sockets in LV, separated by a smooth median bar; marginal areas variable in width, commonly

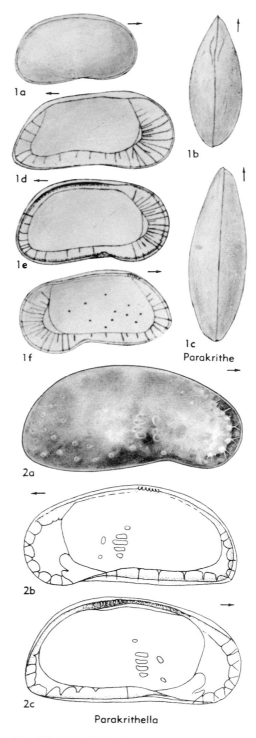

1a

1b

1d

1e

1c

1f

Parakrithe

2a

2b

2c

Parakrithella

FIG. 215. Cytherideidae (Krithinae) (p. Q289).

with vestibule at one or both ends; muscle scars in vertical row of 4 adductors, with 1 or 2 antennal scars and 1 or 2 mandibular scars in front. *Jur.-Rec.*

Neocytherideis PURI, 1952 [*N. elongatus* PURI, 1952 (=*Cytherideis subulata* var. *fasciata* BRADY & ROBERTSON, 1874)] [=*Sahnia* PURI, 1952]. Carapace like that of *Cushmanidea* but anterior end narrower. Duplicature narrow, with large anterior vestibule; radial canals short, straight, widely spaced, few; muscle scars in vertical row of 4, with 6 additional scars in front and above row; hinge weakly lophodont. *Mio.-Rec.*, Eu.——FIG. 213,4a. *N. subulata fasciata,* Rec., Eng.; RV int., ×45 (285).——FIG. 213,4b,c. *N. linearis* (ROEMER), Mio., Fr.(Aquitaine); 4b,c, carapace R, dors., ×75 (42).

Aparchitocythere SWAIN & PETERSON, 1952 [*A. typica*]. Ovate to subquadrate in outline, LV larger than RV; surface smooth except for normal canals. RV hinge with elongate terminal teeth formed by edge of shell, furrow along median part of dorsal margin, dentition not crenulate. *Jur.*, N.Am.—— FIG. 216,3. *A. typica,* U.Jur.(Sundance), Wyo.; 3a,b, ♀ carapace L, dors., ×45; 3c, ♂ LV int., ×45 (358).

Aulocytheridea HOWE, 1951 [*A. margodentata*]. Carapace very solid, like *Cytheridea* in shape. RV hinge with smooth elongate teeth at ends and furrow below middle of dorsal margin, LV with terminal elongate sockets and straight bar between, separated from dorsal margin by accommodation groove; hinge elements smooth; marginal areas regular, with numerous radial canals; muscle scars in vertical row of 4 adductors, with heart-shaped antennal scar in front of top of row and one mandibular scar in front of base of row, 2nd obliquely below and behind near ventral margin (32). *Eoc.*, N.Am.——FIG. 216,2a,b. *A. margodentata,* U.Eoc., Fla.; 2a,b, LV lat., int., ×50 (Howe, n).

Copytus SKOGSBERG, 1939 [*C. caligula*]. Elongate, low, with dorsal edge straighter than in *Cushmanidea;* no sexual dimorphism. Hinge seemingly without teeth; large vestibule at each end, bordered by moderate number of simple radial canals; muscle scars 4, in group near middle, and single scar in front. *Rec.*, Antarct.——FIG. 213,3. *C. caligula; 3a,b,* carapace L, dors., ×37.5 (332).

Cushmanidea BLAKE, 1933 [*Cytheridea seminuda* CUSHMAN, 1906] [=*Cytherideis* AUCTT. (*non* JONES, 1856); ?*Sacculus* NEVIANI, 1928 (*non* GOSSE, 1851); *Pontocythere* DUBOWSKY, 1939; *Hemicytherideis* RUGGIERI, 1952]. Carapace elongate ovate, highest and thickest behind middle; surface smooth or reticulate, with pattern tending to parallel margins. Hinge lophodont, LV with elongate smooth terminal sockets and median projecting flange or bar which may show faint crenulation; anterior with serrate vestibule in

which several radial canals are developed from each serration; muscle scars in vertical row of 4, with one large antennal scar in front and 2 mandibular scars obliquely set in front and below the row (99, 134). *Eoc.-Rec.*, N.Am.-Eu.——FIG. 213, 1a-e. *C. seminuda* (CUSHMAN), Rec., W.Atl.(off R.I.); 1a,b, RV lat., int., ×45; 1c, LV int., ×45; 1d, carapace dors., ×45; 1e, LV hinge, ×110 (Howe, n).——FIG. 213,1f-i. *C. grosjeani* (KEIJ), Eoc.(Lutet.), Belg.; 1f,g, RV lat., int.; 1h, LV lat.; 1i, carapace dors.; all ×75 (42).

Cytherissa SARS, 1925 [*Cythere lacustris* SARS, 1863] [=*Alexandrella* SCHWEYER, 1939 (*non* CHEVREUX, 1911, *nec* TONNOIR, 1926)]. Shape of carapace variable but generally resembling *Cytheridea*, in some species more rectangular both in lateral and end views. Hinge with terminal noncrenulate blades, weak; marginal areas narrow, with radial canals arranged in groups of 4, muscle scars as in *Cytheridea* (18, 68, 220). ?*Oligo.*(Belg.)-*Rec.*(Eu.-Asia), recorded in glacial lakes of Scot., Norway, Swed., Sib.(Baikal).——FIG. 213, 2. *C. lacustris* (SARS), Rec., Norway; 2a,b, carapace L, dors., ×60 (314).

Hulingsina PURI, 1958 [*H. tuberculata*]. Shaped externally like *Cushmanidea* but with reticulate or papillate surface. Internal characters similar, but shell heavier, with strong development of selvage, particularly in RV where flange groove is very wide at posteroventral corner; other internal features as in *Cushmanidea. Mio.-Rec.*, N.Am.—— FIG. 217,1. *H. tuberculata,* Rec., Gulf Mexico; 1a,b, LV lat., int.; 1c,d, RV lat., int.; 1e, carapace dors.; all ×60 (291).

Paracyprideis KLIE, 1929 [*Cytheridea fennica* HIRSCHMANN, 1909]. Carapace like that of *Cytheridea* in shape but hinge teeth of RV and median bar of LV not crenulate; surface smooth except for large normal canals, which in fossil forms are sievelike; large vestibule in front part of interior with few short radial canals diverging from it; muscle scars in vertical row of 4, with usually 2 antennal scars in front of upper end of row, one scar being larger than the other; mandibular scars obliquely below, and additional scars higher in carapace (215, 171). *L.Cret.-Rec.*, Eu.-N.Am.—— FIG. 216,1a,b. *P. fennica* (HIRSCHMANN), Rec., Baltic; 1a,b, ♀ RV int., ♀ LV lat., ×45 (171). ——FIG. 216,1c-e. *P. rarefistulosa* (LIENENKLAUS), Oligo.(Rupel.), Belg.; 1c,d, ♀ RV int., RV lat.; 1e, ♂ LV int., ×75 (42).

Family CYTHERISSINELLIDAE
Kashevarova, 1958

[*nom. transl.* VAN DEN BOLD, herein (*ex* Cytherissinellinae KASHEVAROVA, 1958) (attributed by KASHEVAROVA to SCHNEIDER, 1956, but seemingly in error)] [Materials for this family prepared by W. A. VAN DEN BOLD, Louisiana State University, and R. A. REYMENT, University of Stockholm]

Elongate suboblong, dorsal margin straight, anterior and posterior ends round-

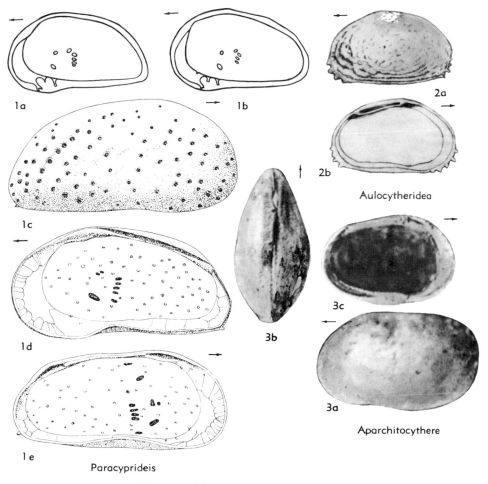

1a   1b   2a

2b

Aulocytheridea

1c

3b   3c

1d

3a

1e

Aparchitocythere

Paracyprideis

FIG. 216. Cytherideidae (Neocytherideidinae) (p. Q290).

ed, with faint to distinct narrow sulcus extending straight downward from mid-dorsal region; surface reticulate and may bear inconspicuous longitudinal ribs. *L.Trias.*

Cytherissinella SCHNEIDER, 1956 [*C. okrajantzi*]. Elongate trapezoidal, length twice height, inflation variable. LV overlapping RV all around but least strongly along straight dorsal margin, ventral margin straight to faintly concave, both margins converging slightly backward; anterior and posterior ends broadly rounded; ornament irregular, composed of reticulations and a flexed, longitudinal, lateral rib running from dorsal third of anterior margin to posterior third of side. LV hinge with terminal subsockets united by sharp, smooth bar. [Brackish water.] *L.Trias.*, USSR(Emba Region). ——FIG. 218,*1.* *C. okrajantzi; 1a,b,* carapace R, dors., ×64 (50).

Lutkevichinella SCHNEIDER, 1956 [*L. bruttanae*]. Carapace rather small, oblong, with greatest height

in posterior half; dorsal margin straight, ventral margin feebly convex. Anterior and posterior margins rounded; midpoint of anterior slightly lower than that of posterior or of equal height. Surface ornamented with roughly concentric reticulations; central sulcus running from middle of dorsal margin vertically to below middle of valve. Hinge elements thin: LV hinge with bar terminating in shallow depressions, RV complementary, with feeble anterior and posterior swellings. *L.Trias.*, USSR(Emba Region).——FIG. 218,*2.* *L. bruttanae; 2a-d,* carapace R, dors., RV and LV hinge, ×64 (50).

Family CYTHERURIDAE G. W. Müller, 1894

[*nom. transl.* REYMENT, herein (*ex* Cytherurinae G.W.MÜLLER, 1894] [=Cytheropterinae HANAI, 1957] [Materials for this family prepared by R. A. REYMENT, University of Stockholm, with assistance on some genera by H. V. HOWE, Louisiana State University, and TETSURO HANAI, University of Tokyo] [Includes Procytheropterinae, Paracytherideinae, Eocytheropterinae MANDELSTAM, 1960]

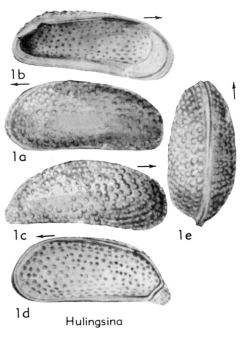

Hulingsina

Fig. 217. Cytherideidae (Neocytherideidinae)
(p. Q290).

Characterized in many genera by small size, presence of caudal processes and flattening of venter, and by occurrence of ventrolateral wing processes in some; surface mostly well ornamented, RV generally slightly larger than LV, especially along dorsal margin. Hinge modified entomodont (as defined by Triebel), typically developed in *Cytheropteron*, which shows crenulated furrow between terminal teeth of protodont valve but altered in *Cytherura* and allied genera by thickening of both ends of hinge bar to form protodont sequence of tooth, socket, median furrow, and socket tooth; eyes present or absent; marginal zones commonly wide, in some genera extending from posterior end far into shell, with small marginal niches in some species; 4 vertically arranged adductor muscle scars and 1 or 2 scars in front. Sexual dimorphism strong in some genera but hardly discernible in others. [Marine and brackish-water environments.] *U.Jur.-Rec.*

Cytherura Sars, 1866 [*Cythere gibba* O.F.Müller, 1785]. Carapace small, bilamellar, anterior margin broadly rounded but pointed in dorsal view, widening backward strongly, posterior margin usually with subdorsal caudal process; RV over-

hanging LV along dorsal margin; ornament variable, irregularly ribbed, reticulate, or punctate; eyes present in living forms; eye tubercles very feeble if developed; males commonly more inflated in dorsal view than females (owing to the bulky copulative organs). Adductor muscle scars in vertical row of 4, with another scar in front of most dorsal and 2 irregular scars in front of this one; 2 elongated mandibular scars at anteroventral angle; anterior and posterior parts of zone of concrescence extremely broad, internal margin and line of concrescence mostly coincident; radial pore canals few, unbranched, long, thin; normal pore canals open; RV hinge with 2 terminal smooth or weakly notched teeth, united by furrow that widens terminally, where it also becomes deeper. Larval forms slightly different, large instars having lower caudal process. Sexual dimorphism relatively strong. [Marine and brackish-water.] *Cret.-Rec.*, cosmop.——Fig. 219,2, 220,4. *C. gibba* (O.F.Müller), Rec., Holl., Kijkduin; 219,2a,b, ♂ carapace R, L, ×60; 220,4a, ♀ carapace dors., ×83; 220,4b, ♂ LV int., ×130 (88, by permission Mouton & Co., The Hague).

Cytheropteron Sars, 1866 [*Cythere latissima* Norman, 1865 (=*C. convexum* Sars, 1866)] (=*Aversovalva* Hornibrook, 1952]. Carapace bilamellar, in lateral view roughly ovoid, with oblique caudal process that points obliquely upward (commonly drawn out strongly) and pointed ventrolateral winglike processes, which may be feebly developed; RV slightly larger than LV, overlapping it distinctly along hinge length; surface smooth or ornamented; no eye tubercles. Inner margin and line of concrescence coincident, apart from presence of narrow vestibule at front middle; RV hinge with terminal notched tooth plates united by crenulated furrow; marginal pore canals few, separated widely, undivided, some grouped in 2's and 3's; normal pore canals open; 4 subvertically arranged, elongated adductor muscle scars with V-shaped scar in front of dorsal end of adductor row, and small round scar inside V. Sexual dimorphism present but not pronounced. *U.Jur.-Rec.*, cosmop.——Fig. 219,1a, 220,5. *C. latissimum* (Norman), Rec., Holl., Kijkduin; 219, 1a, LV lat., ×60; 220,5a,b, RV int., dors., ×115, ×100 (88, by permission of Mouton & Co., The Hague).——Fig. 219,1b-g, *C. aureum* (Hornibrook), Rec., N.Z. (type species of *Aversolvalva*); 1b-d, LV lat., dors., int.; 1e-g, RV lat., dors., int.; all ×75 (32).

Eocytheropteron Alexander, 1933 [*Cytheropteron bilobatum* Alexander, 1929]. Carapace subtrapezoidal in side view, egg-shaped in dorsal view, provided with short caudal process; LV overhanging RV strongly along dorsal margin, ventrally swollen, without side wings. LV hinge with longitudinal crenulated terminal sockets and feebly crenulated ridge between; marginal pore canals

few, straight, some branched. *L.Cret.-Eoc.*, N.Am.
——FIG. 219,9. **E. bilobatum* (ALEXANDER), L.
Cret.(Washita), Tex.; *9a,* carapace R, ×70; *9b,c,*
LV and RV hinge margins, ×100 (89).

Eucytherura G.W.MÜLLER, 1894 [**Cythere complexa* BRADY, 1867; SD ALEXANDER, 1936]. Carapace small, subrhombic to quadrate in side view, inflated, thin-walled; anterior margin broadly rounded, posterior margin with caudal process in dorsal half; surface tuberculate or reticulate; eye tubercle present. Adductor muscle scars 4 in vertical row and another dorsally situated; line of concrescence and inner margin usually coincident, zone of concrescence broad, with a few straight, simple canals, normal pore canals numerous; hinge of same type as in *Cytherura.* Sexual dimorphism slightly reflected by carapace features. *Cret.-Rec.,* N.Am.-Eu.——FIG. 219,4, 220,1. **E. complexa* (BRADY), Rec., Medit.(Gulf of Naples); 219,4, LV lat., ×80 (401); 220,1a,b, carapace L, dors., ×150 (53).

Hemicytherura ELOFSON, 1941 [**Cythere cellulosa* NORMAN, 1865]. Carapace like that of *Cytherura,* RV clearly overhanging LV along dorsal margin; ornament of pits or coarse reticulations; eye tubercles present. Adductor muscle scars 4 in vertical row, 1 anterior to them; zone of concrescence broad anteriorly, narrow backward, triangular anteromarginal niche (if developed), niche in posteroventral section; marginal pore canals simple, few, anterior ones grouped in 3 bundles; RV hinge with 2 terminal tooth plates and intervening furrow. Males longer than females. *Plio.-Rec.,* N.Eu.-S.Eu.——FIG. 219,3, 220,6. **H. cellulosa* (NORMAN), Rec., Holl.; 219,3a,b, LV lat., RV lat., ×60; 220,6a, carapace long. sec., ×85; 220,6b, RV int., ×150 (88, by permission of Mouton & Co., The Hague).——FIG. 219,7. *H. clathrata* (SARS), Rec., N.Atl.; 7a,b, RV lat., carapace dors., ×60 (88, by permission of Mouton & Co.).

Howeina HANAI, 1957 [**H. camptocytheroidea*]. Large, ovate; posterior caudal process indistinct; ventral surface with slight trace of winglike ridges. Hinge like that of *Cytherura* but anterior tooth of RV large and elongate; inner margin with modified S-shape along posterior margin; eye spot indistinct. *Plio.,* E.Asia(Japan).——FIG. 221,1. **H. camptocytheroidea;* 1a, RV lat., ×90 (Hanai, n); 1b,c, LV int., RV int., ×90 (27). [HANAI.]

Kangarina CORYELL & FIELDS, 1937 [**K. quellita*]. Carapace small, thick-walled, with subdorsal caudal process, strong ventral and dorsal ribs with additional lateral ribs, interspaces being pitted. Zone of concrescence broad, line of concrescence and inner margin separated; RV hinge with notched terminal tooth plates and notched median furrow. *Mio.,* C.Am.(Panama).——FIG. 219,5. **K. quellita,* M.Mio.(Gatun F.), Panama C.Z.; 5a-c, RV lat., dors., int., ×100 (126).

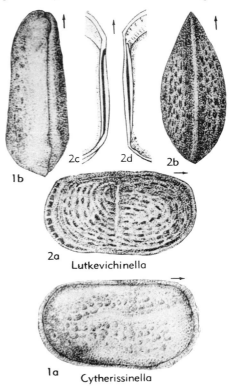

FIG. 218. Cytherissinellidae (p. Q291).

Kobayashiina HANAI, 1957 [**K. hyalinosa*]. Fragile nearly transparent with sharp-pointed alae, caudal process turned downward. Hinge intermediate between *Paijenborchella* and *Cytheropteron;* in RV, anterior tooth bilobate, with steplike projection just below it; median furrow arched and finely crenulate in anterior half, straight and coarsely crenulate in posterior half, and with shallow depression at anterior termination; posterior tooth consisting of single row of elongate knoblike teeth. Other characters similar to *Cytheropteron.* *Plio.,* E.Asia(Japan).——FIG. 222,1. **K. hyalinosa;* 1a, RV lat., ×90 (Hanai, n); 1b,c, RV int., LV int., ×90 (27). [HANAI.]

Mehesella REYMENT, 1960 [**M. paleobiafrensis*] [=*Budaia* MÉHES, 1941 (*non* WELLS, 1933)]. Subequivalved, LV usually overlapping RV slightly along dorsal margin, particularly posterodorsally but RV may overlap LV, ventral region tending to be much swollen; posterior margin drawn out into blunt caudal process; surface with numerous notched longitudinal riblets and/or reticulations, posterior half of ventral margin may be compressed into low but sharp keel; eye spot present, sometimes only discernible from inside. LV hinge with notched terminal sockets and fine uniting crenulated bar (in Eocene species anterior termi-

nation of bar may be thickened to form prominent tooth); moderately broad anterior and posterior marginal zones with 10 or 11 slightly flexed

simple pore canals; inner margin and line of concrescence coincident; adductor muscle field composed of vertical row of 4 spots, also a dorsal field

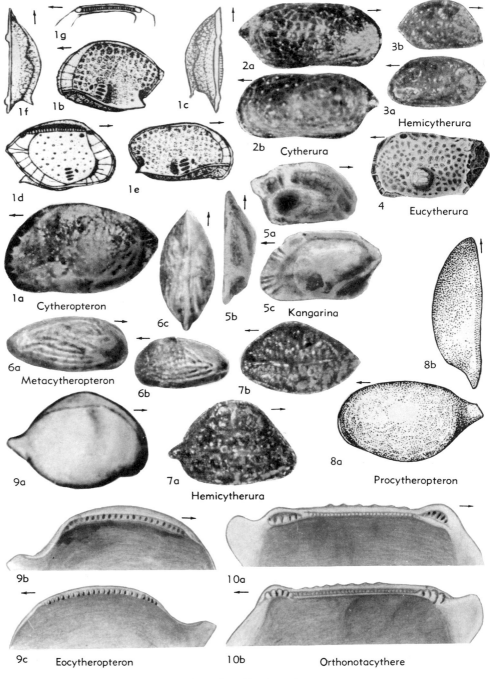

Fig. 219. Cytheruridae (p. Q292-Q299).

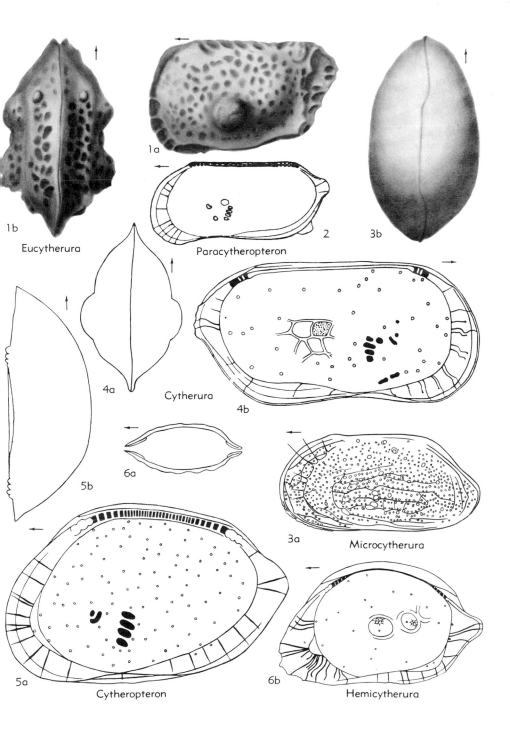

FIG. 220. Cytheruridae (p. *Q*292-*Q*299).

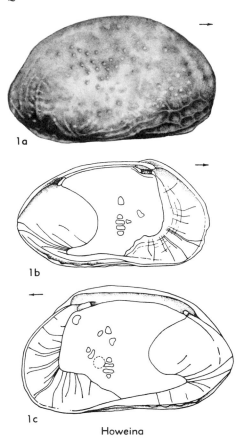

1a

1b

1c

Howeina

Fig. 221. Cytheruridae (p. Q293).

of 3 or 4 spots, and 2 antennal spots. Sexual dimorphism slightly reflected in morphology of carapace. *U.Cret.(U. Maastricht.)-Oligo.,* Eu.(Hung.)-W.Afr.(Nigeria).——Fig. 223,3. *M. biafrensis,* Maastricht., W.Nigeria; *3a,* carapace R, ×60; *3b-d,* LV lat., dors., vent., ×60 (293).——Fig. 224,*1.* *M. paleobiafrensis; 1a,b,* carapace L, dors., ×85; *1c,* RV ant. pore canals, ×135 (293).

Metacytheropteron OERTLI, 1957 [*M. elegans*]. More or less almond-shaped, like *Cytherura* but LV overlapping RV, elongated, especially along dorsal margin, as in *Eocytheropteron;* anterior end broadly rounded, with ventral part of this margin more sharply rounded than dorsal, posterior end bluntly pointed; weak eye tubercle. Inner margin coincident with line of concrescence, surface reticulate and bearing longitudinal ribs, marginal pore canals radial and simple, straight, few; hinge like that of *Cytheropteron.* Slight sexual dimorphism. *U.Jur.-L.Cret.,* Eu.——Fig. 219,6. *M. elegans,* U.Jur.(Kimm.), Fr.; *6a-c,* ♂ carapace R show-

ing dorsal overlap and eye tubercles, ♀ carapace L, ♂ dors., all ×50 (269).

Microcytherura G.W.MÜLLER, 1894 [*M. nigrescens*]. Subquadrate in side view, egg-shaped in dorsal view, fragile, bilamellar; RV slightly larger than LV, overhanging it along dorsal margin; posterior caudal process weakly developed. LV hinge with weakly crenulated anterior furrow, smooth median bar, widened at its extremities, and posterior crenulated furrow with indistinct posterior termination; internal margin and line of concrescence coincident, marginal pore canals few, straight, simple, although some may branch; normal pore canals numerous, with sieve structure in outer opening; oblique row of 4 adductor scars and 2 anterodorsally located scars, 2 mandibular scars. Sexual dimorphism discernible. *Pleist.* (incl. *Rec.),* Eu.(Holl.-Italy).——Fig. 220,3. *M. nigrescens,* Rec., Holl.; *3a,b,* carapace L, dors., ×130 (53).

Orthonotacythere ALEXANDER, 1933 [*Cytheridea? hannai* ISRAELSKY, 1929] [=*Cytheropterina* MAN-

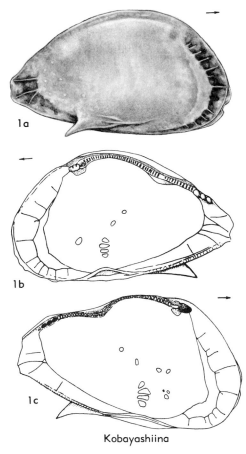

1a

1b

1c

Kobayashiina

Fig. 222. Cytheruridae (p. Q293).

DELSTAM, 1956]. Subquadrangular, dorsal margin almost straight, ventral margin parallel to it, posterodorsal caudal extension present, greatest width ventral; surface with knobs and reticulations. Adductor muscle scars in vertical row of 4; marginal pore canals few, simple, straight, RV hinge with notched terminal tooth plates separated by strongly crenulated furrow. *Jur.-Rec.*, N.Am.——Fig. 219,10. *O. hannai* (ISRAELSKY), Cret., Tex., Ark.; *10a,b,* LV hinge, RV hinge, ×100 (89). ——Fig. 225,1. *O. vegranda* (MANDELSTAM), Cret., USSR(Ukraine) (type species of *Cytheropterina*); *1a-c,* carapace R, RV and LV hinge, ×94 (50).

Otocythere TRIEBEL & KLINGLER, 1959 [*O. callosa*]. Shape in side view resembling that of human ear, small; LV slightly larger than RV, overhanging it along dorsal margin and upper part of broadly rounded anterior margin, greatest height in front of middle, posterior extremity

drawn out in short caudal process, deepened by small furrow on both valves, so that unclosed posterior end looks like a pipe; dorsal outline of inferred females inflated in LV and thereafter straight, ventral margin convex, overhung by ventral inflation; surface with coarse folds and tubercles; strong adductor muscle tubercle in front of middle, bounded anteriorly and posteriorly by broad submarginal folds; no eye structures present. Inner margin and line of concrescence coincident, approximately parallel to outer margin; zone of concrescence rather narrow, marginal pore canals few, well spaced, straight, unbranched, submarginal canals lacking but a few sublateral canals occur, beginning proximally from line of concrescence and cutting marginal fields obliquely outside zone of concrescence; RV hinge with low terminal tooth plates, each with 6 to 9 toothlets, and intervening notched median furrow

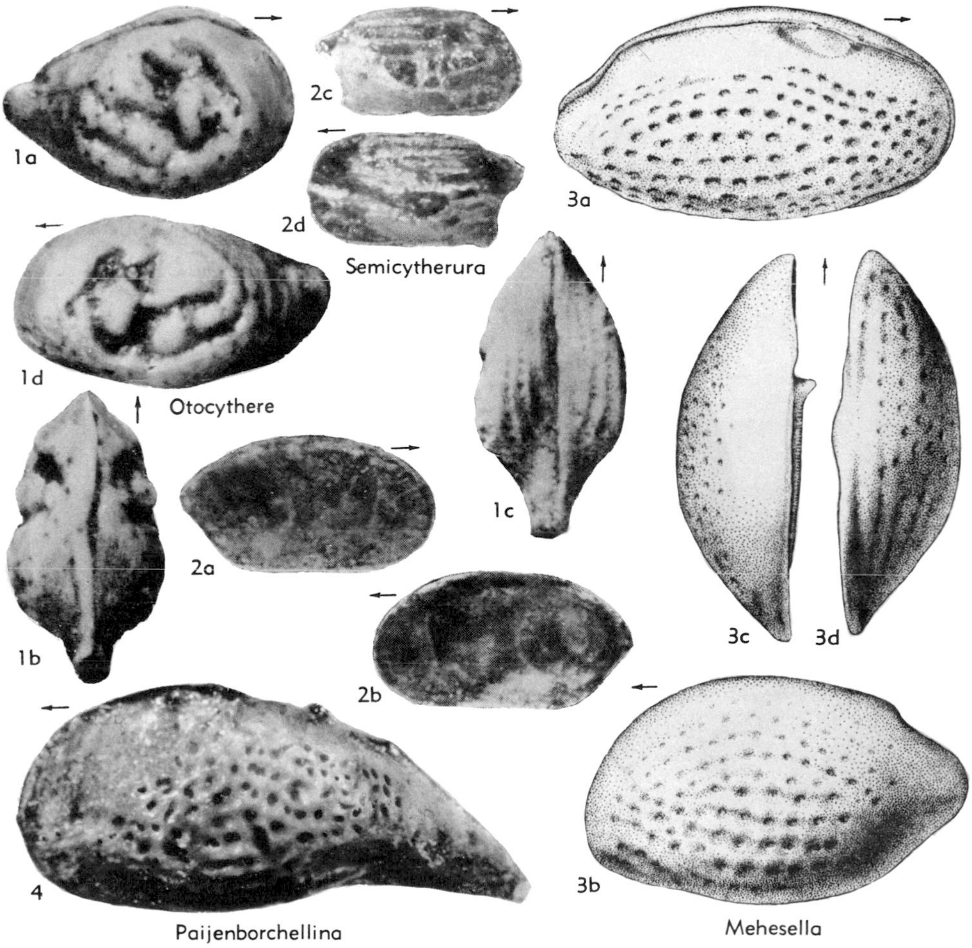

FIG. 223. Cytheruridae (p. Q293-Q300).

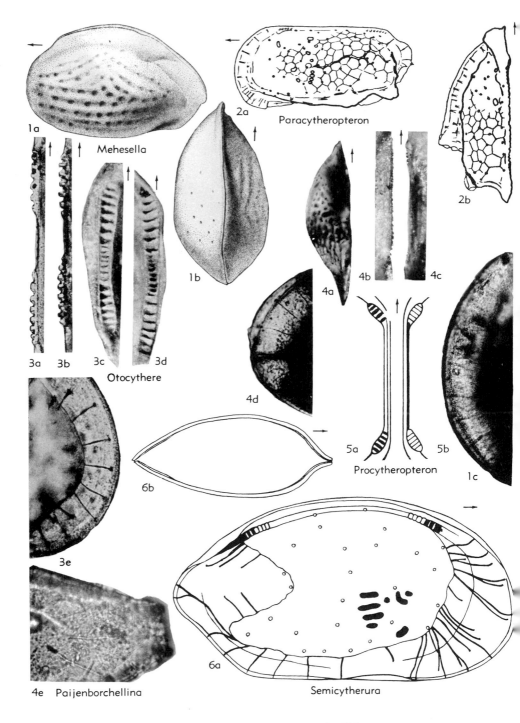

Fig. 224. Cytheruridae (p. Q293-Q300).

(this element shorter than terminal hinge elements in females and shorter than posterior tooth plate in males). [Marine.] *L.Jur.-M.Jur.*, Eu.——FIG. 223,*1*, 224,*3*. **O. callosa*, L.Jur., Ger.(Niedersachsen); 223,*1a-c*, ♀ carapace R, dors., vent., ×80; 223,*1d*, ♂ LV lat., ×80; 224, *3a,b*, ♂ and ♀ RV hinge dors., ×155, ×160; 224,*3c,d*, ♀ LV int., ♀ RV int., showing hinge, ×150; 224,*3e*, ♀ LV int. ant. margin showing radial pore canals, ×240 (382).

Paijenborchellina KUZNETSOVA, 1957 [**P. excelens*]. Carapace rather small, pear-shaped, with strongly produced caudal process, valves subequal but LV generally slightly larger than RV; ventral area gently rounded, on each valve bearing sharp rib that runs to base of posterior caudal projection and even onto it; weakly impressed zone from dorsal margin to middle of side; anterior margin fairly broad, provided with narrow outer lip through which canals do not penetrate, inner margin and line of concrescence coincide; anterior marginal pore canals few, straight, with 3 radially arranged canals in caudal process; selvage well developed, located near outer margin; RV hinge with anterior and posterior tooth plates and crenulated median furrow; eye spots feeble; muscle field and dimorphism unknown. Surface with small pits and weak ribs or reticulations. *L.Cret.(Neocom.)-Eoc.*, Eu.(USSR)-W.Afr.(Nigeria).——FIG. 223,*4*, 224, *4*. *P. ijuensis* REYMENT, *Eoc.*, W.Nigeria; 223,*4*, LV lat., ×125; 224,*4a*, LV dors., ×83; 224,*4b,c*, hinge LV and RV, ×133; 224, *4d,e*, RV int., ant. and post., pore canals, ×430 (Reyment, n).

Paracytheridea G.W.MÜLLER, 1894 [**P. depressa* (=**Cytheropteron bovettensis* SEGUENZA, 1880)] [?=*Mooreina* HARLTON, 1935]. Carapace stout, very much broadened, with rounded anterior and pointed posterior extremities, backwardly directed aliform ridge and sundry swellings on either side. Hinge weak, with faint indications of sockets in left valve, hinge margin or crenulate bar between them; radial canals slender, thickened near middle 53, 54, 327). *?Penn., U.Cret.-Rec.*, Eu.-N.Am.——FIG. 206,*1a-d*. **P. bovettensis* (SEGUENZA), Rec., Italy; *1a,b*, carapace L and vent., ×60 (54); *1c,d*, carapace L and dors., ×75, ×110 (53).——FIG. 206,*1e-h*. *P. brusselensis* KEIJ, Eoc.(Lutet.), Belg.; *1e*, LV ext., ×75; *1f,g*, carapace dors., vent., ×75; *1h*, RV int., ×115 (all 42). [HOWE.]

Paracytheropteron RUGGIERI, 1952 [**Cytheropteron calcaratum* SEGUENZA, 1880, =*Paracytheridea (Paracytheropteron) calcarata* (SEGUENZA) RUGGIERI, 1952]. Shape of carapace intermediate between that of *Paracytheridea* and *Cytherura* but with a cytheropteronoid hinge. Provided with moderately well-developed eye tubercles. Pore canals straight, unbranched, few in number, in posterior part confined to the caudal process. Inner margin and line of concrescence coincide or almost so. Adductor muscle field composed of a

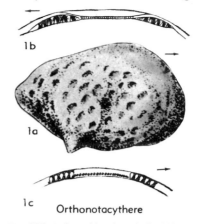

FIG. 225. Cytheruridae (p. Q296-Q297).

slightly inclined vertical row of four elongated spots and larger spot on top of row. At a distance anterior to the dorsalmost thereof a V-shaped spot and posteroventral from it a roundish spot. An elongated antennal spot occurs and there is a dispersed dorsal muscular field. Surface ornament of fine, large reticulations. Sexual dimorphism. *Mio.-Rec.*, Eu.——FIG. 220,*2*, 224,*2*. **P. calcaratum* (SEGUENZA), *Pleist.*, Italy; 220,*2*, RV int., ×70; 224,*2a,b*, LV lat., dors., ×100 (all 301).

Procytheropteron LYUBIMOVA, 1955 [**P. obesum*]. Carapace shaped like that of *Eocytheropteron* and *Mehesella* but with even more pronounced caudal process, LV overlapping RV, ventral margin strongly convex, dorsal less strongly convex; surface with minute pits and feeble ribs. Normal pore canals possibly provided with sieve openings; zone of concrescence narrow, with numerous straight pore canals; RV hinge with terminal teeth (each with 5 notches) and median furrow, LV hinge complementary; muscle field reported to be cytherid in type. [LYUBIMOVA (1955) ascribed genus to MANDELSTAM, designating *P. obesum* LYUBIMOVA, n. sp., as type. MANDELSTAM (1956) published *Procytheropteron* as new genus of his own, designating *Cythere punctatula* var. *virginea* JONES, 1849, as type; neither of these actions is valid.] [Marine.] *U.Jur.(Volg.) - U. Cret. (Maastricht.)*, Eu.(USSR-Eng.)-Asia-N.Am. —— FIG. 219,*8*. **P. obesum*, U.Jur.(Volg.), USSR; *8a,b*, LV lat., dors., ×43 (230).——FIG. 224,*5*. *P. virgineum* (JONES), U.Cret.(Campan.), USSR (Emba); *5a,b*, LV and RV hinge, ×62 (50).

Semicytherura WAGNER, 1957 [**Cythere nigrescens* BAIRD, 1838]. Carapace bilamellar, fragile, surface smooth or distinctly ornamented, RV slightly larger than LV, overhanging it along dorsal margin; with distinct caudal process. LV hinge with notched elongated anterior pit that is indistinctly terminated in front, short bar terminally thickened and crenulated but with middle smooth, notched elongated posterior pit indistinctly terminated at rear; zone of concrescence broad, particularly in posterior part where it extends forward into carapace, in some rather far; line of concrescence and inner margin coincident; normal pore canals open, few; marginal pore canals few, long, commonly grouped, some divided; 4 vertically arranged adductor scars with a group of 3 scars in front, single mandibular scar. Sexual dimorphism pronounced. *Pleist.*(incl. *Rec.*), N.Eu.——Fig. 223,2a,b; 224,6. *S. nigrescens* (Baird), Rec., NE.Atl.; 223,2a,b, ♀ RV lat., ♀ LV lat., ×80; 224,6a, ♀ LV int., ×220; 224,6b, carapace long. sec., ×125 (88, by permission of Mouton & Co., The Hague).——Fig. 223,2c,d. *S. acuticostata* (Sars), Rec., NE.Atl.; 2c,d, RV lat., LV lat., ×60 (88, by permission of Mouton & Co., The Hague).

Family ENTOCYTHERIDAE Hoff, 1942

[*nom. transl.* Howe, herein (*ex* Entocytherinae Hoff, 1942)] [Materials for this family prepared by H. V. Howe, Louisiana State University]

Shell reniform to elliptical, thin, chitinous, laterally compressed, valves nearly equal, surface without protuberances or papillae, but in some species with slight sculpturing and a few hairs. [Commensal or subterranean.] *Rec.*, N.Am.-Eu.-Afr.

Entocythere Marshall, 1903 [*E. cambaria*]. Commensal or subterranean. *Rec.*, N.Am.

E. (Entocythere). Second antenna of ♀ with 3 distal claws. Commensal on crayfish. *Rec.*, N.Am.

E. (Cytherites) Sars, 1926 [*C. insignipes*]. Second antenna of ♀ with 2 distal claws. Commensal on crayfish. *Rec.*, N.Am.——Fig. 226,1. *E. (C.) insignipes; 1a,b*, carapace L, dors., ×100 (315).

E. (Donnaldsoncythere) Rioja, 1942 [*Entocythere donnaldsonensis* Klie, 1938]. Like *E. (Cytherites)* but differs in terminal portion of ♂ antenna. [Subterranean.] *Rec.*, N.Am.——Fig. 226,2. *E. (D.) donnaldsonensis; 2a,b*, ♀ and ♂ LV lat.; *2c,d*, ♀ and ♂ carapace dors., ×100 (220).

Sphaeromicola Paris, 1916 [*S. topsenti*]. Distal podomere of mandibular palp in most species spatulate in shape; respiratory plate of maxilla wanting and masticatory lobes vestigial; penis with shortened base. Commensals of Isopoda and Amphipoda. *Rec.*, S.Eu.-Afr.-N.Am.(Mexico).

Family HEMICYTHERIDAE Puri, 1953

[*nom. transl.* Howe, herein (*ex* Hemicytherinae Puri, 1953)] [Materials for this family prepared by H. V. Howe, Louisiana State University, with contribution on some genera by W. A. van den Bold, Louisiana State University, and R. A. Reyment, University of Stockholm]

Carapace ovate, subrectangular, or somewhat almond-shaped, broadly and obliquely rounded in front, somewhat truncate behind, with angulation at or above junction with ventral margin, in some genera (e.g., *Urocythere, Caudites, Nephokirkos*) with angulation produced into true caudal process; usually somewhat concave above angulation; surface smooth, pitted, reticulate, or longitudinally ribbed, with low keel com-

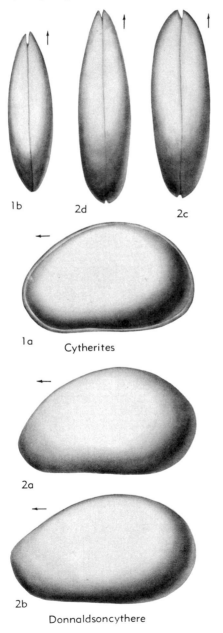

1b   2d   2c

1a   Cytherites

2a

2b   Donnaldsoncythere

Fig. 226. Entocytheridae (p. Q300).

monly present near ventral margin. Hinge merodont in young but holamphidont in adults, except for *Nereina* which remains merodont in adult stage and is tentatively placed here because of its shape, marginal area, and muscle scars. Marginal areas with numerous radial canals, 25 to 80 in anterior region, line of concrescence tending to coincide with inner margin; muscle scars in vertical row of 4 (median scars divided dif-

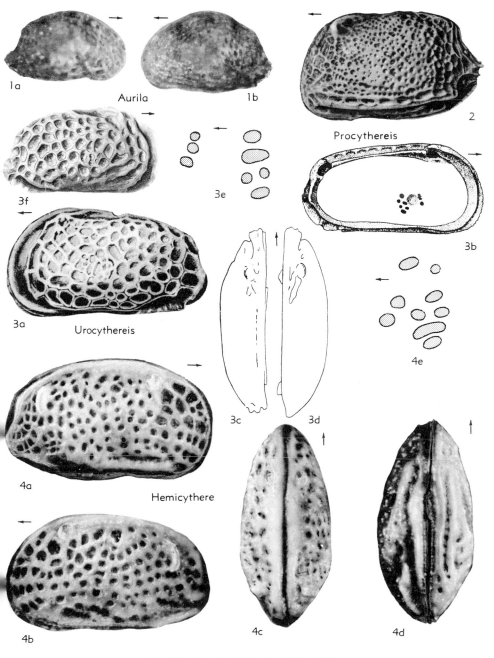

Fig. 227. Hemicytheridae (p. Q302-Q306).

ferently in various genera) with either 2 or 3 (mandibular-antennal) scars obliquely in front of upper part of vertical row (adductors). *Eoc.-Rec.*

Hemicythere SARS, 1925 [*Cythere villosa* SARS, 1866; SD EDWARDS, 1944] [?=*Eucythereis* KLIE, 1940 (*partim*)]. Carapace plump, subrectangular with rather concave posterodorsal margin; surface pitted or reticulate. Hinge holamphidont, heavy; marginal areas with numerous radial canals; muscle scars in vertical row of 4 (next to top element paired) and 2 antennal scars obliquely in front (68, 143, 222, 300). *Eoc.-Rec.*, cosmop.——FIG. 227,4. *H. villosa* (SARS), Rec., NE.Atl. (Norway-Swed.); 4a-d, carapace R, L, dors., vent., ×80; 4e, muscle scars, ×180 (Sylvester-Bradley, n).

Aurila POKORNÝ, 1955 [*pro Auris* NEVIANI, 1928 (*non* SPIX, 1827)] [*Cythere convexa* BAIRD, 1850 (=*Cythere punctata* MÜNSTER, 1830)]. Carapace almond-shaped, rounded in front, pointed behind; surface pitted, eye tubercles distinct. Hinge holamphidont; adductor muscle scars variable, oblique row of 3 antennal scars in front. *Plio.-Rec.*, Eu.——FIG. 227,1a,b, 230,1. *A. convexa* (BAIRD), Rec., Eng.; 227,1a,b, RV lat., LV lat., ×40; 230, 1a, RV int., ×100; 230,1b, hinge, dors., ×45 (all 88, by permission, Mouton & Co., The Hague).——FIG. 228,2a. *A. punctata* (MÜNSTER), Plio., Fr. (Perpignan); LV lat., ×60 (42).——FIG. 228, 2b,c. *A. cicatricosa* (REUSS), Plio., Fr.(Perpignan); 2b,c, LV lat., int., ×60, ×75 (42).

Caudites CORYELL & FIELDS, 1937 [*C. medialis*]. Carapace small, elongate, thick-shelled, more compressed than *Hemicythere*, anterior end rounded,

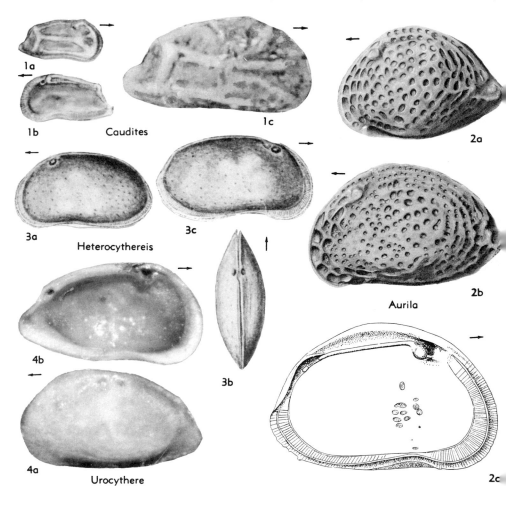

Caudites

Heterocythereis

Aurila

Urocythere

FIG. 228. Hemicytheridae (p. Q302-Q306).

with thickened rim, posterior attenuated and posteroventral corner produced; surface with one or more longitudinal ribs and commonly a transverse rib extending downward from posterocardinal angle. Hinge as in *Hemicythere*. *Eoc.-Rec.*, N. Am.——Fig. 228,*1a,b*. **C. medialis*, Mio., Panama; *1a,b*, RV lat., int., ×60 (126).——Fig. 228, *1c*. *C. jacksonvillensis* SWAIN, M.Eoc., N.Car.; *1c*, RV lat., ×80 (353).

Elofsonella POKORNÝ, 1955 [*pro Paracythereis* ELOF-

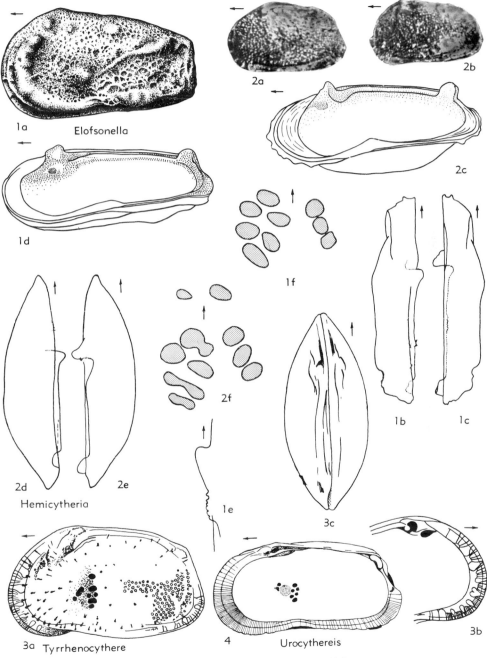

FIG. 229. Hemicytheridae (p. Q304-Q306).

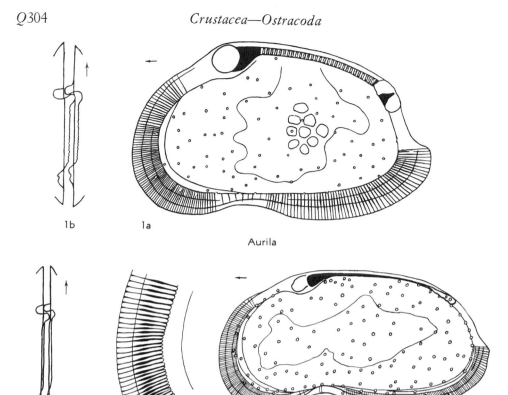

1b   1a

Aurila

2c   2b   Heterocythereis   2a

FIG. 230. Hemicytheridae (p. Q302-Q304).

son, 1941 (*non* DELACHAUX, 1928; *nec* JENNINGS, 1936)] [*Cythere concinna* JONES, 1857]. Carapace oblong, highest at anterior cardinal angle, dorsal and ventral margins straight, converging, posterior end truncate; surface uneven, with marginal rim around anterior and ventral edges. Hinge holamphidont; radial canals numerous; muscle scars in vertical row of 4 (with 2 middle ones paired) and 3 antennal scars forming oblique row in front (38). *Pleist.-Rec.,* Eu.——FIG. 229,1. *E. concinna* (JONES), Rec., Swed.; *1a,b,* ♀ LV ext. and dors.; *1c,d,* ♀ RV dors. and int. oblique vent.; *1e,* juv. RV dors., ant. part of hinge; *1f,* muscle scars; *1a-d,* ×60; *1e-f,* ×200 (276).

Hemicytheria POKORNÝ, 1955 [*Cypridina folliculosa* REUSS, 1850]. Like *Hemicythere* in shape; surface pitted or latticed, with smooth eye tubercle. Hinge holamphidont; muscle scars as in *Hemicythere* except for presence of 3 antennal scars in front instead of 2. *Plio.,* Eu.——FIG. 229,2. *H. folliculosa* (REUSS), Aus.(Vienna basin); *2a,b,* ♂ LV ext., ♀ LV ext., ×35; *2c,* RV int. oblique vent., ×50; *2d,e,* LV and RV dors., ×50; *2f,* muscle scars, ×200 (276).

Heterocythereis ELOFSON, 1941 [*Cythere albo-*

maculata BAIRD, 1838]. Like *Hemicythere* in shape but shell thin and smooth. Hinge holamphidont in adult, merodont in young, anterior teeth strong, posteriors weak; marginal areas moderate in width, line of concrescence and inner margin nearly coincident, radial canals very numerous; muscle scars in vertical row of 4 (upper middle one paired) and row of 3 antennal scars in front. *Rec.,* N.Atl.——FIG. 228,3, 230,2. *H. albomaculata* (BAIRD); 228,3a,b, ♀ carapace L, dors., ×45; 228,3c, ♂ carapace R, ×45; 230,2a,b, ♂ RV int., part of ant. marginal area, ×75, ×200; 230,2c, hinge, dors., ×45; (228,3, 314; 230,2, 88, permission of Mouton & Co., The Hague).

Mutilus NEVIANI, 1928 [*Cythereis (Mutilus) laticancellata* NEVIANI, 1928 (=*Cythere retiformis* TERQUEM, 1878); SD RUGGIERI, 1956] [=*Mutila* NEVIANI, 1929 (*errore*)]. Plump, subrectangular, dorsum arched, ventral margin inturned near middle, anterior obliquely rounded, posterior subtruncate but angulated below middle and somewhat concave above angulation. Surface very coarsely reticulate, with normal canals in reticulations. Hinge holamphidont, front tooth of the RV stepped, back tooth reniform; adductor scars in row of 4 (next to top paired), with oblique row of 3 antennal scars in front. *Mio.-Rec.*, Eu.——Fig. 231,*1*. **M. retiformis* (Terquem), Calabrian(Plio.); *1a-c*, RV lat., int., dors.; *1d*, muscle scars, ×65 (305).

Nephokirkos Howe, 1951 [**N. aquaplanus*]. Resembling *Cytheropteron* in shape of carapace because of well-developed caudal process, but more solidly constructed, with amphidont hinge, and distinct eye sockets and tubercles. Marginal area regular with about 25 radial canals on anterior. Muscle scars a vertical row of 4 (next to top

divided) with 2 oblique antennal scars in front. *Eoc.*, SE.USA(Fla.).——Fig. 190,*5*. **N. aquaplanus; 5a,b*, RV lat., int.; *5c*, carapace, dors.; all ×60 (Howe, n).

Nereina Mandelstam in Mandelstam *et al.*, 1957 [**N. barenzovoensis*]. Carapace elongate, with ventral and smaller dorsal ridge or swelling and low subcentral node; hinge in LV consisting of terminal crenulate sockets, divided into 6 anteriorly and 5 posteriorly [each with 7 to 9 crenulations (van den Bold)] and median smooth ridge [crenulate (van den Bold)]; marginal area not

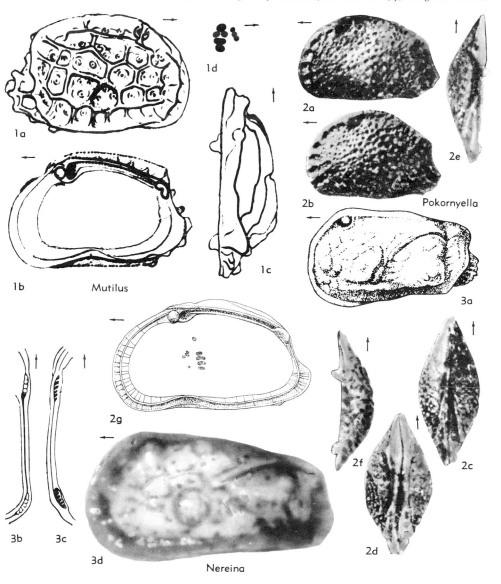

Fig. 231. Hemicytheridae (p. Q304-Q306).

described [line of concrescence and inner margin nearly coinciding in anterior end, pore canals very numerous, thin, some slightly sinuous, mostly apparently straight (VAN DEN BOLD)] muscle scars in posterior row of 4, of which middle 2 are subdivided into 2 scars each, and 3 additional scars in more or less oblique row in front. *Rec.*, N.Asia (Barentz Sea).——FIG. 231,3. **N. barenzovoensis; 3a*, LV lat., ×50; *3b,c*, RV and LV hinge, ×50 (238a); *3d*, LV lat. (Russian Harbor, Novaya Zemlya), ×75 (Howe, n). [BOLD.]

Pokornyella OERTLI, 1956 [**Cythere limbata* BOSQUET, 1852]. Egg- to kidney-shaped in side view, with distinct posteroventral process, ornament of reticulations or small pits, some shells with a posterodorsal swelling; small, glassy eye tubercles. Similar in appearance to *Hemicythere*, but has only about 25 radial canals on anterior, instead of 60 to 80. Inner margin and line of concrescence coincident; marginal pore canals fairly numerous, straight, single or grouped in pairs; adductor muscle field with vertical row of 4 spots, with round depression in front and 2 antennal spots located distally; RV hinge with simple, prominent anterior tooth and simple, narrow posterior tooth, a weak median groove with anterior depression. *Oligo.*, Eu.(Fr.-Belg.-Switz.).——FIG. 231,2. **P. limbata* (BOSQUET), Switz.-Fr.; *2a,b*, ♂ and ♀ LV lat., ×58; *2c,d*, ♂ and ♀ carapace dors., ×58; *2e,f*, ♂ LV dors., ♀ RV dors., ×58; *2g*, RV int., ×75 (*2a-f*, from Switz., 269; *2g*, from France, 42). [HOWE-REYMENT.]

Procythereis SKOGSBERG, 1928 [**Cythereis (Procythereis) torquata*]. Carapace shaped like *Hemicythere;* surface pitted and bearing strong alate ridge near flattened ventral margin. *Rec.*, S.Atl.-Pac.——FIG. 227,2. **P. torquata* (SKOGSBERG), S.Atl.; carapace L, ×50 (72).

Tyrrhenocythere RUGGIERI, 1955 [**T. pignattii*]. Like *Hemicythere* but anterior marginal areas distinguished by peculiar vestibule divided into several rounded pockets that terminate in series of slender, nearly straight radial canals with bulbous distal dilation, line of concrescence well separated from inner margin. *Rec.*, Medit.——FIG. 229,3. **T. pignattii*, Tyrrhen. Sea; *3a,b*, ?♀ LV lat., int., ×60; *3c*, ?♂ carapace, dors., ×60 (304).

Urocythere HOWE, 1951 [**U. attenuata*]. Sexually dimorphic, males resembling *Caudites*, females shorter, similar to *Hemicythere;* surface with little or no ornamentation. Hinge of left valve with deep anterior socket partitioned from ocular sinus, posterior socket ovate, median element consisting of high ridge that fits under dorsal margin of right valve, anterior end of ridge terminated by blunt downwardly directed tooth received by anterior socket of right valve; anterior marginal areas crossed by 8 to 10 pairs of widely spaced radial canals. *Eoc.*, SE.USA.——FIG. 228,4. **U.*

attenuata, Fla.; *4a,b*, ♀ LV ext. and int., ×80 (Howe, n).

Urocythereis RUGGIERI, 1950 [**Cytherina favosa* ROEMER, 1838]. Carapace like *Hemicythere* in shape; surface reticulate. Hinge holamphidont, median groove of right valve deepest just behind elliptical-based anterior tooth, posterior tooth reniform; marginal areas with numerous radial canals; adductor muscle scars in vertical row of 4 (lower middle or both middle scars paired) and 3 antennal scars in front. *Mio.-Rec.*, Eu.——FIG. 227, 3, 229,4. **U. favosa* (ROEMER), Plio., Italy-Fr.; 227,3a*, juv. LV ext., ×75; 227,3b*, ♀ LV int., ×50; 227,3c,d*, LV, RV dors., ×50; 227, 3e*, muscle scars, ×200; 227, 3f*, ♂ RV ext., ×40 (227,3a-e*, Italy, 300; 227,3f*, Fr., 42); 229,4*, ♂ RV int., ×75 (300).

Family KLIELLIDAE Schäfer, 1945

[*nom. transl.* HOWE, herein (*ex* Kliellinae SCHÄFER, 1945)]
[Materials for this family prepared by H. V. HOWE, Louisiana State University]

Shell delicate, small, chitinous or slightly calcareous; smooth or with surface ornamentation; elongate, but narrowed at ends. Subterranean. *Rec.*

Kliella SCHÄFER, 1945 [**K. hyaloderma*]. Elongate-ovate chitinous shells, without definite hinge or surface ornamentation. Subterranean. *Rec.*, Greece.——FIG. 232,2. **K. hyaloderma; 2a,b*, ♂ and ♀ RV ext.; *2c*, carapace dors., ×200 (318).

Nannokliella SCHÄFER, 1945 [**N. dictyoconcha*]. Elongate, slightly calcareous shell, with distinct drop-off from dorsal margin to posterior. Surface ornamented with lines. Subterranean. *Rec.*, Greece.——FIG. 232,1. **N. dictyoconcha; 1a*, RV ext.; *1b*, carapace dors., ×300 (318).

Family LEGUMINOCYTHEREIDIDAE Howe, n. fam.

[Materials for this family prepared by H. V. HOWE, Louisiana State University]

Carapace medium to small in size, elongate ovate in side and dorsal views with greatest thickness generally behind middle, ovate to subcircular in end view; smooth to strongly reticulate, without dorsal and median ribs such as found generally in Trachyleberididae but a rib may be present at junction of lateral and ventral surfaces. External muscle node in some specimens. Hingement somewhat modified holamphidont, with anterior socket of RV triangular and elongated in direction of hinge furrow (in *Triginglymus* and some specimens of *Basslerites* with toothlike thickening of shell wall below and behind anterior hinge element of each valve). Marginal areas nearly regular but usually wider on anterior end, where a vestibule develops; radial canals fairly numerous and straight, normal canals small and rather widely spaced; adductor muscle scars a vertical row of 4, usually with 2 adductors obliquely in front, but in most specimens of *Basslerites* these are fused into a V. [The 2 mandibular scars, set low and forward near the inner margin, are usually overlooked unless the shell is rotated slightly.] *Eoc.-Rec.*

Leguminocythereis Howe, 1936 [**L. scarabaeus* Howe & Law, 1936]. Carapace bean-shaped, with dorsum and venter subparallel, ends rounded, anterior broadest, greatest thickness behind middle; surface reticulate, with a vertical element in upper half and muscle node usually evident. Hinge holamphidont, with sharp anterior tooth in RV, behind which is a triangular socket, furrow, and oblique rounded posterior tooth; marginal areas regular, with small anterior vestibule; adductor muscle scars a vertical row of 4, with 2 antennal scars in front of lower pair and 2 small ovate mandibular scars just above ventral inner margin; normal canals large, widely spaced. *Eoc.-Mio.*, N.Am.-Eu.——Fig. 188,2. **L. scarabaeus*, Oligo., USA(Miss.); *2a,b*, LV lat., int.; *2c*, RV int.; *2d*, carapace, dors.; all ×40 (after 35).——Fig. 191, 2. *L. dumonti* Keij, Eoc.(Led.), Belg.; *2a,b*, ♂ RV lat., dors., ×40; *2c*, ♀ LV lat., ×40; *2d*, ♀ carapace dors., ×40; *2e*, RV int., ×47 (after 42).——Fig. 191,*1*. *L. genappensis* Keij, Eoc. (Lutet.), Belg.; *1a*, ♂ LV lat., ×40; *1b*, ♂ carapace dors., ×40; *1c*, ♀ RV lat., ×40; *1d*, ♀ carapace dors., ×40 (all 42).

Acuticythereis Edwards, 1944 [**A. laevissima*]. Carapace externally rather similar to *Campylocythere* but shorter and RV tending to be subangulate posteriorly. Surface smooth to strongly pitted or finely reticulate. Hinge holamphidont, with prominent pyramidal anterior tooth in RV, behind which triangular socket is less elongate than in *Campylocythere;* adductor scars a vertical row of 4, with 2 rounded antennal scars in front and 2 rounded mandibular scars just above ventral inner margin; normal canals rather large and moderately scattered. *Mio.*, N.Am.——Fig. 233,*1*. **A. laevissima*, USA(N.Car.); *1a-c*, LV lat., int., hinge, ×35; *1d-f*, RV lat., int., hinge, ×35; *1g*, carapace dors., ×35; *1h,i*, LV int., RV int., ×60 (*1a-g*, 143; *1h,i*, Howe, n).

Basslerites Howe in Coryell & Fields, 1937 [*pro Basslerella* Howe, 1935 (*non* Kellett, 1935; *nec* Bouček, 1936)] [*non Basslerites* Teichert, 1937 (=*Rayella* Teichert, 1939] [**Basslerella miocenica* Howe, 1935]. Ovate, small, dorsal and posterior margins forming near right angle; surface smooth or slightly wrinkled posteriorly. Hinge exceptionally strong, holamphidont, in RV with pointed anterior tooth, deep oblique socket drawn

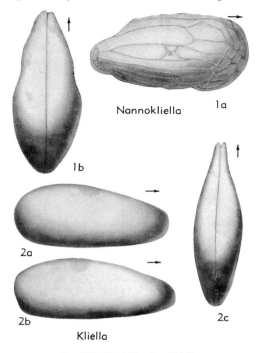

Nannokliella 1a

1b

2a

2b

Kliella 2c

Fig. 232. Kliellidae (p. Q306).

out toward hinge furrow, and oblique ovate posterior tooth; marginal area broadest anteriorly, with distinct vestibule and numerous radial canals; muscle scars a vertical row of 4 adductors, antennal scars generally fused into a V in front, and 2 ovate mandibular scars set obliquely just above ventral inner margin; normal canals thickly spaced on anterior end but widely spaced elsewhere. *Mio.-Rec.*, N.Am.-Eu.——Fig. 233,*3*. **B. miocenicus* (Howe), Mio., USA(Fla.); *3a,b*, carapace R, dors., ×60; *3c,d*, LV int., RV int., ×60 (all 178).

Campylocythere Edwards, 1944 [**C. laeva*]. Carapace elongate ovate, LV larger than RV, overlapping it at anterior cardinal angle and venter; surface smooth or finely pitted. Hinge holamphidont, with sharp triangular anterior tooth in RV, behind which is a very elongate triangular socket, a faint furrow, and ovate posterior tooth. Margin widest anteriorly, with vestibule and numerous straight radial canals; normal canals small, scattered over entire carapace; adductor scars a vertical row of 4 with 2 antennal scars obliquely in front and 2 large ovate mandibular scars just above ventral inner margin. *Mio.*, N.Am.——Fig. 233,2. **C. laeva*, USA(N. Car.); *2a-c*, carapace L, dors., LV hinge, ×35 (143); *2d,e*, LV int., RV int., ×74 (Howe, n).

Triginglymus Blake, 1950 [**T. hyperochus*]. Carapace comparatively delicate, elongate, ovate, with distinct muscle node and tiny eye tubercles; plumpest behind middle. Hinge holamphidont, in RV with sharp anterior tooth, behind which is an elongate V-shaped socket, a narrowing furrow, and an ovate posterior tooth; behind and below anterior elements of each valve thickened shell wall projects in a toothlike manner. Marginal area regular, with a small anterior vestibule and numerous straight radial canals, normal canals small, rather widely spaced; adductor scars in vertical row of 4, with 2 small antennal scars in front of upper part of row and 2 mandibular scars just above ventral inner margin. *Eoc.*, N.Am.——Fig. 233,4. *T. hyperochus*, USA(Ala.); *4a-c*, ♂ LV lat., int., dors.; *4d-f*, ♀ RV lat., int., dors.; all ×50 (6).

Family LEPTOCYTHERIDAE Hanai, 1957

[*nom. transl.* HANAI, herein (*ex* Leptocytherinae HANAI, 1957)] [Materials for this family prepared by TETSURO HANAI, University of Tokyo]

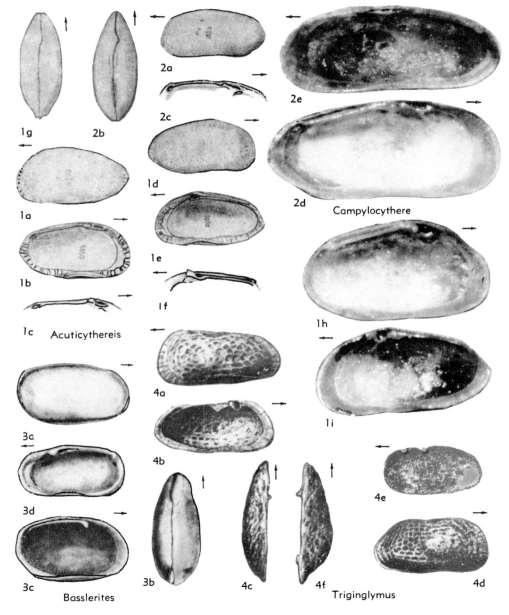

FIG. 233. Leguminocytichereididae (p. *Q307-Q308*).

Carapace comparatively small, elongate to subquadrangular in lateral outline; with distinct posterior cardinal angle. Surface nearly smooth to deeply sculptured, some species with anterior and posteroventral marginal ridges. Marginal area broad with characteristic polyfurcated radial pore canals; hinge modified entomodont; groove below selvage and above median bar of LV receiving or accommodating median element of RV, opening into anterior and posterior sockets (anterior tooth-and-socket structure and two-fold median element varies in different genera); adductor muscle scars in vertical row of 4, with single heart-shaped scar in front. *?Jur., Tert.-Rec.*

Leptocythere SARS, 1925 [*Cythere pellucida* BAIRD, 1850] [=*Leptocythera* SARS, 1925 (*errore*)]. Elongate; surface nearly smooth to punctate. Vestibule moderate. Median hinge element of left valve has 1 tooth at anterior end; corresponding socket of right valve is obscure. *Rec.,* N.Atl.——FIG. 234,*1.* *L. pellucida; 1a-e,* ♀ carapace L, ♂ RV lat., ♀ carapace dors., LV int., RV int., all ×60 (*1a-c,* 314; *1d,e,* 25).

Callistocythere RUGGIERI, 1953 [*Cythere littoralis* MÜLLER, 1894] [=*Cryptocythere* MANDELSTAM, 1958 (obj.)]. Elongate to subquadrangular; surface sculptured by reticulation and undulating ridges. Vestibule poorly developed; more than 2 anterior terminal teeth of median hinge element of LV definitely enlarged, corresponding sockets of RV distinct. Color usually yellow. *?Jur., Tert.-Rec.,* cosmop.——FIG. 234,*2.* *C. littoralis; 2a,b,* LV lat., LV int., ×100 (*2a,* 53; *2b,*302).

Mesocythere HARTMANN, 1956 [*M. foveata*]. Carapace elongate, greatest height in median to post-median position; surface pitted to reticulate; eye spots fused. Hinge ?without teeth, owing to decalcification. *Rec.,* S.Am.(Brazil).——FIG. 234,*3.* *M. foveata; 3a,* ♀ RV lat., ×93; *3b,* ♀ LV lat. outline (surface ornamentation not drawn), ×93; *3c,d,* LV and RV, dors. outline, ×60 (163).

Tanella KINGMA, 1948 [*T. gracilis*]. Elongate, tumid; surface sculptured by reticulations and ridges. Anterior tooth of RV replaced by elongate swelling of dorsal edge, in LV anterior socket lacking; anterior tooth of median bar represented by strong antislip tooth. *Plio.-Rec.,* SE.Asia.——FIG. 234,*4.* *T. gracilis; 4a-d,* LV int., LV lat., RV int. (hinge), carapace dors.; all ×100 (46).

Family LIMNOCYTHERIDAE Klie, 1938

[*nom. transl.* HOWE, herein (*ex* Limnocytherinae KLIE, 1938)] [=Limnicytherinae SARS, 1925] [Materials for this family prepared by H. V. Howe, Louisiana State University]

Valves subequal, shell weakly to strongly calcified; surface smooth, reticulate, or noded. Normal pore canals in general not described; marginal area regular, fairly broad, with tendency to form vestibules at ends and with rather evenly spaced straight radial canals; adductor muscle scars in nearly vertical row of 4, divided in some, antennal scar somewhat crescent-shaped and mandibular scar oval where known (in most genera undescribed); hinge weak, usually adont, but some genera have terminal teeth in RV and sockets in LV. In genera with strong sexual dimorphism, selvage of valves tends to interlock along ventral margin. [Habitat fresh- to brackish-water.] *Jur.-Rec.*

Limnocythere BRADY, 1868 [*Cythere inopinata* BAIRD, 1843; SD BRADY & NORMAN, 1889] [=*Limnicythere* BRADY, 1868 (obj.); *Acanthopus* VERNET, 1878 (*non* KLUG, 1807; *nec* OKEN, 1816; *nec* DAHL, 1823; *nec* LATREILLE, 1829; *nec* DE HAAN, 1835; *nec* MÜNSTER, 1839; *nec* GIEBEL, 1872); *Limnocytheridea* FOREL, 1894 (*nom. nud.*); *Acanthobus* MÜLLER, 1900; *Limnicytheridea* MÜLLER, 1912]. Carapace thin, horny, with reticulate, tuberculate, or spiny surface. Marginal areas broad, with numerous straight radial pore canals; adductor scars 4, with crescentic antennal and oval mandibular scars in front and additional scars above. *Jur.-Rec.,* cosmop.——FIG. 235,*2.* *L. inopinata* (BAIRD), Rec., NW.Eu.; *2a,b,* carapace L, dors., ×50 (54); *2c,d,* RV lat., LV lat., ×40; *2e,* LV int., ×135 (*2c-e,* 88, permission of Mouton & Co., The Hague).

Afrocythere KLIE, 1935 [*A. rostrata*]. Like *Limnocythere* except for different structure of mandibles and maxillae, surface of carapace smooth; only females known. *Rec.,* Afr.——FIG. 235,*4.* *A. rostrata; 4a-c,* ♀ carapace L, R, dors., ×60 (218).

Bisulcocypris PINTO & SANGUINETTI, 1958 [*B. pricei*]. Similar in shape to *Cytheridella,* females pyriform in dorsal view but having 2 sulci in anterior half of carapace instead of one; type species with valves reversed, RV larger than LV. Hinge with terminal sockets and projecting bar along median element; adductor scars in vertical row of 4 below middle, antennal and mandibular scars not described; marginal area and canals not described. *U.Jur.,* Brazil.——FIG. 236,*1a,b.* *B. ventrosa* (SWAIN); *1a,b,* carapace R, dors., ×45 (350).——FIG. 236,*1c-e.* *B. pricei; 1c,d,* ♀ RV lat., ♂ RV int.; *1e,* ♀ carapace dors.,×50 (274).——FIG. 237,*1.* *B. minnekahtensis* (ROTH), Morrison F., USA(S.Dak.); *1a-c,* ♀ carapace, L, dors., vent.; *1d-f,* ♂ carapace L, dors., vent.; all ×36 (200).

Cytheridella DADAY, 1905 [*C. ilosvayi*]. Sexual dimorphism strong, posterior 0.7 of female carapace much inflated. Hinge of RV formed of projecting selvage that fits in flange groove of LV; marginal areas broad, with small anterior vestibule and straight regularly spaced radial canals; adductor scars in row of 4 set low on strong internal ridge. [Differs from *Bisulcocypris* in having only one strong external sulcus in muscle-scar region.] *Rec.*, S.Am.——Fig. 238,1. *C. ilosvayi*, Para.; *1a,b*, RV lat., LV lat.; *1c,d*, ♂ and ♀ carapace dors.; *1e*, RV int.; *1f*, muscle scars; *1a-e*, ×30; *1f*, ×100 (136).

Elpidium F. Müller, 1881 [*E. bromeliarum*]. Resembling *Metacypris* in shape but larger (to 1.3 mm. length) and with differences in appendages; ventral margin with strongly interlocking selvages; muscle scars in oblique row of 4 forward-down-ward from middle; hinge undescribed. [Freshwater.] *Rec.*, S.Am.——Fig. 238,3. *E. bromeliarum*, Brazil; *3a-c*, carapace R, dors., vent., ×40; *3d*, muscle scars, ×100 (*3a*, Howe, n; *3b,c*, 255).

Gomphocythere Sars, 1924 [*Limnicythere obtusata* Sars, 1910]. Sexually strongly dimorphous and from above looks like small *Cytheridella* but venter flattened and outlined by a distinct rim. *Rec.*, Afr.——Fig. 238,2. *G. obtusata* (Sars); *2a,b*, carapace R, dors., ×75 (311).——Fig. 236,2. *G. expansa* (Sars); *2a,b*, ♂ carapace R, vent.; *2c,d*, ♀ carapace dors., vent., ×50 (313).

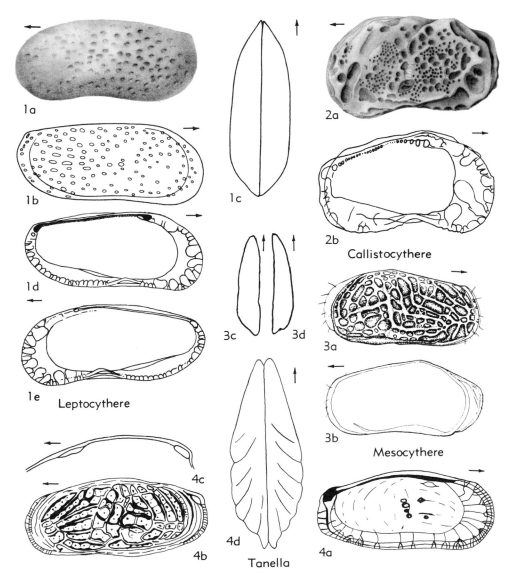

1a

1b

1c

1d

1e   Leptocythere

2a

2b   Callistocythere

3a

3b

3c   3d

Mesocythere

4c

4b   4d   Tanella   4a

Fig. 234. Leptocytheridae (p. Q309).

Leucocythere Kaufmann, 1892 [*L. mirabilis]. Carapace like *Limnocythere* but soft parts of males distinct. *Rec.,* Eu.

Metacypris Brady & Robertson, 1870 [*M. cordata]. Carapace subrhombic from side, female heart-shaped from above, with no sulcus, valve margins incurved except in front. Hinge of RV with laminated angular anterior projection and rectangular strongly produced posterior flange bearing a single sharply cut tooth. [Habitat freshwater.] *Rec.,* Eu.——Figs. 237,2, 238,4. *M. cordata,* Rec., Eng.; 237,2a-c, carapace L, dors., post.; 238,4, carapace vent.; all ×100 (17).

Neolimnocythere Delachaux, 1928 [*N. hexaceros]. Each valve bearing 3 backward-directed hornlike projections. [Genus defined mainly by nature of male copulatory appendage; only type species and one other (covered by short spines)

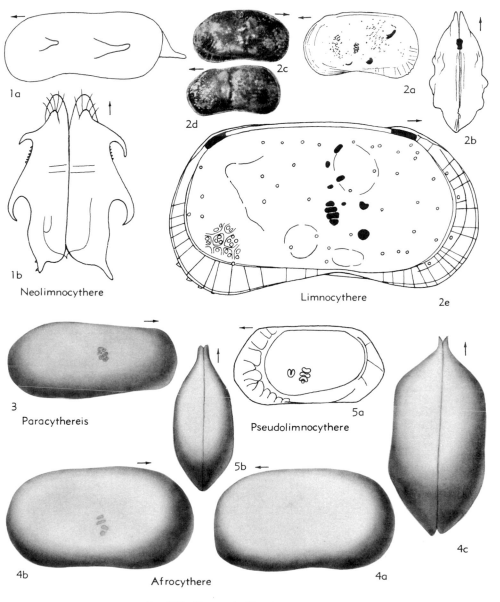

Fig. 235. Limnocytheridae (p. Q309-Q312).

described.] *Rec.*, S.Am.——Fɪɢ. 235,*1*. **N. hexaceros*, Peru; *1a,b*, ♂ carapace L, dors., ×45 (139).

Paracythereis Dᴇʟᴀᴄʜᴀᴜx, 1928 [*non* Jᴇɴɴɪɴɢs, 1936; *nec* Eʟᴏғsᴏɴ, 1941] [**P. impudica*]. Carapace shaped like *Limnocythere;* distinguished by male copulatory appendage and furca, which are chitinized, bearing median and terminal setae.

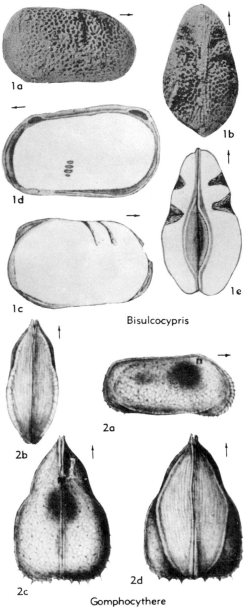

Fɪɢ. 236. Limnocytheridae (p. *Q309-Q310*).

Rec., S.Am.——Fɪɢ. 235,*3*. **P. impudica*, Peru; ♂ carapace R, ×60 (139).

?**Pseudolimnocythere** Kʟɪᴇ, 1938 [**P. hypogaea*]. Carapace elongate ovate, delicate. Marginal areas broad, with few branching radial pore canals; muscle scars in oblique row of 4 with heartshaped scar in front. [Habitat subterranean.] *Rec.*, S.Eu.——Fɪɢ. 235,*5*. **P. hypogaea*, Italy, *5a,b*, ♀ carapace L, dors., ×120 (220).

Theriosynoecum Bʀᴀɴsᴏɴ, 1936 [*pro Morrisonia* Bʀᴀɴsᴏɴ, 1935 (*non* Gʀᴏᴛᴇ, 1874)] [**Morrisonia wyomingensis* Bʀᴀɴsᴏɴ, 1935] [=*Theriosynecum* Mᴀɴᴅᴇʟsᴛᴀᴍ in Gᴀʟᴇᴇᴠᴀ, 1955]. Like *Cytheridella* but with strong nodes on swollen posterior region. Hinge straight and short, ridge of LV fitting groove in RV. [Habitat fresh-water.] *Jur.-Cret.*, N.Am.-Eu.-Asia.——Fɪɢ. 239,*1*. **T. wyomingense* (Bʀᴀɴsᴏɴ), U.Jur. (Morrison F.), USA(Wyo.); *1a-c*, carapace R, dors., vent., ×25 (113).

Family LOXOCONCHIDAE Sars, 1925

[*nom. transl.* Hᴏᴡᴇ, herein (*ex* Loxoconchinae Sᴀʀs, 1925)] [Materials for this family prepared by H. V. Hᴏᴡᴇ, Louisiana State University, with contributions on some genera by R. A. Rᴇʏᴍᴇɴᴛ, University of Stockholm] [Includes Cytheromorphinae Mᴀɴᴅᴇʟsᴛᴀᴍ, 1960]

Small, usually dimorphous carapaces, which may be nearly smooth but mostly are finely to coarsely pitted or reticulate; reniform to rectangularly ovate in lateral view with tendency to develop a posterior caudal process in some genera. Hinge typically gongylodont; in some species with anterior left and posterior right dentition appearing as 2 distinct teeth separated by deep pit but with these teeth normally united above by narrow crescentic ridge that makes horseshoe-shaped structure fitting over single tooth in opposite valve; marginal areas broad, tending to develop vestibules at ends; radial canals few, normal canals widely spaced, large, and in some genera sievelike; adductor scars in slightly oblique row of 4 elongate spots (divided in some), antennal scar U- or C-shaped, mandibular scars 2 oval spots obliquely below and forward. Habitat shallow marine or brackish-water. *Cret.-Rec.*

The genera *Loxoconchella, Phlyctocythere,* and *Elofsonia* resemble typical Loxoconchidae externally and in some of the internal characters but are placed in this family with much doubt. *Loxoconchella* differs from other forms assigned here in having an adont hinge and a lobate line of concrescence. *Phlyctocythere* also has an adont hinge, a high anterior vestibule, and small rounded antennal scar. *Elofsonia* has a hinge that does not fit any described family, possessing in the LV a crenulate anterior ridge, a smooth median bar, and an elongate posterior crenulate socket, though its other characters fit with the loxoconchids very well.

Loxoconcha SARS, 1866 [*Cythere impressa* BAIRD, 1850 (*non* M'COY, 1844) (=*C. rhomboidea* FISCHER, 1855); SD BRADY & NORMAN, 1889 [=*Loxoleberis* SARS, 1866; *Normania* BRADY, 1866 (*non* BOWERBANK, 1869; *nec* BOECK, 1871)]]. Carapace almond-shaped, with straight dorsal margin and sinuous venter; surface pitted or reticulate. Hinge gongylodont, middle element crenulate. Anterior and posterior vestibules with few straight canals; normal canals widely spaced; adductor scars 4, antennal scar crescent-shaped, mandibular scar oval. *Cret.-Rec.*, cosmop.——FIGS. 240,*1e-h*, 241,*3*. *L. rhomboidea* (FISCHER), Rec., NE.Atl.; 240,*1e,f*, ♀ carapace L, dors., ×45; 240,*1g*, ♂ RV lat., ×45; 240,*1h*, RV int. marginal area with radial canals, ×100; 241,*3a,b*, ♀ RV lat., LV lat., ×40; 241,*3c*, ♀ RV int., ×125 (240,*1e-g*, 220; 240,*1h*,379; 241,*3a-c*, 88, by permission of Mouton & Co., The Hague).——FIG. 241,*2*. *L. curryi* KEIJ, Eoc.(Led.), Eng.; *2a-c*, ♂ LV lat., int., dors., ×80 (42).——FIG. 240,*1a-d*. *L. grateloupriana* (BOSQUET), Oligo.-Mio., Eu.; *1a-c*, ♀ LV lat., ♂ carapace R, dors., ×75 (Mio., Fr.); *1d*, ♂ RV int., ×75 (Oligo., Belg.) (all 42).——FIG. 241, *4*. *L. matagordensis* SWAIN, Rec., Gulf of Mexico (off Tex.); *4a,b*, ?♂ LV lat., RV int., ×50 (355).

Cytheromorpha HIRSCHMANN, 1909 [*C. albula* (=*Cythere fuscata* BRADY, 1869); SD SARS, 1925]. Carapace compressed elongate ovate, anterior slightly more broadly rounded than posterior, long margins nearly straight but ventral margin inturned in front of middle; surface smooth, pitted, or faintly reticulate. Hinge gongylodont; adductor scars 4, antennal scar crescent-shaped, mandibular scars oval, oblique; marginal areas wide with anterior and posterior vestibules and few straight radial canals; normal canals widely spaced, irregular (68). *Paleoc., Rec.,* Eu.-N.Am.——FIGS. 240,*3*, 241,*5*. *C. fuscata* (BRADY), Rec., NE.Atl.; 240,*3a*, ♂ RV lat., ×45; 240,*3b,c*, ♀ carapace L, dors., ×45; 241,*5*, ♂ RV int., ×125 (240,*3a-c*, 314; 241,*5*, 88, by permission Mouton & Co., The Hague).

Elofsonia WAGNER, 1957 [*Loxoconcha baltica* HIRSCHMANN, 1909]. Carapace externally like *Loxoconcha*, subequivalved, smooth in type species. Hinge of LV with elongate crenulate anterior tooth, smooth median bar, and elongate crenulate posterior socket; adductor scars in oblique row of 4, antennal scar crescentic, mandibular scars 2, oval, oblique; marginal area broad, with vestibules at ends and few straight radial canals; normal canals large, widely spaced, sievelike. Sexual

dimorphism weakly defined in shell features. *Rec.*, NW.Eu.——FIG. 242,*1*. *E. baltica* (HIRSCHMANN), Rec., Holl.; *1a,b*, LV lat., RV lat., ×60; *1c*, LV int.,×133; *1d*, carapace long. sec., ×80 (88, by permission Mouton & Co., The Hague). [HOWE-REYMENT.]

Hirschmannia ELOFSON, 1941 [*Cythere viridis* O.F. MÜLLER, 1785]. Carapace subreniform; surface smooth except for large normal pore canals. Hinge gongylodont, middle element smooth. Terminal part of antennule 3-jointed. *Rec.*, Eu.——FIG. 241, *1*. *H. viridis* (MÜLLER), NE.Atl.; *1a,b*, ♀ carapace L, dors., ×50 (315); *1c,d*, RV lat., LV lat., ×40; *1e*, RV int., ×115 (*1c-e*, 88, by permission of Mouton & Co., The Hague).

?Loxoconchella TRIEBEL, 1954 [*Loxoconcha honoluliensis* BRADY, 1880]. Carapace externally like

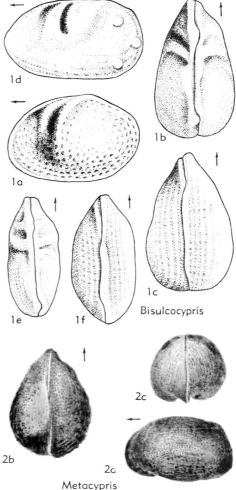

FIG. 237. Limnocytheridae (p. Q309-Q311).

Loxoconcha but with distinct caudal process above middle, valves subequal, surface smooth but with deep pits of normal canals. Hinge adont, furrow in RV and ridge in LV; adductor scars in oblique row of 4, antennal scars of different size, mandibular scars subequal, rounded, oblique; normal canals large, widely spaced. Vestibule ex-

tending from anterior to posterior and very lobate with 2 or 3 short straight radial canals extending from each indentation. *Rec.,* Pacific.——Fig. 240, 2. **L. honoluliensis;* 2a,b, LV lat., RV lat., ×60; 2c, LV int., ×60; 2d,e, RV int. with pore canals, ×300, ×60 (all 379).

?**Phlyctocythere** KEIJ, 1958 [**P. eocaenica*]. Smooth,

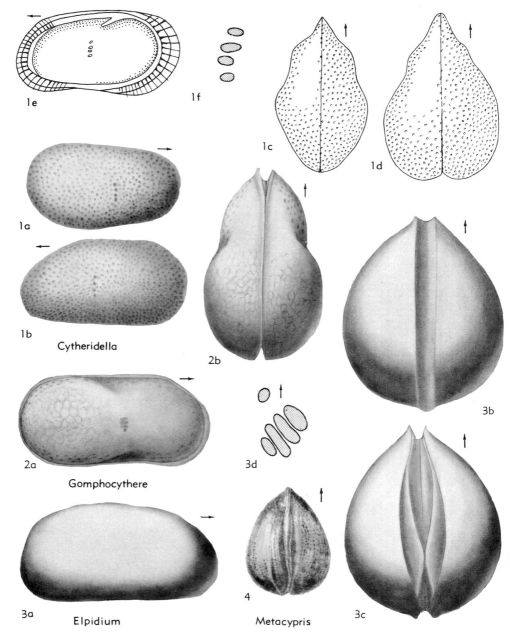

FIG. 238. Limnocytheridae (p. Q310-Q311).

externally much like *Loxoconchella;* internally adont, ridge in RV and furrow in LV; adductor scars in row of 4, antennal scar an oval spot, mandibular scars oval; marginal areas broad, with narrow, high anterior vestibule and shallow posterior vestibule; radial canals straight, numerous, unbranched; normal canals widely spaced. Sexual dimorphism pronounced; ?females shorter and more thick-set than ?males. *Eoc., Eu.*(Fr.).——Fig. 243,*1.* **P. eocaenica; 1a-c,* LV lat., int., dors., ×80 (197).

Family PARADOXOSTOMATIDAE
Brady & Norman, 1889

[Materials for this family prepared by P. C. Sylvester-Bradley, University of Leicester, and H. V. Howe, Louisiana State University]

Carapace elongate, thin-shelled, smooth, more or less compressed. Hinge lophodont (denticulate in some species of *Microcythere,* and in *Pellucistoma*). Adductor muscle leaving 3 to 6 elongate scars that form oblique linear pattern, sloping toward anteroventral margin; duplicature wide; vestibules wide or narrow. [This family includes some of the most abundant present-day ostracodes of the intertidal zone, being particularly common in rock pools, and amongst seaweed, but also abundant in deeper water. Fossils referable to the family seem to be rare, however.] *?Cret., Eoc.-Rec.*

The family can be divided into three subfamilies (Paradoxostomatinae, Microcytherinae, and Cytheromatinae) on the basis of appendages, but no carapace characters have yet been discovered which are diagnostic of them. Even some of the genera are difficult to distinguish on carapace characters alone.

Subfamily PARADOXOSTOMATINAE
Brady & Norman, 1889

[*nom. transl.* G.W.Müller, 1894 (as Paradoxostominae); *nom. correct.* Sylvester-Bradley & Howe, herein (*pro* Paradoxostominae G.W.Müller, 1894)]

Distinguished mainly by features of appendages. *?Cret., Eoc.-Rec.*

Paradoxostoma Fischer, 1855 [**P. dispar*]. Ventral margin sinuous, concave in anterior third, highest point of carapace in posterior third; caudal process absent, or very blunt, though a slight posterodorsal sinuosity is common. Hinge lophodont; vestibule wide, continuous from anterior through venter to posterior end; radial pore canals sparse; muscle-scar pattern with 3 or 4 adductor scars. [Type-species imperfectly known.] *?Cret., Eoc.-Rec.,* cosmop.——Fig. 244,*2.* *P. variabile* (Baird), Rec., Neth.; *2a,b,* RV lat., int., ×60; *2c,* LV lat., ×60 (88).

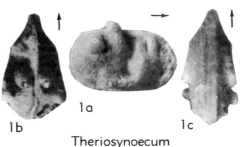

1a

1b   1c

Theriosynoecum

Fig. 239. Limnocytheridae (p. Q312).

?Boldella Keij, 1957 [**B. deldenensis*]. Carapace elongate, with obliquely rounded anterior end and upturned posterior extremity making posterodorsal obtuse angle in outline, posteroventral margin convex; LV slightly overlapping RV ventrally; surface smooth, striate, or faintly reticulate. Marginal areas broad except in front where deep vestibule occurs; radial pore canals numerous, generally bifurcating; adductor muscle scars in oblique row of 4, with scar in front and another anteroventrally. *U.Eoc., ?Mio.,* NW.Eu.——Fig. 244,*1.* **B. deldenensis,* U.Eoc., Belg.; *1a-c,* RV lat., int., dors.; *1d-e,* LV lat., int.; all ×90 (42).

Cytherois Müller, 1884 [**C. virens* (=**Paradoxostoma fischeri* Sars, 1866)]. Like *Paradoxostoma* but ventral margin less sinuous and with no trace of caudal process or posterodorsal sinuosity. *Pleist.-Rec.,* cosmop.——Fig. 244,*3a-c.* **C. fischeri* (Sars), Rec., Neth.; *3a,* LV lat., ×60; *3b,* RV lat., ×60; *3c,* LV int., ×50 (88).

Paracythere G. W. Müller, 1894 [**P. minima*]. Shell thin, fragile, smooth; closure without teeth, some with complete margin. Only females known. [Distinguished largely on nature of appendages.] *Rec.,* Medit. (Gulf of Naples).

Paracytherois Müller, 1894 [**P. striata* (=**Paradoxostoma flexuosum* Brady, 1868); SD Howe, 1955.] Carapace like that of *Paradoxostoma* but anterior vestibule more constricted. *Rec.,* cosmop.——Fig. 244,*7.* **P. flexuosa* (Brady), Rec., Medit.; RV int., ×100 (53).

Sclerochilus Sars, 1866 [**Cythere contorta* Norman, 1862] [=*Sclerochylus* Sohn, 1951 (*errore*)]. Like *Paradoxostoma* but muscle-scar pattern with 5 adductor scars. *?Eoc., Rec.,* cosmop.——Fig. 244,*4.* **S. contortum* (Norman), Rec., Neth.; *4a,* LV lat.; *4b,c,* RV lat., LV int.; all ×60 (88).

Xiphichilus Brady, 1870 [**Bythocythere tenuissima* Norman, 1869; SD Brady & Norman, 1889] [=*Machaerina* Brady & Norman, 1889 (obj.)]. Like *Paradoxostoma* but acuminate or produced at both ends. *Rec.,* Eu.——Fig. 244,*6.* **X. tenuissimum* (Norman), Rec., Scot.; *6a,b,* carapace L, dors., ×40 (15).

Subfamily MICROCYTHERINAE Klie, 1938

Distinguished mainly by features of appendages. *Mio.-Rec.*

Microcythere MÜLLER, 1894 [**M. inflexa;* SD VAN DEN BOLD, 1946]. Carapace small, thin, fragile, flattened ventrally. Hinge seemingly lophodont in type species but denticulate and anomalous in other species hitherto assigned to genus; marginal areas wide at both ends, forming large vestibules; radial pore canals few, distinct; 4 adductor muscle scars, with a single large antennal scar in front. *Mio.-Rec.,* cosmop.——FIG. 245,*1.* **M. inflexa,* Rec., Medit.; *1a,b,* ♂ RV int., ♂ LV int.; *1c,* ♀ LV int.; *1d,* ♀ carapace dors.; all ×180 (53).

Subfamily CYTHEROMATINAE Elofson, 1939

[*nom. correct.* HOWE, herein (*pro* Cytherominae ELOFSON, 1939)]

Distinguished mainly by features of appendages. *Oligo.-Rec.*

Cytheroma G. W. MÜLLER, 1894 [**C. variabilis*].

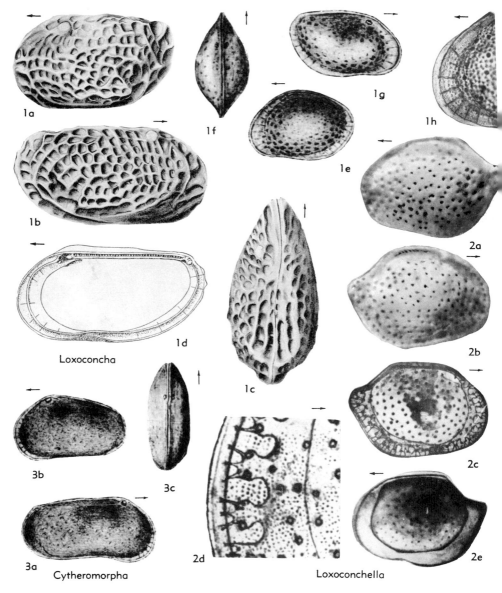

1a   1b   1d   *Loxoconcha*   1f   1c   1e   1g   1h   2a   2b   2c   2d   *Loxoconchella*   2e   3b   3c   3a   *Cytheromorpha*

FIG. 240. Loxoconchidae (p. Q313-Q314).

Carapace elongate, with rounded ends, anterior wider than posterior, lateral outline with venter less sinuous than in *Paradoxostoma*. Hinge lophodont; marginal areas broad except near midventral margin, with large, variably shaped vestibules at each end; radial pore canals (some of which may branch) causing line of concrescence to have lacy pattern. *Oligo.-Rec.,* cosmop.——Fig. 245,2. *C. variabilis*, Rec., Medit.; *2a,b,* RV lat., 2 specimens showing variation; *2c,* carapace dors.; all ×130 (53).

Paracytheroma Juday, 1907 [*P. pedrensis*]. Carapace like that of *Cytheroma. Rec.,* N.Am.

Pellucistoma Coryell & Fields, 1937 [*P. howei*]

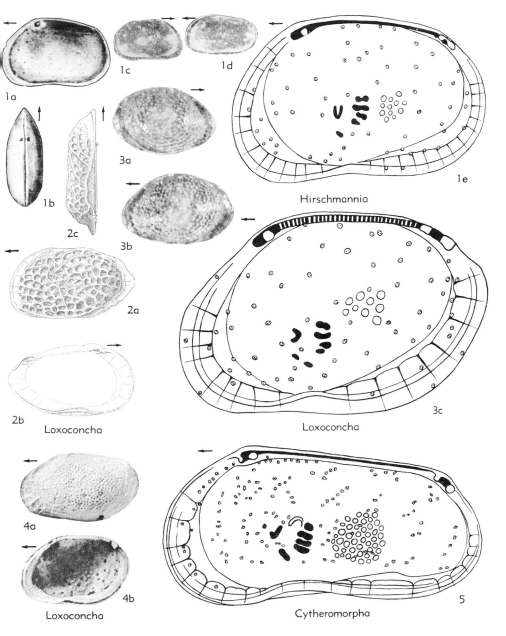

Fig. 241. Loxoconchidae (p. Q313).

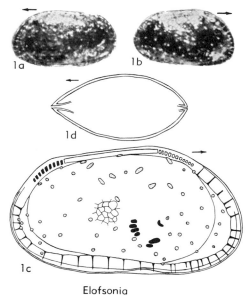

Fig. 242. Loxoconchidae (p. Q313).

[=*Javanella* Kingma, 1948]. Less elongate and more inflated than other genera assigned to family, with marked caudal process, widely spaced radial pore canals (some branching) and merodont hinge. *Mio.,* C.Am.-N.Am.——Fig. 244,5. **P. howei,* Panama; *5a,* RV lat.; *5b,* LV dors.; ×75 (126).

Family PECTOCYTHERIDAE Hanai, 1957

[*nom. transl.* Hanai, herein (*ex* Pectocytherinae Hanai, 1957)] [Materials for this family prepared by Tetsuro Hanai, University of Tokyo, with contributions on some genera by H. V. Howe, Louisiana State University]

Carapace thick, small, subquadrangular in side view, typically narrowing backward from broadly rounded anterior margin; surface coarsely punctate and with marginal ridge accenting periphery. Hinge line straight, of modified merodont (pentodont) type, characterized in LV by anterior and posterior sockets and intervening crenulate bar with terminations that swell into knoblike projections, RV hinge complementary; marginal area broad, vestibule developed anteroventrally; radial pore canals simple, straight, few. *L.Cret.-Rec.*

Pectocythere Hanai, 1957 [**P. quadrangulata*]. Close to *Munseyella,* but very thick, oblong box-shaped; marginal ridge is extremely blunt and bold, having a tendency to circumscribe periphery of shell. Posterior end lacks spines. In LV, lower non-crenulate elements of terminal teeth at each end of median element are larger than upper. Anterior vestibule mostly crescent-shaped. Male slightly elongate. *Plio.-Pleist.,* E.Asia(Japan), N. Am.(Calif.).——Fig. 246,1. **P. quadrangulata,* Plio., Japan; *1a,* RV lat., ×105 (Hanai, n); *1b,c,* RV int., LV int., ×105 (26).

Arcacythere Hornibrook, 1952 [**A. chapmani*]. Carapace small, oblong, narrow, widest posteriorly, dorsal and ventral margins parallel, ends squarely truncate with heavy rims; LV overlapping RV conspicuously at anterodorsal angle; surface reticulate. Four posterior muscle scars and one anterior scar; normal canals not numerous; radial canals few, simple, wide at the bases; line of concrescence deviating slightly from inner margin; RV hinge with 2 terminal simple or crenulate teeth and straight, faintly crenulate, median groove; LV hinge complementary. [Differs from *Leptocythere* in its oblong boxlike form with RV conspicuously overlapping LV at anterodorsal angle.] *U.Cret.-M.Mio.,* N.Z.——Fig. 246,2. **A. chapmani; 2a-c,* LV (holotype) lat., RV int., carapace dors., ×100 (32).

Dolocythere Mertens, 1956 [**D. rara*]. Carapace with rounded quadrangular outline, long margins converging slightly backwards; surface reticulate. LV hinge with terminal smooth sockets open to interior, between them a smooth bar, RV with smooth hinge furrow, upper portion of anterior and posterior margins thickening into false teeth which fit smooth sockets of LV, no accommodation groove; inner margin and line of concrescence coincident; radial canals 8 to 12 in front, about 6 behind, with some pseudoradial canals which reach outer surface; muscle scars in

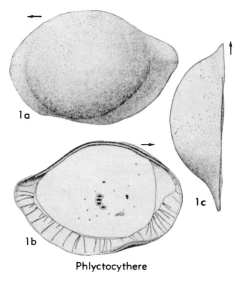

Phlyctocythere

Fig. 243. Loxoconchidae (p. Q315).

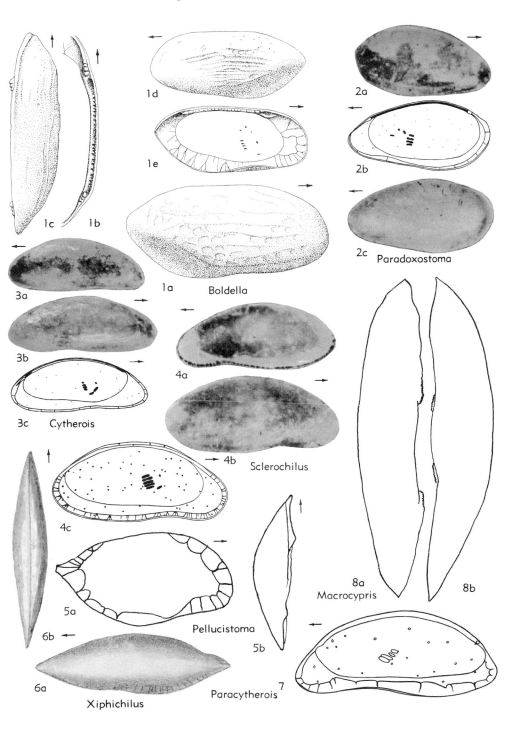

FIG. 244. Macrocyprididae, Paradoxostomatidae (p. Q207, Q315-Q318).

vertical row of 4, V-shaped scar in front opening forward. *L.Cret.(Apt.-Alb.)*, Eu.(Ger.).——Fig. 246,*4*. **D. rara*, L.Alb., NW.Ger.; *4a-d*, ♀ carapace (holotype) R, L, dors., vent., ×75; *4e*, RV lat., ×75; *4f,g*, LV int., RV int., ×100; *4h*, LV and RV from dorsal side, ×100 (250).

Munseyella van den Bold, 1957 [*pro Toulminia* Munsey, 1953 (*non* Zittel, 1878)] [**Toulminia hyalokystis* Munsey, 1953]. Small, compressed, subquadrangular, with nearly straight posterior outline. Surface heavily ornamented; a marginal ridge nearly circumscribes the carapace. In hingement of LV, lower non-crenulate elements of anteromedian and posteromedian teeth are smaller than upper elements. Both upper and lower elements are usually fused together so as to make one knob-like projection. Radial canals straight and few, extending from vestibule. Adductor scars a row of 4 with at least one scar in front. *Paleoc.-Rec.*, N.Am., Japan.——Fig. 246, *3*, 246A,*1*. **M. hyalokystis* (Munsey), Paleoc. (U. Midway), USA(Ala.); *246,3a,b*, RV ext., RV int.,

×140 (259); *246A,1a,b*, RV int., LV int., ×145 (Bold, n).

Family PERMIANIDAE Schneider, 1947

[Materials for this family prepared by W. A. van den Bold, Louisiana State University, and R. A. Reyment, University of Stockholm]

Large, subrectangular, distinct dorsal angles, rounded anteroventrally and posteroventrally, transverse groove extending downward from middle of dorsal margin. Muscle scars in vertical row of 3 with 2 scars in front. *U.Perm.*

Permiana Schneider, 1947 [**P. oblonga* Schneider, 1947]. Carapace elongate rectangular, with almost parallel, slightly concave ventral and dorsal margins and broadly, slightly uneven rounded anterior and posterior margins; RV overlapping LV weakly along dorsal margin; valves swollen ventrally with posteroventral winglike process; surface smooth, pitted or bearing small spines. Hinge simple, LV hinge with median furrow hav-

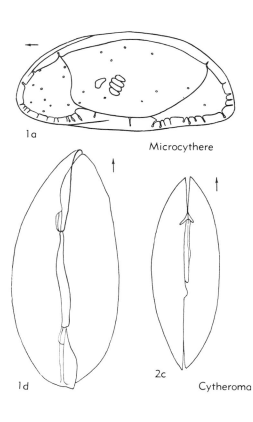

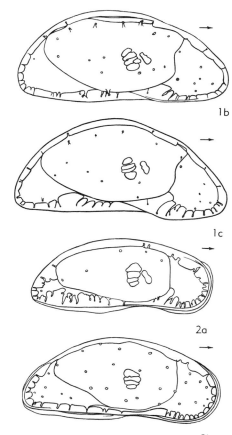

1a Microcythere

1d   2c Cytheroma

1b

1c

2a

2b

Fig. 245. Paradoxostomatidae (Microcytherinae, Cytheromatinae) (p. *Q*316-*Q*317).

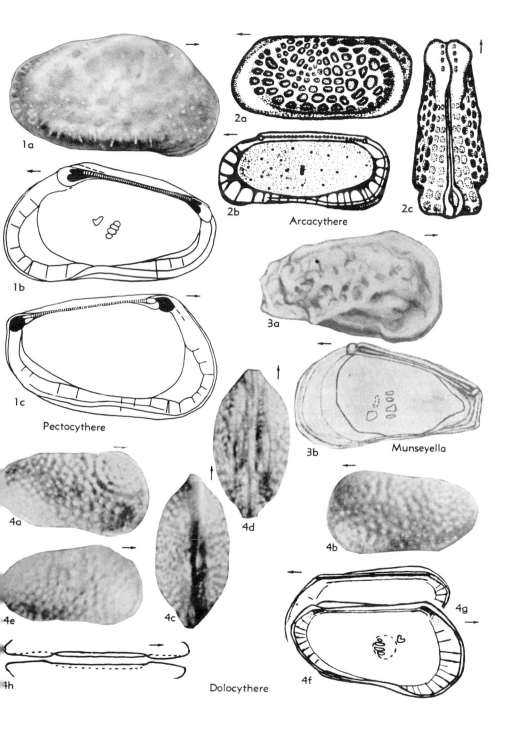

Fig. 246. Pectocytheridae (p. *Q*318-*Q*320).

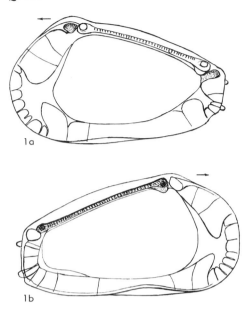

Munseyella

Fig. 246A. Pectocytheridae (p. Q320).

ing steplike sides and weak terminal thickenings corresponding to teeth, RV with shallow terminal impressions. *U.Perm.,* Eu.(USSR).——Fig. 247, *1a-d.* *P. oblonga* Schneider, Volga region; *1a,b,* LV lat., vent., ×45; *1c,d,* LV and RV hinge, ×53 (Schneider, 1947).——Fig. 247,*1e. P. tuberculata* Kashevarova, N.Dvina Basin; LV lat., ×43 (192).——Fig. 247,*1f,g. P. cornuta* Kashevarova, Troitsk-Petchora; *1f,g,* RV lat., vent., ×43 (192).

Family PROGONOCYTHERIDAE
Sylvester-Bradley, 1948

[*nom. transl.* Howe, herein (*ex* Progonocytherinae Sylvester-Bradley, 1948] [Materials for this family prepared by H. V. Howe, Louisiana State University, with contributions from W. A. van den Bold, Louisiana State University, and R. A. Reyment, University of Stockholm]

Ovate to subtriangular in lateral view, rather plump in dorsal view; surface usually pitted or reticulate but may be nearly smooth, some species with nodes or more commonly with longitudinal folds or ribs; eye tubercles prominent in only a few genera. Hingement entomodont or merodont; marginal areas rather wide, radial canals not abundant, tending to be straight, 2 to 20 at anterior end; muscle scars (where described) in nearly vertical row of 4 oval spots for adductors, an oval or crescent-shaped an-

tennal scar in front, and more rarely an oval mandibular scar. [A strictly synthetic family based largely on hingement and external form, since finer details of the carapace belonging to many Jurassic and Lower Cretaceous genera assigned here have not been described.] ?*Penn., Jur.-Rec.*

On the basis of hingement the family may be divided into the subfamilies Progonocytherinae with strictly entomodont hingement, in which the median element is divided into two portions, the anterior one being crenulate; and the Protocytherinae, in which the hingement is merodont, either antimerodont or hemimerodont. In surveying genera assigned to the Progonocytheridae it is appropriate to mention a few forms which are excluded from the family. These are *Cyprideis,* which has entomodont hingement, *Perissocytheridea,* and some species of *Clithrocytheridea,* which may be difficult to distinguish from the Protocytherinae. Following custom these genera are included in the Cytherideidae, a procedure that appears justified, as for nearly a hundred years *Cyprideis* was confused with *Cytheridea,* the type genus of Cytherideidae. The case for *Perissocytheridea,* however, is not so clear, for its hinge is strongly anti-

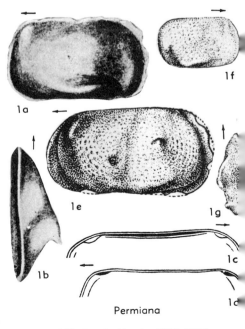

Permiana

Fig. 247. Permianidae (p. Q320-Q322).

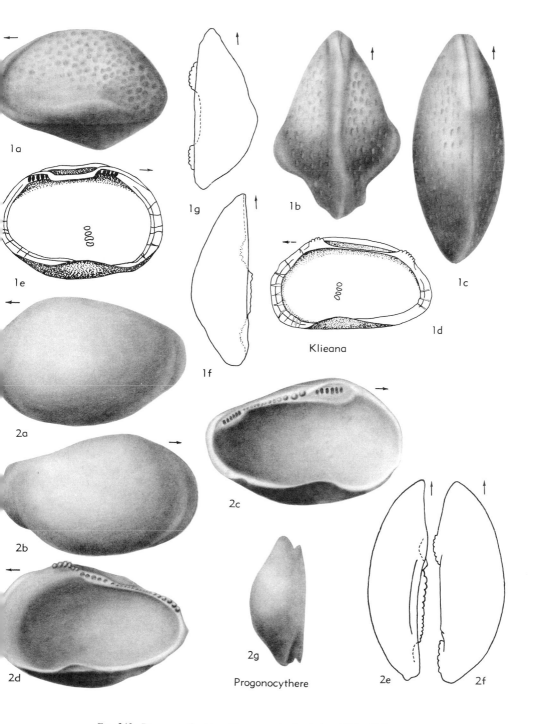

FIG. 248. Progonocytheridae (Progonocytherinae) (p. Q326-Q329).

merodont, its shape is unlike that of *Cytheridea* and close allies, and its ornamentation is much like that of protocytherine genera. The same is true of reticulate species of *Clithrocytheridea* having longitudinal ridges on their surface; these may well belong to a different genus classified as a Tertiary representative of the Progonocytheridae.

The Progonocytherinae appear to have

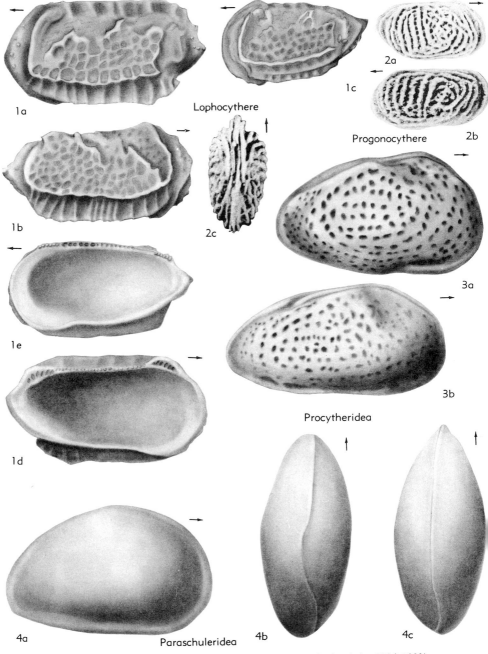

FIG. 249. Progonocytheridae (Progonocytherinae, Protocytherinae) (p. Q326-Q330).

developed from forms included in the Protocytherinae by a strengthening of the anterior crenulations of the middle element of the hinge. Both subfamilies, which are represented in early and middle Mesozoic ostracode assemblages, give evidence of being ancestral to other Upper Cretaceous and Tertiary families. Thus the Xestoleberi-

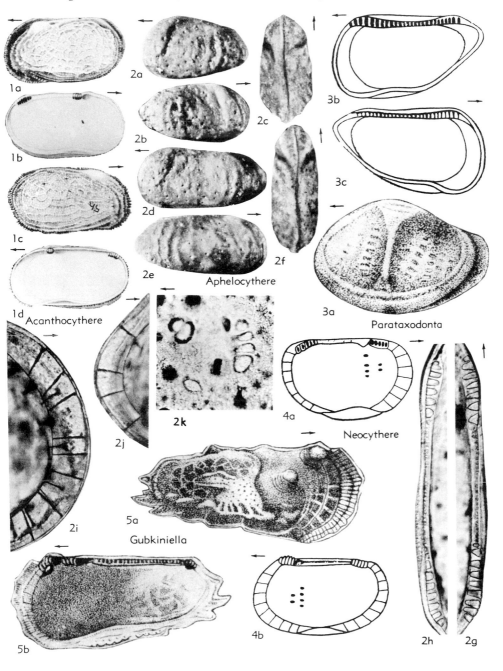

FIG. 250. Progonocytheridae (Progonocytherinae, Protocytherinae) (p. Q326-Q330).

didae, Cytherideidae, and Cytheridae, with their merodont hinges may be thought of as special developments from the Protocytherinae. The Leptocytheridae, Brachycytheridae, Hemicytheridae, Cytherettidae, Trachyleberididae, Campylocytheridae, and Schizocytheridae, with their amphidont and schizodont hinges are clearly developed from the older entomodont hingement of the Protocytherinae.

Subfamily PROGONOCYTHERINAE Sylvester-Bradley, 1948

[Includes Xenocytherinae MANDELSTAM, 1960]

Hinge entomodont. *Jur.-Rec.*

Progonocythere SYLVESTER-BRADLEY, 1948 [*P. stilla*]. Shape similar to *Brachycythere*, ventrally very plump; may be smooth, longitudinally wrinkled, or reticulate; hinge entomodont; muscle scars a vertical row of 4 with 2 in front; marginal area and radial canals not described; no eye tubercle. *Jur.*, Eu.——FIG. 248,2. *P. stilla*, M.Jur., Eng.; *2a,b*, LV lat., RV lat.; *2c,d*, LV int., RV

int.; *2e-g*, LV dors., RV dors., LV post.; all ×75 (364).——FIG. 249,2. *P. hieroglyphica* SWAIN & PETERSON, U.Jur. (Sundance), USA(S.Dak.); *2a,b*, RV and LV ext.; *2c*, ♀ carapace dors., ×50 (357).

Acanthocythere SYLVESTER-BRADLEY, 1956 [*Cythere sphaerulata* JONES & SHERBORN, 1888]. Carapace with shape similar to that of *Echinocythereis* but with lobodont hinge, which is an intermediate stage in evolution of the typical entomodont hinge of the Progonocytheridae to typical amphidont hingement of the Trachyleberididae; carapace medium in size, plump, with straight long margins and rounded ends; surface covered with fine spines and apparently possessing broad eye tubercles. Normal canals large, sparse, about 20 to the valve; marginal area widest anteriorly, with few straight radial canals; muscle scars unknown. *M. Jur.(Bathon.)*, Eng.——FIG. 250,1. *A. sphaerulata* (JONES & SHERBORN); *1a,b*, ♂ LV lat., int.; *1c,d*, ♂ RV lat., int.; all ×47 (367).

Centrocythere MERTENS, 1956 [*C. denticulata*]. Egg-shaped, surface with concentric reticulations and nodes; obliquely rounded in front, narrower

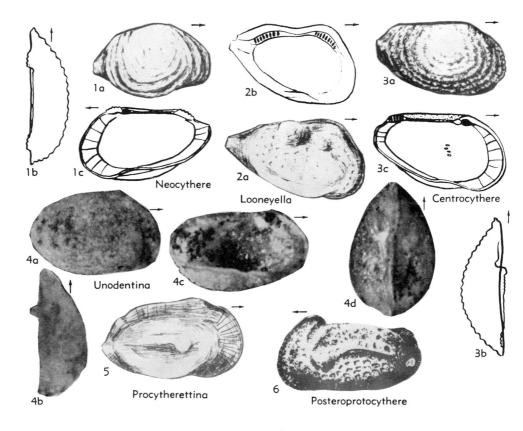

FIG. 251. Progonocytheridae (Progonocytherinae, Protocytherinae) (p. Q326-Q330).

behind. Hinge entomodont, crenulate in front part of median element; with accommodation groove in LV; marginal areas regular, radial canals sparse; muscle scars in near-vertical row of 4; mandibular and antennal scars not known. *L.Cret.*, Ger.——Fig. 251,3. **C. denticulata*, M.Alb.; *3a-c*, RV lat., LV dors., LV int., ×50 (34).

?Gubkiniella Kuznetsova, 1956 [**G. miranda*] [=*Gubkinella* Kuznetsova, 1956 (original variant spel.)]. Elongate-ovate in side view, widest anteriorly; surface reticulate with muscle swelling in front of center, distinct eye spot. Marginal area wide, with about 25 straight radial canals; hinge modified entomodont, consisting in RV from front to back of crenulate tooth, socket, crenulate tooth, another socket, crenulated median furrow, posterior crenulate tooth; muscle scars not known. Sexual dimorphism pronounced, males longer than females, tapering backward, more strongly ornamented and with larger rear spines. *L.Cret.(Barrem.)*, SW. Asia-SE. Eu. (Caucasus-Caspian).——Fig. 250,5. **G. miranda*, SE.Caucasus; *5a,b*, RV lat., RV int., ×80 (50). [Howe-Bold-Reyment.]

Lophocythere Sylvester-Bradley, 1948 [**Cytheridea ostreata* Jones & Sherborn, 1888]. Sexually dimorphous, more or less rectangular, surface reticulate with one or more keels parallel to long margins; eye node distinct; hinge entomodont; other internal details not described. *M.Jur.-U.Jur.*, Eu.——Fig. 249,1. **L. ostreata* (Jones & Sherborn), M.Jur., Eng.; *1a,b*, LV lat., RV lat.; *1c*, juv. LV lat.; *1d,e*, LV int., RV int.; all ×70 (364).

Neocythere Mertens, 1956 [**N. vanveeni*] [=*?Trochiscus* Mandelstam in Mandelstam et al., 1956 (non Heyden, 1826); *?Trochinius* Mandelstam, 1957 (nom. subst. pro *Trochiscus* Mandelstam, 1956)]. Shape and ornamentation similar to that of *Centrocythere* but rear element of middle hinge bar of LV crenulate; other internal characters also similar. *L.Cret.(Barrem.)-U.Cret.(Maastricht.)*, Eu.(Ger.-Holl.-Fr.-?Eng.-?Urals)-?W.Sib.-?C.Asia.——Fig. 251,1. **N. vanveeni*, U.Alb., Ger.; *1a-c*, RV lat., dors., int., ×50 (34). ——Fig. 250,4. *N. punctatula* (Jones) (fide Mandelstam but identity with *Cythere punctatula*, Eng., type species of *Trochinius*, doubtful), Santon., N.Urals; *4a,b*, LV int., RV int., ×43 (50). [Howe-Bold-Reyment.]

Posteroprotocythere Mandelstam in Mandelstam et al., 1958 [**Orthonotacythere proparia* Sharapova, 1939 (?=*O. propria* Sharapova, 1939, p. 43, pl. 3, fig. 32)]. Carapace elongate-ovate, LV larger than RV, anterior end rounded, posterior end produced, dorsal margin straight; reminiscent of *Protocythere* but differs in having hinge subdivided into 4 elements, LV with anterior crenulate socket, postadjacent elongate tooth divided into 5 parts, median crenulate bar, and posterior socket divided into 5 parts. *L.Jur.-U.Cret.*, W.Eu.-SE.Eu.(Caucasus)-SW.Asia(Kazakhstan).—— Fig.

251,6. **P. propria* (Sharapova), L.Cret.(Neocom.), SE.Russia; LV lat., ×50 (34). [Bold.]

Procytherettina Mandelstam in Mandelstam et al., 1958 [**Cythereis solus* Sharapova, 1939]. Carapace kidney-shaped, LV overlapping; valves 3-ribbed, with subcentral tubercle; hinge in LV with anterior socket divided into 5 parts, postadjacent conical tooth with 5 crenulations, median crenulate bar, posterior ovate socket divided into 5 parts; marginal area moderately broad, pore canals moderately numerous, thin, equally spaced, coupled in some. [Differs from *Veenia* in having crenulate hinge teeth and from *Protocythere* in having 4 hinge elements.] *U.Cret.(Cenom.)*, E.Eu.(M. Volga-Ozinki).——Fig. 251,5. **P. sola* (Sharapova); RV lat., ×50 (34). [Bold.]

Unodentina Malz, 1958 [**Macrodentina? spinosa* Schmidt, 1955]. Shape and most other characters similar to *Acanthocythere*; smooth anterior tooth in RV and smooth tooth in anteromedian part of LV hinge. *M.Jur.-U.Jur.*, W.Eu.——Fig. 251,4. **U. spinosa* (Schmidt), U.Jur., Ger.; *4a,b*, RV lat., dors.; *4c*, LV int.; *4d*, carapace dors.; all ×60 (320). [Reyment.]

Xenocythere Sars, 1925 [**Cythere cuneiformis* Brady, 1868]. Carapace wedge-shaped, highest in front, ventral face broad and flattened, valves subequal, moderately strong. Hinge of LV with terminal crenulate pits and between them bar bearing 2 or 3 small teeth at anterior end; marginal areas widest at ends where line of concrescence departs from inner margin; radial canals few; muscle scars in vertical row of 4 with 2 antennal and single mandibular scar in front. (314, 375). *Rec.*, N.Atl.——Fig. 256,8. **X. cuneiformis* (Brady); *8a,b*, carapace L, R, ×75; *8c-e*, RV int. with pore canals, hinge, muscle scars, ×120 (375).

Subfamily PROTOCYTHERINAE Lyubimova in Lyubimova & Khabarova, 1955

[Although cited as a new subfamily introduced by Mandelstam, 1960 (USSR Treatise), the name Protocytherinae was first published by Lyubimova in 1955.] [=?Palaeocytherididae Mandelstam, 1947; includes Pleurocytherinae, Parataxodontinae Mandelstam, 1960]

Hinge merodont. *Jur.-Cret.*

Protocythere Triebel, 1938 [**Cytherina triplicata* Roemer, 1841] [=*Cytherettina* Mandelstam, 1956 (obj.)]. Carapace elongate, LV much larger than RV; surface with 3 longitudinal ridges or swellings. Hinge antimerodont; marginal areas broad, with long radial pore canals upturned in upper part of anterior end; muscle scars in vertical row of 4, with heart-shaped antennal scar in front. *U.Jur.-L.Cret.*, Eu.——Fig. 252,1. **P. triplicata* (Roemer), L.Cret., Ger.; *1a,b*, ♀ LV lat., RV lat.; *1c*, ♀ carapace dors.; *1d,e*, ♂ LV lat., RV lat.; *1f,g*, ♀ LV int., RV int.; *1h,i*, ♀ LV dors., RV dors.; all ×50 (80).——Fig. 253,2. *P. quadricarinata* Swain & Peterson, U.Jur.(Sundance F.), USA(Wyo.); *2a-c,* carapace R, L, vent., ×50 (354).

Aphelocythere Triebel & Klingler, 1959 [**A. undulata*]. Strongly dimorphous, ?males bigger and more elongated than ?females; carapace of medium size, irregularly quadrangular in side view, greatest height just in front of mid-point, LV slightly larger than RV, overhanging it dorsally and in upper part of posterior end; anterior margin broadly rounded, posterior end narrowly rounded to bluntly angled in lower half and almost flat in upper half, dorsal margin of ?females almost straight or weakly convex; surface with broad flat rib near anterior margin beginning at anterodorsal angle and fading below mid-line, central area of valves without coarse ornament, posterior area with few curved vertical riblets; eye spots and internal eye sockets apparently lacking. Inner margin and line of concrescence coincident; marginal pore canals undivided, widely separated, few; true submarginal canals lacking; normal canals large, sieve-shaped; 4 adductor muscle spots in slightly curved vertical row, lowermost scar largest, also frontal pair that may coalesce; RV hinge with terminal low, notched tooth plates and finely crenulated median furrow. [Marine.] *L.Jur.-M.Jur.,* W.Eu.——Fig. 250,2. **A. undulata,* L. Jur., NW.Ger.; *2a-c,* ♀ carapace L, R, dors., ×47; *2d-f,* ♂ carapace L, R, dors., ×47; *2g,h,* RV and LV hinge, int., ×145; *2i,j,* LV int. ant., post., ×195; *2k,* muscle scars, RV int., ×290 (382). [Reyment.]

Hutsonia Swain, 1946 [**H. vulgaris*]. Carapace subpyriform in appearance from side, LV larger than RV; sides pitted or reticulate, venter ribbed, bisulcate above muscle scars (vertical row of 4 with crescent-shaped antennal scar in front); hinge hemimerodont. *Jur.,* N.Am.——Fig. 253,3. **H. vulgaris,* U.Jur., La.; *3a-c,* carapace R, L, and dors.; *3d,* LV int.; *3e,* carapace R; all ×70 (350).

Klieana Martin, 1940 [**K. alata*]. Small, with distinct sexual dimorphism; LV larger than RV, plump females with strongly developed winglike processes on ventral side of both valves, producing broad arrowhead outline in view from above; surface pitted. Hinge hemimerodont; duplicature rather narrow, with widely spaced radial canals in pairs; muscle scars in slightly oblique row of 4.

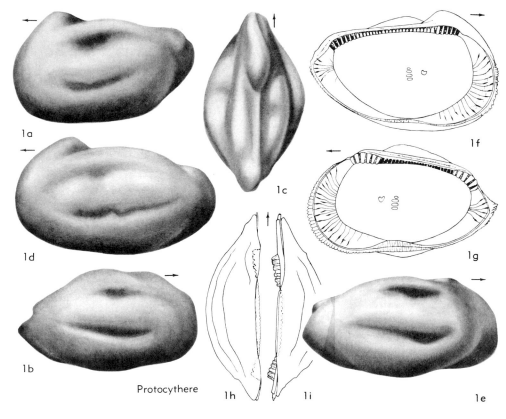

1a 1d 1b

1c

Protocythere 1h 1i

1f 1g 1e

Fig. 252. Progonocytheridae (Protocytherinae) (p. *Q*327).

Jur.-Cret., Eu.——FIG. 248,*1.* *K. alata,* U.Jur., Ger.; *1a,b,* ♀ carapace L, dors.; *1c,* ♂ carapace dors.; *1d,g,* RV int., dors.; *1e,f,* LV int., dors.; all ×80 (51).

Looneyella PECK, 1951 [*Cythere monticula* JONES,

1893]. Subtrigonal to pyriform in side view, dorsal and ventral outlines converging toward well-developed caudal process, anterior end obliquely rounded; alate, with spines on certain species and in known forms reticulate on lateral surface and

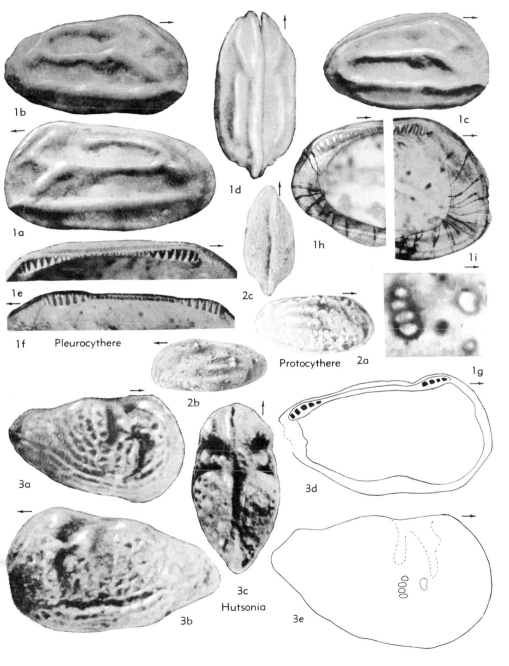

FIG. 253. Progonocytheridae (Protocytherinae) (p. Q327-Q330).

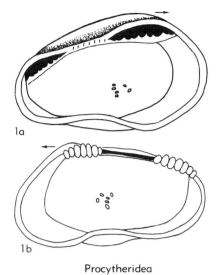

Procytheridea

FIG. 254. Progonocytheridae (Protocytherinae)
(p. Q330).

longitudinally ribbed on ventral surface. Hinge hemimerodont; muscle scars not known; marginal areas regular and radial canals few. *L.Cret.*, N. Am.——FIG. 251,2. *L. monticula* (JONES), Wyo.; *2a,b*, RV lat., LV int., ×50 (34).

Merocythere OERTLI, 1957 [*Clithrocytheridea plena* SCHMIDT, 1954]. Plump, subpyriform in lateral outline, surface pitted, no eye tubercles. Hinge hemimerodont; other internal details lacking. *Jur.*, Ger.

Paraschuleridea SWARTZ & SWAIN, 1946 [*P. antumbonata*]. Shape somewhat like that of the Cytherideidae, but muscle scars in vertical row of 4 with single large scar in front; marginal areas and canals not known. [Based on single specimen. Placed here because of hemimerodont hinge.] *Jur.*, N.Am.——FIG. 249,4. *P. anumbonata*, U.Jur., La.; *4a-c*, carapace R, dors., vent.; all ×60 (77).

?Parataxodonta MANDELSTAM, 1956 [*P. uralensis*]. Carapace elongate oval roughly, with maximum height just behind anterior third, with transverse concavity in side middle; ventral margin broadly rounded, dorsal margin strongly and irregularly arched, posterior half concave, anterior margin broadly and rather irregularly rounded with midpoint lower than posterior; rear margin blunt; LV larger than RV. Anterior zone of concrescence with only few pore canals, no vestibule. Eye spot distinct; surface smooth except for coarse, symmetrically aligned reticulations and weak ribs. LV hinge with crenulated bar with terminations becoming wider and toothlike. *L.Cret.(Apt.-Alb.)*, USSR (Urals-Kazakhstan). —— FIG. 250,3. *P.

uralensis, Urals; *3a*, LV lat., ×63; *3b,c*, RV int., LV int., ×43 (238). [REYMENT.]

Pleurocythere TRIEBEL, 1951 [*P. richteri*] [=?*Annosacythere* KUZNETSOVA, 1957]. Ovate, LV larger than RV and with dorsal keel that is missing on RV; lateral surface with slanting front rib and 3 long ribs connecting anteriorly. Hinge antimerodont; radial canals long, curved, with 2 aberrant canals near posterior hinge terminations; muscle scars in row of 4 with 2 in front. *Jur.*, Eu.—— FIG. 253,1. *P. richteri*, M.Jur., Ger.; *1a*, ♂ LV lat.; *1b*, ♀ RV lat.; *1c,d*, ♀ carapace R, dors.; *1e,f*, LV and RV hinge; *1g*, LV int., muscle scars; *1h,i*, LV marginal areas post., ant.; *1a-d*, ×75; *1e,f,h,i*, ×150; *1g*, ×300 (377).

Procytheridea PETERSON, 1954 [*P. exempla*]. Ovate, highest at anterior cardinal angle, ventral margin sinuous; surface reticulate, with tendency to develop longitudinal ridges. Hinge antimerodont; marginal areas wide, few canals, line of concrescence nearly coinciding with inner margin; muscle scars in vertical row of 4 with several others in front. *Jur.*, N.Am.——FIGS. 249,3; 254,1. *P. exempla*, U.Jur., Wyo.; 249,3a,b, ♀ and ♂ carapace R; 254,1a,b, LV int., RV int.; all ×100 (273).

Family PSAMMOCYTHERIDAE Klie, 1938

[*nom. transl.* HOWE, herein (*ex* Psammocytherinae KLIE, 1938)] [Materials for this family prepared by H. V. HOWE, Louisiana State University]

Shell low, elongate, delicate, transparent. Hinge adont. No eye. [Marine.] *Rec.*, Eu.

Psammocythere KLIE, 1936 [*P. remanei*]. No sexual dimorphism; shells similar; height about 2/5 length; radial pore canals widely spaced in broad anterior marginal area. *Rec.*, Helgoland.—— FIG. 255,1. *P. remanei*; *1a,b*, RV ext. (by reflected and transmitted light); *1c*, RV dors.; all ×100 (219).

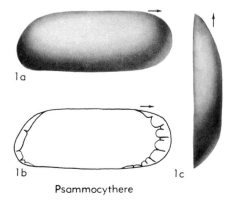

Psammocythere

FIG. 255. Psammocytheridae (p. Q330).

Family SCHIZOCYTHERIDAE Howe, n. fam.

[Materials for this family prepared by H. V. Howe, Louisiana State University, with some contribution from P. C. Sylvester-Bradley, University of Leicester, and R. A. Reyment, University of Stockholm]

Characterized by schizodont dentition; marginal areas relatively broad, crossed by few radial canals; surface rough, commonly reticulate or strongly pitted, tending to develop longitudinal ridge near venter. Muscle scars very difficult to discern because of surface ornamentation but apparently consisting of a vertical or curved row of 4 adductors, in front of which are rounded antennal and mandibular scars. *U.Cret.-Rec.*

Schizocythere Triebel, 1950 [*S. hollandica*]. Carapace small to medium in size, ovate in side view, posterior extremity being slightly angulated near middle, above which it is weakly concave in RV, ovate in dorsal view; surface pitted to strongly reticulate, with distinct eye tubercle and in most species a tendency to have rib separating lateral and ventral surfaces. Hinge schizodont, in RV with split anterior tooth, deep socket divided in 2 parts at end, crenulate hinge furrow and lobed posterior tooth; anterior tooth bifid in LV; marginal areas very broad, with about 5 straight radial canals on anterior part; normal canals rather large, widely spaced, sieve-like; muscle scars in somewhat curved row of 4 adductors, in front of which are single rounded antennal and mandibular scars very difficult to see. *Eoc.-Mio.*, Eu.——Fig. 256, *1a-c. S. batjesi* Keij, Eoc.(Led.-Barton.), Belg.; *1a*, LV lat. (Barton.), ×67; *1b*, carapace dors. (Led.), ×77; *1c*, LV int. (Barton.), ×75 (42). ——Fig. 256,*1d. S. tessellata* (Bosquet), Eoc. (Lutet.), Fr.(Paris Basin); RV lat., ×65 (42).—— Fig. 256,*1e-g. *S. hollandica*, Mio., Holl.; *1e*, RV lat., ×45; *1f*, RV hinge, ×100; *1g*, RV radial pore canals, ×137 (all 376).

Amphicytherura Butler & Jones, 1957 [*Cytherura? dubia* Israelsky, 1929]. Hinge schizodont. Like *Costa* but smaller, less elongate, and with median ridge straight or curved, concave upward, less well defined than in *Costa*. Externally resembles *Eucytherura. U.Cret.*, Eu.-N.Am.——Fig. 257,*1a-c. *A. dubia* (Israelsky), USA(Ark.); *1a-c*, LV lat., RV int., carapace dors.; ×80, ×70, ×77 (34). ——Fig. 257,*1d. A. limburgensis* Howe & Laurencich, Belgium; LV lat.; ×80 (34). [Sylvester-Bradley.]

Neomonoceratina Kingma, 1948 [*N. columbiformis*]. Generally considered to be subgenus of *Paijenborchella* because similar in most features but with upturned caudal process (46). *Mio.-Rec.*, E.Indies-Eu.——Fig. 256,*5. *N. columbiformis*, Rec., Sumatra; *5a,b*, RV int., LV int., ×50; *5c*, RV dors., ×50.——Fig. 256,6. *N.*

mediterranea Ruggieri, Mio., Italy; *6a,b*, LV lat., int., ×50; *6c*, hinge, ×50 (302).

Paijenborchella Kingma, 1948 [*P. iocosa*] [=*Payenborchella* Keij, 1953]. Carapace with obliquely rounded anterior end and long caudal process near lower posterior margin, nearly vertical sulcus marking middle of valves at position of adductor muscles; valves widest near venter where ridge tends to form ala or develop spine, another horizontal ridge crossing sulcus near middle of valves. Hinge heavy, schizodont, with split anterior teeth, crenulate middle element, and lobed posterior tooth in RV; marginal areas broad, with very few radial canals; adductor scars on median ridge, in front of which appear to be single antennal and mandibular scars (148, 220). *U.Cret.-Rec.*, E.Indies-Eu.-N.Am.——Fig. 256,2. *P. iocosa*, Rec., E. Indies(Java), *2a*, carapace dors.; *2b,c*, LV int., RV int.; all ×30 (46).——Fig. 256,3. *P. eocaenica* Triebel, Eoc.(Barton.), Belg.; *3a*, RV lat., ×73 (42); *3b,c*, hinge and radial pore canals, enlarged (375).——Fig. 256,4. *P. longicosta* Keij, Eoc.(Lutet.), Belg.; *4a,b*, ♂ carapace R, dors., ×78; *4c*, ♀ carapace dors., ×78; *4d*, ♀ LV int., ×78 (42).

Palmenella Hirschmann, 1916 [*Cythereis limicola* Norman, 1865] [=*Kyphocythere* Sars, 1925]. Carapace subrectangular, dorsal margin marked by raised cardinal angles, compressed in dorsomedian region, much widened ventrally in form of subalate ridge; surface pitted or reticulate or both. Hinge schizodont; marginal areas broad, with few wavy radial pore canals; muscle scars in subvertical row of 4 with 2 scars in front (314, 375, 172). *Plio.-Rec.*, Eu.-N.Am.——Fig. 256,7. *P. limicola* (Norman); *7a,b*, carapace L, dors., ×40; *7c,d*, hinge and radial pore canals, ×67 (172).

Family SINUSUELLIDAE Kashevarova, 1958

[*nom. transl.* van den Bold, herein (*ex* Sinusuellinae Kashevarova, 1958)] [Materials for this family prepared by W. A. van den Bold, Louisiana State University]

Large, rectangular, with rounded antero- and posteroventral margins; swollen area overhanging mid-ventral part of valves; surface smooth or ornamented. Zone of concrescence forming a flattened fringe. *U. Perm.*

Sinusuella Spizharsky, 1939 [*S. ignota*]. Dorsal outline sinusoid, with process pointed backward; LV overlapping RV only along dorsal margin, entire free border with radially striate fringe or rim; surface smooth or reticulate. LV hinge with crenulate bar, teeth at extremities of bar, corresponding depressions in RV; muscle scars in vertical row of 4 with 2 scars in front. [The muscle-scar pattern of this poorly illustrated, inadequately known genus indicates placement within the

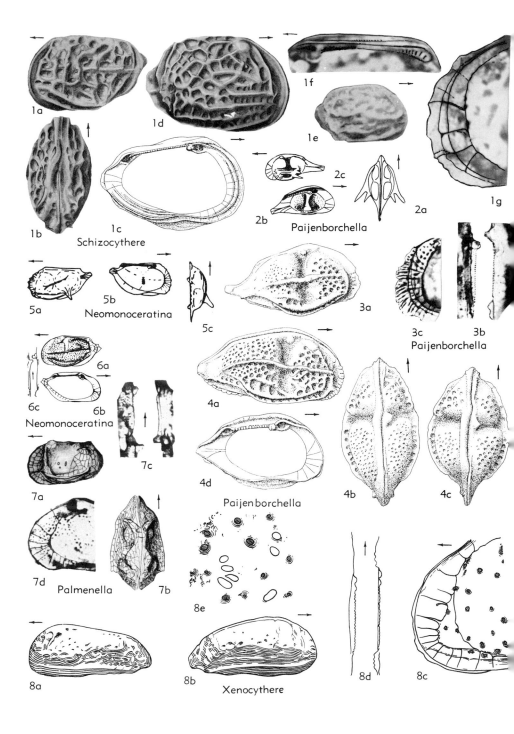

FIG. 256. Progonocytheridae, Schizocytheridae (p. Q327, Q331).

Cytheracea (Howe).] *U.Perm.(Ufim. - Tatar.),* USSR(Sukhon Basin).

Family TOMIELLIDAE Mandelstam in Yavorsky, 1956

[Materials for this family prepared by W. A. van den Bold, Louisiana State University] [Includes Iniellinae Mandelstam, 1960]

Strongly varying in shape, ends rounded, dorsal margin straight, ventral margin straight or concave; valves with submedian transverse groove, commonly curved at posterior side, some genera with subcentral tubercle in front of this groove; surface reticulate, pitted, or smooth. Zone of concrescence well developed, radial pore canals widely spaced, vestibule in anterior end; muscle scars in vertical row of 4 with 2 in front; LV hinge with shallow groove which receives sharpened dorsal margin of RV. *Perm.*

Tomiella Spizharsky, 1937 [**T. yavorskyi* (=**Kirkbya oblongata* Jaishevsky, 1927)]. Anterior end rounded, commonly higher than posterior end, dorsal margin straight, ventral margin concave; valves with submedian transverse groove, usually obliquely rounded toward rear, some species with tubercle in front of this groove. Muscle scars on interior ridge that corresponds to external groove, 4 elongate scars (much like those of *Timiriasevia*) in oblique row; radial pore canals better developed in anterior end, inner lamella weak; LV hinge with narrow, smooth groove, slightly widened anteriorly into which fits sharp dorsal margin of RV. *Perm.,* USSR(Kuznetsk-Tungus region).——Fig. 258,*1.* **T. oblongata* (Jaishevsky), Kuznetsk; *1a,* RV lat., ×20 (337); *1b,* RV lat. (?another specimen), ×43 (238).

Iniella Mandelstam in Yavorsky, 1956 [**Leperditia kuznetskiensis* Spizharsky, 1937]. Elongate, subelliptical, anterior end higher than posterior or equally high, both ends rounded; dorsal margin straight, parallel to ventral margin or converging posteriorly. Muscle scars in vertical row of 4 with 2 in front; LV hinge with weak longitudinal depression in dorsal margin into which fits dorsal margin of RV. *Perm.,* USSR(Kuznetsk-Tungus region).——Fig. 258,*3.* **I. kuznetskiensis,* Kuznetsk; *3a,b,* LV lat., RV lat., ×43 (238).

Kemeroviana Mandelstam in Yavorsky, 1956 [**K. argulata*]. Irregularly elliptical to rounded rectangular, LV larger than RV, ends usually equally high, rounded; dorsal margin straight or sloping toward posterior end, ventral margin obliquely convex; surface with weak ribs that tend to be diagonal. Muscle scars situated in subcentral transverse groove. *Perm.,* USSR(Kuznetsk Basin).——Fig. 258,*5. K. argulata,* Kuznetsk; LV lat., ×43 (238).

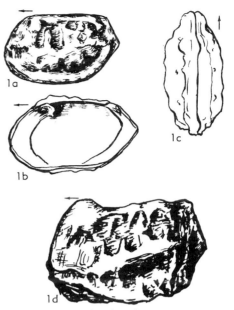

Fig. 257. Schizocytheridae (p. *Q331*).

Suriekovella Mandelstam, 1958 [**Iniella limbata* Mandelstam, 1956]. Carapace elongate, large, elliptical or elongate-ovate, anterior end with broad marginal area, also broad in ventral margin. Surface smooth. [Differs from *Iniella* in its straight dorsal margin and upturned posterior part.] *Perm.,* WC.Asia(Kuznetsk Basin).——Fig. 258,*4. *S. limbata* (Mandelstam), Kuznetsk; RV lat., ×40 (238). [Bold.]

Tomiellina Mandelstam in Yavorsky, 1956 [**T. umbrata*]. Elongate, rectangular, LV larger than RV, anterior end slightly higher than posterior, both rounded; dorsal margin straight, ventral convex, upwardly bent in posterior end, central portion of valve with longitudinal concavity with angular to rounded reticulations. [Differs from *Tomiella* in lacking transverse groove, subcentral tubercle and blunt antero- and posterodorsal angles.] *U.Perm.,* USSR(Kuznetsk Basin).——Fig. 258,*2. T. umbrata,* Kuznetsk; LV lat., ×43 (238).

Family TRACHYLEBERIDIDAE Sylvester-Bradley, 1948

[*nom. correct.* Sylvester-Bradley & Harding, 1954 (*pro* Trachyleberidae Sylvester-Bradley, 1948)] [=Cythereidinae Berousek, 1952 (*nom. correct.* Sylvester-Bradley & Harding, 1954, *pro* Cythereisinae Berousek, 1952)] [Materials for this family prepared by P. C. Sylvester-Bradley, University of Leicester, with aid on some genera by H. V. Howe, Louisiana State University, and R. A. Reyment, University of Stockholm]

Sexual dimorphy common, males longer than females. Carapace subrectangular, dorsal and ventral margins parallel or slightly convergent toward rear, anterior margin broadly rounded, posterior margin subtri-

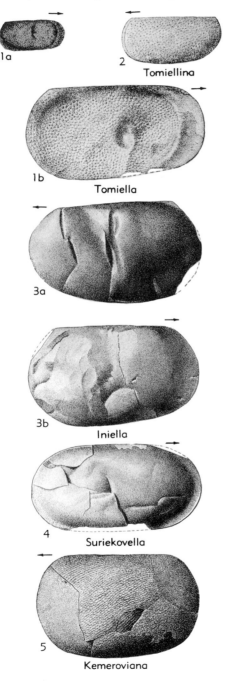

1a

2
Tomiellina

1b
Tomiella

3a

3b
Iniella

4
Suriekovella

5
Kemeroviana

FIG. 258. Tomiellidae (p. Q333).

angular (or caudate in a few genera), posteroventrally more or less produced. Eye tubercles and ocular sinuses well developed (except in *Idiocythere*). Internal muscle-scar pit represented on exterior by an elevated process (subcentral tubercle). Carapace heavily ornamented with spines or ridges or both, or may be reticulate, or (in a few genera) smooth. Details of shape and ornament (coupled with variations in hinge, muscle-scar pattern and duplicature) are used to distinguish genera. Hinge in post-Jurassic adults strongly amphidont, juvenile molts merodont; accommodation groove narrow or absent; duplicature of average width, vestibule narrow or absent; normal pore canals large, of sieve-type, radial pore canals numerous, rather crowded, some crossing each other, in many species majority of canals widening at about middle; muscle-scar pattern mostly with vertical row of 4 just behind muscle-scar pit, and single U-shaped scar or pair of oval antennal scars within pit immediately in front; other genera show break up into greater number of scars. Species of this family are abundant in shallow-water and littoral zones, and extend into deep water. Abundant fossils from Cretaceous onward. Strong and characteristic ornament makes species easy to recognize and valuable for zonal stratigraphy. Apparently descended from *Oligocythereis* (M.Jur.). Caudal process not normally developed. Presence of subcentral tubercle and absence of caudal process serve to distinguish family from Hemicytheridae, but *Quadracythere* and *Orionina* have both caudal process and subcentral tubercle, and are here included in Trachyleberididae. Most Hemicytheridae have one of the adductor muscle scars divided, whereas in most Trachyleberididae they are undivided. ?*L.Jur., M.Jur.-Rec.*

Trachyleberis BRADY, 1898 [*Cythere scabrocuneata BRADY, 1880]. Hinge holamphidont. Ornament of spines, tubercles or blades, dominated by subcentral tubercle, and not arranged in longitudinal lines except on venter; in some species subsidiary reticulations also occur. *Paleoc.-Rec.,* cosmop.——— FIG. 259,1. *T. scabrocuneata (BRADY), Rec.,* Japan; *1a-c,* ♂ LV lat., ♂ RV lat., ♀ RV lat.; *1d,e,* ♂ LV int., ♀ RV int.; *1f,g,* ♂ LV dors., ♀ RV dors.; all ×50 (Sylvester-Bradley).

Actinocythereis PURI, 1953 [*Cythere exanthemata ULRICH & BASSLER, 1904]. Hinge holamphidont. Ornament like *Costa* but all ridges broken up into spines. *Eoc.-Rec.,* N.Am.——FIG. 259,2. **A. exanthemata* (ULRICH & BASSLER), Eoc., USA (Fla.); *2a,b,* carapace R, dors., ×50 (286).

Ambocythere VAN DEN BOLD, 1957 [**A. keiji*]. Carapace subquadrate, with distinct posteroventral projection; subcentral tubercle slight but

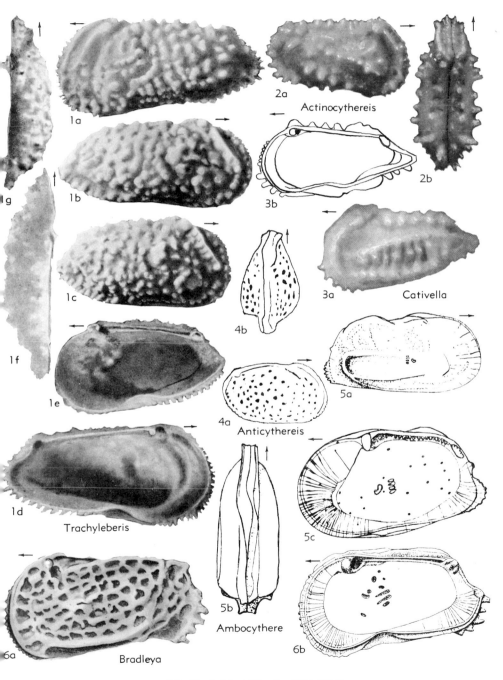

FIG. 259. Trachyleberididae (p. *Q334-Q336*).

definite; ornament of 2 or more thin carinae confined to posterior half of carapace; marginal rims sharply defined. Duplicature wide, with narrow vestibule; hinge hemiamphidont; antennal muscle scar large, crescentic, opening toward front. *Oligo.-Rec.*, Carib.——FIG. 259,5. **A. keiji*, Rec., Carib.; *5a,b*, carapace R, dors.; *5c*, RV int.; all ×90 (101).

Anticythereis VAN DEN BOLD, 1946 [*pro Pseudocythereis* JENNINGS, 1936 (*non* SKOGSBERG, 1928)] [**Pseudocythereis reticulata* JENNINGS, 1936] [=*Stephensonaria* CORYELL in STEPHENSON, 1946 (*nom. subst. pro Pseudocythereis*)]. Carapace plump, like *Echinocythereis*, but surface not spiny and RV larger than LV. *U.Cret.*, N.Am.——FIG. 259,4. **A. reticulata* (JENNINGS), USA(N.J.); *4a,b*, carapace R, dors., ×40 (34).

Archicythereis HOWE, 1936 [**Cythereis yazooensis* HOWE & CHAMBERS, 1934]. Based on a juvenile molt that is generically indeterminable, but certainly belonging to Trachyleberididae; not generally recognized as valid generic name, but useful in the vernacular when immature stages of amphidont hinges are termed the "*Archicythereis* hinge." *Paleog.-Neog.*, cosmop.——FIG. 263,1. *A. holmani* LEROY, Neog., Sumatra; *1a-c*, RV lat., int., LV int., ×24 (200).

Bradleya HORNIBROOK, 1952 [**Cythere arata* BRADY, 1880]. Carapace subquadrate, smooth or reticulate, with distinct dorsal and ventral ridges and subcentral tubercle; no caudal process. Hinge hemiamphidont, with denticulate median element; antennal muscle scar double. Line of concrescence and inner margin coincident, marginal pore canals simple, widened in middle. Sexual dimorphism recorded. *U.Cret.-Rec.*, cosmop.——FIG. 260,4. **B. arata* (BRADY), Rec., N.Z.; *4a*, LV lat.; *4b,c*, RV int., dors.; all ×50 (32).——FIG. 259,6. *B. approximata* (BOSQUET), Eoc. (Lutet.), Fr.; *6a,b*, LV lat., RV int., ×50 (42). [REYMENT.]

Buntonia HOWE, 1935 [**B. shubutaensis*, juvenile (=**Cythereis? israelskyi* HOWE & PYEATT, 1935, adult)] [=*Pyricythereis* HOWE, 1936; *Semicythereis* ELOFSON, 1944]. Hinge holamphidont, with median element smooth or denticulate. Surface smooth, or with longitudinal ridges and furrows. Carapace plump, tapering to narrow rounded posterior; eye tubercle present but not pronounced. [Included with some doubt in Trachyleberididae, as carapace lacks subcentral tubercle.]*?U.Cret., Eoc.-Rec.*, N.|Am.-Eu.-Afr. (Nigeria-Cameroons) ——FIG. 260,1a-d. **B.. shubutaensis*, Eoc., USA (La.); *1a,b*, carapace R, dors.; *1c,d*, RV int., LV int.; all ×80 (178).——FIG. 260,1e. *B. corpulenta* (BRADY & NORMAN), Rec., Swed.; RV int., ×110 (294).

Carinocythereis RUGGIERI, 1956 [**Cytherina carinata* ROEMER, 1838]. Like *Trachyleberis* but subrectangular, dorsal and ventral borders almost parallel; ornament of ridges, spines and reticulations;

subcentral tubercle bladelike, duplicature with narrow but definite anterior and posterior vestibules; hinge holamphidont tending to hemiamphidont (posterior tooth variable). [As here interpreted, the genus includes *Cythere rugipunctata* ULRICH & BASSLER, hitherto assigned to *Puriana* (=*Favella*).] *Eoc.-Rec.*, N.Am.-C.Am.-Eu.—— FIG. 260,3. **C. carinata* (ROEMER), Rec., Italy; *3a,b*, RV lat., int., ×80 (305).

Cativella CORYELL & FIELDS, 1937 [**C. navis*] [=*Navecythere* CORYELL & FIELDS, 1937]. Hinge holamphidont. Like *Costa* but smaller, tapering strongly toward rear, with acuminate posterior end; ridges high, sharply defined, perforate in type species. *Mio.*, N.Am.-C.Am.——FIG. 259,3. **C. navis*, Fla.; *3a*, LV lat.; *3b*, RV int.; ×80 (286).

Costa NEVIANI, 1928 [**Cytherina edwardsi* ROEMER, 1838; SD HOWE, 1955] [=*Rectotrachyleberis* RUGGIERI, 1952]. Hinge holamphidont. Ornament dominated by 3 subparallel ridges, with median ridge slightly diagonal to length, sloping down gently toward anterior end and bent downward rather abruptly at posterior end. Adductor muscle scars 4, in vertical row; antennal scar V-shaped, opening upward. *Mio.-Rec.*, Eu.-Asia-Afr.-?N.Am.——FIG. 261,5. **C. edwardsi* (ROEMER), Plio.-Pleist., Fr.-Italy; *5a,b*, RV lat., int., ×75 (*5a*, 301; *5b*, 42).

Cythereis JONES, 1849 [**Cytherina ciliata* REUSS, 1846 (=**Cytherina ornatissima* REUSS, 1846); SD SUTTON & WILLIAMS, 1939]. Hinge paramphidont. Type-species has ornament of low reticulations, but as currently understood, genus includes all reticulate, costate and spinose members of family with paramphidont hinge and simple muscle scar (4 vertically arranged adductor impressions, with V-shaped antennal scar in front). *L.Cret.-U.Cret.*, cosmop.——FIG. 260,2. **C. ornatissima* (REUSS), U.Cret.(Turon.), Ger., ♂ LV lat., ×50; *2b,c*, ♀ RV lat., ×50, ♂ RV lat., ×60; *2d*, RV dors., ×50 (380).——FIG. 261,1. *C. senckenbergi* TRIEBEL, L.Cret.(Hauteriv.), Ger.; RV int., ×50 (380).

Echinocythereis PURI, 1954 [**Cythereis garretti* HOWE & McGUIRT, 1935]. Like *Trachyleberis* but more inflated, with rounded rather than triangular posterior end; ornament of rounded spines superimposed on reticulations, concentrically arranged in many species; young molts may lack spines. Antennal muscle scar split into 2. *U.Cret.-Rec.*, cosmop.——FIG. 261,3. **E. garretti* (HOWE & McGUIRT), Mio., USA(Fla.); *3a,b*, LV lat., int.; *3c*, RV int.; all ×30 (178).

Henryhowella PURI, 1957 [**Cythere evax* ULRICH & BASSLER, 1904] [*pro Howella* PURI, 1956 (*non* OGILBY, 1899)]. Like *Actinocythereis* but ridges not continuing into anterior half of carapace, where spines are more or less concentrically arranged. *Mio.*, N.Am.——FIG. 261,4. **H. evax* (ULRICH &

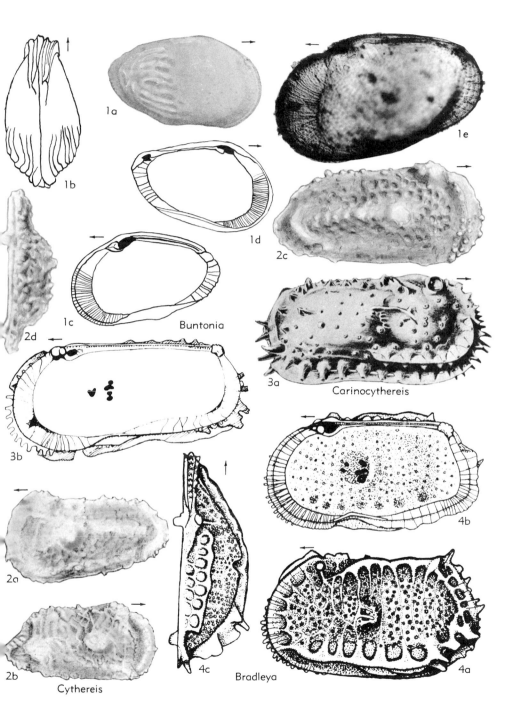

1a, 1b, 1c, 1d, 1e, Buntonia

2a, 2b, 2c, 2d, Cythereis

3a, Carinocythereis, 3b

4a, 4b, 4c, Bradleya

Fig. 260. Trachyleberididae (p. Q336).

BASSLER), USA(Fla.); *4a,b,* LV lat., int., ×40 (290).

Hermanites PURI, 1955 [*pro Hermania* PURI, 1954 (*non* MONTEROSATO, 1844)] [**Hermania reticulata* PURI, 1954]. Carapace subquadrate, reticulate, with pronounced subcentral tubercle, and dorsal and ventral ridges. Hinge holamphidont; antennal muscle scar single, crescentic. *Eoc.-Mio.,* N.Am.-Eu.——FIG. 261,2. **H. reticulatus* (PURI), *Mio.,* USA(Fla.); *2a,b,* RV lat., int.; *2c,* carapace vent.; all ×60 (288).

Hirsutocythere HOWE, 1951 [**H. hornotina*]. Like

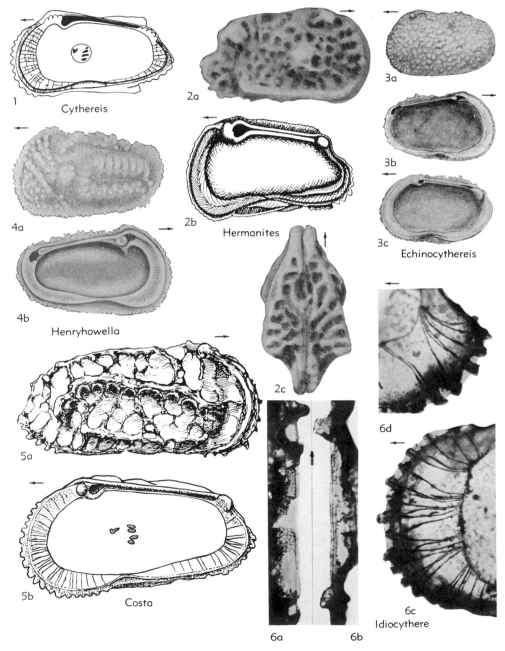

1 Cythereis
2a
3a
3b
3c Echinocythereis
4a
2b Hermanites
4b Henryhowella
2c
5a
5b Costa
6a   6b
6c Idiocythere
6d

FIG. 261. Trachyleberididae (p. Q336-Q339).

Trachyleberis but spines more numerous, duplicature wider, and radial pore canals very numerous and very fine. Typically each marginal spine has 3 radial pore canals leading to it. *Eoc.-Mio.*, USA (Fla.).——Fig. 262,6. **H. hornotina,* Mio., USA (Fla.); LV int., ×60 (Howe, n).

Idiocythere TRIEBEL, 1958 [**I. lutetiana*]. Hinge modified holamphidont, LV with additional strongly projecting conical-pessular tooth in front and above (dorsal to) socket for reception of anterior tooth of RV, with 2nd less strongly developed anterior tooth behind this socket, in usual position for amphidont dentition; hinge tubercles present on exterior of carapace, but eye tubercles not developed. Subcentral tubercle flat but readily discernible. *Eoc.*, Eu.——Figs. 261,6, 266,3. **I. lutetiana,* Lutet., Fr.; 261,6a,b, LV, RV dors. showing hinge, ×100; 261,6c,d, RV int., ant. and post., ×160; 266,3a-d, carapace L, R, dors., vent., ×75 (380).

Isocythereis TRIEBEL, 1940 [**I. fissicostis*]. Like *Cythereis* but hinge hemiamphidont, muscle scar much divided, duplicature wider, and radial pore canals sparser. *L.Cret.(Alb.),* Ger.——Fig. 262, 4a-d. **I. fissicostis;* 4a,b, carapace R, dors., ×120; 4c,d, RV dors., int., ×100 (81).——Fig. 262,4e. *I. fortinodis* TRIEBEL, Ger.; muscle scar of RV int., ×300 (81).

Leniocythere HOWE, 1951 [**L. lebanonensis*]. Subrectangular, dorsal and ventral margins slightly sinuous; surface smooth or faintly reticulate about middle; eye tubercle and subcentral tubercle faint. Hinge holamphidont. *Eoc.*, N.Am.——Fig. 262,1. **L. lebanonensis;* Eoc., USA(Fla.); LV lat., ×30 (177).

Murrayina PURI, 1954 [**M. howei*]. Carapace elongate, reticulate, subrectangular, with well-marked ocular and subcentral tubercles and marginal rims but no longitudinal ridges. Hinge holamphidont. *Mio.*, N.Am.——Fig. 262,3. **M. howei,* Mio., USA(Fla.); 3a,b, RV lat., int.; 3c, LV int.; all ×50 (3a, 287; 3b,c, 177).

?Normanicythere NEALE, 1959 [**Cythere leioderma* NORMAN, 1869]. Carapace- like *Isocythereis* in general form. Hinge line straight, oblique to dorsal margin seen from side; inner margin and line of concrescence well separated anteriorly and at posteroventral angle; radial pore canals simple. *Pleist.* (incl. *Rec.*), NW.Atl.-NE.Atl.-Br.Is.——Fig. 263,2. **N. leioderma* (NORMAN), Rec., Spitzbergen *(2a,b),* Shetland Is. *(2c,d);* 2a,b, ♀ LV int., dors., ×50; 2c,d, ♂ RV int., dors., ×50 (261). [MOORE.]

Occultocythereis HOWE, 1951 [**O. delumbata*]. Hinge holamphidont; broad duplicature traversed by radial pore canals which midway along their length divide into 2 or more branches. *Eoc.*, N. Am.-Eu.——Fig. 267,2. **O. delumbata,* Eoc., USA(Fla.); 2a,b, LV lat., int., ×50 (177).——

FIG. 262,2. *O.* sp., Lutet., Fr.; 2a,b, ♀ carapace R, vent. ×80; 2c, ♂ carapace R, ×80; 2d, RV int., anterior duplicature, ×200 (81).

Oligocythereis SYLVESTER-BRADLEY, 1948 [**Cythere fullonica* JONES & SHERBORN, 1888]. Hinge entomodont. Ornament with prominent subcentral tubercle and well-marked dorsal, ventral, and anterodorsal ridges. *M.Jur.-U.Jur.,* Eu.——Fig. 264, 1. **O. fullonica* (JONES & SHERBORN), Eng.; 1a-c, LV lat., int., dors., 1d-f, RV lat., int., dors.; all ×80 (364).

Orionina PURI, 1954 [**Cythere vaughani* ULRICH & BASSLER, 1904] [=*Jugosocythereis* PURI, 1957]. Reticulate, with sharp, straight median ridge sloping downward from posterodorsal complex and almost reaching anterior margin, posterior end with slight posteroventral extension. Hinge holamphidont. *Eoc.-Rec.,* cosmop.——Fig. 262,5. **O. vaughani* (ULRICH & BASSLER), Mio., USA(Fla.); 5a,b, RV lat., int.; 5c,d, LV lat., int.; all ×50 (5a,c, 287; 5b,d, 177).

?Parexophthalmocythere OERTLI, 1959 [**P. rodewaldensis* BARTENSTEIN & BRAND, 1959]. In lateral aspect like *Cythereis,* although somewhat more extended with rather narrow, sharply triangular posterior extremity, front and rear margins strongly denticulate, with blunt to thorned ventral wing-like process and bulge or strong tooth near dorsal margin; strong, protuberant eye tubercles; rib-like bar of thorns along anterior margin between eye tubercle and uppermost tooth. Surface smooth or reticulate, sides only slightly inflated, very flat near anterior and posterior marginal zones; LV larger than RV in posterodorsal and ventral regions. Marginal zone very broad, proximal section of the inner shell lamella less strongly calcified; numerous threadlike radial pore canals (about 25 anterior); selvage strong; LV hinge with about 5 anterior and posterior small sockets, a median feebly notched bar on dorsal side of sockets (forms a continuation of the selvage); shallow accommodation groove. *L.Cret.(Valangin-Hauteriv.),* Eu.(Ger.-Fr.).——Fig. 264,2. **P. rodewaldensis,* Ger.; 2a,b, LV lat., LV lat., ×58 (270). [REYMENT.]

Phacorhabdotus HOWE & LAURENCICH, 1958 [**P. texanus*]. Hinge holamphidont. Like *Costa* but less elongate, smoother, with ribs less well defined, subcentral tubercle wider, and wider duplicature. Eye tubercle present but indistinct. *U. Cret.*, N.Am.-Eu.——Fig. 264,3. **P. texanus,* Pecan Gap Chalk, USA(Tex.); 3a,b, carapace L, dors.; 3c, RV int.; all ×60 (34).

Platycythereis TRIEBEL, 1940 [**Cythereis excavata* CHAPMAN & SHERBORN, 1893]. Hinge hemiamphidont. Sides flat, much compressed; subcentral tubercle bladelike, deflected backward. *L.Cret.-U. Cret.*, Eu.——Fig. 264,6. **P. excavata* (CHAPMAN & SHERBORN), U.Cret.(Turon.), Ger.; 6a,b, carapace R, dors., ×60; *6c,d,* RV int., dors., ×80 (81).

?**Protobuntonia** GREKOFF, 1954 [**P. numidica*]. Like *Buntonia* but posterior extremity pointed and anterior marginal pore canals straight, unbranched. *U.Cret.-Paleoc.,* Afr.(Tunisia-Algeria-Nigeria).——FIG. 265,2a. **P. numidica;* U.Cret.(Senon.), N.Afr. (Alg.); carapace L, ×65 (294).——FIG. 265,2b-d.

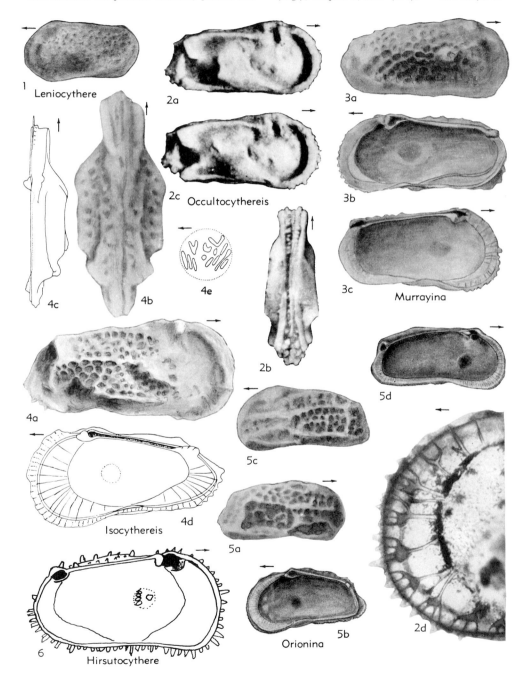

FIG. 262. Trachyleberididae (p. Q338-Q339).

P. ioruba REYMENT, Paleoc., W.Afr.(Nigeria); *2b,* RV lat., ×65; *2c,d,* RV int. (ant., post.), ×240 (294). [REYMENT.]

Puriana CORYELL & FIELDS, 1953 [*pro Favella* CORYELL & FIELDS, 1937 (*non* JORGENSEN, 1925)] [**Favella puella* CORYELL & FIELDS, 1937]. Hinge (according to original description) modified lophodont, with denticulate median element and entire terminal elements. Ornament of ridges and blades dominated by subcentral tubercle. [Genus has usually been interpreted by reference to *Cythere rugipunctata* ULRICH & BASSLER, which differs markedly in hinge and ornament from *P. puella* and is here referred to *Carinocythereis*. Perhaps *P. puella* is based on immature molts and the genus should be in same category as *Archicythereis*.] Mio., C.Am.-N.Am.——FIG. 266,*1.* **P. puella* (CORYELL & FIELDS), Panama; *1a-c,* RV lat., int., dors., ×67 (126).

Quadracythere HORNIBROOK, 1952 [**Cythere truncula* BRADY, 1898]. Like *Bradleya* but even shorter and with distinct caudal process. *Eoc.-Rec.,* cosmop.——FIG. 267,*1.* **Q. truncula* (BRADY), Rec., N.Z.; *1a,* LV lat.; *1b,c,* RV int., dors.; all ×50 (32).——FIG. 264,*5. Q. vermiculata* (BOSQUET), Eoc.(Lutet.), Fr.; *5a,b,* carapace R, RV int., ×50 (42).

?Spongicythere HOWE, 1951 [**S. spissa*]. Hinge holamphidont, carapace inflated, with wide, sinuous anterior duplicature. Ornament of high, spongy reticulations; subcentral tubercle absent or obscure. *Eoc.Oligo.,* N.Am.——FIG. 264,*4.* **S. spissa,* Eoc., USA(Fla.); *4a,b,* LV lat., int., ×40 (177).

?Trachycythere TRIEBEL & KLINGLER, 1959 [**T. tubulosa*]. Carapace of medium size, irregularly quadrangular, LV slightly larger than RV, overhanging only at dorsal ends, front broadly rounded, rear narrowly rounded, greatest height near anterior end, ventral margin weakly convex; distinct eye spot and internal eye socket. Surface reticulate and warty, each wart with coarse sieve-type pore canals. Zone of concrescence fairly broad, inner margin and line of concrescence coincident; radial pore canals rather few, unbranched; both valves with thin bladelike outer list; RV hinge with terminal dentate ridges and fine, crenulate median furrow. *L.Jur.-M.Jur.,* Eu.——FIG. 265,*1.* **T. tubulosa,* Ger.; *1a,b,* RV lat., RV (transmitted light) showing radial pore canals, ×70; *1c,* LV lat., ×70; *1d,e,* carapace dors., vent., ×70; *1f,g,* LV int., RV int., showing hinge, ×165; *1h,i,* RV ant. (int.), RV post. (ext.) showing radial pore canals, ×220; *1j,* two warts, showing sieve-type pore canals, ×450 (382). [REYMENT.]

Trachyleberidea BOWEN, 1953 [**Cythereis prestwichiana* JONES & SHERBORN, 1887]. Like *Costa* but posterior tooth of hinge tending to be lobate (hemiamphidont), antennal muscle scar broken into 2, and median ridge of ornament discontinu-

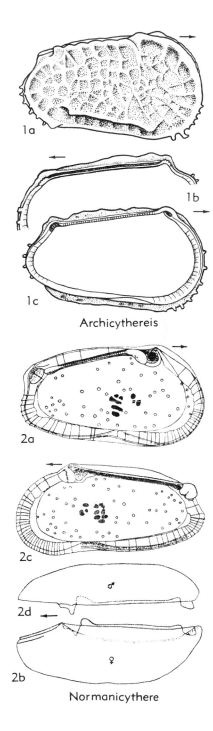

Archicythereis

Normanicythere

FIG. 263. Trachyleberididae (p. Q336-Q339).

ous, dominated by subcentral tubercle, not bent at posterior end. *Paleoc.-Oligo.*, Eu.-N.Am.——FIG. 267,4a. **T. prestwichiana* (JONES & SHERBORN), Eoc.(Ypres.), Belg.; RV lat., ×75 (42).

——FIGS. 266,2, 267,4b. *T. aranea* (JONES & SHERBORN), Eoc.(Ypres.), Belg.; 266,2, LV lat., ×67; 267,4b, RV int., ×75 (42).

Veenia BUTLER & JONES, 1957 [**Cythereis ozanana*

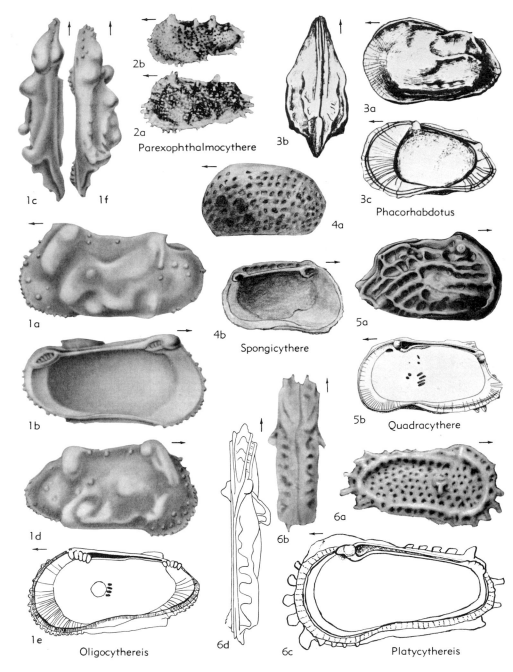

1c   1f

2b

2a

Parexophthalmocythere

3a

3b

3c

Phacorhabdotus

4a

4b

Spongicythere

5a

5b

Quadracythere

1a

1b

1d

1e

Oligocythereis

6a

6b

6c

6d

Platycythereis

FIG. 264. Trachyleberididae (p. Q339-Q341).

ISRAELSKY, 1929]. Hinge holamphidont. Like *Costa* but ribs less well defined, median ridge straight or slightly curved, convex upward; externally may resemble *Protocythere*. Fairly strong sexual dimorphism. *U.Cret.-Paleoc., Eu.-N.Am.*——FIG. 267,*3a*. **V. ozanana* (ISRAELSKY), Ozan(Campan.), USA(Ark.); carapace R, ×50 (89).——FIG. 267,*3b,c*. *V. parallelopora* (ALEXANDER), Navarro(Maastricht.), USA(Tex.);*3b,c*, carapace R, LV int., ×50 (*3a*, 89; *3b,c*, 34).

Family XESTOLEBERIDIDAE Sars, 1928

[*nom. transl. et correct.* HOWE, herein (*ex* Xestoleberinae SARS, 1928)]

Shell stout; with smooth or pitted surface. Characteristically with reniform or arcuate scar below and behind eye region on inside of valves; hingement adont or merodont; adductor muscle scars in vertical row of 4, with more or less arcuate antennal scar in front of upper part of row and may have 2 mandibular scars below and in front of row, marginal areas broad, with vestibule in front; radial canals short. [Habitat marine.] *Cret.-Rec.*

Xestoleberis SARS, 1866 [*Cythere aurantia BAIRD, 1838; SD BRADY & NORMAN, 1889]. Carapace ovate, LV larger than RV. Hinge merodont, with elongate crenulate terminal cusps in RV, separated by somewhat curved to nearly straight, finely crenulate to smooth furrow; marginal areas narrow except in front where vestibule is present; radial canals short, straight; adductor scars in vertical row of 4, with arrowhead-shaped antennal scar in front and 2 mandibular scars below in front, crescent-shaped scar above this group in eye region. *Cret.-Rec.*, cosmop.——FIG. 268,*4*. **X.*

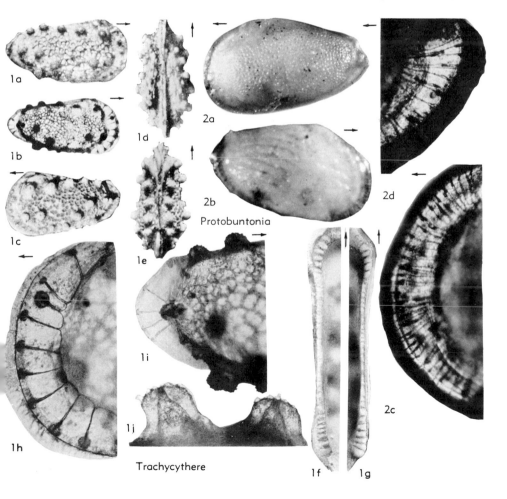

Protobuntonia

Trachycythere

FIG. 265. Trachyleberididae (p. Q340-Q341).

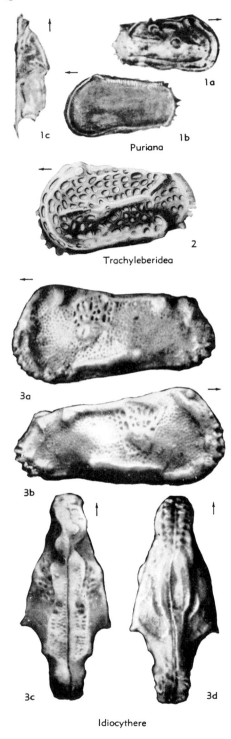

1a

1c   1b

Puriana

2

Trachyleberidea

3a

3b

3c   3d

Idiocythere

FIG. 266. Trachyleberididae (p. Q339-Q343).

aurantia (BAIRD), Rec.; *4a,b*, ♀ carapace R, dors. (Norway), ×100 (65); *4c*, LV and RV dors. (Baja California), ×100 (Sylvester-Bradley, n); *4d*, LV int. (NE.Atl.), ×145 (88, by permission of Mouton & Co., The Hague).——FIG. 269, *1*. *X. subglobosa* (BOSQUET), Eoc.(Lutet.), Fr. (Paris Basin); *1a-c*, LV lat., int., dors., ×75 (42).

Linocheles BRADY, 1907 [*L. vagans*]. Like *Xestoleberis* but distinguished from all Cytheracea in its greatly elongated threadlike 3rd pair of legs. *Rec.*, Antarct.——FIG. 268,*2*. *L. vagans; 2a,b*, carapace L, dors., ×70; *2c*, 3rd leg, ×120 (110).

Microxestoleberis G.W.MÜLLER, 1894 [*M. nana*]. Small, elongate, with projection at posteroventral corner, venter flattened. Hinge adont; reniform muscle scar below eye region, as in *Xestoleberis*. *Eoc.-Rec.*, Eu.——FIG. 268,*1*. *M. nana*, Rec., Medit.; *1a-c*, carapace L, dors., vent., ×130 (53).

Uroleberis TRIEBEL, 1958 [*Eocytheropteron parnensis* APOSTOLESCU, 1955]. Carapace short and high, with strongly arched dorsal outline; posterior end drawn out into caudal process; ventral region very plump; surface smooth or pitted. Shallow eye pits on inside but not external eye tubercles. Hinge of LV with elongate crenulate sockets separated by smooth middle bar above which lies an accommodation groove; marginal area distinct, with anterior vestibule, and straight radial canals; muscle scars in nearly vertical row of 4 with V-shaped antennal scar in front and 2 mandibular scars lower. *Eoc.-Rec.*——FIGS. 268,*3*, 269,*2*. *U. parnensis* (APOSTOLESCU), Eoc.(Lutet.), Fr.; *268,3*, ♀ LV int. eye spot with reniform scar below and behind it, ×165 (380); *269,2a*, RV int., ×90 (42); *269,2b-d*, ♀ carapace L, R, dors., ×80; *269,2e*, RV int. showing eye spot and reniform scar, ×175; *269,2f,g*, ♀ RV and LV hinge, int., ×225; *269,2h*, ♀ LV int. ant., ×355 (269, *2b-h;* 380). [HOWE-REYMENT.]

Family UNCERTAIN

[Includes Speluncellinae, Glorianellinae SCHNEIDER, 1960; Timiriaseviinae, Palaeocytherideinae, Palaeocytherideides, Faluniinae, Mediocytherideisinae (*recte* Mediocytherideidinae) MANDELSTAM, 1960]

Absonocytheropteron PURI, 1957 [*A. carinata*]. Medium in size, with shape like that of *Eucytherura*, oblong, anterior end rounded broadly, posterior compressed, triangular; dorsal and ventral margins slightly concave; valves strongly alate, type species with vertical ribbing, flattened venter tending to be reticulate; eye spot and muscle node present. Hinge amphidont; muscle pattern arranged in a concentric pattern of 4 dorsal scars, 6 ventral scars and single posterior scar; anterior pore canals single, few, straight. *U.Eoc.*, N.Am.——FIG. 270,*1*. *A. carinatum*, USA (Fla.-Miss.); *1a-c*, carapace L, dors., vent.; *1d,e*, RV lat., RV int.; all ×50 (290). [HOWE.]

Aspidoconcha DE VOS, 1953 [*A. limnoriae*]. Carapace with very peculiar shield shape; dorsally arched, laterally inflated, and ventrally flattened. Hingement adont; marginal areas narrow, with short straight canals; other shell features not described. [Commensal on *Limnoria* washed ashore on coast of Holland.] *Rec.*, North Sea.——FIG. 271,1. *A. limnoriae; 1a,* carapaces on abdomen of *Limnoria,* ×48; *1b-e,* "lat.", dors., vent., ×68; *1f,* surface sculpture, ×375 (all 398). [HOWE.]

Atjehella KINGMA, 1948 [*A. semiplicata*]. Elongate ovate, thickest near posterior margin, which appears truncate from side; surface with longitudinal ribbing in rear half. Marginal area very broad around anterior end and venter; radial canals long, few, branching. RV hinge with elongate, notched, terminal teeth, separated by narrow, finely crenulate groove; muscle scars with several small spots near center. *Neog.*, E.Indies(Java-Sumatra).—— FIG. 270,2. *A. semiplicata; 2a,b,* LV lat., int.; *2c,* carapace dors.; *2d,* RV hinge, ×50 (46). [HOWE.]

Bronsteiniana MANDELSTAM, 1956 [*B. galba*]. Oval, smooth, LV larger than RV, greatly inflated posteriorly, both ends well rounded, mid-point of front margin slightly lower than that of rear; dorsal margin convex, ventral margin slightly convex to straight. Internal margin not coincident with line of concrescence, shallow anterior vestibule present; marginal pore canals distinct; muscle field typical of cytherids. *U.Cret.(Cenom.)-Paleoc.,* C.Asia(USSR).——FIG. 271,3. *B. galba,* U.Cret.(Cenom.); *3a,b,* carapace R, dors., ×57 (50). [REYMENT.]

Climacoidea PURI, 1956 [*C. pleurata*]. Carapace laterally subrectangular, with raised ridges roughly parallel to margins; area between ridges crossed by smaller transverse ridges; anterior broadly and obliquely rounded, posterior narrower but rounded. Hinge holamphidont, with well-developed ocular sinus in front of hingement, leading to small external eye tubercles; marginal area broad in front, with widely spaced radial canals; muscle scars in vertical row of adductors with 2 antennal and 2 mandibular scars in front. *Plio.* (or *Pleist.*), N.Am. ——FIG. 270,6. *C. pleurata,* USA(Fla.); *6a,b,* LV lat., RV lat., ×50 (289). [HOWE.]

Cytheralison HORNIBROOK, 1952 1*C. java*]. Large, subquadrate, valves subequal, strongly inflated, some with spines or ridges; anterior margin bearing flange, posterior with blunt caudal process, upper margin of which in LV strongly overlaps that of RV, both processes being convex laterally so as to form caudal chamber with posterior opening. Selvage prominent, isolating caudal chamber when valves are shut; shell thick, built of honeycomb lattice forming hexagonal cells, commonly with slitlike openings; muscle-scar pattern consisting of vertical group of large alternately

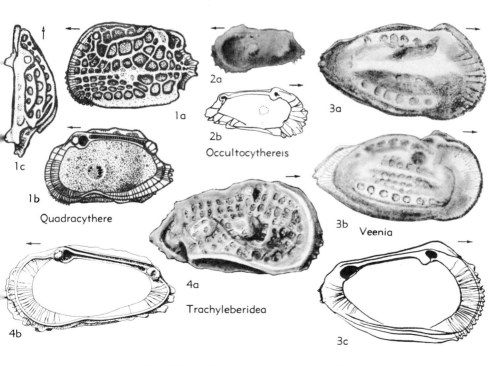

2a

1a

2b

Occultocythereis

3a

1c

1b

Quadracythere

3b Veenia

4a

Trachyleberidea

4b

3c

FIG. 267. Trachyleberididae (p. Q339-Q343).

placed spots, with 2 isolated spots above and in front, adductor muscle group marked externally by deep, elliptical, open chamber; LV hinge with straight simple bar between 2 terminal sockets; marginal pore canals fine and simple; line of concrescence coinciding with inner margin. *Eoc.-Rec.*, N.Z.-Austral.——Fig. 271,*2a-d.* **C. java,* Rec., N.Z.; *2a-c,* LV (holotype) lat., dors., int.;

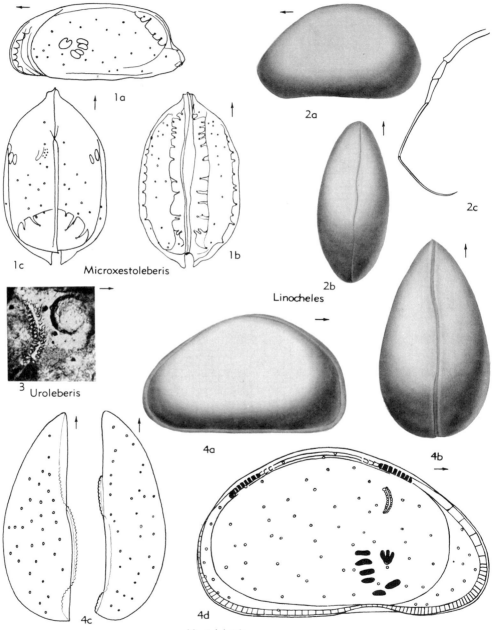

Fig. 268. Xestoleberididae (p. *Q*343-*Q*344).

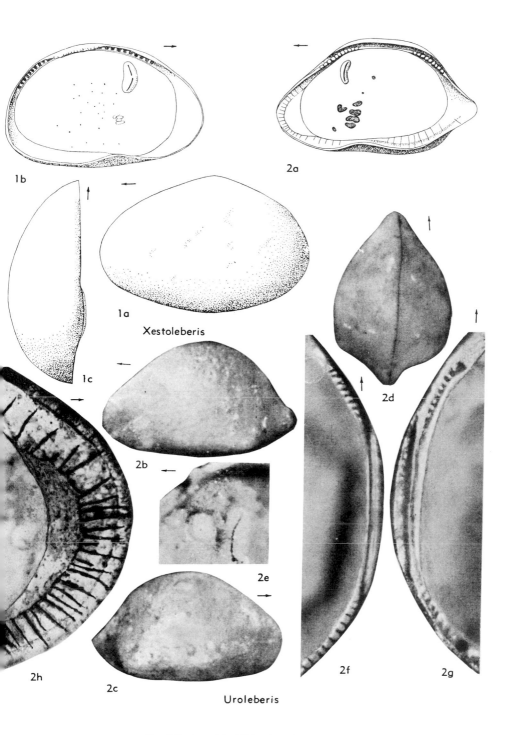

1b

2a

1a

Xestoleberis

1c

2b

2d

2e

2h

2c

2f

2g

Uroleberis

Fig. 269. Xestoleberididae (p. Q343-Q344).

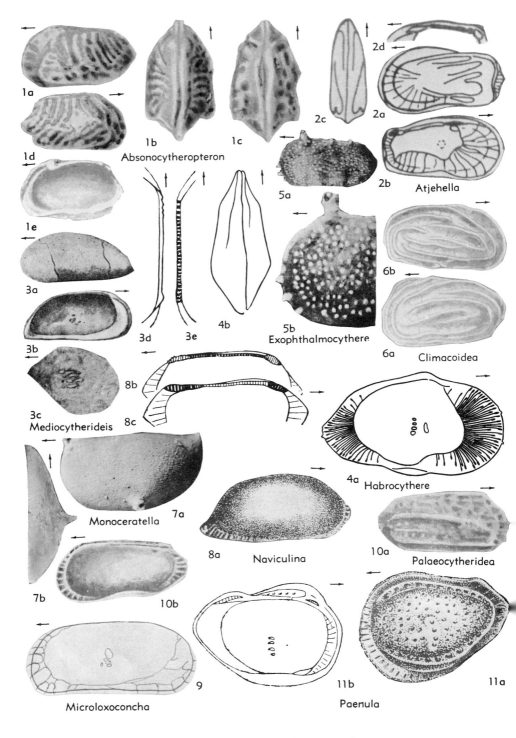

FIG. 270. Family Uncertain (p. Q344-Q353).

2d, carapace (paratype) post.; all ×50 (32).——
Fig. 271,*2e-g. C. parvacauda* Hornibrook, Rec.,
N.Z.; *2e-g,* LV lat., dors., int., ×50 (32). [Rey-
ment.]

Emphasia Mandelstam, 1956 [**E. ceratophaga*].
Inequivalved, slightly inflated, LV much larger

than RV, each valve with 1 or 2 tubercles (including possible adductor muscle tubercle) and strong,
winglike ventrolateral rib; anterior margin rounded, with mid-point lower than that of also
rounded posterior margin, dorsal margin of LV
strongly arched but that of RV almost straight;

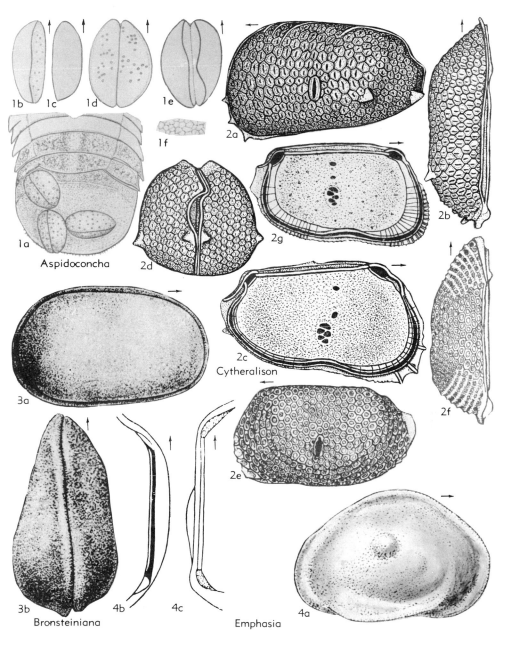

Fig. 271. Family Uncertain (p. Q345-Q350).

surface ornamented with minute pits and tubercles. Internal margin and line of concrescence anteriorly not coincident, making small vestibule; LV hinge with elongated terminal sockets and broad median bar, that of RV complementary, with platelike terminal teeth; muscle field typical of Cytheridae. *M.Jur.*, SW.Asia(USSR).——Fig. 271,4. **E. ceratophaga*, Kazakhstan; *4a*, carapace R, ×65; *4b,c*, RV int., LV int. showing hinge, ×43 (50). [Reyment-Bold.]

Exophthalmocythere Triebel, 1938 [**E. mamillata*]. Differs from other known cytherids in having well-developed eye stalks at anterodorsal angle of subrectangular carapace, which has evenly rounded anterior end somewhat wider than posterior extremity; ornament of small pits, tubercles and spines. Zone of concrescence broad, with few simple pore canals, line of concrescence and inner margin coincident, RV hinge with anterior and posterior teeth (latter notched in some forms) and median furrow with shallow anterior socket be-

hind anterior tooth. *U.Jur.-L.Cret.*, Eu.——Fig. 270,5. **E. mamillata*, U.Jur., Ger.; *5a*, carapace L, ×30; *5b*, same, ant. part showing prominent eye stalk, ×60 (80). [Reyment.]

Falunia Grekoff & Moyes, 1955 [**F. girondica*]. Based on young molts with only partially developed hinge and lacking properly developed marginal areas and radial canals. *Mio.*, Eu.—— Fig. 272,2. **F. girondica*, Fr.; *2a*, LV lat., ×?; *2b*, RV int., ×? (153). [Howe.]

Gemmanella Schneider, 1956 [**G. schweyeri*]. Oval, thin-walled, somewhat inflated in dorsal aspect, LV distinctly overlapping RV all around, dorsal margin straight in middle but curving evenly to join broadly rounded anterior and posterior margins, latter tending to be obtusely angular, ventral margin concave in anterior half, convex in posterior half; surface irregularly pitted, with short broad rib extending toward middle from posterior margin and few feeble ventral riblets, LV less strongly ornamented than RV. Hinge

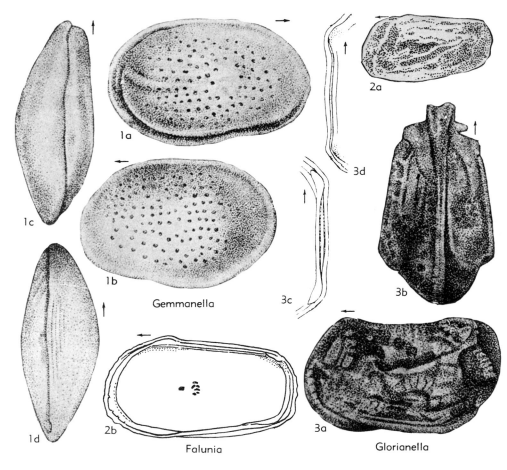

1a
2a
1c
3d
1b
Gemmanella
3c
3b
1d
2b
Falunia
3a
Glorianella

Fig. 272. Family Uncertain (p. Q350-Q351).

simple; LV with groove to accommodate sharp hinge margin of RV. [Marine.] *L.Trias.*, SE.Eu.——Fig. 272,1. *G. schweyeri*, USSR(Emba Region); *1a-d*, carapace R, L, dors., vent., ×64 (50). [BOLD-REYMENT.]

Glorianella SCHNEIDER, 1956 [*G. efforta*]. Small to medium-sized oblong valves differently inflated (LV commonly larger than RV), length twice height, dorsal margin straight or weakly concave, anterior margin rounded, posterior margin blunt and obliquely truncate; surface ornamented with ribs, tubercles and pits. RV hinge with elongated terminal teeth separated by median furrow; muscle pattern of cytherid type. [Nonmarine.] *L.Trias.*, SE.Eu.——Fig. 272,3. *G. efforta*, USSR(Emba Region); *3a,b*, carapace L, dors.; *3c,d*, RV and LV hinge; all ×64 (50). [BOLD-REYMENT.]

Habrocythere TRIEBEL, 1940 [*H. fragilis*]. Small, egg-shaped, smooth; toothless. Marginal areas broad, with long radial pore canals; line of concrescence S-shaped; muscle scars in vertical row of 4 with one in front of middle of row. *L.Cret.*, Ger.——Fig. 270,4. *H. fragilis; 4a,b*, LV int., carapace dors., ×90 (81). [HOWE.]

Hemicytheridea KINGMA, 1948 [*H. reticulata*]. Medium in size, thick-shelled, valves subequal but RV higher than LV; surface reticulate. Marginal area moderately narrow; line of concrescence leaving inner margin at ends; radial pore canals straight, simple; RV hinge with crenulate anterior socket, serrate median groove and heavy crenulate, triangular posterior tooth. Dimorphism pronounced. *Neog.*, E.Indies(Sumatra-Java).——Fig. 273,1. *H. reticulata; 1a,b*, carapace R, dors.; *1c,d*, RV int., LV int.; *1e,g*, LV and RV hinge; all ×50 (46). [HOWE.]

Hemikrithe VAN DEN BOLD, 1950 [*H. occidentalis*]. Elongate ovate, with broadly rounded anterior margin and narrowed at rear which is incised somewhat as in *Krithe;* surface reticulate. RV hinge with elongate faintly notched terminal teeth and crenulate groove between; marginal areas very broad in front, with few long branching radial canals; muscle scars numerous, small, in 2 oblique rows of 4, with additional small spots. *Rec.*, E.Indies(Sumatra).——Fig. 273,3. *H. occidentalis; 3a-c*, LV lat., RV int., carapace dors., ×50 (100). [HOWE.]

Juvenix KUZNETSOVA in MANDELSTAM *et al.,* 1957 [*J. pseudocuspidatus*]. Carapace elongate-ovate, with small transverse sulcus and in some shells a ventral ridge; LV overlapping RV; anterior end higher than posterior; hinge in LV consisting of terminal crenulate sockets (divided into 4), median crenulate bar, crenulations stronger toward anterior end; anterior socket larger than posterior. Radial pore canals widely spaced. *L.Cret.(Barrem.),* SE.Eu.(Caucasus)-SW.Asia(Caspian-Azer-

baidjan).——Fig. 273,5. *J. pseudocuspidatus,* Azerbaidjan; *5a,b*, RV lat., dors.; *5c*, RV int.; all ×50 (238a). [BOLD.]

Laocoonella DEVOS & STOCK, 1956 [*pro Laocoon* DEVOS, 1953 (*non* NIERSTRASZ & ENTZ, 1922)] [*Laocoon commensalis* DEVOS, 1953]. Carapace small, inflated, greatest thickness behind middle, length more than twice height, ovate in dorsal or lateral view, though venter somewhat flattened; eyes confluent. Surface minutely concentrically reticulate. Marginal areas rather broad, with straight radial canals; muscle scars not known. [Commensal on *Limnoria,* Curaçao.] *Rec.*, W.Indies.——Fig. 273,2. *L. commensalis* (DEVOS); *2a*, carapace dors., ×82; *2b*, surface sculpture, enlarged (398).

Mandelstamia LYUBIMOVA, 1955 [*M. facilis*]. [Recorded as new genus by LYUBIMOVA in 1956, but actually already published elsewhere in March, 1955.] Elongate oval in side view, somewhat preplete, dorsal margin straight, anterior and posterior margins broadly rounded, ventral margin with middle concave zone; LV larger than RV; feeble sulcus in dorsal half just in front of midpoint; surface pitted, with or without small tubercles. Muscle field with posterior vertical row of 4 scars and 2 scars in front; anterior zone of concrescence broad, posterior zone narrow, line of concrescence not coincident with inner margin; marginal pore canals straight, widely separated; RV hinge with terminal teeth and median furrow. *U.Jur.,* Eu.(Ger. - USSR) - SW.Asia(Kazakhstan).——Fig. 273,11. *M. facilis*, USSR(Lower Volga); *11a,b*, RV lat., vent., ×94; *11c,d*, RV int., LV int., ×130 (50). [REYMENT.]

Marslatourella MALZ, 1959 [*M. exposita*]. Valves subequal, roughly trapezoidal in side view, ventral area with or without blunt winglike developments; surface smooth. RV hinge with terminal notched teeth united by narrow, shallow notched furrow; inner margin coinciding with line of concrescence, zone of concrescence rather broad; eye tubercles strong, somewhat resembling those of *Exophthalmocythere;* muscle field unknown. *M.Jur.,* Fr.——Fig. 273,4. *M. exposita; 4a-d*, carapace (holotype) R, L, dors., post.; *4e,f*, RV and LV hinge; all ×80 (236). [REYMENT.]

?Mediocytherideis MANDELSTAM, 1956 [*Cytheridea apatoica* SCHWEYER, 1949]. Elongate, low in relation to length, anterior margin well rounded, posterior pointed below mid-point; dorsal margin broadly and evenly arched, ventral margin straight to faintly concave; surface with small rounded pits, tubercles, or thin ribs arranged concentrically especially in anterior and ventral regions. Zone of concrescence moderately broad, with branched radial pore canals, vestibule broadest in anterior part; LV hinge with narrow notched anterior furrow, smooth central furrow and posterior rounded impression. [Muscle-scar pattern and some other features suggest that this genus may be cypridacean, rather than cytheracean, as classed by MANDEL-STAM.] Sexual dimorphism strong, inferred females posteriorly swollen. *Mio.-Plio.,* USSR(Caucasus-

Volga Region).——FIG. 270,3. *M. apatoica* (SCHWEYER); Plio., Caspian Region; *3a,b,* LV lat., int., ×40; *3c,* RV muscle pattern, ×100; *3d,e,* RV and LV hinge, ×70 (50). [REYMENT-BOLD.]

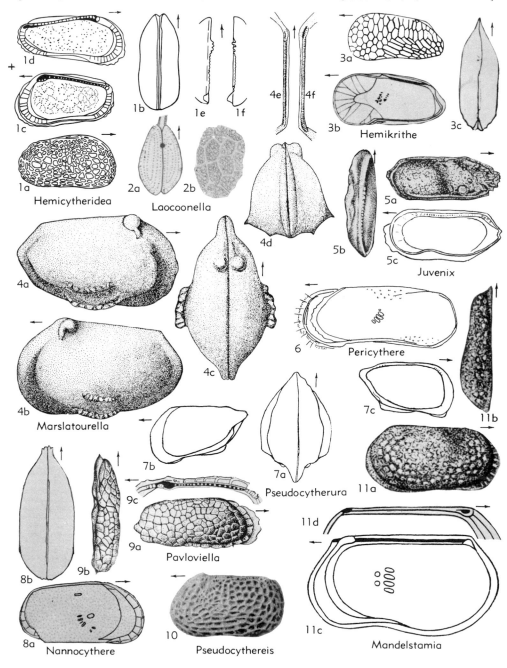

FIG. 273. Family Uncertain (p. *Q351-Q354*).

Microloxoconcha HARTMANN, 1954 [*M. compressa*]. Small, elongate-ovate, very thin and fragile, unornamented. Marginal areas broad at ends, with widely spaced radial canals; line of concrescence not coincident with inner margin, indented near some radial canals, particularly at rear; muscle scars in row of 4 with single small scar in front. *Rec.*, Medit.——FIG. 270,9. *M. compressa;* LV lat., ×155 (162). [HOWE.]

Monoceratella TEICHERT, 1937 [*M. teres*]. Subovate in side view, dorsal border nearly straight (with low hump extending above hinge line toward rear), cardinal angles sharp, nearly 90 degrees; ventral border straight or slightly concave; well-defined sulcus just in front of mid-length; surface smooth; spine projecting outward-backward from ventromedian surface and in some species additional spines in dorsal region. *M.Ord.*, Arct.Can.——FIG. 270,7. *M. teres*, M.Ord.(Edinburg F.), USA(Va.); *7a,b*, LV lat., vent., ×20 (J. C. Kraft, n). [MOORE.]

Nannocythere SCHÄFER, 1953 [*N. remanei*]. Carapace rather delicate, with outswelling in back part of lower shell half, thus producing flat underside. Hinge teeth very small, hinge ledge smooth. Surface covered with numerous sieve-pore fields from which rather long hairs spring out; marginal area broad anteriorly with large vestibule, only few radial canals at each end; muscle scars in curved row of 4 adductors, with single antennal and 2 mandibular scars in front. Dimensions not reported. *Rec.*, Ger.——FIG. 273,8. *N. remanei; 8a,b*, RV lat., carapace dors., ×? (319). [HOWE.]

Naviculina KATZ in MANDELSTAM *et al.*, 1957 [*N. longa* LYUBIMOVA in MANDELSTAM *et al.*, 1957]. Carapace inverse boat-shaped, LV overlapping RV, posterior end pointed, dorsal margin almost straight; hinge in LV with anterior elongate socket, subdivided into 8 parts, median, low, crenulate bar slightly broadened toward ends with coarser crenulations, posterior socket less elongate than anterior one, subdivided into 6 parts; marginal area broad, with widely spaced pore canals. *M.Jur. (Baj.)*, E.Eu.(E.Ukraine).——FIG. 270,8. *N. longa* LYUBIMOVA; *8a-c*, RV lat., RV int., LV int., ×43 (238a).

Paenula KUZNETSOVA in MANDELSTAM *et al.*, 1957 [*P. superba*]. Greatest height in anterior 0.3, longitudinal ribs projecting over dorsal and ventral margin, dorsal margin straight; hinge in LV consisting of terminal slitlike sockets subdivided into 5 parts, median element a thin smooth ridge; muscle scars in vertical row of 4. Surface pitted. *U.Cret.(Cenom.)*, SE.Eu.(Caucasus)-SW.Asia(Caspian-Azerbaidjan).——FIG. 270,11. *P. superba*, Caspian-Azerbaidjan; *11a,b*, LV lat., int., ×70 (238a). [BOLD.]

Palaeocytheridea MANDELSTAM, 1947 [*P. bakirovi* (*non Eucythere denticulata* SHARAPOVA, 1937, invalid SD LYUBIMOVA, 1955)]. Elongate, ovate, higher anteriorly, dorsal margin straight, posterior end of LV angular; LV hinge with elongate, crescent-shaped, denticulate sockets and median smooth bar; line of concrescence and inner margin nearly coincident, pore canals straight, widely spaced; surface smooth, punctate, reticulate, or ridged. [Species assigned to this genus by MANDELSTAM and LYUBIMOVA seem to have only the hinge structure in common and are not all congeneric. *Eucythere denticulata* is judged to be congeneric with *Clithrocytheridea decumana* TRIEBEL but has been assigned by HOWE & LAURENCICH (1958) to *Asciocythere* SWAIN.] *M.Jur.*, SW. Asia(Kazakhstan).——FIG. 270,10. *P. bakirovi; 10a,b*, RV lat., int., ×50 (237). [BOLD.]

Pavloviella KUZNETSOVA in MANDELSTAM *et al.*, 1957 [*P. barremica*]. Elongate-ovate, with laterally much-compressed margin in anterior end; LV hinge with anterior rounded socket bearing postadjacent knob-shaped tooth, median part a crenulate groove ending in a posterior elongate curved socket; surface reticulate, with concentric pattern. [Figures published by the author suggest that the RV (not LV) has the median groove as indicated in the description.] *L.Cret.(Barrem.)*, SE.Eu.(Caucasus)-SW. Asia.(Azerbaidjan-Caspian). —— FIG. 273,9. *P. barremica*, Azerbaidjan; *9a-c*, RV lat., dors., hinge, ×43 (238a). [BOLD.]

Pericythere HARTMANN, 1957 [*P. foveata*]. Shell thin, fragile, very elongated in lateral view, without tubercles or pits, and commonly no ribs; eyes with coalesced pigment. Inner margin parallel to outer margin, selvage strong near the anterior margin, with hairlets, inner margin and line of concrescence apparently not coincident; 5 adductor muscle spots slightly in front of mid-length, hinge untoothed. [Brackish-water, mangrove environment.] *Rec.*, C.Am.-S.Am.——FIG. 273,6. *P. foveata*, El Salvador; LV lat., ×90 (164). [REYMENT-HOWE.]

Pseudocythereis SKOGSBERG, 1928 [*Cythereis (Pseudocythereis) spinifera*] [=*Pseudocytheretta* PURI, 1958]. Moderately large, ovate, resembling *Cytheretta* and *Echinocythereis* in shape, surface entirely covered with reticulations, each reticulum containing small pits. Hinge apparently holamphidont; internal shell characters not described, but from marginal bristles, it appears that radial canals are numerous; 5th appendage not differentiated as in *Cytheretta*, hence family assignment uncertain. *Rec.*, S.Atl.——FIG. 273,10. *P. spinifera* (SKOGSBERG); LV lat., ×30 (72). [HOWE.]

Pseudocytheromorpha PURI, 1957 [*P. elongata*]. Elongate, length nearly twice height, anterior end obliquely rounded, denticulate, posterior end obliquely rounded with tuft of 4 ventral spines; dorsal margin nearly straight, ventral margin concave in middle; surface smooth, pitted or reticulate.

Marginal areas very wide, especially at rear, radial pore canals numerous, long, straight; muscle scars in subcentral pit, arranged in 2 rows of 3 each, middle scar of ventral row showing tendency to divide; RV hinge with anterior crenulate tooth, postjacent socket, and posterior socket connected by median bar. *U.Eoc.*, SE.N.Am.——FIG. 274,1. **P. elongata*, Crystal River F., USA(Fla.); *1a-d*, RV (holotype) lat., int., dors., vent.; *1e,f*, LV (paratype) lat., int.; *1g,h*, carapace dors., vent.; all ×50 (290). [HOWE.]

Pseudocytherura DUBOWSKY, 1939 [**P. pontica*]. Thick-shelled, sculptured, irregularly rhomboidal in side view, with broadly rounded anterior margin and drawn-out process at rear, greatest height 0.25 length from front; winglike processes increase to make greatest breadth 0.7 back from front. Marginal areas regular, radial pore canals undescribed; hinge crenulate at ends but smooth in mid-section. Length of females 1.12 mm., males unknown. *Rec.*, Black Sea.——FIG. 273,7. **P. pontica*; *7a-c*, carapace dors., RV int., LV int., ×27 (142). [HOWE.]

Pseudokrithe MEHES, 1941 [**P. dictyota*]. Rectangular, postplete, with parallel dorsal and ventral margins and almost equally rounded posterior and anterior margins; surface with reticulate ornament. *Paleog.*, Eu.(Hung.). ——FIG. 275,3. **P. dictyota*; *3a,b*, RV lat., LV dors., ×? (249). [REYMENT.]

Pseudoloxoconcha G.W.MÜLLER, 1894 [*P. minima*]. Shell thin, fragile, elongate, very small, highest in front third. Marginal area broad; line of concrescence somewhat far from inner margin at ends but fused below; few radial pore canals, slender, not branched; hinge apparently adont. *Rec.*, Medit. (Gulf of Naples).——FIG. 275,11. **P. minima*; LV lat., ×220 (53). [HOWE.]

Pulviella SCHNEIDER in MANDELSTAM *et al.*, 1957 [**P. ovalis*]. Ovate or egg-shaped, LV larger than RV, ends rounded, dorsal margin arched, greatest width in posterior half; hinge with groove in LV and sharp edge in RV. Surface weakly pitted or smooth, some with fine ribs in ventral portion. [Illustrations given show RV overlapping LV and carapace with a strong ventral ridge not mentioned in the description; also greatest height is located anteriorly.] *L.Trias.*, SE.Eu.(Emba region).——FIG. 274,5. **P. ovalis*; *5a,b*, carapace L, dors., ×40 (238a). [BOLD.]

Rectocythere MALZ, 1958 [*Clithrocytheridea? iuglandiformis* KLINGLER, 1955]. Valves subequal, LV slightly larger than RV, valves only in contact along ventral and both end margins, along dorsal margin LV overlapping RV only around both hinge angles; anterior margin thickened, cut off from strongly ornamented sides by shallow furrow; adductor muscle tubercle present; inner margin and line of concrescence coincident, zone of concrescence fairly broad, with few marginal pore canals, which are straight, unbranched; LV hinge with low, sharp knife-edge that fits into small furrow in RV, definite terminal teeth lacking. *U.Jur.*, W.Eu. —— FIG. 274,9. **R. iuglandiformis* (KLINGLER) M.Kimm.(NW.Ger.); *9a-c*, carapace L, R, dors., ×? (235). [REYMENT.]

Redekea DEVOS, 1953 [**R. perpusilla*]. Small, thin, compressed, elongate; surface marked with widely spaced rounded pits; eyes confluent. Hingement adont; marginal areas rather wide, with vestibules and widely spaced radial canals; muscle scars not known. [Commensal on *Limnoria* washed ashore on coast of Holland.] *Rec.*, North Sea.——FIG. 274,4. **R. perpusilla*; *4a-c*, carapace L, dors., vent., ×84 (398). [HOWE.]

Renngartenella SCHNEIDER in MANDELSTAM *et al.*, 1957 [**R. pennata*]. Elongate-ovate, moderately swollen, with weak transverse sulcus, length twice height, ends compressed, with bladelike ridge in ventral portion of carapace; hinge in LV with narrow terminal sockets and median ridge; surface smooth, but some with spinelike projections. *L.Trias.*, SW.Asia(Astrakan).——FIG. 275,7. **R. pennata*; *7a,b*, RV lat., dors., ×35 (238a). [BOLD.]

Rubracea MANDELSTAM in MANDELSTAM *et al.*, 1957 [**R. artis* LYUBIMOVA in MANDELSTAM *et al.*, 1957]. Elongate, with incised posterior end; hinge in LV with elongate, crenulate terminal sockets and median ridge, in RV with terminal teeth connected to elevated inner margin (?selvage); line of concrescence and inner margin not coinciding at ends. Surface smooth. *M.Jur.*(Callov.), W.Asia Transvolga-Saratov).——FIG. 275,1. **R. artis* LYUBIMOVA, Saratov Region; *1a,b*, carapace R, dors.. ×43; *1c*, RV int., ×60 (238a). [BOLD.]

Ruggieria KEIJ, 1957 [**Cythere micheliniana* BOSQUET, 1852]. Ovate with upturned posterior extremity, marginal spines in front and rear; distinct eye depression located near anterior hinge element; surface ornamented partly or entirely with reticulations or longitudinal ridges and with posteroventral spine. Marginal areas fairly broad, line of concrescence and inner margin coincident; radial pore canals moderately numerous, simple, wavy, widened in middle; muscle field with posterior row of 4 scars, single dorsally open horseshoe-shaped scar in front, and 3 scars above central field; RV hinge with conical anterior tooth and smooth or obscurely lobed posterior tooth, LV hinge with terminal sockets and a crenulated bar with a conical tooth at its anterior termination. Sexual dimorphism pronounced, inferred males more slender than females. [Marine.] *Mio.-Rec.*, Eu.——FIG. 275,9. **R. micheliniana* (BOSQUET), Mio. (Burdigal.), Fr. (Aquitaine Basin); *9a-c*, ♀ LV lat., ♂ RV lat., RV (lectotype) lat.; *9d,e*, ♀ LV int., ♂ RV hinge; all ×40 (42).

Ruttenella VAN DEN BOLD, 1946 [**R. ovata*]. Elongate ovate, smooth, posterior cardinal angle pro-

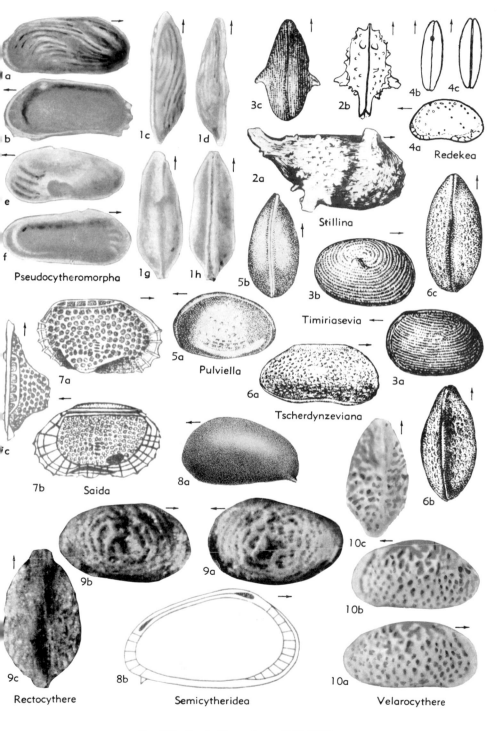

FIG. 274. Family Uncertain (p. Q353-Q358).

jecting slightly in LV. LV hinge with crenulate anterior socket, denticulate dorsal margin, and posterior cardinal area raised to fit over pointed tooth of RV; line of concrescence leaving inner margin at ends; radial pore canals numerous, in anterior region curving upward and downward from middle. *U.Eoc.,* Carib.(Bonaire Is.).——FIG. 275,4. **R. ovata; 4a,b,* carapace R., dors.; *4c,d,* LV int., RV int., ×50 (7). [HOWE.]

Saida HORNIBROOK, 1952 [**S. truncata*]. Minute,

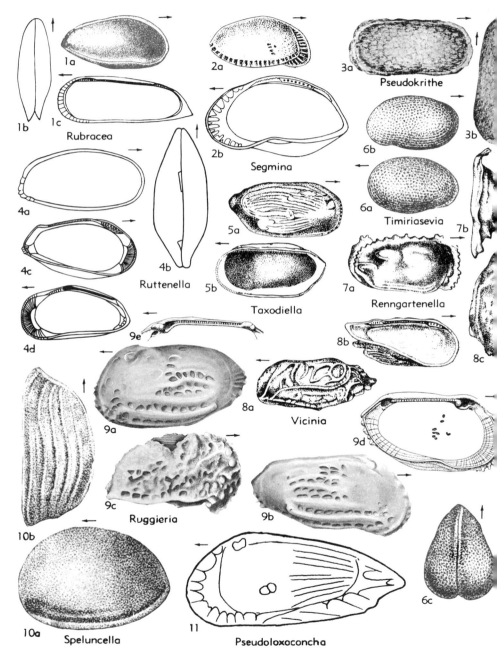

FIG. 275. Family Uncertain (p. Q354-Q358).

subrhomboidal; ventrolateral margins expanded to form blunt processes rather similar to *Cytheropteron;* hinge consisting of 2 terminal simple teeth in RV separated by straight, simple groove; LV with terminal sockets and horizontal, straight, simple bar between; caudal process absent; muscle-scar pattern consisting of 4 elongate scars, one above the other with a single scar in front; radial canals simple, not numerous; line of concrescence coinciding with inner margin; duplicatures wide. *M.Eoc.-Rec.,* N.Z.——Fig. 274,7. **S. truncata,* Rec.; *7a-c,* RV lat., int., dors., ×75 (32). [Howe.]

Segmina Mandelstam in Mandelstam et al., 1957 [**Cytheropteron lunulare* Lienenklaus, 1894]. Carapace small, anterior end obliquely rounded, posterior narrowly rounded or pointed, dorsal margin arched, alate; radial pore canals few, widely spaced, vestibule crescent-shaped; hinge in LV with narrow crenulate ridge and terminal slitlike crenulate grooves, in RV with elongate, terminal, crenulate teeth connected by crenulate groove; muscle scars (*fide* Lienenklaus) in curved posterior row of 4 with 2 additional scars rather far in front. Surface smooth or reticulate. *L.Cret.-U.Oligo.,* W.Eu.-C.Asia-SW.Asia.——Fig. 275,2. *S. obvalla* Kuznetsova; *2a,b,* RV lat., int., ×73, ×100 (238a). [Bold.]

Semicytheridea Mandelstam, 1956 [**Cythere spirifera* Chapman & Sherborn, 1893]. Kidney-shaped, LV larger than RV, anterior broadly rounded, point of maximum height at about anterior 4th; posterior extremity more sharply rounded and with mid-point lower than that of anterior margin; dorsal margin strongly and irregularly arched, ventral margin less strongly convex and more regular, surface smooth except for openings of normal pore canals; strong spine at about posterior 3rd of LV. Anterior zone of concrescence well developed, provided with a few unbranched radial pore canals; muscle field reported to conform to that of Cytheridae; LV hinge with large elongated anterior socket and small elongated posterior socket, median bar thin. *L.Cret.(Apt.-Alb.),* W.Eu.-SE.Eu.(USSR).——Fig. 274,8. **S. spinifera* (Chapman and Sherborn), L.Alb.; *8a,b,* LV lat., int., ×50, ×63 (*8a,* 120; *8b,* 50). [Reyment-Bold.]

Speluncella Schneider, 1956 [**S. spinosa*]. Egg-shaped, inflated ventrally, height about 0.7 of length, dorsal margin strongly arched, mid-point of anterior margin slightly higher than that of posterior, which is pointed; ventral surface with 5 or 6 slightly flexed longitudinal ribs, stretching along entire length, surface otherwise smooth. LV hinge with median groove and terminal tooth-lets. [Fresh-water.] *L.Trias.* USSR(Emba area). ——Fig. 275,10. **S. spinosa; 10a,b,* LV lat., vent., ×86 (50).

Stillina Laurencich, 1957 [**S. asterata*]. Teardrop-shaped in side view, with straight dorsal margin, rather evenly rounded at front and along ventral margin, acuminate at rear by reason of long, drawn-out, upturned posterior caudal process; plump, with greatest thickness below middle; divided by weak median sulcus into subequal broad lobes; eye tubercles strong; surface ornamented with spines and reticulations and marked by strong posteroventral spine. Hinge holomerodont, in RV consisting of crenulate bar with terminal crenulate cusps; marginal areas rather narrow, no vestibule; radial pore canals few. *L.Cret.,* N.Am.——Fig. 274,2. **S. asterata,* Alb.(Goodland F.), USA(Tex.); *2a,b,* carapace R, dors., ×72, ×60 (34).

Taxodiella Kuznetsova in Mandelstam et al., 1957 [**T. fiscellaformis*]. Like *Cytheretta,* but differs in less asymmetry of valves; hinge of LV with denticulate bar, in RV with finely denticulate groove. Shape of carapace indicates relationship to *Cytherurinae* or *Bythocytherinae.* [Since characters of the muscle scars and marginal areas have not been described, no family assignment is possible.] *L.Cret.(Barrem.),* SE.Eu.(Caucasus)-SW. Asia (Caspian-Azerbaidjan).—— Fig. 275,5. **T. fiscellaformis,* Azerbaidjan; *5a,b,* RV lat., int., ×94 (238a). [Bold.]

Thalmannia LeRoy, 1939 [**T. sumatraensis*]. Roundly elongate rectangular, plump, LV larger than RV, overlapping it dorsally and ventrally; surface smooth except for large pits in longitudinal rows. RV hinge with elongate terminal teeth

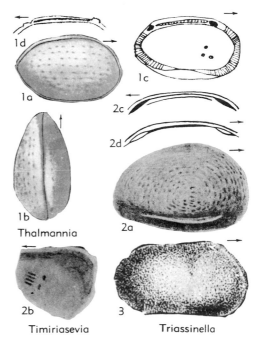

Thalmannia

Timiriasevia

Triassinella

Fig. 276. Family Uncertain (p. *Q358*).

and furrow along dorsal margin, all elements of hinge showing tendency to crenulation, although terminal teeth may be smooth; radial canals numerous, straight; muscle scars 3 or 4, closely spaced. *Mio.*, E.Indies.——Fig. 276,1. **T. sumatraensis*, Sumatra; *1a,b*, carapace R, dors., *1c,d*, LV int., RV hinge; all ×40 (227). [Howe.]

Timiriasevia Mandelstam, 1947 [**T. epidermiformis*]. Carapace roughly kidney-shaped, LV larger than RV, in some species anterior and posterior ends almost equally rounded, but anterior always more broadly rounded than posterior, dorsal margin almost straight to convex, ventral margin weakly convex in anterior third; surface with fine pits, small ribs, or irregular spines (placed usually in a row parallel to and at a short distance from anterior and posterior margins, or dorsocentrally). Zone of concrescence with only a few radial pore canals; inner margin and line of concrescence coincident; no eye spot; RV hinge provided with smooth median furrow and elongated terminal tooth plates; muscle scars in row of 4 adductors with 2 mandibular scars in front. Strong sexual dimorphism, inferred female carapaces being strongly swollen in posterior third. Very like *Metacypris* (syn. *Gomphocythere*) in general features but with different hinge. [Freshwater.] *M.Jur.-L.Cret.*, Eu.-Asia(USSR).——Fig. 276,2. **T. epidermiformis*, M.Jur., SW.Asia; *2a,b*, RV lat., RV muscle field, ×43, ×56 (237); *2c,d*, RV and LV hinge, ×43 (50).——Fig. 275, 6. *T. polymorpha* Mandelstam, L.Cret., SW.Asia; *6a-c*, ♀ carapace L, R, dors., ×? (50).——Fig. 274,3. *T. acuta* Mandelstam, L.Cret., SW.Asia; *3a-c*, ?♂ carapace L, R, dors., ×? (50). [Reyment.]

Triassinella Schneider, 1956 [**T. chramovi*]. Small, equivalved, valves slightly inflated, oblong; dorsal margin straight, in some forms gently rounded; anterior and posterior margins rounded, ventral margin straight to slightly convex; surface ornamented with fine pits, central part of valves with shallow transverse depression, posteroventral part provided with short spine. Hinge simple. [Marine.] *L.Trias.*, USSR.——Fig. 276,3. **T. chramovi*, Emba Region; RV lat., ×64 (50). [Reyment-Bold.]

Tscherdynzeviana Kashevarova, 1958 [**T. busulukensis*]. Elongate oval, with rather truncate posterior end and regularly rounded anterior margin; RV overlapping LV along straight dorsal margin but with reverse overlap along ventral margin, which is gently concave in middle; surface pitted. Zone of concrescence well developed anteriorly but narrow. *U.Perm.* (Tatar.), USSR.——Fig. 274,6. **T. busulukensis*; *6a-c*, carapace R, dors., vent., ×43 (192). [Bold-Reyment.]

Velarocythere Brown, 1957 [**V. scuffeltonensis*]. Medium in size, elongate ovate, with obliquely rounded, rimmed, toothed anterior margin, pos-

terior extremity slightly narrower, rounded in LV and subangulate near middle in RV; dorsal outline slightly arched, with distinct eye spot; ventral outline irregular because of overhang of valves, ventral margin inturned in front of middle; surface rather coarsely reticulate and pitted near middle. Hingement holamphidont, with median elements finely crenulate; marginal areas rather broad at ends, with numerous long paired radial pore canals; muscle scars in vertical row of 3 with a large antennal scar in front. *U.Cret.*, N. Am.——Fig. 274,10. **V. scuffeltonensis*, USA (N.Car.), *10a-c*, carapace R, L, dors., ×50 (117). [Howe.]

Vicinia Kuznetsova in Mandelstam *et al.*, 1957 [**V. sutilis*]. Reminiscent of *Paracytheridea* but LV hinge consisting of anterior socket with 4 crenulations, central denticulate bar with 3 larger teeth at its anterior end [only 2 visible in author's figure] and posterior elongate socket with 5 crenulations. *L.Cret.(Barrem.)*, SE.Eu.(Caucasus)-SW. Asia (Caspian-Azerbaidjan). —— Fig. 275,8. **V. sutilis*; *8a-c*, LV lat., int., dors., ×72 (238a). [Bold.]

Suborder METACOPINA Sylvester-Bradley, n. suborder

[Diagnosis and discussion by P. C. Sylvester-Bradley, University of Leicester]

Hinge distinct, simple to tripartite; muscle scar consisting of secondary scars assembled in a compact group; inner lamella narrow, poorly developed or unknown. *?L.Ord., M. Ord.-L.Cret.*

The Metacopina are podocopids in which the muscle-scar pattern is a circular aggregate of many scars. The duplicature is variable in width. In the Healdiacea it is narrow, and often described as absent. Transverse sections (Fig. 277,*A*) show, however, a calcified inner lamella developed in some genera; this is joined to the outer lamella along a plane of concrescence oblique to the shell surface. This differs only in degree from the situation, common in the Podocopina, in which the plane of concrescence is parallel to the shell surface (Fig. 277,*B*), and in which, therefore, a duplicature is clearly recognizable. Other Podocopina show a plane of commissure which is intermediate in direction (Fig. 277,*C*). The duplicature in the Quasillitacea has not yet been investigated by modern methods; superficial examination suggests that it is intermediate in nature between the conditions described above for the Healdiacea and the Podocopina, respectively. The hinge in the Metacopina is more or less differentiated. Some Healdiacea (?Cavellinidae) have an undifferentiated hinge; others (Healdiidae) have a hinge differentiated into three striated elements. Quasillitacea have a differentiated hinge in which the terminal elements may be denticulate.

The suborder certainly includes the ancestors of the Platycopina (which are related, via *Cavellina,* to the Cavellinidae, and may also be related, via *Hungarella* [*Ogmoconcha*], to the Healdiidae). It may also include ancestors of some of the Podocopina (e.g., some of the Cytheracea may be descended from some of the Quasillitacea); in that case, however, the Podocopina must be polyphyletic, as some of the Cytheracea (e.g., Bythocytheridae) seem to have developed independently from the Palaeocopida. The Healdiacea also show signs of relationship (in the development of ornament) with Paleozoic Bairdiacea.

Superfamily HEALDIACEA Harlton, 1933

[*nom. transl.* MANDELSTAM, 1960 (*ex* Healdiidae HARLTON, 1933)] [Diagnosis and discussion by R. H. SHAVER, Indiana University and Indiana Geological Survey]

Convex-backed, short-hinged ostracodes with hinge and contact margins ridged and grooved in platycopine fashion, LV-over-RV overlap and overreach, adductor muscle scars consisting of numerous aggregate spots. *Dev.-L.Cret.*

Healdiacea have hinges that generally are better differentiated from the contact margin than are those of Cytherellacea, from which the Healdiacea also differ in direction of overlap; Healdiacea have shorter hinges than Quasillitacea, and they lack the separated calcified inner lamellae of Bairdiacea; Healdiacea are externally smooth or ornamented but lack the coarse sculpturing of Thlipsuracea.

Genera classified in families of the Healdiacea are predominantly middle to late Paleozoic in distribution, occurring especially in Devonian and Carboniferous formations (Fig. 278). Only two, or possibly three, genera range above the Permian.

Family HEALDIIDAE Harlton, 1933

[Materials for this family prepared by R. H. SHAVER, Indiana University and Indiana Geological Survey, with some additions by W. A. VAN DEN BOLD, Louisiana State University]

Carapaces convex-backed, with suboblong

to subtriangular outlines and nearly straight venters in lateral view, commonly with posterior, and less commonly with anterior sculpturing in ridges and spines; valves hinged posterodorsally; LV larger than RV, overlap and overreach (where present) LV over RV. Without separated calcified inner lamellae but with hinges and contact margins simply ridged and grooved or shouldered in platycopine fashion; adductor muscle scar circular, consisting of numerous spots generally arranged in concentric rings or rows. Most genera probably exhibit moderate sexual dimorphism. *Dev.-L.Cret.*

Healdiidae have distinctive marginal areas which serve to separate them from other families. A duplicature is either lacking or is completely fused with the outer lamella so that a vestibule and other associated structures are lacking around the entire margin.

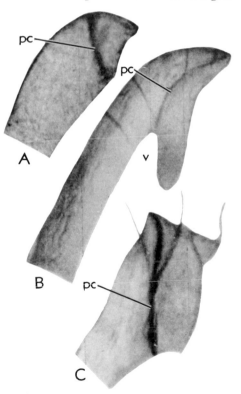

FIG. 277. Transverse sections through the anterior border of the left valves of (*A*) *Ogmoconcha contractula* TRIEBEL (Healdiacea, Metacopina) ×240; (*B*) *Pelocypris zilchi* TRIEBEL (Cypridacea, Podocopina) ×350; (*C*) *Cyprideis torosa* (JONES) (Cytheracea, Podocopina) ×370. [Explanation: *pc,* plane of concrescence; *v, vestibule*] (Triebel).

The LV is larger than RV and with a heavy selvage and prominent inner selvage groove around the entire margin, overlaps it except commonly posterodorsally and anteriorly, where the flange groove and flange of the RV variably have conspicuous development

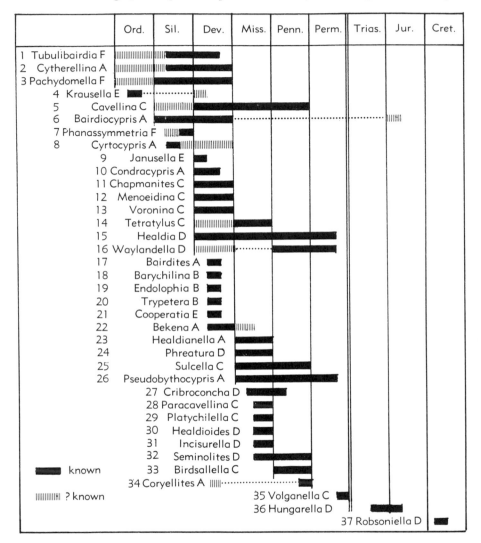

Fig. 278. Stratigraphic distribution of healdiacean ostracodes (Moore, n). Classification of genera in families is indicated by letter symbols (A—Bairdiocyprididae, B—Barychilinidae, C—Cavellinidae, D—Healdiidae, E—Krausellidae, F—Pachydomellidae). An alphabetical list of genera with index numbers furnishes cross reference to the serially arranged numbers on the diagram.

Generic Names with Index Numbers

so that overlap may be mutual; the LV commonly overreaches the RV even in areas of mutual overlap. Closure of the valves is accomplished by the selvage of the RV articulating in the selvage groove of the LV and additionally by outer grooving of the RV at the hinge for reception of the selvage of the LV. The hinge and adjacent contact margins are commonly further modified by transverse microteeth-like crenulations in the selvage groove and selvage of the LV and RV, respectively.

Several authors have noted considerable specific variation among some healdiid genera, and others have erected species of a genus on minute differences. Commonly, the variants are of two forms, one with greater height and, in dorsal view, with more symmetrically convex outlines; the other is relatively lower, and, in dorsal view, the greatest thickness is near the posterior end, producing a lateral outline with acuminate anterior and blunt posterior extremities. The two forms are interpreted as male and female, respectively, although either choice presents evidence contradictory to the normal aspect of sexual dimorphism in other ostracodes.

The classification of healdiid genera has been determined by re-evaluation of the relative importance of taxonomic characters, the most essential of which are thought to be marginal structures, muscle scar, direction of overlap, and shape. While generally diagnostic, shape is not considered all-important, for reliance on shape often has resulted in indiscriminate grouping of dissimilar genera. Direction of overlap is emphasized as a valuable taxonomic character even above the generic level in this group, previous objections and exceptions to this being without force when orientation is correct and when other evidence is considered.

Features serving to differentiate genera of the Healdiidae from Cytherellidae and Cavellinidae are LV-over-RV overlap and overreach, associated generally with somewhat more complex articulation and sculpture. Healdiidae are distinguished from Bairdiidae by muscle-scar patterns and simpler metacopine margins which lack separated calcified inner lamellae. Healdiidae differ from Bairdiocyprididae in their characteristic healdiid shape as opposed to bythocypridid shape and in the presence of ridges and spines on the carapace.

Healdia ROUNDY, 1926 [**H. simplex*]. Carapace generally subtriangular in lateral view, with angularly arched dorsum, venter nearly straight, anterior border mostly broadly rounded and posterior end commonly truncate; greatest thickness posterior, producing cuneate appearance in dorsal view; surface of each valve generally smooth and posteriorly sculptured with either a sickle-shaped to straight, vertical ridge or shoulder, or 1 or 2 backward-pointing spines, or both; LV larger than RV, with overlap and overreach of LV over RV but overlap commonly reduced or lacking in hinge area located posterodorsally. Hinge and contact margins consisting of groove or depressed shoulder in LV into which fits edge of RV, which also is grooved or beveled marginally; dorsal elements of articulation of each valve commonly crenulated transversely in numerous, minute toothlets and grooves extending through and beyond hinge area; adductor muscle scar circular, with numerous aggregate spots arranged in concentric rings or rows (121, 299, 82). *Dev.-Perm.*, cosmop.——FIG. 279,*1a-h*. *H. cara* BRADFIELD, Penn., USA(Ill.); *1a,b*, ?♂ carapace R, dors.; *1c,d*, ?♀ carapace R, dors. (=*H. aspinosa* COOPER), ×100 (Shaver, n); *1e,f*, RV int., LV int., showing marginal areas (Shaver, n); *1g,h*, long. and transv. secs.; all ×100 (Shaver, n).——FIG. 279,*1i*. *H.* sp., Penn., USA(Tex.); muscle scar from RV int., ×200 (82).

Cribroconcha COOPER, 1941 [**C. costata*]. Resembles *Healdia* but variably with coarsely pitted surface anterior to ridge; imperfectly known. *M.Miss.-L.Penn.*, N.Am.(Ill.-Ark.).——FIG. 280, *1*. **C. costata*, Ill.; *1a,b*, carapace R, dors., ×80 (Shaver, n).

Healdioides CORYELL & ROZANSKI, 1942 [**H. diversus*]. Like *Seminolites*, with anterior ridge but without pits; commonly with spines. *U.Miss.*, N. Am.——FIG. 280,*3*. **H. diversus*, USA(Ill.); *3a,b*, carapace R, dors., ×100 (130).

Hungarella MÉHES, 1911 [**Bairdia? problematica* DADAY, 1911] [=*Ogmoconcha* TRIEBEL, 1941]. Resembling *Healdia* but carapace subovate with rounded extremities, relatively higher and shorter in lateral view; lacks cuneate appearance in dorsal view and without posterior ridges or spines; LV overlaps and overreaches RV and is conspicuously larger. Muscle scar with fewer spots; hinge and contact margins mostly similar but with more prominent transverse crenulation of hinge (82). *U.Trias.-L.Jur.*, Eu.——FIG. 281,*1a-d*. *H.* sp., Ger.; *1a-c*, RV lat., dors., LV int., ×85 (Shaver, n); *1d*, transv. sec. through hinge, ×240 (376).—— FIG. 281,*1e*. *H. contractula* (type species of *Ogmoconcha*), Ger.; long. sec. through ant. margin, ×180 (376). [SHAVER prepared the above diagnosis and figures for *Ogmoconcha* on the assumption that this name is not a synonym of *Hungarella*. He has not seen the specimens representing the type species of *Hungarella*, but the type figures and description suggest to him the presence of radial pore canals and a duplicature. He therefore has no basis for agreement on the suggested synonymy or on assignment of *Hungarella* to Healdiidae, as have other *Treatise* authors. He appreciates, however, the late editorial necessity that has left this taxon unsatisfactorily described.]

Incisurella COOPER, 1941 [**I. prima*]. Resembles *Healdia* but sculpture consists of vertically subovate area impressed into shell near posterior border of each valve; imperfectly known. *U.Miss.*, N.Am.(Ill.-Okla.).——FIG. 282,1. **I. prima*, USA (Ill.); *1a,b*, carapace R, dors., ×90 (Shaver, n).

Phreatura JONES & KIRKBY, 1886 [**P. concinna*]. Resembles *Incisurella* but with smaller impressed area near anterior border of each valve; imperfectly known. *L.Carb.*, Eu.(Eng.).

Robsoniella KUSNETSOVA in MANDELSTAM *et al.*, 1956 [**R. obovata*]. Carapace elongate-ovate reniform, LV projecting, anterior end narrower than posterior and bent downward, dorsal margin arched; surface smooth or bearing weak striae in central region. Duplicature broad, with relatively few radial pore canals, which are straight and thin; muscle scars 4 to 5, not arranged in cypridid manner; LV hinge with terminal elongate crenulate teeth and median narrow bar, denticulate on dorsal side. [Differs from *Ogmoconcha* in

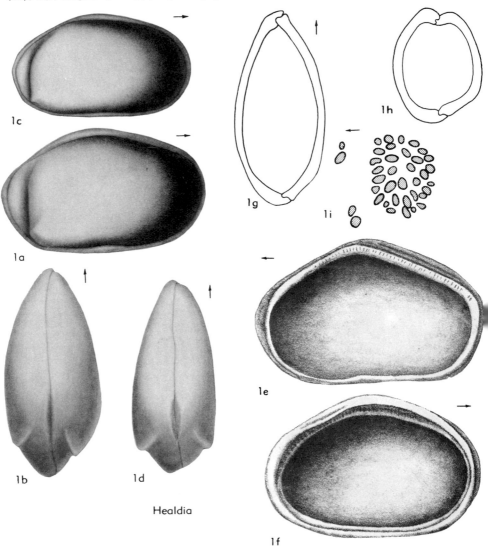

1c

1a

1b   1d

Healdia

1h

1g   1i

1e

1f

FIG. 279. Healdiidae (p. Q361).

stronger differentiation of hinge features.] *L.Cret.* (*Apt.-Alb.*), SE.Eu.(Caucasus).——Fig. 283,*1.* **R. obovata;* *1a,b,* RV lat., int., ×70; *1c,d,* LV int., dors., ×70; *1e,f,* LV and RV hinge, ×195 (50). [Bold.] [Shaver disagrees strongly with inclusion of *Robsoniella* in the Healdiidae. He believes especially that the duplicature and associated morphology and muscle scars of that genus, and to a lesser extent the shape and hinge structures, would render the family diagnosis meaningless if this assignment should be considered as a valid one. The genus is assigned here on the basis of the original author's judgment; a more important consideration to Shaver, except for late editorial neces-

sity, is that *Treatise* authors are urged to improve, if they can, upon taxonomic concepts.]

Seminolites Coryell, 1928 [**S. truncatus*]. Like *Healdia* but mostly with sickle-shaped ridge nearly paralleling anterior border and shallow depression commonly paralleling both ridges on concave side; known species without spines; mostly coarse-pitted like *Cribroconcha* (21, 121). *U.Miss.-Penn.,* N.Am.——Fig. 280,*2.* **S. truncatus,* Penn., USA (Ill.); *2a,b,* ?♂ carapace R, dors.; *2c,* ?♀ carapace dors. (=*S. elongatus* Coryell), ×90 (Shaver, n).

Waylandella Coryell & Billings, 1932 [**W. spinosa*] [=*Harltonella* Bradfield, 1935]. Differs

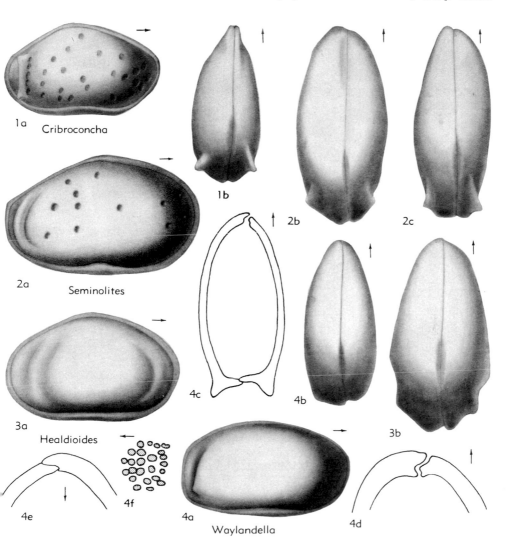

1a Cribroconcha
1b
2b
2c
2a Seminolites
3a Healdioides
3b
4a Waylandella
4b
4c
4d
4e
4f

Fig. 280. Healdiidae (p. Q361-Q364).

from *Healdia* in generally more elongate outline in lateral view, dorsal angulation mostly lacking; posterior terminated as in *Healdia* to somewhat pointed ventrally; sculpture generally weaker but variably consisting of 1 or 2 spines, or ridge or both, on each valve posteriorly (21, 123). *?Dev., Penn.-Perm.*, N.Am.-Eu.——Fig. 280,*4a,b*. *W. ardmorensis* (BRADFIELD), Penn., USA(Ill.); *4a,b*, carapace R, dors., ×60 (Shaver, n).——Fig. 280, *4c-f*. *W. obesa* COOPER, Penn., USA(Ill.); *4c*, long. sec., ×60 (Shaver, n); *4d,e*, transv. secs., through hinge and venter, ×100 (Shaver, n); *4f*, muscle scar from LV ext., ×100 (Shaver, n).

Family BAIRDIOCYPRIDIDAE Shaver, n.fam.

[Materials for this family prepared by R. H. SHAVER, Indiana University and Indiana Geological Survey]

Convex-backed ostracodes of bythocyprididid shapes, mostly without ornamentation and sculpturing; LV larger, with overlap and overreach of LV over RV; lacking separated calcified inner lamellae but with short hinges and contact margins simply ridged and grooved in platycopine fashion; adductor muscle scars circular and consisting of numerous closely grouped spots; sexual dimorphism slight or unknown in most genera. *?Ord., Sil.-Perm., ?Jur.*

Bairdiocyprididae are intermediate morphologically between Healdiidae and Bairdiidae but are easily distinguished from other families. They resemble Healdiidae in having similar muscle scars and configuration of the contact margins; true duplicatures either are lacking or are mostly fused with the outer lamellae, so that vestibules are lacking; the contact margins and hinges of

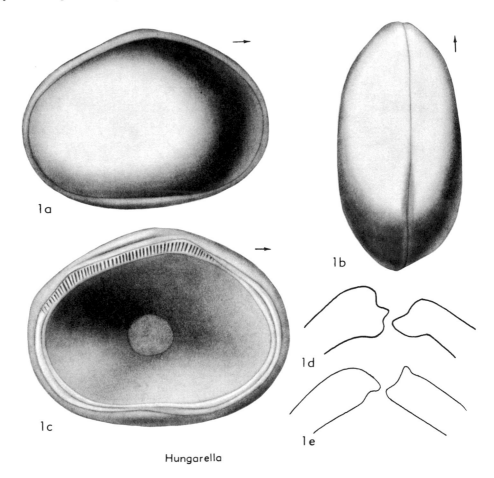

Hungarella

FIG. 281. Healdiidae (p. Q361).

LV are considered to consist of selvage, and proximally from it, selvage groove and list, whereas the contact margins of RV consist principally of the selvage. Bairdiocyprididae differ from Healdiidae by mostly lacking the distinctive dorsal angularity and posterior sculpturing in ridges and spines of that family.

The family mostly resembles Bairdiidae in shape and lack of sculpture but has adductor muscle scars and marginal structures that are foreign to true bairdiids. Bairdiocyprididae are differentiated from Cytherellidae and Cavellinidae by LV over RV overlap and overreach; they are much less tumid than *Pachydomella* and *Phanassymetria;* they have simpler hinges and contact margins than *Ponderodictya, Quasillites, Menoeidina,* and related genera, in which reversal of overlap occurs dorsally.

Possibly, *Bythocypris* does not exist in Paleozoic strata. Re-examinations of several of the many-score Paleozoic species referred to it have resulted in assignments to other genera. Part of these genera and remaining supposed Paleozoic species of *Bythocypris* are better assigned to the Bairdiocyprididae; nevertheless, the critical morphology of some genera included here is unknown.

Bairdiocypris KEGEL, 1932 [*Bythocypris (Bairdiocypris) gerolsteinensis* KEGEL, 1932]. Carapace large, heavy-shelled, mostly subtriangular in lateral view, with considerable dorsal convexity and nearly straight ventral border; lacking the great tumidity of *Pachydomella;* LV conspicuously overreaching RV; contact marginal structures as described for family; adductor muscle scar with 70 or more spots; commonly with posteroventral furrow on RV. Surface smooth or punctate. *Sil.-Dev., ?Jur.,* cosmop.——FIG. 284,2a,b. *B. gerolsteinensis* (KEGEL), M.Dev., Ger.; 2a,b, carapace (holotype) R, dors., ×20 (224).——FIG. 284,2c,d. *B. uexheimensis* (KEGEL), M.Dev., Ger.; 2c, transv. vent. edge, ×40; 2d, adductor muscle scar, ×60 (224).

Bairdites CORYELL & MALKIN, 1936 [*B. deltasulcata*]. Externally, nearly like *Bairdiocypris* but differs in having posterior crescentic ridge in each valve with depression in front. *M.Dev.,* N.Am.—— FIG. 284,1. *B. deltasulcata; 1a,b,* RV lat., dors., ×25 (Shaver, n).

Bekena GIBSON, 1955 [*B. diaphrovalvis*]. Nearly like *Bairdiocypris,* but type species without ventral furrow and having crescentic, compressed areas near anterior and posterior borders of RV (both features being of doubtful generic value); contact marginal structures apparently conforming to fam-

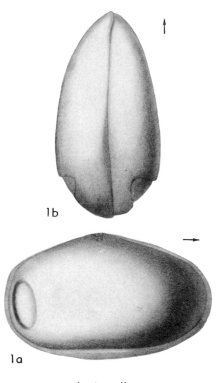

1b

1a

Incisurella

FIG. 282. Healdiidae (p. Q362).

ily diagnosis; adductor muscle scar unknown. *M. Dev.-U.Dev., ?L.Miss.,* N.Am.-Eu.(Ger.).——FIG. 285,3. *B. diaphrovalvis,* U.Dev., USA(Iowa); 3a,b, carapace R, dors., ×38 (155).

Condracypris ROTH, 1929 [*C. binoda*]. Large bairdiocypridid with 2 crescentic ridges located antero- and posterocentrally on surface of each valve; adductor muscle scars and marginal structures of type species (from USA) unknown; European specimens doubtfully referred to genus having greater overreach than type species but typical bairdiocypridid muscle scars and marginal structures. *L.Dev.-M.Dev.,* N.Am.-?Eu. ——FIG. 284,3a,b. *C. binoda,* L.Dev., USA (Okla.); 3a,b, carapace R, dors., ×40 (297).—— FIGS. 284,3c, 285,4. *C. ?circumvallata* (KUMMEROW), M.Dev., Ger.; 284,3c, transv. sec., LV at right, ×35; 285,4a,b, carapace R, dors., ×25; 285, 4c, adductor muscle scar, ×70 (226).

Coryellites KELLETT, 1936 [*pro Coryellina* KELLETT, 1935 (*non* BRADFIELD, 1935)] [*Coryellina firma* KELLETT, 1935]. Externally almost like *Pseudobythocypris* but has blunt node with depression in front near posteroventral angle of LV; contact marginal structures and muscle scars unknown. *?M.Dev.,* Eu.(Pol.), *U.Penn.,* N.Am.(Kans.).

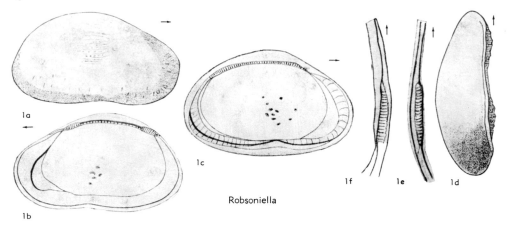

la
lb
lc
ld
le
lf

Robsoniella

FIG. 283. Healdiidae (p. Q362-Q363).

?Cyrtocyprus CORYELL & WILLIAMSON, 1936 [*C. subovata] [=Cyrtocypris CORYELL & WILLIAMSON, 1936 (nom. van.)]. Nearly like Cytherellina in external appearance and doubtfully distinguished from that genus by truncation of posterior border in lateral view; adductor muscle scars and contact marginal structures unknown; like some middle Paleozoic species assigned to Cavellina. M.Sil., ?Dev., N.Am.-?Eu.——FIG. 285,2a,b. *C. subovata, M.Sil., USA(Ind.); 2a,b, carapace (holotype) R, dors., ×23 (131).——FIG. 285,2c,d. C.? chvorostanensis POLENOVA), Dev., Russia; 2c,d, carapace R, dors., ×50 (278).

Cytherellina JONES & HOLL, 1869 [*Beyrichia siliqua JONES, 1855] [=Orthocypris KUMMEROW, 1953]. Carapace without external sculpture, smooth, elongate bythocypridid in side view, with point of greatest height in posterior half, terminal borders rounded, ventral border nearly straight; adductor muscle scars and contact marginal structures unknown, but many Paleozoic species assigned to Bythocypris are similar to Cytherellina. [Type specimens figured by JONES and JONES & HOLL consist partly of internal molds with clinging shell material that show 2 elongate, nearly vertical, shallow depressions without external expression, one located centrally and the other anteriorly.] ?Ord., M.Sil.-Dev., cosmop.——FIG. 285,1. *C. siliqua (JONES), M.Sil., Eng.; 1a-c, carapace R, vent., post.; 1d, LV int. with clinging shell, lat.; all ×20 (188).

Healdianella POSNER, 1951 [*H. darwinulinoides]. Nearly like Healdia in shape and small size but lacking ridges, shoulders, and spines; contact-marginal structures and adductor muscle scars as described for family. [The Devonian species from Russia referred to the genus apparently should be assigned to Cytherellina and other genera; some of POSNER's original species appear to differ from Cytherellina only in their smaller size.] L.Carb.,

E.Eu.——FIG. 285,5. *H. darwinulinoides; 5a,b, carapace R, vent., ×72 (281).——FIG. 284,4. H. darwinulinoides?; adductor muscle scar, ×70 (281).

?Longiscula NECKAJA, 1958 [*L. arcuaris]. Nonsulcate, outline depressed subtriangular, dorsal margin gently arcuate, ventral somewhat concave in central part, carapace broadest in dorsal part, LV overlapping RV along free margin, most distinct in central part of ventral margin (overlap may be indistinct or not developed along anterior and posterior ends), RV overlapping LV along dorsal margin, particularly in central part; no adventral structures or internal cavellinoid lamellae present. Dimorphism not observed. Surface smooth, tubercles developed in posterior part. M.Ord.-M.Sil., NW.Eu.(Eng.-Baltoscandia). —— FIG. 143,3. *L. arcuaris, M.Ord., NW.USSR(Pskow area); 3a,b, carapace (holotype) R lat., dors. (post. end up), ×28 (264). [HESSLAND.]

Pseudobythocypris SHAVER, 1958 [*Bythocypris pediformis KNIGHT, 1928]. Carapace small, smooth, thin-shelled; most species differing from other bythocypridids in showing laterally a short posteroventral slope or upsweep that meets posterodorsal border in sharp angle; marginal structures, undifferentiated hinge, and adductor muscle scars as described for family; dimorphism slight. [Differs from Waylandella in lacking posterior ridges, shoulders, and spines, and from Cytherellina in its smaller size and posteroventral slope of most species.] Miss.-Perm., N.Am.-Eu. ——FIG. 286,1. *P. pediformis (KNIGHT), Penn., USA(Ill.); 1a,b, ? ♂ carapace R, dors.; 1c,d, RV int., LV int.; all ×42 (Shaver, n).

?Reversocypris PŘIBYL, 1955 [*R. regularis]. Outline resembling Bythocypris in lateral view, with greatest height in posterior portion; stated to have reversal of overlap along venter, but duplicature-like structure along ventral contact margin of RV and contact margins, as well as hingement and adductor muscle scar, are not well understood. [The orientation adopted by Přibyl is reversed here, so that the genus resembles *Bairdiocypris* with overreach, if not overlap, mostly LV over RV.] *L.Dev.,* Czech.——Fig. 287,*1a-d.* *R. regularis; 1a-d,* carapace (holotype) ?L, ?R, dors., vent., ×28 (61).——Fig. 287,*1e-g.* *R. klukovicensis* Přibyl; *1e,f,* ?RV lat., vent., ×28; *1g,* ?RV lat., int., ×40 (61).

?**Silenis** Neckaja, 1958 [*S. subtriangulatus*]. Outline subtriangular, short straight hinge, ventral margin generally concave in central part, nonsulcate, LV overlapping RV along entire free margin and if developed, RV overlap along dorsal margin, no adventral or interior structures. Dimorphism not observed. Surface smooth. *L.Sil.-M.Sil.,* NW.Eu.——Fig. 143,*2.* *S. subtriangulatus,* Wenlock., Est.; carapace (holotype) R lat., ×18 (264). [Hessland.]

?Family BARYCHILINIDAE Ulrich, 1894

[Materials for this family prepared by R. V. Kesling, University of Michigan, with addition by H. W. Scott, University of Illinois]

RV much larger than LV, overlapping it around free border and with corners conspicuously projecting beyond those of LV. Hinge line short, straight, depressed, with groove in RV for accommodation of edge of LV, interior of both valves with elongate vertical ridge marking position of external sulcus. *M.Dev., ?L.Miss.*

Barychilina Ulrich, 1891 [*B. punctostriata*]. Each valve with distinct S_2. Surface with pattern of ridges in many species, outermost ridges nearly concentric in outline but central ones sloping posterodorsal-anteroventrally; some species with punctae in furrows between ridges. *M.Dev.,* N.Am.——Fig. 288,*2a.* *B. punctostriata,* M.Dev.(Onondaga), USA(Ind., Falls of Ohio); ×20 (385).——Fig. 288,*2b,c.* *B. embrithes* Kesling & Kilgore, M.Dev., USA(Mich.); *2b,c,* carapace L, dors., ×30 (209).

Endolophia Kesling, 1954 [*E. chariessa*]. Like *Barychilina* but lacking sulci; internal ridges without external sulcate counterparts. *M.Dev.,* N.Am.——Fig. 288,*3.* *E. chariessa,* USA(Ohio); *3a,b,* carapace L, dors., ×30; *3c,d,* LV int., RV int., ×30 (203).

Trypetera Kesling, 1954 [*T. barathrota*]. Valves

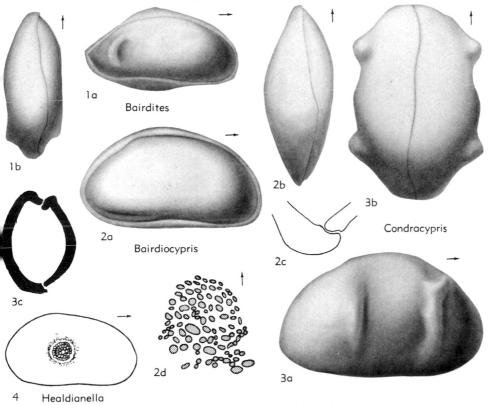

1a Bairdites
1b

2a Bairdiocypris
3c

2d Healdianella
4

2b
2c
3b Condracypris
3a

Fig. 284. Bairdiocyprididae (p. Q365-Q366).

with large pit in place of S_2; surface ornamented with smaller pits. *M.Dev.*, N.Am.——Fig. 288,1. *T. barathrota*, USA(Ohio); *1a,b*, carapace L, dors., ×30 (203).

?**Venula** COOPER, 1941 [*Primitiopsis? striatus* CRONEIS & FUNKHOUSER, 1938]. Small, valves unequal, RV overlapping LV; dorsal margin straight, ventral margin subparallel to dorsal; greatest width near posterior end, wedge-shaped in dorsal view; surface covered with thin ribs connected at various points; S_2 faintly developed; probable dimorphism indicated by swelling of posterior portion. *L.Miss.*, USA(Ill.). [SCOTT.]

?**Family CAVELLINIDAE Egorov, 1950**

[*nom. transl.* POLENOVA, 1960 (*ex* Cavellininae EGOROV, 1950)] [=?Volganellidae MANDELSTAM in MANDELSTAM *et al.*, 1956; ?*Volganellacea* (MANDELSTAM) KASHEVAROVA, 1959] [Materials for this family prepared by R. H. BENSON, University of Kansas, with contributions from W. A. VAN DEN BOLD, Louisiana State University, R. H. SHAVER, Indiana University and Indiana Geological Survey, and R. A. REYMENT, University of Stockholm]

Carapace subovate to subelliptical in lateral view, subelliptical in dorsal view with noticeable posterior swelling in female. Dorsal border convex, commonly with slight anterodorsal slope; ventral border almost straight, varying to slightly convex or con-

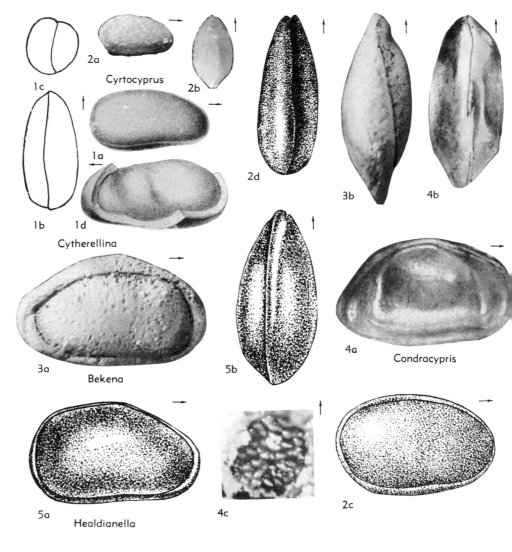

FIG. 285. Bairdiocyprididae (p. Q365-Q366).

cave; ends round or subround. Valves unequal, right larger, overreaching left completely around margin, emphasized most along dorsum and venter, less conspicuously on ends. Surface smooth, genera distinguished by slight muscle scar pit or development of posterior rim and possibly anterior rim. Left valve with duplicature fused or lacking, right valve with continuous groove dipping dorso-anteriorly and very well developed along venter; inner lip forms depression suggesting analogy with left valve of ostracodes with tripartite hinge, duplicature very narrow as in the Cytherellidae; no vestibule. Hinge lacking or at least without differentiation of marginal structure along dorsum. Aggregate circular muscle scar slightly anterodorsal to mid-carapace located on a low ridge extending downward and normal to dorsal margin, commonly more strongly developed in the female. *?Sil., Dev.-Penn., ?L.Perm.-?U.Perm.*

In this family the duplicature and details of overreach are intermediate in character between those typical of the Healdiidae and Cytherellidae. The muscle scar and tendency toward arching of the dorsum are typical of the Healdiidae, the development of a posterior rim being possibly a relict structure derived from healdiid ancestors. The Cavel-

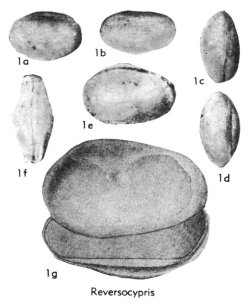

Reversocypris

Fig. 287. Bairdiocyprididae (p. Q366-Q367).

linidae resemble Cytherellidae in size and in nature of the valve contacts but differ in muscle-scar structure.

Cavellina CORYELL, 1928 [*C. puchella*] [=?*Cavellinella* BRADFIELD, 1935; *Alvenus* HAMILTON, 1942 (erroneous original spelling), *Alveus* HAMILTON, 1942]. Carapace oblong to ovate in lateral view, dorsum moderately arched, venter slightly concave to convex, ends rounded, with posteroventer and anterodorsal extremities slightly to moderately truncated. Subovate in dorsal view, posterior end thicker than anterior, especially pronounced in female. Surface smooth; contact margin of right (larger) valve grooved along inner edge so as to receive edge of smaller left valve which may be more subangular than right valve marginally. Sexual dimorphism expressed by shorter, higher, and thicker carapace of females especially in posterior part, where inner body cavity is more fully developed than in male. Some species show tendency to develop shallow muscle-scar pit (as in *Sulcella*) and posterior rim or ridge (as in *Birdsallella, Paracavellina*). *?Sil., Dev.-Penn., ?Perm.,* N.Am.-Eu.-Austral.——FIG. 289,4. *C. puchella,* Penn., USA(Okla.-Tex.); *4a-e,* carapace L, dors., vent., post., ant., ×30 (21); *4g,* muscle scar, ×350; *4f,* RV int., ×60 (82).

Birdsallella CORYELL & BOOTH, 1933 [*B. simplex*]. Carapace small, cavellinoid with only slight right-over-left overreach; posterior swelling develops into an accentuated crest or beveled ridge giving carapace a wedge-shaped outline in dorsal view. Differs from *Sulcella* only in lack of muscle-scar

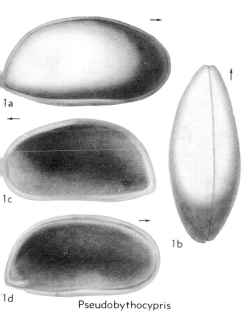

Pseudobythocypris

Fig. 286. Bairdiocyprididae (p. Q366).

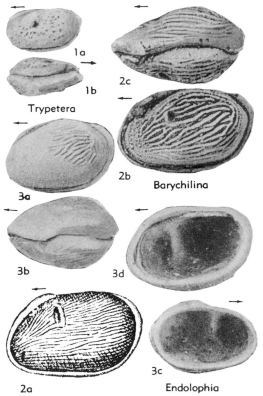

1a

1b

2c

Trypetera

2b

Barychilina

3a

3b

3d

3c

2a

Endolophia

Fig. 288. Barychilinidae (p. Q367-Q368).

pit and shape of posterior ridge. *Penn.*, N.Am.——
Fig. 289,2. **B. simplex*, U.Penn., USA(Tex.);
2a,b, carapace L, dors., ×50 (124).

Chapmanites Krömmelbein, 1954 [**C. crassus*].
Carapace moderately large, less elongate than in
Cavellina. Pronounced marginal right-over-left
overreach, culminating in a high comblike dorsal
flange which tends to be directed toward posterior
end in some species. Surface smooth. *Dev.*, Austral.
——Fig. 289,1. **C. crassus*, M.Dev., SE.Austral.;
1a-c, carapace R, L, dors., ×60 (225).

?Menoeidina Stewart, 1936 [**M. subreniformis*].
Carapace small, elongate-ovate to beanlike in lateral view; very similar to *Birdsallella* but smaller,
with reversed overreach. Probably transitional between representatives of Healdiidae and Cavellinidae. Surface smooth but with posterior median
crescentic ridge. *Dev.*, N.Am.——Fig. 290,2. **M.
subreniformis*, M.Dev., USA(Ohio); *2a,b*, carapace R, dors., ×65 (345).

Paracavellina Cooper, 1941 [**P. elliptica*]. Carapace
like *Cavellina* in general shape, outline, and valve
relationships; differs from other cavellinids in that
both posterior and anterior ends bear a spineless
ridge close to and parallel with margins so as to

form a furrow just inside the ridge. Surface usually
smooth but may be minutely pitted. *U.Miss.*, N.
Am.——Fig. 290,3. **P. elliptica*, U.Miss., USA
(Ill.); *3a,b*, carapace L, dors., ×50 (20).

Platychilella Cooper, 1942 [*pro Platychilus* Cooper,
1941 (*non* Jakolev, 1874)] [**Platychilus ovoides*
Cooper, 1941]. Like *Cavellina* but much smaller
and has a narrow overlap terminally, and with
shallow sulcate area centrodorsally; these features
of doubtful generic significance. *U.Miss.*, Ill.——
Fig. 289,3. *P. sp. ovoides*; *3a,b*, carapace L, dors.,
×40 (20).

Sulcella Coryell & Sample, 1932 (May) [**S. sulcata*] [=?*Sansabelloides* Harris & Lalicker, 1932
(June)]. Carapace like *Cavellina* in shape and
contact of valves but somewhat smaller; posterodorsal cardinal angle slightly sharper than anterodorsal in some forms; posterior margin bordered
by distinct ridge. Surface smooth, except for shallow sulcus which extends from dorsal margin to a
pronounced submedian pit. *Miss.-Penn.*, N.Am.
——Fig. 290,1. **S. sulcata*, Penn., USA(Tex.);
1a-d, ♂ carapace L, dors., vent., ant.; *1e-i*, ♀
carapace L, dors., vent., post., ant.; all ×50 (21).

Tetratylus Cooper, 1941 [**T. elliptica*]. Carapace
like *Cavellina* in general shape. It differs from
other cavellinids in that overreach (right-over-
left) is less emphasized; anterior and posterior
ridges which run parallel to margins like *Paracavellina* terminate dorsally and ventrally in round,
knoblike spines of variable length. A shallow
muscle-scar pit indicates a close relationship to
Sulcella. Surface smooth to finely punctate. *?Dev.,
Miss.*, N.Am.-Eu.——Fig. 291,1. **T. elliptica*, U.
Miss., USA(Ill.); *1a-c*, ♂ carapace L, dors., post.;
1d-f, ♀ carapace L, dors., post.; all ×85 (20).

?Volganella Sharapova & Mandelstam, 1956
[**Volganella magna* Spizharsky, 1956]. Usually
large, irregularly oval, with broadly rounded anterior, posteriorly convergent dorsal and ventral
margins and narrowly rounded posterior margin,
dorsal margin almost straight or faintly convex,
ventral margin straight to faintly concave; surface
smooth, moderately convex, greatest convexity
posteroventral from center; RV slightly overlapping LV. Zone of concrescence narrow, line of
concrescence coinciding with inner margin; vestibule and eye spots lacking; RV hinge with thin,
knifelike bar that fits below dorsal margin of LV.
[Fresh-water.] *U.Perm.*, USSR(Sukhon Basin).
——Fig. 291,2. *V. magna* Spizharsky; LV lat.,
×40 (50).——Fig. 292,1. *V. spizharskyi*; RV lat.,
×35 (50). [Reyment-Bold.]

?Voronina Polenova, 1952 [**V. voronensis*]. Possibly equivalent to *Cavellina*, for reversal of Polenova's orientation provides close resemblance to
C. mesodevonica Pokorný. *Dev.*, Russia.——Fig.
292,2. **V. voronensis*; *2a,b*, ♀ carapace (holotype) L, vent.; *2c,d*, ♂ carapace L, vent.; all
×45 (277). [Shaver.]

?Family KRAUSELLIDAE Berdan, n. fam.

[Materials for this family prepared by JEAN BERDAN, United States Geological Survey]

Asymmetrical, smooth ostracodes, LV distinctly different from RV, overlapping it ventrally and in some forms overreaching it dorsally; posterior end of either valve produced as spine. Hinge line straight, shorter than greatest length, hinge simple; muscle scar unknown; duplicature absent. *M.Ord.-M.Dev.*

Krausella ULRICH, 1894 [*K. inaequalis*] [=*Rayella* TEICHERT, 1939 (*pro Basslerites* TEICHERT,

Fig. 289. Cavellinidae (p. Q369-Q370).

1937, *non* Howe, 1937)]. LV suboval in outline, overlapping RV ventrally, RV being produced posteriorly in short, blunt spine. Hinge line may coincide with dorsal margin, may be entrenched below dorsal margin of both valves, or LV may overreach RV along hinge line. *M.Ord., ?L.Dev.,* N.Am.-Eu.——Figs. 293,1, 294,1*a,b.* *K. inaequalis,* M.Ord., USA (Ill.); 293,1*a,* carapace (holotype) R, ×15; 293,1*b,* carapace (topotype) vent., ×15; 294,1*a,* carapace dors., ×20; 294,1*b,* cross-

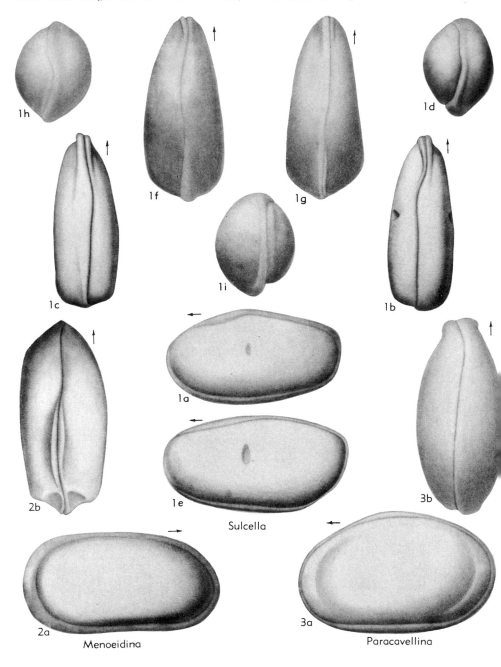

Fig. 290. Cavellinidae (p. *Q*370).

section showing overlap, ×33 (Berdan, n).——Fig. 294,2a-c. *K.* sp., M.Ord. (Edinburg F.), USA(Va.); *2a-c,* carapace R, L, vent., ×20 (J. C. Kraft, n).

?**Cooperatia** TOLMACHOFF, 1937 [*pro Cooperia* TOLMACHOFF, 1926 (*non* RANSOM, 1907)] [*Cooperia granum* TOLMACHOFF, 1926]. Suboval in lateral outline, LV acuminate at rear, rounded acuminate in front; LV overlapping RV all round, forming both anterior and posterior ends of carapace, dorsal margin curved. *M.Dev.,* Arct. N.Am.——Fig. 293,3. *C. granum* (TOLMACHOFF); *3a,b,* carapace L, dors.; *3c,* carapace R; all ×28 (370).

Janusella ROTH, 1929 [*J. biceratina*]. LV subtriangular, with spine at apex of dorsum, RV like that of *Krausella. L.Dev.,* N.Am.——Fig. 293,2. *J. biceratina,* USA(Okla.); *2a,b,* carapace (topotype) R, dors., ×30 (Berdan, n).

Family PACHYDOMELLIDAE
Berdan & Sohn, n. fam.

[Materials for this family prepared by JEAN BERDAN and I. G. SOHN, United States Geological Survey]

Inequivalved, subovate, asymmetrical, medium-sized, with thick shells; tubules normal to shell surface, wider on interior surface, invisible on exterior of well-pre-

served specimens; hinge simple, hinge line straight, shorter than greatest length of carapace. Surface smooth, punctate or rugulose. *?Ord., Sil.-Dev.*

Pachydomella ULRICH, 1891 [*P. tumida*] [=?*Senescella* STEWART & HENDRIX, 1945]. Round cross-section, prominent shoulder on larger valve separated from dorsum by deep groove parallel to hinge line. Surface rugulose or smooth. *?L.Dev., M.Dev.,* C.N.Am.——Fig. 295,1. *P. tumida;* M. Dev., USA(Ky.); *1a-d,* carapace (holotype) R, dors., vent., post.; *1e,* LV (topotype) int. showing tubules; all ×30 (Sohn, n).

Tubulibairdia SWARTZ, 1936 [*T. tubulifera*]. Round cross-section, lacking shoulders or well-defined grooves. Surface smooth or finely punctate. *?Ord., M.Sil.-M.Dev.,* E.N.Am.-C.Eu.——Fig. 295,2a,b. *T. tubulifera,* L.Dev., Pa.; *2a,b,* LV (holotype) artificial cast, lat., dors., ×30.——Fig. 295,2c,d, *T. punctulata* (ULRICH), M.Dev., USA(Ky); *2c,d,* LV (topotype) post., LV radiograph showing internal tubules (Sohn, n).

Phanassymetria ROTH, 1929 [*P. triserrata*] [=*Phanassymetrica* NEAVE, 1940 (*errore*); *Phanasymmetrica* VAN DEN BOLD, 1946 (*errore*)]. Angu-

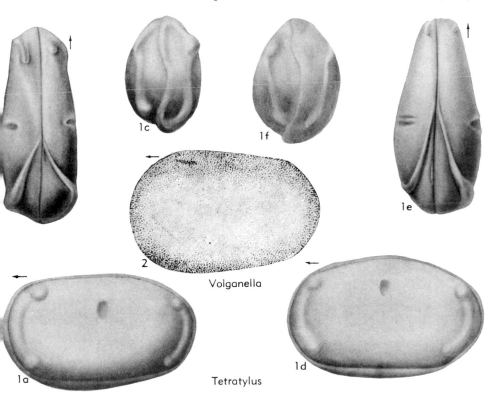

1c

1f

1e

2

Volganella

1a

Tetratylus

1d

FIG. 291. Cavellinidae (p. Q370).

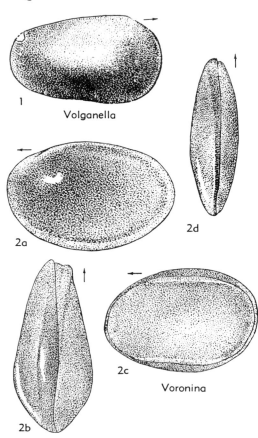

Volganella

2a

2d

2c

Voronina

2b

Fig. 292. Cavellinidae (p. Q370).

lar cross section, groove parallel to hinge line may be present on LV. Surface smooth. ?M.Sil., L.Dev., C.N.Am.——Fig. 295,3a-c. *P. triserrata, L.Dev., USA(Okla.); 3a-c, carapace (holotype) L, dors., ant., ×30 (Sohn, n).——Fig. 295,3d. P. sp., L. Dev., USA(Okla.); LV int., showing tubules, ×30 (335).

Superfamily QUASILLITACEA Coryell & Malkin, 1936

[nom. transl. SYLVESTER-BRADLEY, 1956 (ex Quasillitidae CORYELL & MALKIN, 1936)] [Diagnosis and discussion by H. W. SCOTT, University of Illinois]

Ostracodes with a primitive tripartite hinge; mostly straight-backed, some arched; muscle-scar circular, composed of many small secondary scars; inner calcareous lamella absent or poorly developed; surface ornamented. Dev.-Miss., ?Penn.

The Quasillitacea are included in the Metacopina because of the presence of an inner calcareous lamella in some forms, general outline and ornamentation of the carapace, and the tripartite hinge. However, a duplicature has not been demonstrated to exist in all forms and the details of hinge structure are inadequately known in many closely related groups. The muscle-scar pattern is suggestive of that found in Healdiidae, and, in fact, is very unlike anything known in the Podocopina. Hingement, ornamentation, and outline are somewhat similar to such features in the Cytheracea. In the final allocation of the quasillitids, it becomes a question of relative values. Is hingement more important than muscle-scar pattern; or is the muscle-scar pattern of less value than a poorly developed duplicature and general outline?

The stratigraphic distribution of quasillitacean genera of ostracodes is shown graphically in Figure 296.

Family QUASILLITIDAE Coryell & Malkin, 1936 (November)

[=Graphiodactylidae KELLETT, 1936 (December)] [Materials for this family prepared by I. G. SOHN, United States Geological Survey, and L. E. STOVER, Tulsa, Oklahoma]

Straight-hinged, subquadrate ostracodes with well-developed to vestigial marginal flange on end margins. Hinge (merodont) with terminal crenulated teeth and sockets. Surface ornamented by ridges and grooves, usually with posterior spines. Muscle scar round, aggregate consisting of circular spots; inner lamella narrow (KELLETT, 1936), usually not preserved. Dev.-Miss., ?Penn.

Quasillites CORYELL & MALKIN, 1936 [*G. obliquus] [=?Burlella CORYELL & BOOTH, 1933; Spinovina CORYELL & MALKIN, 1936; ?Lucasella STEWART, 1936; Allostracites PŘIBYL, 1953 (pro Paracythere BASSLER, 1932, non MÜLLER, 1894]. Surface ornamented with grooves and ridges, posterior spine and ridge in front of posterior margin. Dev.-Miss. (L.Carb.), ?Penn., N.Am.-Eu.——Fig. 297,1. *Q. obliquus, Dev., Can.(Ont.); 1a-d, carapace (holotype) R, L, dors., vent., ×30 (Sohn, n).——Fig. 298,2. Q. sp. cf. Q. obliquus, Dev., Can.(Ont.); RV int. (converted to fluoride, anterodorsal portion broken), ×60 (Sohn, n).

Costatia POLENOVA, 1952 [*C. posneri]. Differs from Jenningsina in vertical trend of ornamental ribs, groove joining dental sockets, and crenulated opposing ridge. Muscle spot not surrounded by rim. M.Dev., Eu.(USSR).——Fig. 299,1a-d. *C. posneri, 1a,b, carapace (holotype) L, dors.; 1c,d, LV int., RV int.; ×60 (277).——Fig. 299, 1e,f. C. cavernosa POLENOVA; 1e,f, carapace (holotype) R, dors., ×60 (277).

?Eriella STEWART & HENDRIX, 1945 [*E. robusta*]. Differs from *Jenningsina* in concentric ornamentation, absence of rim around muscle-scar pit, and presence of posteroventral spine. Hinge unknown. *Dev.*, N.Am.——FIG. 300,*3.* *E. robusta*, USA (Ohio); *3a-d*, carapace (holotype) R, L, dors., vent., ×70 (Sohn, n.).

?Euglyphella WARTHIN, 1934 [*Strepula sigmoidalis* JONES, 1890]. Small, with high or low bifurcating ridges; accommodation groove above dental sockets. *Dev.*, N.Am.——FIG. 298,*1.* *E. sigmoidalis*, Dev., N.Y.; *1a*, LV (holotype) lat., ×60; *1b*, LV lat., ×60 (Sohn, n).

Graphiadactyllis ROTH, 1929 [*Kirkbya lindahli arkansana* GIRTY, 1910] [=*Graphiodactylus* ROTH, 1929; *Bassleria* HARLTON, 1929]. Like *Quasillites* but differs in presence of well-developed marginal flanges and lack of posterior ridge and spine. *Miss.*, N.Am.——FIG. 300,*1.* *G. arkansana*, USA (Ark.); *1a*, carapace (lectotype) R, ×40; *1b*, dors., ×30; *1c,d*, RV int., LV int., ×40 (Sohn, n.).

Jenningsina CORYELL & MALKIN, 1936 [*Graphiodactylus catenulatus* VAN PELT, 1933]. Differs from *Quasillites* in muscle-scar pit surrounded by rim,

FIG. 294. Krausellidae (p. Q371-Q373).

Krausella

Janusella

Cooperatia

FIG. 293. Krausellidae (p. Q371-Q373).

steep posterior shoulder, and absence of spines. *Dev.*, N.Am.——FIG. 300,*2.* *J. catenulata;* *2a-d*, carapace L, R, dors., vent; *2e*, LV int., ×80 (Stover, n).

?Scalptina STOVER, 1956 [*S. incuda*]. Differs from *Jenningsina* in having reticulations, a smooth central spot and lacking an anteroventral flange. *M. Dev.*, N.Am.

Family BUFINIDAE Sohn & Stover, n. fam.

[Materials for this family prepared by I. G. SOHN, United States Geological Survey, and L. E. STOVER, Tulsa, Oklahoma, and with contribution from R. H. SHAVER, Indiana University and Indiana Geological Survey]

Straight-hinged, subovate, smooth, striate or reticulate, with terminal ridges subconcentric to end margins, and usually beaded end margins. Hinge and muscle scar like *Graphiadactyllis.* Each curved ridge may be reduced to one or more spines, or one ridge may be missing. *Dev., ?Penn.*

Bufina CORYELL & MALKIN, 1936 [*B. elata* CORYELL & MALKIN (=?*Moorea bicornuta* ULRICH, 1891)] [=*Parabufina* SMITH, 1956]. Well-developed terminal ridges, beaded end margins; posterior ridge may be reduced to 2 thick spines. *Dev.*, E.N.Am.——FIG. 301,*1a.* *B. elata*, Dev., Can.; carapace (holotype) R, ×40 (Sohn, n).——FIG. 301,*1b-e.* B. ?*bicornuta* (ULRICH), Dev., USA (N.Y.); *1b-d*, carapaces (*1b*, holotype) R, L, dors.; *1e*, LV int.; all ×40 (Sohn, n, 348).

?Aurigerites ROUNDY, 1926 [*A. texanus*]. Posterior ridge horseshoe-shaped, ? no anterior ridge or spines. Based on corroded carapaces so that hingement and muscle scar unknown. *Penn.*, SW.USA.——FIG. 301,*2.* *A. texanus*, USA(Tex.); *2a,b,* carapace (lectotype, herein designated) R, dors., ×40 (Sohn, n).

?**Bythocyproidea** STEWART & HENDRIX, 1945 [*B. sanduskyensis*]. Anterior margin narrower than posterior; posterior ridge on RV only, no anterior ridge or spines. Hinge and muscle scar unknown. *M.Dev.*, E.USA.——FIG. 302,1. *B. sanduskyensis*, USA(Ohio); 1a,b, carapace (lectotype, herein designated) R, dors., ×36 (Shaver, n.).——FIG. 301,3. *B. eriensis* STEWART & HENDRIX, USA (Ohio); 3a,b, carapace (lectotype, herein designated) R, dors., ×40 (Sohn, n).

Punctomosea STOVER, 1956 [*Thrallella cristata* SWARTZ & ORIEL, 1948]. Differs from *Bythocyproidea* only in having posterior ridges on both valves. *M.Dev.*, N.Am.——FIG. 302,2. *P. cristata; 2a,b*, carapace R, dors.; 2c, RV int.; all ×25 (348).

Family ROPOLONELLIDAE
Coryell & Malkin, 1936

[Materials for this family prepared by I. G. SOHN, United States Geological Survey]

Straight-hinged, subtriangular, minute, with terminal ridges that may be spinose or single spines. Ridge-and-groove hinge known in one genus; muscle scar and surface ornament unknown. *Dev.*

Ropolonellus VAN PELT, 1933 [*B. papillatus*]. Rows of spinelets near end margins. Hinge line 0.7 or more of greatest length. *Dev.*, E.USA.——303,3. *R. papillatus*, N.Y.; 3a,b, carapace R, dors., ×60; 3c,d, carapace (another specimen) R, dors., ×40 (Sohn, n).

Rudderina CORYELL & MALKIN, 1936 [*R. extensa*]. Single anterior and posterior backwardly-directed

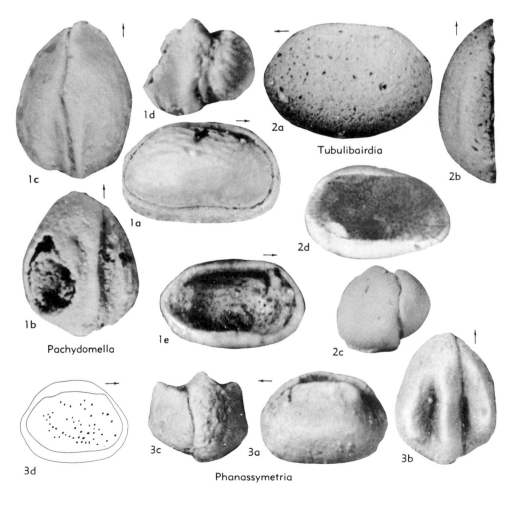

1c 1d 1a 1b Pachydomella 1e

2a Tubulibairdia 2b 2d 2c

3c 3a 3b 3d Phanassymetria

FIG. 295. Pachydomellidae (p. Q373-Q374).

spines extending from valve surface above contact of valves, located in ventral 0.3 of valve height. *M.Dev.*, N.Am.——Fɪɢ. 303,*1*. *R. extensa*, Can.; *1a-d*, LV lat., int., dors., vent., ×40 (Sohn, n).

Varicobairdia Pokorný, 1950 [**Bairdia (Varicobairdia) ķettneri*]. Terminal ridges with tubercles or spines. Hinge line 0.7 or less of greatest length. *Dev.*, C.Eu.——Fɪɢ. 303,*2*. **V. ķettneri*, Czech.; *2a-c*, carapace (topotype) R, L, dors., ×40 (Sohn, n).

Family UNCERTAIN

Unicornites Pokorný, 1950 [**U. homeomorphus*]. Carapace elongate, subquadrate in lateral view, fusiform in dorsal view, with posteroventral spine on RV, with variable LV over RV overlap, except none or reversed overlap along hinge. [These features suggest Quasillitidae, but the surface is smooth and the hinge, contact margins, and adductor muscle scar are unknown. *M.Dev.*, Eu.——Fɪɢ. 298,*3*. **U. homeomorphus*, Czech.; *3a,b*, carapace R, vent., ×25 (275). [Shaver.]

?Superfamily THLIPSURACEA Ulrich, 1894

[*nom. transl.* Kesling, herein (*ex* Thlipsuridae Ulrich, 1894)] [Diagnosis by R. V. Kesling, University of Michigan]

Carapace subovate to elongate-subelliptical in lateral view; dorsal border slightly convex, ventral border nearly straight, ends subround; valves unequal, LV overreaching RV and with slight left-over-right overlap; posterior end compressed, rest of lateral surface convex except for pits, furrows, or depressed areas. Insofar as known, hinge consisting of groove in one valve and corresponding ridge in the other (some genera have groove in LV, others in RV); some species with elongate anterior and posterior sockets below ridge; function of sockets unknown, since opposite valve has no elements fitting into them. Details of muscle scars unknown. *?Ord., Sil.-Dev.*

The stratigraphic occurrences of thlipsuracean genera are indicated diagrammatically in Figure 296.

Family THLIPSURIDAE Ulrich, 1894

[Materials for this family prepared by R. V. Kesling, University of Michigan]

Carapace subovate to elongate-subelliptical in lateral view, subelliptical to subquadrate in dorsal view; dorsal border convex; ventral border nearly straight or, in some species, slightly concave; ends round or subround; valves unequal, in known species, LV overreaching RV on all margins, over-

lap slight; lateral surfaces convex except for pits, furrows, or depressed areas and posterior end of carapace, which is characteristically compressed. Hingement, development of ?duplicature, and muscle scars unknown in most genera, hence taxonomic position of the family remains in doubt. *?Ord., Sil.-Dev.*

	Ord.	Sil.	Dev.	Miss.	Penn.
1 Octonaria E					
2 Thlipsuroides E					
3 Thlipsura E					
4 Thrallella E					
5 Thlipsurella E					
6 Thlipsuropsis E					
7 Eucraterellina E					
8 Rothella E					
9 Bufina A					
10 Costatia B					
11 Eriella B					
12 Euglyphella B					
13 Jenningsina B					
14 Ropolonellus C					
15 Varicobairdia C					
16 Quasillites B					
17 Bythocyproidea A					
18 Punctomosea A					
19 Scalptina B					
20 Rudderina C					
21 Unicornites D					
22 Eustephanella E					
23 Favulella E					
24 Hyphasmaphora E					
25 Ponderodictya E					
26 Stibus E					
27 Strepulites E					
28 Thlipsurina E					
29 Graphiadactyllis B					
⁞ ? known		30 Aurigerites A			

Fɪɢ. 296. Stratigraphic distribution of quasillitacean and thlipsuracean ostracode genera (Moore, n). Classification in families is indicated by letter symbols (A—Bufinidae, B—Quasillitidae, C—Ropolonellidae, D—Quasillitacea, family uncertain, E—Thlipsuridae). An alphabetical list of genera furnishes cross reference to the serially arranged numbers given on the diagram.

Generic Names with Index Numbers

Aurigerites—30	*Quasillites*—16
Bufina—9	*Ropolonellus*—14
Bythocyproides—17	*Rothella*—8
Costatia—10	*Rudderina*—20
Eriella—11	*Scalptina*—19
Eucraterellina—7	*Stibus*—26
Euglyphella—12	*Strepulites*—27
Eustephanella—22	*Thlipsura*—3
Favulella—23	*Thlipsurella*—5
Graphiadactyllis—29	*Thlipsurina*—28
Hyphasmaphora—24	*Thlipsuroides*—2
Jenningsina—13	*Thlipsuropsis*—6
Octonaria—1	*Thrallella*—4
Ponderodictya—25	*Unicornites*—21
Punctomosea—18	*Varicobairdia*—15

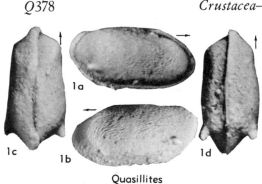

la

lc   lb   ld

Quasillites

Fig. 297. Quasillitidae (p. Q374).

Thlipsura Jones & Holl, 1869 [*T. corpulenta] [=Craterellina Ulrich & Bassler, 1913]. Lateral surfaces strongly convex except for flat depressed area at posterior; contact of flat area with convex part of valve very abrupt; in some species, one or more furrows form re-entrants extending forward into convex part of valve. Sil.-Dev., Eu.-N.Am. ——Fig. 304,2d. *T. corpulenta, Sil., Eng.; carapace R, ×50 (85).——Fig. 304,2a,b. T. confluens Swartz, L.Dev., Pa.; 2a,b, carapace R, dors., ×60 (297).——Fig. 304,2c. T. robusta (Ulrich & Bassler), L.Dev., Pa.; carapace R, ×35 (74).

Eucraterellina Wilson, 1935 [*E. randolphi] Like Thlipsura, except posterior depression completely surrounded by distinct rim at level of rest of valve to form crater, which in some species has nodes within that rise to general level of valve but in other species has prolongations of lateral surface into it, so that crater becomes deep C-shaped furrow. L.Dev., N.Am.——Fig. 305,1a. *E. randolphi, Helderberg, Tenn.; carapace L, ×50 (403).——Fig. 305,1b. E. oblonga (Ulrich & Bassler), USA(Pa.); carapace L, ×50 (74).—— Fig. 305,1c. E. crateriformis Swartz, USA(Pa.); carapace R, ×50 (74).

Eustephanella Swartz & Swain, 1942 [*Eustephanus catastephanes Swartz & Swain, 1941] [=Eustephanus Swartz & Swain, 1941 (non Reichenbach, 1849)]. Surface of each valve broadly convex, with pits tending to lie in curving furrows that more or less parallel dorsal border, much as in Octonaria; posterior slope concave; crest at summit of this slope bearing 2 posteriorly directed short spines. [Differs from Octonaria in lacking a marginal flange, and from Stibus in being more elongate and having pits more evenly distributed and in grooves.] M.Dev., N.Am.—— Fig. 304,3. *E. catastephanes (Swartz & Swain), Onond., USA(Pa.); LV lat., ×40 (76).

?Favulella Swartz & Swain, 1941 [*Bythocypris favulosa Jones, 1889]. Submarginal ridge in each valve, indistinct dorsally in some species, setting off marginal flange, which is low and indistinct ventrally in both valves and dorsally also in RV;

posterior part of submarginal ridge extended into 2 well-developed posteriorly-directed spines; area enclosed by ridge honeycombed with coarse, close-set puncta of various shapes and sizes, except on subcentral smooth spot. Hinge of LV with main longitudinal groove, long anterior accessory groove somewhat oblique to main groove, and possibly short posterior accessory groove; these grooves receive slightly modified dorsal edge of RV; LV apparently overlapping RV except at ends. [This genus appears closely related to Ponderodictya.] M.Dev., N.Am.——Fig. 306,6. *F. favulosa, Onond., USA(Pa.); 6a, juv. LV ext., ×35; 6b, RV ext., ×35 (76).

Hyphasmaphora Van Pelt, 1933 [*H. textiligera]. Each valve divided into sloping smooth marginal area and central reticulate area; conspicuous circular pit in center of each valve. Posterior part of marginal area sloping sharply away from reticulate area to posterior border; LV hinge with long groove; narrow flange along border of each end of valve. M.Dev., N.Am.——Fig. 304,4. *H. textiligera, Hamilton, USA(Mich.); LV lat., ×70 (400).

Octonaria Jones, 1887 [*O. octoformis]. Lateral surface of each valve divisible into large, nearly flat central area and convex marginal area extending out to free edge and hinge line; central part of flat area marked by pits or pitlike depressions, in some species large and few, in others small and numerous; edge of flat area forms rim around pitted area; LV in some species has fewer pits than RV, and in one species has none. ?Ord., Sil.-Dev., Eu.-N.Am.——Fig. 304,1d. *O. octoformis, Sil., Eng.; RV lat., ×40 (185).——Fig. 304,1a-c. O. laevilatata Kesling & Kilgore, M.Dev.(Hamilton), USA(Mich.); 1a-c, carapace L, R, dors., ×50 (209).——Fig. 307,1. O. bifurcata Stover, M.Dev. (Hamilton), USA(N.Y.); 1a,b, carapace R, L, ×30 (348).

?Ponderodictya Coryell & Malkin, 1936 [*Cytherella(?) bispinulatus Stewart, 1927 (=*Leperditia punctulifera Hall, 1860)] [=Hamiltonella Stewart, 1936]. Dorsal margin sinuous; surface closely punctate; carapace swollen posteriorly; some specimens ornamented with one or more spines on posterior and rarely on anterior; crescentic swollen ridge on anterior; intraspecific variation very great; LV overlapping selvage of RV along venter. Hinge hemimerodont; muscle scar an oval aggregate clump, marked on outer surface by lack of puncta. [Included in Thlipsuridae with doubt; may be related to Quasillitidae (Metacopina) or even Healdiidae (Metacopina).] M.Dev., N.Am.——Fig. 306,3. *P. punctulifera, M.Dev. (Hamilton), USA(Mich.), Ont.-N.Y.; 3a,b, RV lat., vent.; 3c, hinge (RV int.), all ×35; 3d, muscle scar (RV int.), ×100 (82).

Rothella Wilson, 1935 [*Dizygopleura recta Roth, 1929]. Carapace subrectangular in lateral and dorsal views; dorsal border nearly straight or slightly convex; greatest length through middle of valve, greatest height posterior; surface with anterior and posterior craters, which are vertically elongate in some species, round in others; some species with rims around craters, others with ridge or lip at one side of crater, and some with no raised structures. L.Dev., N.Am.——Fɪɢ. 305,2. *R. recta, Helderberg., USA(Tenn.); 2a,b, carapace R, post., ×40 (297).

Stibus Swartz & Swain, 1941 [*S. kothornostibus]. Surface of each valve convex, with steep, concave posterior slope; crest at top of slope bearing 2 small, posteriorly directed spines, as in *Eustephanella*; one or more furrows, in some species only a row of pits, near to and approximately parallel with anterior and anterodorsal borders; in some species posterior part of each valve with numerous small pits but in others this part is smooth. M. Dev., N.Am.——Fɪɢ. 306,2. *S. kothornostibus, Onond., USA(Pa.-W.Va.); LV lat., ×60 (76).

Strepulites Coryell & Malkin, 1936 [*S. mooki] [=Octonariella Bassler, 1941]. Like *Octonaria* but with more strongly developed narrow ridges, somewhat parallel to borders; many species with short, low ridges connected to main ridges and surrounding pitlike areas. Dev., N.Am.——Fɪɢ. 305,4a. *S. mooki, M.Dev.(Hamilton); Can. (Ont.); carapace R, ×50 (22).——Fɪɢ. 305,4b. S. quadricostata (Van Pelt), M.Dev.(Hamilton), USA(Mich.); LV lat., ×60 (391).——Fɪɢ. 305, 4c. S. crescentiformis (Van Pelt), M.Dev.(Hamilton), USA(Mich.); carapace R, ×70 (391).—— Fɪɢ. 307,2. S. divectus Stover, M.Dev.(Hamilton), USA(N.Y.); 2a,b, carapace L, R, ×25 (348).

Thlipsurella Swartz, 1932 [*T. ellipsoclefta]. Like *Thlipsura* except for depressed areas; distinct ridge rising from posterior border; 2 sublongitudinal furrows in rear part of each valve, separated from posterior border by ridge; short subvertical furrow near middle of valve or shallow pit in anterior half, or both. M.Sil.-U.Sil., Eu.; L.Dev., N.Am. ——Fɪɢ. 305,3. *T. ellipsoclefta, L.Dev., N.Am.; 3a,b, LV lat., dors., ×60 (74).

Thlipsurina Bassler, 1941 [*non* Kummerow, 1953] [*T. elongata]. Like *Thlipsura*, but with broad, shallow, transverse depression near middle of each valve; narrow concave area forming posterior slope, bounded anteriorly by distinct ridge, immediately in front of which is elongate, nearly vertical sulcus widest at bottom, extending from dorsal border almost to ventral. M.Dev., N.Am. ——Fɪɢ. 306,5. *T. elongata, Onond., USA (Tenn.); 5a,b, LV lat., RV lat., ×40 (4).

Thlipsuroides Morris & Hill, 1951 [*T. thlipsuroides]. Overlap confined to narrow area; each valve with 2 long, subparallel, nearly longitudinal grooves (in type species with pits along bottom); in dorsal view, posterior slope of each valve nearly flat, set at angle of about 45° to hinge.

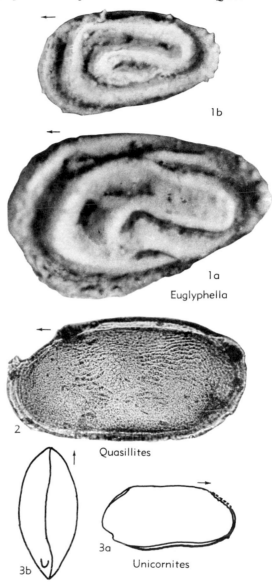

Euglyphella

Quasillites

Unicornites

Fɪɢ. 298. Quasillitidae, Family Uncertain (p. Q374-Q377).

Sil., N.Am.——Fɪɢ. 306,1. *T. thlipsuroides, M. Sil., USA(Tenn.); 1a,b, carapace R, dors., ×25 (253).

Thlipsuropsis Swartz & Whitmore, 1956 [*T. diploglyptulis]. Markedly inequivalved, LV overlapping RV; LV with posterior depressed area confluent with anteriorly projecting furrows, RV lacking depressed area; narrow furrow parallel to anterodorsal border. U.Sil.-L.Dev., N.Am.——Fɪɢ.

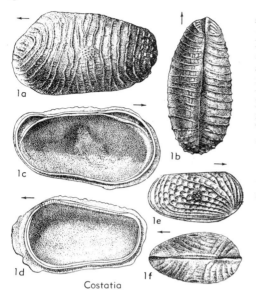

Costatia

Fig. 299. Quasillitidae (p. *Q*374).

307,3. **T. diploglyptulis,* U.Sil., USA(N.Y.); *3a,b,* carapace L, R, ×45 (78).

Thrallella STEWART & HENDRIX, 1945 [**T. phaseolina*]. Small; crescentic ridge in posterior part of each valve, subparallel to posterior border; shallow pits in area immediately in front of crescentic ridge, rear row of pits regularly spaced along concave edge of ridge. *M.Dev.,* N.Am.——FIG. 306,*4.* **T. phaseolina,* Hamilton., USA(Ohio); RV lat., ×70 (347).

Suborder PLATYCOPINA Sars, 1866

[*nom. correct.* SYLVESTER-BRADLEY, herein (*pro* Platycopa SARS, 1866)] [Type genus: *Cytherella* JONES, 1859; SD SARS, 1866] [Diagnosis and discussion by H. W. SCOTT, University of Illinois, and P. C. SYLVESTER-BRADLEY, University of Leicester]

Podocopida in which muscle-scar pattern is a biserial aggregate of small scars. Dorsal margin convex, ends round; valves unequal, larger (typically RV) overlapping smaller; uninterrupted contact furrow in larger RV receiving selvage edge of smaller valve; duplicature narrow or wanting; dimorphism by posterior swelling. *Jur.-Rec.*

SARS based definition of this assemblage mainly on characters of the appendages, as indicated by his description, which follows:

Lower antennae [antennae proper] biramous, equal, similar to the feet of Copepoda, the basal part biarticulate and geniculate, with numerous setae attached to both margins. Upper antennae [antennules] very large and strong, multiarticulate, geniculate at the base, with short spines. Mandibles

small and feeble, with a large palp. There are only three pairs of thoracic limbs, all maxilliform. Mandibular palp and first maxillae [maxillulae] are provided with a pair of combs with large bristles attached to the inner surfaces. First and second maxillae [maxillulae and third post-oral limbs] provided with large branchial plates; the third maxillae [fourth post-oral limb] rudimentary in the female, in the sea they are evolute and prehensile. The postabdominal rami [furcal rami] are small and narrow, distinctly separate, spinose at their apices (65, translated by SCOTT and WAINWRIGHT).

The Platycopina are a suborder composed of only one family. The lack of a well-defined inner calcareous lamella separates them from the Podocopina; lack of lobes, sulci, and ventral frills and the convex outline of the back in side view distinguish them from most palaeocopids, the uninterrupted contact furrow separating them from all other ostracodes, and RV-over-LV overlap further differentiating them from most Metacopina.

They are readily separated from the thlipsurids by absence of pits and furrows and from the Healdiacea by their RV-over-LV overlap and unmodified hinge. The hinges of the Healdiacea are better defined than those in the cytherellids.

Among the Cytherellidae are two very common subgenera that possess very different external characteristics; these are *Cytherella (Cytherella),* with a smooth unornamented carapace, and *C. (Cytherelloidea),* with a carapace modified by ribs and in some species pits, tubercles, and a muscle-scar depression. Dimorphism in the family is recognized by the posterior swelling of the female carapace. In *Cytherelloidea* dimorphism may be expressed by greater posterior tumidity of the female and a modification of the ribs. In some females the ribs are more pronounced than in males, in others more subdued.

The Cavellinidae are very similar to the Cytherellidae and at one time were considered to be the same. They differ in muscle-scar structure, the cavellinids possessing numerous small secondary scars set in a circular to ovate cluster and the cytherellids possessing a double row of about 10 or more individual scars. In all other respects the cavellinids are like *Cytherella;* in fact, they are so similar that once they were considered merely to be dimorphs. The family has been questionably reported from the Silurian but is essentially a Devonian-to-Middle Permian group, with greatest development in the Pennsylvanian, whereas the cytherellids range from Jurassic to Recent.

Cavellina is considered to be the central stock from which the cytherellids developed. The many Paleozoic forms which have been called *Cytherella* probably should be transferred to *Cavellina*. If we had no knowledge of the muscle scars we would be forced to rely on the single factor of direction of overlap in separating cytherellids from cavellinids. Though the muscle scars are distinct, biserial, and few in *Cytherella,* and irregularly arranged and many in *Cavellina,* we cannot overlook the fact that one of the trends in ostracode evolution was reduction in the number of adductor muscle fibers. The cytherellids appear to be a direct de-

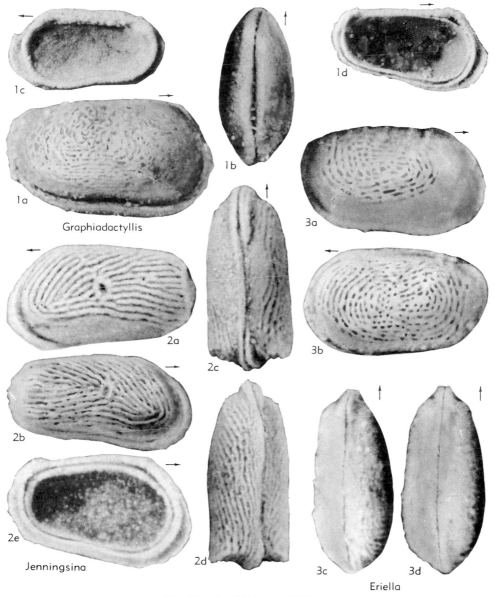

FIG. 300. Quasillitidae (p. Q375).

FIG. 301. Bufinidae (p. Q375-Q376).

velopment from the cavellinids by the simple process of reducing the number of muscle scars. During this reduction, the duplicature, contact groove, and general outline of the carapace remained unchanged, although the dorsum tended to become straighter in the cytherellids. The only basis for separating cavellinids from cytherellids is difference in muscle-scar patterns.

The stratigraphic distribution of platycopine genera (and subgenera) is shown diagrammatically in Figure 308.

Family CYTHERELLIDAE Sars, 1866

[Materials for this family prepared by R. A. REYMENT, University of Stockholm]

Carapace mostly oval in side view, anterior and posterior ends generally subequal in height; RV typically larger than LV. Marginal pore canals lacking or (as in *Cytherella*) represented by short simple canals of normal type; contact furrow entire, hinge undifferentiated; selvage of smaller valve forming valve edge; Mesozoic and Cenozoic (including Recent) representatives with adductor-muscle field composed of 2 curved parallel rows, each with 5 to 9 elongated spots. Dimorphic, with brood section in posterior part of carapace. *Jur.-Rec.*

Cytherella JONES, 1849 [*Cytherina ovata* ROEMER, 1840; SD ULRICH, 1894] [=*Morrowina* LOETTERLE, 1937]. Carapace small to moderately large, thick-shelled; surface smooth or ornamented with

pits or ribs; RV larger than LV, its margin being grooved all around; shape and ornament of opposite valves commonly different. LV hinge with dorsal ridge, RV with corresponding furrow; no marginal pore canals, although normal canals in marginal area may resemble them. *Jur.-Rec.,* world-wide.

C. (Cytherella). Surface smooth to faintly ribbed concentrically, posterior weakly denticulated in some species, carapace of Recent forms milky white; egg-shaped in side view; anterior end rather compressed, rear end more inflated, especially in females, most species having equally rounded ends but some with slightly sharper rear ends; Recent species with hairs, particularly on posterior part of valves. Adductor-muscle fields with 2 usually slightly bent rows, each with 5 to 9 longitudinal to subrectangular spots, muscle field commonly on slight inner elevation that appears externally as a depression. Sexual dimorphism prominent, females being larger than

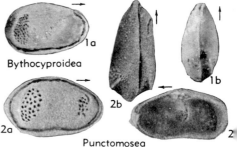

FIG. 302. Bufinidae (p. Q376).

males, with internal swellings that may divide valves into anterior and posterior parts, but such ridge being poorly developed or absent in males; extremities of Recent species greatly specialized. *Jur.-Rec.,* cosmop.——Fig. 309,1. *C. (C.) abyssorum* Sars, Neog., Br. Is.; *1a,b,* ♀ carapace L, dors.; *1c,d,* ♂ carapace R, dors.; all ×38 (312).

C. (Cytherelloidea) Alexander, 1929 [*Cytherella williamsoniana* Jones, 1849]. Differs from *C. (Cytherella)* in generally stronger ornament, especially ribbing, and generally more compressed shell form, but because variation within a single species may range from strongly ornamented to entirely smooth, distinction from *C. (Cytherella)* is provisional (97); forms regarded as females have 2 round impressions on inner surface. *Jur.-Rec.,* cosmop.——Fig. 310,2a,b. *C. (C.) williamsoniana* Jones, L.Cret., Eng.; *2a,b,* RV lat., LV lat., ×50 (328).——Fig. 310,2c. *C. (C.) ouachitensis* Howe, U.Eoc. (Jackson.), USA(La.); LV lat., ×50 (328).——Fig. 310,2d. *C. (C.) alabamensis* Howe, Oligo., USA(Ala.); LV lat., ×50 (328).

C. (Staringia) Van Veen, 1936 [*Terquemia falcoburgensis;* SD Howe & Laurencich, 1958] [=*Terquemia* Van Veen, 1932 (*non* Tate, 1868)]. Thick-shelled, with smooth surface, carapace tending to be drawn out strongly; forms regarded as females commonly with single posterior impression in each valve and more inflated posteriorly; ventral and dorsal margins usually but not invariably weakly convex, showing tendency to be almost straight. *U.Cret.(Maastricht.),* Holl.——Fig. 310,1. *S. falcoburgensis; 1a-e,* ♀

carapace R, L, dors., vent., post.; *1f,g,* ♂ carapace R, dors.; *1h,* ♂ RV int.; all ×25 (395).

?Ankumia Van Veen, 1932 [*A. bosqueti*]. Carapace thick-walled, RV larger than LV, ventral margin strongly concave; surface with smooth concentric rings. Hinge with anterior and posterior teeth. Sexual dimorphism present. *U.Cret.,* Eu. (Holl.).——Fig. 310,3. *A. bosqueti; 3a-e,* ♀ RV lat., vent., dors., ant., int., ×25 (395).

?Platella Coryell & Fields, 1937 [*P. gatunensis*]. Subquadrate, thin-shelled, ornamented with numerous pits and a median, subdorsal, shallow sulcus; RV larger than LV, overlapping it along dorsal and ventral margins, receiving LV in shallow groove. Muscle scars reported to form irregular groups on interior surface of sulcus. *Mio.,* Panama.——Fig. 310,4. *P. gatunensis; 4a,b,* LV lat., dors., ×65 (126).

PODOCOPIDA, Suborder and Family UNCERTAIN

[Materials for this section prepared by authors as severally recorded at end of generic descriptions]

Abursus Loranger, 1954 [*A. beaumontensis*]. Apparently differing from *Quasillites* and *Graphiadactyllis* only in having surface reticulated in polygonal pattern. *U.Dev.,* Alba.——Fig. 310A,1. *A. beaumontensis; 1a,b,* carapace R, dors., ×30 (232). [Shaver.]

?Altha Neckaja, 1958 [*A. modesta*]. Nonsulcate, elongate and straight-backed, somewhat preplete, ventral margin mainly parallel to dorsal, RV overlapping LV along entire free margin, left overlap generally along anterior part of dorsal

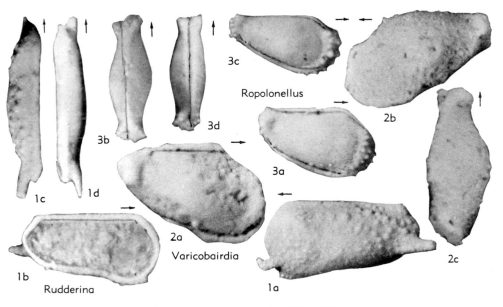

3c

Ropolonellus

2b

3d

3a

2a

Varicobairdia

1c   1d

1b

Rudderina

1a

2c

Fig. 303. Ropolonellidae (p. Q376-Q377).

margin; posterior part of hinge in depression; adventral structures and interior cavellinoid partition lacking. Dimorphism not observed. Surface smooth or perforate. *L.Sil.-M.Sil.,* NW.Russia.——Fig. 143,*1*. **A. modesta,* Llandov., Lithuania, *1a-d,* carapace (holotype), R lat., L lat., dors., ant., ×35; *1e,* juv. L lat., ×35 (264). [HESSLAND.] [Earlier assigned questionably to Bairdiidae but emphatically rejected from this family by SHAVER.]

Anchistrocheles BRADY & NORMAN, 1889 [**A. fumata* BRADY, 1890]. Reniform in lateral view, very

narrow in dorsal view; diagnosis based on the arrangement of setae on thoracic legs and furcae and on mandibular and maxillar parts which are more slender than in *Bairdia* and *Bythocypris.* [Marine; assigned by BRADY to both *Bairdiidae* and *Cyprididae.*] *Rec.,* C.Pac.(Samoa-Fiji)-Antarctica-?N.Atl.-?Ceylon.——Fig. 310A,*2. A. ?acerosa* (BRADY), N.Atl.; *2a,b,* carapace R, dors., ×40 (108). [SHAVER.]

Artifactella CORYELL & BOOTH, 1933 [**A. tomahawki*]. Small, egg-shaped in lateral view; shal-

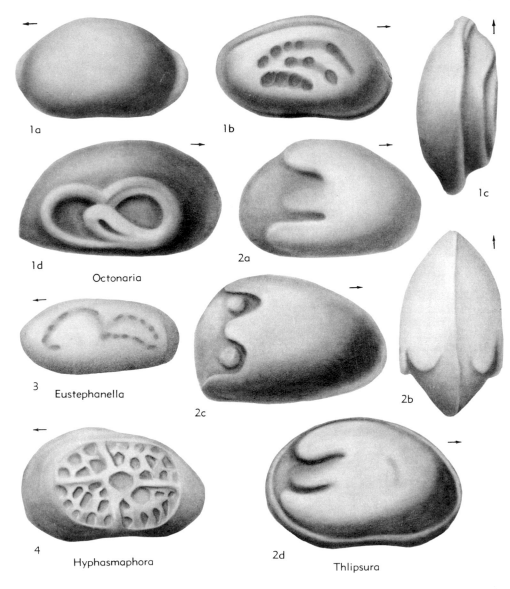

1a   *1b*   *1c*

Octonaria

1d   *2a*

3   Eustephanella

2c   *2b*

Hyphasmaphora

4   *2d*   Thlipsura

FIG. 304. Thlipsuridae (p. Q378).

low, vertically aligned depression near center of valves; LV larger than RV, with sinuate line of commissure; surface smooth. [Probably belongs in Cypridacea, but hinge, contact marginal structures, and adductor muscle scar unknown.] *U.Penn.*, N.Am.——Fig. 310A,*11*. **A. tomahawki*, USA (Tex.); *11a,b*, carapace R, dors., ×60 (124). [Shaver.]

Bosquetia Brady, Crosskey, & Robertson, 1874 [**B. robusta*]. Rather regularly ovate in side view,

with greatest height slightly in front of mid-length; margins well rounded, anterior broadly, posterior more narrowly, dorsal somewhat flattened in middle, ventral strongly convex; ovoid, in dorsal view widest near mid-point; surface smooth. Muscle field, hinge, marginal structures and dimorphism unknown. *Pleist.*, NW.Eu.(Scot.).—— Fig. 310A,*4*. **B. robusta; 4a-d*, carapace L, dors., vent., ant., ×27 (14). [Reyment.]

Daleiella Bouček, 1937 [**Cythere corbuloides*

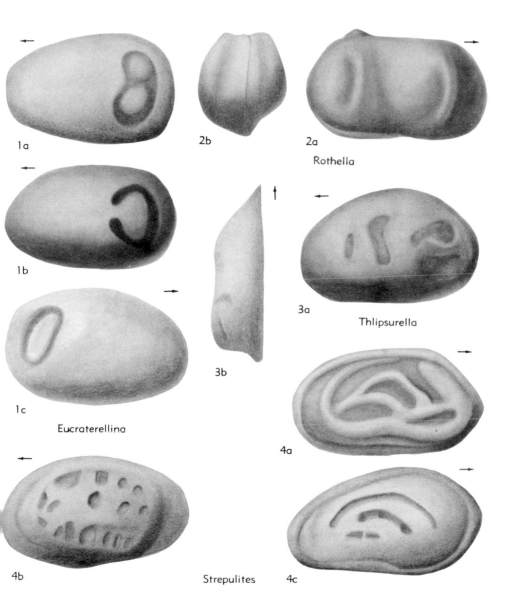

1a

2b

2a

Rothella

1b

3a

Thlipsurella

1c

Eucraterellina

3b

4a

4b

Strepulites   4c

Fig. 305. Thlipsuridae (p. Q378-Q379).

JONES & HOLL, 1869]. Subtriangular in dorsal and lateral views, relatively very tall and extremely tumid; LV commonly larger than RV, with extreme dorsal overreach that differentiates genus from *Microcheilinella;* surface smooth. Hinge, contact marginal structures and adductor muscle scar unknown. *M.Sil.,* Eu.——FIG. 310A,*5. *D. corbuloides* (JONES & HOLL); *5a,b,* carapace R, dors., ×20 (188). [SHAVER.]

Ellesmerina GLEBOVSKAJA & ZASPELOVA in GLEBOV-

SKAJA, 1948 [*E. incognita*] [=*Mossolovella* EGOROV, 1953 (obj.)]. Apparently related to *Bairdiocypris,* having mostly similar shape and contact-marginal structures, but differing in its more complex hingement with additional ridge and groove in each valve (278). [According to GLEBOVSKAJA & ZASPELOVA in POLENOVA (1953), *Ellesmerina* and designation of *E. incognita* as its type species was published in 1948 (not seen by SHAVER). EGOROV (145) mistakenly concluded that *E. incognita* had

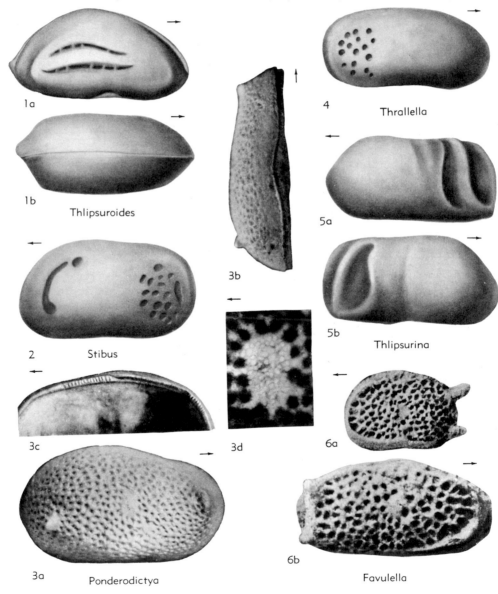

1a

1b
      Thlipsuroides

2   Stibus

3a   Ponderodictya

3b

3c   3d

4   Thrallella

5a

5b   Thlipsurina

6a

6b   Favulella

FIG. 306. Thlipsuridae (p. Q378-Q380).

been assigned by its authors to *Ellesmeria* TOLMACHOFF and therefore introduced for it the new name *Mossolovella*.] *Dev.*, Russia.——FIG. 310A,*6*. *E. incognita*; *6a,b*, carapace R, vent.; *6c*, LV int.; all ×50 (145). [SHAVER.]

Healdiacypris BRADFIELD, 1935 [*H. perplexa*]. Nearly like *Bairdia* in shape, but with reversal of overlap and overreach along short depressed hinge; contact marginal structures including calcified inner lamella; adductor muscle scar unknown. *M. Penn.*, N.Am.(Okla.-Ill.).——FIG. 310A,*8*. *H. perplexa*, Okla.; *8a,b*, carapace R, dors., ×55 (11).——FIG. 310B,*4*. *H. acuminata* COOPER, Ill.; *4a*, long. sec., LV at left; *4b*, transv. sec. through hinge, LV at left; both ×60 (Shaver, n). [SHAVER.]

Macrocyproides SPIVEY, 1939 [*M. clermontensis*]. Short, high, compressed; dorsum strongly arched, highest postmedially, venter nearly straight, front end narrowly rounded, extended below, posterior margin much broader; RV overlapping LV except posteroventrally where LV may slightly overlap; surface smooth. [Marine.] *M.Ord.-U.Ord.*, N.Am.——FIG. 310A,*9*. *M. clermontensis*, U.Ord. (Maquoketa F.), USA(Minn.); *9a,b*, RV lat., int., ×50 (J. R. Cornell, n). [SWAIN.]

Microcheilinella GEIS, 1933 [*pro Microcheilus* GEIS, 1932 (*non* KITTL, 1894)] [*Microcheilus distortus* GEIS, 1932]. Nearly like *Daleiella* but lacking great height and extreme dorsal overreach of type species of that genus; carapace elongate-oval in lateral view; overlap and overreach strong except along hinge where overreach remains LV-over-RV but overlap is reversed; contact marginal structures consist of duplicature and vestibule with related structures reminiscent of Cypridacea; adductor muscle scar unknown. *Sil.-L.Perm.*, N.Am.-Eu.——FIGS. 310A,*3*, 310B,*3*. *M. distorta*, U.Miss.

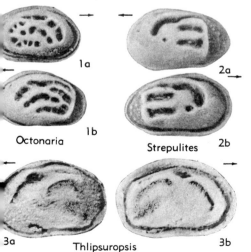

FIG. 307. Thlipsuridae (p. Q378-Q379).

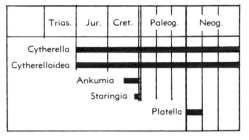

FIG. 308. Stratigraphic distribution of cytherellid genera and subgenera.

(Salem Ls.), USA(Ind.); 310A,*3a,b*, carapace R, dors., ×27 (Shaver, n); 310B,*3a,b*, carapace, long. and transv. secs., LV at left, ×57 (Shaver, n). [SHAVER.]

Punctaparchites KAY, 1934 [*Cytheropsis rugosus* JONES, 1858]. Small, compressed, valves subequal; hinge margin convex; venter nearly straight, anterior end broadly rounded, much wider than posterior extremity; surface coarsely pitted. [Marine.] *M.Ord.*, N.Am.——FIGS. 310A,*10*, 310B,*1*. *P. rugosus* (JONES); 310A,*10a,b*, LV lat., int. (M.Ord., Minn.), ×30 (J. R. Cornell, n); 310B,*1a-c*, carapace L, R, dors., ×40 (193). [SWAIN.]

Silenites CORYELL & BOOTH, 1933 [*S. silenus* (=*Carbonia? lenticularis* KNIGHT, 1928]. Bythocypridid ostracodes with great height, dorsal convexity, and LV-over-RV overreach; lacking great tumidity of *Pachydomella*; hinge and contact marginal structures ridged and grooved as in *Bairdiocypris* but differing in its adductor muscle scar that consists of 2 vertical rows of spots in pinnate pattern. *?Sil.*, *U.Penn.-L.Perm.*, cosmop.——FIG. 310A,*7*. *S. lenticularis* (KNIGHT), U.Penn., USA (Tex.); *7a,b*, carapace R, dors., ×35 (124). [SHAVER.]

Order MYODOCOPIDA Sars, 1866

[*nom. correct.* POKORNÝ, 1953 (*pro* Myodocopa SARS, 1866] [Type genus: *Cypridina* MILNE EDWARDS, 1840; SD SYLVESTERBRADLEY, herein] [Diagnosis by P. C. SYLVESTER-BRADLEY, University of Leicester]

Valves subequal, ornamented or smooth. Anterior rostrum and incisure may or may not be developed. Antennae (="second antennae") modified as swimming organs. Dimorphic. [Marine.] *Ord.-Rec.*

The order includes most planktonic ostracodes.

Suborder MYODOCOPINA Sars, 1866

[Type genus: *Cypridina* MILNE EDWARDS, 1840; SD SYLVESTERBRADLEY, herein] [Diagnosis and discussion by P. C. SYLVESTER-BRADLEY, University of Leicester]

Dorsal margin straight or curved; an-

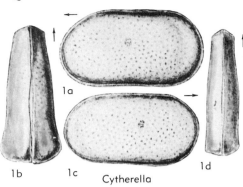

FIG. 309. Cytherellidae (p. Q382-Q383).

terior margin usually with rostral incisure, rostrum and rostral sinus; caudal siphon developed in some. Nuchal furrow may be well defined, especially in Paleozoic forms, lacking rostrum. Size highly variable, including macroscopic forms 2 or 3 cm. in diameter. Shell may lack calcification. One or both sexes usually free-swimming during at least part of life cycle. Antennules (="first antennae") not modified for swimming. [Marine.] *Ord.-Rec.*

The suborder can be divided conveniently into two artificial groups, in one of which no rostral incisure is developed, the other bearing a rostral incisure and usually also a rostrum and rostral sinus. The first group (without a rostral incisure) includes the superfamilies Entomozoacea, Entomoconchacea, and Thaumatocypridacea. The second group (with rostral incisure) includes the Cypridinacea and Halocypridacea. The assemblages are artificial in the sense that the three superfamilies without rostral incisure are not necessarily more closely related to each other than to the superfamilies of the second group. The Entomozoacea may be ancestral to the Thaumatocypridacea, but evidence of this is slight; indeed, the Entomozoacea and Entomoconchacea may not even belong to the Myodocopina.

A majority of the known myodocopid genera are restricted to rocks of Paleozoic age (30 genera); only two genera are identified definitely in post-Paleozoic deposits other than Recent (Fig. 311).

?Superfamily ENTOMOZOACEA (Jones, 1873) Přibyl, 1951

[Having priority from 1873, since family-group name Entomididae JONES, 1873, is based on a generic name which is

a junior homonym] [=Entomidacea JONES, 1873, *nom. transl.* SCHMIDT, 1941 (*ex* Entomidae JONES, 1873)] [Diagnosis and discussion by P. C. SYLVESTER-BRADLEY, University of Leicester]

Large, usually over 1 mm., without rostrum or rostral sinus, with long, commonly deep nuchal furrow (slight or absent in some genera). Muscle-scar details unknown. Classed in Myodocopida with some doubt; several authors have maintained that *Bolbozoe* and some other genera are not true ostracodes. *Ord.-Perm.*

The nuchal furrow is by most authors regarded as deciding factor for orientation within superfamily, anterior being considered to lie on its concave side.

Family ENTOMOZOIDAE Přibyl, 1951

[=Entomidae JONES, 1873] [Materials for this family prepared by P. C. SYLVESTER-BRADLEY, University of Leicester]

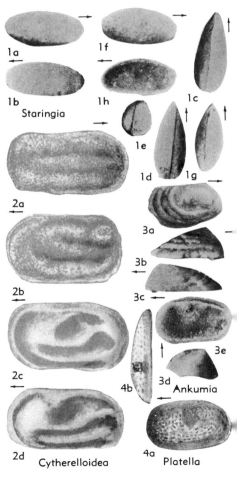

FIG. 310. Cytherellidae (p. Q383).

Like Bolbozoidae, but lacking anterodorsal swelling. Surface striate in many forms. *Ord.-Perm.*

Subfamily ENTOMOZOINAE Přibyl, 1951

Surface smooth, punctate, or variously striate. Nuchal furrow in most genera long, narrow, usually deep, more or less in center of dorsum. *Ord.-Perm.*

Entomozoe PŘIBYL, 1951 [*pro Entomis* JONES, 1861 (*non* HERRICH-SCHAEFFER, 1856)] [**Entomis tuberosa* JONES, 1861]. Like *Richteria*, but surface smooth or punctate, not striate. *Sil.-Perm.*, cosmop. ——FIG. 312,2a-e. *E. zoppii* (CANAVARI), Sil., Sardinia; *2a-e*, carapace R, L, dors., vent., post., ×20 (119).——FIG. 312,2f. **E. tuberosa* (JONES), U. Sil., Scot.; RV lat., ×8 (182a).——FIG. 312,2g-j.

E. meneghinii (CANAVARI), Sil., Sardinia; *2g-j*, carapace R, L, dors., vent., ×25 (119).

Bertillonella STEWART & HENDRIX, 1945 [**B. subcircularis*] [=*Waldeckella* RABIEN, 1954]. Like *Entomoprimitia* but muscle-scar pit absent. *Dev.*, Eu.-N.Am.——FIG. 313,4a. **B. subcircularis*, Dev., Ohio; carapace R, ×60 (Sylvester-Bradley, n).—— FIGS. 313,4b, 314,5. *B. erecta* (RABIEN) (type species of *Waldeckella*), U.Dev., Ger.; *313,4b*, RV lat. (external impression), ×80; *314,5a-c*, RV (steinkern) lat., dors., ant., ×30 (292).

Entomoprimitia KUMMEROW, 1939 [**Primitia hattungensis* MATERN, 1929 (=**Cypridina nitida* ROEMER, 1850)] [=*Omphalentomis* KUMMEROW, 1953]. Like *Nehdentomis* but nuchal furrow reduced to inconspicuous groove or, in many species, to dimple-like depression just anterior to midpoint of dorsal margin; surface striations con-

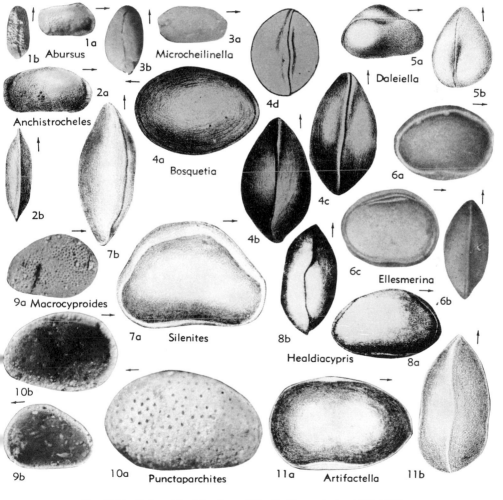

FIG. 310A. Podocopida, Suborder and Family Uncertain (p. Q383-Q387).

1a, 1b Abursus
3a, 3b Microcheilinella
2a Anchistrocheles
2b
4a Bosquetia
4b 4c 4d
5a Daleiella
5b
6a
6b, 6c Ellesmerina
7a Silenites
7b
8a, 8b Healdiacypris
9a Macrocyproides
9b
10a Punctaparchites
10b
11a, 11b Artifactella

centrically arranged round muscle-scar pit. *Dev.,*
cosmop.——FIG. 315,*1a-c.* *E. nitida* (ROEMER)
(steinkern), U.Dev., Ger.; *1a-c,* RV lat., dors.,
post., ×30 (292).——FIG. 315,*1d,e.* *E. splendens*
(WALDSCHMIDT, 1885), U.Dev., Ger.; *1d,e,* carapace R, dors., ×15 (292).

Nehdentomis MATERN, 1929 [*Entomis (Nehdento-

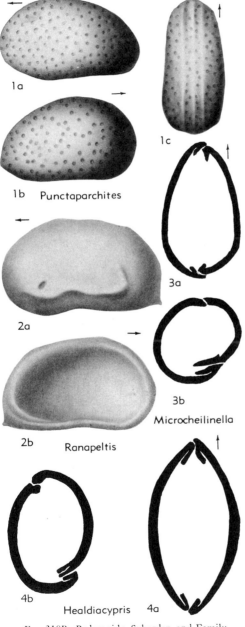

1a

1b   Punctaparchites

2a

2b   Ranapeltis

3a

3b

Microcheilinella

4b

4a

1c

Healdiacypris

FIG. 310B. Podocopida, Suborder, and Family
Uncertain (p. Q252-Q253, Q387).

mis) nehdensis]. Like *Richteria,* but with nuchal
furrow terminating in a well-defined external
muscle-scar pit. *Dev.,* Ger.——FIG. 315,*4.* *N.
nehdensis* (MATERN), U.Dev., Ger.; *4a,b,* LV lat.,
dors., ×50 (244).

Pseudoentomozoe PŘIBYL, 1951 [*Entomis pteroides*
CANAVARI, 1899]. Like *Rhomboentomozoe* but
without ventral spine, and with a posteroventral
carina running diagonally upwards from mid-
point of ventral margin. *U.Sil.,* Sardinia.——FIG.
315,*3.* *P. pteroides* (CANAVARI), L.Ludlov., Sardinia; *3a,b,* LV lat., vent., ×20; *3c,d,* RV lat.,
vent., ×20 (119).

Rhomboentomozoe PŘIBYL, 1951 [*Cryptocaris?
rhomboidea* BARRANDE, 1872]. Carapace subtriangular, striate, with pronounced ventral spine.
Ord.-Sil., Eu.-N.Am.——FIG. 315,*2.* *R. rhomboidea* (BARRANDE), Sil., Boh.; RV lat., ×20
(284).

Richteria JONES, 1874 [*Cypridina serratostriata*
SANDBERGER, 1842; SD KEGEL, 1933] [=*Ento-
minella* LIVENTAL, 1945]. Carapace oblong, with
nuchal furrow extending downwards more than
halfway. Dorsal margin nearly straight, cardinal
angles curved. Surface ornamented with striations
which are usually longitudinal or concentric. With
or without small anterodorsal tubercle. *Sil.-Perm.,*
cosmop. (284, 292).——FIG. 312,*1.* *R. serratostriata* (SANDBERGER), U.Dev., Ger.; RV lat., ×30
(244).——FIG. 313,*1.* *R. lamarmorai* (CANAVARI),
Sil., Sardinia; *1a-e,* carapace R, L, dors., vent.,
post., ×15 (119).

Ungerella LIVENTAL, 1948 [*Cypridina calcarata*
RICHTER, 1856] [=*Franklinella* STEWART &
HENDRIX, 1945 (*non* NELSON, 1937)]. Like
Rhomboentomozoe, but with posterodorsal spine
in addition to mid-ventral spine (284). *U.Dev.-L.
Carb.,* Eu.-N.Am.——FIG. 313,*3a,b.* *U. novecosta*
(STEWART & HENDRIX) (type species of *Franklin-
ella*), *3a,b,* carapace L, R, ×40 (Sylvester-Bradley,
n).——FIG. 313,*3c.* *U. calcarata* (RICHTER), U.
Dev., Ger.; RV lat., ×40 (284).

Vltavina BOUČEK, 1936 [*V. bohemica*]. Carapace
striate, elongate, dorsal margin straight with cardinal angles produced into horizontal spines directed anteriorly and posteriorly. Nuchal furrow
rather shallow. *U.Dev.,* Boh.——FIG. 313,*2.* *V.
bohemica;* RV lat., ×50 (10).

Subfamily BOUCIINAE Přibyl, 1951

Like Entomozoinae, but with posteroventral sulcus in addition to nuchal furrow.
U.Sil.

Boucia AGNEW, 1942 [*pro Basslerella* BOUČEK, 1936
(*non* KELLETT, 1935; *nec* HOWE, 1935)] [*Basslerella ornatissima* BOUČEK, 1936]. Carapace bean-
shaped, with concave dorsal margin. Ornamented
with fine transverse striations. *U.Sil.,* Bohemia.——
FIG. 314,*6.* *B. ornatissima* (BOUČEK), M.Ludlov.,
Bohemia; *6a,b,* RV lat., dors., ×30 (10).

Subfamily RICHTERININAE Sylvester-Bradley, nov.

Carapace accurately elliptical or circular in side view, and thus equilateral; ornamented with fine or strong raised striae. Nuchal furrow slight or absent. *Dev.*

Richterina GÜRICH, 1896 [*Cytherina costata* RICHTER, 1869]. Elliptical; no nuchal furrow; ornament uninterrupted by muscle-scar pit. *Dev.*, Eu. ——FIG. 314,3a. *R. costata*, U.Dev., Ger.; ?L or RV, lat., ×30 (244).——FIG. 314,3b,c. R. *viltata* (GÜRICH), U.Dev., Ger.; 3b,c, RV lat., vent., ×60 (244).

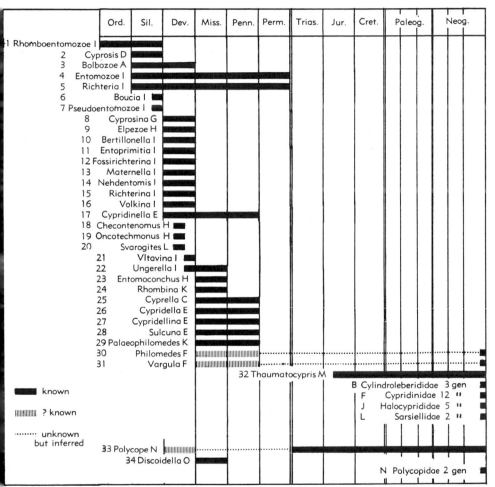

FIG. 311. Stratigraphic distribution of myodocopid ostracode genera (Moore, n). Classification in families is indicated by letter symbols (A—Bolbozoidae, B—Cylindroleberididae, C—Cyprellidae, D—Cypridinacea, Family Uncertain, E—Cypridinellidae, F—Cypridinidae, G—Cyprosinidae, H—Entomoconchidae, I—Entomozoidae, J—Halocyprididae, K—Rhombinidae, L—Sarsiellidae, M—Thaumatocyprididae, N—Polycopidae, O—Cladocopina, Family Uncertain). An alphabetical list of genera furnishes cross reference to the serially arranged numbers on the diagram.

Generic Names with Index Numbers

Bertillonella—10	*Cyprosis*—2	*Oncotechmonus*—19	*Richterina*—15
Bolbozoe—3	*Discoidella*—34	*Palaeophilomedes*—29	*Sulcuna*—28
Boucia—6	*Elpezoe*—9	*Philomedes*—30	*Svarogites*—20
Checontonomus—18	*Entomoconchus*—23	*Polycope*—33	*Thaumatocypris*—32
Cyprella—25	*Entomozoe*—4	*Pseudoentomozoe*—7	*Ungerella*—22
Cypridella—26	*Entoprimitia*—11	*Rhombina*—24	*Vargula*—31
Cypridellina—27	*Fossirichterina*—12	*Rhomboentomozoe*—1	*Vltavina*—21
Cypridinella—17	*Maternella*—13	*Richteria*—5	*Volkina*—16
Cyprosina—8	*Nehdentomis*—14		

Fossirichterina MATERN, 1929 [**Richterina (Fossirichterina) intercostata*]. Like *Richterina*, but ornament interrupted by muscle scar pit. *Dev., Eu.*——FIG. 314,2. **F. intercostata* (MATERN), U. Dev., Ger.; *2a,b,* ?RV lat., ?vent., ×30 (244).

Maternella RABIEN, 1954 [**Richterina (?) costata* (RICHTER) var. *dichotoma* PAEKELMANN, 1913]. Like *Richterina* but subcircular and slightly asymmetrical, with anterior margin more sharply curved than posterior; in dorsal view anterior slope less

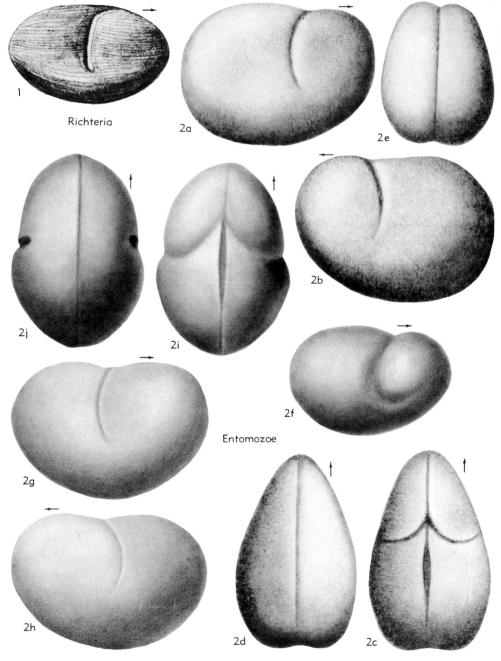

Richteria

Entomozoe

FIG. 312. Entomozoidae (Entomozoinae) (p. Q389-Q390).

abrupt than posterior. Dev., Eu.——Fig. 314,*1*. *M. dichotoma* (RABIEN), U.Dev., Ger.; *1a*, LV (external impression); *1b*, RV (external impression); *1c-e*, (steinkern), LV lat., dors., post.; all ×30 (292).

Volkina RABIEN, 1954 [*Entomis (Nehdentomis)

zimmermanni VOLK, 1939]. Like *Richterina*, but with distinct dimple just anterior to mid-point of dorsal margin representing nuchal furrow. Dev., Eu.——Fig. 314,*4*. *V. zimmermanni* (VOLK), U.Dev., Ger.; *4a,b*, LV lat., dors., ×60, ×90 (292).

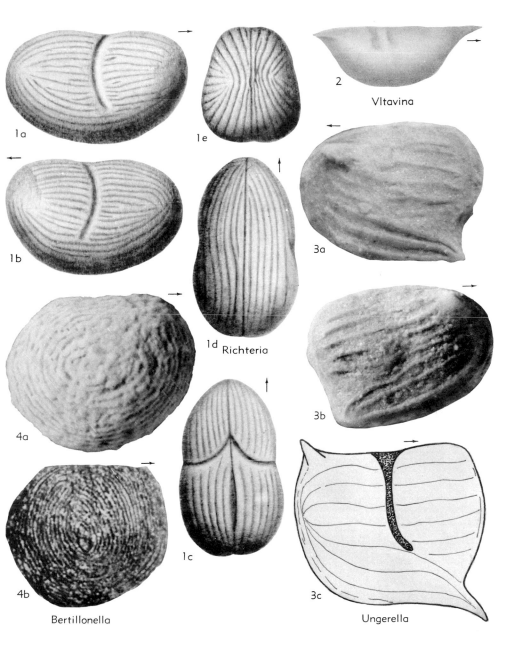

Fig. 313. Entomozoidae (Entomozoinae) (p. Q389-Q390).

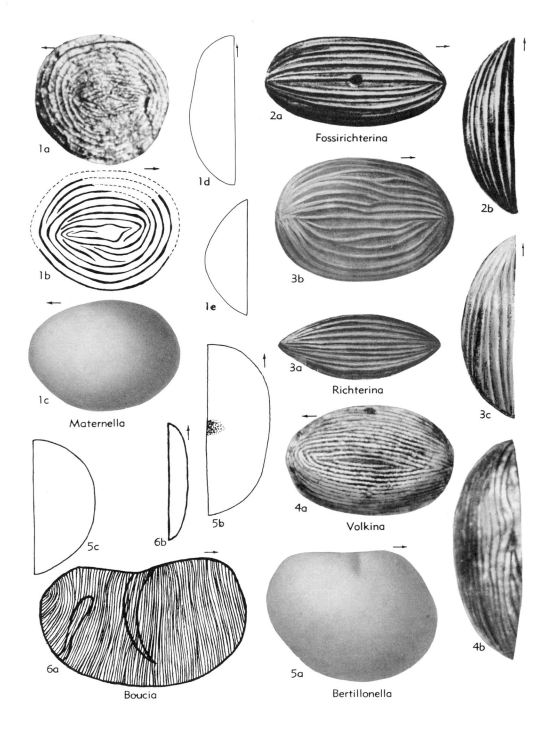

FIG. 314. Entomozoidae (Entomozoinae, Bouciinae, Richterininae) (p. Q389-Q393).

Family BOLBOZOIDAE Bouček, 1936

[Materials for this family prepared by P. C. SYLVESTER-BRADLEY, University of Leicester]

Hemispherical, anterodorsal swelling developed immediately in front of nuchal fur-row. Subsidiary ventral sulcus present in some forms. Surface smooth, punctate or reticulate. *Sil.-Dev.,* Eu.

Bolbozoe BARRANDE, 1872 [**B. anomala* SD BASSLER & KELLETT, 1934]. With characters of family.

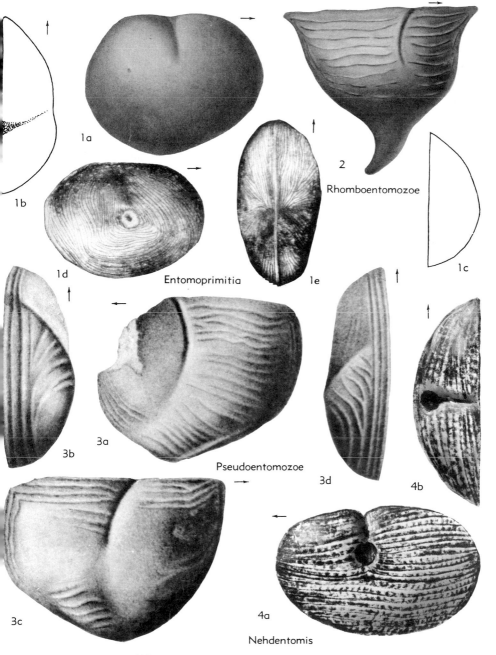

FIG. 315. Entomozoidae (Entomozoinae) (p. Q389-Q390).

Sil.-Dev., Eu.——FIG. 316,*1c-f*. **B. anomala*, internal cast; Sil., Boh.; *1c,d*, LV lat., dors.; *1e,f*, RV lat., vent.; all ×7 (95).——FIG. 316,*1a,b*. *B. bohemica* BARRANDE; Sil., Boh.; *1a,b*, RV lat., LV lat., ×3 (95).——FIG. 316,*1g-i*. *B. bohemica?*, Sil., Sardinia; *1g-i*, LV lat., dors., vent., ×5 (119).

Superfamily ENTOMOCONCHACEA Brady, 1868

[*nom. transl.* SYLVESTER-BRADLEY, 1953 (*ex* Entomoconchidae BRADY, 1868)] [Diagnosis by P. C. SYLVESTER-BRADLEY, University of Leicester]

Macroscopic, without rostrum or rostral sinus, usually with a posterior siphon. Muscle-scar pattern oval, composed of radiate linear scars. *Dev.-Carb.*

Family ENTOMOCONCHIDAE Brady, 1868

[Materials for this family prepared by P. C. SYLVESTER-BRADLEY, University of Leicester]

Anterior margin vertical, almost straight. Left valve usually larger than right, overreaching it particularly at anterodorsal and anteroventral angles. Caudal siphon usually developed. *Dev.-Carb.*

Subfamily ENTOMOCONCHINAE Brady, 1868

[*nom. transl.* SYLVESTER-BRADLEY, herein (*ex* Entomoconchidae BRADY, 1868)]

Carapace smooth, unornamented, usually rather tumid. *Dev.-Carb.*

Entomoconchus M'COY, 1839 [**E. scouleri*]. Subspheroidal; anterior margin flanked by shallow or deep furrows; anterodorsal and anteroventral angles slightly protuberant. Siphon at posteroventral corner or in middle of posterior margin, but lacking in some. *L.Carb.*, Eu.——FIG. 317,*1*. **E. scouleri*, L.Carb., Eire; *1a-d*, carapace L, ant., vent., post., ×2 (366a).

Elpezoe PŘIBYL, 1950 [*pro Elpe* BARRANDE, 1872 (*non* ROBINEAU-DESVOIDY, 1863)] [*Elpe inchoata

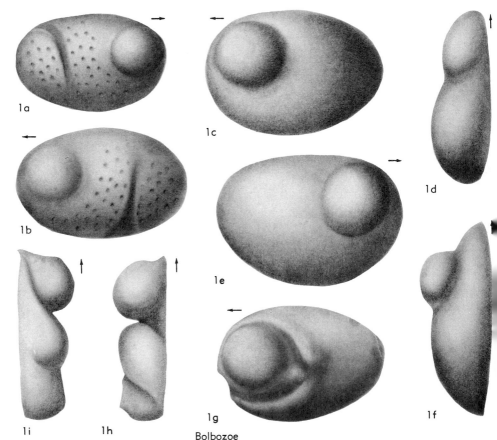

1a

1c

1d

1b

1e

1i   1h

1g

1f

Bolbozoe

FIG. 316. Bolbozoidae (p. Q395-Q396).

BARRANDE, 1872]. Like *Entomoconchus* but more compressed anteriorly, compressed area being limited posterodorsally by a pronounced "cheek." *Dev.,* Eu.——FIG. 317,2. **E. inchoata,* L.Dev., Bohemia; carapace R, ×6 (283).

Subfamily ONCOTECHMONINAE Kesling, 1954

Carapace coarsely punctate, ornamented with small ridges more or less concentric round anterior. *M.Dev.*

Oncotechmonus KESLING, 1954 [**O. chemotus*]. Carapace oval in side view, compressed, tapering sharply to posterior. Ornamented ridges crowded anteriorly, more widely spaced towards posterior. Posteroventral gape wide, long. *M.Dev.,* N.Am.——FIG. 318,1. **O. chemotus,* M.Dev., Lake Erie; *1a-d,* carapace (holotype) R, L, dors., vent., ×4.5 (202).

Checontonomus KESLING, 1954 [**C. cophus*]. Like *Oncotechmonus,* but posteroventral gape narrow or absent, and ornamental ridges restricted to anterior. *M.Dev.,* N.Am.——FIG. 318,2. **C. cophus,* M.Dev., Lake Erie; *2a-e,* carapace R, L, dors., vent., ant., ×4.5 (202).

Family CYPROSINIDAE Whidborne, 1890

[Materials for this family prepared by P. C. SYLVESTER-BRADLEY, University of Leicester]

Siphon produced as caudal process directed upward (or siphon may be interpreted as rostrum when orientation is reversed, dorsal becoming ventral, anterior becoming posterior); shallow nuchal furrow present. *Dev.*

Cyprosina JONES, 1881 [**C. whidbornei*]. Ovoid, widest in posterior 3rd, tapering forward; anterior margin receding toward venter. *Dev.,* Eng.—— FIG. 319,1. **C. whidbornei;* LV lat., ×2 (366a).

Superfamily THAUMATOCYPRIDACEA G. W. Müller, 1906

[*nom. transl. et correct.* SYLVESTER-BRADLEY, herein (*ex* Thaumatocyprinae G. W. MÜLLER, 1906)] [Diagnosis by P. C. SYLVESTER-BRADLEY, University of Leicester]

Carapace subcircular in outline, with no rostrum, sinus or incisure, but with projecting spines or processes developed close to plane of commissure. *M.Jur.-Rec.*

The recognition of a superfamily containing a single known family and genus seems anomalous, even though such monotypical taxa are not unique. In the case of Thaumatocypridacea such classification is well justified, because *Thaumatocypris* is a very distinctive, long-ranging genus with soft parts (observed in Recent specimens) that confirm correct placement in the Myo-

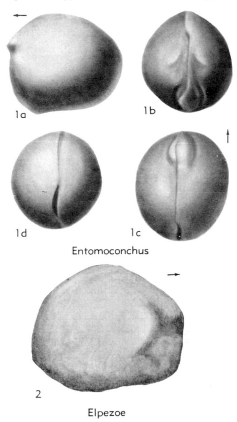

FIG. 317. Entomoconchidae (Entomoconchinae) (p. Q396-Q397).

docopina; at the same time it is far removed from other superfamilies of the suborder. New fossil species of *Thaumatocypris* from post-Paleozoic strata keep coming to light.

Family THAUMATOCYPRIDIDAE G. W. Müller, 1906

[*nom. transl. et correct.* SYLVESTER-BRADLEY, herein (*ex* Thaumatocyprinae G. W. MÜLLER, 1906] [Materials for this family prepared by P. C. SYLVESTER-BRADLEY, University of Leicester]

Characters of superfamily. *M.Jur.-Rec.*

Thaumatocypris G. W. MÜLLER, 1906 [**T. echinata*]. Anomalous rare genus with unique method of swimming; known from one Recent and several Jurassic species. *M.Jur.-Rec.,* cosmop.——FIG. 320,1. **T. echinata,* Rec., Ind.O.; *1a-c,* carapace L, R (juv.), dors., ×30 (258a).

Superfamily CYPRIDINACEA Baird, 1850

[*nom. transl.* SYLVESTER-BRADLEY, herein (*ex* Cypridinidae BRADY, 1868, *nom. correct. pro* Cypridinadae BAIRD, 1850)] [=Cypridiniformes SKOGSBERG, 1920] [Diagnosis by P. C. SYLVESTER-BRADLEY, University of Leicester]

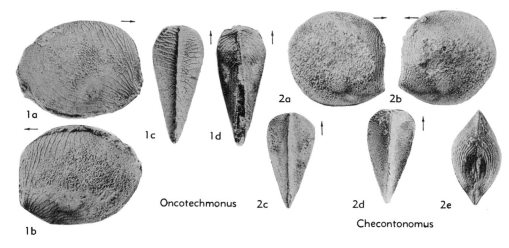

1a

1c   1d

1b   Oncotechmonus   2c

2a   2b

2d   2e

Checontonomus

FIG. 318. Entomoconchidae (Oncotechmoninae) (p. Q397).

Carapace usually calcified, strongly in some; rostrum downcurved usually below line of dorsal border, overhanging an anterior incisure (or gape) through which antennae of living animal protrude. Specialized hinge structure may be developed or lacking, but if present, it rarely consists of more than terminal teeth in RV, with long narrow smooth groove between and corresponding elements in LV. *Sil.-Rec.*

Family CYPRIDINIDAE Baird, 1850

[*nom. correct.* BRADY, 1868 (*pro* Cypridinadae BAIRD, 1850); *nom. transl.* BRADY, 1868 (*ex* Cypridiniae DANA, 1852, *nom. transl. et correct. ex* Cypridinadae BAIRD, 1850)] [Materials for this family prepared by P. C. SYLVESTER-BRADLEY, University of Leicester]

Rostrum down-curved, overhanging well-marked sinus; incisure more or less cruciform. *?Carb., Rec.*

A description of cypridinid appendages and eye structures given by SARS (65), translated by SCOTT & WAINWRIGHT, follows:

Only one pair of feet that are unique in shape, the appendages being elongate and curved above, flexible, annular, vermiform, forming spines toward their apices. Upper antennae [antennules] large, distinctly articulate, geniculate at the base. The characteristic part of the mandibles is absent. Second maxillae [third post-oral limb] are provided with large branchial plates. Composite eyes stalked, widely separated; anteriorly between these is a large, simple and tentaculate eye on the small forehead.

Subfamily CYPRIDININAE Baird, 1850

[*nom. transl. et correct.* DANA, 1852 (*ex* Cypridinadae BAIRD, 1850)]

Carapace more or less strongly calcified, usually smooth; dorsal border arched, anterior margin of rostrum evenly curved or sinuous. Mostly rather large forms (more than 2 mm.); sexual dimorphism weak or absent. *?Carb., Rec.*

Cypridina MILNE EDWARDS, 1840 [**C. reynaudi*] [=*Daphnia* M'COY, 1844 (*non* MÜLLER, 1776); *Pyrocypris* G. W. MÜLLER, 1890; *Eupathistoma* BRADY, 1898]. Pronounced anterodorsal angle leading to downwardly directed rostrum, posteroventral extremity produced; carapace not calcified; animal phosphorescent. *Rec.*, trop.——FIG. 321,*1. C. inermis* (G. W. MÜLLER) (=**C. reynaudi*), Malaya; ♂ carapace L, ×30 (53, 109).

Azygocypridina SYLVESTER-BRADLEY, 1950 [*pro Crossophorus* BRADY, 1880 (*non* HEMPRICH & EHRENBERG, 1828)] [**Crossophorus imperator* BRADY, 1880]. Carapace like *Gigantocypris* but smaller, with rostrum proportionally larger. *Rec.*, cosmop.——FIG. 321,*8. A. gibber* (G.W.MÜLLER), E.Indies; carapace L, ×15 (Sylvester-Bradley, n).

Codonocera BRADY, 1902 [**C. cruenta*]. Carapace strongly calcified, oval, with or without anterior cardinal angle, bearing a blunt caudal process, which perhaps forms a siphon. *Rec.*, E.Indies-Ind. O.——FIG. 321,*2. *C. cruenta*, E.Indies; *2a,* ♂ carapace L, ×20; *2b,* muscle scar, ×60 (Brady, 1902).

Gigantocypris G. W. MÜLLER, 1895 [**G. agassizi;* SD SYLVESTER-BRADLEY, herein]. Carapace large (more than 10 mm.), globular, not calcified; rostrum small. *Rec.*, cosmop.——FIG. 321,*4. *G. agassizi,* Pac.; ♂ carapace L, ×2 (255a).

Heterodesmus BRADY, 1865 [**H. adamsi* BRADY, 1866] [=*Siphonostra* SKOGSBERG, 1920]. Carapace produced posteriorly into a siphon; hinge anterior

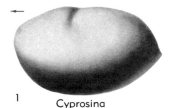

1 Cyprosina

Fig. 319. Cyprosinidae (p. Q397).

with projecting tooth in R valve. *Rec.*, Japan-Austral.——Fig. 321,3. *H. spinifer* (Skogsberg), Austral.; ♀ carapace L, ×20 (71).

Macrocypridina Skogsberg, 1920 [*Cypridina castanea* Brady, 1897]. Carapace weakly calcified, tapering backward; rostrum sinuous. *Rec.*, Atl.O.-Ind.O.——Fig. 321,5. *M. castanea* (Brady), Atl.; ♀ carapace L, ×8 (16).

Monopia Claus, 1873 [*Cypridina monopia* Claus, 1873] [=*Eumonopia* Claus, 1891; *Cypridinodes* Brady, 1902]. Shell heavily calcified, smooth or sculptured, oval, produced posteriorly as acuminate or truncate caudal process; rostrum sinuous, incurved. *Rec.*, Pac.——Fig. 321,6. *M. acuminata* Skogsberg; ♂ carapace L, ×8 (71).

Vargula Skogsberg, 1920 [*Cypridina norvegica* Baird, 1860]. Carapace smooth, LV slightly larger than RV with little overlap except at posteroventral corner; dorsal border evenly arched, continuous with rostrum, posteroventral corner little produced; hinge without teeth. *?Carb., Rec.,* cosmop.

V. (Vargula). *?Carb., Rec.,* cosmop.——Fig. 321, 7. *V. norvegica* (Baird), off Norway; *7a,b,* ♀ carapace L, ant., ×15 (Sylvester-Bradley, n); *7c,* ♂ carapace R, ×15 (365); *7d,* muscle scar, RV int., ×40 (71).

V. (Doloria) Skogsberg, 1920 [*Cypridina (D.) levis*]. Carapace as in *V. (Vargula)*. *Rec.*, Antarct.-S.Am.

Subfamily PHILOMEDINAE G. W. Müller, 1912

Shell strongly calcareous, ornamented (except smooth in *Philomedes*); dorsal border straight or arched; rostrum truncate, rounded or pointed; posteroventral corner angular, or produced as slight caudal process. Sexual dimorphism commonly very marked. *?Carb., Rec.*

Philomedes Liljeborg, 1853 [*P. longicornis*] [=*Bradycinetus* Sars, 1866]. Shell calcified, smooth; LV very slightly larger than RV; rostrum abruptly truncated; hinge very weak but an elongated tooth may be developed at posterior end of hinge line in RV with corresponding socket in LV. Pronounced sexual dimorphism in both shape of shell (males longer) and rostral sinus (males wider and shallower). *?Carb., Rec.,* cosmop.——

Fig. 322,3. *P. brenda* (Baird) (?=*P. longicornis*), Rec., off Norway; *3a,b,* ♀ carapace L, ant., ×20; *3c,* ♂ carapace L; all ×20 (65).

Pleoschisma Brady, 1890 [*P. moroides;* SD Sylvester-Bradley, herein]. Shell ornamented with puncta, reticula, or tubercles; rostral sinus absent or very slight. *Rec.*, Pac.——Fig. 322,4. *P. moroides; 4a,b,* carapace R, dors., ×40 (108).

Pseudophilomedes G. W. Müller, 1894 [*P. foveolata*] [=*Paramekodon* Brady & Norman, 1896]. Relatively small (less than 2 mm.), ornamented with puncta and in some with ridges and other projections; rostrum blunt, caudal process short, ill-defined; hinge probably as in *Tetragonodon*. *Rec.*, Medit.-Atl.——Fig. 322,2. *P. foveolatus,* Medit.; ♀ carapace L, ×60 (53).

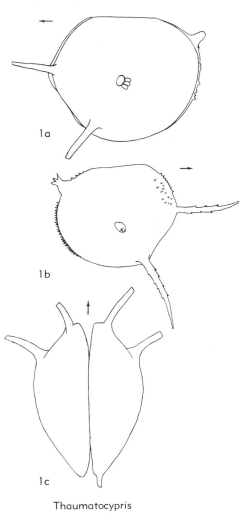

1a

1b

1c

Thaumatocypris

Fig. 320. Thaumatocyprididae (p. Q397).

Streptoleberis Brady, 1890 [*S. crenulata] [=Scleroconcha Skogsberg, 1920]. Relatively large (more than 2 mm.); surface highly sculptured; like Pseudophilomedes but rostrum and caudal process more pronounced. Rec., cosmop.——Fig. 322,1. S. appellöfi (Skogsberg), Antarct.; ♀ carapace L, ×15 (71).

Tetragonodon Brady & Norman, 1896 [*T. cten-

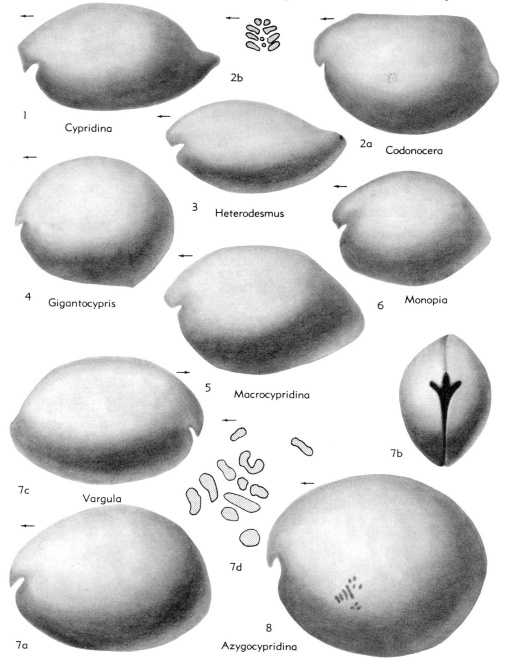

FIG. 321. Cypridinidae (Cypridininae) (p. Q398-Q399).

orhynchus; SD Sylvester-Bradley, herein]. Shell ornamented with spines or puncta; LV larger; rostrum long, pointed, projecting forward at about 45 degrees; posteroventral caudal process directed diagonally upward in continuation of ventral margin. Hinge straight, median bar with terminal sockets in LV, very narrow median groove with terminal teeth in RV. *Rec.,* Atl.——Fig. 322,5. *T.

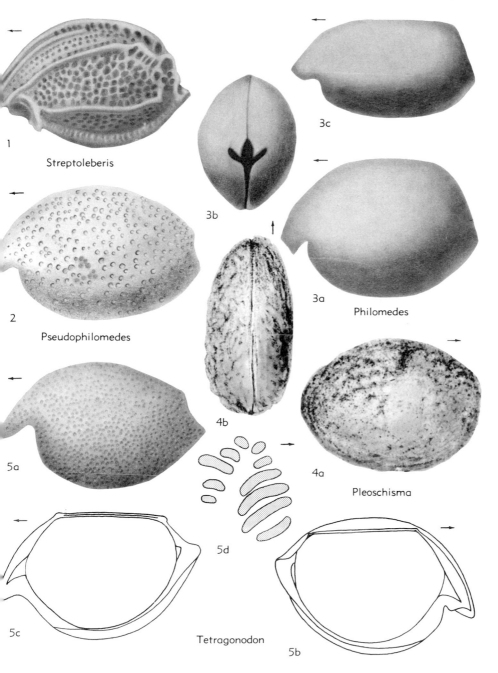

FIG. 322. Cypridinidae (Philomedinae) (p. Q399-Q402).

ctenorhynchus; 5a, ♀ carapace L, ×25 (16); 5b,c, ♀ LV int., ♀ RV int., ×25; 5d, muscle scar, RV ext., ×50 (Sylvester-Bradley, n).

Family CYLINDROLEBERIDIDAE G. W. Müller, 1906

[*nom. transl. et correct.* SYLVESTER-BRADLEY, herein (*ex* Cylindroleberinae G. W. MÜLLER, 1906)] [=Asteropidae BRADY, 1874] [Materials for this family prepared by P. C. SYLVESTER-BRADLEY, University of Leicester]

Rostrum down-curved, overhanging rostral incisure and almost overlapping rostral sinus; muscle-scar pattern spiral. *Rec.*

Cylindroleberis BRADY, 1867 [*pro Asterope* PHILIPPI, 1840 (*non* HÜBNER, 1816; *nec* MÜLLER & TROSCHEL, 1840)] [*Asterope mariae BAIRD, 1840; SD SYLVESTER-BRADLEY, herein] [=*Copechaete* HESSE, 1878; *Copechaeta* CARUS, 1880 (*nom. van. pro Copechaete*); *Asteropina* STRAND, 1928 (*nom. van. pro Asterope*)]. Shell calcareous, smooth, more or less elongate, posterior evenly rounded. Males longer than females, with sinuous dorsal margin making anterior higher than posterior. Muscle-scar pattern consisting of less than 20 rounded scars arranged in a loose spiral. *Rec.,* cosmop.——FIG. 323,3. **C. mariae* (BAIRD), off Norway; *3a,b,* ♀ carapace L, ant., ×20; *3c,* ♂ carapace R, ×20; *3d,* muscle scar, RV ext., ×75 (107).

Cyclasterope BRADY, 1897 [**C. hendersoni;* SD SKOGSBERG, 1920]. Like *Cylindroleberis* but posterior acuminate in some, or outline may be subcircular; muscle-scar pattern consisting of many elongate scars (30 to 40) arranged in a close spiral. *Rec.,* cosmop.

C. (**Cyclasterope**). *Rec.,* cosmop.——FIG. 323,2. *C. fascigera* BRADY, E.Indies; *2a,* ♂ carapace L, ×8; *2b,* muscle scar, RV int., ×20 (107).

C. (**Cycloleberis**) SKOGSBERG, 1920 [**Cylindroleberis lobiancoi* G. W. MÜLLER, 1895], *Rec.,* cosmop.

Asteropteron SKOGSBERG, 1920 [*Asterope fusca G. W. MÜLLER, 1894]. Shell highly sculptured with strongly projecting ridges or with lateral winglike expansions. *Rec.,* cosmop.——FIG. 323,1. **A. fuscum* (G.W.MÜLLER), Japan; *1a,b,* carapace L, dors., both × 15 (53).

Family CYPRELLIDAE Sylvester-Bradley, n. fam.

[Materials for this family prepared by P. C. SYLVESTER-BRADLEY, University of Leicester]

Carapace annulate; rostrum down-curved; incisure horizontal; posterior produced into caudal siphon. *Carb.*

Cyprella DEKONINCK, 1841 [**C. chrysalidea*]. With low subcentral tubercle and rather deep and narrow sinuate nuchal furrow behind. Venter commonly inflated, particularly toward anterior. *Carb., Eu.*——FIG. 324,1. **C. chrysalidea,* L.Carb.(Visé.), Belg.; *1a,b,* carapace R, ant.; *1c,* dorsal view of partly opened carapace; all ×10 (Sylvester-Bradley, n).

Family CYPRIDINELLIDAE Sylvester-Bradley, n. fam.

[Materials for this family prepared by P. C. SYLVESTER-BRADLEY, University of Leicester]

Rostral incisure transverse, with antero-

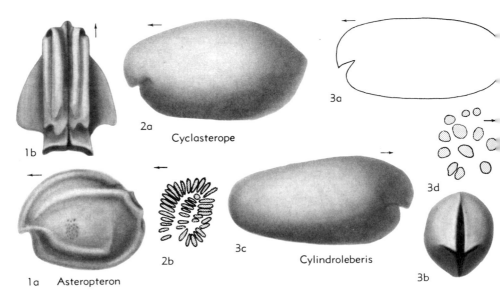

3a

2a

Cyclasterope

1b

3d

2b

3c

Cylindroleberis

1a   Asteropteron

3b

FIG. 323. Cylindroleberididae (p. *Q402*).

ventral prow projecting at least as far forward as rostrum. Caudal siphon invariably present, although not developed uniformly as a projecting process. *Dev.-Carb.*

Cypridinella JONES & KIRKBY, 1874 [**C. cummingi;* SD BASSLER & KELLETT, 1934] [=*Offa* JONES & KIRKBY, 1874]. Size medium (2 to 10 mm.); prow more or less produced, more prominent than rostrum; rostral incisure a narrow, horizontal slit; caudal process blunt or acuminate. *Dev.-Carb.,* Eu.——FIG. 325,*1. C. monitor* JONES & KIRKBY, L. Carb.(Visé.), Belg.; *1a-c,* carapace R, dors., post., ×5 (Sylvester-Bradley, n).

Cypridella DEKONINCK, 1841 [*non* VÁVRA, 1895] [**C. cruciata*]. Like *Cypridellina,* but with subcentral swelling replaced by pronounced backwardly directed tubercle, curved nuchal furrow developed behind tubercle; other tubercles may be present; prow not so pronounced, commonly extending no farther than rostrum; caudal siphon well developed. *Carb.,* Eu.——FIG. 325,*2. C.* sp., Carb.(Visé.), Belg.; *2a-d,* carapace L, dors., ant., post. (rostrum reconstr.), ×10 (365).

Cypridellina JONES & KIRKBY, 1874 [**C. clausa;* SD BASSLER & KELLETT, 1934]. Like *Cypridinella* but with subcentral swelling slightly above center on each valve. *Carb.,* Eu.——FIG. 325,*3. C. galea* JONES & KIRKBY, Eire; carapace L, ×7 (39).

Sulcuna JONES & KIRKBY, 1874 [**S. lepus;* SD BASSLER & KELLETT, 1934]. Like *Cypridellina* but subcentral swelling replaced by backwardly directed dorsal protuberance and defined posteriorly by shallow nuchal furrow. *Carb.,* Eu.——FIG. 325, *4. S. cuniculus* JONES & KIRKBY, Eire; LV lat. (reconstr., rostrum and sinus hypothetical), ×8 (Sylvester-Bradley, n).

Family RHOMBINIDAE
Sylvester-Bradley, 1951

[Materials for this family prepared by P. C. SYLVESTER-BRADLEY, University of Leicester]

Rostrum truncate, down-curved; anteroventral border receding; marginal rim developed more or less strongly along ventral border. *Carb.*

Rhombina JONES & KIRKBY, 1874 [**R. hibernica;* SD BASSLER & KELLETT, 1934]. Posterior tumid, posterior margin evenly curved; no nuchal furrow. *L.Carb.,* Eu.——FIG. 326,*2. R. oblonga* (JONES & KIRKBY), Eire; *2a,b,* LV lat., dors., ×10; *2c,* muscle scar, ×20 (366).

Palaeophilomedes SYLVESTER-BRADLEY, 1951 [**Philomedes bairdiana* JONES & KIRKBY, 1874]. Posterior margin triangular, possibly with siphon; short nuchal furrow pointing toward posteroventral corner. *Carb.,* Eu.——FIG. 326,*1. *P. bairdianus* (JONES & KIRKBY), L.Carb., Eire.; RV lat., ×10 (366).

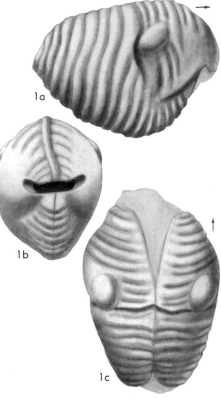

Cyprella

FIG. 324. Cyprellidae (p. Q402).

Family SARSIELLIDAE
Brady & Norman, 1896

[=Rutidermatidae BRADY & NORMAN, 1896] [Materials for this family prepared by P. C. SYLVESTER-BRADLEY, University of Leicester, with addition by I. G. SOHN, U.S. Geological Survey]

Carapace strongly calcified, heavily sculptured and ornamented, subcircular or oval in lateral outline, with pronounced caudal process; rostrum and sinus present or absent. Sexual dimorphism extreme. *?M.Dev., Rec.*

Sarsiella NORMAN, 1869 [**S. capsula*] [=*Eurypylus* BRADY, 1869; *Nematohamma* BRADY & NORMAN, 1896]. Prominent, more or less acuminate caudal process somewhat below mid-line, and usually with less prominent posterodorsal process; carapace of female subcircular, without rostrum or sinus, that of male with pronounced overhanging blunt rostrum, as in *Streptoleberis. Rec.,* cosmop. ——FIG. 327,*1. *S. capsula,* Medit.; *1a,* ♂ carapace L; *1b,c,* ♀ carapace L, dors.; all ×40 (♂.)

Rutiderma BRADY & NORMAN, 1896 [**R. compress..*]. Males unknown, females resembling males of

Sarsiella but longer and with less pronounced rostrum and caudal process. *Rec.*, cosmop.——Fig. 328,*1*. **R. compressa*, Atl.; ♀ carapace L, ×40 (258a).

?**Svarogites** PŘIBYL, 1951 [**S. spinosus*]. Differs from *Sarsiella* in large size (3 mm.-6 mm.), presence of an anterodorsal node and posterodorsal spine, and subcentral circular impression. *M.Dev.*, *C.Eu.*——Fig. 327,*2*. **S. spinosus*, Czech.; *2a,b*, carapace (holotype) R, post., ×50 (248a). [SOHN.]

Family UNCERTAIN

Cyprosis JONES, 1881 [**C. haswelli*]. Genus based on single specimen, subsequently lost; characters doubtful. *Sil.*, Scot.

Superfamily HALOCYPRIDACEA Dana, 1852

[*nom. transl. et correct.* SYLVESTER-BRADLEY, herein (*ex* Halocypridae DANA, 1852)] [=Halocypriformes SKOGSBERG, 1920] [Diagnosis by P. C. SYLVESTER-BRADLEY, University of Leicester]

Carapace almost or entirely uncalcified; rostrum projecting in continuation of more or less straight dorsal border. *Rec.*

Family HALOCYPRIDIDAE Dana, 1852

[*nom. correct.* SYLVESTER-BRADLEY, 1956 (*pro* Halocypridae DANA, 1852)] [Materials for this family prepared by P. C. SYLVESTER-BRADLEY, University of Leicester]

Characters of superfamily. *Rec.*

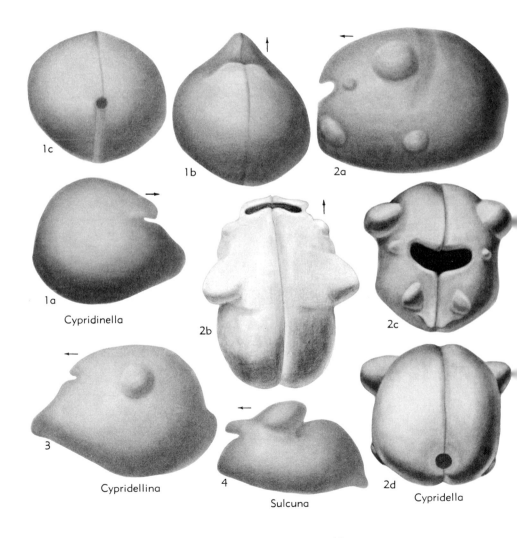

1c

1b

2a

1a

Cypridinella

2b

2c

3

Cypridellina

4

Sulcuna

2d

Cypridella

FIG. 325. Cypridinellidae (p. *Q*403).

Halocypris DANA, 1852 [*Conchaecia inflata* DANA, 1849; SD SYLVESTER-BRADLEY, 1956] =[*Halocypria* CLAUS, 1874]. Carapace short, rostrum slight. *Rec.*, cosmop.——FIG. 329,2. *H. inflata* (DANA), Rec., ♂ carapace L, ×30 (Dana, 1852).

Conchoecia DANA, 1849 [*C. magna* CLAUS, 1874 (ICZN pend.)] [=*Conchaecia* DANA, 1849 (ICZN pend.); *Conchoecetta, Conchoecilla, Conchoecissa, Mikroconchoecia, Pseudoconchoecia, Paraconchoecia* CLAUS, 1890; *Microconchoecia* G.W.MÜLLER, 1894; *Metaconchoecia* GRANATA & CAPORIACCO, 1949]. Carapace elongate, rostrum pronounced, pointed. *Rec.*, cosmop.——FIG. 329,1. *C. spinirostris* (CLAUS), Atl.; ♂ carapace L, ×50 (Sylvester-Bradley, n).

Macroconchoecia GRANATA & CAPORIACCO, 1949 [*Conchoecia reticulata* MÜLLER, 1906]. Like *Conchoecia*, but shell reticulate. *Rec.*, Atl., Ind.O.——FIG. 329,3. *M. reticulata*, Rec., Atl.; carapace L, ×60 (258a).

Euconchoecia G. W. MÜLLER, 1890 [*E. chierchiae*]. Carapace as in *Conchoecia* except that so-called asymmetrical glands are symmetrical. *Rec.*, cosmop.

Archiconchoecia G. W. MÜLLER, 1894 [*A. striata*]. Shell as in *Conchoecia. Rec.*, cosmop.

Suborder CLADOCOPINA Sars, 1866

[*nom. correct.* SYLVESTER-BRADLEY, herein (*pro* Cladocopa SARS, 1866)] [Type: =*Polycope* SARS, 1866; SD SYLVESTER-BRADLEY, herein] [=Polycopiformes SKOGSBERG, 1920] [Diagnosis by P. C. SYLVESTER-BRADLEY, University of Leicester]

Carapace subcircular in lateral outline, without gape of any kind; muscle-scar pattern composed of 3 closely juxtaposed scars. Described species (except doubtful Paleozoic fossils) all rather small (less than 0.8 mm.). Both antennules (="first antennae") and antennae (="second antennae") modified for swimming. *?Dev., Miss.-Rec.*

The description of this division by SARS (1866) is as follows:

"Lower antennae [antennae proper] biramous, both rami fully evolute, mobile and natatory. Upper antennae [antennules] also natatory, not geniculate, terminated by a small bundle of long setae. Mandibles distinguished by a short and very small, pediform palp that serves as a branchial appendage. There are only two pairs of thoracic limbs, the anterior large, bifid, natatory, the posterior membraneous and branchial. No eyes. Post abdomen divided into two short plates [furcal rami] that are pointed posteriorly." [Transl. by SCOTT & WAINWRIGHT.]

Family POLYCOPIDAE Sars, 1866

[Materials for this family prepared by P. C. SYLVESTER-BRADLEY, University of Leicester]

Characters of order. *?Dev., L.Jur.-Rec.*

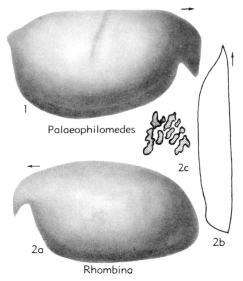

FIG. 326. Rhombinidae (p. Q403).

Polycope SARS, 1866 [*P. orbicularis*] [=*Cypridinopsis* ARMSTRONG, 1871]. Carapace entire in lateral outline except for minute downward-directed spine developed on anterior margin in some, perfect circle of outline broken by faint indication of cardinal angles and slightly protuberant anterior end; surface smooth, punctate or reticulate. Narrow vestibules developed in both valves in posteroventral area; hinge line straight, sunk in slight dorsal furrow; in type species anterior and posterior elements of LV hinge with short ridge, above groove, above projecting ridge and RV with ridge above groove; median element simple, in some other species (as *P. sublaevis*) this structure is reversed. *?Dev., L.Jur.-Rec.*, cosmop.——FIG. 330,2a-c. *P. orbicularis*, Rec., off Norway; 2a,b, ♂ carapace L, dors., ×70 (65); 2c, LV int., ×70 (Sylvester-Bradley, n).——FIG. 330,2d,e. *P. sublaevis* SARS, Rec., Atl.; 2d,e, LV int., RV int. (showing reversed hinge), ×70 (Sylvester-Bradley, n).

Polycopsis G. W. MÜLLER, 1894 [*Polycope compressa* BRADY & ROBERTSON, 1869; SD SYLVESTER-BRADLEY, herein]. Like *Polycope* but more compressed and with anterior border serrate. *Rec.*, cosmop.——FIG. 330,1. *P. compressa*, off Norway; 1a,b, carapace L, dors., ×70 (65).

Parapolycope KLIE, 1936 [*P. germanica*]. Shell as in *Polycopsis. Rec.*, Balt.

Family UNCERTAIN

?Discoidella CRONEIS & GALE, 1938 [*D. simplex*]. Carapace small, subcircular, coarsely reticulate, widest dorsally. Hinge line short, straight, lying below dorsal border in conspicuous hinge channel.

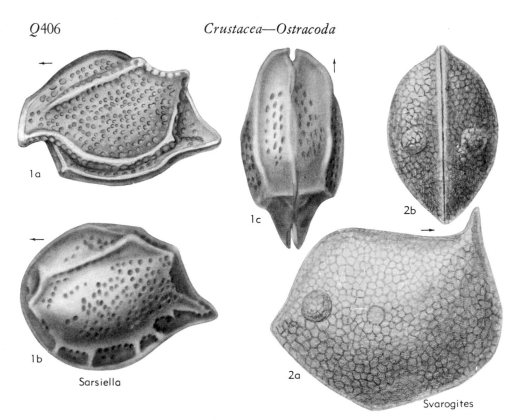

1a

1c

2b

1b

Sarsiella

2a

Svarogites

FIG. 327. Sarsiellidae (p. Q403-Q404).

This genus is referred to the Cladocopina with considerable doubt. *Miss., N.Am.*——FIG. 330,*3a.* *D. simplex,* Miss. (Chester), Ill.; carapace L, ×100 (132).——FIG. 330,*3b. D. ampla* COOPER, Miss. (Chester); carapace post., ×100 (Sylvester-Bradley, n).

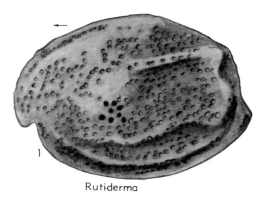

1

Rutiderma

FIG. 328. Sarsiellidae (p. Q403-Q404).

OSTRACODA, Order and Suborder UNCERTAIN

Family BUREGIIDAE Polenova, 1953

[Materials for this family prepared by R. H. SHAVER, Indiana University and Indiana Geological Survey]

Carapace subovate, with straight dorsal border, high, rounded at extremity, RV commonly overlapping LV, some with ventral projections or spines. *Dev.*

Buregia ZASPELOVA in POLENOVA, 1953 [**B. bispinosa*]. Characters of family. *Dev.,* Russ.——FIG. 331,*1a-d.* **B. bispinosa; 1a-d,* RV lat., dors., LV lat., dors., ×33 (278).——FIG. 331,*1e-g.* *B. krestovnikovi* POLENOVA; *1e-g,* carapace (holotype) L, dors., vent., ×30 (278).

OSTRACODA, Order, Suborder, and Family UNCERTAIN

[Materials for this section prepared by authors as severally recorded at end of generic descriptions. Included also by the editor are names of genera published in the USSR 1960 Treatise which are not contained in preceding sections of this Treatise.]

Acrossula KUMMEROW, 1953 [**A. u-scripta*]. Straight-backed, with long hinge line; quadrilobate or trilobate, with large L_2 and sharp-ridged

L_1 joined with L_3 by connecting structure; velar structure seemingly developed and some with carinal ridge; subvelar area stated to be channeled. Dimorphism not oberved. *Dev., Eu.*——Fig. 332, *1.* **A. u-scripta,* M.Dev., Ger.; *1a,* RV (holotype), ×12; *1b,c,* carapace vent., post., ×12 (47). [Hessland.]

Ampuloides Polenova, 1952 [**A. verrucosa*]. Straight-hinged, somewhat preplete, nonsulcate, LV slightly larger than RV; adult specimens generally very gibbous, with furrow along free margin corresponding to interior septum in anterior part; dimorphism possibly indicated by difference in gibbosity; surface warty. *M.Dev.-U.Dev.* USSR. ——Fig. 333,*1.* **A. verrucosa,* M.Dev. (U. Givet.); *1a,* carapace (holotype) L; *1b,* vent. (post. end up), ×45 (60). [Hessland.]

Arcuaria Neckaja, 1958 [**A. sineclivula*]. Generally high, lateral outline subtriangular, anterior margin sloping more steeply than posterior, ventral concave in central part; valves globose (most in dorsal region), LV overlapping RV along free margin and RV overlapping LV along central or posterocentral part of dorsal margin; no adventral or interior structures or dimorphism reported; surface smooth but some shells provided with tubercles or fine spines in posterior part. *M.Ord.* Eu.(NW.Russian Platform).——Fig. 333,4. **A.*

sineclivula, Lithuania; *4a,b,* carapace (holotype) R, dors.; *4c,* RV lat. (juv. instar), all ×15 (264). [Hessland.]

Ballardina Harris, 1957 [**B. concentrica*]. Hinge line straight, carapace less than 2 mm. long in known species, compressed, elongate-oval, S_2 a centrodorsal depression or pit, elongate, sloping posteroventrally, constricted medially, lying below and behind prominent dorsal ridge that curves downward in anterodorsal part of valve; velar ridge, apparently restricted. *M.Ord.,* N.Am.—— Fig. 333,*2.* **B. concentrica,* USA(Okla.); RV lat., ×25 (161). [Kesling.]

Balticella Thorslund, 1940 [**B. oblonga*]. Subquadrate, long, straight-backed, bisulcate; LV overlapping RV, especially along ventral margin; deep median sulcus extending more than half of valve height; anterior sulcus subdued; well-defined large ovate lobe at anterior margin of median sulcus not reaching dorsal margin. *Ord.,* Eu.-N.Am.——Fig. 332,*3. B. deckeri* Harris, M.Ord. (Edinburg F.), USA(Va.); *3a-c,* RV lat., int., LV lat., ×20 (J. C. Kraft, n). [Moore.]

Boucekites Přibyl, 1951 [**B. devonicus*]. Nonsulcate, equivalved, without overlap, smooth, amplete or somewhat preplete; dorsal margin slightly convex, hinge straight; cardinal angles ending in short rounded spines; no adventral structures un-

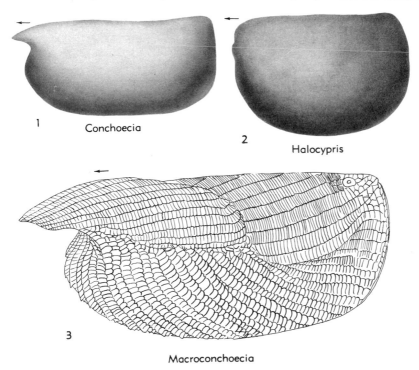

1   Conchoecia

2   Halocypris

3   Macroconchoecia

Fig. 329. Halocyprididae (p. Q405).

less represented by permanent posteroventral tubercles one on each valve. Dimorphism not observed. *M.Dev.* Eu.——FIG. 332,5. **B. devonicus,* Czech.; LV lat., ×35 (284). [HESSLAND.]

Brachycytheropteron KUZNETSOVA, 1960 [**Cytheropteron bicornutum* ALEXANDER, 1933]. *L.Cret.,* Eu. (Caucasus), N.Am.(Tex.).

Celechovites POKORNÝ, 1950 [**C. cultratus*]. Straight-backed, elongate, anterior margin rounded, posterior end pointed at about mid-height, ventral margin parallel to dorsal, anterior and ventral part of posterior areas pinched; inequivalved (LV larger than RV); no adventral structures, dimorphism not observed. *M.Dev.,* Eu.——

FIG. 332,2. **C. cultratus,* Czech.; *2a,b,* carapace (holotype) R, vent., ×50 (275). [HESSLAND-SHAVER.]

Ceratocypris POULSEN, 1934 [**C. symmetrica*]. Nonsulcate, nonsculptured, without adventral structure except possibly a rounded adventral bend, area between bend and ventral margin being slightly channeled; ventral region swollen and extended backward into hollow spine; surface smooth; possibly dimorphic, as indicated by differences in gibbosity. *Ord.-Sil.,* Eu.(Baltoscandia)-Greenl.—— FIG. 332,9. **C. longispina* HESSLAND, L.Ord. (Llanvirn.), Swed.; *9a-c,* LV lat., vent., ant., ×45 (30). [HESSLAND.]

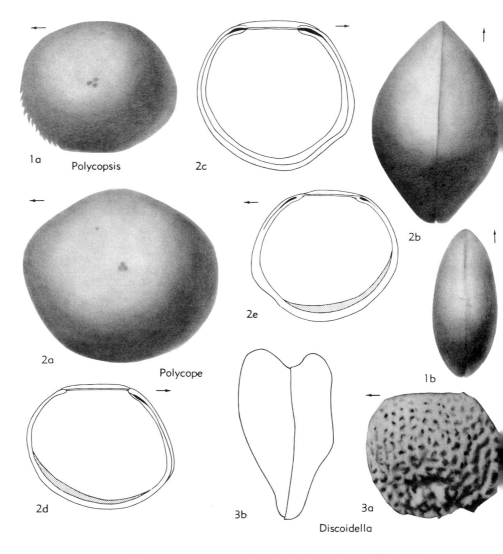

1a   Polycopsis   2c

2a
2b

2e

Polycope

2d   1b

3b   3a

Discoidella

FIG. 330. Polycopidae, Cladocopina, Family Uncertain (p. *Q405-Q406*).

Coeloenellina POLENOVA, 1952 [**C. parva*]. Dev., USSR.

Craspedopyxion JAANUSSON, 1957 [**Primitia undulosa* ÖPIK, 1937]. Outline subcircular, straight-backed, postplete, strongly convex; with single deep sulcus almost at mid-length of valve, and prominent node behind sulcus; flattened marginal border widest at front and rear. *M.Ord.*, NW.Eu. (Est.).——FIG. 334,*1*. **C. undulosum* (ÖPIK); LV lat.(reconstr.), ×35 (36). [MOORE.]

Ginella V. IVANOVA, 1960 [(*pro Pinella* IVANOVA, 1955) (*non* STEPHENSON, 1941)] [**Pinella tenuispina* IVANOVA, 1955]. *M.Ord.*, USSR.

Hemiaechminoides MORRIS & HILL, 1952 [**H. monospinosus*]. Nonsulcate, smooth, slightly postplete, hinge straight, dorsal margin somewhat convex; valves unequal, RV overlapping LV along entire free border, LV provided with dorsal spine directed laterally-backward; no adventral structures. Dimorphism not observed. *M.Sil.*, N.Am.——FIG. 332,*8*. **H. monospinosus*, USA(Tenn.); *8a,b*, carapace (reconstr.) L, post., ×40 (254). [HESSLAND.]

Hupehella HOU, 1955 [**H. lunata*]. Outline subovate, nonsulcate (or possibly with slight sulcal depression), swollen umbo extending above hinge which is straight and provided with terminal sockets and slitlike central furrow in LV (no structures are observed in hinge of RV); ventral margin and ventral parts of terminal margins surrounded by broad flange which may be concave ventrally. Dimorphism not reported. Surface smooth or reticulate. *U.Dev.*, E.Asia.——FIG. 333, *3*. **H. lunata*, Hsiehkingsu F., China(Changyang Distr., Hupeh Prov.); *3a*, RV (syntype) lat.; *3b*, LV (syntype) int.; ×37 (176). [HESSLAND.]

Ilmenoindivisia EGOROV, 1954 [**I. wjadica*]. Dev., USSR.

Indivisia ZASPELOVA, 1954 [**I. indistincta*]. *Carb.*, USSR. (Type genus of Indivisiinae EGOROV, 1954).

Ivaria NECKAJA, 1960 [(*pro Glossopsis* NECKAJA, 1953) (*non* BUSH, 1904; *nec* HESSLAND, 1949)] [**Glossopsis robusta* HESSLAND, 1949]. *Ord.*, USSR.

Moorea JONES & KIRKBY in JONES & HOLL, 1869 [**M. silurica* JONES & HOLL, 1869; SD S. A. MILLER, 1892]. Small, straight-backed, subhemicircular, with raised marginal rim continuing along dorsum below contact for about 0.7 of length of dorsal margin, then turning sharply down and ?posteroventrally to merge with convexity of valve; surface pitted. *Sil.*, Eng. [SOHN.]

Paleocythere TOLMACHOFF, 1926 [**P. typa*]. Reminiscent of male cavellinids, with ovate outline in lateral view, moderately blunt posterior extremity in dorsal view, and strong RV-over-LV overreach; narrow anterior sulcus and differentiated ridge-and-groove hinge with terminal teeth in LV and sockets in RV may denote placement in Palaeocopida;

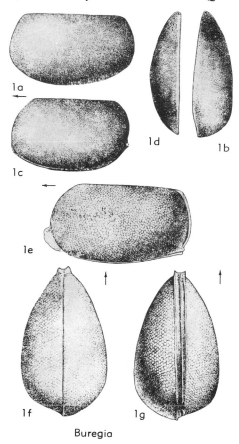

Buregia

FIG. 331. Ostracoda, Order and Suborder Uncertain (p. Q406).

seemingly differs from *Ellesmeria* only by its smooth surface. Adductor muscle scar and contact-marginal structures unknown. *M.?Dev.*, Ellesm.——FIG. 332,6. **P. typa; 6a,b*, carapace L, dors., ×13 (370). [SHAVER.]

Parapyxion JAANUSSON, 1957 [**Primitia subovata* THORSLUND, 1948]. Straight-backed, nonsulcate or unisulcate, dorsal region steeply sloping but not protruding above hinge line, peripheral area along free margins gently sloping, without marginal depression; adductor muscle impression well defined on lateral surface, rounded, moderately large. *M.Ord.-U.Ord.*, Eu.(Swed.-Czech.).——FIG. 334, *2*. **P. subovatum* (THORSLUND), M.Ord., Swed.; LV lat. (reconstr.), ×35 (36). [MOORE.]

Platyrhomboides HARRIS, 1957 [**P. quadratus*]. Trapezoidal in side view, straight-hinged, equivalved, triangular in transverse section, thickest through edge of flattened venter; surface smooth, punctate, reticulate or spinose. *Ord.*, N.Am.—— FIG. 332,7. *P.* sp., M.Ord.(Edinburg F.), USA

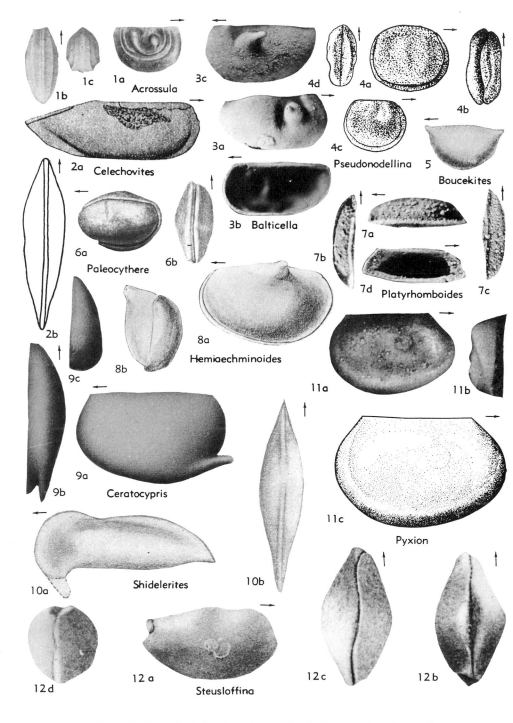

FIG. 332. Ostracoda, Order, Suborder, and Family Uncertain (p. *Q*406-*Q*412).

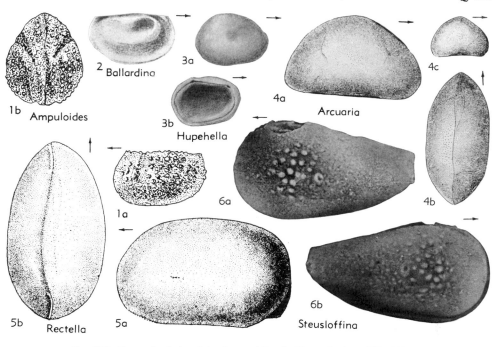

FIG. 333. Ostracoda, Order, Suborder, and Family Uncertain (p. Q407-Q412).

(Va.); *7a-d,* LV lat., dors., vent., int., ×30 (J. C. Kraft, n). [MOORE.]

Pseudonodellina POLENOVA, 1955 [**Nodella? parvula* POLENOVA, 1953]. Small, very high in relation to length, almost circular in lateral aspect; LV larger than RV, overlapping it along free margin; with elongate umbos, extending above hinge to cause convexity of dorsal margin; with 3 feeble dorsoventral lobes, no adventral structures. No dimorphism reported. *Dev.,* USSR.——FIG. 332,*4a,b. P. rotundata* POLENOVA; *4a,b,* carapace R, dors., ×45 (279).——FIG. 332,*4c,d. P. strelniensis* POLENOVA; *4c,d,* carapace R, dors., ×45 (279). [HESSLAND-REYMENT.]

Pseudoperissocytheridea MANDELSTAM, 1960 [**Protocythere crassula* MANDELSTAM, 1947]. *Jur.,* Asia (Kazakhstan), S.Am.

Pyxion THORSLUND, 1948 [**Primitia carinata* HADDING, 1913]. Postplete, nonsulcate to unisulcate (broad depression), presulcal node consisting of fairly large oblong knob, some shells also with postsulcal dorsal inflation; adventral structure constituting ridge or bend along entire free margin. Dimorphism not observed. *M.Ord.,* Eu.(Baltoscandia).——FIG. 332,*11. *P. carinatum* (HADDING), M.Ord.(Carodoc.), Swed.; *11a,b,* RV (lectotype, THORSLUND, 1948) lat., ant., ×25; *11c,* RV lat. (reconstr.), ×35 (36). [HESSLAND.]

Rectella NECKAJA, 1958 [*nom. subst. pro Mica* NECKAJA, 1952 (*non* BUDDE-LUND, 1908; *nec* PETRUNKEVITCH, 1925)] [**Mica inaequalis* NECK-

AJA, 1952]. Nonsulcate, lengthened, preplete, dorsal and ventral margins approximately parallel, dorsal corners rounded; inequivalved (LV larger); hinge depressed (dorsum epicline); dimorphism not observed; spines may be developed in posterior part. *Ord., ?Dev.,* Eu.(Baltoscandia).——FIG. 333,*5. *M. inaequalis,* M.Ord.(Caradoc), Lithuania; *5a,b,* carapace L, vent., ×53 (262).

Reginacypris SCHNEIDER, 1960 [**Cytherina abcissa* REUSS, 1850]. *Mio.,* Czech.-USSR(Urals).

Shidelerites MORRIS & HILL, 1951 [**S. typus*]. Subequivalved, smooth, slender, attenuate in dorsal and lateral views but with rounded anterior border in lateral view; anterior beaklike projection reminiscent of Cypridinidae but lacking slit; re-entrants in 2 long borders offer mechanical difficulty to either choice of hinge border; contact marginal structures and adductor muscle scar unknown. *M.Sil.,* Ind.——FIG. 332,*10. *S. typus; 10a,b,* carapace (holotype) L, dors., ×28 (253). [SHAVER.]

Sigillium KUZNETSOVA, 1960 [**S. procerum*]. [Type genus of Sigilliuminae MANDELSTAM, 1960 (*recte* Sigilliinae)]. *Paleog.,* USSR(Azerbaijan).

Steusloffina TEICHERT, 1937 [**S. ulrichi*]. Dorsal margin long, hinge depressed (dorsum epicline); outline of free margin asymmetrical, posterodorsal angle acute; nonsulcate; lateral prominence (bulb or spine) generally present, tending to be broken at base leaving characteristic crater-like depression; no adventral structure. Dimorphism not ob-

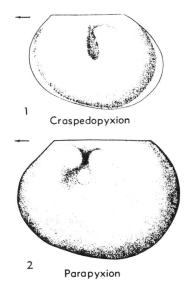

1

Craspedopyxion

2

Parapyxion

FIG. 334. Ostracoda, Order, Suborder, and Family Uncertain (p. Q409).

served. *M.Ord.-U.Ord.,* NW.Eu.(Baltoscandia)-Greenl.-N.Am.(Arct. Can.).——FIG. 332,*12.* **S. ulrichi,** M.Ord.(Trenton.), Can.(Arct.); *12a-d,* carapace (holotype) R lat., dors., vent., ant., ×30 (368).——FIG. 333,*6. S. papillosa* HESSLAND, U. Ord. (Ashgill.), Est.; *6a,b,* carapace (holotype) L, R, ×27 (30). [HESSLAND.]

Timanella EGOROV, 1950 [**T. typica*]. *Dev.,* USSR (Timan).

NOMINA DUBIA

Generic names published for ostracodes which in the course of collaborative work by authors contributing to the *Treatise* have not been placed in some category given in previous pages are gathered together here with such annotations as are available.

Allostraca ULRICH & BASSLER, 1932 [**A. fimbriata*]. Name given as explanation of a figured specimen of some representative of the Hollinacea, possibly *Apatobolbina* or *Chilobolbina;* no published description. Unrecognizable. *L.Miss.,* USA(Tenn.). [SCOTT.]

Antitomis GÜRICH, 1896 [**A. bisulcata*]. Assigned to Leperditiidae by author but may belong to Entomozoidae (5); description and illustrations insufficient for recognition. *Sil.,* Eu.(Pol.). [MOORE.]

Aparchitellina POLENOVA, 1955 [**A. decorata*]. Referred by author to Leperditellidae. *Dev.,* USSR.

Bernix JONES, 1884 [**Beyrichia tatei* JONES, 1864]. Based on poorly preserved specimens; seemingly differs from *Hypotetragona* in that dorsum is not incised. Unrecognizable. *Carb.,* Eng. [SHAVER.]

Bohemia SNAJDR, 1951.

Bryocypris RØEN, 1956.

Budnianella BOUČEK, 1936 [**B. caroli*]. Referred to Kirkbyidae by author; classified with Beyrichiacea, family uncertain by Henningsmoen (29). *Sil.,* Eu.(Czech.). [MOORE.]

Bursulella JONES, 1887 [**B. triangularis*]. Type species consists of small triangular shells with 2 spines projecting from ? ventral border; described as bivalved with crenulate contact margins. [Probably not an ostracode; species assigned to the genus by later authors not considered to be congeneric.] *U.Sil.,* Swed.(Gotl.). [SHAVER.]

Callizoe BARRANDE, 1872. Broken fragments of unidentifiable ostracode related to Isochilinidae. [SCOTT.]

Carnarvonopsis SWARTZ, 1954 [no type species]. Undescribed and unknown; *nom. nud.* [SCOTT.]

Caryon BARRANDE, 1872 [**C. bohemicum*] [=*Caryon* BARRANDE in BIGSBY, 1868 *(nom. nud.)*]. Two large (length up to 30 mm.), dissimilar valves described from molds; cannot be classed as representing Ostracoda; probably cephalic parts of trilobites. *U.Ord.,* Czech. [SHAVER.]

Colpos MOBERG, 1895 [**C. insignis*]. Unidentifiable. *U.Sil.,* Swed. [SHAVER.]

Cornia LUTKEVICH, 1939 [**C. papillaria* LUTKEVICH, 1937] [=*Cornia* LUTKEVICH, 1937, *nom. nud.*]. Holotype of type species is oval fossil with growth lines and length of 3.8 mm., poorly preserved in matrix; classified by LUTKEVICH and others as Phyllopoda but listed by AGNEW as Ostracoda. *U.Perm.,* Russ. [SHAVER.]

Ctenentoma SCHMIDT, 1941 [**Entomis umbonata* STEUSLOFF, 1894]. Type species with frill and small node anterior to long median sulcus. The one known specimen and holotype possibly is an internal mold and is thought by some authors to belong to *Steusloffia.* Many species assigned to *Ctenentoma* are referable to other genera (29). *?L.Ord.* (glacial erratic), Ger. [SHAVER.]

Cuselina AMMON in REIS, SCHUSTER, KOEHNE & AMMON, 1910 [**C. impressa*]. Type material consists of molds and broken carapaces in matrix; identified by author as Phyllopoda or preferably Ostracoda; lectotype selected by GUTHÖRL has length of 4 to 5 mm. and is unidentifiable. *L. Perm.,* Ger. [SHAVER.]

Cyclocytheridea MANDELSTAM in LYUBIMOVA, 1955 [**?Type species*]. Carapace ovate with rounded ends, valves reticulate; hinge with terminal rounded teeth and median groove in RV. Name printed on plate as designation of *Camptocythere nordvikensis* MANDELSTAM (attributed to SHARAPOVA but a *nomen nudem*). [REYMENT-BOLD.]

Cytheropsis M'COY, 1849 [*non* SARS, 1866]. No species mentioned; first species assigned to genus is *Cytheropsis aldensis* M'COY, 1851. ?Senior synonym of *Pontocypris* BASSLER & KELLETT, 1934. *Cam.-Ord.,* Scot.

Cytherurina MANDELSTAM, 1958 [**Hemicytherura*

cellulosa Hornibrook (*non* Norman, ?date)]. [Genus currently invalid because based on junior homonym.] [Bold.]

Diagonella Swartz, 1945 [no type species]. Undescribed and unknown; *nom. nud.* [Scott.]

Dithyrocaris Portlock, 1843 [(*pro Argas* Scouler, 1835) (*non* Latreille, 1795)] [*Dithyrocaris colei* Portlock, 1843; SD Shaver, herein]. Since Portlock states: "Not having yet received Dr. Scouler's generic characters," it is evident that the first valid publication and description of *Dithyrocaris* in Portlock is solely by him, although he attributed the name to Scouler. Not Ostracoda, although the 1850 examples of Jones probably are referable to Kirkbyidae. [Shaver.]

Donellina Egorov, 1950 [*D. grandis*]. Assigned by author to Kloedenellidae. *U.Dev.*, USSR. [Levinson.]

Ellesmeria Tolmachov, 1926 [*E. ovata*]. Subovate, RV overlapping LV all around; surface punctate (5). *Dev.*, Arct. N.Am.(Ellesmereland). [Moore.]

Elpinella Přibyl & Šnajdr, 1950 [*Leperditia radiata* 1879]. Doubtfully referred to Entomoconchidae. *Ord.*, USA(Ohio). [Howe.]

Entomidella Jones, 1873 [*Entomis buprestis* Jones in Hicks, 1872; SD Jones, 1884] [=*Leperditia buprestis* Salter in Hicks, 1865 (*nom. nud.*)]. Holotype of type species is an elongate bivalved carapace 9 mm. long, with transverse fractures erroneously thought by Jones comparable to furrows of *Entomis*; possibly a branchiopod. *?L. Cam.*, *M.Cam.*, *?L.Ord.*, ?Can.(N.B.). [Shaver.]

Eocytherella Bonnema, 1933 [No type species designated]. Provisional name to include *Cytherella smithi* Jones, 1887, and *C. troedssoni* Bonnema, 1933 (*nom. subst. pro Primitia tenera* Troedsson, 1918). *Sil.*, Eu.(Swed.). [Howe.]

Famenella Polenova, 1953 [*F. incondita*].

Fidelitella V. Ivanova, 1960 [*pro Trilobella* V. Ivanova, 1955 (*non* Woodward, 1924)] [*Trilobella unica* V. Ivanova, 1955]. Referred to Tetradellidae by author. *M.Ord.*, USSR. [Levinson.]

Geffenites Coryell & Sohn, 1938 [*G. jungae*]. Based on inadequate material that probably is related to Hollinidae. *Miss.*, USA(W.Va.). [Scott.]

Gibba Fuchs, 1920 [*Beyrichia (gibba) spinosa*]. Type species is invalid junior homonym of *B. spinosa* (Hall, 1852) Hall, 1859. [=*Paraechmina spinosa*]. *Dev.*, Eu.(Ger.). [Howe.]

Gipsella Egorov, 1950 [*G. polkvoii*]. Illustrations inadequate for recognition. *U.Dev.*, USSR. [Scott.]

Glyptolichwinella Posner in Samoilova, 1951.

Golcondella Croneis & Gale, 1938 [*G. sulcata*]. Type specimens examined by Scott, who considered them to be molts of unknown ostracodes. *Miss.*, USA(Ill.). [Scott.]

Goniocypris Brady & Robertson, 1870 [*G. mitra*]. Originally assigned to Ostracoda but consisting of tiny, smooth, triangular, bivalved shells later identified by Brady & Norman as "fry of *Anodonta cygnaea*" (Mollusca) (Brady & Norman). *Rec.*, Eng. [Shaver.]

Gravia Polenova, 1952 [*G. aculata*]. *Dev.*, USSR. Type genus of Graviidae Polenova, 1952.

Hesvechilus Brady, 1875. In list of Ostracoda by Brady as *Hesvechilus contortus* (Norman); original source, if any, and combination unknown; (*nom. nud.*) *Rec.*, Fr.-Eng. [Shaver.]

Hlubocepina Přibyl, 1955.

Huarpina Rusconi, 1954.

Isocythere Terquem, 1885 [*I. nova*]. P. C. Sylvester-Bradley reports (19 February, 1960) that his examination of Terquem's type specimens establishes that they are "clearly not as described." *Jur.*, Eu.(Fr.). [Moore.]

Jonesina Ulrich & Bassler, 1908 [*Beyrichia fastigiata* Jones and Kirkby, 1867]. Straight-backed small trilobate ostracodes, hinge and overlap unknown (type lost, *fide* Johnson, 1936). [Most species referred to this genus belong to *Geisina* and *Hypotetragona*.] *L.Carb.*, Eng.

Junctocytheretta Anonymous, 1956, in Mandelstam *et al*. Printed as name on plate for species *J. signata* Mandelstam; in descriptive text referred to as *Eucytherura signata* M. The name *Junctocytheretta* is crossed over in Mandelstam's personal copy of the paper. [Reyment.]

Kelletella Delo, 1930 [*K. naviculata*]. Only known specimen is believed to be a steinkern; indeterminable. *U.Penn.*, Tex. [Scott.]

Leioditia Ulrich in Jones, 1891 (*nom. nud.*). ?Equivalent to *Elpezoe* Přibyl, 1950.

Leioprimitia Kummerow, 1939 [*L. punctata*]. Referred to Primitiidae [Leperditellidae] by author. *L.Carb.*, Eu.(Ger.). [Moore.]

Leptoprimitia Kummerow, 1953 [*L. compressa*]. Referred to Primitiidae [Leperditellidae] by author. *Dev.*, Eu.(Pol.). [Moore.]

Lichwinella Posner, 1948.

Lucasella Stewart, 1936 [*L. mundula*]. [Holotype a badly corroded steinkern of undeterminable affinities. Species described by Stewart & Hendrix belong to *Graphiadactyllis*.] *Dev.*, Ohio.

Monoculus Linné, 1758. Apparently, no type species designated subsequently among 9 referred to genus by Linné, of which 6 have been assigned by later authors to other genera in several orders of non-ostracode Crustacea and 3 are considered to be unidentifiable ostracodes; at least *M. conchaceus* has had fresh-water cypridids assigned to it by later authors. *Rec.* [Shaver.]

Neochilina Matern, 1929 [*N. binsenbachensis*]. Referred to Eurychilininae by author and Beyrichiacea, family uncertain, by Henningsmoen, 1953. *Dev.*, Eu.(Ger.). [Howe.]

Nezamyslia Přibyl, 1955.

Nothozoe Barrande, 1872 [*N. pollens*] [=*Nothozoe* Barrande in Bigsby, 1868 (*nom. nud.*)]. Nearly featureless internal molds shaped like some ostracodes, but excessively large (length up to 65 mm.; possibly a leperditiid (BARRANDE). ?*Cam.*, ?*Sil.*, Eu.-?N.Am. [SHAVER.]

Novakina BOUČEK, 1936 [*N. applanans*]. Referred to Aechmininae by author and to Beyrichiacea, family uncertain by HENNINGSMOEN (29). *Sil.*, Eu.(Czech.). [HOWE.]

Opisthoplax KUMMEROW, 1943 [*O. compressa*]. Referred to Primitiidae [Leperditellidae] by author. *Sil.*, Eu.(E.Ger.). [HOWE.]

Orozoe BARRANDE, 1872. Based on molt stages of form possibly similar to Aristozoe, unidentifiable. [SCOTT.]

Orthoconchoecia GRANATA & CAPORIACCO, 1949 [no type species designated]. Referred to Halocyprididae by authors. *Rec.*, Medit. [HOWE.]

Paradoxorhyncha CHAPMAN, 1904 [*P. foveolata*]. Based on misinterpreted type specimen obscured by matrix; unrecognizable. *Jur.*, W.Austral. [SYLVESTER-BRADLEY.]

Paragraphylus CORYELL & ROZANSKY, 1942 [*P. reticulatus*]. [Holotype a broken and corroded specimen; unrecognizable.] *Miss.*, Ill.

Parajonesites IVANOVA, 1955 [*P. notabilis*]. Referred to Primitiidae [Leperditellidae] by author. *U.Ord.*, USSR. [LEVINSON.]

Parenthetica SWARTZ, 1945. Undescribed; name only given in list. *L.Dev.*, N.Y.-Md.-W.Va. [SWARTZ (January, 1959, letter to SHAVER) states name is not a misprint or *lapsus calami* for *Parenthatia* KAY, as suggested by HOWE (1955), but a *nomen nudum*.] [SHAVER.]

Plagionephrodes MOREY, 1935 [*P. uninodosus*]. [Types lost, according to R. E. PECK (October, 1955). Illustrations and descriptions inadequate for determining affinities.] *Dev.*, ?*L.Miss.*, Mo.

Pulvillites ÖPIK, 1937 [*P. triangulum*]. Referred to Cypridacea by author. *Ord.*, Eu.(Est.). [HOWE.]

Pyxiprimitia SWARTZ, 1936 [*P. ventriclefta*]. Referred to Primitiidae [Leperditellidae] by author, Drepanellidae by SCHMIDT (69), and Beyrichiacea family uncertain, by HENNINGSMOEN (29). *Dev.*, USA(Pa.). [HOWE.]

Quadricollina ÖPIK, 1953 [*Q. initialis*]. Referred to Drepanellidae by author. *L.Sil.*, Austral. [LEVINSON.]

Russia POLENOVA, 1952 [*Gravia (Russia) unicostata*]. Referred to Acronotellidae by author. *Dev.*, USSR. [HOWE.]

Sacculus NEVIANI, 1928 [*non* GOSSE, 1851; *nec* HIRASE, 1927]. RUGGIERI (305) reports that name was published for 2 species; *Cythereis (Sacculus)*

trigibbosa and *C. (S.) tetragibbosa*, the first classed by RUGGIERI as synonym of *Cypridina haidingeri* REUSS (1849), which he thinks probably belongs to *Bradleya*. The second species is really not identifiable, but probably is some representative of *Gaudites*. *Tert.*, Italy. [HOWE.]

Selebratina POLENOVA, 1952 [*Gravia (S.) dentata*].

Semilukiella POLENOVA, 1952 [*S. zaspelovae*]. ?Hor., USSR.

Spinosa LORANGER, 1954 [No type species]. *Nom. nud.* probably intended as subgenus of Cytheridea. [HOWE.]

Sulcatia POLENOVA, 1952 [*?*]. ?Hor., USSR.

Sulcocavellina POLENOVA, 1952 [*S. incognita*]. ?Hor., USSR.

Sulcoindivisa EGOROV, 1954 [*S. svinordensis*].

Synaphe CHAPMAN, 1914 [*non* HUEBNER, 1825; *nec* THOMSON, 1864; *nec* JONES & KIRKBY, 1896] [*S. mesozoica*]. Invalid junior homonym. *Trias.*, Austral.

Tetrasulcata MATERN, 1929 [*T. fluens*]. Based on partly exfoliated right valve that differs from *Jonesina* in having 4 instead of 3 lobes, rearmost compressing a node. *U.Dev.*, Eu.(NE.Ger.). [SOHN.]

Tricornina BOUČEK, 1936 [*T. navicula*]. Referred to Primitiidae [Leperditellidae] by author and Alanellidae [Beecherellidae] by HENNINGSMOEN (29). *Sil.*, Eu.(Czech.). [HOWE.]

Trubinella PŘIBYL, 1949 [*Hippa latens* BARRANDE, 1872 (*non Hippa* FABRICIUS, 1787)]. Probably an ostracode instar; unidentifiable. *Ord.*, Czech. [SCOTT.]

Zaborovia POLENOVA, 1952 [*Z. obscura*]. Referred to Leperditellidae by author. *Dev.*, USSR. [HOWE.]

Generic Names Incorrectly Classed by Authors as Belonging to Ostracoda

Cryptocaris BARRANDE, 1872. Referred by the author to Phyllopoda; 8 species, one of which (*C. ?rhomboidea*) has been defined by PŘIBYL (1951) as type species of *Rhomboentomozoe*, a myodocopid ostracode. *Sil.*, Czech.

Lepidilla MATTHEW, 1886 [*L. anomala*]. Classed by author as an ostracode but regarded as conchostracan by ULRICH & BASSLER (87). *Cam.*, Can. (N.B.).

Lepiditta MATTHEW, 1886 [*L. alata*]. Same entry as for *Lepidilla*. *Cam.*, Can.(N.B.).

Zonozoe BARRANDE, 1872 [*Z. drabowiensis*]. Not an ostracode. *Sil.*, Czech.

REFERENCES

Agnew, A. F.
(1) 1942, *Bibliographic index of new genera and families of Paleozoic Ostracoda since 1934:* Jour. Paleont., v. 16, no. 6, p. 756-763 (Tulsa).

Apostolescu, Vespasian
(2) 1955a, *Description de quelques ostracodes de Lutétien du Bassin de Paris:* Cahiers Géologiques, no. 28-29, p. 241-279, 7 pl. (Paris).

Baird, William
(3) 1837-1838, *The natural history of the British Entomostraca:* Mag. Zool. & Bot., Edinburgh, v. 1, 1837, p. 35-41, 309-333, 514-526, pl. 8-10; v. 2, 1838, p. 132-144, 400-412, pl. 16.

Bassler, R. S.
(4) 1941, *Ostracoda from the Devonian (Onondaga) chert of West Tennessee:* Washington Acad. Sci., Jour., v. 31, no. 1, p. 21-27, 37 fig.

——, **& Kellett, Betty**
(5) 1934, *Bibliographic index of Paleozoic Ostracoda:* Geol. Soc. America, Spec. Paper 1, 500 p., pl. (New York).

Blake, D. B.
(6) 1950, *Gosport Eocene Ostracoda from Little Stave Creek, Alabama:* Jour. Paleont., v. 24, no. 2, p. 174-184, pl. 29, 30 (Tulsa).

Bold, W. A. van den
(7) 1946, *Contribution to the study of Ostracoda with special reference to the Tertiary and Cretaceous microfauna of the Caribbean region:* Amsterdam, July 11, 1946, 167 p., 18 pl., fig.

Bonnema, J. H.
(8) 1930, *Orientation of the carapaces of Paleozoic Ostracoda:* Jour. Paleont., v. 4, no. 2, p. 109-20, 14 fig. (Tulsa).

Bosquet, J. A. H.
(9) 1852, *Description des entomostracés fossiles des terrains Tertiares de la France et de la Belgique:* Acad. Roy. Sci. Belg., Mém. Sav., v. 24, 142 p., 6 pl. (Bruxelles).

Bouček, Bedřich
(10) 1936, *Die ostracoden des Böhmischen Ludlows (Stufe):* Neues Jahrb., Beil.-Band 76, Abt. B, Heft 1, p. 31-98, pl. 206 (Stuttgart).

Bradfield, H. H.
(11) 1935, *Pennsylvanian ostracods of the Ardmore Basin, Oklahoma:* Bull. Am. Paleont., v. 22, no. 73, 145 p., 13 pl. (Ithaca).

Brady, G. S.
(12) 1868, *A monograph of the Recent British Ostracoda:* Linn. Soc. London, Trans., v. 26, p. 353-495, pl. 23-41.
(13) 1880, *Report on the Ostracoda dredged by H.M.S. Challenger during the years 1873-1876:* Rept. Voyage Challenger, Zool., v. 1, pt. 3, p. 1-184, pl. 1-44.

——, **Crosskey, H. W., &
Robertson, David**
(14) 1874, *A monograph of the Post-Tertiary Entomostraca of Scotland including species from England and Ireland:* Paleontogr. Soc., Mon., p. i-v, 1-274, pl. 1-16 (London).

——, **& Norman, A. M.**
(15) 1889, *A monograph of the marine and freshwater Ostracoda of the North Atlantic and of Northwestern Europe. Section I. Podocopa:* Roy. Dublin Soc., Sci. Trans., ser. 2, v. 4, p. 63-270, pl. 8-23.
(16) 1896, *A monograph of the marine and freshwater Ostracoda of the North Atlantic and of Northwestern Europe. Sections 2-4. Myodocopa, Cladocopa, and Platycopa:* Same, ser. 2, v. 5, p. 621-746, pl. 50-68.

——, **& Robertson, David**
(17) 1870, *Notes of a week's dredging in the west of Ireland:* Ann. & Mag. Nat. History, ser. 4, v. 3, p. 353-374, pl. 18-22 (London).

Bronstein, Z. S.
(18) 1947, *Faune de L'URSS. Crustacés: v. II, no. 1. Ostracodes des eaux douces:* Inst. Zool. Acad. Sci. L'URSS, new ser., no. 31, 339 p., 206 fig. (Moscow-Leningrad).

Claus, Carl
(19) 1893, *Beiträge zur kenntniss der Süsswasser-Ostracoden:* Arbeit. Zool. Inst. Univ. Wien, v. 10, p. 147-216, pl. 1-12.

Cooper, C. L.
(20) 1941, *Chester ostracodes of Illinois:* Illinois State Geol. Survey, Rept. Inv. No. 77, 101 p., 14 pl. (Urbana).
(21) 1946, *Pennsylvanian ostracodes of Illinois:* Illinois State Geol. Survey, Bull. 70, 177 p., 21 pl. (Urbana).

Coryell, H. N., & Malkin, D. S.
(22) 1936, *Some Hamilton ostracodes from Arkona, Ontario:* Am. Mus. Nov., no. 891, 20 p,. 2 pl. (New York).

Elofson, Olof
(23) 1941, *Sur Kenntnis der marinen Ostracoden Schwedens mit besonderer Berücksichtigung des Skageraks:* Zool. Bidr. f. Uppsala, v. 19, p. 215-534, 52 fig., 42 maps.

Furtos, N. C.
(24) 1933, *The Ostracoda of Ohio:* Ohio Biol. Survey, Bull., v. 5, no. 6 (Bull. 29), p. 412-524, pl. 1-16 (Columbus).

Hanai, Tetsuro
(25) 1957a, *Studies on the Ostracoda from Japan I. Subfamily Leptocytherinae n. subfam.:*

Fac. Sci. Univ. Tokyo, Jour., sec. 2, v. 10, pt. 3, p. 431-468, pl. 7-10.

(26) 1957b, *Studies on the Ostracoda from Japan II. Subfamily Pectocytherinae n. subfam.:* Same, sec. 2, v. 10, pt. 3, p. 469-482, pl. 11.

(27) 1957c, *Studies on the Ostracoda from Japan III. Subfamilies Cytherurinae G. W. Müller (emend. G. O. Sars, 1925) and Cytheropterinae n. subfam.:* Same, sec. 2, v. 11, pt. 1, p. 11-36, pl. 2-4.

(28) 1959a, *Studies on the Ostracoda from Japan IV. Family Cytherideidae Sars, 1925:* Same, sec. 2, v. 11, pt. 3, p. 291-308, pl. 16-18.

Henningsmoen, Gunnar

(29) 1953a, *Classification of Paleozoic straight-hinged ostracods:* Norsk. Geol. Tidsskr., v. 31, p. 185-288, pl. 1-2 (Oslo).

Hessland, Ivar

(30) 1949, *Investigations of the Lower Ordovician of the Siljan District, Sweden. I. Lower Ordovician ostracods of the Siljan District, Sweden:* Uppsala Univ. Min-geol. Inst. Bull., v. 33, p. 97-408, pl. 1-18.

Hoff, C. C.

(31) 1942a, *The ostracods of Illinois. Their biology and taxonomy:* Illinois Biol. Mon., v. 19, p. 5-196, pl. 1-9 (Urbana).

Hornibrook, N. de B.

(32) 1952, *Tertiary and Recent marine Ostracoda of New Zealand:* New Zealand Geol. Survey, Paleont. Bull. 18, 82 p., 18 pl. (Wellington).

Howe, H. V., & Chambers, Jack

(33) 1935, *Louisiana Jackson Eocene Ostracoda:* Dept. Conserv., Louisiana Geol. Survey, Geol. Bull., no. 7, 96 p., 6 pl. (Baton Rouge).

——, **& Laurencich, Laura**

(34) 1958, *Introduction to the Study of Cretaceous Ostracoda:* Louisiana State Univ. Press, 536 p., fig. (Baton Rouge).

——, **& Law, John**

(35) 1936, *Louisiana Vicksburg Oligocene Ostracoda:* Dept. Conserv., Louisiana Geol. Survey, Geol. Bull., no. 7, 96 p., 6 pl. (Baton Rouge).

Jaanusson, Valdar

(36) 1957, *Middle Ordovician ostracodes of central and southern Sweden:* Geol. Inst. Univ. Uppsala, Bull., v. 37, p. 176-442, 15 pl., 46 fig.

Jones, T. R.

(37) 1849, *A monograph of the Entomostraca of the Cretaceous Formation of England:* Palaeontogr. Soc., Mon., 1849, 40 p., 7 pl. (London).

(38) 1857, *A monograph of the Tertiary Entomostraca of England:* Same, (1856) 1857, p. 1-68, pl. 1-6.

——, **& Kirkby, J. W.**

(39) 1874, *A monograph of the British fossil bivalved Entomostraca from the Carboniferous formations. Part I. The Cypridinadae and their allies:* Same, p. 1-56, pl. 1-5.

Kaufmann, Alfred

(40) 1900a, *Zur systematik der Cypriden:* Naturf. Gesell. Bern, Mitteil., p. 103-109.

(41) 1900b, *Cypriden und Darwinuliden der Schweiz:* Revue Suisse Zool., v. 8, p. 209-423, pl. 15-31 (Geneva).

Keij, A. J.

(42) 1957, *Eocene and Oligocene Ostracoda of Belgium:* Verh. kon. Belg. Inst. Natuurwetensch., no. 136, 210 p., 33 pl., 6 fig. (Bruxelles).

Kellett, Betty

(43) 1934, *Ostracodes from the Upper Pennsylvanian and the Lower Permian strata of Kansas: II. The genus Bairdia:* Jour. Paleont., v. 8, no. 2, p. 120-138 (Tulsa).

Kesling, R. V.

(44) 1951b, *Terminology of ostracod carapaces:* Univ. Michigan Mus. Paleont. Contr., v. 9, no. 4, p. 93-171, 18 pl., 7 fig., 5 charts (Ann Arbor).

——, **& McMillan, G. W.**

(45) 1951, *Ostracods of the family Hollinidae from the Bell Shale of Michigan:* Same, v. 9, no. 2, p. 45-81, pl. 1-7, fig.

Kingma, J. T.

(46) 1948, *Contributions to the knowledge of the Young-Caenozoic Ostracoda from the Malayan region:* 118 p., 11 pl. (Utrecht).

Kummerow, E. H. E.

(47) 1953, *Über oberkarbonische und devonische Ostracoden in Deutschland und in der Volksrepublik Polen:* Geologie, Beiheft Zeitschr., no. 7, p. 1-75, pl. 1-7 (Berlin).

Levinson, S. A.

(48) 1950B, *The hingement of Paleozoic Ostracoda and its bearing on orientation:* Jour. Paleont., v. 24, p. 63-75 (Tulsa).

(49) 1951, *Thin sections of Paleozoic Ostracoda and their bearing on taxonomy and morphology:* Same, v. 25, no. 5, p. 553-560, pl. 77.

Mandelstam, M. I., et al.

(50) 1956, *Ostracoda:* All-Union Sci. Res. Geol. Inst. (VSEGEI), Contr. Paleont., new ser., no. 12, Ostracodes, p. 87-144, pl. 19-27, fig. 16-53 (Moscow).

Martin, G. R. P.

(51) 1940, *Ostracoden des norddeutschen Purbeck und Wealden:* Senckenberg., v. 22, p. 275-361, pl. 1-13 (Frankfurt a-M.).

Martinsson, Anders

(52) 1955, *Studies in the ostracode family Primitiopsidae:* Geol. Inst. Univ. Uppsala, Bull., v. 35.

Müller, G. W.

(53) 1894, *Die Ostracoden des Golfes von Neapel*

und der angrenzenden Meeres-abschnitte: Fauna u. Flora Neapel, Mon. 21, p. i-viii, 1-404, pl. 1-40 (Berlin).

(54) 1912, *Ostracoda in das Tierreich. Eine Zusammenstellung und Kennzeichnung der rezenten Tierformen:* Im Auftrage der Königl. Preuss. Akad. Wiss., Lief. 31, p. k-xxxiii, 1-434, 92 fig. (Berlin).

Müller, O. F.
(55) 1776, *Zoologiae Danicae Prodromus, seu animalium Daniae et Norvegiae indigenarum characters, nomina, et synonyma imprimis popularium:* Havniae, v. 8, p. l-xxxii, 1-282. 1-282.

(56) 1785, *Entomostraca seu insecta testacea, quae in aquis Daniae et Norvegiae reperit, descripsit et iconibus illustravit:* Lipsiae et Havniae, 135 p., 21 pl.

Oertli, H. J.
(57) 1956, *Ostrakoden aus der oligozänen und miozänen Molasse der Schweiz:* Abh. Schweiz. Paläont. Gesell., v. 74, 119 p., 16 pl. (Zurich).

Öpik, A. A.
(58) 1937, *Ostracoda from the Ordovician Uhaku and Kukruse Formations of Estonia:* Ann. Soc. Univ. Tartu, v. 43, no. 1-2, p. 65-138, 15 pl., 8 fig.

Pokorný, Vladimir
(59) 1953, *A contribution to the taxonomy of the Paleozoic ostracods:* Sborník Ústředního Ustavu Geologického, Svazek XX, oddíl paleont., p. 213-232 (Praha).

(60) 1958, *Grundzüge der Zoologischen Mikropalaeontologie:* v. 2, p. 66-453 (Berlin).

Přibyl, Alois
(61) 1955, *A contribution to the study of ostracods of the Bohemian Devonian and their stratigraphical significance: Sborník Ústředního Ustavu Geologického,* Svazek 21, oddíl paleont., 141 p. (Praha).

Raymond, P. E.
(62) 1935, *Leanchoilia and other mid-Cambrian Arthropoda:* Harvard Univ., Mus. Comp. Zoology, Bull., v. 76, no. 6, p. 205-230 (Cambridge).

(63) 1946, *The genera of fossil Conchostraca—an order of bivalved Crustacea:* Same, Bull., v. 96, no. 3, p. 217-307.

Roemer, F. A.
(64) 1840, *Die Versteinerungen des Norddeutschen Kreidegebirges:* Hannover, 1840, 145 p., 16 pl.

Sars, G. O.
(65) 1866, *Oversigt af Norges marine Ostracoder:* Norske Vidensk.-Akad. Forhandl. (1865), p. 1-130.

(66) 1888, *Nye Bidrag til Kundskaben om Middlehavets Invertebratfauna. 4. Ostracoda*

Mediterranea: Arch. Math. Naturvidenskab, v. 12, p. 173-324, pl. 1-20 (Oslo).

(67) 1894, *Contributions to the knowledge of the fresh-water Entomostraca of New Zealand as shown by artificial hatching from dried mud:* Norske Vidensk.—akad., Videnskabs-Selskabets Skr.. I. Math-naturv. Kl., no. 5, pl. 1-8 (Oslo).

(68) 1922-1928, *An account of the Crustacea of Norway. Vol. 9, Crustacea:* Bergen Mus., pt. XV-XVI, 277 p., 119 pl. (Oslo).

Schmidt, E. A.
(69) 1941, *Studien im böhmischen Caradoc (Zahořan-Stufe). I. Ostrakoden aus den Bohdalecschichten und über die Taxonomie der Beyrichiacea:* Senckenb. Naturf. Gesell. Abh. 454, 87 p., 5 pl., 2 fig. (Frankfurt a-M.).

Scott, H. W.
(70) 1951, *Instars and shell morphology of Eoleperditia fabulites:* Jour. Paleont., v. 25, no. 3, p. 321-326, pl. 51, 3 fig. (Tulsa).

Skogsberg, Tage
(71) 1920, *Studies on marine Ostracods. Part I. (Cypridinids, halocyprids, and polycopids):* Zool. Bidr. f. Uppsala, suppl. ed. i, 784 p., 153 fig.

(72) 1928, *Studies on marine ostracods. Part II. External morphology of the genus Cythereis with descriptions of twenty-one new species:* California Acad. Sci., Occas. Papers, no. XV, p. 1-155, pl. 1-6 (San Francisco).

Sohn, I. G.
(73) 1954, *Ostracoda from the Permian of the Glass Mountains, Texas:* U. S. Geol. Survey, Prof. Paper 264-A, 24 p., 5 pl. (Washington).

Swartz, F. M.
(74) 1936, *Revision of the Primitiidae and Beyrichiidae with new Ostracoda from the Lower Devonian of Pennsylvania:* Jour. Paleont., v. 10, p. 541-586, pl. 78-89 (Tulsa).

(75) 1949, *Muscle marks, hinge and overlap features and classification of some Leperditiidae:* Same, v. 23, no. 3, p. 306-327, pl. 65-67, 2 fig.

————, & Swain, F. M.
(76) 1941, *Ostracodes of the Middle Devonian Onondaga beds of central Pennsylvania:* Geol. Soc. America, Bull., v. 52, p. 381-458, pl. 1-8 (New York).

(77) 1946, *Ostracoda from the Upper Jurassic Cotton Valley Group of Louisiana and Arkansas:* Jour. Paleont., v. 20, no. 4, p. 362-373, pl. 52, 53 (Tulsa).

————, & Whitmore, F. C., Jr.
(78) 1956, *Ostracoda of the Silurian Decker and Manlius limestones in New Jersey and East-*

ern New York: Same, v. 30, no. 5, pl. 103-110, fig.

Sylvester-Bradley, P. C.

(79) 1949, *The ostracod genus Cypridea and the zones of the upper and middle Purbeckian:* Geol. Assoc., Proc., v. 60, p. 125-153, pl. 3, 4, fig. 17-25 (London).

Triebel, Erich

(80) 1938a, *Die Ostrakoden der deutschen Kreide. 2. Die Cytheridea-Arten der unteren Kreide:* Senckenberg. 20, p. 471-501, pl. 1-6 (Frankfurt a-M.).

(81) 1940, *Die Ostracoden der Deutschen Kreide. 3. Cytherideinae und Cytherinae aus der unteren Kreide:* Same, v. 22, p. 160-227, 10 pl.

(82) 1941, *Zur Morphologie und Ökologie der fossilen Ostracoden, mit Beschreibung einiger neuer Gattungen und Arten:* Same, v. 23, p. 294-400, 15 pl.

Ulrich, E. O.

(83) 1894(1897), *The Lower Silurian Ostracoda of Minnesota:* Minnesota Geol. Nat. History Survey, Rept., v. 3, pt. 2, ch. 7, p. 629-693, pl. 43-46 (Minneapolis).

(84) 1900, *New American Paleozoic Ostracoda. No. I. Ctenobolbina and Kirkbya:* Cincinnati Soc. Nat. History, Jour., v. 19, p. 179-186, pl. 8.

———, & Bassler, R. S.

(85) 1908, *New American Paleozoic Ostracoda. Preliminary revision of the Beyrichiidae, with descriptions of new genera:* U. S. Natl. Mus., Proc., v. 35, p. 277-340, fig. 1-61, pl. 37-44 (Washington).

(86) 1923a, *Paleozoic Ostracoda: Their morphology, classification, and occurrence:* Maryland Geol. Survey, Silurian vol., p. 271-391, fig. 11-26 (Baltimore).

(87) 1931, *Cambrian bivalved Crustacea of the order Conchostraca:* U.S. Natl. Mus., Proc., v. 78, art. 4, p. 1-130, pl. 1-10 (Washington).

Wagner, C. W.

(88) 1957, *Sur les ostracodes du Quaternaire Recent des Pays-Bas et leur utilisation dans l'étude géologique des dépôts holocénes:* Mouton & Co., 259 p., 50 pl. ('s-Gravenhage).

SOURCES OF ILLUSTRATIONS

(89) Alexander, C. I. (1933)
(90) Almeida, F. M. de (1950)
(91) Anderson, F. W. (1939)
(92) Apostolescu, Vespasian (1955b)
(93) ——— (1956)
(94) Baird, William (1850)
(95) Barrande, Joachim (1872)
(96) Bartenstein, Helmut, & Brand, Erich (1959)
(97) Bettenstaedt, Franz (1958)
(98) Blake, Charles (1931)
(99) ——— (1933)
(100) Bold, W. A. van den (1950)
(101) ——— (1957)
(102) ——— (1958)
(103) ——— (1959)
(104) Bonnema, J. H. (1909)
(105) ——— (1913)
(106) ——— (1938)
(107) Brady, G. S. (1868)
(108) ——— (1890)
(109) ——— (1898)
(110) ——— (1907)
(111) ——— (1910)
(112) ——— (1913)
(113) Branson, C. C. (1935)
(114) ——— (1936)
(115) Brongniart, Charles (1876)
(116) Bronstein, Z. S. (1930)
(117) Brown, P. M. (1957)
(118) Burr, J. H., Jr.
(119) Canavari, Mario (1899)

(120) Chapman, Frederick, & Sherborn, C. D. (1893)
(121) Coryell, H. N. (1928)
(122) ——— (1930)
(123) ———, & Billings, G. D. (1932)
(124) ———, & Booth, R. T. (1933)
(125) ———, & Cuskley, V. A. (1934)
(126) ———, & Fields, Suzanne (1937)
(127) ———, & Johnson, S. C. (1939)
(128) ———, & Rogatz, Henry (1932)
(129) ———, Sample, C. H., & Jennings, P. H. (1935)
(130) ———, & Rozanski, George (1942)
(131) ———, & Williamson, Marjorie (1936)
(132) Croneis, Carey, & Gale, A. S., Jr. (1939)
(133) ———, & Gutke, R. L. (1939)
(134) Cushman, J. A. (1906)
(135) Daday, Eugen von (1895)
(136) ——— (1905)
(137) ——— (1910)
(138) Delo, David (1930)
(139) Delachaux, Theodore (1928)
(140) Dobbin, C. N. (1941)
(141) Dons, J. A., & Henningsmoen, Gunnar (1949)
(142) Dubowsky, N. F. (1939)
(143) Edwards, R. A. (1944)
(144) Egorov, V. G. (1950)
(145) ——— (1953)
(146) ——— (1954)
(147) Elofson, Olof (1939)

(148) Forel, F. A. (1894)
(149) Frederickson, E. A. (1946)
(150) Geis, H. L. (1932)
(151) Gibson, L. B. (1955)
(152) Goerlich, Franz (1952)
(153) Grekoff, Nicolas (1951)
(154) ———, & Moyes, Jean (1955)
(155) Groom, Theodore (1902)
(156) Gürich, George (1896)
(157) Hall, James (1852)
(158) Hanai, Tetsuro (1959b)
(159) Harlton, B. H. (1929)
(160) Harris, R. W. (1931)
(161) ——— (1957)
(162) Hartmann, Gerd (1954)
(162a) ——— (1955)
(163) ——— (1956)
(164) ——— (1957)
(165) Henningsmoen, Gunnar (1948)
(166) ——— (1953b)
(167) ——— (1954)
(168) Herbst, H. V. (1957)
(169) Hill, B. L. (1954)
(170) ——— (1955)
(171) Hirschmann, Nikolaj (1909)
(172) ——— (1916)
(173) Hoff, C. C. (1942b)
(174) Hornibrook, N. de B. (1949)
(175) Hou, Y. T. (1953)
(176) ——— (1955)
(177) Howe, H. V. (1951)
(178) ———, et al. (1935)
(179) Ivanova, V. A. (1955)
(180) Jaanusson, Valdar, & Martinsson, Anders (1956)
(181) Jennings, P. H. (1936)
(182) Jones, T. R. (1850)
(182a) ——— (1861)
(183) ——— (1870)
(184) ——— (1885)
(185) ——— (1888)
(186) ——— (1889)
(187) ———, & Brady, G. S. (1874)
(188) ———, & Holl, H. B. (1869)
(189) ——— (1886)
(190) ———, & Kirkby, J. W. (1886)
(191) ———, & Sherborn, C. D. (1888)
(192) Kashevarova, N. P. (1958)
(193) Kay, Marshall (1934)
(194) ——— (1940)
(195) Kegel, Wilhelm (1932)
(196) ——— (1933)
(197) Keij, A. J. (1958)
(198) Kellett, Betty (1935)
(199) ——— (1936)
(200) Kesling, R. V. (1951a)
(201) ——— (1952)
(202) ——— (1953)
(203) ——— (1954)
(204) ——— (1957)
(205) ——— (1958)
(206) ———, et al. (1958)
(207) ———, & Copeland, M. J. (1954)
(208) ———, & Hussey, R. C. (1953)
(209) ———, & Kilgore, J. E. (1952)
(210) ———, & Peterson, R. M. (1958)
(211) ———, & Tabor, N. R. (1953)
(212) ———, & Wagner, P. L. (1956)
(213) ———, & Weis, Martin (1953)
(214) King, R. L. (1855)
(215) Klie, Walter (1929)
(215a) ——— (1930)
(216) ——— (1931)
(217) ——— (1932)
(217a) ——— (1933)
(218) ——— (1935)
(219) ——— (1936)
(220) ——— (1938)
(221) ——— (1939)
(222) ——— (1940)
(223) Krause, Aurel (1889)
(224) Krommelbein, Karl (1952)
(225) ——— (1954)
(226) ——— (1955)
(227) LeRoy, L. W. (1939)
(228) Lienenklaus, E. (1894)
(229) Lyubimova, P. S. (1952)
(230) ——— (1955)
(231) ——— (1956)
(231a) ——— (1959)
(232) Loranger, D. M. (1954)
(233) Lowndes, A. G. (1932)
(234) Lutkevich, E. C. (1939)
(235) Malz, Heinz (1958)
(236) ——— (1959)
(237) Mandelstam, M. I. (1947)
(238) ——— (1956)
(238a) ——— (1957)
(239) ——— et al. (1958)
(240) Marple, M. F. (1952)
(241) Marshall, W. S. (1903)
(242) Martin, G. P. R. (1958)
(243) Martinsson, Anders (1956)
(244) Matern, Hans (1929)
(245) Matthew, G. F. (1886)
(246) ——— (1896)
(247) ——— (1899)
(248) ——— (1902)
(249) Méhes, Gyula (1941)
(250) Mertens, von Erwin (1956)
(251) Moberg, J. C. (1895)
(252) ———, & Segerberg, C. O. (1906)
(253) Morris, R. W., & Hill, B. L. (1951)
(254) ——— (1952)
(255) Müller, Fritz (1881)
(255a) Müller, G. W. (1895)
(256) ——— (1898)
(257) ——— (1900)
(258) ——— (1901)
(258a) ——— (1906)

(259) Munsey, G. C., Jr. (1953)
(260) Münster, Georg Graf von (1830)
(261) Neale, J. W. (1959)
(262) Neckaja, A. I. (1952)
(263) ———— (1953)
(264) ———— (1958)
(265) ————, & Ivanova, V. A. (1956)
(266) Neviani, A. (1928)
(267) Norman, A. M. (1865)
(268) Oehlert, D. P. (1877)
(269) Oertli, H. J. (1957)
(270) ———— (1959)
(271) Öpik, A. A. (1953)
(272) Paris, P. (1916)
(273) Peterson, J. A. (1954)
(274) Pinto, I. P., & Sanguinetti, Y. T. (1958)
(275) Pokorný, Vladimir (1950)
(276) ———— (1955)
(277) Polenova, E. N. (1952)
(278) ———— (1953)
(279) ———— (1955)
(280) Portlock, J. E. (1843)
(281) Posner, V. M. (1951)
(282) Poulsen, Christian (1937)
(283) Přibyl, Alois (1950)
(284) ———— (1951)
(284a)———— (1953)
(285) Puri, H. S. (1952)
(286) ———— (1953)
(287) ———— (1954)
(288) ———— (1955)
(289) ———— (1956)
(290) ———— (1957)
(291) ———— (1958)
(292) Rabien, Arnold (1954)
(293) Reyment, R. A. (1959)
(294) ————, & Elofson, Olof (1959)
(295) Rioja, Enrique (1942)
(296) Roemer, F. A. (1838)
(297) Roth, R. I. (1929)
(298) ————, & Skinner, J. M. (1930)
(299) Roundy, P. V. (1926)
(300) Ruggieri, Giuliano (1950)
(301) ———— (1952)
(302) ———— (1953)
(303) ———— (1954)
(304) ———— (1955)
(305) ———— (1956)
(305a) Sars, G. O. (1889)
(306) ———— (1895)
(307) ———— (1898)
(308) ———— (1899)
(309) ———— (1903)
(310) ———— (1905)
(311) ———— (1910)
(312) ———— (1923)
(313) ———— (1924)
(314) ———— (1925)
(315) ———— (1926)
(316) ———— (1935)

(317) Sarv, L. I. (1959)
(318) Schäfer, H. W. (1945)
(319) ———— (1953)
(320) Schmidt, E. A. (1955)
(321) Schneider, G. F. (1939)
(322) ———— (1948)
(323) Scott, H. W. (1944)
(324) ———— (1959)
(325) ————, & Smith, W. H. (1951)
(326) ————, & Summerson, C. H. (1943)
(327) Seguenza, Giuseppe (1880)
(328) Sexton, J. V. (1935)
(329) Sharapova, E. G. (1937)
(330) Sharpe, R. W. (1903)
(331) Shaver, R. H. (1958)
(332) Skogsberg, Tage (1939)
(333) Sohn, I. G. (1950)
(334) ———— (1953)
(335) ————, & Berdan, Jean (1952)
(336) Spivey, R. C. (1939)
(337) Spizharsky, T. H. (1937)
(338) ———— (1939)
(339) ———— (1956)
(340) Stephenson, M. B. (1935)
(341) ———— (1936)
(342) ———— (1938)
(343) ———— (1946)
(344) Steusloff, A. (1894)
(345) Stewart, G. A. (1936)
(346) ————, & Hendrix, W. E. (1939)
(347) ———— (1945)
(348) Stover, L. E. (1956)
(349) Swain, F. M. (1944)
(350) ———— (1946)
(351) ———— (1947)
(352) ———— (1949)
(353) ———— (1951)
(354) ———— (1952)
(355) ———— (1955)
(356) ———— (1957)
(357) ————, & Peterson, J. A. (1951)
(358) ————, & ———— (1952)
(359) Swartz, F. M. (1956)
(360) ————, & Hendrix, W. E. (1945)
(361) ————, & Oriel, S. S. (1948)
(362) Sylvester-Bradley, P. C. (1941)
(363) ———— (1947)
(364) ———— (1948)
(365) ———— (1950)
(366) ———— (1951)
(366a)———— (1953)
(367) ———— (1956)
(368) Teichert, Curt (1937)
(369) Thorslund, Per (1940)
(369a)———— (1948)
(370) Tolmachoff, I. P. (1926)
(371) Tressler, W. L. (1939)
(372) ————, & Smith, E. M. (1948)
(373) Triebel, Erich (1938b)
(374) ———— (1939)

(375) ——— (1949)
(376) ——— (1950)
(377) ——— (1951)
(378) ——— (1952)
(379) ——— (1954)
(380) ——— (1958)
(381) ——— (1959)
(382) ———, & Klingler, W. (1959)
(383) Ulrich, E. O. (1879)
(384) ——— (1890)
(385) ——— (1891)
(386) ——— (1892)
(387) ——— (1906)
(388) ——— (1916)
(389) ———, & Bassler, R. S. (1913)
(390) Upson, M. E. (1933)
(391) VanPelt, H. L. (1933)

(392) Vávra, Vélav (1891)
(393) ——— (1895)
(394) ——— (1901)
(395) Veen, J. E. van (1932)
(396) ——— (1936)
(397) Vernet, H. (1878)
(398) Vos, A. P. C. de (1953)
(399) Wainwright, John (1959)
(400) Warthin, A. S., Jr. (1948)
(401) Weingeist, Leo (1949)
(402) Wetherby, A. G. (1881)
(403) Wilson, C. W., Jr. (1935)
(404) Wiman, Carl (1902)
(405) Zalányi, Béla (1929)
(406) ——— (1944)
(407) Zaspelova, V. S. (1952)

ADDENDUM

By S. A. Levinson and H. V. Howe

The following ostracode taxa which are found neither in the foregoing parts of this *Treatise* nor in the USSR *Treatise* are listed in this section together with references to publications in which the taxa appear.

Actinoseta Kornicker, 1958 [*A. chelisparsa*]. Rec., BahamaI. [Assigned to Asteropidae.]

Alternochelata Kornicker, 1958 [*Rutiderma (Alternochelata) polychelata*]. Rec., BahamaI. [Assigned to Rutidermatidae.]

Bodenia Ivanova, 1959 [*B. aspera*]. M.Ord., Russ. [Assigned to Beyrichiidae.]

Chelicopia Kornicker, 1958 [*C. arostrata*]. Rec., BahamaI. [Assigned to Sarsiellidae.]

Citrella Oertli, 1959 [*C. nitida*]. M.Jur., Fr. [Assigned to Cytheridae (Cytheropterinae).]

Cyprideidae Martin, 1940 [nom. transl. Martin, 1948 (ex Cyprideinae Martin, 1940).]

Discoidellidae Přibyl, 1958. New family, order uncertain, to include *Discoidella* Croneis & Gale, 1938.

Dolborella Ivanova, 1959 [*D. plana*]. U.Ord., Russ. [Assigned to Drepanellidae.]

Egorovella Ivanova, 1959 [*E. compacta*]. M.Ord., Russ. [Assigned to Tetradellidae.]

Eocypridina Kesling & Ploch, 1960 [*E. campbelli*]. U.Dev., Ind. [Assigned to Cypridinidae (Cypridinae).]

Fissocythere Malz, 1959 [*F. calloglypta*]. M.Jur., Ger. [Assigned to Cytheridae (Trachyleberidinae).]

Gesoriacula Oertli, 1959 [*G. plana*]. M.Jur., Fr. [Assigned to Cytheridae.]

Incongruellina Ruggieri, 1958 [*I. semispinescens*]. U.Plio.-Quat. [Assigned to Cytheridae (Brachycytherinae).]

Isobythocypris Apostolescu, 1959 [*I. unispinata*]. L.Jur., Fr. [Assigned to Bairdiidae.]

Lophodentina Apostloescu, 1959 [*L. lacunosa*]. L.Jur., Fr. [Assigned to Cytheridae (Cytherinae).]

Parariscus Oertli, 1959 [*P. bathonicus*]. M.Jur., Fr. [Assigned to Cytheridae (Cytherurinae).]

Pichottia Oertli, 1959 [*P. muris*]. M.Jur., Fr. [Assigned to Cytheridae (Cytherideinae).]

Planusella Ivanova, 1959 [*P. bicornis*]. M.Ord., Russ. [Assigned to Drepanellidae.]

Platylophocythere Oertli, 1959 [*P. hessi*]. U.Jur., Switz. [Assigned to Cytheridae (Cytherinae).]

Pribylina Ivanova, 1959 [*P. levis*]. M.Ord., Russ. [Assigned to Graviidae.]

Pseudocypris Herbst, 1958 (non Daday, 1908) [*P. dietzi*]. Fresh water. Rec., Austral.

Pteroleperditia Hamada, 1959 [*Herrmannina ehlersi* Kesling, 1958]. Ord.-Sil., Mich. [Assigned to Leperditiidae.]

Quasibuntonia Ruggieri, 1958 [*Cythere radiatopora* Seguenza, 1880]. Plio., Italy. [Assigned to Cytheridae (Trachyleberinae).]

Rhysomagis Kesling, Kavary, Takagi, Tillman, & Wulf, 1959 [*R. dichelomota*]. M.Dev., N.Y. [Assigned to Alanellidae.]

Tanycypris Triebel, 1959 [*Cypris madagascarensis* G. W. Müller, 1898]. Rec., Madagascar. [Assigned to Cyprididae (Cypridinae).]

REFERENCES

Apostolescu, Vespasian, 1959. *Ostracodes du Lias du Bassin de Paris:* Rev. Inst. Fr. Pétrole, v. 14, no. 6, p. 795-826, pl. 1-4.

Hamada, Takashi, 1959, *Discovery of a Devonian ostracod in the Fukuji District, Gifu Prefecture, West Japan:* Japan. Jour. Geol. Geog., v. 30, p. 39-51, 1 text-fig.

Hanai, Tetsuro, 1959, *Studies on the Ostracoda from Japan; 4-family Cytherideidae Sars, 1925:* Fac.

Sci. Univ. Tokyo, Jour., sec. 2, v. 11, pt. 3, p. 291-308, pl. 16-18, 4 text-fig.

Herbst, H. V., 1958, *Neue Cypridae (Crustacea Ostracoda) aus Australien II*: Zool. Anzeiger, v. 160, no. 9/10, p. 177-192, 52 text-fig.

Ivanova, V. A., 1959a, *Some Ordovician ostracodes of the Siberian platform*: Acad. Nauk U.S.S.R., Pal. Jour., no. 4, p. 130-142, pl. 8.

———, 1959b, *Ostracode genera from Ordovician deposits of the Siberian platform which are new or previously unknown in the U.S.S.R.: Material for "Fundamental Paleontology"*: Acad. Nauk U.S.S.R., Pal. Inst., no. 3, p. 71-83, 9 text-fig.

Kesling, R. V., Kavary, E., Takagi, R. S., Tillman, J. R., & Wulf, G. R., 1959, *Quasillitid and alanellid ostracods from the Centerfield limestone of western New York*: Univ. Michigan Mus. Paleont. Contr., v. 15, no. 2, p. 15-31, pl. 1-4, 4 text-fig.

———, & Ploch, R. A., 1960, *New Upper Devonian cypridinacean ostracod from southern Indiana*: Same, v. 15, no. 12, p. 281-292, 3 pl.

Kornicker, L. S., 1958, *Ecology and taxonomy of Recent marine ostracodes in the Bimini Area, Great Bahama Bank*: Inst. Marine Science, Univ. Texas, v. 5, p. 194-300, 89 text-fig.

Malz, Heinz, 1959, *Ostracoden-studien im Dogger,*
2: *Fissocythere n. g.*: Senckenberg. lethaea, v. 40, no. 3/4, p. 317-331, pl. 1-2.

Martin, G. P. R., 1958, *Uber die systematische Stellung der Gattung Cypridea Bosquet (Ostracoda), nebst Beschreibung der Wealden-Basis-Ostracode C. buxtorfi n. sp.*: Neues Jahrb. Geol. Min., v. 7, p. 312-320, 1 text-fig.

Oertli, H. J., 1959a, *Platylophocythere, eine neue Ostrakoden-Gattung aus dem untern Malm des schweizer Juras*: Ecologae Geol. Helv., v. 52, no. 2, p. 953-957, pl. 1-2, 3 text-fig.

———, 1959b, *Les ostracodes du Bathonien du Boulonnais 1. Les "microstracodes"*: Rev. Micropaléontologie, v. 2, no. 3, p. 115-226, pl. 1-3, 5 text-fig.

Přibyl, Alois, 1957 [1958], *The ostracodes of the Upper Carboniferous (Namurian A) of Czechoslovakia (Porubá Beds) and its importance for the Ostrava-Karviná coal district*: Sborník Ustr. Ust. Geol., v. 24, p. 7-52 (in Czech.), p. 53-93 (in Russian), p. 94-95 (English abstr.), pl. 1-16.

Ruggieri, Giuliano, 1958, *Alcuni ostracodi del Neogene italiano*: Soc. Ital. Sci. Nat. Milano, Atti, v. 97, no. 2, p. 127-146, 30 text-fig.

Triebel, Erich, 1959, *Zur Kenntnis der Ostracoden-Gattungen Isocypris und Dolerocypris*: Senck. Biologica, v. 40, no. 3/4, p. 155-170, pl. 12-18.

SUPPLEMENT—USSR TREATISE ON OSTRACODA

By Raymond C. Moore

The publication during the first half of 1960 of a volume of the Russian paleontological treatise containing descriptions and illustrations of Ostracoda is of special interest to workers on this group located in countries outside of the Soviet Union because (1) systematics of the entire subclass are indicated according to the views of Russian specialists and (2) information on many generic and suprageneric taxa defined in recent years on the basis of fossils collected in European and Asiatic parts of the USSR is here brought together for the first time. The treatise, prepared under direction of Yu. A. Orlov, has added importance in that most of the papers containing original descriptions of the ostracode taxa recognized by Russian paleontologists are unavailable generally beyond borders of the USSR or at least are very difficult to obtain. Accordingly, it seems desirable to summarize the contents and arrangement of systematic parts of the Russian treatise on Ostracoda for comparison with treatment given on preceding pages of this volume.

All of the type-setting for Part Q (Ostracoda) of the *Treatise on Invertebrate Paleontology* had been completed before I learned about the Russian volume concerned with ostracodes, and therefore it has been possible only to make changes in headings needed to take account of priority in publication by various Russian authors of some family-group taxa in our *Treatise* that were being designated as new. It is fortunate that changes of this sort have been possible in proof, for it avoids the need for corrections in future. Also, it has been possible to include this supplement. My acknowledgment of the kindness of Dr. Wolfgang Struve, Forschungs-Institut Senckenberg, Frankfurt am-Main, and thanks to him are expressed for loan of his copy of the Russian treatise volume. Instructions for changes of our proof and addition of this supplement could then be sent by me from Europe to the United States.

It should be noted that the USSR treatise is restricted essentially to fossils known from the Soviet Union, whereas ours undertakes to take account of the entire world. Effort has been made to include both generic and higher-rank ostracode taxa reported from the USSR in our *Treatise* but omissions, incomplete records of occurrence, and even misinterpretations are possible in view of dependence of *Treatise* authors on literature alone, with the added impediment of language barriers.

The following summary of ostracode taxonomic units found in the Russian treatise is designed to focus attention on differences between classification and nomenclature given by our Soviet colleagues and those adopted by contributors to our *Treatise*. It is difficult to make comparisons adequately in very limited space, particularly as regards classificatory arrangement, and so a reader must study this for himself, aided by a few explanations. (1) Unless the name of a taxon is accompanied by an annotation (generally citing data given in the Russian text), no significant discrepancy of recorded author, date, type species, and the like is noted between the USSR and our versions. (2) Names of taxa adopted in the Russian text that are lacking in ours (by omission or rejection) are printed in boldface type, generally followed by an explanation enclosed by brackets. (3) Erroneously formed names included in the Russian text are printed in italics, those of family-group taxa being in capital letters. (4) Classificatory placement of genera in subfamilies according to Russian authors is ignored for purposes of comparison with our *Treatise,* but genera differently classified at the family level are marked by a dagger (†). Family-group taxa cited in the Russian systematic arrangement are also indicated by a dagger if they differ in placement with respect to next higher-rank taxa (e.g., †Kirkbyidae, which is assigned to Drepanellacea by Russian authors but classified in Kirkbyacea in our *Treatise*).

Classification of Ostracoda in USSR Treatise

Palaeocopida (By A. F. Abushik, A. I. Neckaja, E. N. Polenova, & I. E. Zanina)

Leperditiida POKORNÝ, 1953 (suborder) [=Leperditicopida] (By Abushik, Polenova, & Zanina)

†**Leperditacea** [*recte* Leperditiacea] JONES, 1856 (superfam.) (By Abushik, Polenova, & Zanina)
Leperditiidae
Leperditiinae JONES, 1856

Leperditia; Briartina; **Kiaeria** GLEBOWSKAJA, 1949 [*Leperditia kiaeri GLEBOWSKAJA, 1949]; Schrenckia, Sibiritia. Non-USSR: Heterochilina.

Herrmannininae ABUSHIK, 1960 (USSR Treatise) [*recte* 1957, p. 237]
Hermannina [=Hermannella]; Eoleperditia; *Mölleritia* [*recte* Moelleritia]; †Paenaequina. Non-USSR: Anisochilina; Chevroleperditia. Possibly assignable here: †Cambria NECKAJA & V. IVANOVA, 1956.

Isochilinidae SWARTZ, 1949 [*nom. transl.* ABUSHIK, 1960 (USSR Treatise) (*ex* Isochilininae SWARTZ, 1949)]
Isochilina; †Gibberella; Hogmochilina. Non-USSR: Teichochilina; Dihogmochilina; †Saffordellina [=†Saffordella ULRICH & BASSLER, 1923]. Probably assignable here: Holtedahlites [=Holtedahlina SOLLE, 1935].

†**Aparchitacea** [*nom. transl.* SWARTZ, 1945 (*ex* Aparchitidae JONES, 1901) (By Polenova)
Aparchitidae [=Leperditellidae ULRICH & BASSLER, 1906]
Aparchites; †Aparchitella; †Aparchitellina; **Coeloenellina** POLENOVA, 1952 [*C. parva*]; †Conchoprimites; †Conchoprimitia; **Ginella** V. IVANOVA, 1960 (USSR Treatise) [*pro* †Pinella V. IVANOVA, 1955 (*non* STEPHENSON, 1941)] [*Pinella tenuispina V. IVANOVA, 1955]; †Leperditella [=†Punctaparchites]; Macronotella; †Microcoelonella; †Paraparchites]; †Paraschmidtella; †Pseudoparaparchites; †Schmidtella; †Zaborovia. Non-USSR: †Ardmorea; †Paraparchitella; †Proparaparchites. Probably in addition: †Bertillonella; †Bonneprimites; †*Coeloenella* [*recte* Coelonella]; Cyathus; †Microparaparchites.

†Aechminidae
Aechmina. Non-USSR: Aechminaria; Paraechmina; Sigynus. Probably assignable here: †Leightonella.

Graviidae POLENOVA, 1952
†Gravia; †*Boučekites* [*recte* Boucekites]; †Monoceratella; †*Přibylites (Přibylites)* [*recte* Pribylites] [=Gravia (Russia) POLENOVA, 1952]; †*P. (Parapřibylites)* [*recte* P. (Parapribylites)]; †Saccelatia [=†Sphenocibysis]; †Selebratina. Non-USSR: †Coryellina. Probably assignable here: †Acronotella [=†?Acrotonella]; †Ceratocypris; †Pinnatulites; †Tricornina.

Beyrichiida POKORNÝ, 1954 [=Beyrichicopina + Kloedenellocopina] (By A. I. Neckaja, E. N. Polenova, I. E. Zanina, & V. S. Zaspelova)
Beyrichiacea JONES, 1954 [Matthew, 1886]
†**Primitiidae** ULRICH & BASSLER, 1923
†**Primitiinae** ULRICH & BASSLER, 1923
†Primitia; †Bromidella; †Chilobolbina; †Haploprimitia.

†**Eurychilininae** [see Eurychilinidae]

†Eurychilina; †Coelochilina; †Laccochilina; †Mirochilina; †*Öpickella* [sic] [*recte* Oepikella]. Non-USSR: †Apatochilina; †Platybolbina [=†Platychilina]. Probably assignable here: †Novakina; †Trubinella [=Hippa].

†Euprimitiinae HESSLAND, 1949

†Euprimitia; †Euprimites; †Hallatia; †Laccoprimitia; †Primitiella; †Punctoprimitia. Non-USSR: †Halliella; †Pyxiprimitia. Probably assignable here: †Hlubocepina; †Pseudoleperditia.

†Primitiopsidae

Primitiopsis; *Clavafabella* [*recte* Clavofabella]; **Primitiopsella** POLENOVA, 1960 (USSR Treatise) [*pro* Leperditellina POLENOVA, 1955 (*non* NECKAJA, 1955)]. Non-USSR: Amygdalella; Leiocyamus; Sulcicuneus. Probably assignable here: Limbinaria.

†Tetradellidae

Tetradellinae SWARTZ, 1936

Tetradella; †Ctenobolbina; †Ctenonotella; **Fidelitella** V. IVANOVA, 1960 (USSR Treatise) [*pro* Trilobella V. IVANOVA, 1955 (*non* WOODWARD, 1924)] [*Trilobella unica V. IVANOVA, 1955]; †Hesslandella [=†Ordovicia; †Sigmobolbina]; †*Öpikium* [*recte* Oepikium] [=†Biflabellum]; †Piretella [=†?Duhmbergia]; †Polyceratella; †Pseudostrepula; †Quadrilobella; †Rigidella; †Steusloffia; †Tallinella; †Tetrada NECKAJA, 1960 (USSR Treatise) [*?recte* NECKAJA, 1958]. Non-USSR: †Adelphobolbina.

†**Ceratopsinae** NECKAJA, 1958

†Ceratopsis [=†Kiesowia; † Sigmoopsis]; †Aulacopsis; †Ctenoloculina; Dilobella; †Hollina; **Ivaria** NECKAJA, 1960 (USSR Treatise) [*Glossopsis robusta HESSLAND, 1949)] [=Glossopsis NECKAJA, 1953 (*non* HESSLAND, 1949)]; †Ogmoopsis; †Winchellatia. Non-USSR: †Quadrijugator. Probably assignable here: †Bolbina; †Piretopsis; †Rakverella; †Ullerella [=†Ullia]; †Zygobolboides.

Bassleratiidae

Non-USSR: Bassleratia; Bellornatia; †*Öpikatia* [*recte* Opikatia]; Raymondatia; Thomasatia.

Beyrichiidae JONES, 1854 [*recte* MATTHEW, 1886]

Beyrichiinae JONES, 1854

Beyrichia; Apatobolbina; †Craspedobolbina; Dibolbina. Non-USSR: †Balticella; Bolbibollia.

Kloedeniinae ULRICH & BASSLER, 1923

Kloedenia; Cornikloedenia; *Kyammodes* [*recte* Kyamodes]; †Plethobolbina; Welleria. Non-USSR: Lophokloedenia; Myomphalus; Welleriopsis; Zygobeyrichia.

Zygobolbinae ULRICH & BASSLER [see Zygobolbidae]

Drepanellina. Non-USSR: †Zygobolba; †Bonnemaia; †Mastigobolbina; †Zygobolbina; †Zygosella.

Treposellinae HENNINGSMOEN, 1954

Bolbiprimitia. Non-USSR: Hibbardia; Phlyctiscapha; Saccarchites; Treposella. Probably assignable here: Mesomphalus.

†Hollinidae [note exclusion of Hollina, which is classed in Tetradellidae!]

Hollinella [=Basslerina; Hollites]; Parabolbina; Tetrasacculus [=Pterocodella; Workmanella]. Non-USSR: Abditoloculina; Bisaccalus; Falsipollex; Proplectrum; Ruptivelum; Subligaculum; Triemilomatella. Probably assignable here; Janischewskya.

Drepanellacea ULRICH & BASSLER, 1923 [*nom. transl.* POLENOVA & ZANINA, 1960 (USSR Treatise) (*ex* Drepanellidae ULRICH & BASSLER, 1923)]

Drepanellidae

Drepanellinae ULRICH & BASSLER, 1923 [=Neodrepanellinae ZASPELOVA, 1952]

Drepanella [=Drepanala]; · †Bicornellina; †Limbatula; Neodrepanella; Tetracornella. Non-USRR: †Aechminella [=†Mammoides]; †Crescentilla; †Parenthatia, †Parulrichia; Scofieldia. Probably assignable here: †Bicornella.

Nodellinae ZASPELOVA, 1952

Nodella; Acantonodella; †Pseudonodella; †Schweyerina; Subtella. Non-USSR: †Lindsayella; †Waldronites [pro †Cornulina]. Probably assignable here: †Balantoides; †Cornigella; †Polyzygia; †Pseudonodellina.

Bolliinae BOUČEK, 1936 [see Bolliidae]

†Bollia; †Ulrichia. Non-USSR: †Jonesella [=†Vogdesella]; †Kayatia; Kinnekullea; †Maratia; †Melanella; †Pseudulrichia; †Signetopsis. Probably assignable here: †Jonesites; [=†Placentula] †Parajonesites; †Richina; †Boursella; †Tetrastorthynx, †Tmemolophus; †Xystinotus.

†Kirkbyidae

Kirkbya; †Amphissites [=†Scytia; †Albanella; †Kindlella]; †Editia; †Kellettina; Knightina; Tenebrion. Non-USSR: †Amphizona; †Arcyzona; Aurikirkbya; †Brillius; †Cardiniferella; †Chironiptrum; Coronakirkbya; †Kirkbyites; †Mauryella; †Reticestus; †Semipetasus; †Strepula. Probably assignable here: †Roundyella [=†Amphissella; †Scaberina]; †Svarogites.

†Youngiellidae

Youngiella [=Youngia]; †Moorea; Moorites. Non-USSR: Hardinia; †Kellettella.

Kloedenellacea

Kloedenellidae

Kloedenella [=†Kloedenellina]; Dizygopleura; Eukloedenella; †?Prosopeionum. Non-USSR: Neokloedenella.

Mennerellidae Polenova, 1960 (USSR Treatise) †Mennerella; †Gipsella; †Semilukiella; †Uchtovia.

Knoxidae Egorov, 1950 [*nom. transl.* Polenova, 1960 (USSR Treatise) (*ex* Knoxinae Egorov, 1950)] [Invalid because family contains no nominal type genus from which name could be derived (see Beyrichiopsidae).] †Carboprimitia; †Chesterella; †Evanovia; †Kirkbyina; †Knoxiella; †Knoxites; †Marginia; †Milanovskya; †Plavskella. Non-USSR: †Knoxina [=Lokius]; †Beyrichiella [=Synaphe]; †Geffenina [=†Geffenites]; †Hastifaba; †Lamarella; †Reversabella; †Sansabella [=†Persansabella], †Sargentina. Probably assignable here: †Oliganisus.

Perprimitiidae Egorov, 1950 [*nom. transl.* Polenova, 1960 (USSR Treatise) (*ex* Perprimitiinae Egorov, 1950)] †Perprimitia; †Jonesina [=†Coryella; †Nuferella]; †Kloedenellitina; †Mennerites. Non-USSR; †Geisina. Probably assignable here: †Limnoprimitia.

Glyptopleuridae Glyptopleura [=Ceratopleurina; Glyptopleurites; Idiomorpha; Idiomorphina]; †Beyrichiopsis; †Tambovia. Non-USSR: †Deloia; †Denisonella [*pro* †Denisonia]; Glyptopleurina; †Glyptopleuroides; Mesoglypha; †Venula.

Lichwinidae Posner, 1950 [*recte* ?Lichviniidae] [*nom. transl.* Polenova, 1960 (USSR Treatise) (*ex* Lichwininae Posner, 1950)] [see Lichviniidae] Lichwinia [see Lichvinia] Evlanella; Kalugia. gia.

Indivisiidae Egorov, 1954 [*nom. transl.* Polenova, 1960 (USSR Treatise) (*ex* Indivisiinae Egorov, 1954)] Indivisia Zaspelova, 1954 [*I. indistincta]; Ilmenoindivisia Egorov, 1954 [*I. wjadica]; †Sulcoindivisia. Non-USSR; †Lochriella. Probably assignable here: †Ellipsella.

†Buregiidae Buregia.

Myodocopida Müller, 1894 [*recte* Sars, 1866] (By E. N. Polenova & I. E. Zanina) Myodocopa Sars, 1865 [see Myodocopina] (By Zanina) Cypridinidae Cypridina; †Cypridella; †Cypridellina. Non-USSR; †Cyprella; †Cypridinella; †Cyprosina; †Cyprosis: †Palaeophilomedes; Philomedes [=Bradycinetus]; †Rhombina; †Sulcuna.

Entomoconchidae Entomoconchus; Elpezoe [=Elpe]; †Leioditia. Non-USSR: Checontonomus; †Offa; Oncotechmonus.

Halocypridae [*recte* Halocyprididae]

Conchoeciidae Sars, 1866

Cladocopa Sars, 1865 [see Cladocopina] (By Polenova & Zanina) Polycopidae Polycope. Non-USSR: Polycopsis.

†Entomozoidae [=Entomididae] Entomozoinae Entomozoe [=Entomis]; Franklinella; Richterina; Rhomboentomozoe. Non-USSR: *Entoprimitia* [*recte* Entomoprimitia]; Pseudoentomozoe.

Boučiinae [*recte* Bouciinae] *Bouča* [*recte* Boucia] [=Basslerella Bouček, 1936].

Podocopida (By N. P. Kashevarova, P. S. Lyubimova, M. I. Mandelstam, E. N. Polenova, G. F. Schneider, & I. E. Zanina)

Platycopa Sars, 1865 [see Platycopina] (By Mandelstam & Polenova) †Cavellinidae Egorov, 1950 [*nom. transl.* Polenova, 1960 (USSR Treatise) (*ex* Cavellininae Egorov, 1950)] Cavellina; †Donellina; Sulcella [=Sansabelloides]; †Sulcocavellina; Timanella Egorov, 1950 [*T. typica]. Non-USSR: *Birdsalella* [*recte* Birdsallella]; Cavellinella; †Ellesmeria; Paracavellina; Platychilella [=Platychilus].

†Barychilinidae Non-USSR: Barychilina; Endolophia; †Pachydomella; †Palaeocythera; Trypetera.

Cytherellidae Cytherella; *Cytherelloides* [*recte* Cytherelloidea]. Non-USSR; Staringia [=Terquemia]. Probably assignable here: Ankumia.

Podocopa Sars, 1865 [see Podocopina] (By Kashevarova, Lyubimova, Mandelstam, Polenova, Schneider & Zanina)

Thlipsuridacea Jones, 1869 [*recte* Thlipsuracea] [*nom. transl.* Polenova & Zanina, 1960 (USSR Treatise (*ex* Thlipsuridae Jones, 1869—*recte* Ulrich, 1894)]

Thlipsuridae Jones, 1869 [*recte* Ulrich, 1894] Thlipsura [=Craterellina]; Octonaria; †Poloniella; Thrallella. Non-USSR: Eucraterellina; Eustephanella [=Eustephanus]; Hyphasmophora; Octonariella; Rothella; Stibus; Strepulites; Thlipsurella; Thlipsurina; Thlipsuroides; Thlipsuropsis. Probably assignable here: Favulella; †Phanassymetria; †Ranapeltis.

†Quasillitidae [=Graphiodactylidae] †Bufina; †Janetina; Jenningsina [=Quasillites; Costatia; ?Scalptina; †Svantovites]. Non-USSR: *Graphiodactyllis* [*recte* Graphiadactyllis] [=Graphiodactylus; Basslearia; Eriella; †Savagellites (*pro* Savagella)]; †Parabufina; †Paragraphylus. Probably assignable here: Allostracites [=†Allostraca]; Paracythere; Spinovina.

†Ropolonellidae
Non-USSR: Ropolonellus; †Euglyphella; Rudderina; *Varis* (*recte* Varix). Probably assignable here: †Plagionephrodes.

†Healdiacea HARLTON, 1933 [*nom. transl.* MANDELSTAM, 1960 (USSR Treatise) (*ex* Healdiidae HARLTON, 1933)] (By Kashevarova, Polenova, & Zanina)

Healdiidae
Healdia; †Bythocyproidea [=†Punctomosea]; Cribroconcha; †Healdianella; †Microcheilinella [=†Microcheilus; †Daleiella]; †Ponderodictya [=†Hamiltonella]; Robsoniella; Waylandella. Non-USSR: †Alveus; †Coryellites [*pro* Coryellina]; †Healdiacypris; Healdioides; Incisurella; †Lucasella; †Menoeidina; Phreatura; †Reversocypris; Seminolites; †Tetratylus. Probably assignable here: †Carbonita [*pro* Carbonia] [=†?Hilboldtina].

†Darwinulidae
Darwinula [=Darwinella; Cyprione; Polycheles; †Suchonellina]; Darwinuloides; †Suchonella. Probably assignable here: †Pruvostina; †Whipplella.

†Scrobiculidae
Scrobicula.

Bairdiacea (By Polenova & Zanina)
Bairdiidae [=Nesideidae]
Bairdia [=Nesidea; Bairdiacypris]; Acratia; Acratina; Acratinella; Actuaria; Bairdianella; Bairdoppilata; †Basslerella; †Burlella; †Celechovites; Fabalicypris, †Famenella; †Mossolovella [=Ellesmerina]; †Silenites; †Steusloffina; Triebelina. Non-USSR; †Artifactella;Bairdiolites; †Bairdites; †Camdenidea; Ceratobairdia; †Condracypris; †Elpinella; †Harltonella; †Hastacypris; †Haworthina; †Shidelerites; †Tubulibairdia. Probably assignable here: †Bairdiocypris; †Macrocypris.

Beecherellidae
Beecherella; †Krausella; †Rayella [*pro* Basslerites]. Non-USSR: Acanthoscapha; †Cooperatia [*pro* Cooperia]. Probably assignable here: †Janusella; Ulrichella.

Volganellacea MANDELSTAM, 1956 [*nom. transl.* MANDELSTAM, 1960 (USSR Treatise) [*recte* KASHEVAROVA, 1958, p. 337) (*ex* Volganellidae MANDELSTAM, 1956] (By Mandelstam & Schneider)

Volganellidae MANDELSTAM, 1956
†Volganella.

†Placideidae
Placidea.

Cypracea SYLVESTER-BRADLEY, 1949 [=Cypridacea Dana, 1849] [*recte* Cypridacea Baird, 1845] (By Lyubimova, Mandelstam & Schneider)

Cypridae [*recte* Cyprididae]

Pontocyprinae [*recte* Pontocypridinae] [see Pontocyprididae]
†Pontocypris [=Erythrocypris]; †Bythocypris; †Paracypris; †Pontocyprella; †Propontocypris.

Aglaiocyprinae SCHNEIDER, 1960 [*recte* Aglaiocypridinae] (USSR Treatise)
†Aglaiocypris [=†Aglaia]

Argilloeciinae MANDELSTAM, 1960 (USSR Treatise)
†Argilloecia [=†Argillaecia; †Protargilloecia]. Non-USSR: †Bosquetia.

Disopontocyprinae [*recte* Disopontocypridinae]
Disopontocypris; Amplocypris [=†Thaminocypris], †Bakunella; Caspiocypris; **Caspiolla** MANDELSTAM, 1960 (USSR Treatise) [*pro* Caspiella MANDELSTAM, 1956 (*non* THIELE, 1928)]; †Caspiollina; †Liventalina; **Pontoniella** MANDELSTAM, 1960 (USSR Treatise) [*pro* Pontonella MANDELSTAM, 1956 (*non* HELLER, 1856)]; Rectocypris. **Reginacypris** SCHNEIDER, 1960 (USSR Treatise) [*Cytherina abcissa* REUSS, 1850].

†Cyprideinae [=*Rostrocyprinae* (*recte* Rostrocypridinae)] [See Ilyocyprididae]
†Cypridea [=†Cytherideis (partim); †Pseudocypridina; †Cyamocypris; †Morinina; †Langtonia; †Paracypridea; †Ulwellia]; †Cyprideamorphella; †Ilyocyprimorpha; †Limnocypridea; Mongolianella.

Ilyocyprinae [*recte* Ilyocypridinae] [*see* Ilyocyprididae]
†Ilyocypris [=†Ilyocyprella]; †Origoilyocypris.

Cyclocyprinae [*recte* Cyclocypridinae] [see Cyclocyprididae]
†Cyclocypris; †Cypria; Exuocypris. Non-USSR: †Cyclocypria; †Cytherites; †Physocypria.

Lineocyprinae SCHNEIDER, 1960 (USSR Treatise) [*recte* Lineocypridinae]
†Lineocypris; †Fergania.

Clinocyprinae MANDELSTAM, 1960 (USSR Treatise) [*recte* Clinocypridinae]
†Clinocypris; †Subulacypris.

Candoninae
Candona; †Candoniella; †Cryptocandona. Non-USSR: Candonella; Candonopsis; †Eucandona; †Nannocandona; †Paracypria; †Pontoparta; †Pseudocandona; †Riocypris; †Typhlocypris.

Paracandoninae SCHNEIDER, 1960 (USSR Treatise)
†Paracandona.

Eucyprinae SARS, 1925 [*nom. transl.* SCHNEIDER, 1960 (USSR Treatise) (*ex* Eucyprides SARS, 1925)] [*recte* Eucypridinae] [Note that if Cyprididae is divided into subfamilies, one of these must be Cyprinae, and this subfamily must include Cypris; therefore, "Eucyprinae" of USSR Treatise is invalid.]

Eucypris [=Microcypris]; Cypris [=†Cyprisia]; Dolerocypris; Lycopterocypris; Paraeucypris; Pseudoeucypris.

Mediocyprinae SCHNEIDER, 1960 (USSR Treatise) [*recte* Mediocypridinae]

Mediocypris. Non-USSR: Amphicypris; Cypriconcha; Strandesia.

Herpetocyprinae SCHNEIDER, 1960 (USSR Treatise) [=Herpetocypridinae KAUFMANN, 1900] [If this subfamily is recognized, its valid name is that first published by KAUFMANN]

Herpetocypris [=Erpetocypris]; Prionocypris; Stenocypris. Non-USSR: *Arcocypris* [*recte* Acocypris]; Astenocypris [=Leptocypris]; Ilyodromus; Isocypris; Megalocypris; Stenocypris.

Cyprinae [*recte* Cypridinae] [See note given with Eucyprinae, to which USSR Treatise assigns Cypris.]

Cypricercus; †Cyprois; **Kassinina** MANDELSTAM, 1960 [*pro* Kassinia MANDELSTAM, 1956 (*non* Khabakov, 1937)]; Ussuriocypris. Non-USSR: Candonocypris; *Chlamidotheca* [*recte* Chlamydotheca].

Cyprinotinae BRONSTEIN, 1947 [*nom. transl.* SCHNEIDER, 1960 (USSR Treatise) (*ex* Cyprinotini BRONSTEIN, 1947)]

Cyprinotus [=Cypridonotus; Hemicypris]; Dogelinella; Hemicyprinotus; Paracyprinotus. Non-USSR: Heterocypris.

Baturinellinae SCHNEIDER, 1960 (USSR Treatise)

Baturinella.

Herpetocyprellinae BRONSTEIN, 1947 [*nom. transl.* SCHNEIDER, 1960 (USSR Treatise) (*ex* Herpetocyprellini BRONSTEIN, 1947)]

Herpetocyprella. Non-USSR: †Tuberocypris; †Tuberocyproides.

Cypridopsinae

Cypridopsis [=Pionocypris; Proteocypris]; Potamocypris [=Paracypridopsis]; Zonocypris. Non-USSR: Cypretta; Cypridopsella; Cyprilla; Oncocypria; Pseudocypretta.

Advenocyprinae SCHNEIDER, 1960 (USSR Treatise) [*recte* Advenocypridinae]

Advenocypris.

Cytheracea (By Kashevarova, Mandelstam, & Schneider)

Permianidae

Permiana.

Cytheridae

Tomiellinae [see Tomiellidae]

†Tomiella; †Kemeroviana; †Tomiellina.

Sinusuellinae [see Sinusuellidae]

†Sinusuella.

Iniellinae MANDELSTAM, 1960 (USSR Treatise)

†Iniella; †Suriekovella.

Speluncellinae SCHNEIDER, 1960 (USSR Treatise)

†Speluncella; †Gemmanella; †Pulvella.

Glorianellinae SCHNEIDER, 1960 (USSR Treatise)

†Glorianella; †Renngartenella.

Cytherissinellinae SCHNEIDER, 1960 (USSR Treatise) [see Cytherissinellidae] [*recte* KASHEVAROVA, 1958]

†Cytherissinella; †Lutkevichinella; †Triassinella.

Schulerideinae MANDELSTAM, 1960 (USSR Treatise) [*recte* 1959, p. 472]

†Schuleridea [=†Paraschuleridea; †Aequacytheridea]; †Apatocythere [=†Habrocythere]. Non-USSR: †Dordoniella; †Palaeomonsmirabilia.

Cuneocytherinae MANDELSTAM, 1960 (USSR Treatise) [see Cytherideidae]

†Cuneocythere; †Archeocuneocythere.

Pyrocytherideinae MANDELSTAM, 1960 (USSR Treatise)

†Pyrocytheridea

Timiriaseviinae MANDELSTAM, 1960 (USSR Treatise) [*recte,* 1959, p. 471]

†Timiriasevia; †Bronsteinia; †Theriosynoecum [=†Gomphocythere; †Morrisonia]. Non-USSR: †Metacypris. Probably assignable here: †Emphasia.

Palaeocytherideinae MANDELSTAM, 1960 (USSR Treatise)

Palaeocytherideides MANDELSTAM, 1960 (USSR Treatise)

†Palaeocytheridea; †Hutsonia; †Klieana; †Looneyella; †Palaeocytheridella; †Procytheridea; †Rubracea.

Camptocytherides MANDELSTAM, 1960 (USSR Treatise)

Camptocythere; †Aparchitocythere; †Mandelstamia; †Semicytheridea.

Eucytherinae [see Cytherideidae]

†Eucythere; †Cytherissa [=†Alexandrella]; †Rotundracythere.

Krithinae MANDELSTAM, 1960 (USSR Treatise) [see Cytherideidae] [*recte* MANDELSTAM in BUBIKAN, 1958]

†Krithe [=†Ilyobates; †Paracytherideis]; †Suzinia. Non-USSR: †Pseudokrithe. Probably assignable here: †Hemikrithe.

Pontocytherinae MANDELSTAM, 1960 (USSR Treatise)

†Pontocythere [=†Cytherideis *(partim)*; †Hemicytherideis; †Neocytherideis; †Sahnia]. Probably assignable here: †*Copitus* [*recte* Copytus].

Mediocytherideisinae MANDELSTAM, 1960 (USSR Treatise) [*recte* Mediocytherideidinae]

†Mediocytherideis.

Leptocytherinae [see Leptocytheridae]

†Leptocythere [=†Leptocythera] †Callistocy-

there. Probably assignable here: †Ento-cythere.

Limnocytherinae [=Limnicytherinae] [see Limnocytheridae]
†Limnocythere [=†Limnicythere; †Leuco-cythere]. Non-USSR: †Gomphocythere.

Procytheropterinae MANDELSTAM, 1960 (USSR Treatise)
†Procytheropteron; †Paenula.

Dolocytherideinae MANDELSTAM, 1960 (USSR Treatise)
†Dolocytheridea; **Aenigma** KUZNETSOVA, 1956; †Asciocythere.

Clithrocytherideinae MANDELSTAM, 1960 (USSR Treatise) [*recte* 1959, p. 460]
†Clithrocytheridea [=†Heterocyprideis]; †Pseudocytheridea; †Ruttenella. Non-USSR: †Perissocytheridea.

Cytherideinae [see Cytherideidae]
†Cytheridea [=†Haplocytheridea; †Leptocy-theridea; †Phractocytheridea]; †Cyprideis [=†Anomocytheridea; †Amonocythere; †Paracyprideis]. Non-USSR: †Aulocytheri-dea; †Eucytheridea; †Toulminia. Probably assignable here: †Hemicytheridea.

Macrodentinae MANDELSTAM, 1960 (USSR Treatise) [Invalid because subfamily contains no nominal genus from which name could be derived.]
†Macrodentina; †Amphicythere; †Exophthal-mocythere; †Oligocythereis. Non-USSR: †*Acantocythere* [*recte* †Acanthocythere]; †Dictyocythere.

Trachyleberinae [*recte* Trachyleberidinae] [see Trachyleberididae]

Trachyleberides MANDELSTAM, 1960 (USSR Treatise [*recte* Trachyleberidides SYLVES-TER-BRADLEY, 1948 (*nom. transl.* MANDEL-STAM, 1960)]
†Trachyleberis [=†Actinocythereis; †Anticy-thereis; †Archicythereis (*partim*); †Aurila; †Bradleya; †Echinocythereis; †Hemicy-theria; †Leguminocythereis; †Orionina; †Pseudocythereis; †Rectotrachyleberis; †Trachyleberidea; †Trachyleberina; †Tri-ginglymus; †Tyrrhenocythere] †Brachy-cthere [=†Digmopteron] †Cythereis [=†Cativella; †Navecythere; †Platy-cythereis (*partim*)]; †Digmocythere. Non-USSR: †Hirsutocythere; †Leniocythere; †Occultocythere; †Protobuntonia; †Spongi-cythere.

Hemicytherides MANDELSTAM, 1960 (USSR Treatise) [see Hemicytheridae]
†Hemicythere[=†Elofsonella; †Eucythereis; †Heterocythereis; †Paracythereis; †Procy-thereis]. Non-USSR: †Hermanites [*pro* Hermania].

Pterygocytherides MANDELSTAM, 1960 (USSR Treatise) [*recte* Pterygocythereidides]

†Pterygocythereis [=†Alatacythere; †Archi-cythereis (*partim*)]; †Caudites [=†Isocy-thereis; †Platycythereis (*partim*)]; †Quad-racythere; †Puriana [=†Carinocythereis; †Favella]. Probably assignable here: †Cli-macoidea; †Howella.

Schizocytherides MANDELSTAM, 1960 (USSR Treatise)
†Schizocythere; Cnestocythere; †Gubkiniella; Paijenborchella.

Cytherinae [see Cytheridae]
Cythere [=Hirschmannia]; †Urocythereis. Probably assignable here: †Arcacythere; Nereina.

Cytheromorphinae MANDELSTAM, 1960 (USSR Treatise)
†Cytheromorpha.

Loxoconchinae [see Loxoconchidae]
†Loxoconcha [=†Normania]; †Loxoconch-ella; **Loxoella** KUZNETSOVA, 1956 [*L. in-violata*]. Non-USSR: †Pseudoloxoconcha. Probably assignable here; Loxocythere.

Protocytherinae MANDELSTAM, 1960 (USSR Treatise) [*recte* Protocytherinae LYUBIMOVA in LYUBIMOVA & KHABAROVA, 1955]
†Protocythere; †Posteroprotocythere; †Procy-therettina; **Pseudoperissocytheridea** MAN-DELSTAM, 1960 (USSR Treatise) [*Proto-cythere crassula* MANDELSTAM, 1947]

Cytherettinae [see Cytherettidae]
†Cytheretta [=†Acuticythereis; †Basslerella; Basslerites; †Buntonia; †Campylocythere; †Cylindrus; †Cytherettina; †Javanella; †Paracytheretta; Pellucistama (*sic*) [=†Pel-lucistoma]; †Prionocytheretta; †Pseudocy-theretta; †Pyricythereis]. Non-USSR: †Monsmirabilia.

Sigilliuminae MANDELSTAM, 1960 (USSR Treatise) [*recte* Sigilliinae]
Sigillium KUZNETSOVA, 1960 (USSR Treatise) [*S. procerum*].

Faluniinae MANDELSTAM, 1960 (USSR Treatise)
†Falunia; †Taxodiella. Non-USSR: †Tanella; †Thalmannia.

Progonocytherinae [see Progonocytheridae]
†Progonocythere [=†Lophocythere]; †Juve-nix; †Pavloviella; †Trochinius [=†Trochis-cus]; †Vicinia.

Paracytherideinae MANDELSTAM, 1960 (USSR Treatise) [*recte* 1959, p. 477]
†Paracytheridea.

Pleurocytherinae MANDELSTAM, 1960 (USSR Treatise)
†Pleurocythere; †Annosacythere; †Naviculina; †Orthonotacythere.

Parataxodontinae MANDELSTAM, 1960 (USSR Treatise)
†Parataxodonta.

Eocytheropterinae MANDELSTAM, 1960 (USSR Treatise) [*recte* 1959, p. 474]

†Eocytheropteron [=†Budia]

Cytherurinae [see Cytheruridae]

†Cytherura [=Tetracytherura]; †Cytherurina. Non-USSR: †Pseudocytherura.

Xenocytherinae MANDELSTAM, 1960 (USSR Treatise)

†Xenocythere.

Cytheropterinae HANAI, 1957

†Cytheropteron [=†Aversovalva; †Kangarina]; **Brachycytheropteron** KUZNETSOVA, 1960 (USSR) Treatise) [*Cytheropteron bicornutum ALEXANDER, 1933]; †Cytheropterina; †Eucytherura; †Hemicytherura; †Paijenborchellina; †Segmina.

Xestoleberinae [*recte* Xestoleberidinae] [see Xestoleberididae]

†Xestoleberis. Non-USSR: †Linocheles; †Microcytherura; †Microxestoleberis. Probably assignable here: †Atjehella; Cushmanoidea [*recte* †Cushmanidea]; †Ovocytheridea; †Scabriculocypris.

Paradoxostomidae [*recte* Paradoxostomatidae]

Paradoxostominae [*recte* Paradoxostomatinae]

†Paradoxostoma; †Cytherois; †Paracytherois; †Sclerochilus. Non-USSR: †Cytheroma; †Paracythere; †Paracytheroma.

Pseudocytherinae SCHNEIDER, 1960 (USSR Treatise)

†Pseudocythere; †Jonesia [=†Macrocythere]. Non-USSR: †Xiphichilus.

Bythocytherinae [see Bythocytheridae]

†Bythocythere; †Bythoceratina; †Bythocytheremorpha [=†Monoceratina ALEXANDER, 1934]; †Cytheralison; †Miracythere. Non-USSR: †Luvula; †Macrocytherina.

Incertae Sedis

Altha; Ampuloides; Discoidella; †Monoceratina ROTH, 1928; Rectella; †Samarella; Serenida; †Voronina. Non-USSR: †Alanella; †Aurigerites; Bernix; †Berounella; †Bideirella; Budnianella; Dicranella; †Doraclatum; †Eoconchoecia; Golcondella; †Hesperidella; †Hypotetragona; †Karlsteinella; †Kegelites [=Girtyites]; †Kirkbyella; †Kirkbyellina; †Milleratia; †Miltonella; †Mooreina; †Murrayina; Nanopsis; Nezamyslia; †Parahealdia; †Placentella; Pseudophanasymmetria; Pyxion; †Sanniolus; Tetrasulcata; †Tribolbina [=?Beyrichiana]; Unicornites; †Verrucosella; Vltavina; †Waldeckella.

Unrecognizable

Antitomis; †Binodella; †Bolbozoe; Bursulella; †Carbinobolbina; Ctenentoma; †Cyrtocypris; †Gillina; †Glossomorphites [*pro* Glossopsis]; †Kuleschowkia; *Ladella* [*recte* †Laddella]; Leioprimitia; Macrocyproides; †Neobeyrichiopsis; Pulvillites; Quadricollina; †Senescella; †Warthinia.

EDITORIAL NOTE

After the completion of typesetting for the *Treatise* volume on Ostracoda, Dr. W. D. IAN ROLFE, University of Birmingham (England), who has been working at Harvard University to prepare descriptions of phyllocarid and other crustaceans for publication in the *Treatise,* raised a question about possible disagreement in classification of some genera considered by him to belong in the area of his assignment. Especially, this might relate to forms included by other workers in Ostracoda or Branchiopoda, or classed as *nomina dubia*.

Examination has shown that four nominal genera included in Part Q are claimed by Dr. ROLFE to belong among phyllocarids, rather than among ostracodes. These are *Aristozoe* BARRANDE, 1868 (p. Q109), *Callizoe* BARRANDE, 1872 (p. Q412), *Dithyrocaris* PORTLOCK, 1843 (p. Q413), and *Orozoe* BARRANDE, 1872 (p. Q414); the first-mentioned is here classed in the Isochilinidae and the other three as *nomina dubia*.

In the opinion of Dr. ROLFE, *Dithyrocaris* is without doubt a phyllocarid, *Callizoe* is almost surely so, but *Aristozoe* and *Orozoe* are more doubtfully classed with the phyllocarids. He states: "It is clear from PORTLOCK, 1843 (p. 313), that SCOULER is responsible for the name *Dithyrocaris,* and adequate 'indication' of the homonymic genus *Argas* was provided by SCOULER, if only because the specimen was well illustrated by SCOULER's 1835 paper (fig. 3)." ROLFE's study has led to submittal of revised nomenclatural data concerning *Dithyrocaris,* as follows.

Dithyrocaris SCOULER in PORTLOCK, 1843 (p. 313) [*pro Argas* SCOULER, 1835 (*non* LATREILLE, 1795, *nec* OKEN, 1815)] [*Argas testudineus* SCOULER, 1835; SD VAN STRAELEN & SCHMITZ, 1934] [=*Dithyrocarus* M'COY, 1842 (*nom. nud.*); *Rhachura* SCUDDER, 1878; *Argus* PACKARD, 1879 (*nom. null.*); *Rachura* HALL & CLARKE, 1888 (*nom. null.*)].

R. C. MOORE

INDEX

Names included in the following index are classified typographically as follows: (1) Roman capital letters are used for suprafamilial taxonomic units which are recognized as valid in classification; (2) italic capital letters are employed for suprafamilial categories which are considered to be junior synonyms of valid names; (3) morphological terms and generic family names accepted as valid are printed in roman type; and (4) generic and family names classed as invalid, including junior homonyms and synonyms, are printed in italics. Page numbers printed in boldface type as (**Q234**) indicate the location of systematic descriptions or definitions of morphological terms.

ARLIS

Alaska Resources
Library & Information Services
Anchorage, AK